Selected Titles in This Series

196 **Yu. N. Lin′kov,** Asymptotic statistical methods for stochastic processes, 2001

195 **Minoru Wakimoto,** Infinite-dimensional Lie algebras, 2001

194 **Valery B. Nevzorov,** Records: Mathematical theory, 2001

193 **Toshio Nishino,** Function theory in several complex variables, 2001

192 **Yu. P. Solovyov and E. V. Troitsky,** C^*-algebras and elliptic operators in differential topology, 2001

191 **Shun-ichi Amari and Hiroshi Nagaoka,** Methods of information geometry, 2000

190 **Alexander N. Starkov,** Dynamical systems on homogeneous spaces, 2000

189 **Mitsuru Ikawa,** Hyperbolic partial differential equations and wave phenomena, 2000

188 **V. V. Buldygin and Yu. V. Kozachenko,** Metric characterization of random variables and random processes, 2000

187 **A. V. Fursikov,** Optimal control of distributed systems. Theory and applications, 2000

186 **Kazuya Kato, Nobushige Kurokawa, and Takeshi Saito,** Number theory 1: Fermat's dream, 2000

185 **Kenji Ueno,** Algebraic Geometry 1: From algebraic varieties to schemes, 1999

184 **A. V. Mel′nikov,** Financial markets, 1999

183 **Hajime Sato,** Algebraic topology: an intuitive approach, 1999

182 **I. S. Krasil′shchik and A. M. Vinogradov, Editors,** Symmetries and conservation laws for differential equations of mathematical physics, 1999

181 **Ya. G. Berkovich and E. M. Zhmud′,** Characters of finite groups. Part 2, 1999

180 **A. A. Milyutin and N. P. Osmolovskii,** Calculus of variations and optimal control, 1998

179 **V. E. Voskresenskiĭ,** Algebraic groups and their birational invariants, 1998

178 **Mitsuo Morimoto,** Analytic functionals on the sphere, 1998

177 **Satoru Igari,** Real analysis—with an introduction to wavelet theory, 1998

176 **L. M. Lerman and Ya. L. Umanskiy,** Four-dimensional integrable Hamiltonian systems with simple singular points (topological aspects), 1998

175 **S. K. Godunov,** Modern aspects of linear algebra, 1998

174 **Ya-Zhe Chen and Lan-Cheng Wu,** Second order elliptic equations and elliptic systems, 1998

173 **Yu. A. Davydov, M. A. Lifshits, and N. V. Smorodina,** Local properties of distributions of stochastic functionals, 1998

172 **Ya. G. Berkovich and E. M. Zhmud′,** Characters of finite groups. Part 1, 1998

171 **E. M. Landis,** Second order equations of elliptic and parabolic type, 1998

170 **Viktor Prasolov and Yuri Solovyev,** Elliptic functions and elliptic integrals, 1997

169 **S. K. Godunov,** Ordinary differential equations with constant coefficient, 1997

168 **Junjiro Noguchi,** Introduction to complex analysis, 1998

167 **Masaya Yamaguti, Masayoshi Hata, and Jun Kigami,** Mathematics of fractals, 1997

166 **Kenji Ueno,** An introduction to algebraic geometry, 1997

165 **V. V. Ishkhanov, B. B. Lur′e, and D. K. Faddeev,** The embedding problem in Galois theory, 1997

164 **E. I. Gordon,** Nonstandard methods in commutative harmonic analysis, 1997

163 **A. Ya. Dorogovtsev, D. S. Silvestrov, A. V. Skorokhod, and M. I. Yadrenko,** Probability theory: Collection of problems, 1997

162 **M. V. Boldin, G. I. Simonova, and Yu. N. Tyurin,** Sign-based methods in linear statistical models, 1997

161 **Michael Blank,** Discreteness and continuity in problems of chaotic dynamics, 1997

160 **V. G. Osmolovskiĭ,** Linear and nonlinear perturbations of the operator div, 1997

(Continued in the back of this publication)

Function Theory in Several Complex Variables

Contents

Preface

"When you ask around and persist fondly in the Way, you can never find the Truth. Maintain your belief. Devote yourself to the Way. It will turn out to be the mighty Truth."

from "the 42 teachings of Buddha"

One of the subjects in mathematical nature is created by unifying three notions: complex numbers, coordinate systems, and the related notions of differentiation and integration. The space $\mathbf{C}^n$ of n-tuples of complex numbers is ruled by the coordinate system $(z_1, \ldots, z_n)$. If a complex-valued function in a domain in $\mathbf{C}^n$ is differentiable in each variable, it can be represented locally as a convergent power series. It has a natural domain of existence, in which it behaves in its own characteristic way, i.e., it creates its own mathematical world. We call such a function an analytic function.

In the case of one complex variable, an analytic function has a distinguishing property. In this case its real and imaginary parts are harmonic functions which are conjugate to each other. Namely, if one is considered as a potential, then the other is the flow which the potential induces. A harmonic function is uniquely determined by its boundary values; we can construct a harmonic function with prescribed boundary values and construct locally its conjugate harmonic function, which is unique up to an additive constant. This makes it easy to construct analytic functions of one complex variable. The main properties of analytic functions of one complex variable can be explained from this observation.

When we want to describe concepts in nature by using analytic functions, it is not enough to use only those of one complex variable. The theory of analytic functions of several complex variables is quite difficult to treat, compared to the theory in one complex variable. One reason for this is the freedom of the form of domains in $\mathbf{C}^n$ due to the increase in the dimension. Another reason is that both the real and the imaginary parts of an analytic function are now pluriharmonic functions, which imposes a stronger restriction than being merely harmonic. For example, in some cases, a pluriharmonic function is uniquely determined by its boundary values on some proper subset of the boundary, and we cannot always construct a pluriharmonic function with prescribed boundary values on a given portion of the boundary. Therefore, it is difficult to construct analytic functions of several complex variables. Since function theory in one complex variable generally proceeds by constructing analytic functions, we cannot simply use the one-variable approach in the case of several complex variables.

The most particular phenomena in the study of analytic functions in several complex variables which does not appear in the case of one complex variable is the fact that the natural domain of an analytic function is not arbitrary, i.e., it is not true that any domain in $\mathbf{C}^n$ is a natural domain of existence of some analytic function. This fact is important. We call a domain in $\mathbf{C}^n$ which is the natural domain of existence of some analytic function a *domain of holomorphy*. The principal problem in function theory in several complex variables is to study which domains are domains of holomorphy, and to determine which objects we can construct in a domain of holomorphy.

This book is an attempt to explain results in the theory of functions of several complex variables which were mostly established from the late 19th century through the middle of the 20th century. The focus is to introduce the mathematical world which was created by my advisor, Kiyoshi Oka (1901-1978). I have attempted to remain as close as possible to Oka's original work.

Kiyoshi Oka, at the beginning of his research, regarded the collection of problems which he encountered in the study of domains of holomorphy as large mountains which separate today and tomorrow. Thus, he believed that there could be no essential progress in analysis without climbing over these mountains.

The work of Oka can be divided into two parts. The first is the study of analytic functions in univalent domains in $\mathbf{C}^n$. Here he proved that three concepts: domains of holomorphy, holomorphically convex domains, and pseudoconvex domains, are equivalent; and, moreover, that the Poincaré problem, the Cousin problems, and the Runge problem – when stated properly – can be solved in domains of holomorphy satisfying the appropriate conditions. The second part was to establish a method by which we can study analytic functions defined in a ramified domain over $\mathbf{C}^n$ in which the branch points are considered as interior points of the domain. He proceeded in this later work under the assumption that the results valid in univalent domains in $\mathbf{C}^n$ should similarly hold in a ramified domain over $\mathbf{C}^n$. However, the true situation was contrary to his intuition, i.e., a ramified domain of holomorphy is not always a holomorphically convex domain.

Oka's establishment of his method to treat analytic functions in a ramified domain has proved to be indispensable not only in analysis but also in other fields of mathematics.

This book consists of parts I and II, according to Oka's earlier and later work mentioned above. In part I we treat analytic functions in a univalent domain in $\mathbf{C}^n$. In part II we treat analytic functions in an analytic space; this is a slight generalization of a ramified domain over $\mathbf{C}^n$. The one exception to our adherence to Oka's program is that the fact that a pseudoconvex univalent domain is a domain of holomorphy will be proved in part II in a more general setting by modifying Oka's original ideas.

A mathematical object is abstract and is described by use of words and notation. We should note that the words and the notation themselves are not really mathematics. Mathematics can be realized as a flow of the consciousness which is really creating mathematical nature. After such a process, mathematical nature lives individually in the mind of each person who has studied it. He seems to hear a voice coming from the bottom of his mind, or to feel the glow of a living object within his mind. This process is essential when we study the established works of the pioneers of a field. If mathematical nature lives correctly within a person's mind, then when he encounters a certain problem, he may not recall the knowledge to solve it immediately, but he will be able to understand the problem itself in order to solve it.

The difficulty in studying mathematics is the procedure for giving life and meaning to the mathematics. The first step is to organize and expand upon the material written by use of words and notation in a concrete form, so that we can proceed with further steps.

I hope that this book is a worthwhile initial step for the reader in order to understand the mathematical world which was created by Kiyoshi Oka.

Toshio Nishino
June 22, 1996 at Kyoto

Preface to the English Edition

This book was written, after long consideration, with the intent to make Oka's original ideas easier to understand. One of the main reasons to pursue this project was the recommendation of Professor John Wermer. During the time while I was writing the original version of the book in Japanese, Professor Katsumi Nomizu had already started urging the AMS to publish an English translation.

Oka's original papers may appear to be difficult to read. However, when we truly understand his original thoughts, we gain much more than simply mathematical results. I hope that this book helps the reader to better comprehend Oka's work.

As for the English translation, Professors Norman Levenberg (Auckland University) and Hiroshi Yamaguchi (Nara Women's University) devoted much time and effort to translating the Japanese version; they had to overcome the difficulties caused by the many differences between Western and Japanese culture. I greatly appreciate their effort. Also, many thanks to the people at the AMS, particularly Ralph Sizer, for their patience and understanding.

Toshio Nishino
March 3, 2000 at Kyoto

Part 1

Fundamental Theory

CHAPTER 1

Holomorphic Functions and Domains of Holomorphy

1.1. Complex Euclidean Space

1.1.1. Complex Euclidean Space. We let $\mathbf{C}$ denote the Euclidean plane of one complex variable. To emphasize the variable used, e.g., w, we use the notation $\mathbf{C}_w$. For a positive integer n, we let $\mathbf{C}^n$ denote the n-dimensional complex Euclidean space generated by the n complex variables $z_1, \dots, z_n$. Given a point $z = (z_1, \dots, z_n) \in \mathbf{C}^n$, we call z_j the j-th coordinate of z and we call $\mathbf{C}_{z_j}$ the j-th coordinate plane of $\mathbf{C}^n$. Then $\mathbf{C}^n$ is the product of the n complex planes $\mathbf{C}_{z_j}$ $(j = 1, \dots, n)$. It is sometimes convenient to use the two real-dimensional plane to model $\mathbf{C}_{z_j}$. To visualize a point $z' = (z'_1, \dots, z'_n)$ of $\mathbf{C}^n$, we imagine n coordinate planes $\mathbf{C}_{z_j}$ $(j = 1, \dots, n)$ lying on the same plane $\mathbf{C}$; we take the point z'_j on $\mathbf{C}_{z_j}$ for $j = 1, \dots, n$ and regard their combination as the point z' in $\mathbf{C}^n$ (see Figure 1).

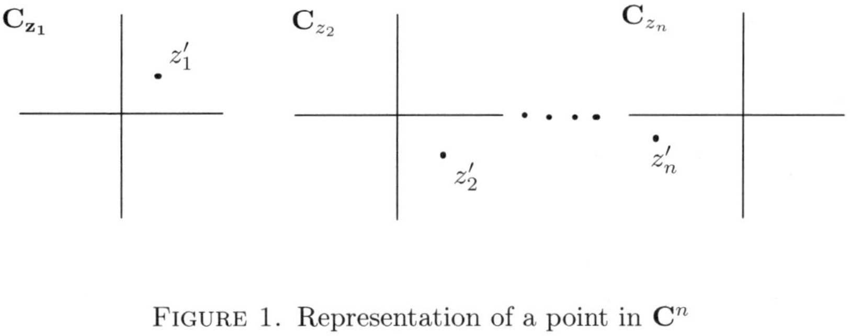

FIGURE 1. Representation of a point in $\mathbf{C}^n$

By a **linear coordinate transformation** we mean a linear transformation of $\mathbf{C}^n$:

$$\mathcal{L}: \ w_i = b_i + a_{i1}z_1 + \dots + a_{in}z_n \qquad (i = 1, \dots, n)$$

where b_i, a_{ij} $(i, j = 1, \dots, n)$ are complex numbers with $\det(a_{ij}) \neq 0$. We refer to $(w_1, \dots, w_n)$ as the new coordinate system of $\mathbf{C}^n$; thus if a point $P_0 \in \mathbf{C}^n$ has coordinates $z^0 = (z_1^0, \dots, z_n^0)$ in the (old) coordinate system $(z_1, \dots, z_n)$, then it has coordinates $w^0 = (w_1^0, \dots, w_n^0)$ in the new coordinate system $(w_1, \dots, w_n)$ where $w^0 = \mathcal{L}(z^0)$.

1.1.2. Projections, Product Spaces, and Sections. Let r and s be positive integers and set $n = r + s$. The space $\mathbf{C}^n$ of the n variables $z_1, \dots, z_n$ is the product of $\mathbf{C}^r$ with variables $z_1, \dots, z_r$ and $\mathbf{C}^s$ with variables $z_{r+1}, \dots, z_n$. For a point $z' = (z'_1, \dots, z'_n) \in \mathbf{C}^n$, we call $(z'_1, \dots, z'_r)$ the **projection of z' to $\mathbf{C}^r$** and we call the map sending $z' = (z'_1, \dots, z'_n)$ to $(z'_1, \dots, z'_r)$ the **projection from $\mathbf{C}^n$ to $\mathbf{C}^r$**. For a subset E of $\mathbf{C}^n$, the set consisting of the projections of all points z in E is called the **projection of E to $\mathbf{C}^r$** and will be denoted $\underline{E}$. We define in a similar manner the projection from $\mathbf{C}^n$ to $\mathbf{C}^s$.

Let $E_1 \subset \mathbf{C}^r$ and $E_2 \subset \mathbf{C}^s$. For any $z' = (z'_1, \dots, z'_r)$ in $\mathbf{C}^r$ and $z'' = (z''_{r+1}, \dots, z''_n)$ in $\mathbf{C}^s$, we consider the ordered pair

$$(z', z'') = (z'_1, \dots, z'_r, z''_{r+1}, \dots, z''_n)$$

as a point of $\mathbf{C}^n$. We denote the set of all such pairs by $E_1 \times E_2$, called the **product set of E_1 and E_2**. In a similar manner we can define the product set of more than two sets. Let $\mathbf{C}^n = \mathbf{C}^r \times \mathbf{C}^s$ and let $E \subset \mathbf{C}^n$. For $a = (a_1, \dots, a_r) \in \mathbf{C}^r$ we let $E(a)$ denote the set of all points of E whose projection to $\mathbf{C}^r$ is a, and we call $E(a)$ the **section** (or **fiber**, sometimes) of E over $z_j = a_j$ $(j = 1, \dots, r)$. Note that the projection of $E(a)$ to the space $\mathbf{C}^s$ is one-to-one. Thus we often identify $E(a) \subset \mathbf{C}^n$ with the projection of $E(a)$ to $\mathbf{C}^s$, and we consider E as a variation of the sets $E(a)$ in $\mathbf{C}^s$ varying with the parameter $a \in \mathbf{C}^r$. In the special case where $E = E_1 \times E_2$, we identify $E(a)$ with E_2 for $a \in E_1$ and with $\emptyset$ for $a \notin E_1$.

Let $E \subset \mathbf{C}^n$ and $F \subset \mathbf{C}^r$. We let $E(F) \subset \mathbf{C}^n$ denote the set of all points of E whose projection to $\mathbf{C}^r$ is contained in F; i.e.,

$$E(F) = \bigcup_{a \in F} E(a).$$

1.1.3. Domains and Product Domains. By a **domain** in $\mathbf{C}^n$ we will mean an open and connected subset of $\mathbf{C}^n$, although we will have occasion to drop the connectivity assumption. We let ∂D denote the **boundary** of the domain D. In general, for any subset E of $\mathbf{C}^n$, we let $\overline{E}$ denote the **closure** of E. For a bounded domain D, we call the closure $\overline{D} = D \cup \partial D$ of D a **closed domain**.

To represent a domain D in $\mathbf{C}^n$ more concretely, as described in 1.1.1 we consider n coordinate planes $\mathbf{C}_{z_j}$ $(j = 1, \dots, n)$ on the same plane $\mathbf{C}$. Taking a point z'_j $(j = 1, \dots, n-1)$ on each coordinate plane $\mathbf{C}_{z_j}$, we set $z' = (z'_1, \dots, z'_{n-1}) \in \mathbf{C}^{n-1}$. On the coordinate plane $\mathbf{C}_{z_n}$, we draw the section $D(z')$ of D over $z_j = z'_j$ $(j = 1, \dots, n-1)$ (identifying this section with its projection to $\mathbf{C}_{z_n}$), so that $D(z')$, which is a domain in $\mathbf{C}_{z_n}$, varies with the parameter $z' \in \mathbf{C}^{n-1}$. The totality of these sections $D(z')$, varying in the complex plane $\mathbf{C}_{z_n}$, gives a realization of the domain D in $\mathbf{C}^n$. We remark that even if D is connected and simply connected in $\mathbf{C}^n$, a section $D(z')$ in $\mathbf{C}_{z_n}$ is not necessarily connected.

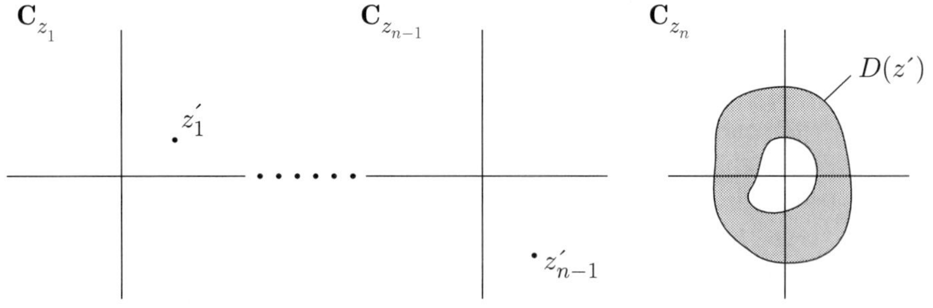

FIGURE 2. Representation of a domain

Let Δ_j be a domain in the coordinate plane $\mathbf{C}_{z_j}$ $(j = 1, \ldots, n)$. The product set $\Delta = \Delta_1 \times \ldots \times \Delta_n$ in $\mathbf{C}^n$ is called a **product domain** in $\mathbf{C}^n$, and $\Delta_j \subset \mathbf{C}_{z_j}$ is called the z_j **component set** of Δ. When $n \geq 2$ and each Δ_j is connected (but not necessarily simply connected), the boundary $\partial\Delta$ of the product domain Δ consists of one connected component. The product domain Δ is connected and simply connected if and only if each z_j component set Δ_j $(j = 1, \ldots, n)$ is connected and simply connected. In general, there is no easy way to describe, geometrically or analytically, a domain in $\mathbf{C}^n$. However, in the case of product domains Δ, we can represent the component sets of Δ on n separate coordinate planes. Often when we need to choose a neighborhood of a point z in $\mathbf{C}^n$, we will take a product domain consisting of one simply connected component.

1.1.4. Complex Hyperplanes, Polydisks, and Balls. Let n and m $(0 < m < n)$ be positive integers. We consider the space $\mathbf{C}^n$ of n complex variables $z_1, \ldots, z_n$, and the space $\mathbf{C}^m$ of m complex variables $t_1, \ldots, t_m$. Let a_{jk} $(j = 1, \ldots, n;\ k = 1, \ldots, m)$ be nm complex numbers such that the rank of the matrix (a_{jk}) is equal to m, and let b_j $(j = 1, \ldots, n)$ be any n complex numbers. Consider the mapping T from $\mathbf{C}^m$ to $\mathbf{C}^n$ sending $(t_1, \ldots, t_m)$ to $(z_1, \ldots, z_n)$ by the rule

$$z_j = a_{j1}t_1 + \cdots + a_{jm}t_m + b_j \qquad (j = 1, \ldots, n).$$

We call the image set $L := T(\mathbf{C}^m)$ an **m-dimensional complex hyperplane** in $\mathbf{C}^n$. In particlar, when $m = 1$, L is called a **complex line**. An m-dimensional complex hyperplane can also be given as the solution set of a finite number of simultaneous linear equations for n unknown complex numbers $z_1, \ldots, z_n$.

Let $E \subset \mathbf{C}^n$ and let $L = T(\mathbf{C}^m)$ be an m-dimensional complex hyperplane in $\mathbf{C}^n$. Then $E \cap L$ is called the **section of E in L** in $\mathbf{C}^n$. We often identify $E \cap L$ with its pre-image $T^{-1}(E \cap L)$ in $\mathbf{C}^m$.

REMARK 1.1. We can identify $\mathbf{C}^n$ with real $2n$-dimensional Euclidean space $\mathbf{R}^{2n}$. Under this identification, an m-dimensional hyperplane L in $\mathbf{C}^n$ is always a real $2m$-dimensional hyperplane. However, not all real $2m$-dimensional hyperplanes can be regarded as complex m-dimensional hyperplanes. For example, given two distinct points p and q in $\mathbf{C}^n$, the family of all real two-dimensional hyperplanes passing through p and q is a $(2n-2)$-dimensional real-parameter family, among which exactly one plane is a complex line.

Let $a = (a_1, \ldots, a_n) \in \mathbf{C}^n$ and $r_j > 0$ $(j = 1, \ldots, n)$. We call the subset of $\mathbf{C}^n$ given by

$$\Delta \ : \ |z_j - a_j| < r_j \qquad (j = 1, \ldots, n),$$

the (open) **polydisk** centered at a with **polyradius** r_j $(j = 1, \ldots, n)$. This is a special type of product domain. In particular, when $r_j = r$ $(j = 1, \ldots, n)$, we call Δ a **polydisk with radius** r. We allow $r_j = +\infty$. The closure $\overline{\Delta}$ of Δ is called a **closed polydisk**. If $n = 2$, Δ is called a **bidisk**.

For a polydisk Δ centered at a with polyradius r_j $(j = 1, \ldots, n)$, we call

$$\mathcal{E} \ : \ |z_j - a_j| = r_j \qquad (j = 1, \ldots, n)$$

the **distinguished boundary** of Δ. The topological boundary $\partial\Delta$ of Δ in $\mathbf{C}^n$ is a real $(2n-1)$-dimensional set which contains the real n-dimensional distinguished boundary $\mathcal{E}$.

Let $a = (a_1, \dots, a_n) \in \mathbf{C}^n$ and let $r > 0$. We call the subset of $\mathbf{C}^n$ given by

$$B : |z_1 - a_1|^2 + \cdots + |z_n - a_n|^2 < r^2$$

the **open** (Euclidean) **ball** centered at a with radius r and

$$\mathcal{Q} : |z_1 - a_1|^2 + \cdots + |z_n - a_n|^2 \leq r^2$$

the **closed ball** centered at a with radius r. We often use the simple notation:

$$\|z - a\| := (|z_1 - a_1|^2 + \cdots + |z_n - a_n|^2)^{1/2}$$

for any $z, a \in \mathbf{C}^n$.

REMARK 1.2. Let $\mathcal{Q}$ be the closed ball centered at the origin with radius r in $\mathbf{C}^n$ and let L be a complex line passing through the origin in $\mathbf{C}^n$. Define $\mathcal{Q}^0 = \mathcal{Q} \cap L$. Then the projection C_j of $\mathcal{Q}^0$ to each coordinate plane $\mathbf{C}_{z_j}$ is a disk centered at $z_j = 0$ (possibly of radius 0). Furthermore, if we let r_j $(j = 1, \dots, n)$ denote the radius of the disk C_j $(j = 1, \dots, n)$, then $r^2 = \sum_{j=1}^n r_j^2$, so that the Euclidean area πr^2 of $\mathcal{Q}^0$ is equal to the sum $\pi \sum_{j=1}^n r_j^2$ of the Euclidean areas of these disks.

1.1.5. Boundary Distance. Let D be a domain in $\mathbf{C}^n$. For $z' = (z'_1, \dots, z'_n) \in D$, the supremum of $r > 0$ such that the closed ball

$$\mathcal{Q}_r : |z_1 - z'_1|^2 + \cdots + |z_n - z'_n|^2 \leq r^2$$

centered at z' with radius r is contained in D is called the (Euclidean) **boundary distance from** z' **to** ∂D and is denoted by $d_D(z')$. Note that $|d_D(z') - d_D(z'')| \leq \|z' - z''\|$ for $z', z'' \in D$. The boundary distance thus defines a positive-valued, continuous function $d_D(z)$ on D called the **boundary distance function on** D. For $E \subset D$,

$$d_D(E) := \inf_{z \in E} d_D(z)$$

is called the **boundary distance from** E **to** ∂D.

We also consider another kind of boundary distance. For $z' \in D$, the supremum of $r > 0$ such that the closed polydisk

$$\overline{\Delta}_r : |z_j - z'_j| \leq r \qquad (j = 1, \dots, n)$$

centered at z' with radius r is contained in D is called the **polydisk boundary distance from** z' **to** ∂D and is denoted by $\delta_D(z')$. For $E \subset D$,

$$\delta_D(E) := \inf_{z \in E} \delta_D(z)$$

is called the **polydisk boundary distance from** E **to** ∂D.

For $D \subset \mathbf{C}^n$ and $E \subset D$, $\overline{E} \cap D$ is called the **closure of** E **in** D. If $\overline{E} \cap D$ is compact, we say E is **relatively compact in** D and we write $E \subset\subset D$. For example, if D is a bounded domain in $\mathbf{C}^n$, and if the Euclidean or polydisk boundary distance from E to ∂D is positive, then E is relatively compact in D. However, this is not true in general if D is unbounded.

1.1.6. Compactification. We often want to add ideal boundary points to $\mathbf{C}^n$ in such a way that the new space becomes compact. If $n = 1$, there is a unique compactification obtained by adding one ideal boundary point to $\mathbf{C}$; this gives the **Riemann sphere** $\widehat{\mathbf{C}}$. If $n \geq 2$, there are several possible compactifications of $\mathbf{C}^n$; we discuss two standard ones.

1. Osgood Space. In $\mathbf{C}^n$ with coordinates $z_1, \ldots, z_n$, let $\widehat{\mathbf{C}}_{z_j}$ $(j = 1, \ldots, n)$ be the Riemann sphere of the coordinate plane $\mathbf{C}_{z_j}$. The product space

$$\widehat{\mathbf{C}}^n = \widehat{\mathbf{C}}_{z_1} \times \ldots \times \widehat{\mathbf{C}}_{z_n}$$

is called n-dimensional **Osgood space**. Thus, in a sense, n-dimensional Osgood space $\widehat{\mathbf{C}}^n$ is constructed by adding n copies of $(n-1)$-dimensional Osgood space $\widehat{\mathbf{C}}^{n-1}$ to $\mathbf{C}^n$.

The mapping $\Phi(z_1, \ldots, z_n) = (w_1, \ldots, w_n)$, where

$$w_j = \frac{a_j z_j + b_j}{c_j z_j + d_j} \qquad (a_j d_j - b_j c_j \neq 0, \ j = 1, \ldots, n)$$

are linear fractional transformations, defines an analytic bijection from $\widehat{\mathbf{C}}^n$ to $\widehat{\mathbf{C}}^n$. These transformations are transitive on $\widehat{\mathbf{C}}^n$; i.e., given any point $(z'_1, \ldots, z'_n) \in \widehat{\mathbf{C}}^n$, there exists Φ as above with $\Phi(z'_1, \ldots, z'_n) = (0, \ldots, 0)$.

2. Projective Space. Let $z' = (z'_0, z'_1, \ldots, z'_n)$ and $z'' = (z''_0, z''_1, \ldots, z''_n)$ be points in $\mathbf{C}^{n+1} \setminus \{0\}$. We call these points equivalent if there exists $c \in \mathbf{C} \setminus \{0\}$ such that $z''_j = cz'_j$ $(j = 0, 1, \ldots, n)$. The equivalence classes in $\mathbf{C}^{n+1} \setminus \{0\}$ form an n-dimensional space called **complex projective space**, which we denote by $\mathbf{P}^n$. The coordinates $z = (z_0, z_1, \ldots, z_n)$ are called **homogeneous coordinates** for $\mathbf{P}^n$ and will be denoted by $[z_0 : z_1 : \ldots : z_n]$. If $n = 1$, $\mathbf{P}^1 := \mathbf{P}$ is equal to the Riemann sphere $\widehat{\mathbf{C}}$.

We get a bijective correspondence by sending the point $z = [z_0 : z_1 : \ldots : z_n]$ in $\mathbf{P}^n$ with $z_0 \neq 0$ to the point $w = (w_1, \ldots, w_n)$ in $\mathbf{C}^n$, where

$$w_j = z_j / z_0 \qquad (j = 1, \ldots, n).$$

The set of all points $z = [z_0 : z_1 : \ldots : z_n] \in \mathbf{P}^n$ with $z_0 = 0$ can be identified with the $(n-1)$-dimensional complex space consisting of all points with homogeneous coordinates $(z_1, \ldots, z_n)$ in $\mathbf{C}^n \setminus \{0\}$. Thus $\mathbf{P}^n$ is a compactification of $\mathbf{C}^n$ obtained by adding the space $\mathbf{P}^{n-1}$ as the set of ideal boundary points to $\mathbf{C}^n$. We call $\mathbf{C}^n$ the **finite part** of $\mathbf{P}^n$ and $\mathbf{P}^{n-1}$ the **hyperplane at infinity**; the coordinates $w = (w_1, \ldots, w_n)$ are called **inhomogeneous coordinates** for $\mathbf{P}^n \setminus \mathbf{P}^{n-1}$.

Let $m < n$ be a positive integer and let

$$L_k(z) = a_{0k} z_0 + a_{1k} z_1 + \cdots + a_{nk} z_n \qquad (k = 1, \ldots, m)$$

be linear functions of $z = [z_0 : z_1 : \ldots : z_n]$. The set H_L of all points $z \in \mathbf{P}^n$ which satisfy the equations $L_1(z) = \cdots = L_m(z) = 0$ is called a **complex hyperplane** in $\mathbf{P}^n$. When the functions $L_k(z)$ $(k = 1, \ldots, m)$ are linearly independent, H_L is $(n-m)$-dimensional and may be considered as an $(n-m)$-dimensional projective space.

Let $A = (a_{ij})$ be an $(n+1, n+1)$ matrix with non-zero determinant. The linear transformation $\Psi([z_0 : z_1 : \ldots : z_n]) = ([w_0 : w_1 : \ldots : w_n])$, where

$$w_j = a_{0j} z_0 + a_{1j} z_1 + \cdots + a_{nj} z_n \qquad (j = 0, 1, \ldots, n),$$

is an analytic bijection between $\mathbf{P}^n$ with homogeneous coordinates z and $\mathbf{P}^n$ with homogeneous coordinates w. We call Ψ a **projective transformation** of $\mathbf{P}^n$. These transformations are transitive on $\mathbf{P}^n$; given any point with homogeneous coordinates $[z_0 : z_1 : \ldots : z_n] \in \mathbf{P}^n$, there exists Φ as above with $\Phi([z_0 : z_1 : \ldots : z_n]) = [1 : 0 : \ldots : 0]$.

1.2. Analytic Functions

1.2.1. Power Series. Fix $a = (a_1, \ldots, a_n) \in \mathbf{C}^n$ and consider a power series centered at a in the n complex variables $z_1, \ldots, z_n$:

$$\mathcal{P}(z) = \sum_{j_1, \ldots, j_n \geq 0} \alpha_{j_1, \ldots, j_n} (z_1 - a_1)^{j_1} \cdots (z_n - a_n)^{j_n}.$$

The set of all points z' in $\mathbf{C}^n$ such that $\mathcal{P}(z)$ converges uniformly in some neighborhood of z' is called the **domain of convergence** of $\mathcal{P}(z)$ and is denoted by $\mathcal{D}_{\mathcal{P}}$. Clearly $\mathcal{D}_{\mathcal{P}}$ is open.

REMARK 1.3. If $n \geq 2$, there may exist points $z' \notin \overline{\mathcal{D}_{\mathcal{P}}}$ for which $\mathcal{P}(z')$ converges. For example, in $\mathbf{C}^2$ with coordinates (z_1, z_2), consider the power series

$$\mathcal{P}(z_1, z_2) = z_1 + z_1 z_2 + z_1 z_2^2 + \cdots$$

centered at $(0, 0)$. Then $\mathcal{D}_{\mathcal{P}}$ is the bidisk $(|z_1| < \infty) \times (|z_2| < 1)$, while $\mathcal{P}(z_1, z_2)$ converges at any point on the complex line $z_1 = 0$.

If the domain of convergence $\mathcal{D}_{\mathcal{P}}$ of a power series $\mathcal{P}(z)$ is not empty, then $\mathcal{P}(z)$ defines a continuous function on $\mathcal{D}_{\mathcal{P}}$ that has partial derivatives $\partial\mathcal{P}/\partial z_j$, $j = 1, \ldots, n$, which are obtained by termwise differentiation of $\mathcal{P}(z)$ with respect to z_j, $j = 1, \ldots, n$. Here we define $\partial/\partial z_j$ in the usual calculus sense; for more on these differential operators, see Remark 1.6 in section 1.3.2. A complex-valued function $f(z)$ defined in a domain D in $\mathbf{C}^n$ is called **analytic in** D if $f(z)$ can be represented by a convergent power series in a neighborhood of each point in D.

Let D be a domain in $\mathbf{C}^n$ and let $a = (a_1, \ldots, a_n) \in \mathbf{C}^n$. If whenever $z' = (z'_1, \ldots, z'_n)$ lies in D, the entire distinguished boundary

$$|z_j - a_j| = |z'_j - a_j| \qquad (j = 1, \ldots, n)$$

of the polydisk $|z_j - a_j| < |z'_j - a_j|$ $(j = 1, \ldots, n)$ is contained in D, then D is called a **Reinhardt domain centered at** a. If, moreover, $z' = (z'_1, \ldots, z'_n) \in D$ implies that the entire closed polydisk

$$|z_j - a_j| \leq |z'_j - a_j| \qquad (j = 1, \ldots, n)$$

is contained in D, then the Reinhardt domain D centered at a is said to be **complete**.

PROPOSITION 1.1. *The domain of convergence $\mathcal{D}_{\mathcal{P}}$ of a power series $\mathcal{P}(z)$ centered at a in $\mathbf{C}^n$ is a complete Reinhardt domain centered at a.*

PROOF. If $z' = (z'_1, \ldots, z'_n) \in \mathcal{D}_{\mathcal{P}}$, then the terms

$$|\alpha_{j_1, \ldots, j_n} (z'_1 - a_1)^{j_1} \cdots (z'_n - a_n)^{j_n}|$$

in the series $\mathcal{P}(z')$ converge to 0 as $j_1 + \ldots + j_n \to \infty$. Hence it suffices to prove that if the terms in the series $\mathcal{P}(z')$ are bounded, then $\mathcal{P}(z)$ converges uniformly on each compact subset of the polydisk

$$\Delta : \ |z_j - a_j| < |z'_j - a_j| \qquad (j = 1, \ldots, n).$$

Thus we assume there exists an $M > 0$ such that

$$|\alpha_{j_1,\dots,j_n}(z'_1 - a_1)^{j_1} \cdots (z'_n - a_n)^{j_n}| \le M$$

for all $j_1, \dots, j_n$. Let $0 < \rho < 1$. In the closed polydisk

$$|z_j - a_j| \le \rho\,|z'_j - a_j| \qquad (j = 1, \dots, n),$$

we have

$$\begin{aligned} &\sum_{j_1,\dots,j_n \ge 0} |\alpha_{j_1,\dots,j_n}(z_1 - a_1)^{j_1} \cdots (z_n - a_n)^{j_n}| \\ &\le M \sum_{j_1,\dots,j_n \ge 0} \rho^{j_1+\dots+j_n} = \frac{M}{(1-\rho)^n}. \end{aligned}$$

Thus $\mathcal{P}(z)$ converges absolutely and uniformly on any compact subset of Δ. Since z' was an arbitrary point of $\mathcal{D}_{\mathcal{P}}$, it follows that $\mathcal{D}_{\mathcal{P}}$ is a complete Reinhardt domain centered at a. □

1.2.2. Associated Multiradius of Convergence. Let $r = (r_1, \dots, r_n)$ be an n-tuple of positive numbers. Let $\mathcal{P}(z)$ be a power series centered at $a = (a_1, \dots, a_n)$ in $\mathbf{C}^n$. If $\mathcal{P}(z)$ is convergent in the polydisk

$$\Delta\ :\ |z_j - a_j| < r_j \qquad (j = 1, \dots, n)$$

and is divergent in the product domain

$$|z_j - a_j| > r_j \qquad (j = 1, \dots, n),$$

then r is called an **associated multiradius of convergence** of $\mathcal{P}(z)$. Note that we make no assumptions for points on the topological boundary of Δ.

An associated multiradius of convergence can be determined by the following formula.

THEOREM 1.1 (Hadamard). *If r is an associated multiradius of convergence of $\mathcal{P}(z)$, then*

$$\overline{\lim}_{j_1+\dots+j_n \to \infty} \sqrt[j_1+\cdots+j_n]{|\alpha_{j_1,\dots,j_n}|\, r_1^{j_1} \cdots r_n^{j_n}} = 1. \tag{1.1}$$

PROOF. Let $r = (r_1, \dots, r_n)$ be an associated multiradius of convergence of $\mathcal{P}(z)$ and let $\rho := \overline{\lim}_{j_1+\dots+j_n \to \infty} \sqrt[j_1+\cdots+j_n]{|\alpha_{j_1,\dots,j_n}|\, r_1^{j_1} \cdots r_n^{j_n}}$. We prove that $\rho = 1$ by contradiction. If $\rho < 1$, fix ρ' with $\rho < \rho' < 1$. Then

$$\sqrt[j_1+\cdots+j_n]{|\ \alpha_{j_1,\dots,j_n}\ |\, r_1^{j_1} \cdots r_n^{j_n}} \le \rho'$$

for all but a finite number of n-tuples $(j_1, \dots, j_n)$. Then for any z' satisfying

$$|\ z'_j - a_j\ | = \frac{r_j}{\rho'} \qquad (j = 1, \dots, n),$$

we have

$$|\alpha_{j_1,\dots,j_n}(z'_1 - a_1)^{j_1} \cdots (z'_n - a_n)^{j_n}| \le 1$$

for all but finitely many terms. Thus, as noted in the proof of Proposition 1.1, $\mathcal{P}(z)$ is convergent in the polydisk $|z_j - a_j| < r_j/\rho'$ $(j = 1, \dots, n)$. Since $0 < \rho' < 1$, this contradicts the fact that r is an associated multiradius of convergence of $\mathcal{P}(z)$.

On the other hand, if $\rho > 1$, fix ρ' with $\rho > \rho' > 1$. Then

$$\sqrt[j_1+\cdots+j_n]{|\alpha_{j_1,\dots,j_n}|\, r_1^{j_1} \cdots r_n^{j_n}} \ge \rho'$$

for infinitely many n-tuples $(j_1, \ldots, j_n)$. Then for any z' satisfying

$$|z'_j - a_j| = \frac{r_j}{\rho'} \qquad (j = 1, \ldots, n),$$

we have

$$|\alpha_{j_1,\ldots,j_n}(z'_1 - a_1)^{j_1} \cdots (z'_n - a_n)^{j_n}| \geq 1$$

for infinitely many terms. Since $\rho' > 1$, this again contradicts our assumption that r is an associated multiradius of convergence of $\mathcal{P}(z)$. □

The theorem implies that the power series $\partial\mathcal{P}/\partial z_j$ obtained by differentiating each term of $\mathcal{P}$ with respect to z_j is a power series centered at a having the same associated multiradius of convergence as $\mathcal{P}$.

REMARK 1.4. We have $z^0 \in \mathcal{D}_{\mathcal{P}}$ if and only if there exist a neighborhood δ of z^0 in $\mathbf{C}^n$ and constants $M > 0$ and $0 < \rho < 1$ such that

$$|\alpha_{j_1,\ldots,j_n}(z_1 - a_1)^{j_1} \cdots (z_n - a_n)^{j_n}| \leq M\rho^{j_1+\ldots+j_n}$$

for all $j = (j_1, \ldots, j_n)$ and $z \in \delta$.

1.2.3. Convexity of Domains of Convergence. Thus far, the theory of power series of several complex variables has not differed significantly from the theory in one complex variable. In this section, we will study a type of convexity occurring in all domains of convergence $\mathcal{D}_{\mathcal{P}}$.

Let D be a complete Reinhardt domain centered at $a = (a_1, \ldots, a_n)$ in $\mathbf{C}^n$ and let $\mathcal{L}(z_1, \ldots, z_n) := (u_1, \ldots, u_n)$, where

$$u_j := \log \mid z_j - a_j \mid \qquad (j = 1, \ldots, n);$$

thus $\mathcal{L}$ is a mapping from $\mathbf{C}^n \setminus \{a\}$ into $\mathbf{R}^n$. We let $\tilde{D}$ denote the image of D under $\mathcal{L}$. If $\tilde{D}$ is geometrically convex as a subset of $\mathbf{R}^n$, we say that D is **logarithmically convex** in $\mathbf{C}^n$.

THEOREM 1.2 (Fabry). *The domain of convergence $\mathcal{D}_{\mathcal{P}}$ of a power series $\mathcal{P}(z)$ centered at a in $\mathbf{C}^n$ is logarithmically convex in $\mathbf{C}^n$.*

PROOF. Let $\widetilde{\mathcal{D}}_{\mathcal{P}} \subset \mathbf{R}^n$ be the image of $\mathcal{D}_{\mathcal{P}}$ under the mapping $\mathcal{L}$. We will use $u_1, \ldots, u_n$ for coordinates in $\mathbf{R}^n$. Associated to each term

$$\alpha_{j_1,\ldots,j_n}(z_1 - a_1)^{j_1} \cdots (z_n - a_n)^{j_n}$$

of the power series $\mathcal{P}(z)$, we let $H_{(j)}$ be the half-space in $\mathbf{R}^n$ defined by

$$H_{(j)} \ : \ j_1 u_1 + \cdots + j_n u_n + \log |\alpha_{j_1,\ldots,j_n}| < 0.$$

A point $u' \in \mathbf{R}^n$ belongs to $\widetilde{\mathcal{D}}_{\mathcal{P}}$ if and only if there is a neighborhood V of u' in $\mathbf{R}^n$ such that $V \subset H_{(j)}$ for all but finitely many $j = (j_1, \ldots, j_n)$ (this follows from Remark 1.4). Since $H_{(j)}$ is a half-space, it follows that if u' and u'' are contained in $\widetilde{\mathcal{D}}_{\mathcal{P}}$, then the segment $[u', u'']$ in $\mathbf{R}^n$ is also contained in $\widetilde{\mathcal{D}}_{\mathcal{P}}$. Thus $\widetilde{\mathcal{D}}_{\mathcal{P}}$ is geometrically convex in $\mathbf{R}^n$. □

This fact was discovered in 1902 by Fabry [**18**]; it shows that the domain of convergence of a power series in several complex variables has very special properties. The theorem implies that the zero set of a holomorphic function of $n \geq 2$ complex variables does not contain isolated points.

EXAMPLE 1.1. In $\mathbf{C}^2$ with coordinates z_1 and z_2, let

$$\mathcal{P}(z_1, z_2) = \sum_{j,k\geq 0} a_{j,k} z_1^j z_2^k$$

be a power series about $(0,0)$. If the domain of convergence $\mathcal{D}_{\mathcal{P}}$ of $\mathcal{P}(z_1, z_2)$ contains the polydisks

$$|z_1| < 1, \quad |z_2| < \infty \qquad \text{and} \qquad |z_1| < \infty, \quad |z_2| < 1,$$

then $\mathcal{D}_{\mathcal{P}}$ is all of $\mathbf{C}^2$.

1.2.4. Estimation of Coefficients. We next study the Cauchy estimates for coefficients of power series. Let

$$\mathcal{P}(z) = \sum_{j_1,\dots,j_n \geq 0} \alpha_{j_1,\dots,j_n} (z_1 - a_1)^{j_1} \cdots (z_n - a_n)^{j_n}$$

be a power series centered at $a \in \mathbf{C}^n$ and let $\mathcal{D}_{\mathcal{P}}$ be the domain of convergence of $\mathcal{P}(z)$. Let

$$\overline{\Delta} : |z_j - a_j| \leq r_j \qquad (j = 1, \dots, n)$$

be a closed polydisk contained in $\mathcal{D}_{\mathcal{P}}$, and fix $M > 0$ such that

$$|\mathcal{P}(z)| \leq M \quad \text{in } \overline{\Delta}.$$

THEOREM 1.3 (Cauchy Estimates). *The coefficients of $\mathcal{P}(z)$ satisfy*

$$|\alpha_{j_1,\dots,j_n}| \leq \frac{M}{r_1^{j_1} \cdots r_n^{j_n}}.$$

PROOF. Let $\nu = (\nu_1, \dots, \nu_n)$ be an n-tuple of integers. Then

$$\int_0^{2\pi} \cdots \int_0^{2\pi} (e^{i\theta_1})^{\nu_1} \cdots (e^{i\theta_n})^{\nu_n} \, d\theta_1 \cdots d\theta_n = \begin{cases} 0, & \nu \neq (0, \dots, 0), \\ (2\pi)^n, & \nu = (0, \dots, 0), \end{cases}$$

where $i^2 = -1$. We let $\mathcal{E}$ denote the distinguished boundary of Δ, i.e., $\mathcal{E} : |z_j - a_j| = r_j$ $(j = 1, \dots, n)$, and we form the integral

$$I = \int \cdots \int_{\mathcal{E}} \frac{\mathcal{P}(z)}{(z_1 - a_1)^{j_1+1} \cdots (z_n - a_n)^{j_n+1}} \, dz_1 \cdots dz_n.$$

By integrating term by term, we obtain

$$I = (2\pi)^n \alpha_{j_1,\dots,j_n}.$$

On the other hand, standard estimates for the integral yield

$$| I | \leq \frac{(2\pi)^n M}{r_1^{j_1} \cdots r_n^{j_n}},$$

and the result follows. □

COROLLARY 1.1. *If two power series $\mathcal{P}_1(z)$ and $\mathcal{P}_2(z)$ centered at a in $\mathbf{C}^n$ agree in a neighborhood of a, then they are identical.*

1.3. Holomorphic Functions

1.3.1. Definition. Let $f(z)$ be a complex-valued function defined on a domain D in $\mathbf{C}^n$. If $f(z)$ satisfies the following two conditions:

1. $f(z)$ is continuous in D, and
2. $f(z)$ has partial derivatives $\partial f/\partial z_j$ $(j = 1, \ldots, n)$ in D,

then we say that $f(z)$ is a **holomorphic function on** D. For a closed subset E of $\mathbf{C}^n$, we say $f(z)$ is **holomorphic on** E if $f(z)$ is holomorphic in a neighborhood of E. In particular, we often use the terminology that $f(z)$ is **holomorphic at a point** a if $f(z)$ is defined and is holomorphic in a neighborhood of a in $\mathbf{C}^n$.

By this definition, a holomorphic function $f(z)$ is necessarily holomorphic in each variable z_j $(j = 1, \ldots, n)$ separately. Thus, many propeties for holomorphic functions of one complex variable remain valid for holomorphic functions of several complex variables.

One of the most important properties is the Cauchy integral representation of holomorphic functions on polydisks. Let D be a domain in $\mathbf{C}^n$ and let $a = (a_1, \ldots, a_n)$ be a point in D. Let

$$\overline{\Delta} \, : \, | \, z_j - a_j \, | \leq r_j \qquad (j = 1, \ldots, n)$$

be a closed polydisk centered at a which is contained in D; as usual we let $\mathcal{E}$ be the distinguished boundary of Δ.

THEOREM 1.4 (Cauchy Integral Formula). *If $f(z)$ is holomorphic on D, then $f(z)$ has the following integral representation in Δ:*

$$f(z_1, \ldots, z_n) = \frac{1}{(2\pi i)^n} \int \cdots \int_{\mathcal{E}} \frac{f(\zeta_1, \ldots, \zeta_n)}{(\zeta_1 - z_1) \cdots (\zeta_n - z_n)} \, d\zeta_1 \cdots d\zeta_n.$$

PROOF. The proof is by induction on the dimension n. For $n = 1$ this is the classical Cauchy integral formula. We now assume the result is true in dimension $n - 1$. Fix any point $z = (z_1, \ldots, z_n)$ in Δ. Since $f(z)$ is holomorphic in the complex variable z_1, we have from the one-variable case that

$$f(z_1, \ldots, z_n) = \frac{1}{2\pi i} \int_{\mathcal{E}_1} \frac{f(\zeta_1, z_2, \ldots, z_n)}{\zeta_1 - z_1} \, d\zeta_1, \tag{1.2}$$

where $\mathcal{E}_1 = \{|z_1 - a_1| = r_1\}$. Now fix any point ζ_1 on the circle $\mathcal{E}_1$. Then $f(\zeta_1, z_2, \ldots, z_n)$ is a holomorphic function of the $n-1$ complex variables $z_2, \ldots, z_n$ on the closed polydisk $\overline{\Delta}' : |z_j - a_j| \leq r_j$ $(j = 2, \ldots, n)$ in $\mathbf{C}^{n-1}$. It follows from the inductive hypothesis that

$$f(\zeta_1, z_2, \ldots, z_n) = \frac{1}{(2\pi i)^{n-1}} \int \cdots \int_{\mathcal{E}'} \frac{f(\zeta_1, \zeta_2, \ldots, \zeta_n)}{(\zeta_2 - z_2) \cdots (\zeta_n - z_n)} \, d\zeta_2 \cdots d\zeta_n,$$

where $\mathcal{E}'$ denotes the distinguished boundary of Δ'. We substitute this formula into (1.2) to obtain an iterated integral. Since $f(z)$ is continuous in D, the iterated integral can be replaced by the desired integral formula. □

REMARK 1.5. This proof also gives a Cauchy integral formula for holomorphic functions $f(z)$ in D when the polydisk is replaced by any product domain $\Delta = \Delta_1 \times \ldots \times \Delta_n$ contained in D having boundary component sets $\partial \Delta_j$ which consist of smooth curves in the plane $\mathbf{C}_{z_j}$. Here, the integration takes place over the n real-dimensional set $\partial \Delta_1 \times \cdots \times \partial \Delta_n$.

It follows from Theorem 1.4 that any holomorphic function $f(z)$ in D has partial derivatives of all orders with respect to each variable z_j $(j = 1, \dots, n)$ at any point of D and the resulting functions are also holomorphic in D. Furthermore, in any closed polydisk $\overline{\Delta} \subset D$ centered at a point a in D, $f(z)$ can be expanded into an absolutely and uniformly convergent power series $\mathcal{P}(z)$; hence $f(z)$ is analytic in D. Thus the holomorphic functions of several complex variables are also analytic, just as in the case of one complex variable. By the Cauchy integral formula, we can write any partial derivative

$$\frac{\partial^{j_1+\cdots+j_n} f}{\partial z_1^{j_1} \cdots \partial z_n^{j_n}}(z)$$

of $f(z)$ as

$$\frac{j_1! \cdots j_n!}{(2\pi i)^n} \int \cdots \int_{\mathcal{E}} \frac{f(\zeta_1, \dots, \zeta_n)}{(\zeta_1 - z_1)^{j_1+1} \cdots (\zeta_n - z_n)^{j_n+1}} \, d\zeta_1 \cdots d\zeta_n.$$

It follows that if we write

$$f(z) = \mathcal{P}(z) = \sum_{j_1, \dots, j_n \geq 0} \alpha_{j_1, \dots, j_n} (z_1 - a_1)^{j_1} \cdots (z_n - a_n)^{j_n}$$

in Δ, then the coefficient $\alpha_{j_1, \dots, j_n}$ is given by

$$\alpha_{j_1, \dots, j_n} = \frac{1}{(2\pi i)^n} \int \cdots \int_{\mathcal{E}} \frac{f(\zeta_1, \dots, \zeta_n)}{(\zeta_1 - a_1)^{j_1+1} \cdots (\zeta_n - a_n)^{j_n+1}} \, d\zeta_1 \cdots d\zeta_n.$$

1.3.2. Cauchy-Riemann Equations. In this section we study the real and imaginary parts of a holomorphic function of n complex variables $z = (z_1, \dots, z_n)$. We write

$$z_j = x_j + i y_j \qquad (i^2 = -1;\ j = 1, \dots, n),$$

where x_j and y_j are real numbers. For a holomorphic function $f(z)$, we set

$$f(z) = u(x, y) + i v(x, y),$$

where $u(x, y)$ and $v(x, y)$ are the real and imaginary parts of $f(z)$; $x = (x_1, \dots, x_n)$ and $y = (y_1, \dots, y_n)$. From the one-variable Cauchy-Riemann equations for each z_j $(j = 1, \dots, n)$, we have

$$\frac{\partial u}{\partial x_j} = \frac{\partial v}{\partial y_j}, \quad \frac{\partial u}{\partial y_j} = -\frac{\partial v}{\partial x_j} \qquad (j = 1, \dots, n). \tag{1.3}$$

By differentiating these equations with respect to x_k and y_k, we see that both the real and the imaginary parts of a holomorphic function satisfy the following system of partial differential equations of second order:

$$\frac{\partial^2 \varphi}{\partial x_j \partial x_k} + \frac{\partial^2 \varphi}{\partial y_j \partial y_k} = 0, \quad \frac{\partial^2 \varphi}{\partial x_j \partial y_k} - \frac{\partial^2 \varphi}{\partial x_k \partial y_j} = 0 \qquad (j, k = 1, \dots, n). \tag{1.4}$$

A function $\varphi(x, y)$ satisfying (1.4) is called **pluriharmonic**. If u and v satisfy (1.3), we call v a **pluriharmonic conjugate** of u.

In general, a real- or complex-valued function $\varphi(x, y)$ defined on a domain D in $\mathbf{C}^n$ is said to be of **class** C^2 if it is of class C^2 with respect to the $2n$ real variables x_j and y_j.

For a real-valued function $u(x, y)$ of class C^2 on a domain D in $\mathbf{C}^n$, we define the 1-form

$$\omega = -\sum_{j=1}^{n} \frac{\partial u}{\partial y_j} dx_j + \sum_{j=1}^{n} \frac{\partial u}{\partial x_j} dy_j.$$

Condition (1.4) for the function u is equivalent to the condition that ω is locally exact in D. In this case, if we take v with $dv = \omega$, then u and v satisfy condition (1.3). If D is simply connected and we set

$$f(z) = u(x, y) + iv(x, y),$$

then $f(z)$ is a holomorphic function of $z_1, \dots, z_n$.

REMARK 1.6. For a complex variable $z_j = x_j + iy_j$, we define

$$\frac{\partial}{\partial z_j} = \frac{1}{2}\left(\frac{\partial}{\partial x_j} - i\frac{\partial}{\partial y_j}\right), \qquad \frac{\partial}{\partial \overline{z}_j} = \frac{1}{2}\left(\frac{\partial}{\partial x_j} + i\frac{\partial}{\partial y_j}\right).$$

Then condition (1.3) that the complex-valued function $f(z_1, \dots, z_n)$ is differentiable with respect to the variable z_j becomes

$$\frac{\partial f}{\partial \overline{z}_j} = 0.$$

Similarly, condition (1.4) that a real-valued function $u(z_1, \dots, z_n)$ is pluriharmonic becomes

$$\frac{\partial^2 u}{\partial z_j \partial \overline{z}_k} = 0 \qquad (j, k = 1, \dots, n).$$

We will use these conditions for the rest of the book.

1.3.3. Pluriharmonic Functions. From the definition in the previous section, it follows that a pluriharmonic function $u(z)$ is a harmonic function with respect to the $2n$ real variables $x_1, \dots, x_n$ and $y_1, \dots, y_n$, namely, $\sum_{j=1}^{n}(\partial^2 u/\partial x_j^2 + \partial^2 u/\partial y_j^2) = 0$. Moreover, such a function is harmonic with respect to each complex variable $z_j = x_j + iy_j$:

$$\frac{\partial^2 u}{\partial x_j^2} + \frac{\partial^2 u}{\partial y_j^2} = 0 \qquad (j = 1, \dots, n).$$

Indeed, the following stronger condition is valid.

PROPOSITION 1.2. *Let $\varphi(z)$ be a real-valued function of class C^2 in a domain $D \subset \mathbf{C}^n$. Then $\varphi(z)$ is pluriharmonic in D if and only if for any complex line L, the restriction of $\varphi(z)$ to $L \cap D$ is harmonic as a function of one complex variable on each component of $L \cap D$.*

PROOF. Let $L : t \to ct + b$ be a complex line which passes though a point $b = (b_1, \dots, b_n) \in D$ and has a direction given by $c = (c_1, \dots, c_n) \in \mathbf{C}^n \setminus \{0\}$. For any $t \in \mathbf{C}$ such that $ct + b \in D$, we set

$$\Phi(t) := \varphi(c_1 t + b_1, \dots, c_n t + b_n).$$

Then

$$\frac{\partial^2 \Phi}{\partial t \partial \overline{t}}(t) = \sum_{j,k=1}^{n} \frac{\partial^2 \varphi}{\partial z_j \partial \overline{z}_k}(ct + b) c_j \overline{c}_k,$$

and the result follows. □

As with Cauchy's integral formula, Poisson's formula in one complex variable can be generalized to the case of n complex variables. Let D be a domain in $\mathbf{C}^n$ and fix $a = (a_1, \dots, a_n) \in D$. Let $\overline{\Delta}$ be a closed polydisk centered at a and contained in D:

$$\overline{\Delta} \,:\, |z_j - a_j| \leq r_j \qquad (j = 1, \dots, n).$$

We set

$$P_j(r_j, \rho_j, \theta_j, \vartheta_j) = \frac{r_j^2 - \rho_j^2}{r_j^2 + \rho_j^2 - 2r_j\rho_j\cos(\theta_j - \vartheta_j)},$$

the one-variable Poisson kernel for $|z_j - a_j| < r_j$, where $z_j = a_j + \rho_j e^{i\vartheta_j}$, and define

$$P(r, \rho, \theta, \vartheta) = \prod_{j=1}^{n} P_j(r_j, \rho_j, \theta_j, \vartheta_j),$$

the **Poisson kernel** for $\Delta \subset \mathbf{C}^n$. For any $z = (z_1, \dots, z_n)$ in Δ; i.e., $z_j = a_j + \rho_j e^{i\vartheta_j}$ with $\rho_j < r_j$, a pluriharmonic function $\varphi(z)$ in D can be represented at z using the **Poisson formula**

$$\varphi(z_1, \dots, z_n) = \frac{1}{(2\pi)^n}\int_0^{2\pi} \cdots \int_0^{2\pi} P(r, \rho, \theta, \vartheta)\, \varphi(re^{i\theta} + a)\, d\theta_1 \cdots d\theta_n, \tag{1.5}$$

where we use the notation $re^{i\theta} + a = (r_1 e^{i\theta_1} + a_1, \dots, r_n e^{i\theta_n} + a_n)$.

REMARK 1.7. Poisson's formula (1.5) is valid for any C^2 function $\varphi(z)$ which is harmonic in each complex variable z_j $(j = 1, \dots, n)$. However, a function which is harmonic in each complex variable z_j $(j = 1, \dots, n)$ is not necessarily pluriharmonic in the n complex variables $z = (z_1, \dots, z_n)$. For example, in $\mathbf{C}^2$ with coordinates $z_1 = x_1 + iy_1$, $z_2 = x_2 + iy_2$, consider the function

$$\varphi(z_1, z_2) = x_1 y_1 x_2 y_2.$$

This function is harmonic in each complex variable z_1 and z_2, but it is not pluriharmonic in $z = (z_1, z_2)$.

If $\varphi(\zeta)$ is any real-valued function of class C^2 on the distinguished boundary of the polydisk Δ in $\mathbf{C}^n$, $n \geq 2$, the function $\varphi(z)$ in Δ defined by the Poisson integral formula (1.5) is harmonic in each complex variable z_j but is not necessarily pluriharmonic in $z = (z_1, \dots, z_n)$.

1.3.4. Elementary Properties of Holomorphic Functions. We list some elementary properties of holomorphic functions of several complex variables which are proved by the same methods as in the case of one complex variable.

1. Liouville's theorem. Let $f(z)$ be an entire function in $\mathbf{C}^n$; i.e., a holomorphic function in all of $\mathbf{C}^n$. If $|f(z)|$ is bounded in $\mathbf{C}^n$, then $f(z)$ is a constant in $\mathbf{C}^n$. More generally, let $\Delta_r \,:\, |z_j| < r$ $(j = 1, \dots, n)$ and let $M(r) = \mathrm{Max}\{|f(z)| \mid z \in \overline{\Delta_r}\}$. If there exists an integer $\nu \geq 1$ such that

$$\lim_{r \to \infty} M(r)/r^\nu = 0,$$

then $f(z)$ is a polynomial of degree at most $\nu - 1$ in $\mathbf{C}^n$.

Contrary to the case of one complex variable, there exist domains D in $\mathbf{C}^n$, $n > 1$, with $\mathbf{C}^n \setminus D$ having non-empty interior but such that every bounded holomorphic function in D is constant. For example, we will soon see (as a consequence of Osgood's theorem in section 1.5.2) that the complement of a ball has this property.

2. Identity theorem. Let $f(z)$ and $g(z)$ be holomorphic functions in a domain D in $\mathbf{C}^n$. If $f(z) = g(z)$ for all z in a non-empty open set δ in D, then $f(z) \equiv g(z)$ in D. Hence, analytic continuation of holomorphic functions in several complex variables can be performed as in the case of one complex variable.

Contrary to the case of one complex variable, the zero set of a holomorphic function in a domain $D \subset \mathbf{C}^n$, $n \geq 2$, contains no isolated points. Thus, even if $f(z) = g(z)$ in a set with accumulation points in D, it does not necessarily follow that $f(z) = g(z)$ in D. For example, in $\mathbf{C}^2$ with variables z and w, we can take $f(z, w) = z$ and $g(z, w) = z^2$.

3. Maximum principle. Let $f(z)$ be a holomorphic function in a domain D in $\mathbf{C}^n$. If $|f(z)|$ attains its maximum at a point of D, then $f(z)$ is constant in D.

Contrary to the case of one complex variable, in some domains D in $\mathbf{C}^n$, $n > 1$, there exists a proper closed subset $\boldsymbol{e}$ of ∂D such that any holomorphic function $f(z)$ in D with continuous boundary values attains its maximum modulus at a point of $\boldsymbol{e}$. Given $D \subset \mathbf{C}^n$, the smallest set $\boldsymbol{e} \subset \partial D$ with this property is called the **Shilov boundary of** D. For example, the Shilov boundary of a polydisk $|z_j| < r_j$ $(j = 1, \ldots, n)$ is the distinguished boundary $|z_j| = r_j$ $(j = 1, \ldots, n)$; on the other hand, the Shilov boundary of an open ball B is the topological boundary, the sphere ∂B.

4. Weierstrass' theorem. Let $\{f_j\}_{j=1,2,\ldots}$ be a sequence of holomorphic functions in a domain D in $\mathbf{C}^n$. If $\{f_j\}$ converges uniformly on each compact set in D, then the limit function $f(z)$ is a holomorphic function in D.

Let $\{f_j\}$ be a sequence of holomorphic functions in D which are uniformly bounded in D; i.e., there exists $M > 0$ such that $|f_j(z)| \leq M$ $(j = 1, 2, \ldots)$ in D. Then Stieltjes' theorem holds: if $\{f_j\}$ converges uniformly on a non-empty open set δ in D, then $\{f_j\}$ converges uniformly on each compact set in D. However, Vitali's theorem does not necessarily hold: if we replace δ by a set with accumulation points in D, $\{f_j\}$ might not converge uniformly on each compact set in D. For example, take D to be the unit bidisk centered at the origin in $\mathbf{C}^2$ with variables z and w, and take $f_j(z, w) := (-1)^j z^j$, $j = 1, 2, \ldots$.

5. Montel's theorem. Let $\mathcal{F}$ be a family of holomorphic functions in D. Assume that there exists an $M > 0$ such that $|f(z)| \leq M$ in D for all $f \in \mathcal{F}$. Then $\mathcal{F}$ is uniformly equicontinuous in D and hence is a normal family. By Picard's theorem we can replace the uniform boundedness by the condition that there exist two different complex values a and b such that each $f \in \mathcal{F}$ omits the values a and b in D.

6. Rado's theorem. Let $f(z)$ be a complex-valued continuous function in D and let e be the zero set of $f(z)$, i.e., $e = \{z \in D | f(z) = 0\}$. If $f(z)$ is holomorphic in $D \setminus e$, then $f(z)$ is holomorphic in all of D.

REMARK 1.8. Although Rado's theorem is important in the theory of functions in one and several complex variables, its proof is not often given in standard textbooks. Below we give the proof in the case where D is the unit disk in one complex variable.

PROOF. Let $\Delta : |z| < 1$ in $\mathbf{C}$, and let $f(z) \not\equiv 0$ be a continuous function in $\overline{\Delta}$ with $|f(z)| < 1$. We let e denote the zero set of $f(z)$ in Δ, and we let ω denote the interior of $\Delta \setminus e$. Let $u(z) = \Re f(z)$ on Δ. We form the harmonic function $\widehat{u}(z)$ on Δ, where $\widehat{u}(z) = u(z)$ on $\partial\Delta$, by use of the Poisson integral formula. It follows from the maximum principle for harmonic functions that for any $\gamma > 0$,

$$\gamma \log |f(z)| \le \widehat{u}(z) - u(z) \le -\gamma \log |f(z)|, \quad z \in \Delta.$$

Hence $\widehat{u}(z) = u(z)$ on ω. Similarly, we have $\widehat{v}(z) = v(z)$ in ω for $v(z) = \Im f(z)$. We set $\widehat{f}(z) := \widehat{u}(z) + i\widehat{v}(z)$ in Δ. Since $\widehat{f}(z) = f(z)$ in ω, it follows that $\widehat{f}(z)$ is holomorphic in Δ. On the other hand, both $\widehat{f}(z)$ and $f(z)$ are continuous on Δ and the zero set of $\widehat{f}(z)$ is isolated in Δ; hence, e is isolated and $\widehat{f}(z) \equiv f(z)$ in Δ. □

1.3.5. Holomorphic Mappings. We let $z = (z_1, \dots, z_n)$ denote the variables in $\mathbf{C}^n$ and $w = (w_1, \dots, w_m)$ those for $\mathbf{C}^m$. Let $D \subset \mathbf{C}^n$ be a domain and let $f_k(z)$ $(k = 1, \dots, m)$ be holomorphic functions in D. We call

$$T : w_k = f_k(z) \qquad (k = 1, \dots, m)$$

a **holomorphic mapping from** D **into** $\mathbf{C}^m$. If $T(D) \subset D' \subset \mathbf{C}^m$, then T is called a **holomorphic mapping from** D **into** D'.

Let $T : w_k = f_k(z)$ $(k = 1, \dots, m)$ be a holomorphic mapping from D into D', and let $g(w_1, \dots, w_m)$ be a holomorphic function in D'. Then

$$G(z) = g(f_1(z), \dots, f_m(z))$$

is a holomorphic function in D which satisfies

$$\frac{\partial G}{\partial z_j} = \frac{\partial g}{\partial w_1}\frac{\partial f_1}{\partial z_j} + \cdots + \frac{\partial g}{\partial w_n}\frac{\partial f_m}{\partial z_j} \qquad (j = 1, \dots, m).$$

For a holomorphic mapping $T : w_k = f_k(z)$ $(k = 1, \dots, m)$, we call the matrix

$$\frac{\partial(f_1, \dots, f_m)}{\partial(z_1, \dots, z_n)} = \left(\frac{\partial f_k}{\partial z_j}\right) \qquad (j = 1, \dots, n;\ k = 1, \dots, m)$$

the (complex) **Jacobian matrix of** T. In the case $m = n$, the determinant

$$\left|\frac{\partial(f_1, \dots, f_n)}{\partial(z_1, \dots, z_n)}\right| = \left|\left(\frac{\partial f_k}{\partial z_j}\right)\right| \qquad (j, k = 1, \dots, n)$$

is called the (complex) **Jacobian determinant of** T.

Let $T_1 : w_k = f_k(z)$ $(k = 1, \dots, m)$ be a holomorphic mapping from $D_1 \subset \mathbf{C}^n$ into $D_2 \subset \mathbf{C}^m$ and let $T_2 : v_k = g_k(w)$ $(k = 1, \dots, l)$ be a holomorphic mapping from D_2 into $D_3 \subset \mathbf{C}^l$. Then the composition $T = T_2 \circ T_1$ is a holomorphic mapping from D_1 into D_3. If we write $T : v_k = h_k(z)$ $(k = 1, \dots, l)$, then we have

$$\frac{\partial(h_1, \dots, h_l)}{\partial(z_1, \dots, z_n)} = \frac{\partial(g_1, \dots, g_l)}{\partial(w_1, \dots, w_m)} \cdot \frac{\partial(f_1, \dots, f_m)}{\partial(z_1, \dots, z_n)}. \tag{1.6}$$

In the case $n = m = l$,

$$\left|\frac{\partial(h_1, \dots, h_n)}{\partial(z_1, \dots, z_n)}\right| = \left|\frac{\partial(g_1, \dots, g_n)}{\partial(w_1, \dots, w_n)}\right| \cdot \left|\frac{\partial(f_1, \dots, f_n)}{\partial(z_1, \dots, z_n)}\right|.$$

We prove the following.

PROPOSITION 1.3. *Let* $T : w_k = f_k(z)$ $(k = 1, \dots, n)$ *be a holomorphic mapping from* $D \subset \mathbf{C}^n$ *into* $\mathbf{C}^n$. *Suppose there exist* $z_0 \in D$ *and* $w_0 = T(z_0)$ *such that*

$$\left|\frac{\partial(f_1, \dots, f_n)}{\partial(z_1, \dots, z_n)}\right| \ne 0 \quad \text{at } z = z_0.$$

Then T is a bijection from a neighborhood δ of z_0 onto a neighborhood δ' of w_0, and the inverse mapping T^{-1} is a holomorphic mapping from δ' onto δ.

We call T a **biholomorphic mapping** between δ and δ', and we say that δ and δ' are **biholomorphically equivalent**.

PROOF of Proposition 1.3. Let $z_j := x_j + iy_j$ and $w_k := u_k + iv_k$. From the Cauchy-Riemann equations,

$$\begin{aligned}\frac{\partial(u_1, v_1, \dots, u_n, v_n)}{\partial(x_1, y_1, \dots, x_n, y_n)} &= \frac{\partial(f_1, \overline{f}_1, \dots, f_n, \overline{f}_n)}{\partial(z_1, \overline{z}_1, \dots, z_n, \overline{z}_n)} \\ &= \frac{\partial(f_1, \dots, f_n, \overline{f}_1, \dots, \overline{f}_n)}{\partial(z_1, \dots, z_n, \overline{z}_1, \dots, \overline{z}_n)} = \left|\frac{\partial(f_1, \dots, f_n)}{\partial(z_1, \dots, z_n)}\right|^2 \neq 0 \quad \text{at} \quad z = z_0.\end{aligned}$$

It follows that T is a bijection from a neighborhood δ of z_0 onto a neighborhood δ' of w_0.

We write $T^{-1}: \; z_j = g_j(w_1, \overline{w}_1, \dots, w_n, \overline{w}_n)$ $(j = 1, \dots, n)$, so that

$$z_j = g_j(f_1(z), \overline{f_1(z)}, \dots, f_n(z), \overline{f_n(z)}) \qquad (j = 1, \dots, n).$$

Thus, for each j, k $(j, k = 1, \dots, n)$,

$$0 = \frac{\partial z_j}{\partial \overline{z}_k} = \frac{\partial g_j}{\partial \overline{w}_1}\overline{\left(\frac{\partial f_1}{\partial z_k}\right)} + \cdots + \frac{\partial g_j}{\partial \overline{w}_n}\overline{\left(\frac{\partial f_n}{\partial z_k}\right)} \quad \text{in } \delta.$$

By taking a smaller neighborhood δ of z_0 if necessary, we may assume

$$\frac{\partial(f_1, \dots, f_n)}{\partial(z_1, \dots, z_n)} \neq 0 \quad \text{in} \quad \delta.$$

Then we have $\partial g_j / \partial \overline{w}_k = 0$ $(k = 1, \dots, n)$ in δ'. Hence $g_j(w)$ $(j = 1, \dots, n)$ are holomorphic functions in δ'. □

The converse of Proposition 1.3 is also valid; this may be seen using (1.6): if T is a biholomorphic mapping from $\delta \subset \mathbf{C}^n$ onto $\delta' \subset \mathbf{C}^n$, then the Jacobian determinant of T does not vanish in δ. Indeed, using arguments from the next chapter, we will see that the conclusion is true without the assumption that T^{-1} is a holomorphic mapping (see Remark 2.8).

1.3.6. Plurisubharmonic Functions. In the theory of functions of one complex variable, the study of both harmonic and subharmonic functions is important. In the theory of functions of several complex variables, the study of plurisubharmonic and pluriharmonic functions plays a much more important role than the study of subharmonic and harmonic functions in the underlying $2n$ real variables.

Let $\varphi(z)$ be an uppersemicontinuous function defined on a domain D in $\mathbf{C}^n$ with $-\infty \leq \varphi(z) < +\infty$. If the restriction $\varphi_{|L \cap D}$ of $\varphi(z)$ to any complex line L in D is a subharmonic function of one complex variable on each component of $L \cap D$, then $\varphi(z)$ is called **plurisubharmonic** in D. For convenience, the function $\varphi(z) \equiv -\infty$ on D is considered to be plurisubharmonic in D.

If $-\varphi(z)$ is plurisubharmonic in D, $\varphi(z)$ is called **plurisuperharmonic** in D. If both $\varphi(z)$ and $-\varphi(z)$ are plurisubharmonic in D, then $\varphi(z)$ is pluriharmonic in D. This is clear from Proposition 1.2 if $\varphi(z)$ is of class C^2 in D.

If $f(z)$ is holomorphic in D, then $|f(z)|$ and $\log |f(z)|$ are plurisubharmonic in D.

For functions of class C^2 in D, we have the following criterion for plurisubharmonicity.

PROPOSITION 1.4. *Let $\varphi(z)$ be a real-valued function of class C^2 on a domain D in $\mathbf{C}^n$. Then $\varphi(z)$ is plurisubharmonic in D if and only if the* **complex Hessian matrix** *of $\varphi(z)$,*

$$\left(\frac{\partial^2 \varphi}{\partial z_j \partial \overline{z}_k}\right)_{j,k=1,\dots,n}, \tag{1.7}$$

is positive semidefinite at each point z of D.

PROOF. Fix $z = (z_1, \dots, z_n) \in D$ and let $c = (c_1, \dots, c_n) \in \mathbf{C}^n$ satisfy $\|c\|^2 := |c_1|^2 + \cdots + |c_n|^2 = 1$. For $t \in \mathbf{C}$ with $|t| \ll 1$, we let

$$\Phi(t) := \varphi(c_1 t + z_1, \dots, c_n t + z_n)$$

be the restriction of φ to a small disk centered at z and in the direction of c. Then

$$\frac{\partial^2 \Phi}{\partial t \partial \overline{t}}(0) = \sum_{j,k=1}^{n} \frac{\partial^2 \varphi}{\partial z_j \partial \overline{z}_k}(z)\, c_j \overline{c}_k,$$

which proves the proposition. □

If the complex Hessian matrix (1.7) of $\varphi(z)$ is positive definite at $z_0 \in D$, then $\varphi(z)$ is said to be **strictly plurisubharmonic** at z_0. If $\varphi(z)$ is strictly plurisubharmonic at all points of D, we call $\varphi(z)$ a **strictly plurisubharmonic function** in D.

The following properties of plurisubharmonic functions follow from the analogous properties of subharmonic functions of one complex variable.

1. Let $\varphi(z)$ be a plurisubharmonic function in D. Let $\overline{\Delta} : |z_j - a_j| \le r_j$ $(j = 1, \dots, n)$ be a closed polydisk in D and let $P(r, \rho, \theta, \vartheta) = \prod_{j=1,\dots,n} P_j(r_j, \rho_j, \theta_j, \vartheta_j)$ be the Poisson kernel for Δ, where $z_j = a_j + \rho_j e^{i\vartheta_j}$ and

$$P_j(r_j, \rho_j, \theta_j, \vartheta_j) := \frac{r_j^2 - \rho_j^2}{r_j^2 + \rho_j^2 - 2 r_j \rho_j \cos(\theta_j - \vartheta_j)}.$$

From the subharmonicity in each variable, we obtain

$$\begin{aligned}&\varphi(z_1, \dots, z_n)\\ &\quad\le \frac{1}{(2\pi)^n} \int_0^{2\pi} \cdots \int_0^{2\pi} P(r, \rho, \theta, \vartheta)\, \varphi(a_1 + r_1 e^{i\theta_1}, \dots, a_n + r_n e^{i\theta_n})\, d\theta_1 \cdots d\theta_n.\end{aligned}$$

In particular, setting $\rho_j = 0$ $(j = 1, \dots, n)$, we have

$$\varphi(a) \le \frac{1}{(2\pi)^n} \int_0^{2\pi} \cdots \int_0^{2\pi} \varphi(a_1 + r_1 e^{i\theta_1}, \dots, a_n + r_n e^{i\theta_n}) d\theta_1 \cdots d\theta_n.$$

Multiplying each side of this inequality by $r_1 \cdots r_n$ and integrating from $r_j = 0$ to r_j $(j = 1, \dots, n)$, we obtain

$$\varphi(a) \le \frac{1}{V} \int \cdots \int_{\Delta} \varphi(z_1, \dots, z_n) dv,$$

where V is the Euclidean volume of Δ and dv denotes the volume element in $\mathbf{C}^n$.

2. If $\varphi_1(z)$ and $\varphi_2(z)$ are plurisubharmonic in D, then so is

$$\varphi(z) = \max(\varphi_1(z), \varphi_2(z)).$$

Furthermore, let $\{\varphi_\iota\}_{\iota\in I}$ be a family of plurisubharmonic functions in D which are locally uniformly bounded above. Then the uppersemicontinuous regularization

$$\Phi(z) := \overline{\lim_{z'\to z}}\,[\sup_{\iota\in I}\varphi_\iota(z')]$$

of the upper envelope $\sup_{\iota\in I}\varphi_\iota(z)$ is plurisubharmonic in D.

3. Let $\varphi(z)$ be plurisubharmonic in D and let $\xi(t)$ be a (real-valued) convex increasing function on $-\infty < t < \infty$. Then $\Psi(z) := \xi(\varphi(z))$ is plurisubharmonic in D.

4. Let $\{\varphi_n\}_{n=1,2,\dots}$ be a sequence of plurisubharmonic functions in D. If $\{\varphi_n\}$ converges uniformly on compact subsets of D, or if $\{\varphi_n\}$ is monotonically decreasing in D, then the limit function is plurisubharmonic in D. This last fact, combined with 2, implies that if $\{\varphi_n\}_{n=1,2,\dots}$ is a sequence of plurisubharmonic functions in D which are locally bounded above, then

$$\overline{\lim_{z'\to z}}\,[\overline{\lim_{n\to\infty}}\,\varphi_n(z')], \qquad z\in D,$$

is a plurisubharmonic function in D.

5. (Invariance under holomorphic mappings) Let $\varphi(z)$ be plurisubharmonic in D and let

$$T:\ z_j = g_j(w) \qquad (j=1,\dots,n)$$

be a holomorphic mapping from a domain D' in $\mathbf{C}^m$ with coordinates $w=(w_1,\dots,w_m)$ into D. Then

$$G(w) := \varphi(g_1(w),\dots,g_n(w))$$

is plurisubharmonic in D'.

1.3.7. Hartogs Series. In this section we describe another type of series representation for holomorphic functions. To simplify the discussion we consider the product space $\mathbf{C}^{n+1} = \mathbf{C}^n\times\mathbf{C}$ of the $n+1$ variables $z_1,\dots,z_n,w$, where $(z_1,\dots,z_n)\in\mathbf{C}^n$ and $w\in\mathbf{C}$. Let D be a domain in $\mathbf{C}^n$ and let a be a point in $\mathbf{C}$. We consider a power series

$$\mathcal{H}(z,w) = \sum_{j=0}^{\infty}\alpha_j(z)(w-a)^j \tag{1.8}$$

in the single variable w centered at a, where the coefficents $\alpha_j(z)$ $(j=0,1,2,\dots)$ are holomorphic functions in D. We call such a power series a **Hartogs series** in w centered at a.

Let $D\subset\mathbf{C}^n$, $\Delta=\{w\in\mathbf{C} : |w-a|<r\}$, and set $G := D\times\Delta$. Then any holomorphic function $f(z,w)$ in G can be represented by a Hartogs series (1.8) in w centered at a. Each coefficient $\alpha_j(z)$ can be obtained as follows: if we fix a radius r_0 $(0<r_0<r)$ and a circle $\gamma_0 :\ |w-a| = r_0$ centered at a, then

$$\alpha_j(z) = \frac{1}{2\pi i}\int_{\gamma_0}\frac{f(z,\zeta)}{(\zeta-a)^{j+1}}\,d\zeta \qquad (j=0,1,2,\dots).$$

Given a Hartogs series $\mathcal{H}(z,w)$, let $\mathcal{D}_{\mathcal{H}}$ be the set of points $(z',w')\in\mathbf{C}^{n+1}$ such that $\mathcal{H}(z,w)$ converges uniformly in a neighborhood of (z',w'). We call $\mathcal{D}_{\mathcal{H}}$ the **domain of convergence** of $\mathcal{H}(z,w)$. As we now show, the domain of convergence of a Hartogs series is convex in a sense similar to the logarithmic convexity of the domain of convergence of a power series.

Let $\mathcal{D}$ be a domain in the product space $\mathbf{C}^{n+1}$ of the n complex variables $z_1, \dots, z_n$ and the one complex variable w. If $(z', w') \in \mathcal{D}$ implies

$$\{z'\} \times \{w \in \mathbf{C} : |w-a| = |w'-a|\} \subset \mathcal{D},$$

then $\mathcal{D}$ is called a **Hartogs domain** centered at a. Moreover, if $(z', w') \in \mathcal{D}$ implies

$$\{z'\} \times \{w \in \mathbf{C} : |w-a| \leq |w'-a|\} \subset \mathcal{D},$$

then the Hartogs domain $\mathcal{D}$ is said to be **complete**. In this case the projection of $\mathcal{D}$ to $\mathbf{C}^n$ is called the **base** of $\mathcal{D}$ and will be denoted by D.

Let $\mathcal{D}$ be a complete Hartogs domain in $\mathbf{C}^{n+1}$ centered at a, and let D be the base of $\mathcal{D}$. For any $z' \in D$, the section $\mathcal{D}(z')$ of $\mathcal{D}$ over $z_j = z'_j$ $(j = 1, \dots, n)$ may be identified with an open disk centered at a of radius $\mathcal{R}(z')$. Thus $\mathcal{R}(z)$ defines a positive-valued function on D (which may attain the value $+\infty$). We call $\mathcal{R}(z)$ the **Hartogs radius** of $\mathcal{D}$ with respect to a. Since $\mathcal{D}$ is open, $\mathcal{R}(z)$ is a lowersemicontinuous function on D.

If the function $-\log \mathcal{R}(z)$ associated to the complete Hartogs domain $\mathcal{D}$ is plurisubharmonic on D, then $\mathcal{D}$ is said to be **logarithmically convex**.

THEOREM 1.5. *Let $\mathcal{H}(z,w)$ be a Hartogs series centered at a such that $\mathcal{H}(z,w)$ is holomorphic for (z,w) in $D \times \Delta \subset \mathbf{C}^n \times \mathbf{C}$, where $D \subset \mathbf{C}^n$ and $\Delta = \{w \in \mathbf{C} : |w-a| < r\}$. Then the domain of convergence $\mathcal{D}_{\mathcal{H}}$ of $\mathcal{H}(z,w)$ is a logarithmically convex and complete Hartogs domain centered at a.*

PROOF. We may assume that $a = 0$ and $\mathcal{H}(z,w) = \sum_{j=0}^{\infty} \alpha_j(z) w^j$. We fix $D' \times \Delta_0 \subset\subset D \times \Delta$, where $D' \subset\subset D$ and $\Delta_0 = \{w \in \mathbf{C} : |w| < r_0\}$ with $r_0 < r$. By our assumption, $\mathcal{H}(z,w)$ is a bounded, holomorphic function on $\overline{D}' \times \overline{\Delta_0}$, so that there exists an $M > 0$ such that $|\alpha_j(z)| \leq M/r_0^j$ for all $j = 0, 1, \dots$ and $z \in D'$. Therefore, $\frac{1}{j} \log |\alpha_j(z)|$ $(j = 0, 1, \dots)$ is a plurisubharmonic function in D' with

$$\frac{1}{j} \log |\alpha_j(z)| \leq \frac{1}{j} \log M - \log r_0 \qquad (j = 0, 1, \dots), \quad z \in D'. \tag{1.9}$$

For $z \in D'$, we let $R(z)$ denote the radius of convergence of the Taylor series $\mathcal{H}(z,w)$ in w, i.e., $1/R(z) = \overline{\lim}_{j \to \infty} \sqrt[j]{|\alpha_j(z)|}$. We set

$$1/\tilde{R}(z) := \overline{\lim_{z' \to z}} \left(\overline{\lim_{j \to \infty}} \sqrt[j]{|\alpha_j(z)|} \right), \qquad z \in D',$$

and

$$\tilde{\mathcal{D}} := \bigcup_{z \in D'} (z, \tilde{\Gamma}(z)), \qquad \text{where } \tilde{\Gamma}(z) = \{w \in \mathbf{C} : |w| < \tilde{R}(z)\}.$$

Using property 4 of plurisubharmonic functions from the last section, under condition (1.9) we see that $-\log \tilde{R}(z)$ is a plurisubharmonic function in D'. It follows that $\tilde{\mathcal{D}}$ is a logarithmically convex and complete Hartogs domain centered at 0.

To prove the theorem, since $D' \subset\subset D$ was arbitrary, it suffices to show that $\tilde{\mathcal{D}} = \mathcal{D}'_{\mathcal{H}} := \mathcal{D}_{\mathcal{H}} \cap (D' \times \mathbf{C})$. Clearly $\mathcal{D}'_{\mathcal{H}}$ is contained in $\tilde{\mathcal{D}}$; thus we fix a point $(z', w') \in \tilde{\mathcal{D}}$ and proceed to show that (z', w') lies in $\mathcal{D}'_{\mathcal{H}}$. Since $\tilde{\mathcal{D}}$ is a complete Hartogs domain centered at $w = 0$, we can find a product domain $\delta' \times \Gamma' \subset\subset \tilde{\mathcal{D}}$, where $\Gamma' = \{w \in \mathbf{C} : |w| < \rho'\}$, which contains the point (z', w'). Clearly $\rho' \leq \tilde{R}(z) \leq R(z)$ for any $z \in \delta'$. Then $\mathcal{H}(z,w)$ is a holomorphic function in $\delta' \times \Delta_0$, and, for any fixed $z \in \delta'$, the radius of convergence of the Taylor series $\mathcal{H}(z,w)$ in w is greater than or equal to ρ'. From Remark 1.11 at the end of section 1.4.3, it follows that $\mathcal{H}(z,w)$ is a holomorphic function in $\delta' \times \Gamma'$; hence (z',w') belongs to $\mathcal{D}'_{\mathcal{H}}$. □

REMARK 1.9. The Hartogs radius $\mathcal{R}(z)$ of the domain of convergence of the Hartogs series $\mathcal{H}(z,w)$ is not always equal to the radius of convergence of the power series $\mathcal{H}(z,w)$ with respect to w for fixed $z \in D$. As an example, in $\mathbf{C}^2$ with variables (z,w), consider the Hartogs series centered at $w=0$ given by

$$\mathcal{H}(z,w) = z + zw + zw^2 + \cdots .$$

Then $\mathcal{D}_{\mathcal{H}} = \mathbf{C} \times \{|w| < 1\}$. Hence $\mathcal{R}(z) \equiv 1$ on $\mathbf{C}$, while the radius of convergence $R(0)$ of $\mathcal{H}(0,w)$ is $+\infty$.

We can also consider a **Hartogs-Laurent series** centered at a in the w-plane; i.e., a Laurent series in w of the form

$$\mathcal{L}(z,w) = \sum_{j=-\infty}^{\infty} \alpha_j(z)(w-a)^j, \tag{1.10}$$

where each $\alpha_j(z)$ is a holomorphic function on a domain D in $\mathbf{C}^n$.

Let D be a domain in $\mathbf{C}^n$ and let

$$\Delta^* \ : \ r_1 < |w-a| < r_2$$

be an annulus centered at a in $\mathbf{C}$. Set $G^* = D \times \Delta^*$. Then any holomorphic function $f(z,w)$ in G^* can be represented by a Hartogs-Laurent series (1.10). Furthermore, if we take a circle $\gamma_0 : \ |w-a| = r_0$ where $r_1 < r_0 < r_2$, then we have

$$\alpha_j(z) = \frac{1}{2\pi i}\int_{\gamma_0} \frac{f(z,\zeta)}{(\zeta-a)^{j+1}}\, d\zeta \qquad (j = 0, \pm 1, \pm 2, \dots).$$

As in the case of Hartogs series, we can define the domain of convergence $\mathcal{D}_{\mathcal{L}}$ of a Hartogs-Laurent series $\mathcal{L}(z,w)$; $\mathcal{D}_{\mathcal{L}}$ is a Hartogs domain centered at a. Given $z' \in D$, the section $\mathcal{D}_{\mathcal{L}}(z')$ over $z_j = z'_j$ $(j = 1, \dots, n)$ of $\mathcal{D}_{\mathcal{L}}$ is an annulus centered at a, which may be the entire w-plane $\mathbf{C}$ or all of $\mathbf{C} \setminus \{a\}$. If we let $\mathcal{R}_e(z)$ $(\mathcal{R}_i(z))$ denote the outer (inner) radius of $\mathcal{D}_{\mathcal{L}}(z)$ for $z \in D$, then $\log \mathcal{R}_i(z)$ and $-\log \mathcal{R}_e(z)$ define plurisubharmonic functions on D.

1.3.8. Riemann's Removable Singularity Theorem. In this last section of 1.3 we discuss Riemann's theorem concerning removable singularities for holomorphic functions. Let $f(z)$ be a non-constant holomorphic function on a domain D and let Σ be the zero set of $f(z)$ in D:

$$\Sigma = \{z \in D \mid f(z) = 0\}.$$

Such sets will be discussed in Chapter 2.

THEOREM 1.6 (Riemann). *Let $g(z)$ be a holomorphic function on $D \setminus \Sigma$. If $g(z)$ is bounded in $D \setminus \Sigma$, then $g(z)$ may be holomorphically extended to all of D.*

PROOF. Fix $a \in \Sigma$. It suffices to prove that $g(z)$ has a holomorphic extension to a neighborhood of a in D. By using a suitable linear transformation in $\mathbf{C}^n$, we may assume that the section $\Sigma' \subset \mathbf{C}_{z_n}$ of Σ over $z_j = a_j$ $(j = 1, \dots, n-1)$ consists of only one point a_n for z_n near a_n in $\mathbf{C}_{z_n}$. Furthermore, since Σ is a closed subset of D, we can find a closed polydisk $\Lambda = \overline{\Delta} \times \Gamma$ in D, where

$$\begin{aligned} \overline{\Delta} &: \quad |z_j - a_j| \le r \qquad (j = 1, \dots, n-1), \\ \Gamma &: \quad |z_n - a_n| \le \rho, \end{aligned}$$

with the property that $(\overline{\Delta} \times \partial\Gamma) \cap \Sigma = \emptyset$. Thus $g(z)$ is holomorphic in a neighborhood of $\overline{\Delta} \times \partial\Gamma$; hence $g(z)$ can be represented by a Hartogs-Laurent series centered at a_n:

$$g(z) = \sum_{j=-\infty}^{\infty} \alpha_j(z_1, \dots, z_{n-1})(z_n - a_n)^j,$$

where $\alpha_j(z_1, \dots, z_{n-1})$ $(j = 0, \pm 1, \pm 2, \dots)$ are holomorphic functions on Δ. To prove the theorem, it suffices to prove that this series reduces to a Hartogs series; i.e.,

$$\alpha_j(z_1, \dots, z_{n-1}) \equiv 0 \quad \text{for } j < 0.$$

To verify this, fix $a' \in \Delta$, and let $\Sigma(a')$ be the section of Σ over $z_j = a'_j$ $(j = 1, \dots, n-1)$. Since $f(a', z_n) \neq 0$ for $z_n \in \partial\Gamma$, we see that $\Gamma \cap \Sigma(a')$ consists of a finite number of points in Γ. Using the fact that $g(a', z_n)$ is bounded and holomorphic as a function of the single variable z_n in $\Gamma \setminus \Sigma(a')$, it follows from Riemann's removable singularity theorem for holomorphic functions of one complex variable that $g(a', z_n)$ extends to be holomorphic in Γ. Thus $\alpha_j(a') = 0$ for $j < 0$, and the theorem is proved. □

1.4. Separate Analyticity Theorem

To obtain Cauchy's integral formula in section 1.3.1 we assumed that a holomorphic function of n complex variables $z = (z_1, \dots, z_n)$ was continuous in z and had first partial derivatives with respect to each variable z_j $(j = 1, \dots, n)$. We now show that the continuity is implied by the existence of these first partial derivatives.

THEOREM 1.7 (Separate Analyticity Theorem). *A complex-valued function of n complex variables $(z_1, \dots, z_n)$ which has first partial derivatives with respect to each variable z_j $(j = 1, \dots, n)$ is holomorphic as a function of n complex variables.*

This theorem was discovered in 1906 by Hartogs. We give the proof by induction on the dimension n, our primary inductive argument. In the case $n = 1$ the theorem is trivial. Thus, assuming the theorem is true for n complex variables, we prove it for $n+1$ complex variables (z, w) in $\mathbf{C}^{n+1}$, where $z \in \mathbf{C}^n$ and $w \in \mathbf{C}$. Since the argument is local, we let Λ be a closed polydisk with center at the origin in $\mathbf{C}^{n+1}$,

$$\begin{aligned} &\Lambda = \overline{\Delta} \times \Gamma, \\ &\overline{\Delta} : \ |z_j| \le r_j \quad (j = 1, \dots, n), \qquad \Gamma : \ |w| \le \rho, \end{aligned}$$

and we assume that $f(z, w)$ is a complex-valued function defined on Λ which has first partial derivatives with respect to each variable $z_1, \dots, z_n$ and w. In this setting we shall show that $f(z, w)$ is a holomorphic function on Λ.

1.4.1. Bounded Case. First we show that $f(z,w)$ is holomorphic in Λ under the assumption that $f(z,w)$ is bounded on Λ; i.e., we assume there exists an $M>0$ such that

$$|f(z,w)| \leq M$$

for $(z,w)\in\Lambda$. Since $f(z,w)$ is holomorphic as a function of $w\in\Gamma$ for each fixed $z\in\overline{\Delta}$, it follows from the Cauchy estimates for one complex variable that if

$$f(z,w)=\sum_{j=0}^{\infty}\alpha_j(z)\,w^j,$$

then

$$|\alpha_j(z)| \leq \frac{M}{\rho^j} \qquad (j=0,1,2,\dots). \tag{1.11}$$

By Weierstrass' theorem on locally uniformly convergent sequences of analytic functions, it suffices to show that each $\alpha_j(z)$ $(j=0,1,\dots)$ is holomorphic for $z\in\overline{\Delta}$.

We prove this by induction on j, our secondary induction. For $j=0$, we have $\alpha_0(z)=f(z,0)$. Since $f(z,0)$ is holomorphic in $\overline{\Delta}$ by the primary inductive assumption, $\alpha_0(z)$ is holomorphic in $\overline{\Delta}$. Now let l be any nonnegative integer and assume that each $\alpha_j(z)$ $(j=0,1,\dots,l)$ is holomorphic on $\overline{\Delta}$. To prove that $\alpha_{l+1}(z)$ is holomorphic in $\overline{\Delta}$, we consider the following family of holomorphic functions $\{F_w(z)\}_{w\in\Gamma}$ for $z\in\overline{\Delta}$:

$$F_w(z)=\frac{f(z,w)-\sum_{j=0}^{l}\alpha_j(z)w^j}{w^{l+1}}.$$

By inequality (1.11) we have

$$|F_w(z)| \leq \sum_{k=1}^{\infty}|\alpha_{l+k}(z)|\,|w|^{k-1} \leq \frac{M}{\rho^l(1-\frac{|w|}{\rho})}.$$

Thus $\{F_w(z)\}$ is uniformly bounded on $\overline{\Delta}$ for $|w|\leq\rho_0<\rho$. Since $\lim_{w\to 0}F_w(z)=\alpha_{l+1}(z)$ pointwise for $z\in\overline{\Delta}$, it follows from Weierstrass' theorem that $\alpha_{l+1}(z)$ is holomorphic in $\overline{\Delta}$. Hence $f(z,w)$ is holomorphic in Λ.

REMARK 1.10. We see from the proof that if, under the boundedness assumption, we assume only that $f(z,w)$ is holomorphic as a function of $w\in\Gamma$ for any fixed $z\in\overline{\Delta}$, and, in addition, we assume that there exists a sequence $\{w_j\}$ in Γ with $w_j\neq 0$ and $\lim_{j\to\infty}w_j=0$ such that each function $f(z,w_j)$ is holomorphic as a function of $z\in\overline{\Delta}$, then we can conclude that $f(z,w)$ is holomorphic for $(z,w)\in\Lambda$.

1.4.2. Use of Baire's Theorem. To prove the general case we will first use the Baire Category Theorem to show that there exists an open set γ in Γ such that $f(z,w)$ is holomorphic in $\overline{\Delta}\times\gamma$. For each positive integer ν we define

$$e_\nu=\{w'\in\Gamma \mid |f(z,w')|\leq\nu \quad \text{for each } z \text{ in } \overline{\Delta}\}.$$

Since $f(z,w)$ is holomorphic for z in $\overline{\Delta}$ if $w\in\Gamma$ is fixed (by the primary inductive hypothesis), we have

$$e_\nu\subset e_{\nu+1} \quad (\nu=1,2,\dots), \qquad \bigcup_{\nu=1}^{\infty}e_\nu=\Gamma.$$

Furthermore, e_ν is a closed subset of Γ. To see this, let w_j $(j = 1, 2, \dots)$ be a sequence in e_ν which converges to a point w_0 in Γ and suppose, for the sake of obtaining a contradiction, that there exists a point $z' \in \overline{\Delta}$ such that $|f(z', w_0)| > \nu$. By assumption, $f(z', w)$ is holomorphic with respect to w in Γ and hence is continuous at w_0. Thus

$$\lim_{j\to\infty} f(z', w_j) = f(z', w_0),$$

which contradicts $|f(z', w_j)| \le \nu$ $(j = 1, 2, \dots)$.

From Baire's theorem, we deduce that at least one of the sets e_ν contains an interior point. If we let γ denote the interior of such a set e_ν, then $|f(z, w)|$ is bounded in $\overline{\Delta} \times \gamma$. Using the result in the previous section, we get that $f(z, w)$ is holomorphic in $\overline{\Delta} \times \gamma$.

1.4.3. General Case. In order to prove that $f(z, w)$ is holomorphic in Λ in the general case, we may assume from the previous result that there exists a positive number $\rho_0 < \rho$ such that if we set $\Gamma' = \{w : |w| \le \rho_0\}$, then $f(z, w)$ is holomorphic and bounded in $\overline{\Delta} \times \Gamma'$. Thus we can develop $f(z, w)$ into a Hartogs series centered at $w = 0$:

$$f(z, w) = \sum_{j=0}^{\infty} \alpha_j(z)\, w^j, \tag{1.12}$$

where $\alpha_j(z)$ $(j = 0, 1, \dots)$ is holomorphic in $\overline{\Delta}$ and, from the boundedness of $f(z, w)$ on $\overline{\Delta} \times \Gamma'$ and the Cauchy estimates, there exists an $M > 0$ such that for $z \in \overline{\Delta}$,

$$|\alpha_j(z)| \le \frac{M}{\rho_0^j} \qquad (j = 0, 1, 2, \dots).$$

Moreover, for any fixed z in $\overline{\Delta}$, $f(z, w)$ is a holomorphic function of $w \in \Gamma$; hence the radius of convergence of the power series (1.12) in w is greater than or equal to ρ.

We will need the following lemma of Hartogs.

LEMMA 1.1. *Let $\overline{\Delta}$: $|z_j| \le r_j$ $(j = 1, \dots, n)$ be a closed polydisk in $\mathbf{C}^n$ with distinguished boundary $\mathcal{E}$ and let $\{u_k\}_{k=1,2,\dots}$ be a sequence of plurisubharmonic functions on $\overline{\Delta}$. Assume that there exist two positive constants l and L with $l < L$ such that*

$$\sup_k u_k(z) \le L$$

for all $z \in \overline{\Delta}$ and

$$\overline{\lim_{k\to\infty}}\, u_k(z') \le l$$

for each $z' \in \mathcal{E}$. Given positive numbers $r'_j < r_j$ and $l' > l$, if we set $\overline{\Delta}'$: $|z_j| \le r'_j$ $(j = 1, \dots, n)$, then there exists an integer N such that

$$u_k(z) \le l' \quad \text{on } \overline{\Delta}'$$

for all $k \ge N$.

PROOF. We set $l'' := (l' + l)/2$ and $\varepsilon := (l' - l)/2$. For each integer $\nu \ge 1$, we define

$$e_\nu := \{z' \in \mathcal{E} :\ u_k(z') \le l'' \ \text{ for all } k \ge \nu\}.$$

By assumption we have

$$e_\nu \subset e_{\nu+1} \quad (\nu = 1, 2, \dots), \qquad \bigcup_{\nu=1}^{\infty} e_\nu = \mathcal{E}.$$

If we let $e_\nu^c := \mathcal{E} - e_\nu$ and we let $m(e_\nu^c)$ be the (real) n-dimensional measure of the set $e_\nu^c \subset \mathcal{E}$, then

$$\lim_{\nu \to \infty} m(e_\nu^c) = 0.$$

We can thus find a positive integer N such that

$$\prod_{j=1}^{n} (\frac{r_j + r_j'}{r_j - r_j'}) L\, m(e_\nu^c) < \varepsilon$$

for all $\nu \geq N$. Let $P(r, \rho, \theta, \vartheta)$ be the Poisson kernel for Δ. Since

$$\begin{aligned} & u_\nu(\rho_1 e^{i\vartheta_1}, \dots, \rho_n e^{i\vartheta_n}) \\ \leq \; & \frac{1}{(2\pi)^n} \int_0^{2\pi} \cdots \int_0^{2\pi} P(r, \rho, \theta, \vartheta)\, u_\nu(r_1 e^{i\theta_1}, \dots, r_n e^{i\theta_n})\, d\theta_1 \cdots d\theta_n, \end{aligned}$$

it follows that for any positive $\rho_j < r_j'$ and all $\nu \geq N$,

$$u_\nu(\rho_1 e^{i\vartheta_1}, \dots, \rho_n e^{i\vartheta_n}) \leq l'' + \prod_{j=1}^{n} \left(\frac{r_j + r_j'}{r_j - r_j'} \right) L\, m(e_\nu^c) < l',$$

which proves the lemma. □

We return to the proof of Theorem 1.7 in the general case. We put

$$u_k(z) = \frac{1}{k} \log |\alpha_k(z)| \qquad (k = 1, 2, \dots).$$

Then each $u_k(z)$ is a plurisubharmonic function on $\overline{\Delta}$ which satisfies

$$u_k(z) \leq -\log \rho_0 + \log M \quad \text{on } \overline{\Delta} \qquad (k = 1, 2, \dots).$$

Furthermore, for any fixed $z' \in \overline{\Delta}$, since the radius of convergence of the power series (1.12) in w is at least ρ,

$$\overline{\lim_{k \to \infty}}\, u_k(z') \leq -\log \rho.$$

We now apply Lemma 1.1 to the sequence $\{u_k\}_{k=1,2,\dots}$. Given positive numbers $r_j' < r_j$ $(j = 1, \dots, n)$ and $\rho' < \rho$, we set

$$\overline{\Delta}' \; : \; |z_j| \leq r_j' \quad (j = 1, \dots, n) \qquad \text{and} \qquad \Gamma' \; : \; |w| \leq \rho'.$$

Then there exists a positive integer N such that

$$u_k(z) \leq -\log \rho' \quad \text{on } \overline{\Delta}'$$

for all $k \geq N$. In other words,

$$|\alpha_k(z)| \rho'^k \leq 1 \quad \text{on } \overline{\Delta}' \qquad (k \geq N).$$

It follows that the Hartogs series (1.12) converges absolutely and uniformly on any compact set in the interior $(\Lambda')^o$ of $\Lambda' = \overline{\Delta}' \times \Gamma'$. Thus, again applying the Weierstrass theorem on power series, we see that $f(z, w)$ is holomorphic with respect to the $n+1$ complex variables $(z, w) \in (\Lambda')^o$. Since $r_j' < r_j$ and $\rho' < \rho$ were arbitrary, $f(z, w)$ is holomorphic in Λ. □

REMARK 1.11. From the general case of Hartogs theorem, we have the following result:

Let $f(z,w) = \sum_{j=0}^{\infty} \alpha_j(z)w^j$ *be a holomorphic function for* $(z,w) \in \Delta \times \gamma$, *where* $\Delta \subset \mathbf{C}^n$ *and* $\gamma = \{w \in \mathbf{C} : |w| < r\} \subset \mathbf{C}$. *Suppose that for any* $z \in \Delta$, *the Taylor series of* $f(z,w)$ *in* w *has radius of convergence greater than or equal to* ρ *(independent of* $z \in \Delta$*). Then* $f(z,w)$ *is holomorphic for* (z,w) *in* $\Delta \times \Gamma$, *where* $\Gamma = \{w \in \mathbf{C} : |w| < \rho\}$.

This fact remains valid under weaker conditions than those stated (cf. T. Terada [**72**]). To simplify the description we consider $\mathbf{C}^2$ with variables z and w. Let $\Lambda = \overline{\Delta} \times \Gamma$ be a closed bidisk centered at the origin and let $f(z,w)$ be a complex-valued function on Λ. Let $e \subset \Gamma$ and assume that $f(z,w)$ satisfies the following two conditions:

1. For any fixed $z' \in \overline{\Delta}$, $f(z',w)$ is holomorphic with respect to $w \in \Gamma$.
2. For any fixed $w' \in e$, $f(z,w')$ is holomorphic with respect to $z \in \overline{\Delta}$.

Then $f(z,w)$ is holomorphic with respect to the variables $(z,w) \in \Lambda$ if the logarithmic capacity of e is positive.

1.5. Domains of Holomorphy

1.5.1. Analytic Continuation. A holomorphic function of several complex variables can be locally represented by a power series. Hence its analytic continuation is unique as in the case of one complex variable. Indeed, following the ideas of Weierstrass, there are no qualitative differences between the theory of analytic continuation in the case of several complex variables and in the case of one complex variable, as the following two important theorems illustrate.

THEOREM 1.8 (Monodromy Theorem). *Let* D *be a simply connected domain in* $\mathbf{C}^n$ *and let* $\mathcal{P}(z)$ *be a power series centered at a point* p *in* D *whose domain of convergence is non-empty. If* $\mathcal{P}(z)$ *can be analytically continued to any point* $q \in D$ *along any continuous arc* l *in* D *joining* p *to* q, *then the function* $f(z)$ *obtained by this continuation is a single-valued holomorphic function on* D.

THEOREM 1.9 (Countable Valency Theorem). *Let* $\mathcal{P}(z)$ *be a power series centered at a point* p *in* $\mathbf{C}^n$ *whose domain of convegence is non-empty. If we analytically continue* $\mathcal{P}(z)$ *along all arcs starting from* p *for which a continuation is possible, then for any point* q *in* $\mathbf{C}^n$, *the function* $f(z)$ *obtained by this continuation has at most countably many branches over* q.

1.5.2. Domains of Holomorphy. Let $f(z)$ be a holomorphic function in a domain D in $\mathbf{C}^n$. We analytically continue $f(z)$ to as many points in $\mathbf{C}^n$ as possible. This gives us a canonical domain $\widetilde{D}$ such that $f(z)$ is holomorphic in $\widetilde{D}$ but $f(z)$ cannot be analytically continued beyond any boundary point of $\widetilde{D}$. We say that $\widetilde{D}$ is the **natural domain of** $f(z)$ or the **domain of holomorphy of** $f(z)$. In general, we say that D is a **domain of holomorphy** if there exists at least one holomorphic function whose domain of holomorphy coincides with D. Given a domain D in $\mathbf{C}^n$, the maximal domain $\widetilde{D}$ such that any holomorphic function on D is necessarily holomorphic on $\widetilde{D}$ is called the **envelope of holomorphy** of D.

REMARK 1.12. In studying analytic continuation, there are problems regarding multiple-valuedness (multivalency), branch points, and points at infinity. In Part

I, we will not discuss these problems; hence the term *domain* refers to a *univalent* (or schlicht) domain in $\mathbf{C}^n$ and the term *envelope of holomorphy* refers to a one-sheeted envelope of holomorphy. With respect to points at infinity, we mention that the analytic continuation to a point at infinity p_∞ in the Osgood space $\widehat{\mathbf{C}}^n$ or in complex projective space $\mathbf{P}^n$ may be treated as in the case of a point p in $\mathbf{C}^n$ by transforming p_∞ to the origin O with a linear coordinate transformation (cf., section 1.1.6).

In the case of one complex variable, every domain in $\mathbf{C}$ is a domain of holomorphy. On the other hand, in the case of $\mathbf{C}^n$ with $n > 1$, determining which domains are domains of holomorphy is an area of reseach which will be discussed in the forthcoming chapters. Here we mention a theorem which illustrates the distinguished character of domains of holomorphy.[1]

THEOREM 1.10 (Osgood). [2] *Let D be a domain in $\mathbf{C}^n$ and let E be a compact set in D. Assume that $D \setminus E$ is connected. Then any holomorphic function $f(z)$ on $D \setminus E$ can be analytically continued to a (single-valued) holomorphic function on all of D.*

PROOF. We consider $\mathbf{C}^n$ as the product of $\mathbf{C}^{n-1}$ with variables $z' := (z_1, \dots, z_{n-1})$ and the complex plane $\mathbf{C}_{z_n}$ with variable z_n. Given a set $S \subset \mathbf{C}^n$ and a set $\sigma \subset \mathbf{C}^{n-1}$, we use the following notation: $\underline{S}$ is the projection of S to $\mathbf{C}^{n-1}$; $S(\sigma)$ is the set of all points $(z', z_n) \in S$ such that $z' \in \sigma$; and S^0 and $S^0(\sigma)$ will denote the interiors of S and $S(\sigma)$. We note that $S(\sigma)$ may be empty. In the case where σ consists of a single point z' in $\mathbf{C}^{n-1}$, we write $S(\sigma) = S(z')$. We will identify the fiber $S(z')$ with the set in $\mathbf{C}_{z_n}$ consisting of those points z_n with $(z', z_n) \in S$.

By assumption, we have $\underline{E} \subset\subset \underline{D}$ and $E(z') \subset\subset D(z')$ for any $z' \in \underline{E}$. Fix such a z'. We take a finite number of smooth, closed Jordan curves L in $D(z')$ such that if U is the closed domain bounded by L, i.e., $L = \partial U$, then $E(z') \subset\subset U^0 \subset\subset D(z')$. We next take a sufficiently small neighborhood v of z' in $\underline{D}$ such that if $V := v \times U$, then $E(v) \subset\subset V^0 \subset\subset D(v)$, where $V^0 = v \times U^0$. We note that $f(z)$ is defined and holomorphic on $D(v) \setminus V^0$.

For any $z = (z_1, \dots, z_n) \in V^0$, we consider the integral

$$g(z_1, \dots, z_{n-1}, z_n) = \frac{1}{2\pi i} \int_L \frac{f(z_1, \dots, z_{n-1}, \zeta)}{\zeta - z_n} \, d\zeta.$$

Then $g(z)$ defines a holomorphic function in V^0. If we can find a non-empty open set δ_0 in v such that $f(z)$ is holomorphic in $V(\delta_0)$ (for example, if $E(\delta_0) = \emptyset$), then, by applying Cauchy's theorem for the complex variable z_n, we get that $g(z) = f(z)$ on $V(\delta_0)$; hence $f(z)$ has an analytic continuation to the single-valued function $g(z)$ on V^0.

---

[1]In the introduction of Oka's paper [IX] there appears the following sentence: "La théorie générale du prolongement analytique à une seule variable est semblable à la plaine campagne; là, on n'a pu trouver, malgré les nombreux efforts, aucun fait en dehors des prévisions de la logique formelle. Au contraire, le cas de plusieurs variables nous apparaît comme un pays montagneux, très escarpé."

[2]This theorem is essentially due to Hartogs. In the textbook of Osgood [**56**] there is a proof which appears to be incomplete. A complete proof may be found in A. B. Brown [**4**]. The proof given here is due to the author.

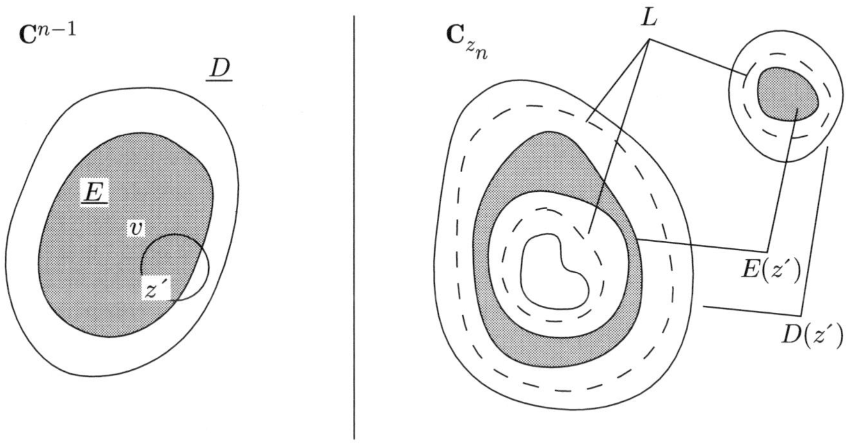

FIGURE 3. Osgood's theorem

Since $\underline{E}$ is compact in $\underline{D}$, we can find a finite number of points $\{z_\nu\}_{\nu=1,\dots,l}$ in $\underline{E}$ such that if we construct the corresponding sets

$$L_\nu = \partial U_\nu, \qquad V_\nu := v_\nu \times U_\nu,$$

and the corresponding functions $g_\nu(z)$ for each z_ν, and then we set $\Omega := \bigcup_{\nu=1}^{l} V_\nu^0$, we have $E \subset\subset \Omega \subset\subset D$. We may assume that the Jordan curves L_ν intersect the curves L_μ with $\nu \neq \mu$ in at most finitely many points.

The theorem will be a consequence of the following lemma in one complex variable.

LEMMA 1.2. *Let U_j $(j = 1, 2)$ be closed domains (not necessarily connected) in the complex plane $\mathbf{C}_w$, each bounded by a finite number of smooth Jordan curves L_j, i.e., $\partial U_j = L_j$. Suppose that $L_1 \cap L_2$ is a finite set of points. Given a holomorphic function $f(z)$ on the closed set*

$$U_1 \cup U_2 - (U_1^0 \cap U_2^0)$$

(here U_j^0 denotes the interior of U_j), we define

$$g_j(w) := \frac{1}{2\pi i} \int_{L_j} \frac{f(\zeta)}{\zeta - w}\, d\zeta$$

for $w \in U_j$ $(j = 1, 2)$. Then $g_1(w) = g_2(w)$ on $U_1^0 \cap U_2^0$.

PROOF. We set $U_1^e := U_1 \setminus U_2^0$, $U_2^e := U_2 \setminus U_1^0$ and

$$L_1^e := U_1^e \cap L_1, \quad L_2^i := U_1^e \cap L_2, \quad L_2^e := U_2^e \cap L_2, \quad L_1^i := U_2^e \cap L_1.$$

From the relation $U_1 \cup U_2 - (U_1^0 \cap U_2^0) = U_1^e \cup U_2^e$, $f(z)$ is defined on U_j^e $(j = 1, 2)$. Since $\partial U_1^e = L_1^e \cup (-L_2^i)$ and $\partial U_2^e = L_2^e \cup (-L_1^i)$, it follows from Cauchy's theorem that, for any $w \in U_1^0 \cap U_2^0$,

$$\frac{1}{2\pi i} \int_{L_1^e} \frac{f(\zeta)}{\zeta - w}\, d\zeta = \frac{1}{2\pi i} \int_{L_2^i} \frac{f(\zeta)}{\zeta - w}\, d\zeta$$

and

$$\frac{1}{2\pi i}\int_{L_1^i}\frac{f(\zeta)}{\zeta-w}\,d\zeta=\frac{1}{2\pi i}\int_{L_2^e}\frac{f(\zeta)}{\zeta-w}\,d\zeta.$$

Since $L_j = L_j^e \cup L_j^i$ $(j=1,2)$, we have $g_1(w)\equiv g_2(w)$ on $U_1^0\cap U_2^0$. □

To finish the proof of Osgood's theorem, we assume that $v_\nu \cap v_\mu \neq \emptyset$ $(1\leq \nu,\mu\leq l)$ and set $\delta := v_\nu\cap v_\mu$. By assumption, $f(z)$ is defined on the closed domain

$$V_\nu(\delta)\cup V_\mu(\delta)-(V_\nu^0(\delta)\cap V_\mu^0(\delta)) \quad \subset\subset (D\setminus E)^0(\delta).$$

It follows from Lemma 1.2 that for any fixed $z'\in\delta$,

$$g_\nu(z',z_n)=g_\mu(z',z_n) \quad \text{for all } z_n \text{ with } z_n\in V_\nu^0(z')\cap V_\mu^0(z');$$

i.e., $g_\nu(z)=g_\mu(z)$ in $V_\nu^0(\delta)\cap V_\mu^0(\delta)$. This means that $g_\nu(z)$ is an analytic continuation of $g_\mu(z)$. Therefore, setting $g(z)=g_\nu(z)$ in V_ν^0 $(\nu=1,\dots,l)$, we get a (single-valued) holomorphic function $g(z)$ on Ω. On the other hand, some v_ν contains a non-empty open set δ_0 such that $E(\delta_0)=\emptyset$; thus $g(z)$ is also the analytic continuation of $f(z)$ to Ω. Since D is connected, $D\cap\Omega\neq\emptyset$, and $E\subset\Omega$, $f(z)$ can thus be analytically continued to the entire domain D.[3] □

We deduce from this theorem that any bounded domain of holomorphy in $\mathbf{C}^n$ for $n\geq 2$ must have only one boundary component. In general, when we study holomorphic functions, we consider them to be defined on their domains of holomorphy.

REMARK 1.13. In $\mathbf{C}^n$ with $n>1$, not all domains are domains of holomorphy, and it is an important problem to determine the envelope of holomorphy of a domain. We give two interesting examples.

1. There exists a univalent domain D in $\mathbf{C}^n$ whose envelope of holomorphy has infinitely many sheets. For example, in $\mathbf{C}^2$ with variables z and w we consider the sets

$$\Sigma_1 : z=1,\ |w|\leq 1 \quad\text{and}\quad \Sigma_2 : z=e^{it},\ |w|=e^t \ (0\leq t<\infty).$$

We set $\Sigma=\Sigma_1\cup\Sigma_2$, and construct a univalent domain D in $\mathbf{C}^2$ which contains Σ and does not intersect $\{e^{it}\}\times\{|w|=e^{t+(2k+1)\pi}\}$ $(0\leq t\leq\infty,\ k=0,1,\dots)$. Then the envelope of holomorphy $\widehat{D}$ of D contains the set

$$\Sigma^* : z=e^{it},\quad |w|\leq e^t \qquad (0\leq t<\infty)$$

in $\mathcal{R}\times\mathbf{C}_w$, where $\mathcal{R}$ is the Riemann surface of $\log z$ over the z-plane, while $\widehat{D}$ is itself contained in the product set $\mathcal{R}\times\mathbf{C}_w$. To verify that $\Sigma^*\subset\widehat{D}$, we need to appeal to a result of Hartogs (Theorem 4.1) which will be proved later. Let $\mathcal{F}$ be the family of all holomorphic functions on D. Given $t\geq 0$, we define the following subsets of $\mathbf{C}^2$:

$$\sigma(t):=\{e^{it}\}\times(|w|=e^t),\qquad [\sigma](t):=\{e^{it}\}\times(|w|<e^t),$$

and

$$\Sigma^*(t):=\bigcup_{0<t'<t}[\sigma](t).$$

[3]In the case when $D\setminus E$ consists of several connected components $G_1,\dots,G_m$ we set $f(z)=f_k(z)$ on G_k $(k=1,\dots,m)$. Then the function $g(z)$ obtained by the above procedure using $f(z)$ is the analytic continuation of $f_{k_0}(z)$ on the component G_{k_0}, where ∂G_{k_0} contains the outer boundary of D.

Define

$$T := \sup\{t \geq 0 \mid \text{ each } f \in \mathcal{F} \text{ is holomorphic on } \Sigma^*(t') \text{ for all } t' \leq t\}.$$

Since $\Sigma_1 \subset D$, we have $T > 0$. The claim will follow if we can show $T = +\infty$. We prove this by contradiction; hence we assume that $T < +\infty$. Choose a neighborhood $U \times V$ of $\sigma(T)$ in D, where $U = \{|z - e^{iT}| < \delta\}$, $V = \{e^{T'} < |w| < e^{T''}\}$ with $\delta > 0$, $0 < T' < T < T''$, and $e^{it} \in U$ $(T' < t < T'')$. Let $f \in \mathcal{F}$ and take t_0 with $T' < t_0 < T$. Then $\sigma(t_0) \subset U \times V$, and $f(z)$ is holomorphic on $[\sigma](t_0)$ by the definition of T. Since f is holomorphic on $U \times V$, Theorem 4.1 implies that f is holomorphic on $U \times \{|w| < e^{t_0}\}$, and hence on $U \times \{|w| < e^{T''}\}$. It follows that f is holomorphic on $\Sigma^*(T'')$. Since $f \in \mathcal{F}$ was arbitrary, we have $T'' \leq T$, which is a contradiction.

2. Conversely, there exists a multivalent domain whose envelope of holomorphy is a univalent domain. For example, in $\mathbf{C}^2$ with variables z and w we consider the three domains

$$\Gamma \, : \, 1 < |z|^2 + |w|^2 < 3,$$

$$\Delta_1 \, : \, |z-1|^2 + |w|^2 < r, \qquad |z|^2 + |w|^2 < 1$$

and

$$\Delta_2 \, : \, |z+1|^2 + |w|^2 < r, \qquad |z|^2 + |w|^2 < 1,$$

where r is a real number satisfying $1 < r < \sqrt{2}$. By gluing Δ_1 and Δ_2 to Γ along the sphere $|z|^2 + |w|^2 = 1$, we obtain a domain D which is two-sheeted over a neighborhood of the origin. By Theorem 1.10 any holomorphic function on D can be analytically continued to a single-valued holomorphic function on the (univalent) ball $B = \{|z|^2 + |w|^2 < 3\}$; hence the envelope of holomorphy $\widehat{D}$ of D contains B. Since B is a domain of holomorphy, we have $\widehat{D} = B$.

1.5.3. Holomorphically Convex Domains. In this section we give an analytic characterization of domains of holomorphy which is due to Cartan and Thullen.

Let D be a domain in $\mathbf{C}^n$ with variables $z_1, \ldots, z_n$. Let E be a compact set in D and let $r = \delta_D(E) > 0$ denote the polydisk boundary distance from E to ∂D. Given ρ $(0 < \rho < r)$ and $z' = (z'_1, \ldots, z'_n) \in E$, we consider the polydisk $\Delta^{\rho}_{z'} : |z_j - z'_j| < \rho$ $(j = 1, \ldots, n)$ centered at z' with radius ρ. We set

$$E^{\rho} := \bigcup_{z' \in E} \Delta^{\rho}_{z'},$$

so that $E \subset E^{\rho} \subset\subset D$. The sets E^{ρ} will occur in the Thullen lemma below.

Following Cartan and Thullen, we consider a class $\mathcal{K}$ of holomorphic functions in D which satisfies the following properties:

1. $f \in \mathcal{K}$ implies $\partial f / \partial z_j \in \mathcal{K}$, $j = 1, \ldots, n$.
2. For any complex number c and any integer $l \geq 1$, $f \in \mathcal{K}$ implies $cf^l \in \mathcal{K}$.

We call $\mathcal{K}$ a **regular class** of holomorphic functions in D.

Standard examples are the class of all polynomials in $\mathbf{C}^n$; the class of all holomorphic functions in D; and the class of all functions which are holomorphic in a given domain D' which contains D.

We have the following lemma concerning these classes.

LEMMA 1.3 (Thullen [**73**]). *Let $\mathcal{K}$ be a regular class of holomorphic functions in D and let E be a compact set in D. We let $r = \delta_D(E) > 0$ denote the polydisk boundary distance from E to ∂D. If z^0 is a point in D at which we have the inequality*

$$|f(z^0)| \le \max_{z\in E} |f(z)| \quad \textit{for all } f \in \mathcal{K},$$

then every $f \in \mathcal{K}$ can be analytically continued to the polydisk $\Delta^r_{z^0} : |z_j - z^0_j| < r$ $(j = 1, \dots, n)$ centered at z^0 with radius r.

PROOF. Fix $f \in \mathcal{K}$. For any ρ $(0 < \rho < r)$, we set

$$A(\rho) := \max_{z\in E^\rho} |f(z)| < \infty.$$

It follows from the Cauchy estimates that for any $z \in E$

$$\frac{1}{j_1! \cdots j_n!} \left| \frac{\partial^{j_1+\cdots+j_n} f}{\partial^{j_1} z_1 \cdots \partial^{j_n} z_n}(z) \right| \le \frac{A(\rho)}{\rho^{j_1+\cdots+j_n}}.$$

Since $z^0 \in D$, we can form the Taylor expansion of $f(z)$ centered at z^0:

$$f(z) = \sum \alpha_{j_1,\dots,j_n} (z_1 - z^0_1)^{j_1} \cdots (z_n - z^0_n)^{j_n}. \tag{1.13}$$

By the hypothesis and condition 1 in the definition of a regular class $\mathcal{K}$ we obtain

$$|\alpha_{j_1,\dots,j_n}| = \left| \frac{1}{j_1! \cdots j_n!} \frac{\partial^{j_1+\cdots+j_n} f}{\partial^{j_1} z_1 \cdots \partial^{j_n} z_n}(z_0) \right| \le \frac{A(\rho)}{\rho^{j_1+\cdots+j_n}}.$$

Therefore, the right-hand-side of (1.13) converges absolutely and uniformly on any compact set in the polydisk $\Delta^\rho_{z^0}$ centered at z^0 with radius ρ; hence f can be analytically continued to $\Delta^\rho_{z^0}$. Since $0 < \rho < r$ was arbitrary, the lemma is proved. □

REMARK 1.14. This lemma has meaning for any point $z^0 \in D$ such that the polydisk boundary distance $\delta_D(z^0)$ from z^0 to ∂D is less than r; i.e., even in the case when $\Delta^r_{z^0}$ contains points outside of D. The lemma then implies that any holomorphic function belonging to $\mathcal{K}$ extends analytically to these points.

REMARK 1.15. In the proof of the lemma, condition 2 in the definition of a regular class $\mathcal{K}$ was not used. However, using condition 2 we can show that, given any $0 < \rho < r$, every $f \in \mathcal{K}$ satisfies

$$\max_{z\in\Delta^\rho_{z^0}} |f(z)| \le \max_{z\in E^\rho} |f(z)|.$$

This inequality follows by applying the same method as in the proof of the lemma to the functions $f^l \in \mathcal{K}$.

Lemma 1.3 yields an analytic characterization of a domain of holomorphy, as we will show in the Cartan-Thullen theorem below.

Let D be a domain in $\mathbf{C}^n$ and let $\mathcal{K}$ be a regular class of holomorphic functions in D. For a compact set E in D, we define the following closed subset of D:

$$\widehat{E}_{\mathcal{K}} := \{z' \in D \mid |f(z')| \le \max_{z\in E} |f(z)| \text{ for all } f \in \mathcal{K}\}. \tag{1.14}$$

We call the set $\widehat{E}_{\mathcal{K}}$ the **$\mathcal{K}$-convex hull** of E. In particular, in the case when $\mathcal{K}$ coincides with the class of all polynomials, $\widehat{E}_{\mathcal{K}}$ is called the **polynomial hull** of E. When $\mathcal{K}$ is the class of all holomorphic functions in D, $\widehat{E}_{\mathcal{K}}$ is called the **holomorphic hull** of E relative to D.

We have the following.

COROLLARY 1.2. *Let D be a bounded domain of holomorphy in $\mathbf{C}^n$. For $r > 0$, let $D^{[r]} = \{z \in D : \delta_D(z) > r\}$. Then the holomorphic hull of $\overline{D^{[r]}}$ relative to D is equal to $\overline{D^{[r]}}$. The same is true for the set $\overline{D^{(r)}}$ which is obtained by replacing the polydisk boundary distance $\delta_D(z)$ by the Euclidean boundary distance $d_D(z)$.*

PROOF. The first assertion follows directly from Lemma 1.3. For the second assertion, let A be an $n \times n$ unitary matrix so that $z^A := Az$ defines a new Euclidean coordinate system for $\mathbf{C}^n$. For $p \in D$, we let $\delta_D^A(p)$ denote the polydisk boundary distance from p to ∂D measured in these new coordinates. For $r > 0$, we set $D^{[A,r]} = \{p \in D : \delta_D^A(p) > r\}$. Taking $r' = r/\sqrt{2n}$, we note that $\overline{D^{(r)}} = \bigcap_A \overline{D^{[A,r']}}$, where the intersection is taken over all $n \times n$ unitary matrices A. From the first assertion, the holomorphic hull of each $\overline{D^{[A,r']}}$ relative to D is equal to $\overline{D^{[A,r']}}$; thus the same is true for $\overline{D^{(r)}}$. $\square$

If a domain D in $\mathbf{C}^n$ satisfies the condition that $\widehat{E}_{\mathcal{K}} \subset\subset D$ for any compact set E in D, then D is called a **$\mathcal{K}$-convex domain**. In particular, when $\mathcal{K}$ is the class of all polynomials (resp., all holomorphic functions in D), D is called a **polynomially convex** (resp., **holomorphically convex**) **domain**.

Let $\mathcal{K}$ be the class of all monomials in the variables $z_1, \dots, z_n$. We see that a domain D in $\mathbf{C}^n$ is $\mathcal{K}$-convex if and only if D is a logarithmically convex complete Reinhardt domain centered at the origin in $\mathbf{C}^n$.

Using these notions, we have the following theorem.

THEOREM 1.11 (Cartan-Thullen [**13**]). *Let D be a domain of holomorphy in $\mathbf{C}^n$ and let $\mathcal{K}$ be a regular class of holomorphic functions in D. Assume that $\mathcal{K}$ contains a holomorphic function f whose domain of holomorphy is equal to D, and that $\mathcal{K}$ also contains the coordinate functions $z_1, \dots, z_n$. Then D is $\mathcal{K}$-convex.*

PROOF. Since $\mathcal{K}$ contains the coordinate functions, it suffices to prove the theorem for bounded domains D in $\mathbf{C}^n$. Let E be any compact set in D. We let $r = \delta_D(E) > 0$ denote the polydisk boundary distance from E to ∂D. Let z^0 be any point in D such that $\delta_D(z^0) < r$. By Lemma 1.3 and the hypothesis that $f \in \mathcal{K}$, there exists a function $\varphi \in \mathcal{K}$ which satisfies the inequality

$$\max_{z \in E} |\varphi(z)| < |\varphi(z^0)|.$$

It follows from the Heine-Borel theorem that D is $\mathcal{K}$-convex. $\square$

As a special case of the theorem when $\mathcal{K}$ is the class of all holomorphic functions in D, we have the following corollary.

COROLLARY 1.3. *A domain of holomorphy is holomorphically convex.*

1.5.4. Analytic Polyhedra. Let D be a domain in $\mathbf{C}^n$ and let f_j $(j = 1, \dots, m)$ be a finite collection of holomorphic functions in D. We consider the following closed subset Λ of D defined by the inequalities:

$$\Lambda = \{z \in D \mid |f_j(z)| \leq 1, \ j = 1, \dots, m\}.$$

If a (closed) connected component Λ_0 of Λ is compact in D, we call Λ_0 an **analytic polyhedron** in D. A finite union of compact, connected components of Λ in D will also be called an analytic polyhedron in $\mathbf{C}^n$.[4]

[4]The notion of analytic polyhedron is due to A. Weil [**77**].

We have the following proposition.

PROPOSITION 1.5. *Let D be a domain in $\mathbf{C}^n$ and let $\mathcal{K}$ be a regular class of holomorphic functions in D. Assume that D is $\mathcal{K}$-convex. Then there exists a sequence of analytic polyhedra $\mathcal{P}_j$ $(j = 1, 2, \dots)$ defined by holomorphic functions in $\mathcal{K}$ such that*

$$\mathcal{P}_j \subset\subset \mathcal{P}^0_{j+1} \qquad (j = 1, 2, \dots), \quad \text{and} \quad D = \bigcup_{j=1}^{\infty} \mathcal{P}_j.$$

PROOF. It suffices to prove that, given any compact set E in D, we can find an analytic polyhedron $\mathcal{P}$ determined by holomorphic functions in $\mathcal{K}$ such that $E \subset \mathcal{P} \subset\subset D$. By assumption, the $\mathcal{K}$-convex hull $\widehat{E}_{\mathcal{K}}$ of E is compact in D. Let $r > 0$ be the polydisk boundary distance from $\widehat{E}_{\mathcal{K}}$ to ∂D, and fix a positive number $\rho < r$. We form the compact subset

$$\widehat{E}^{\rho}_{\mathcal{K}} = \bigcup_{z \in \hat{E}_{\mathcal{K}}} \Delta^{\rho}_{z}$$

of D. Given any $z' \in \partial \widehat{E}^{\rho}_{\mathcal{K}}$ (so that $z' \notin \widehat{E}_{\mathcal{K}}$), we can find a holomorphic function $g \in \mathcal{K}$ such that

$$\max_{z \in E} |g(z)| < 1 < |g(z')|.$$

Thus there exists a neighborhood $\delta_{z'} \subset\subset D$ of z' such that the set $\{z \in D \mid |g(z)| \leq 1\} \cap \delta_{z'} = \emptyset$. Since $\partial \widehat{E}^{\rho}_{\mathcal{K}}$ is compact in D, we can find a finite number of these neighborhoods δ_k $(k = 1, \dots, l)$ and holomorphic functions g_k $(k = 1, \dots, l)$ associated to δ_k such that if we set

$$\Lambda = \{z \in D \mid |g_k(z)| \leq 1, \ k = 1, \dots, l\},$$

then $E \subset\subset \Lambda$ and $\Lambda \cap \partial \widehat{E}^{\rho}_{\mathcal{K}} = \emptyset$. It follows that a finite union Λ_0 of connected components of Λ satisfies $E \subset\subset \Lambda_0 \subset\subset \widehat{E}^{\rho}_{\mathcal{K}}$. Hence Λ_0 is the desired analytic polyhedron. □

This proposition yields the converse of Theorem 1.11.

THEOREM 1.12. *Let D be a domain in $\mathbf{C}^n$ and let $\mathcal{K}$ be a regular class of holomorphic functions in D. If D is $\mathcal{K}$-convex, then D is a domain of holomorphy.*

PROOF. From Proposition 1.5 we can find a sequence of analytic polyhedra $\mathcal{P}_j$ $(j = 1, 2, \dots)$ in D, each defined by holomorphic functions in $\mathcal{K}$, such that

$$\mathcal{P}_j \subset\subset \mathcal{P}^0_{j+1} \qquad (j = 1, 2, \dots), \quad \text{and} \quad D = \bigcup_{j=1}^{\infty} \mathcal{P}_j.$$

Furthermore, we can find a sequence of points $\{z^j\}$ in D such that $z^j \in \partial \mathcal{P}_j$ $(j = 1, 2, \dots)$ and whose set of accumulation points in $\mathbf{C}^n$ is ∂D. Then for each z^j we can find a function $f_j \in \mathcal{K}$ such that

$$|f_j(z)| < 1 \ \text{ in } \ \mathcal{P}_j, \qquad f_j(z^j) = 1.$$

We next take sequences of positive numbers ε_j $(j = 1, 2 \dots)$ and positive integers l_j $(j = 1, 2 \dots)$ such that

$$\sum_{j=1}^{\infty} \varepsilon_j < \infty, \qquad |f_j(z)|^{l_j} < \varepsilon_j \ \text{ in } \ \mathcal{P}_j \ \ (j = 1, 2 \dots).$$

The infinite product of holomorphic functions

$$F(z) := \prod_{j=1}^{\infty} [1 - (f_j(z))^{l_j}]$$

converges uniformly to a holomorphic function on each $\mathcal{P}_j$; hence F defines a holomorphic function in D. Clearly, $F \not\equiv 0$ in D, while $F(z^j) = 0$ $(j = 1, 2 \ldots)$. If F could be analytically continued to a point $p \in \partial D$ (i.e., if there exist a neighborhood $V \subset \mathbf{C}^n$ of p and a holomorphic function f in V with $F|_V = f$), then $\partial D \cap V$ would be contained in the set $\Sigma := \{z \in V : f(z) = 0\}$. From the results of the next chapter, we will see that this forces $\Sigma = V$, i.e., $f \equiv 0$ in V, which leads to a contradiction. Thus F cannot be analytically continued to any point of ∂D; hence D is the domain of holomorphy of F. □

This theorem, together with the corollary of Theorem 1.11, implies the following result.

THEOREM 1.13. *A domain D in $\mathbf{C}^n$ is a domain of holomorphy if and only if D is holomorphically convex.*

CHAPTER 2

Implicit Functions and Analytic Sets

2.1. Implicit Functions

As shown in the previous section, the set of zeros of a holomorphic function $f(z)$ of $n \geq 2$ complex variables does not contain isolated points. Furthermore, Osgood's theorem implies that this set is not relatively compact in the domain of definition of f. In this section we will make a more detailed study of the zero sets of holomorphic functions of n complex variables.

2.1.1. Zero Sets of Holomorphic Functions. For convenience, we consider $\mathbf{C}^{n+1} = \mathbf{C}^n \times \mathbf{C}$, where $\mathbf{C}^n$ is the space of the n complex variables $z_1, \dots, z_n$ and $\mathbf{C}$ is the complex plane of the variable w. Let D be a domain in $\mathbf{C}^{n+1}$ and let $f(z,w)$ be a holomorphic function on D. Fix a point (a,b) in D. If f satisfies the two conditions

$$(1)\ f(a,b) = 0, \qquad (2)\ f(a,w) \not\equiv 0,$$

then we say that $f(z,w)$ satisfies the **Weierstrass condition** at (a,b) for the coordinates (z,w). Clearly if $f(z,w) \not\equiv 0$ satisfies condition (1), then we can find a linear coordinate transformation yielding new coordinates (z,w) for which $f(z,w)$ satisfies the Weierstrass condition at (a,b).

Now assume that $f(z,w)$ satisfies the Weierstrass condition at (a,b) in the coordinates (z,w). We consider the section $D(a)$ in the w-plane over $z_j = a_j$ $(j = 1, \dots, n)$. Then $f(a,w)$ is holomorphic in the variable w in $D(a)$. We let $\nu \geq 1$ denote the order of the zero of $f(a,w)$ at b in $D(a)$; thus

$$f(a,w) = A_0(a)(w-b)^\nu + A_1(a)(w-b)^{\nu+1} + \dots,$$

where $A_0(a) \neq 0$. In the domain $D(a)$ we let $\Gamma : |w-b| \leq \rho$ be a closed disk centered at b with radius $\rho > 0$ sufficiently small so that $f(a,w) \neq 0$ at any point w in the punctured disk $\Gamma \setminus \{b\}$. We then fix $r > 0$ (depending on ρ) such that the closed polydisk $\Lambda = \overline{\Delta} \times \Gamma$, where

$$\overline{\Delta} : \ |z_j - a_j| \leq r \quad (j = 1, \dots, n),$$

lies in D and $f(z,w) \neq 0$ on $\overline{\Delta} \times \partial\Gamma$.

LEMMA 2.1. *For any fixed z' in Δ, the holomorphic function $f(z',w)$ of the variable w has ν zeros in Γ (counted with multiplicity).*

PROOF. For each $z' \in \Delta$ we let $\nu(z')$ denote the number of zeros (counted with multiplicity) of the holomorphic function $w \to f(z',w)$ in Γ. By the argument principle, we have

$$\nu(z') = \frac{1}{2\pi i} \int_{\partial\Gamma} \frac{\partial f(z',w)/\partial w}{f(z',w)} dw.$$

Since $f(z, w) \neq 0$ on $\Delta \times \partial\Gamma$, the integral on the right-hand side is continuous for z' in Δ. Hence $\nu(z') \equiv$ const. in Δ. □

In Lemma 2.1 we let $\eta_j(z')$ $(j = 1, \ldots, \nu)$ denote the zeros of $f(z', w)$ in Γ for fixed $z' \in \Delta$. Note that we may have $\eta_i(z') = \eta_j(z')$ for some i, j $(1 \leq i, j \leq \nu)$. We remark that

$$\lim_{z' \to a} \eta_j(z') = b \qquad (j = 1, \ldots, \nu). \tag{2.1}$$

To see this, fix ρ' $(0 < \rho' < \rho)$. We then choose r' $(0 < r' < r)$ such that if we let $\Gamma' \subset \Gamma$ denote the closed disk centered at b with radius ρ', and we let $\Delta' \subset \Delta$ denote the polydisk centered at a with radius r', then $f(z, w) \neq 0$ on $\Delta' \times \partial\Gamma'$. The same argument used in the proof of Lemma 2.1 shows that $\eta_j(z') \in \Gamma'$ $(j = 1, \ldots, \nu)$ for each $z' \in \Delta'$. Since $\rho' > 0$ $(\rho' < \rho)$ was arbitrary, we have (2.1).

In Lemma 2.1, suppose that $f(z, w) = 0$ has only one zero $\eta(z)$ for each $z \in \Delta$, that is, $\eta_1(z) = \cdots = \eta_\nu(z) \equiv \eta(z)$, and the order of the zero $w = \eta(z)$ of $f(z, w)$ equals ν for all $z \in \Delta$. Then we have the following result.

LEMMA 2.2. *$\eta(z)$ is a holomorphic function on Δ.*

PROOF. Given any $z \in \Delta$, applying the residue theorem we get

$$\nu \cdot \eta(z) = \frac{1}{2\pi i} \int_{\partial\Gamma} w \, \frac{\partial f(z, w)/\partial w}{f(z, w)} \, dw.$$

For $w \in \partial\Gamma$, the function under the integral in the right-hand side is a holomorphic function of z in Δ. Thus the integral is a holomorphic function for z in Δ; hence so is $\eta(z)$. □

From these two lemmas we see that the zero set of a holomorphic function of $n+1$ complex variables is a complex n-dimensional set which is analytic in a certain sense. We call the zero set of a holomorphic function an **analytic hypersurface**.

2.1.2. Representations of Analytic Hypersurfaces. Let D be a domain in $\mathbf{C}^{n+1}$ with the variables $z_1, \ldots, z_n$ and w. We let $f(z, w)$ be a holomorphic function in D, and we let $\mathcal{S}$ denote the zero set of $f(z, w)$ in D:

$$\mathcal{S} = \{(z, w) \in D \mid f(z, w) = 0\}.$$

Fix $(a, b) \in \mathcal{S}$ and let $\Lambda = \overline{\Delta} \times \Gamma$ be a closed polydisk in D, where

$$\overline{\Delta} : |z_j - a_j| \leq r \quad (j = 1, \ldots, n), \qquad \Gamma : |w - b| \leq \rho.$$

We assume that Λ has been chosen so that we are in the situation of the previous section:

1. $f(z, w) \neq 0$ for any $(z, w) \in \overline{\Delta} \times \partial\Gamma$.
2. The holomorphic function $f(a, w)$ of $w \in \Gamma$ does not vanish at any $w \in \Gamma \setminus \{b\}$. Let $\nu \geq 1$ denote the order of the zero of $f(a, w)$ at b.
3. For each $z' \in \overline{\Delta}$, let $m(z')$ denote the number of distinct zeros of the holomorphic function $f(z', w)$ of w in Γ. We set

$$l = \max\{m(z') \mid z' \in \overline{\Delta}\}, \tag{2.2}$$

so that $l \leq \nu$.

Let

$$\mathcal{S}_0 = \mathcal{S} \cap \Lambda.$$

From the proof of Lemma 2.1 we see that for any $z' \in \Delta$, the equation $f(z', w) = 0$ has ν zeros counted with multiplicity in Γ.

We have the following proposition.

PROPOSITION 2.1. *The set* $\mathcal{S}_0$ *can be written in the form*

$$P(z, w) := (w - b)^l + A_1(z)(w - b)^{l-1} + \cdots + A_l(z) = 0, \tag{2.3}$$

where each $A_j(z)$ $(j = 1, \ldots, l)$ *is a holomorphic function of* z *in* Δ *satisfying* $A_j(a) = 0$.

PROOF. Let c be a point in Δ such that the equation $f(c, w) = 0$ has l distinct solutions $w = b_k$ $(k = 1, \ldots, l)$ in Γ. For each b_k we let

$$\gamma_k \,:\, |w - b_k| \le \rho'$$

be a closed disk with radius ρ' sufficiently small such that $\gamma_k \subset \Gamma$ and $\gamma_k \cap \gamma_h = \emptyset$ for $k \ne h,\ 1 \le k, h \le l$. We then let

$$\overline{\delta} \,:\, |z_j - c_j| \le r' \qquad (j = 1, \ldots, n)$$

be a closed polydisk centered at c with radius $r' > 0$ chosen small enough to insure that $f(z, w) \ne 0$ for any $(z, w) \in \delta \times \partial\gamma_j$ (see Figure 1).

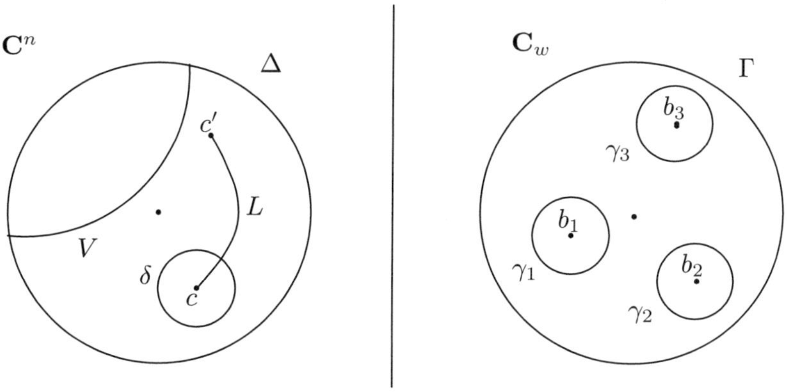

FIGURE 1. Representations of analytic hypersurfaces

From condition 3 and Lemma 2.1 we see that for any $z' \in \delta$ and for each $k = 1, \ldots, l$, there exists precisely one zero of the holomorphic function $f(z', w)$ for w in γ_k, which we denote by $\eta_k(z')$. Lemma 2.2 implies that each $\eta_k(z)$ $(k = 1, \ldots, l)$ is a holomorphic function of z in δ.

Next we consider the set $\mathcal{V}$ of all points z' in Δ such that the equation $f(z', w) = 0$ has l distinct zeros in Γ. By the above argument, $\mathcal{V}$ is an open subset of Δ. We let V denote the connected component of $\mathcal{V}$ which contains the point c above.

Given any point $c' \in V$, we can join c to c' by an arc L in Δ. Then each holomorphic function $\eta_k(z)$ $(k = 1, \ldots, l)$ defined on δ can be analytically continued along the arc L to a neighborhood δ' of the point c'. We use the same notation $\eta_k(z)$ to denote the function obtained by this continuation, which is thus defined in a neighborhood of L in V. The theorem on invariance of analytic relations under analytic continuation implies that the set $\{(z, \eta_k(z)) \mid z \in L\}$ is contained in $\mathcal{S}_0$. Therefore, for any $z' \in V$, the values $\eta_k(z')$ $(k = 1, \dots, l)$ are the l distinct zeros of the equation $f(z, w) = 0$ for w in Γ.

Thus, in the case when $V = \Delta$, each $\eta_k(z)$ defines a single-valued holomorphic function in Δ. Hence the set $\mathcal{S}_0$ is given by the equation

$$P(z, w) := \prod_{k=1}^{l} [w - \eta_k(z)] = 0.$$

By expanding $P(z, w)$ as a polynomial in w and using $\eta_k(a) = 0$ $(k = 1, \dots, l)$, we have the desired representation of $\mathcal{S}_0$.

We next treat the case when $V \neq \Delta$. Take any closed curve L in V containing the point c; we consider c as the initial and terminal point of L. Each holomorphic function $\eta_k(z)$ $(k = 1, \dots, l)$ defined on δ (so that $\eta_k(z) \subset \gamma_k$) can be analytically continued along the curve L, and the resulting function, which is now defined on a neighborhood of the terminal point c, must be identical with one of the functions $\eta_{j_k}(z) \subset \gamma_{j_k}$, where $j_k \in \{1, \dots, l\}$. Since $j_k \neq j_h$ for $k \neq h$, it follows that $(j_1, \dots, j_l)$ is a permutation of $(1, \dots, l)$. Thus for any $z \in V$, if we let $\eta_k(z)$ $(k = 1, \dots, l)$ denote the l distinct zeros of the equation $f(z, w) = 0$ for w in Γ, the square of the product of the differences

$$d(z) := \prod_{h<k} [\eta_h(z) - \eta_k(z)]^2$$

defines a single-valued, holomorphic function on V.

Now fix a boundary point $z' \in \partial V$ in Δ. Since the number of distinct zeros of the equation $f(z', w) = 0$ is less than l, it follows from (2.1) that

$$\lim_{z \to z'} \min_{k \neq h} |\eta_k(z) - \eta_h(z)| = 0;$$

hence

$$\lim_{z \to z'} d(z) = 0.$$

By setting $d(z) \equiv 0$ in $\Delta \setminus V$, we obtain that $d(z)$ is a continuous function on Δ which is holomorphic at all points z where $d(z) \neq 0$. By Radó's theorem, we conclude that $d(z)$ is holomorphic on Δ.

Now let σ denote the zero set of $d(z)$ in Δ, and let $\Delta^0 = \Delta \setminus \sigma = V$. Define

$$P(z, w) := \prod_{k=1}^{l} [w - \eta_k(z)]$$

for $(z, w) \in \delta \times \mathbf{C}$. Expanding this expression with respect to w, we obtain

$$\begin{aligned} P(z, w) &= w^l + a_1(z) w^{l-1} + \cdots + a_l(z) \\ &= (w - b)^l + A_1(z)(w - b)^{l-1} + \cdots + A_l(z). \end{aligned} \tag{2.4}$$

Each coefficient function $a_h(z)$, and hence each $A_h(z)$ $(h = 1, \dots, l)$, is holomorphic on δ and can be analytically continued from δ to any point $z \in \Delta^0$ along an arc connecting c and z. Since each $a_h(z)$, and hence each $A_h(z)$, is clearly a symmetric function of the zeros $\eta_k(z)$ $(k = 1, \dots, l)$, it follows that each $A_h(z)$ is a single-valued, holomorphic function on Δ^0.

Note that each $\eta_k(z)$ is contained in Γ, so that $\eta_k(z)$ is bounded in Δ^0; hence $A_h(z)$ is bounded in Δ^0. From Riemann's removable singularity theorem, it follows that each $A_h(z)$ is a holomorphic function on all of Δ. By (2.1), it is easy to check that the analytic hypersurface defined by $P(z,w) = 0$ in Λ coincides with the hypersurface $\mathcal{S}^0$. Furthermore, since the equation $f(a,w) = 0$ for $w \in \Gamma$ has only one zero $w = b$ (by condition 2), we see that each coefficient $A_k(z)$ $(k = 1, \ldots, l)$ vanishes at a. This completes the proof of Proposition 2.1. □

REMARK 2.1. The holomorphic function $d(z)$ on V defined above is equal to the **discriminant** of the polynomial $P(z,w)$ in (2.4) with respect to w, i.e., $d(z)$ is the determinant of the following $(2l+1) \times (2l+1)$ square matrix:

$$d(z) = \begin{vmatrix} 1 & a_1(z) & \cdots & \cdots & & a_l(z) & & \\ & & \ddots & & \ddots & & & \\ & & 1 & a_1(z) & \cdots & & \cdots & a_l(z) \\ l & (l-1)a_1(z) & \cdots & \cdots & a_{l-1}(z) & & & \\ & & \ddots & & \ddots & & & \\ & & & l & (l-1)a_1(z) & \cdots & \cdots & a_{l-1}(z) \end{vmatrix},$$

where all non-indicated entries are 0. This is the same as the **resultant** of $P(z,w)$ and $(\partial P/\partial w)(z,w)$.

In general, a polynomial $P(z,w)$ in w whose coefficients $A_h(z)$ are holomorphic functions in a domain D in $\mathbf{C}^n$ is called a **pseudopolynomial** in w. In particular, when D is a polydisk Δ in $\mathbf{C}^n$, the coefficient of the term of highest degree in w is identically equal to 1, and each $A_h(z)$ vanishes at the center of Δ, we call $P(z,w)$ a **distinguished pseudopolynomial** in w.

Let $P(z,w)$ be a distinguished pseudopolynomial in w of degree l. The discriminant $d(z)$, constructed as in the proof of the previous proposition, defines a holomorphic function on Δ. We let σ denote the zero set of $d(z)$ in Δ. Then the hypersurface $P(z,w) = 0$ in $\Delta \times \mathbf{C}_w$ is the graph of the multivalued holomorphic function $w = \eta(z)$ over $\Delta \setminus \sigma$ composed of l distinct branches $\{\eta_k(z)\}$ $(k = 1, \ldots, l)$ with the property that if a point z in $\Delta \setminus \sigma$ approaches a point $\zeta \in \sigma$, then each branch $\eta_k(z)$ tends to a point $w_k \in \mathbf{C}_w$ with $w_k = w_h$ for some $k \neq h$. We call the multivalued holomorphic function $\eta(z)$ on Δ the **implicit function** or the **algebraic function** determined by the equation $P(z,w) = 0$.

2.1.3. Weierstrass Preparation Theorem. In the previous section we studied the structure of analytic hypersurfaces as subsets of $\mathbf{C}^{n+1}$. We will now make a more systematic study involving the notions of multiplicity and irreducibility.

Let $f(z,w)$ be a holomorphic function in a domain D in $\mathbf{C}^{n+1}$. Fix (a,b) in $\mathcal{S} = \{(z,w) \in D \mid f(z,w) = 0\}$ and let $\Lambda = \overline{\Delta} \times \Gamma$ be a closed polydisk in D chosen so that we are in the situation described in 2.1.2. We take a point c in Δ such that $f(c,w)$ has exactly l distinct zeros $w = b_k$ in Γ, where l is defined in (2.2). We use the same notation as in the proof of Proposition 2.1: b_k $(k = 1, \ldots, l)$, γ_k, δ, $\eta_k(z)$ and $\Delta^0 = \Delta \setminus \sigma$, where $\sigma = \{z \in \Delta : d(z) = 0\}$.

For each $k = 1, \ldots, l$, we let $\nu_k \geq 1$ denote the order of the zero of $f(c,w)$ at the point $b_k := \eta_k(c)$; hence $\sum_k \nu_k = \nu$. Fix $c' \in \Delta^0$ and let L be an arc in Δ^0 joining c and c'. Then the holomorphic function $\eta_k(z)$, which is defined on the polydisk δ centered at c, can be analytically continued along L to c'; the values of this continuation, which we continue to denote by $\eta_k(z)$, lie in Γ. If we set $\eta_k(c') := b'_k$, then b'_k is one of the zeros of the function $f(c',w)$ for w in Γ.

Furthermore, the order of the zero of $f(c', w)$ at $w = b'_k$ is equal to ν_k, the order of the zero of $f(c, w)$ at the point b_k. This follows since the number of distinct zeros of $f(c, w)$ for w in Γ is the maximal number l.

We divide the family C consisting of the l holomorphic functions $\{\eta_k(z) \mid k = 1, \dots, l\}$ on δ into subclasses as follows: identify $\eta_{k_1}(z)$ with $\eta_{k_2}(z)$ if there exists a closed curve L in Δ^0 with initial and terminal point c such that the function element $\eta_{k_1}(z)$ at the initial point c is analytically continued along L to the function element $\eta_{k_2}(z)$ at the terminal point c. Clearly this gives a stratification of the family C into subclasses C_h $(h = 1, \dots, m)$ of equivalent function elements, which we write as $\eta_{h,k}$, $k = 1, \dots, l_h$; i.e.,

$$C = \bigcup_{h=1}^{m} C_h, \qquad \text{where } C_h := \{\eta_{h,k}(z) \mid k = 1, \dots, l_h\}.$$

For convenience, for each function element $\eta_{h,k}$, we use the notation $\gamma_{h,k}$ to denote the disk with center $b_{h,k}$ which corresponds to the disk γ_k with center b_k associated to the function η_k described previously. For fixed h $(h = 1, \dots, m)$, the order of the zero of $f(z, w)$ in w at each $\eta_{h,k}(z)$ $(k = 1, \dots, l_h)$ will be denoted by ν_h; this notation is consistent since this number is independent of $k = 1, \dots, l_h$. Thus we have $\nu = \sum_{h=1}^{m} l_h \nu_h$.

REMARK 2.2. If we shrink the radius r of the polydisk Δ centered at a, then the number m of subclasses C_h $(h = 1, \dots, m)$ may increase but cannot decrease. Since m is always less than or equal to l, it follows that if $r > 0$ is sufficiently small, the number of classes $\{C_h\}$ of the family C obtained by the above stratification on the polydisk Δ centered at a is independent of r.

For a fixed subclass C_h $(h = 1, \dots, m)$, we set

$$P_h(z, w) := \prod_{k=1}^{l_h} [w - \eta_{h,k}(z)] \qquad \text{in } \Delta \times \mathbf{C}_w.$$

Expanding this expression with respect to w, we obtain

$$P_h(z, w) = (w - b)^{l_h} + A_1^h(z)(w - b)^{l_h - 1} + \cdots + A_{l_h}^h(z).$$

Each coefficient $A_k^h(z)$ $(k = 1, \dots, l_h)$ is a single-valued holomorphic function of z in Δ satisfying $A_k^h(a) = 0$. We let $\mathcal{S}_h$ denote the zero set of $P_h(z, w)$ in Λ, so that $\mathcal{S}_0 = \mathcal{S} \cap \Lambda = \bigcup_{h=1}^{m} \mathcal{S}_h$.

We will need the following lemma, which follows directly from the invariance of analytic relations under analytic continuation.

LEMMA 2.3. *Let $F(z, w)$ be a holomorphic function in Λ. If $F(z, w)$ vanishes identically on an analytic hypersurface $w = \eta_{h,k}(z)$ in $\delta \times \gamma_{h,k}$, then $F(z, w)$ vanishes identically on the analytic hypersurface $\mathcal{S}_h$.*

In particular, $P_h(z, w)$ is a distinguished pseudopolynomial for w centered at (a, b) which satisfies the hypothesis of the lemma. Furthermore, we see from the above remark that if the radius r of the polydisk Δ centered at a is sufficiently small, then $P_h(z, w)$ is **irreducible at** (a, b), meaning that $P_h(z, w)$ cannot be written as a product of two non-constant distinguished pseudopolynomials in w

centered at (a, b). We define

$$P^*(z, w) := \prod_{h=1}^{m} [P_h(z, w)]^{\nu_h} \qquad \text{in } \Lambda,$$

and we have the following lemma.

LEMMA 2.4. *Let* $\omega(z, w) := f(z, w)/P^*(z, w)$ *for* (z, w) *in* Λ. *Then* $\omega(z, w)$ *is a non-vanishing holomorphic function on* Λ.

PROOF. By construction, $P_h(z, w) \neq 0$ on $\Delta \times \partial\Gamma$ for $h = 1, \dots, m$, so that $\omega(z, w)$ is holomorphic near $\Delta \times \partial\Gamma$. Thus we can develop $\omega(z, w)$ into a Hartogs-Laurent series of the form

$$\omega(z, w) = \sum_{j=-\infty}^{\infty} \beta_j(z)(w - b)^j,$$

where $\beta_j(z)$ is holomorphic in Δ for each $j = 0, \pm 1, \pm 2, \dots$. From the construction of $P^*(z, w)$, it follows that for any fixed $z' \in \Delta^0 = \Delta \setminus \sigma$, the holomorphic functions $f(z', w)$ and $P^*(z', w)$ of w in Γ have the same zeros with the same multiplicities. Hence the ratio $\omega(z', w)$ is a non-vanishing holomorphic function for w in Γ, so that $\beta_j(z') = 0$ for $j < 0$. Thus $\beta_j(z) = 0$ in all of Δ for $j < 0$. Therefore $\omega(z, w)$ is a holomorphic function in Λ and $\omega(z, w) \neq 0$ in $\Delta^0 \times \Gamma$. Since $\omega(z, w) \neq 0$ on $\Delta \times \partial\Gamma$, we conclude from Proposition 2.1 that $\omega(z, w) \neq 0$ on Λ. □

Summarizing these results, we have the following theorem.

THEOREM 2.1 (Weierstrass Preparation Theorem). *Let* $f(z, w)$ *be a holomorphic function on a domain* D *in* $\mathbf{C}^{n+1}$. *Assume that* $f(z, w)$ *satisfies the Weierstrass condition at a point* (a, b) *in* D *for the coordinates* (z, w). *Then there exists a closed polydisk* $\Lambda = \overline{\Delta} \times \Gamma \subset D$ *centered at* (a, b) *such that on* Λ, $f(z, w)$ *can be written in the form*

$$f(z, w) = \omega(z, w) \prod_{h=1}^{m} [P_h(z, w)]^{\nu_h}, \tag{2.5}$$

where each $P_h(z, w)$ *is an irreducible distinguished pseudopolynomial in* w *at the point* (a, b) *whose coefficients are holomorphic functions of* z *in* Δ, *and* $\omega(z, w)$ *is a non-vanishing holomorphic function for* (z, w) *in* Λ.

In the two-dimensional case, we get more information from the Weierstrass condition. Let $f(z, w)$ be a non-constant holomorphic function in a domain D in $\mathbf{C}^2$. Suppose that $f(a, b) = 0$ and $f(a, w) \equiv 0$ near $w = b$ in the w-plane. From the Taylor expansion of $f(z, w)$ about (a, b), there exist a positive integer μ and a neighborhood D' of (a, b) in D such that

$$f(z, w) = (z - a)^{\mu} f^0(z, w)$$

in D', where $f^0(z, w)$ is a holomorphic function of (z, w) in D' with $f^0(a, w) \not\equiv 0$. In particular, if $f^0(a, b) = 0$, then $f^0(z, w)$ satisfies the Weierstrass condition at (a, b) for the coordinates (z, w) in D. Thus without using a preliminary linear coordinate transformation we get the irreducible decomposition of $f(z, w)$ in a closed polydisk $\Lambda = \Delta \times \Gamma$ centered at (a, b):

$$f(z, w) = \omega(z, w)(z - a)^{\mu} \prod_{h=1}^{m} [P_h(z, w)]^{\nu_h}.$$

REMARK 2.3. From the proofs of Proposition 2.1, Lemma 2.3, and Lemma 2.4 we see that a global version of the Weierstrass preparation theorem holds:

Let D be a domain in $\mathbf{C}^n$ with variables $z = (z_1, \dots, z_n)$ and let U be a domain in the complex plane $\mathbf{C}$ with variable w. Consider the product domain $G = D \times U$ in $\mathbf{C}^{n+1}$. Suppose that $f(z,w)$ is a holomorphic function of (z,w) in $\overline{G}$ such that $f(z,w) \neq 0$ for (z,w) in $\overline{D} \times \partial U$. Then $f(z,w)$ can be written in the following form on all of G:

$$f(z,w) = \omega(z,w) \prod_{h=1}^{m} [P_h(z,w)]^{\nu_h},$$

where $P_h(z,w)$ $(h = 1, \dots, m)$ are pseudopolynomials which are monic in w with coefficients that are holomorphic functions of z in D, and $\omega(z,w)$ is a non-vanishing holomorphic function in G.

2.2. Analytic Sets (Local)

2.2.1. Definition. Let D be a domain in $\mathbf{C}^n$. A subset $\mathcal{E}$ of D is called an **analytic set** in D if $\mathcal{E}$ is defined locally as the common zero set of a finite number of holomorphic functions. To be precise, this means that for any point p in D there exists an open neighborhood U of p in D and a finite number of holomorphic functions $f_j(z)$ $(j = 1, \dots, l)$ in U such that

$$\mathcal{E} \cap U = \bigcap_{j=1}^{l} \{z \in U \mid f_j(z) = 0\}.$$

This is an equality of sets, i.e., we do not take multiplicity into account. Thus we may assume that each f_j has no repeated factors. By definition, an analytic set $\mathcal{E}$ in D is a closed subset of D. For the sake of convenience, the empty set and the whole domain D are considered to be analytic sets in D. If $\mathcal{E} \neq D$, then $\mathcal{E}$ is nowhere dense in D and does not separate D. An analytic hypersurface in D (i.e., the zero set of a single holomorphic function in D) is a particular type of analytic set in D.

Let E be a closed set in $\mathbf{C}^n$. Then we say that $\mathcal{E}$ is an analytic set in the closed set E if there exists an open neighborhood D of E in $\mathbf{C}^n$ such that $\mathcal{E}$ is an analytic set in D.

We note that a non-empty analytic set $\mathcal{E}$ in a closed disk $\overline{\Delta}$ in the complex plane $\mathbf{C}$ is either a finite set of points or coincides with $\overline{\Delta}$. This follows from the identity theorem for holomorphic functions of one complex variable.

We begin with the following proposition.

PROPOSITION 2.2. *Let $\mathcal{E}$ and $\mathcal{F}$ be analytic sets in a domain D in $\mathbf{C}^n$. Then the union $\mathcal{E} \cup \mathcal{F}$ and the intersection $\mathcal{E} \cap \mathcal{F}$ are also analytic sets in D.*

PROOF. Let $p \in D$. There exist a neighborhood U of p in D and a finite number of holomorphic functions $f_j(z)$ $(j = 1, \dots, l)$ and $g_k(z)$ $(k = 1, \dots, m)$ such that $\mathcal{E} \cap U = \bigcap_{j=1}^{l} \{z \in U \mid f_j(z) = 0\}$ and $\mathcal{F} \cap U = \bigcap_{k=1}^{m} \{z \in U \mid g_k(z) = 0\}$.

Then we have

$$\begin{aligned}(\mathcal{E}\cup\mathcal{F})\cap U &= \bigcap_{\substack{j=1,\dots,l\\ k=1,\dots,m}} \{z\in U \mid f_j(z)\cdot g_k(z)=0\};\\ (\mathcal{E}\cap\mathcal{F})\cap U &= \bigcap_{\substack{j=1,\dots,l\\ k=1,\dots,m}} \{z\in U \mid f_j(z)=0,\quad g_k(z)=0\}.\end{aligned}$$

It follows that both $\mathcal{E}\cup\mathcal{F}$ and $\mathcal{E}\cap\mathcal{F}$ are analytic sets in D. □

Let $\mathcal{E}$ be an analytic set in a domain D in $\mathbf{C}^n$. If $\mathcal{E}$ can be decomposed in the form

$$\mathcal{E}=\mathcal{E}_1\cup\mathcal{E}_2 \qquad \text{in } D,$$

where $\mathcal{E}_1$ and $\mathcal{E}_2$ are analytic sets in D such that neither $\mathcal{E}_i$ contains the other, then we say that $\mathcal{E}$ is **reducible** in D. Otherwise we say that $\mathcal{E}$ is **irreducible** in D.

Let $p\in\mathcal{E}$. If for all sufficiently small polydisks U centered at p, the analytic set $\mathcal{E}\cap U$ in U is reducible in U, we say that $\mathcal{E}$ is **reducible** at the point p. Otherwise $\mathcal{E}$ is said to be **irreducible** at p.

REMARK 2.4. An analytic set $\mathcal{E}$ in D may be irreducible at a point $p\in\mathcal{E}$ but reducible at a point $q\in\mathcal{E}$ which is arbitrarily close to p in D. For example, in $\mathbf{C}^3$ with variables x, y, and z, consider the analytic hypersurface $\mathcal{E}$ defined by

$$z^2-xy^2=0.$$

Then $\mathcal{E}$ is irreducible at the origin in $\mathbf{C}^3$, but $\mathcal{E}$ is reducible at any point $q=(x,0,0)$ in $\mathcal{E}$ with $x\neq 0$.

We now define the dimension of an analytic set at a point. Let $\mathcal{E}$ be an analytic set in a domain D in $\mathbf{C}^n$. Let $p\in\mathcal{E}$. Take a complex hyperplane L of dimension r_L $(0\le r_L\le n)$ passing through p such that in a neighborhood of p in D, $\mathcal{E}\cap L$ consists of only the point p. Here, by a complex hyperplane of dimension 0 we mean an isolated point, while a complex hyperplane of dimension n means the entire space $\mathbf{C}^n$. We let r $(0\le r\le n)$ denote the maximal such r_L with this property. Then $n-r$ is called the **dimension** of the analytic set $\mathcal{E}$ at the point p, and r is called the **codimension** of $\mathcal{E}$ at p. Furthermore, the maximum of the dimensions of $\mathcal{E}$ at all points q in $\mathcal{E}$ is called the **dimension** of the analytic set $\mathcal{E}$ in D. If $\mathcal{E}$ has the same dimension at each point of $\mathcal{E}$ in D, then $\mathcal{E}$ is said to be of **pure dimension** in D; if this dimension is l, we say that $\mathcal{E}$ is a **pure l-dimensional** analytic set in D.

In particular, a pure one-dimensional analytic set is often called an **analytic curve**. Clearly an analytic hypersurface $\mathcal{S}$ in D is a pure $(n-1)$-dimensional analytic set in D. A set of isolated points in D with no accumulation point in D is an analytic set of dimension 0, while the domain D itself is an analytic set of dimension n in D. For the sake of convenience the empty set $\emptyset$ is considered as an analytic set of dimension -1.

Let $\mathcal{E}_1$ and $\mathcal{E}_2$ be analytic sets in D which have dimension ν_1 and ν_2 at a point p in D. Then the dimension of the analytic set $\mathcal{E}_1\cup\mathcal{E}_2$ is equal to $\max(\nu_1,\nu_2)$, while that of $\mathcal{E}_1\cap\mathcal{E}_2$ is at most $\min(\nu_1,\nu_2)$.

2.2.2. Projections of Analytic Sets. Let $\mathcal{E}$ be an analytic set in a domain D in $\mathbf{C}^n$. We set $n = r+s$ where r and s are positive integers, and we consider $\mathbf{C}^n$ as the product of $\mathbf{C}^r$ with variables $z_1, \dots, z_r$ and $\mathbf{C}^s$ with variables $w_1, \dots, w_s$. Let $\mathcal{E}$ contain the origin 0 in $\mathbf{C}^n$. We let $D(0)$ and $\mathcal{E}(0)$ denote the sections of D and $\mathcal{E}$ over the s-dimensional hyperplane $z_j = 0$ $(j = 1, \dots, r)$. Then $0 \in \mathcal{E}(0) \subset D(0) \subset \mathbf{C}^s$, and the section $\mathcal{E}(0)$ is an analytic set in $D(0)$.

Suppose that in some neighborhood δ of the origin in $\mathbf{C}^s$ the set $\mathcal{E}(0)$ consists of the single point 0 (we use the same notation for the origin in $\mathbf{C}^r$, $\mathbf{C}^s$ and $\mathbf{C}^n$). Then the dimension of $\mathcal{E}$ at the origin in $D \subset \mathbf{C}^n$ is at most r.

Let Γ be a closed polydisk centered at the origin 0 in $\mathbf{C}^s$ with radius ρ_k $(k = 1, \dots, s)$,

$$\Gamma \; : \; |w_k| \le \rho_k \qquad (k = 1, \dots, s),$$

with ρ_k chosen so that $\Gamma \subset D(0)$ and $\Gamma \cap \mathcal{E}(0) = \{0\}$.

Then we let $\overline{\Delta}$ be a closed polydisk centered at the origin 0 in $\mathbf{C}^r$ with radius r_j $(j = 1, \dots, r)$,

$$\overline{\Delta} \; : \; |z_j| \le r_j \qquad (j = 1, \dots, r),$$

with r_j chosen so that $\Lambda := \overline{\Delta} \times \Gamma \subset D$ and $(\overline{\Delta} \times \partial\Gamma) \cap \mathcal{E} = \emptyset$. This is possible because $\mathcal{E}$ is a closed subset of D in $\mathbf{C}^n$. We define

$$\mathcal{E}^0 := \mathcal{E} \cap \Lambda,$$

which is an analytic set in Λ.

We let π_r denote the projection mapping of Λ onto $\overline{\Delta}$. Then we have the following proposition, which is indispensable for inductive arguments on dimension.

PROPOSITION 2.3. *The projection $\pi_r(\mathcal{E}^0)$ of $\mathcal{E}^0$ onto $\overline{\Delta}$ is an analytic set in $\overline{\Delta}$.*

PROOF. We prove the proposition by induction on $s = n - r$. To begin, we assume $s = 1$, so that $r = n - 1$. We let π_{n-1} denote the projection from $\mathbf{C}^n = \mathbf{C}^{n-1} \times \mathbf{C}_w$ onto $\mathbf{C}^{n-1}$. Fix z' in $\pi_{n-1}(\mathcal{E}^0)$. Since $(\overline{\Delta} \times \partial\Gamma) \cap \mathcal{E} = \emptyset$ and the section $\mathcal{E}^0(z')$ is an analytic set in the closed disk $\Gamma \subset \mathbf{C}_w$, it follows that $\mathcal{E}^0(z')$ in Λ consists of a finite number of points $p_i = (z', w_i)$ $(i = 1, \dots, \mu)$. For each point p_i in $\mathcal{E}^0(z')$ we choose a polydisk $\Lambda_i = \Delta_i \times \Gamma_i$, where Δ_i is a polydisk centered at z' and Γ_i is a disk centered at w_i in Γ, such that

$$\Lambda_i \subset \Lambda, \quad \Lambda_i \cap \Lambda_j = \emptyset \; (i \ne j), \quad (\Delta_i \times \partial\Gamma_i) \cap \mathcal{E}^0 = \emptyset.$$

By Proposition 2.2, it suffices to prove that the projection $\pi_{n-1}(\mathcal{E}^0 \cap \Lambda_i)$ of each analytic set $\mathcal{E}^0 \cap \Lambda_i$ $(i = 1, \dots, \mu)$ is an analytic set in Δ_i. For simplicity, we write $z' = 0$, $p_i = 0$, $\Delta_i = \Delta$, $\Gamma_i = \Gamma$, and $\Lambda_i = \Lambda$ as described above.

Note that Λ is a polydisk centered at 0 in $\mathbf{C}^n$; by taking a smaller polydisk if necessary, we can write

$$\mathcal{E}^0 = \bigcap_{j=1}^{l} \{(z, w) \in \Lambda \mid f_j(z, w) = 0\},$$

where each $f_j(z, w)$ $(j = 1, \dots, l)$ is a holomorphic function on Λ. Furthermore, from the condition that $(\Delta \cap \partial\Gamma) \cap \mathcal{E}^0 = \emptyset$, we have $\emptyset = (\{0\} \cap \partial\Gamma) \cap \mathcal{E}^0 = \mathcal{E}^0(0) \cap \partial\Gamma$, so that one of the $f_j(z, w)$ $(j = 1, \dots, l)$, say $f_1(z, w)$, satisfies $f_1(0, 0) = 0$ and $\{f_1(0, w) = 0\} \not\subset \partial\Gamma$; i.e., the one-variable holomorphic function $f_1(0, w)$ is not identically zero in Γ. By taking a smaller disk $\Gamma' \subset \Gamma$ centered at $w = 0$ with the same property that $(\overline{\Delta} \cap \partial\Gamma') \cap \mathcal{E}^0 = \emptyset$ (if necessary), we can assume $f_1(0, w) \ne 0$ for any $w \in \Gamma' \setminus \{0\}$. For simplicity, we use the same notation $\Gamma' = \Gamma$. By Proposition 2.1, the zero set of $f_1(z, w)$ in Λ, denoted $\mathcal{S}$, can be written in the following form:

$$P(z,w) := w^\nu + A_1(z)w^{\nu-1} + \ldots + A_\nu(z) = 0 \qquad \text{in } \Lambda,$$

where each $A_k(z)$ $(k = 1, \ldots, \nu)$ is a holomorphic function in Δ satisfying $A_k(0) = 0$. Note that $P(z, w)$ may be reducible but, from the construction in Proposition 2.1, it cannot have repeated factors; thus the discriminant $d(z)$ of $P(z, w)$ with respect to w does not vanish identically on Δ. We let σ denote the zero set of $d(z)$ in Δ, and we set $\Delta' := \Delta \setminus \sigma$.

Let $c = (c_1, \ldots, c_{n-1}) \in \Delta'$. We let $\mathcal{S}(c) \subset \Gamma$ denote the section of $\mathcal{S}$ over the hyperplane $z_j = c_j$ $(j = 1, \ldots, n-1)$. The set $\mathcal{S}(c)$ consists of ν distinct points $w = b_k$ $(k = 1, \ldots, \nu)$. For each $k = 1, \ldots, \nu$, let γ_k be a closed disk in Γ centered at b_k and with radius $\rho' > 0$,

$$\gamma_k \ : \ |w - b_k| \leq \rho',$$

with ρ' chosen so that $\gamma_k \cap \gamma_h = \emptyset$ $(k \neq h)$. Next, let δ be a small closed polydisk in Δ centered at c with radius $r' > 0$,

$$\delta \ : \ |z_j - c_j| \leq r' \qquad (j = 1, \ldots, r),$$

with r' chosen so that $\mathcal{S} \cap (\delta \times \partial\gamma_k) = \emptyset$ $(k = 1, \ldots, \nu)$. By Lemma 2.2, in the polydisk $\lambda_k := \delta \times \gamma_k \subset \Lambda$, the analytic set $\mathcal{S}$ can be written in the form

$$w = \eta_k(z) \qquad (k = 1, \ldots, \nu),$$

where each $\eta_k(z)$ is a single-valued holomorphic function in δ satisfying $\eta_k(c) = b_k$.

We introduce the $l - 1$ complex variables u_j $(j = 2, \ldots, l)$, and construct the following holomorphic function on $\delta \times \mathbf{C}^{l-1}$:

$$H(z,u) := \prod_{k=1}^{\nu} \left[f_2(z, \eta_k(z))u_2 + \cdots + f_l(z, \eta_k(z))u_l \right].$$

Expanding this function as a homogeneous polynomial in u_j $(j = 2, \ldots, l)$, we can write

$$H(z,u) = \sum g_{j_1 \ldots j_{l-1}}(z) u_{j_1} \ldots u_{j_{l-1}},$$

where each $g_{j_1 \ldots j_{l-1}}(z)$ is a holomorphic function in δ. To avoid multiple indices, we write $g_J(z) := g_{j_1 \ldots j_{l-1}}(z)$ for $1 \leq J \leq l' := (l-1)^{\nu-1}$. Each $\eta_k(z)$ $(k = 1, \ldots, \nu)$ can be analytically continued to any point $z' \in \Delta'$ along any arc in Δ' connecting c and z'. Moreover, the function element at z' obtained by this continuation coincides with one of the functions $\eta_{j_k}(z')$ determined by the equation $P(z', w) = 0$. Since each $g_J(z)$ $(J = 1, \ldots, l')$ is symmetric with respect to $\eta_k(z)$ $(k = 1, \ldots, \nu)$, we conclude that $g_J(z)$ can be analytically continued along any arc in Δ' and thus defines a single-valued holomorphic function in Δ'. Since each $\eta_k(z)$ $(k = 1, \ldots, \nu)$ is bounded in Δ', the same is true of each $g_J(z)$ $(J = 1, \ldots, l')$. It follows from Riemann's removable singularity theorem that $g_J(z)$ can be analytically extended across the analytic set σ to all of Δ. We use the same notation $g_J(z)$ to denote this holomorphic extension in Δ.

We set

$$\mathcal{E}' := \bigcap_{J=1}^{l'} \{z \in \Delta \mid g_J(z) = 0\},$$

which defines an analytic set in Δ. Note that we may have $g_J(z) \equiv 0\ (J = 1, \dots, l')$, i.e., $\Delta = \mathcal{E}'$.

To finish the proof of Proposition 2.3 for $s = 1$, we will show that $\mathcal{E}'$ coincides with the projection of $\mathcal{E}^0$ onto Δ; i.e.,

$$\mathcal{E}' = \pi_{n-1}(\mathcal{E}^0) \qquad \text{in} \quad \Delta. \tag{2.6}$$

In fact, we note from (2.1) that each $\eta_k(z)$ $(k = 1, \dots, \nu)$ is defined and continuous at all points of σ. From the construction of $w = \eta_k(z)$ using $f_1(z, w)$, and from the representation of $\mathcal{E}^0$,

$$\mathcal{E}^0 = \bigcap_{j=1}^{l} \{(z, w) \in \Lambda \mid f_j(z, w) = 0\},$$

we easily obtain the following equivalent representation of $\mathcal{E}^0$:

$$\mathcal{E}^0 = \bigcup_{k=1}^{\nu} \{(z, \eta_k(z)) \in \Lambda \mid f_j(z, \eta_k(z)) = 0\ (j = 2, \dots, l)\}.$$

To prove (2.6), first fix $z' \in \mathcal{E}'$; i.e., let $g_J(z') = 0$ $(J = 1, \dots, l')$. It follows that $H(z', u) \equiv 0$ for $u \in \mathbf{C}^{l-1}$. Hence for some $k \in \{1, \dots, \nu\}$

$$f_2(z', \eta_k(z'))\, u_2 + \cdots + f_l(z', \eta_k(z'))\, u_l \equiv 0$$

for all $u \in \mathbf{C}^{l-1}$. Therefore,

$$f_j(z', \eta_k(z')) = 0 \qquad (j = 2, \dots, l).$$

It follows from the description of $\mathcal{E}^0$ above that $(z', \eta_k(z')) \in \mathcal{E}^0$.

Conversely, fix $(z', w') \in \mathcal{E}^0$. From the description of $\mathcal{E}^0$, we have $w' = \eta_k(z')$ for some $k \in \{1, \dots, \nu\}$ such that $f_j(z', \eta_k(z')) = 0$ $(j = 2, \dots, l)$. It follows that $H(z', u) \equiv 0$ for $u \in \mathbf{C}^{l-1}$, and hence that each coefficient $g_J(z') = 0$ $(J = 1, \dots, l')$; i.e., $z' \in \mathcal{E}'$. Thus (2.6) is valid and Proposition 2.3 for the case $s = 1$ is proved.[1]

Now we prove Proposition 2.3 for the case $s \geq 2$ under the assumption that it is true for the case $s - 1$. Using $\Delta \subset \mathbf{C}^r$ and $\Gamma \subset \mathbf{C}^s$ as described in the beginning of the section, prior to the statement of the proposition, we set $\Lambda' = \Delta' \times \Gamma' \subset \mathbf{C}^{n-1} \times \mathbf{C}$, where

$$\begin{aligned} \Delta' &: \quad |z_j| \leq r_j \quad (j = 1, \dots, r), \qquad |w_k| \leq \rho_k \quad (k = 1, \dots, s-1), \\ \Gamma' &: \quad |w_s| \leq \rho_s. \end{aligned}$$

By assumption $\mathcal{E}^0 \cap (\Delta \times \partial\Gamma) = \emptyset$, we see that $\mathcal{E}^0 \cap (\Delta' \times \partial\Gamma') = \emptyset$ in Λ'. Since the proposition has already been proved in the case $s = 1$, the projection $\mathcal{E}'$ of the analytic set $\mathcal{E}^0 \cap \Lambda'$ onto Δ' is thus an analytic set in Δ'.

Note that if we let Γ_{s-1} denote the polydisk

$$\Gamma_{s-1} : |w_k| \leq \rho_k \qquad (k = 1, \dots, s-1),$$

then we have $\mathcal{E}' \cap (\Delta \times \partial\Gamma_{s-1}) = \emptyset$. It follows from the inductive hypothesis that the projection $\mathcal{E}''$ of $\mathcal{E}'$ onto Δ is also an analytic set in Δ. Since $\mathcal{E}''$ is clearly identical with the projection of $\mathcal{E}^0$ onto Δ, it follows that the projection of $\mathcal{E}^0 \subset \mathbf{C}^n$ onto $\Delta \subset \mathbf{C}^r$ is an analytic set in Δ. We thus conclude that the projection $\pi_r(\mathcal{E}^0)$ of the analytic set $\mathcal{E}^0$ in Λ onto Δ is an analytic set in Δ. □

---

[1] This elegant method of introducing the $l-1$ indeterminants $u_2, \dots, u_l$ is due to R. Remmert and K. Stein [**69**]. However, this technique is not essential to verify the proposition.

REMARK 2.5. Under the same hypothesis as in Proposition 2.3, it follows that the dimension of $\mathcal{E}^0$ at the origin 0 is at most r. Furthermore, we see from the proof that $\dim \pi_r(\mathcal{E}^0)$ at $z=0$ in Δ is equal to $\dim \mathcal{E}^0$ at the origin 0 in Λ. Thus, $\pi_r(\mathcal{E}^0) = \Delta$ if and only if $\dim \mathcal{E}^0 = r$ at the origin 0.

2.2.3. Locally Algebraic Analytic Sets. In this section we consider analytic sets with a specific structure; the so-called complete locally algebraic analytic sets. Our goal is to develop the machinery needed to study irreducible decompositions of analytic sets in the next section. As in the previous section we set $n = r+s$ and write $\mathbf{C}^n = \mathbf{C}^r \times \mathbf{C}^s$, where $\mathbf{C}^r$ and $\mathbf{C}^s$ are the spaces of the r complex variables $z_1, \ldots, z_r$ and of the s complex variables $w_1, \ldots, w_s$. We consider a polydisk Δ in $\mathbf{C}^r$ centered at the origin 0 with radius r_j $(j = 1, \ldots, r)$,

$$\Delta : \ |z_j| < r_j \qquad (j = 1, \ldots, r).$$

For each variable w_k $(k = 1, \ldots, s)$, we consider a monic distinguished pseudopolynomial $P_k(z, w_k)$ in w_k of degree l_k,

$$P_k(z, w_k) = w_k^{l_k} + \alpha_1^k(z) w_k^{l_k - 1} + \cdots + \alpha_{l_k}^k(z) \qquad (k = 1, \ldots, s),$$

where each coefficient $\alpha_j^k(z)$ $(j = 1, \ldots, l_k)$ is a holomorphic function on Δ satisfying $\alpha_j^k(0) = 0$. We assume that $P_k(z, w_k)$ has no repeated factors, although we allow it to be reducible. In each coordinate plane $\mathbf{C}_{w_k}$ we take a closed disk Γ_k centered at the origin with radius ρ_k,

$$\Gamma_k : \ |w_k| \le \rho_k \qquad (k = 1, \ldots, s),$$

where ρ_k is chosen sufficiently large so that for any fixed $z \in \Delta$, all l_k solutions of the equation $P_k(z, w_k) = 0$ for w_k in $\mathbf{C}_{w_k}$ are contained in Γ_k. We set

$$\Gamma := \Gamma_1 \times \cdots \times \Gamma_s \subset \mathbf{C}^s \qquad \text{and} \qquad \Lambda := \Delta \times \Gamma \subset \mathbf{C}^n.$$

We consider the pseudopolynomials $P_k(z, w_k)$ as functions on Λ which are independent of $s-1$ variables $w_1, \ldots, \widehat{w_k}, \ldots, w_s$. Then we get an analytic set Σ in Λ given by the following s equations:

$$\Sigma : \ P_k(z, w_k) = 0 \qquad (k = 1, \ldots, s).$$

This analytic set is called a **complete locally algebraic analytic set** with parameters $z_1, \ldots, z_r$. Note that Σ is a pure r-dimensional analytic set in the product space $\Delta \times \mathbf{C}^s$ as well as being a pure r-dimensional analytic set in the polydisk $\Lambda = \Delta \times \Gamma$.

We now consider in greater detail complete locally algebraic analytic sets. To simplify the exposition, we recall the notion of holomorphic mappings (or vector-valued functions) introduced in section 1.3.5. Let D be a domain in $\mathbf{C}^n$, and let $f_j(z)$ $(j = 1, \ldots, m)$ be holomorphic functions in D. Then the m-tuple of holomorphic functions,

$$f(z) := (f_1(z), \ldots, f_m(z)),$$

is called a **holomorphic mapping from $D \subset \mathbf{C}^n$ into $\mathbf{C}^m$**. In other words, $f(z)$ is a $\mathbf{C}^m$-valued holomorphic function on D. As in the case of complex-valued holomorphic functions, we can consider the analytic continuation of $f(z)$ along an arc l in $\mathbf{C}^n$ starting from a point of D; this is merely the **simultaneous analytic continuation** of all the functions $f_j(z)$ $(j = 1, \ldots, m)$ along the arc l.

Let $\Sigma : \ P_k(z, w_k) = 0$ $(k = 1, \ldots, s)$ be a complete locally algebraic analytic set in $\Lambda = \Delta \times \Gamma$. We let $d_k(z)$ denote the discriminant of $P_k(z, w_k)$ with respect to w_k and we let σ_k denote the zero set of $d_k(z)$ in Δ. Note that $d_k(z) \not\equiv 0$. We also set

$$\sigma = \bigcup_{k=1}^{s} \sigma_k \qquad \text{and} \qquad \Delta' = \Delta \setminus \sigma.$$

Let $c = (c_1, \dots, c_r)$ be a fixed point in Δ'. For each pseudopolynomial $P_k(z, w_k)$ ($k = 1, \dots, s$) of degree l_k in w_k, we denote by $b_{k,j}$ ($j = 1, \dots, l_k$) the set of all complex numbers w_k satisfying the equation $P_k(c, w_k) = 0$. These l_k solutions are distinct and simple (order 1). For each $j = 1, \dots, l_k$, we take a closed disk $\gamma_{k,j}$ in $\mathbf{C}_{w_k}$,

$$\gamma_{k,j} \ : \ |w_k - b_{k,j}| \le \rho' \qquad (j = 1, \dots, l_k),$$

centered at $b_{k,j}$ and with radius ρ' sufficiently small so that $\gamma_{k,j} \subset \Gamma_k$ and $\gamma_{k,j} \cap \gamma_{k,i} = \emptyset$ if $i \ne j$. Next we take a closed polydisk δ in Δ',

$$\delta \ : \ |z_j - c_j| \le r' \qquad (j = 1, \dots, r),$$

centered at c and with radius r' chosen so that $P_k(z, w_k) \ne 0$ for any $(z, w_k) \in \delta \times \partial\gamma_{k,j}$ ($j = 1, \dots, l_k$). We let $H_{k,j}$ denote the zero set of $P_k(z, w_k)$ in the polydisk $\lambda_{k,j} := \delta \times \gamma_{k,j}$. From Lemma 2.2, the analytic hypersurface $H_{k,j}$ can be described as

$$H_{k,j} : \ w_k = \eta_{k,j}(z) \qquad \text{in } \lambda_{k,j},$$

where $\eta_{k,j}(z)$ is a single-valued holomorphic function in δ (see Figure 2).

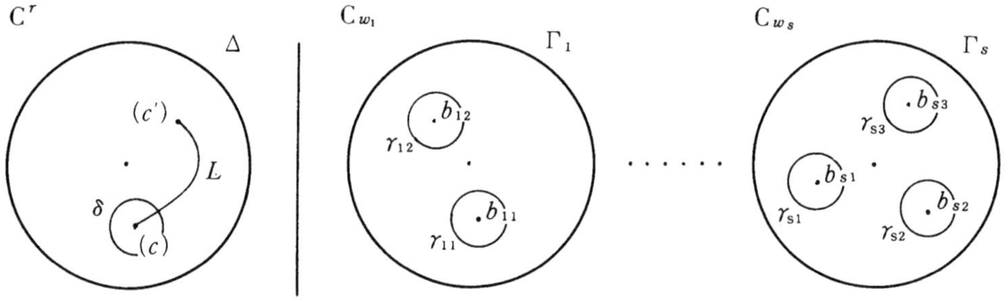

FIGURE 2. Representation of analytic set

For each $k = 1, \dots, s$, we choose a number j_k ($1 \le j_k \le l_k$), and form the s-tuple of integers $j := (j_1, \dots, j_s)$. Then we construct an associated holomorphic mapping on δ, the holomorphic $\mathbf{C}^s$-valued function

$$\eta_j(z) := (\eta_{1,j_1}(z), \dots, \eta_{s,j_s}(z)).$$

The total number of such mappings is $N := l_1 \cdots l_s$. We set

$$\gamma_j := \gamma_{1,j_1} \times \cdots \times \gamma_{s,j_s} \subset \Gamma \qquad \text{and} \qquad \lambda_j := \delta \times \gamma_j \subset \Lambda.$$

In the closed polydisk λ_j centered at $(c, b_{1,j_1}, \dots, b_{s,j_s})$ in $\mathbf{C}^n$ we set

$$\Sigma_j \ : \ w = \eta_j(z), \qquad z \in \delta,$$

so that Σ_j is a pure r-dimensional analytic set in λ_j. From the definition of the analytic set Σ in Λ, we see that $\Sigma \cap (\delta \times \Gamma)$ coincides with the union of the N analytic sets Σ_j.

The holomorphic mapping $\eta_j(z)$ can be analytically continued to any point $c' \in \Delta'$ along any arc L in Δ' connecting the points c and c', i.e., simultaneous analytic continuation of all $\eta_{1,j_1}(z), \ldots, \eta_{s,j_s}(z)$ along L.

From the theorem on invariance of analytic relations under analytic continuation, we see that the arc

$$w = \eta_j(z), \qquad z \in L,$$

is contained in Σ. Therefore, similar to our procedure in section 2.1.3, we can classify the N holomorphic mappings $\eta_j(z)$ defined on δ into subclasses K^h ($h = 1, \ldots, m$) as follows: η_j^h and $\eta_{j'}^h$ belong to the same class if and only if η_j^h can be analytically continued to $\eta_{j'}^h$ along a closed curve L in Δ' with initial and terminal point c. Here we use the notation $\eta_j^h(z)$ to denote a holomorphic mapping belonging to the subclass K^h.

REMARK 2.6. If we shrink the polydisk Δ, the number m of distinct classes K^h may increase but m is always bounded above by N. Thus this number m is invariant for sufficiently small Δ; for the rest of the section, we fix such a Δ.

For each subclass K^h ($h = 1, \ldots, m$), we analytically continue the mappings $\eta_j^h(z)$ on δ along all possible arcs L in Δ' for which the simultaneous continuation is possible. We continue to use the same notation $\eta_j^h(z)$ for the holomorphic mappings now defined on Δ'. We then define the r-dimensional analytic set Σ_h' in $\Lambda' := \Delta' \times \Gamma$ by

$$\Sigma_h' : \; w = \eta_j^h(z), \qquad z \in \Delta', \qquad \eta_j^h \in K^h,$$

and set

$$\Sigma_h := \overline{\Sigma_h'} \qquad \text{in } \Lambda = \Delta \times \Gamma.$$

We note that the set Σ_h is uniquely determined by the subclass K^h ($h = 1, \ldots, m$).

We have the following two lemmas.

LEMMA 2.5. *Each Σ_h ($h = 1, 2, \ldots, m$) is an analytic set in the polydisk Λ.*

PROOF. We fix $h = 1, \ldots, m$. We introduce s new complex variables v_k ($k = 1, \ldots, s$) and construct the following function of (z, w, v) in $\delta \times \mathbf{C}^s \times \mathbf{C}^s$:

$$Q(z, w, v) := \prod_{(j)} \left[\{w_1 - \eta_{1,j_1}^h(z)\} v_1 + \cdots + \{w_s - \eta_{s,j_s}^h(z)\} v_s \right],$$

where $\eta_{k,j_k}^h(z)$ ($k = 1, \ldots, s$) are the component functions of the holomorphic mapping $\eta_j^h(z)$, and the product is taken over all holomorphic mappings $\eta_j^h(z)$ from the class K^h. Expanding $Q(z, w, v)$ into a homogeneous polynomial of v_k ($k = 1, \ldots, s$), we see that the coefficients $g_J(z, w)$ ($J = 1, \ldots, l'' := s^{\#K^h}$) are polynomials in $w_1, \ldots, w_s$ whose coefficients $\alpha_J^i(z)$ are holomorphic functions of z in δ.

Let $c' \in \Delta'$ and connect c and c' by an arc L in Δ'. Since all the holomorphic mappings $\eta_j^h(z)$ in K^h can be analytically continued from c to c' along L, the function $Q(z, w, v)$, as a function of z, can also be analytically continued. Note that in the w and v variables, $Q(z, w, v)$ is a polynomial.

On the other hand, from the explicit form of $Q(z, w, v)$, it is clearly symmetric with respect to all the holomorphic mappings $\eta_j^h(z)$ belonging to K^h. It follows that $Q(z, w, v)$ defines a single-valued, holomorphic function in $\Delta' \times \mathbf{C}^s \times \mathbf{C}^s$. In particular, all of the coefficient functions $\alpha_J^i(z)$ of $g_J(z, w)$ are single-valued, holomorphic functions in Δ'. Since all the mappings $\eta_j^h(z)$ are bounded in Δ', the same is true of each $\alpha^i_J(z)$. It now follows from Riemann's removable singularity theorem that each $\alpha^i_J(z)$ can be analytically extended across the analytic hypersurfaces σ to all of Δ, and thus $g_J(z,w)$ can be analytically continued to the domain $\Delta \times \mathbf{C}^s$. Thus, using the same notation $g_J(z,w)$ $(J = 1, \ldots, l'')$ for this single-valued holomorphic extension, $g_J(z,w)$ is a polynomial in $w_1, \ldots, w_s$ whose coefficients are single-valued holomorphic functions of z in Δ.

We now consider the following analytic set in Λ:

$$\Sigma''_h : \; g_J(z,w) = 0 \qquad (J = 1, \ldots, l'').$$

From the definition of Σ'_h given before the statement of the lemma, and using the definition of $Q(z,w,v)$, it is easy to verify that Σ''_h coincides with the set $\Sigma_h := \overline{\Sigma'_h}$. Hence, Σ_h is an analytic set in Λ. □

From the proof we have the following remark.

REMARK 2.7. The analytic set Σ_h can be written as $\Sigma_h = \{(z,w) \in \Lambda \mid f_j(z,w) = 0 \; (j = 1, \ldots, \nu)\}$, where each $f_j(z,w)$ $(j = 1, \ldots, \nu)$ is a holomorphic function on Λ.

Finally we have the following lemma.

LEMMA 2.6. *Each Σ_h $(h = 1, \ldots, m)$ is an irreducible analytic set at the origin 0 in $\mathbf{C}^n$.*

PROOF. Let $F(z,w)$ be a holomorphic function in Λ such that $F(z,w) = 0$ on one of the analytic sets $\Sigma^h_j : \; w = \eta^h_j(z)$ in $\delta \times \gamma_j$. Then the theorem on invariance of analytic relations under analytic continuation implies that $F(z,w) \equiv 0$ on all of Σ_h. Thus Σ_h is irreducible at the origin 0 in $\mathbf{C}^n$. □

Indeed, we conclude that the complete locally algebraic analytic set Σ in Λ can be represented as the union of a finite number of irreducible analytic sets Σ_h in Λ,

$$\Sigma = \bigcup_{h=1}^{m} \Sigma_h, \tag{2.7}$$

and this union is the irreducible decomposition of Σ at the origin 0 in $\mathbf{C}^n$.

We call each irreducible component Σ_h $(h = 1, \ldots, m)$ a **locally algebraic analytic component** of the complete locally algebraic analytic set Σ.

To represent Σ'_h in $\Delta' \times \Gamma$, we used the $\mathbf{C}^s$-valued holomorphic functions $\eta^h_j(z)$ on Δ'. If $z \in \Delta'$ approaches a point $p \in \sigma$, we see from (2.1) that each branch of $\eta^h_j(z)$ tends to a certain point P in Γ. We thus get a single multiply $\mathbf{C}^s$-valued function $\eta^h_j(z)$ on Δ, which we call a **locally algebraic holomorphic mapping** (or a **locally vector-valued algebraic function**) on Δ. Moreover, the analytic set Σ_h in Λ is called the **graph of** $\eta^h_j(z)$ or the **analytic set determined by** $w = \eta^h_j(z)$.

2.2.4. Irreducible Decompositions of Analytic Sets. We return to the study of general analytic sets. Let $\mathcal{E}$ be an analytic set in a domain D in $\mathbf{C}^n$. We assume that $\mathcal{E}$ contains the origin 0 and that the dimension of $\mathcal{E}$ at 0 is less than or equal to r. We then choose suitable coordinates $z_1, \ldots, z_r$ and $w_1, \ldots, w_s$ (where $n = r + s$) so that the section $\mathcal{E}(0)$ of $\mathcal{E}$ over the hyperplane $z_j = 0$ $(j = 1, \ldots, r)$ consists of the single point 0 in $\mathbf{C}^s$ in a neighborhood of the origin.

We have the following lemma.

LEMMA 2.7. *For each variable w_k $(k = 1, \dots, s)$ there exists a distinguished pseudopolynomial $P_k(z, w_k)$ in w_k with the following property: the complete locally algebraic analytic set Σ with parameters $z_1, \dots, z_r$ defined by*

$$\Sigma \ : \ P_k(z, w_k) = 0 \qquad (k = 1, \dots, s)$$

contains the analytic set $\mathcal{E}$ in a neighborhood of the origin 0 in $\mathbf{C}^n$.

PROOF. For a fixed integer k $(1 \leq k \leq s)$, consider the space $\mathbf{C}^{r+1}$ of the complex variables $z_1, \dots, z_r$ and w_k. We take a closed polydisk Λ_k centered at 0 in $\mathbf{C}^{r+1}$, where

$$\Lambda_k := \overline{\Delta} \times \Gamma_k, \qquad \overline{\Delta} \ : \ |z_j| \leq r_j \quad (j = 1, \dots, r), \qquad \Gamma_k \ : \ |w_k| \leq \rho_k.$$

Set

$$\Gamma := \Gamma_1 \times \cdots \times \Gamma_s, \qquad \Lambda = \overline{\Delta} \times \Gamma.$$

By assumption we can choose $r_j > 0$ $(j = 1, \dots, r)$ and $\rho_k > 0$ $(k = 1, \dots, s)$ sufficiently small so that $\mathcal{E} \cap (\overline{\Delta} \times \partial\Gamma) = \emptyset$. We set

$$\mathcal{E}^0 := \mathcal{E} \cap \Lambda, \qquad \mathcal{E}_k := \pi_k(\mathcal{E}^0) \quad (k = 1, \dots, s),$$

where π_k is the projection map from Λ onto Λ_k. By Proposition 2.3 and $\mathcal{E} \cap (\overline{\Delta} \times \partial\Gamma) = \emptyset$, each $\mathcal{E}_k$ is an analytic set in Λ_k $(k = 1, \dots, s)$. Thus we can assume $\mathcal{E}_k$ can be written in a neighborhood δ_k of the origin 0 in Λ_k in the form

$$g_j^k(z, w_k) = 0 \qquad (j = 1, \dots, m_k),$$

where each $g_j^k(z, w_k)$ is a holomorphic function in δ_k.

We note that the section $\mathcal{E}_k(0)$ of $\mathcal{E}_k$ over the hyperplane $z_j = 0$ $(j = 1, \dots, r)$ consists of the single point 0 in a neighborhood of the origin in the disk Γ_k of the w_k plane. It follows that at least one of the functions $g_j^k(z, w_k)$ $(j = 1, \dots, m_k)$, say $g_1^k(z, w_k)$, satisfies the Weierstrass condition at the origin 0 in the coordinates (z, w_k). By taking a smaller polydisk if necessary, we can assume that $g_1^k(z, w_k) \neq 0$ for $(z, w_k) \in \overline{\Delta} \times \partial\Gamma_k$. We let $\widetilde{\mathcal{E}}_k$ denote the zero set of the holomorphic function $g_1^k(z, w_k)$ in Λ_k, so that $\mathcal{E}_k \subset \widetilde{\mathcal{E}}_k$. We see from Theorem 2.1 that $\widetilde{\mathcal{E}}_k$ coincides with the zero set of a distinguished pseudopolynomial $P_k(z, w_k)$ in w_k in the polydisk Λ_k. Note we may assume that for all $k = 1, \dots, s$, the same polydisk $\overline{\Delta}$ centered at 0 in $\mathbf{C}^r$ is taken in the construction of $\Lambda_k = \overline{\Delta} \times \Gamma_k$. It then follows that the analytic set $\mathcal{E} \cap \Lambda$ is contained in the complete locally algebraic analytic set Σ in Λ defined as

$$\Sigma : \ P_k(z, w_k) = 0 \qquad (k = 1, \dots, s).$$

Lemma 2.7 is thus proved. □

The complete locally algebraic analytic set Σ in Lemma 2.7 can be decomposed into the union of irreducible components $\Sigma = \bigcup_{h=1}^{l} \Sigma_h$ in a polydisk Λ centered at the origin 0 in $\mathbf{C}^n$. Here we may need to take a smaller polydisk Λ than in the proof of Lemma 2.7. Since $\mathcal{E}^0 \subset \Sigma$, it follows that $\mathcal{E}^0$ can be decomposed into the following (not necessarily irreducible) analytic sets:

$$\mathcal{E}^0 = \bigcup_{h=1}^{l} \mathcal{E}_h \qquad \text{in } \Lambda,$$

where

$$\mathcal{E}_h = \mathcal{E} \cap \Sigma_h \qquad (h = 1, \dots, l).$$

If we let r_h denote the dimension of $\mathcal{E}_h$ $(h = 1, \ldots, l)$ at the origin 0, then $0 \leq r_h \leq r$. Furthermore, $\mathcal{E}_h = \Sigma_h$ in Λ if and only if $r_h = r$, by the irreducibility of Σ_h. If $r_h \leq r - 1$, then $\mathcal{E}_h$ is an r_h-dimensional analytic set which is contained in Σ_h. The union of the analytic sets $\mathcal{E}_h$ which are of dimension r at the origin 0 will be denoted $\mathcal{E}^r$. The union of the remaining sets $\mathcal{E}_h$ will be denoted by $\mathcal{E}^*$. Hence we have the decomposition $\mathcal{E}^0 = \mathcal{E}^r \cup \mathcal{E}^*$, where $\mathcal{E}^r$ consists of all components of $\mathcal{E}^0$ which are pure r-dimensional irreducible analytic sets in Λ, while $\mathcal{E}^*$ is an analytic set of dimension at most $r - 1$ in Λ.

From Proposition 2.3 and Remark 2.5, $\pi_r(\mathcal{E}^*)$ is an analytic set in Δ of dimension at most $r - 1$. Thus, by taking a linear transformation of $z_1, \ldots, z_r$, if necessary, we may assume that the section $\mathcal{E}^*(0)$ of $\mathcal{E}^*$ over the hyperplane $z_1 = 0, \ldots, z_{r-1} = 0$ in a neighborhood of the origin 0 in $\mathbf{C}^{s+1}$ either consists of the single point 0 in $\mathbf{C}^{s+1}$ or is empty. We now repeat the above procedure for the analytic set $\mathcal{E}^*$ and obtain the following theorem.

THEOREM 2.2. *Let $\mathcal{E}$ be an analytic set in a domain D in $\mathbf{C}^n$, and let $a = (a_1, \ldots, a_n)$ be a point in $\mathcal{E}$ at which the dimension of $\mathcal{E}$ is r. Then there exist coordinates $(z_1, \ldots, z_n)$ and a polydisk Δ centered at the point a in D,*

$$\Delta \ : \quad |z_j - a_j| < r_j \qquad (j = 1, \ldots, n),$$

such that the analytic set $\mathcal{E}^0 = \Delta \cap \mathcal{E}$ can be decomposed into a finite number of irreducible analytic sets in Δ. Moreover, each irreducible component is pure dimensional in Δ, and each pure s-dimensional irreducible analytic set (with $s \leq r$) coincides with a locally algebraic analytic component in Δ with parameters $z_1, \ldots, z_s$.

We note from the proof that the coordinates $(z_1, \ldots, z_n)$ for $\mathbf{C}^n$ satisfy the conditions of Theorem 2.2 at the point $a = (a_1, \ldots, a_n)$ in such a way that, for $s \leq r$, the intersection $\mathcal{E}_s \cap \{z_1 = a_1, \ldots, z_s = a_s\}$ consists of the single point a in a neighborhood of a in $\mathbf{C}^n$, where $\mathcal{E}_s$ denotes the s-dimensional irreducible components of $\mathcal{E}$ at a.

Given an analytic set $\mathcal{E}$ in a neighborhood of a point a in $\mathbf{C}^n$, the coordinates $z = (z_1, \ldots, z_n)$ of $\mathbf{C}^n$ are said to satisfy the **Weierstrass condition for $\mathcal{E}$ at a** if the conclusion of Theorem 2.2 holds for $\mathcal{E}$ and a in these coordinates. We then call the closed polydisk $\overline{\Delta}$ centered at a in Theorem 2.2 a **Weierstrass canonical neighborhood of $\mathcal{E}$ at the point a** in $\mathbf{C}^n$.

REMARK 2.8. Let D be a domain in $\mathbf{C}^r_z$ and let $f_k(z)$ $(k = 1, \ldots, s)$ be s single-valued, complex-valued functions on D. We set $f(z) = (f_1(z), \ldots, f_s(z))$, $z \in D$, and consider the following subset $\mathcal{E}$ of $D \times \mathbf{C}^s_w$:

$$\mathcal{E} \ : \ w = f(z), \quad z \in D.$$

If either (i) $\mathcal{E}$ is an analytic set in $D \times \mathbf{C}^s_w$ or (ii) $f(z)$ is continuous on D and there exists an r-dimensional analytic set Σ in $D \times \mathbf{C}^s_w$ such that $\mathcal{E} \subset \Sigma$, then $f(z)$ is holomorphic on D.

Since the proofs in each case are similar, we will only give the proof for case (ii). Let $p_0 = (z_0, w_0) \in \mathcal{E}$. Using a linear coordinate transformation which is sufficiently close to the identity, if necessary, we may assume that the coordinates (z, w) satisfy the Weierstrass condition for Σ at p_0. Then we can find a polydisk $\Lambda := \Delta \times \Delta' \subset D \times \mathbf{C}^s_w$ centered at p_0 such that $(\mathcal{E} \bigcup \mathcal{S}) \cap (\Delta \times \partial \Delta') = \emptyset$ and such that $\Sigma \cap \Lambda$ is the union of components of a complete locally algebraic analytic set $\tilde{\Sigma}$,

$$\tilde{\Sigma} \,:\, P_k(z, w_k) = 0 \quad \text{in } \Lambda.$$

Equivalently,

$$\tilde{\Sigma} \,:\, w = \eta^l(z) \quad (l = 1, \ldots m), \quad z \in \Delta,$$

where $\eta^l(z) = (\eta_1^l(z), \ldots, \eta_s^l(z))$ $(l = 1, \ldots, m)$ is a locally algebraic vector-valued analytic function on Δ. We let $\sigma \subset \Delta$ be the union of the zero set of the discriminant $d_k(z)$ of $P_k(z, w_k)$ with respect to w_k $(k = 1, \ldots, s)$.

Fix $z' \in \Delta \setminus \sigma$. Since $\mathcal{E} \subset \Sigma \subset \tilde{\Sigma}$, we have

$$f(z') = \eta^l(z') \quad \text{for some } l = 1, \ldots, m.$$

Since $\eta^k(z) \neq \eta^l(z)$ $(k \neq l)$ for each point in $\Delta \setminus \sigma$, it follows from the continuity of $f(z)$ and $\eta^l(z)$ that $f(z) = \eta^l(z)$ on $\Delta \setminus \sigma$. Since σ is an analytic set in Δ of dimension at most $r - 1$, we conclude that $\eta^l(z)$ and $f(z)$ are single-valued on Δ; and, indeed, that $f(z) = \eta^l(z)$ on Δ. In particular, $f(z)$ is holomorphic on Δ.

This remark immediately implies the following fact. If $T : \delta \to \delta'$ is a holomorphic map between two domains δ and δ' in $\mathbf{C}^n$ which is one-to-one and onto, then T^{-1} is holomorphic.

Now let $\mathcal{E}$ be an analytic set in a domain D in $\mathbf{C}^n$. Let $p \in \mathcal{E}$ and let $\mathcal{E}$ be of pure dimension r at p. If there exists a closed polydisk $\Lambda = \overline{\Delta} \times \Gamma$ centered at $p = (a, b)$ in the coordinates (z, w) of $\mathbf{C}^n = \mathbf{C}^r \times \mathbf{C}^s$,

$$\begin{aligned} \overline{\Delta} \;&:\; |z_j - a_j| \leq r_j \qquad (j = 1, \ldots, r), \\ \Gamma \;&:\; |w_k - b_k| \leq \rho_k \qquad (k = 1, \ldots, s), \end{aligned}$$

such that $\mathcal{E} \cap \Lambda$ can be described in the form

$$w_k = \eta_k(z_1, \ldots, z_r) \qquad (k = 1, \ldots, s),$$

where $\eta_k(z_1, \ldots, z_r)$ $(k = 1, \ldots, s)$ are single-valued holomorphic functions in $\overline{\Delta}$, then the point p is called a **nonsingular** point of $\mathcal{E}$. Otherwise p is called a **singular** point of $\mathcal{E}$. The set of all nonsingular points of $\mathcal{E}$ in D is called the **nonsingular part** of $\mathcal{E}$. Clearly the closure of the nonsingular part of $\mathcal{E}$ equals $\mathcal{E}$. If $\mathcal{E}$ is irreducible in D, then the nonsingular part of $\mathcal{E}$ is connected.

REMARK 2.9. Let $\mathcal{E}$ be a pure r-dimensional analytic set in a domain $D \subset \mathbf{C}^n$. Then the set S of singular points of $\mathcal{E}$ in D is an analytic set of dimension at most $r - 1$.

PROOF. We maintain the notations $\Lambda = \Delta \times \Gamma \subset \mathbf{C}_z^r \times \mathbf{C}_w^s$ and $P_k(z, w_k) (k = 1, \ldots, s)$ used to verify formula (2.7) in Λ. Since the statement is local, we may assume that $D = \Lambda$ and $\mathcal{E} = \bigcup_{h=1}^{m'} \Sigma_h$ in Λ, where Σ_h is given in (2.7) and $m' \leq m$. Then the set S of singular points of $\mathcal{E}$ in Λ is of the form $S = S_1 \cup S_2$ where

$$\begin{aligned} S_1 &= \bigcup_{1 \leq k, l \leq m';\, k \neq l} (\Sigma_k \cap \Sigma_l), \\ S_2 &= \bigcap_{k=1}^{s} \{p \in \mathcal{E} \mid \frac{\partial P_k}{\partial w_k}(p) = \frac{\partial P_k}{\partial z_j}(p) = 0 \ (j = 1, \ldots, r)\}. \end{aligned}$$

Thus, S is an analytic set of dimension at most $r - 1$ in Λ. $\square$

REMARK 2.10. Let $\mathcal{E}$ and $\mathcal{F}$ be two irreducible analytic sets in a domain D in $\mathbf{C}^n$, with dimensions ν and μ. Then either the analytic set $\mathcal{E} \cap \mathcal{F}$ in $\mathbf{C}^n$ coincides with one of $\mathcal{E}$ or $\mathcal{F}$, or the dimension of $\mathcal{E} \cap \mathcal{F}$ at any point is strictly less than $\min\{\nu, \mu\}$. In the latter case, $\mathcal{E} \cap \mathcal{F}$ can be locally decomposed into a finite number of r_j-dimensional irreducible analytic sets, where $r_j < \min\{\nu, \mu\}$.

This may be proved by taking coordinates $(z_1, \dots, z_n)$ which satisfy the Weierstrass condition for both $\mathcal{E}$ and $\mathcal{F}$. This remark yields the following useful corollary.

COROLLARY 2.1. *The intersection of an infinite number of analytic sets in a domain D in $\mathbf{C}^n$ is an analytic set in D.*

2.3. Weierstrass Condition

Let $\mathcal{E}$ be an analytic set in a domain D in $\mathbf{C}^n$ whose dimension is at most $n-1$; i.e., $\mathcal{E} \neq D$. Let p be any point of $\mathcal{E}$ and let $z = (z_1, \dots, z_n)$ be coordinates of $\mathbf{C}^n$. From Proposition 2.3, we can easily find uncountably many systems of coordinates $z' = (z'_1, \dots, z'_n)$ of $\mathbf{C}^n$ which are sufficiently close to the given coordinates $z = (z_1, \dots, z_n)$ and such that $\mathcal{E}$ satisfies the Weierstrass condition at the point p in the z' coordinates. By "sufficiently close" we mean that $z'_1, \dots, z'_n$ are obtained from $z_1, \dots, z_n$ by a linear transformation whose transformation matrix is arbitrarily close to the identity matrix. Therefore we can find a dense subset K of $\mathcal{E}$ and coordinates $w = (w_1, \dots, w_n)$ of $\mathbf{C}^n$ for which $\mathcal{E}$ satisfies the Weierstrass condition at each point q of K. However, this does *not* yield the existence of coordinates for $\mathbf{C}^n$ such that $\mathcal{E}$ satisfies the Weierstrass condition at *every* point of the analytic set $\mathcal{E}$.

Thus the following theorem will be important, not only in this section, but especially in Part II of this book.

THEOREM 2.3. *Let D_j $(j = 1, 2, \dots)$ be a countable collection of domains in $\mathbf{C}^n$ and let $\mathcal{E}_j$ be an analytic set in D_j $(j = 1, 2, \dots)$ whose dimension is at most $n - 1$. Let $z = (z_1, \dots, z_n)$ be coordinates for $\mathbf{C}^n$. Then there exist coordinates $w = (w_1, \dots, w_n)$ for $\mathbf{C}^n$ sufficiently close to the coordinates $(z_1, \dots, z_n)$ such that* **every** *analytic set $\mathcal{E}_j$ $(j = 1, 2, \dots)$ satisfies the Weierstrass condition at* **each** *point of $\mathcal{E}_j$ $(j = 1, 2, \dots)$ in the w coordinates.*

Note we do not assume that $D_i \cap D_j \neq \emptyset$ for $i \neq j$. This section will be devoted to the proof of the theorem.[2]

2.3.1. Complex Lines Contained in an Analytic Hypersurface. As a first step towards the proof of the theorem, we study the question of determining when a complex line is contained in the zero set of a given holomorphic function. Let $f(z)$ be a non-constant holomorphic function in a domain D in $\mathbf{C}^n$. We let S denote the zero set of $f(z)$ in D. Given $z = (z_1, \dots, z_n) \in S$, we fix a complex line L passing through the point z in the (complex) direction $w = (w_1, \dots, w_n)$, i.e., $L: \ t \in \mathbf{C} \to wt + z = (w_1 t + z_1, \dots, w_n t + z_n) \in \mathbf{C}^n$, and we consider the restriction of $f(z)$ to L:

$$F(t) := f(wt + z).$$

The function $F(t)$ is defined and holomorphic for t in a neighborhood γ of $t = 0$ in the complex t-plane. A necessary and sufficient condition that the hypersurface

[2]This theorem was first proved by H. Grauert [**21**]. The proof given here is due to the author [**39**].

S contains a portion of the complex line L lying in a neighborhood of the point z in $\mathbf{C}^n$ is that $F(t) \equiv 0$ on γ. Clearly if w satisfies this condition, then so does any multiple $sw = (sw_1, \dots, sw_n)$ for $s \in \mathbf{C}$. For convenience, we include the case when the direction is $0 = (0, \dots, 0)$; clearly the 0 direction satisfies the above condition.

Let Γ denote the closed polydisk centered at the origin with radius 1 in $\mathbf{C}^n$, and let γ_ρ denote the closed disk centered at the origin with radius ρ in the complex t-plane; i.e.,

$$\Gamma : |w_j| \leq 1 \quad (j = 1, \dots, n), \qquad \gamma_\rho : |t| \leq \rho.$$

Fix a point $z^0 = (z_1^0, \dots, z_n^0)$ in the domain D, and choose a closed polydisk $\overline{\Delta}$ centered at z^0 with radius $r > 0$ sufficiently small so that $\overline{\Delta}$ lies in D,

$$\overline{\Delta} \ : \ |z_j - z_j^0| \leq r \qquad (j = 1, \dots, n).$$

We let $\rho > 0$ denote the polydisk distance from $\overline{\Delta}$ to ∂D. For any $(z, w, t) \in \overline{\Delta} \times \Gamma \times \gamma_\rho$, we set, using the same notation from above,

$$F(z, w, t) := f(wt + z),$$

which defines a holomorphic function in $\overline{\Delta} \times \Gamma \times \gamma_\rho$.

We have the following lemma.

LEMMA 2.8. *Let σ be the set of all points w in Γ with the following property: there exist a point $z \in \overline{\Delta} \cap S$ and a neighborhood V of the point z in $\mathbf{C}_z^n$ such that S contains the portion of a complex line L passing through the point z in the direction w which lies in V. Then σ is a closed, nowhere dense subset of Γ.*

PROOF. We develop $F(z, w, t)$ into a power series with respect to t,

$$F(z, w, t) = A_0(z, w) + A_1(z, w)\, t + A_2(z, w)\, t^2 + \cdots,$$

so that each coefficient $A_j(z, w)$ $(j = 0, 1, 2, \dots)$ is a holomorphic function of (z, w) in the closed polydisk $\Lambda := \overline{\Delta} \times \Gamma$. Consider the subset Σ of Λ defined by the countable number of equations

$$\Sigma := \{(z, w) \in \Lambda : \ A_j(z, w) = 0, \ j = 0, 1, 2, \dots\}.$$

It follows from Corollary 2.1 that Σ is an analytic set of dimension at most $2n - 1$ in Λ. From the necessary and sufficient condition that the hypersurface S contains the portion of a complex line L lying in a neighborhood V of the point z, we see that

$$\sigma = \pi_w(\Sigma) \quad \text{in } \Gamma,$$

where π_w is the projection from Λ onto Γ. Since Λ is closed, σ is a closed subset of Γ.

We now show that $\pi_w(\Sigma)$ contains no non-empty open set. Suppose, for the sake of obtaining a contradiction, that $\pi_w(\Sigma)$ contains a non-empty open set U. For any $a = (a_1, \dots, a_n) \in U$, we consider the section $\Sigma(a)$ of Σ over the hyperplane $w_j = a_j$ $(j = 1, \dots, n)$. For each $p \in \Sigma(a)$, we can find a sufficiently small neighborhood λ_p of the point p in Λ such that $\lambda_p \cap \Sigma$ can be decomposed into a finite number of irreducible components at p. Since $\Sigma(a)$ is compact in Λ, we can find a finite number of these neighborhoods λ_{p_j} $(j = 1, \dots, l)$ such that $\Sigma(a) \subset \bigcup_{j=1}^{l} \lambda_{p_j}$. We let Σ_j^k $(k = 1, \dots, m_j)$ denote the irreducible components of each analytic set $\lambda_{p_j} \cap \Sigma$ $(j = 1, \dots, l)$ in λ_{p_j}. By assumption, the union $\bigcup_{j=1}^{l} \bigcup_{k=1}^{m_j} \pi_w(\Sigma_j^k) \subset \Gamma$

contains a as an interior point. Consequently, one of the sets $\{\pi_w(\Sigma_j^k)\}_{j,k} \subset \Gamma$, say $\pi_w(\Sigma_j^k)$, contains the point a in its interior.

For simplicity, we write $\Sigma_j^k = \Sigma_0$, $\lambda_j = \lambda_0$, and $p_j = p_0$. Thus $\pi_w(\Sigma_0)$ contains $a = \pi_w(p_0)$ in its interior. We can find a sufficiently small polydisk $\lambda_1 := \delta_1 \times \gamma_1 \subset \lambda_0$ centered at p_0 so that $\Sigma_1 := \lambda_1 \cap \Sigma_0$ is an irreducible analytic set in λ_1 with $\pi_w(\Sigma_1) = \gamma_1$. Clearly $\dim \Sigma_1 = n + r$ for some $r \geq 0$. For each $w' \in \gamma_1$, we let $\Sigma_1(w')$ be the section of Σ_1 over the hyperplane $w = w'$. Since $\pi_w(\Sigma_1) = \gamma_1$, we can use Proposition 2.3 to prove that $\dim \Sigma_1(w')$ is always greater than or equal to r and that there exist $w^0 \in \gamma_1$ with $\dim \Sigma_1(w^0) = r$. Let $z^0 = (z_1^0, \ldots, z_n^0)$ be a point in $\Sigma(w^0)$ at which $\dim \Sigma_1(w^0)$ equals r. Then we can find coordinates $z_1, \ldots, z_n$ of $\mathbf{C}^n$ such that the $(n-r)$-dimensional hyperplane H in $\mathbf{C}^{2n}$ defined by

$$H: \ w = w^0, \ z_1 = z_1^0, \ldots, z_r = z_r^0$$

satisfies the condition that in a neighborhood of (z^0, w^0) on H, the intersection $\Sigma_1 \cap H$ consists of the single point (z^0, w^0). We can therefore find a sufficiently small polydisk $\lambda := \delta \times \gamma$, where δ is an $(n+r)$-dimensional polydisk centered at $(w^0, z_1^0, \ldots, z_r^0)$ and γ is an $(n-r)$-dimensional polydisk centered at $(z_{r+1}^0, \ldots, z_n^0)$, such that $\Sigma_1 \cap \lambda$ can be described in the form

$$z_k = \xi_k(w, z_1, \ldots, z_r) \qquad (k = r+1, \ldots, n)$$

for $(w, z_1, \ldots, z_r) \in \delta$, where each $\xi_k(w, z_1, \ldots, z_r)$ is an algebraic function in δ. If we fix a nonsingular point $(w, z) = (\alpha, \beta)$ in $\Sigma_1 \cap \lambda$, then we can find a smaller polydisk $\delta^* \subset \delta$ centered at $(\alpha, \beta_1, \ldots, \beta_r)$ in which each $\xi_k(w, z_1, \ldots, z_r)$ $(k = r+1, \ldots, n)$ is a single-valued holomorphic function.

We fix $|t| << 1$ and construct the holomorphic mapping $T_t: \ z' = \Xi(t, w)$ from the n variables w near $w = \alpha$ to the n variables $z' = (z_1', \ldots, z_n')$ near $z' = \alpha t + \beta$ by the formula

$$\begin{aligned} z_j' &= w_j t + \beta_j && (j = 1, \ldots, r), \\ z_k' &= w_k t + \xi_k(w, \beta_1, \ldots, \beta_r) && (k = r+1, \ldots, n). \end{aligned}$$

By the construction of Σ_0, we see that $f(z') = f(\Xi(t, w)) \equiv 0$ for all (t, w) sufficiently close to $(0, \alpha)$.

On the other hand, we see that the determinant of the Jacobian matrix of T_t,

$$\frac{\partial(z_1', \ldots, z_n')}{\partial(w_1, \ldots, w_n)},$$

is a monic polynomial in t whose coefficients are holomorphic functions of w near $w = \alpha$. Therefore, for some $|t| << 1$, T_t is a bijective map from a neighborhood of α in $\mathbf{C}_w^n$ onto a neighborhood ω of $\alpha t + \beta$ in $\mathbf{C}_{z'}^n$. Hence $f(z') \equiv 0$ in ω, which contradicts the hypothesis that $f(z)$ is non-constant in D. □

2.3.2. Hypersurface Case. Using Lemma 2.8, we can prove Theorem 2.3 in the hypersurface case.

LEMMA 2.9. *Let D_j $(j = 1, 2, \ldots)$ be a countable collection of domains in $\mathbf{C}^n$ and let $f_j(z)$ be a holomorphic function in D_j $(j = 1, 2, \ldots)$. Let S_j $(j = 1, 2, \ldots)$ denote the zero set of $f_j(z)$ in D_j. Given coordinates $(z_1, \ldots, z_n)$ of $\mathbf{C}^n$, there exist coordinates $(w_1, \ldots, w_n)$ sufficiently close to $(z_1, \ldots, z_n)$ such that each hypersurface S_j $(j = 1, 2, \ldots)$ satisfies the Weierstrass condition in the coordinates $(w_1, \ldots, w_n)$ at each of its points.*

PROOF. Let $\mathcal{L}$ be the set of all directions $w \in \Gamma$ such that for some S_j ($j = 1, 2, \ldots$) and for some point $p \in S_j$, there is a neighborhood of p in D_j such that the portion of the complex line L with direction w passing through p lying in this neighborhood is contained in S_j.

Given any point p in one of the sets S_j, we consider a closed polydisk $\delta_{j,p}$ centered at p and contained in the domain D_j. We let $\mathcal{L}(\delta_{j,p})$ denote the set of all directions $w \in \Gamma$ such that for some point $q \in \delta_{j,p} \cap S_j$, there is a neighborhood of q in D_j such that the portion of the complex line L with direction w passing through q lying in this neighborhood is contained in S_j. By Lemma 2.8, the set $\mathcal{L}(\delta_{j,p})$ is a closed, nowhere dense subset of Γ. Note that $\bigcup_{j=1}^{\infty} S_j$ can be covered by a countable number of these sets $\delta_{j,p}$, say $\{\delta_\nu\}_{\nu=1,2,\ldots}$, and thus $\mathcal{L} = \bigcup_{\nu=1}^{\infty} \mathcal{L}(\delta_\nu)$. By Baire's theorem, $\Gamma \setminus \mathcal{L}$ is dense in Γ; hence we can take a direction $w^0 \in \Gamma \setminus \mathcal{L}$ which is as close as we please to the direction $(0, \ldots, 0, 1)$, say $w^0 = (\varepsilon_1, \ldots, \varepsilon_{n-1}, 1)$. If we consider the coordinate transformations $w_i = z_i - \varepsilon_i z_n$ ($i = 1, \ldots, n-1$), $w_n = z_n$, then the coordinates $(w_1, \ldots, w_n)$ satisfy the conditions of the lemma; i.e., each S_j ($j = 1, 2, \ldots$) satisfies the Weierstrass condition in these coordinates at every point of S_j. □

2.3.3. General Case of Analytic Sets. To prove Theorem 2.3 for general analytic sets we use induction on the dimension n of $\mathbf{C}^n$. For $n = 1$ there is nothing to prove. We thus prove Theorem 2.3 in $\mathbf{C}^n$ under the assumption that it is true in $\mathbf{C}^{n-1}$.

Fix one of the domains D_j ($j = 1, 2, \ldots$). For each point $p \in \mathcal{E}_j$, we can find a neighborhood δ_p of p in $\mathbf{C}^n$ and a finite number of holomorphic functions $f_k^j(z)$ ($k = 1, \ldots, \nu_p$) such that $\delta_p \cap \mathcal{E}_j$ is given by the ν_p equations $f_k^j(z) = 0$ ($k = 1, \ldots, \nu_p$) in δ_p. Therefore we can find a countable number of such neighborhoods δ_i ($i = 1, 2, \ldots$) and holomorphic functions $f_k^j(z)$ ($k = 1, \ldots, \nu_i$) in δ_i such that the sets δ_i cover $\mathcal{E}_j$, i.e., $\mathcal{E}_j \subset \bigcup_{i=1}^{\infty} \delta_i$, and $\delta_i \cap \mathcal{E}_j = \{z \in \delta_j \mid f_k^j(z) = 0 \ (k = 1, \ldots, \nu_i)\}$. For simplicity in notation, we write $\delta_j = D_j$; in other words, it suffices to prove the theorem under the assumption that each $\mathcal{E}_j$ ($j = 1, 2, \ldots$) is an analytic set in a domain $D_j \subset \mathbf{C}^n$ described by global functions in D_j; i.e.,

$$\mathcal{E}_j : \ f_k^j(z) = 0 \quad (k = 1, \ldots, \nu_j) \qquad \text{in } D_j,$$

where each $f_k^j(z)$ ($k = 1, \ldots, \nu_j$) is holomorphic in D_j.

For each $j = 1, 2, \ldots$, we choose one of the functions $f_k^j(z)$ ($k = 1, \ldots, \nu_j$), say $f_1^j(z)$, and we let S_j denote the zero set of $f_1^j(z)$ in D_j. Note that $\mathcal{E}_j \subset S_j$. From Lemma 2.9, we find coordinates $w = (w_1, \ldots, w_n)$ sufficiently close to the original coordinates $z = (z_1, \ldots, z_n)$ such that each S_j ($j = 1, 2, \ldots$) satisfies the Weierstrass condition for the coordinates w at any point p of S_j.

Fix $a \in S_j$. We can find a closed polydisk $\lambda = \delta \times \gamma$ centered at a in $\mathbf{C}^n$, where

$$\delta : \ |w_j - a_j| \le r_j \quad (j = 1, \ldots, n-1), \qquad \gamma : \ |w_n - a_n| \le r_n,$$

such that $S_j \cap (\delta \times \partial\gamma) = \emptyset$. We let $S_{j,0} := S_j \cap \lambda$ denote the zero set of $f_1^j(z)$ in λ, and we decompose $S_{j,0}$ into irreducible components $S_{j,0} := \bigcup_{l=1}^{n_j} S_{j,0}^l$ in λ. Setting $\mathcal{E}_{j,0} = \mathcal{E}_j \cap \lambda$, since $\mathcal{E}_{j,0} \subset S_{j,0}$, we have $\mathcal{E}_{j,0} \cap (\delta \times \partial\gamma) = \emptyset$. From Proposition 2.3 it follows that the projection $\mathcal{E}_{j,0}^*$ of $\mathcal{E}_{j,0}$ onto δ is an analytic set in $\delta \subset \mathbf{C}^{n-1}$.

We note that

$$\mathcal{E}_{j,0} = \bigcup_{l=1}^{n_j} \mathcal{E}_{j,0}^l, \qquad \text{where} \qquad \mathcal{E}_{j,0}^l = \mathcal{E}_{j,0} \cap S_{j,0}^l.$$

Thus, if $\mathcal{E}_{j,0}^* = \delta$, we have $\mathcal{E}_{j,0}^l = S_{j,0}^l$ for some l by the irreducibility of $S_{j,0}^l$. We collect all such sets $\mathcal{E}_{j,0}^l$ and denote their union by $\mathcal{E}'_{j,0}$; i.e., $\mathcal{E}'_{j,0}$ is the union of all the $(n-1)$-dimensional components (i.e., complex hypersurfaces) of the analytic set $\mathcal{E}_{j,0}$ in λ. Since $\mathcal{E}'_{j,0} \subset S_{j,0}$, it follows from the construction of the coordinates $w = (w_1, \ldots, w_n)$ that $\mathcal{E}'_{j,0}$ satisfies the Weierstrass condition in these coordinates at any point $p \in \mathcal{E}'_{j,0}$. We let $\mathcal{F}_{j,0}$ denote the union of the other sets $\mathcal{E}_{j,0}^l$, so that $\mathcal{E}_{j,0} = \mathcal{E}'_{j,0} \cup \mathcal{F}_{j,0}$ and $\mathcal{F}_{j,0}$ is an analytic set in λ of dimension at most $n-2$. Thus the projection $\mathcal{F}_{j,0}^*$ of $\mathcal{F}_{j,0}$ onto δ is an analytic set in δ of dimension at most $n-2$.

Each S_j $(j = 1, 2, \ldots)$ can be covered by a countable number of closed polydisks λ as above; we denote these polydisks by

$$\lambda_{j,k} = \delta_{j,k} \times \gamma_{j,k} \ \subset \mathbf{C}^{n-1} \times \mathbf{C}_{w_n} \qquad (j, k = 1, 2, \ldots).$$

We set $\mathcal{E}_{j,k} := \mathcal{E}_j \cap \lambda_{j,k}$ and $S_j \cap \lambda_{j,k} := \bigcup_{l=1}^{n_{j,k}} S_{j,k}^l$, the decomposition of S_j into irreducible components. Then we have

$$\mathcal{E}_{j,k} = \bigcup_{l=1}^{n_{j,k}} (\mathcal{E}_{j,k} \cap S_{j,k}^l) \equiv \bigcup_{l=1}^{n_{j,k}} \mathcal{E}_{j,k}^l.$$

We let $\mathcal{F}_{j,k}$ denote the union of the analytic sets $\mathcal{E}_{j,k}^l$ having dimension at most $n-2$. Then the projection $\mathcal{F}_{j,k}^*$ of $\mathcal{F}_{j,k}$ onto $\delta_{j,k}$ is an analytic set in $\delta_{j,k} \subset \mathbf{C}^{n-1}$ of dimension at most $n-2$. .

Thus in $\mathbf{C}^{n-1}$ with the $n-1$ variables $w_1, \ldots, w_{n-1}$, we have a countable collection of polydisks $\delta_{j,k}$ $(j, k = 1, 2, \ldots)$ and analytic sets $\mathcal{F}_{j,k}^*$ in each $\delta_{j,k}$ having dimension at most $n-2$. It follows from the inductive hypothesis that there exist coordinates $u' = (u_1, \ldots, u_{n-1})$ obtained by a linear transformation of $w' = (w_1, \ldots, w_{n-1})$ and sufficiently close to w' such that each $\mathcal{F}_{j,k}^*$ $(j, k = 1, 2, \ldots)$ satisfies the Weierstrass condition in the u' coordinates at any point q of $\mathcal{F}_{j,k}^*$. Thus, since $\mathcal{F}_{j,k} \cap (\delta_{j,k} \times \partial\gamma_{j,k}) = \emptyset$, each $\mathcal{F}_{j,k}$ itself necessarily satisfies the Weierstrass condition in the coordinates $u = (u', w_n)$ for $\mathbf{C}^n$ at each point $p \in \mathcal{F}_{j,k}$. □

Theorem 2.3 will be used in the next section to investigate the global structure of analytic sets.

2.4. Analytic Sets (Global)

2.4.1. Global Irreducible Decomposition of Analytic Sets. Let $\mathcal{E}$ be an analytic set in a domain D in $\mathbf{C}^n$. From Theorem 2.3, we can find coordinates $z = (z_1, \ldots, z_n)$ for which the analytic set $\mathcal{E}$ satisfies the Weierstrass condition at each of its points. At each point $a = (a_1, \ldots, a_n) \in \mathcal{E}$, we take a Weierstrass canonical neighborhood δ_a of $\mathcal{E}$ in these coordinates. Then $\mathcal{E}$ can be covered by a countable number of such neighborhoods, which we denote by

$$\delta_k \ : \ |z_j - a_j^k| \le r_j^k \qquad (j = 1, \ldots, n; \ k = 1, 2, \ldots).$$

We set $\mathcal{E}_k := \mathcal{E} \cap \delta_k$ $(k = 1, 2, \ldots)$, and we consider the irreducible decomposition of $\mathcal{E}_k$ in δ_k:

$$\mathcal{E}_k = \bigcup_{\nu=1}^{l_k} \mathcal{E}_{k,\nu}.$$

$\mathcal{F}$ satisfy the Weierstrass condition at each point. Thus $\mathcal{E}$ and $\mathcal{F}$ can be written locally as

$$\mathcal{E} : z_j = \eta_j(z_1, \dots, z_d) \qquad (j = d+1, \dots\dots, n),$$
$$\mathcal{F} : z_k = \xi_k(z_1, \dots\dots, z_r) \qquad (k = r+1, \dots, n).$$

We fix a point $a = (a_1, \dots, a_n)$ of $\mathcal{E}$; then we set $a' := (a_1, \dots, a_r)$ and we denote by $D(a'), \mathcal{E}(a'), \mathcal{F}(a') \subset \mathbf{C}^{n-r}$ the sections of $D, \mathcal{E}, \mathcal{F}$ over the hyperplane $z_j = a_j$ $(j = 1, \dots, r)$ in $\mathbf{C}^n$. We note that $\mathcal{E}(a')$ is an isolated set in $D(a')$. Since $\mathcal{F}(a')$ is analytic in $D(a') \setminus \mathcal{E}(a')$, $\mathcal{F}(a')$ is an isolated set in $D(a') \setminus \mathcal{E}(a')$, but it may have accumulation points in $\mathcal{E}(a')$.

We can thus find a neighborhood V' of $a'' = (a_{r+1}, \dots, a_n)$ in $D(a')$ such that $V' \cap \mathcal{E}(a')$ consists of the single point a''; and we take a closed polydisk Γ in $\mathbf{C}^{n-r}$ (with coordinates $z_{r+1}, \dots, z_n$) centered at a'' and with radius ρ_k $(k = r+1, \dots, n)$,

$$\Gamma : |z_k - a_k| \leq \rho_k \qquad (k = r+1, \dots, n),$$

where the ρ_k are chosen sufficiently small so that $\partial\Gamma \cap (\mathcal{E}(a') \cup \mathcal{F}(a')) = \emptyset$. We next take a closed polydisk Δ in $\mathbf{C}^r$ (with coordinates $(z_1, \dots, z_r)$) centered at a' with radius ρ_j $(j = 1, \dots, r)$,

$$\Delta : |z_j - a_j| \leq \rho_j \qquad (j = 1, \dots, r),$$

where the ρ_j are chosen sufficiently small so that $\Lambda := \Delta \times \Gamma \subset D$ and $(\Delta \times \partial\Gamma) \cap (\mathcal{E} \cup \mathcal{F}) = \emptyset$. These choices are possible because $\mathcal{E}$ is analytic at a and $\mathcal{F}$ is closed in $D \setminus \mathcal{E}$. We set $\mathcal{E}^0 := \mathcal{E} \cap \Lambda$ and $\mathcal{F}^0 := \mathcal{F} \cap \Lambda$. Then Theorem 2.3 implies that the projection $\mathcal{E}'$ of $\mathcal{E}^0$ onto Δ is an analytic set in Δ and the dimension of $\mathcal{E}'$ is d. From the assumption that $r > d$, it follows that the set $\Delta' := \Delta - \mathcal{E}'$ is a non-empty domain in $\mathbf{C}^r$ and that $\mathcal{F} \cap (\Delta' \times \Gamma)$ is an analytic set in $\Delta' \times \Gamma$.

Now let $z' = (z_1', \dots, z_r') \in \Delta'$. The section $\mathcal{F}^0(z')$ of $\mathcal{F}^0$ over the hyperplane $z_j = z_j'$ $(j = 1, \dots, r)$ is a finite set; we denote its cardinality by $\zeta(z')$. The nonnegative integer-valued function $\zeta(z')$ is easily seen to be a lowersemicontinuous function of z' in Δ'.

Let ν be any nonnegative integer, and let e_ν be the set of all points $z' \in \Delta'$ such that $\zeta(z') \leq \nu$. By the lowersemicontinuity of $\zeta(z)$, e_ν is a closed set in Δ', and clearly

$$e_\nu \subset e_{\nu+1} \quad (\nu = 1, 2, \dots), \qquad \Delta' = \bigcup_{\nu=1}^{\infty} e_\nu.$$

It follows from Baire's theorem that some e_ν contains interior points in Δ'. We let ν_0 be the smallest integer ν such that e_ν contains at least one interior point. Let $e_{\nu_0}^i$ denote the interior of e_{ν_0} in Δ'; then we shall prove that

$$\Delta' = e_{\nu_0}. \tag{2.8}$$

We prove (2.8) by contradiction. If (2.8) is false, then there exists a point $b' \in \Delta' \cap \partial e_{\nu_0}^i$, where $\partial e_{\nu_0}^i$ denotes the boundary of $e_{\nu_0}^i$ in Δ'. Since e_{ν_0} is a closed set in Δ', the section $\mathcal{F}^0(b')$ of $\mathcal{F}^0$ over $z' = b'$ consists of at most ν_0 points in Γ. Since $\dim \mathcal{F} = r$, we can thus find a neighborhood δ of b' in Δ' such that $\mathcal{F}^0 \cap (\delta \times \Gamma)$ coincides with a finite number of locally algebraic analytic sets in $\delta \times \Gamma$ with parameters $z_1, \dots, z_r$ in δ: $z_k = \eta_k(z_1, \dots, z_r)$ $(k = r+1, \dots, n)$. Consequently, given any point $c' \in \delta$, the number $\zeta(c')$ of points of the section $\mathcal{F}^0(c') \subset \Gamma$ of $\mathcal{F}^0$ over the hyperplane $z' = c'$ remains constant, say ζ_0, except perhaps for points c' belonging to an (at most) $(r-1)$-dimensional analytic set σ in

δ. Since $(\delta \setminus \sigma) \cap e^i_{\nu_0} \neq \emptyset$, we have $\zeta_0 \leq \nu_0$. Hence $\delta \subset e_{\nu_0}$, which is a contradiction; and (2.8) is proved.

Since $\mathcal{F} \cap (\Delta' \times \Gamma)$ is an analytic set in $\Delta' \times \Gamma$ with $\Delta' \cap \partial\Gamma = \emptyset$, it follows that $\zeta(z') = \nu_0$ for each $z' \in \Delta'$ except perhaps for an analytic set in Δ' of dimension at most $r - 1$.

Using the same technique as in the proof of Proposition 2.1, we see that $\mathcal{F}^0 \cap (\Delta' \times \Gamma)$ consists of a finite number of irreducible components of complete, algebraic analytic sets Σ' defined by

$$z_k^{\nu_k} + A_1^k(z') z_k^{\nu_k - 1} + \ldots + A_{\nu_k}^k(z') = 0 \quad (k = r+1, \ldots, n), \tag{2.9}$$

where $\nu_k \leq \nu_0$ and each $A_l^k(z')$ $(l = 1, \ldots, \nu_k)$ is a bounded, single-valued holomorphic function in Δ'. By the Riemann removable singularity theorem for holomorphic functions, each $A_k^j(z')$ can be holomorphically extended to all of Δ. Thus, (2.9) defines a complete, algebraic analytic set Σ in Λ, and Σ equals the closure $\overline{\Sigma'}$ of Σ' in Λ. Therefore, the closure $\overline{\mathcal{F}^0}$ of $\mathcal{F}^0$ in Λ is an analytic set in Λ, and the theorem is proved. □

2.5. Projections of Analytic Sets in Projective Space

Since the notion of analytic sets is local, we can define an analytic set in a domain G of n-dimensional complex projective space $\mathbf{P}^n$ or in a product space of the form $D \times \mathbf{P}^n$, where D is a domain in $\mathbf{C}^m$. The dimension of such an analytic set $\mathcal{E}$ in G and the irreducible decomposition of $\mathcal{E}$ in G are defined as in the case of an analytic set in a domain of $\mathbf{C}^n$.

Let D be a domain in $\mathbf{C}^m$ and consider the product domain $\Omega := D \times \mathbf{P}^n$. In this section we study analytic sets $\mathcal{E}$ in Ω.

We take coordinates $u = (u_1, \ldots, u_m)$ of $\mathbf{C}^m$ and homogeneous coordinates $[z] = [z_0 : z_1 : \ldots : z_n]$ of $\mathbf{P}^n$. We let π_1 and π_2 denote the projections from $\Omega = D \times \mathbf{P}^n$ onto D and $\mathbf{P}^n$. Let $\mathcal{E}$ be an analytic set in Ω and let e be a subset of D. We set

$$\mathcal{E}(e) := \pi_1^{-1}(e) \cap \mathcal{E}.$$

In the special case when e is a single point u' of D, the set $\mathcal{E}(u')$ can be regarded as an analytic set in $\mathbf{P}^n$. Then $\mathcal{E}(u')$ consists of a finite number of irreducible *compact* analytic sets in $\mathbf{P}^n$. We use the notation

$$d(u') := \dim \mathcal{E}(u')$$

for the maximal dimension of the irreducible components of $\mathcal{E}(u')$.

2.5.1. Chow's Theorem. We begin by proving a slight generalization of Chow's theorem, which says that an analytic set in $\mathbf{P}^n$ must be algebraic. To define this notion, let $[z] = [z_0 : z_1 : \ldots : z_n]$ be homogeneous coordinates of $\mathbf{P}^n$ and let $P_k(z)$ $(k = 1, \ldots, \mu)$ be a homogeneous polynomial in the coordinates $z = (z_0, z_1, \ldots, z_n)$ of $\mathbf{C}^{n+1}$. Then the set of all points z of $\mathbf{C}^{n+1}$ defined by the m equations

$$P_k(z) = 0 \qquad (k = 1, \ldots, \mu)$$

canonically defines an analytic set in $\mathbf{P}^n$ which we call an **algebraic set** in $\mathbf{P}^n$.[3]

[3] The fact that **any** analytic set in $\mathbf{P}^n$ must be an algebraic set in $\mathbf{P}^n$ was first proved by L. Chow [**14**]. The idea of the proof given here is due to W. Rothstein [**64**] (see also R. Remmert and K. Stein [**69**]). In Part II we shall show that it is possible in Theorem 2.6 to take the neighborhood δ of u^0 in D to be an analytic polyhedron in D.

THEOREM 2.6. *Let $\mathcal{E}$ be an analytic set of dimension $\rho \geq 1$ in $\Omega = D \times \mathbf{P}^n$ whose projection to D is non-empty. Let u^0 be any point in D. Then there exists a neighborhood δ of u^0 in D such that $\mathcal{E}(\delta) = \mathcal{E} \cap (\delta \times \mathbf{P}^n)$ can be written as the common zero set of a finite number of holomorphic functions $P_k(u,z)$ $(k = 1, \ldots, \mu)$, where each $P_k(u,z)$ is a homogeneous polynomial of degree m_k in the coordinates z of $\mathbf{C}^{n+1}$ whose coefficients are holomorphic functions of u in δ:*

$$P_k(u,z) = \sum_{|\mathbf{j}|=m_k} A_k^{(\mathbf{j})}(u) z_0^{j_0} z_1^{j_1} \cdots z_n^{j_n}, \tag{2.10}$$

where $\mathbf{j} = (j_0, j_1, \ldots, j_n)$ and $|\mathbf{j}| = \sum_{l=0}^n j_l$.

PROOF. We may assume that $\mathcal{E}$ is a ρ-dimensional irreducible analytic set in Ω. Let $z = (z_0, z_1, \ldots, z_n)$ be coordinates for $\mathbf{C}^{n+1}$. For convenience we write $(\mathbf{C}^{n+1})^* = \mathbf{C}^{n+1} \setminus \{0\}$. Any point $z \in (\mathbf{C}^{n+1})^*$ corresponds to a point $[z] \in \mathbf{P}^n$. To the analytic set $\mathcal{E}$ in Ω, we associate the set $\mathcal{E}_0$ in the product space $D \times (\mathbf{C}^{n+1})^*$ defined as

$$\mathcal{E}_0 := \{(u,z) \in D \times (\mathbf{C}^{n+1})^* \mid (u,[z]) \in \mathcal{E}\}, \tag{2.11}$$

which will be called the **associated set** for $\mathcal{E}$. Since $[z'] = [z]$ in $\mathbf{P}^n$ if and only if $z' = tz$ for some $t \neq 0$ in $\mathbf{C}$, it follows that the set $\mathcal{E}_0$ is a cone and is a $(\rho+1)$-dimensional analytic set in $D \times (\mathbf{C}^{n+1})^*$. We first show that $\mathcal{E}_0$ is analytically extendable to the the set $D \times \{0\}$, i.e., the closure $\overline{\mathcal{E}_0}$ of $\mathcal{E}_0$ in the product space $\widetilde{\Omega} := D \times \mathbf{C}^{n+1}$ is a $(\rho+1)$-dimensional analytic set in $\widetilde{\Omega}$.

Case 1: $\rho \geq m$. Since $\dim \mathcal{E}_0 = \rho + 1 > m = \dim(D \times \{0\})$, it follows from Theorem 2.5 that $\overline{\mathcal{E}_0}$ is analytic in $\widetilde{\Omega}$. In this case, let $u_0 \in D$. Since $\overline{\mathcal{E}_0}$ is analytic at the point $(u_0, 0)$, we can find a polydisk $\lambda := \delta \times \gamma$ centered at $(u_0, 0)$ in $D \times \mathbf{C}^{n+1}$ and a finite number of holomorphic functions $g_h(u,z)$ $(h = 1, \ldots, \nu)$ in λ such that

$$\overline{\mathcal{E}_0} \cap \lambda = \{(u,z) \in \lambda \mid g_h(u,z) = 0 \ (h = 1, \ldots, \nu)\}.$$

We develop each $g_h(u,z)$ into a Taylor series with respect to the variables $z \in \gamma$ and we rearrange this series into a series of homogeneous polynomials in $z_1, \ldots, z_n$,

$$g_h(u,z) := \sum_{k=0}^{\infty} \left\{ \sum_{l_0+\cdots+l_n=k} (A_h)_k^l(u) z_0^{l_0} z_1^{l_1} \cdots z_n^{l_n} \right\},$$

where each $(A_h)_k^l(u)$ $(1 \leq l \leq \nu_k := \binom{n+k}{k})$ is a holomorphic function in δ. Fix (u,z) in $\overline{\mathcal{E}_0}$. Since $(u,tz) \in \overline{\mathcal{E}_0}$ for all $t \in \mathbf{C}$, in particular, we have

$$\sum_{k=0}^{\infty} t^k \left\{ \sum_{l_0+\cdots+l_n=k} (A_h)_k^l(u) z_0^{l_0} z_1^{l_1} \cdots z_n^{l_n} \right\} \equiv 0 \qquad \text{in } |t| < 1. \tag{2.12}$$

We let $\sum_{k=0}^{\infty} t^k (P_h)_k(u,z)$ denote the function on the left-hand side of (2.12); thus $(P_h)_k(u,z)$ $(h = 1, \ldots, \nu; k = 0, 1, \ldots)$ is a homogeneous polynomial of degree k in $z \in \mathbf{C}^{n+1}$ whose coefficients are holomorphic functions of $u \in \delta$. It follows from (2.12) that

$$\overline{\mathcal{E}_0} \cap (\delta \times \mathbf{C}^{n+1}) = \bigcap_{h=1}^{\nu} \bigcap_{k=0}^{\infty} \{(u,z) \in \delta \times \mathbf{C}^{n+1} \mid (P_h)_k(u,z) = 0\}.$$

Using Corollary 2.1, the theorem is proved in case 1.

Case 2: $\rho < m$. We set $q = m - \rho \geq 1$ and we use variables $w = (w_1, \ldots, w_q) \in \mathbf{C}^q$ and $z = (z_0, z_1, \ldots, z_n) \in \mathbf{C}^{n+1}$. We form the set

$$\mathcal{F} = \{(u, [z : w]) \in D \times \mathbf{P}^{n+q} \mid (u, [z]) \in \mathcal{E}, \quad w \in \mathbf{C}^q\},$$

so that $\mathcal{F}$ is an $m = (\rho + q)$-dimensional irreducible analytic set in $D \times \mathbf{P}^{n+q}$. Just as $\mathcal{E}$ gave rise to $\mathcal{E}_0$ via (2.11), $\mathcal{F}$ gives rise to the $(m+1)$-dimensional analytic set $\mathcal{F}_0$ in $D \times (\mathbf{C}^{n+q+1})^*$.

Let $u_0 \in D$. Applying case 1 for $\mathcal{F}$, we can find a neighborhood δ of u_0 in D and a finite number of homogeneous polynomials $G_h(u, z, w)$ $(h = 1, \ldots, \nu)$ of degree m_h in $(z, w) \in \mathbf{C}^{n+q+1}$ whose coefficients are holomorphic functions of $u \in \delta$,

$$G_h(u, z, w) = \sum_{|\mathbf{j}_1| + |\mathbf{j}_2| = m_h} A_h^{(\mathbf{j}_1, \mathbf{j}_2)}(u) z^{\mathbf{j}_1} w^{\mathbf{j}_2} \qquad (h = 1, \ldots, \nu),$$

such that

$$\overline{\mathcal{F}_0}(\delta) = \{(u, z, w) \in \delta \times \mathbf{C}^{n+q+1} \mid G_h(u, z, w) = 0 \ (h = 1, \ldots, \nu)\}.$$

If we rearrange the sum

$$G_h(u, z, w) = \sum_{s=0}^{m_h} \left(\sum_{|\mathbf{k}| = s} (B_h)_s^{(\mathbf{k})}(u, z) w_1^{k_1} \cdots w_q^{k_q} \right),$$

where $\mathbf{k} = (k_1, \ldots, k_q)$, then each $(B_h)_s^{(\mathbf{k})}(u, z)$ is a homogeneous polynomial of degree $m_h - s$ in the coordinates $z \in \mathbf{C}^{n+1}$ whose coefficients are holomorphic functions of $u \in \delta$. Since $(u, [z]) \in \mathcal{E}(\delta)$ if and only if $(u, [z : w]) \in \mathcal{F}(\delta)$ for all $w \in \mathbf{C}^q$ (or, equivalently, $(u, z) \in \overline{\mathcal{E}_0}(\delta)$ if and only if $(u, z, w) \in \overline{\mathcal{F}_0}(\delta)$), it follows that

$$\overline{\mathcal{E}_0}(\delta) = \bigcap_{h=1}^{\nu} \bigcap_{s=0}^{m_h} \{(u, z) \in \delta \times \mathbf{C}^{n+1} \mid (B_h)_s^{(\mathbf{k})}(u, z) = 0, \quad \text{where } |\mathbf{k}| = s\}.$$

This proves the theorem in case 2. □

COROLLARY 2.2. *Under the same notation as in Theorem 2.6, the set* $e_n^0 := \{u \in \delta \mid d(u) = n\}$ *is an analytic set in* δ.

Indeed, using the notation in (2.10), we have

$$e_n^0 = \{u \in \delta \mid A_k^{(\mathbf{j})}(u) = 0 \text{ for all } k \text{ and } \mathbf{j}\},$$

which proves the corollary.

2.5.2. Projection. Given an integer s with $0 \leq s \leq n$, we consider the following two projective subspaces of $\mathbf{P}^n$:

$$\begin{aligned} \mathcal{K}^s \quad &: \quad z_k = 0 \qquad (k = s+1, \ldots, n), \\ \mathcal{H}_{n-s-1} \quad &: \quad z_h = 0 \qquad (h = 0, 1, \ldots, s), \end{aligned}$$

so that $\mathcal{K}^s \cap \mathcal{H}_{n-s-1} = \emptyset$. For convenience, we set $\mathcal{H}_{-1} = \emptyset$. To each point $[z] = [z_0 : \ldots : z_n] \in \mathbf{P}^n \setminus H_{n-s-1}$ we associate the point $[z]_s = [z_0 : \ldots : z_s : 0 : \ldots : 0] \in \mathcal{K}^s$. We write $[z]_s := [z_0 : \ldots : z_s]$ and canonically identify the subspace $\mathcal{K}^s$ with $\mathbf{P}^s$. The mapping $\varphi_s([z]) := [z]_s$ from $\mathbf{P}^n$ to $\mathcal{K}^s$ is called the **projection** from $\mathbf{P}^n$ to $\mathcal{K}^s = \mathbf{P}^s$. We also define the associated projection $\Psi_s(u, [z]) := (u, [z]_s)$ from $D \times \mathbf{P}^n$ to $D \times \mathcal{K}^s = D \times \mathbf{P}^s$.

We shall prove a lemma which corresponds to the case $r = n-1$ of Proposition 2.3 for analytic sets in a domain of $\mathbf{C}^n$. We use the notation

$$[\mathbf{e}_i] := [0 : \ldots : 1 : \ldots : 0] \in \mathbf{P}^n \qquad (i = 0, 1, \ldots, n),$$

where the "1" occurs in the $(i+1)$-th slot.

LEMMA 2.10. *Let $\mathcal{E}$ be an analytic set in Ω and let u^0 be a point in D. Assume that $[\mathbf{e}_n] \notin \mathcal{E}(u^0)$; i.e., $\pi_2(\mathcal{E}(u^0)) \cap \mathcal{H}_0 = \emptyset$. Then we can find a neighborhood δ of u^0 in D such that*

(1) *The projection $\mathcal{F}(\delta) := \Psi_{n-1}(\mathcal{E}(\delta))$ of $\mathcal{E}(\delta)$ onto $\delta \times \mathcal{K}^{n-1}$ is an analytic set in $\delta \times \mathcal{K}^{n-1}$, and*
(2) *$\dim \mathcal{E}(u) = \dim \mathcal{F}(u)$ for all $u \in \delta$.*

PROOF. By Theorem 2.6, we can find a neighborhood δ of u^0 in D such that $\mathcal{E}(\delta)$ gives rise to the analytic set $\overline{\mathcal{E}(\delta)_0}$ in $\delta \times \mathbf{C}^{n+1}$ defined by $P_k(u, z) = 0$ $(k = 1, \ldots, \mu)$, where each $P_k(u, z)$ is a homogeneous polynomial in $\mathbf{C}^{n+1}$ whose coefficients are holomorphic functions of $u \in \delta$.

Since $[\mathbf{e}_n] \notin \mathcal{E}(u^0)$, we have $P_k(u^0, 0, \ldots, 0, 1) \neq 0$ for some k $(1 \leq k \leq \mu)$. For simplicity, let $k = 1$ and

$$P_1(u, z) = \sum_{|\mathbf{j}|=m} a^{(\mathbf{j})}(u) z_0^{j_0} z_1^{j_1} \cdots z_n^{j_n},$$

where $\mathbf{j} = (j_0, j_1, \ldots, j_n)$. By taking a smaller neighborhood δ of u^0 in D, if necessary, we may assume that

$$P_1(u, 0, \ldots, 0, 1) \neq 0 \quad \text{for all} \quad u \in \overline{\delta}. \tag{2.13}$$

We shall show that this δ satisfies the conclusion of the theorem.

For simplicity, we set

$$\mathcal{E} = \mathcal{E}(\delta), \qquad \mathcal{F} = \Psi_{n-1}(\mathcal{E}(\delta)).$$

We write $z' = (z_0, z_1, \ldots, z_{n-1}) \in \mathbf{C}^n$ and $[z'] = [z_0 : z_1 : \ldots : z_{n-1}] \in \mathbf{P}^{n-1}$. We note that $(u, [z']) \in \mathcal{F}$ if and only if there exists at least one point $z_n \in \mathbf{C}$ such that $(u, [z]) = (u, [z' : z_n]) \in \mathcal{E}$. Equivalently, if we let $\mathcal{F}_0 \subset \delta \times \mathbf{C}^n$ and $\mathcal{E}_0 \subset \delta \times \mathbf{C}^{n+1}$ denote the associated sets for $\mathcal{F}$ and $\mathcal{E}$, then $(u, z') \in \overline{\mathcal{F}_0}$ if and only if there exists at least one point $z_n \in \mathbf{C}$ such that $(u, z) = (u, z', z_n) \in \overline{\mathcal{E}_0}$. It thus suffices to show that there exist a finite number of homogeneous polynomials $h_\alpha(u, z')$ $(\alpha = 1, \ldots, M)$ of $z' \in \mathbf{C}^n$ whose coefficients are holomorphic functions of $u \in \delta$ such that $\overline{\mathcal{F}_0}$ consists of the common zero set of $h_\alpha(u, z')$ $(\alpha = 1, \ldots, M)$ in $\delta \times \mathbf{C}^n$.

To show this, we note from (2.13) that

$$P_1(u, z) = A_0(u) z_n^m + Q_1(u, z') z_n^{m-1} + \cdots + Q_m(u, z'),$$

where $A_0(u) \neq 0$ for $u \in \overline{\delta}$ and where $Q_j(u, z')$ is a homogeneous polynomial in $\mathbf{C}^n$ of degree $m - j$ whose coefficients are holomorphic functions of $u \in \delta$. It follows that

$$P_1(u, z) = A_0(u) \cdot (z_n - \xi_1(u, z')) \cdots (z_n - \xi_m(u, z'))$$

with

$$|\xi_j(u, z')| \leq K(1 + \|z'\|^{m-1}), \quad \xi_j(u, \lambda z') = \lambda \xi_j(u, z') \ (\lambda \in \mathbf{C}), \tag{2.14}$$

where $K > 0$ is a constant independent of $u \in \delta$ and $j = 1, \ldots, m$.

Following the idea of Remmert-Stein in the proof of Proposition 2.3, we introduce $\mu - 1$ complex variables $X_2, \dots, X_\mu$ and set

$$H(u, z', X) := \prod_{j=1}^{m} [X_2 P_2(u, z', \xi_j(u, z')) + \cdots + X_\mu P_\mu(u, z', \xi_j(u, z'))] .$$

Then $H(u, z', X)$ is a holomorphic function in $\delta \times \mathbf{C}^n \times \mathbf{C}^{\mu-1}$ which can be written as

$$H(u, z', X) = \sum_{|\mathbf{a}|=m} h^{(\mathbf{a})}(u, z') X_2^{a_2} \cdots X_\mu^{a_\mu},$$

where $\mathbf{a} = (a_2, \dots, a_\mu)$. We also have the following:

(i) $(u, z') \in \overline{\mathcal{F}_0}$ if and only if (u, z') belongs to the common zero set of $h^{(\mathbf{a})}(u, z') = 0$ for all $|\mathbf{a}| = m$.

(ii) From (2.14), each $h^{(\mathbf{a})}(u, z')$ with $|\mathbf{a}| = m$ is a homogeneous polynomial of $z' \in \mathbf{C}^n$ whose coefficients are holomorphic functions of $u \in \delta$.

This proves (1).

Furthermore, since $\mathcal{E} \cap (\delta \times [\mathbf{e}_n]) = \emptyset$, for any $u \in \delta$ and $[a'] = [a_0 : \dots : a_{n-1}] \in \mathcal{F}(u)$, there exist at most a finite number of points $a_n \in \mathbf{C}$ such that $[a' : a_n] \in \mathcal{E}(u)$. It follows that $\dim \mathcal{E}(u) = \dim \mathcal{F}(u)$. We thus obtain (2). □

We remark that for any given $[a] \in \mathbf{P}^n$, there exists a linear transformation L of $\mathbf{P}^n$ such that $L([a]) = [\mathbf{e}_n]$; and for any given $[a] \in \mathbf{P}^n$ such that $[a] \notin H_{n-r-1}$ $(0 \le r \le n-1)$ (i.e., $[a]$ is not contained in the subspace of $\mathbf{P}^n$ spanned by $[\mathbf{e}_j]$ $(j = n, n-1, \dots, n-r)$) there exists a linear transformation L of $\mathbf{P}^n$ such that $L([\mathbf{e}_j]) = [\mathbf{e}_j]$ $(j = n, n-1, \dots, n-r)$ and $L([a]) = [\mathbf{e}_{n-r-1}]$.

Using the lemma, and an induction argument on the dimension of $\mathbf{P}^n$ together with this remark, we obtain the following corollary.

COROLLARY 2.3. *Let $\mathcal{E}$ be an analytic set in Ω and let u^0 be a point in D such that $d(u^0) = \dim \mathcal{E}(u^0) = r$ $(0 \le r \le n)$. Then there exist homogeneous coordinates $[z] = [z_0 : \dots : z_n]$ of $\mathbf{P}^n$ and a neighborhood δ of u^0 in D such that*

(1) *$\mathcal{H}_{n-r-1} \cap \mathcal{E}(\delta) = \emptyset$;*

(2) *$d(u') \le r$ for any $u' \in \delta$, i.e., $d(u)$ is a lowersemicontinuous function on δ; and*

(3) *$\mathcal{E}^{(r)}(\delta) := \Psi_r(\mathcal{E}(\delta))$ is an analytic set in $\delta \times \mathcal{K}^r$ such that $\dim \mathcal{E}(u) = \dim \mathcal{E}^{(r)}(u)$ for all $u \in \delta$, and hence $\mathcal{E}^{(r)}(u^0) = \mathcal{K}^r = \mathbf{P}^r$.*

We use these results to prove the following proposition.

PROPOSITION 2.4. *Let $\mathcal{E}$ be an analytic set in Ω and let $u^0 \in D$. If $d(u^0) = r$, then there exists a neighborhood δ of u^0 in D such that*

$$e_r^0 := \{u \in \delta \mid d(u) = r\}$$

is an analytic set in δ.

PROOF. From (3) in Corollary 2.3, we can find a neighborhood δ of u^0 in D such that $\mathcal{E}^{(r)}(\delta)$ is an analytic set in $\delta \times \mathbf{P}^r$. By Theorem 2.6, if we form the associated set $\mathcal{E}_0^{(r)}$ for $\mathcal{E}^{(r)}$ in $\delta \times (\mathbf{C}^{r+1})^*$, there exist a finite number of homogeneous polynomials $Q_k(u, z)$ $(k = 1, \dots, \nu)$ for z in $\mathbf{C}^{r+1}$ whose coefficients are holomorphic functions of u in δ:

$$Q_k(u,z) = \sum_{|\mathbf{j}|=m_k} B_k^{(\mathbf{j})}(u) z_0^{j_0} z_1^{j_1} \cdots z_r^{j_r},$$

such that

$$\overline{\mathcal{E}_0^{(r)}}(\delta) = \{(u,z) \in \delta \times \mathbf{C}^{r+1} \mid Q_k(u,z) = 0 \ (k = 1, \dots, \nu)\}.$$

Since $\dim \mathcal{E}(u) = \dim \mathcal{E}^{(r)}(u)$ for $u \in \delta$ and since $\dim \mathcal{E}^{(r)}(u) = r$ if and only if $\mathcal{E}^{(r)}(u) = \mathbf{P}^r$, or equivalently, $\overline{\mathcal{E}_0^{(r)}}(u) = \mathbf{C}^{r+1}$, it follows that

$$e_r^0 = \{u \in \delta \mid B_k^{(\mathbf{j})}(u) = 0 \ \ 1 \le k \le \nu; \ \ |\mathbf{j}| = m_k\},$$

so that e_r^0 is an analytic set in δ. □

COROLLARY 2.4. *Under the same hypotheses as in Proposition 2.4, let* $d(u^0) = r$ *and* $\dim e_r^0 = s$. *Then* $\mathcal{E}(\delta)$ *contains an analytic set of dimension* $s + r$ *in* $\delta \times \mathbf{P}^n$, *so that* $\dim \mathcal{E}(\delta) \ge s + r$.

REMARK 2.13. Even in the case when $\mathcal{E}$ is irreducible in $\delta \times \mathbf{P}^n$, we do not necessarily have $\dim \mathcal{E}(\delta) = s + r$. Let $D = \mathbf{C}^2$ with variables u_1 and u_2, and let $\Omega = \mathbf{C}^2 \times \mathbf{P}^1$. We use homogeneous coordinates $[x : y]$ in $\mathbf{P}^1$. Let $\mathcal{E}$ be the analytic set in Ω defined by the single homogeneous linear equation

$$\mathcal{E} : u_1 x + u_2 y = 0.$$

Then $\mathcal{E}$ is of dimension 2 in Ω. The set of points $u = (u_1, u_2)$ in $\mathbf{C}^2$ such that $\dim \mathcal{E}(u) = 1$, i.e., such that $\mathcal{E}(u) = \mathbf{P}^1$, consists of only one point, $(0,0)$. Thus, $r + s = 1 + 0 = 1$. However, for any neighborhood δ of $(0,0)$ in $\mathbf{C}^2$ we have $\dim \mathcal{E}(\delta) = 2$.

In particular, if $D = \emptyset$, in a manner similar to Lemma 2.10, we obtain the following corollary for $r = 0, \dots, n-1$ (the case $r = n-1$ being an immediate consequence of the lemma).

COROLLARY 2.5. *Let* $\mathcal{F}$ *be an analytic set in* $\mathbf{P}^n$ *and suppose that* $\mathcal{F} \cap \mathcal{H}_{n-r-1} = \emptyset$. *Then the projection* $\varphi_r(\mathcal{F})$ *is an analytic set in* $\mathcal{K}^r$.

We deduce the following property of analytic sets in $\mathbf{P}^n$, which will be used later.

COROLLARY 2.6. *Let* $\mathcal{F}$ *be an analytic set in* $\mathbf{P}^n$ *of dimension* r *where* $0 \le r \le n$. *Then for* **each** $(n-r)$*-dimensional hyperplane* L^{n-r} *of* $\mathbf{P}^n$, *the intersection* $\mathcal{F} \cap L^{n-r}$ *is non-empty; while for* **some** $(n-r-1)$*-dimensional hyperplane* L^{n-r-1}, $\mathcal{F} \cap L^{n-r-1} = \emptyset$.

PROOF. The result is trivial if $r = n$; thus we assume $0 \le r \le n-1$. The second part of the corollary follows from the definition of dimension; without loss of generality, we can assume that $\mathcal{F} \cap \mathcal{H}_{n-r-1} = \emptyset$. For the first part, we can choose coordinates so that $L^{n-r} = \mathcal{H}_{n-r}$. From the previous corollary, together with the assumption that $\mathcal{F}$ has dimension r, we conclude that $\varphi_r(\mathcal{F}) = \mathcal{K}^r$. But $\mathcal{H}_{n-r} \cap \mathcal{K}^r$ contains $[\mathbf{e}_r]$. We note that $[\mathbf{e}_r] \in \varphi_r(\mathcal{F})$ means that there exists at least one point $[a] \in \mathcal{F}$ of the form $[a] = [0 : \dots : 0 : 1 : a_{r+1} : \dots : a_n]$. Since $[a] \in \mathcal{H}_{n-r}$, we have the corollary. □

2.5.3. Analyticity of e_r. Let $\mathcal{E}$ be an analytic set in Ω. For an integer r with $-1 \le r \le n$, we define

$$e_r := \{u \in D \mid d(u) = r\},$$

where $d(u) = -1$ if $\mathcal{E}(u) = \emptyset$. We consider all r such that $e_r \neq \emptyset$ and arrange them in increasing order; using the notation r_j, $j = 0, 1, \ldots, \mu$, we have $r_j < r_{j+1}$. We set

$$E_{r_j} := \bigcup_{k=j}^{\mu} e_{r_k} \qquad (j = 0, 1, \ldots, \mu).$$

Note that $E_{r_0} \supset E_{r_1} \supset \ldots \supset E_{r_\mu} = e_{r_\mu}$. Applying Proposition 2.4 inductively, we obtain the following facts:

(1) *E_{r_μ} is an analytic set in D.*
(2) *For each r_j $(j = 1, \ldots, \mu - 1)$, the subset e_{r_j} of D is an analytic set in $D \setminus E_{r_{j+1}}$; we set $s_j := \dim e_{r_j}$.*
(3) *e_{r_0} is a dense, open subset of D.*

Our goal in this section is to show that the closure $\overline{e_{r_j}}$ of each e_{r_j} in D is an analytic set in D. To achieve this goal, we require two lemmas.

LEMMA 2.11. *Assume that $\mathcal{E}$ is an irreducible analytic set in Ω of dimension ρ. Then $\pi_1(\mathcal{E})$ is an irreducible analytic set in D. In case $r_0 = -1$, $\pi_1(\mathcal{E})$ coincides with $\overline{e_{r_1}}$ (the closure of e_{r_1} in D) and is of dimension $s_1 = \rho - r_1$.*

PROOF. In case when $r_0 \geq 0$, we have $\pi_1(\mathcal{E}) = D$, which proves the lemma. We consider the case $r_0 = -1$, i.e., there exists a non-empty open set G in D such that $\mathcal{E}(u) = \emptyset$ for all $u \in G$. Since $\dim \mathcal{E} = \rho$ in Ω, it follows from Corollary 2.4 that $\rho \geq s_j + r_j$ $(j = 1, \ldots, \mu)$.

If $j = 1$, we have the relation $\rho = s_1 + r_1$. To see this, we take a nonsingular point $(u^0, [a^0])$ of the analytic set $\mathcal{E}(e_{r_1})$ in $(D \setminus E_{r_2}) \times \mathbf{P}^n$ such that u^0 and $[a^0]$ are non-singular points of e_{r_1} and $\mathcal{E}(u^0)$, respectively. Since $u^0 \notin E_{r_2}$ and $r_0 = -1$, we can find a neighborhood δ of u^0 in D such that $\delta \cap \pi(\Sigma) = \delta \cap e_{r_1} \neq \delta$. Fix a neighborhood τ of $[a^0]$ in $\mathbf{P}^n$. If δ and τ are sufficiently small, then we have $\mathcal{E} \cap (\delta \times \tau) = \mathcal{E}(e_{r_1}) \cap (\delta \times \tau)$. The latter set has dimension at least $s_1 + r_1$. On the other hand, from the irreducibility of $\mathcal{E}$ it follows that the former set and $\mathcal{E}$ are of the same dimension ρ. From Corollary 2.4 we conclude that $\rho = s_1 + r_1$.

Thus $r_j > r_1$ $(j = 2, \ldots, \mu)$ implies that

$$s_1 > s_j \qquad (j = 2, \ldots, \mu). \tag{2.15}$$

From (2), e_{r_1} is an analytic set in $D \setminus E_{r_2}$. Since e_{r_1} is pure s_1-dimensional and e_{r_2} is an analytic set of dimension s_2 in $D \setminus E_{r_3}$, it follows from Theorem 2.5 and (2.15) that the closure of e_{r_1} in $D \setminus E_{r_3}$, which is equal to $\overline{e_{r_1}} \cap (D \setminus E_{r_3})$, is an analytic set in $D \setminus E_{r_3}$. It is also pure s_1-dimensional in $D \setminus E_{r_3}$. By repeating this procedure, we conclude that the closure $\overline{e_{r_1}}$ in D is a pure s_1-dimensional analytic set in D.

We thus see that $\mathcal{E}(\overline{e_{r_1}})$ is a ρ-dimensional analytic set in Ω. It follows from the irreducibility of $\mathcal{E}$ in Ω and the inclusion $\mathcal{E}(\overline{e_{r_1}}) \subset \mathcal{E}$ that $\mathcal{E} = \mathcal{E}(\overline{e_{r_1}})$, which proves the lemma. □

This immediately implies

COROLLARY 2.7. *For any analytic set $\mathcal{E}$ in Ω, the projection $\pi_1(\mathcal{E})$ is an analytic set in D.*

LEMMA 2.12. *Each E_{r_j} $(j = 0, \dots, \mu)$ is an analytic set in D.*

PROOF. We let $\mathcal{L}^{n-r_j}$ denote the set of all $(n-r_j)$-dimensional hyperplanes in $\mathbf{P}^n$. Fix $u^0 \notin E_{r_j}$. Then there exists $L \in \mathcal{L}^{n-r_j}$ such that $\mathcal{E}(u^0) \cap L = \emptyset$. On the other hand, if $u \in E_{r_j}$, then $\mathcal{E}(u) \cap L \neq \emptyset$ for each $L \in \mathcal{L}^{n-r_j}$.

For each $L \in \mathcal{L}^{n-r_j}$, we set $L_0 := D \times L$, which is an analytic set in Ω. We thus have

$$E_{r_j} \;=\; \bigcap_{L \in \mathcal{L}^{n-r_j}} \pi_1(L_0 \cap \mathcal{E}).$$

To apply Corollary 2.7, for each fixed $L \in \mathcal{L}^{n-r_j}$ we can consider L_0 as Ω in Corollary 2.7. Thus, each projection $\pi_1(L_0 \cap \mathcal{E})$ is an analytic set in D. Thus Corollary 2.1 yields that E_{r_j} is an analytic set in D. □

From this lemma we obtain the following theorem.

THEOREM 2.7. *For each r_j $(j = 0, 1 \dots, \mu)$, the closure $\overline{e_{r_j}}$ of e_{r_j} in D is an analytic set in D.*

PROOF. Since both e_{r_μ} and $E_{r_{\mu-1}} = e_{r_\mu} \cup e_{r_{\mu-1}}$ are analytic sets in D (from (2) and Lemma 2.12), it follows from the local irreducible decomposition theorem of analytic sets (Theorem 2.2) that the closure $\overline{e_{r_{\mu-1}}}$ of $e_{r_{\mu-1}}$ in D is an analytic set in D. Repeating this argument, we obtain the theorem. □

These results on analytic sets $\mathcal{E}$ in $\Omega = D \times \mathbf{P}^n$ can be modified (Corollary 2.8); this will be useful in Chapter 6.

Let $D \subset \mathbf{C}^m_u$ be a domain and set $\Omega' := D \times \mathbf{C}^n_w$. Let $P_j(u,w)$ $(j = 1, \dots, \nu)$ be a polynomial in $w = (w_1, \dots, w_n)$ whose coefficients are holomorphic functions of u on D. Let $\Sigma :\ P_j(u,w) = 0$ $(j = 1, \dots, \nu)$ be an analytic set in Ω', and let $\Sigma(u)$ be the section of Σ over $u \in D$, i.e., $\Sigma(u) = \{w \in \mathbf{C}^n_w \mid (u,w) \in \Sigma\}$. We assume that there exists a polydisk $\lambda := \delta \times \Gamma \subset \mathbf{C}^m_u \times \mathbf{C}^n_w$ in Ω' such that, if we set $\sigma := \Sigma \cap \lambda$, then the section $\sigma(u)$ of σ over each point $u \in \delta$ consists of a finite number of points in $\mathbf{C}^n_w$. We let Σ_0 denote the irreducible component of Σ containing σ (thus $\dim \Sigma_0 = m$).

COROLLARY 2.8. *Under the above setting, there exists an analytic set e in D of dimension at most $m-1$ such that for each $u \in D \setminus e$, $\Sigma_0(u)$ consists of $l \geq 1$ distinct points in $\mathbf{C}^n_w$, where l is an integer independent of $u \in D \setminus e$.*

PROOF. Let $w_i = \zeta_i/\zeta_0$ $(i = 1, \dots, n)$ and let k_j $(j = 1, \dots, \nu)$ be the degree of $P_j(u,w)$ in w. Then we can form the homogeneous polynomial $\widehat{P}_j(u,\zeta)$ in $\zeta = (\zeta_0, \zeta_1, \dots, \zeta_n)$, where $P_j(u,w)\zeta_0^{k_j} = \widehat{P}_j(u,\zeta)$. We set $[\zeta] = [\zeta_0 : \zeta_1 : \dots : \zeta_n]$ and $[\zeta'] = [\zeta_1 : \dots : \zeta_n]$ and identify

$$\mathbf{C}^n_w \;\cong\; \{[\zeta] \in \mathbf{P}^n_\zeta \mid \zeta_0 \neq 0\}, \quad \mathcal{H}_{n-1} \;= \{[\zeta] \in \mathbf{P}^n_\zeta \mid \zeta_0 = 0\} \;\cong\; \mathbf{P}^{n-1}_{\zeta'},$$

so that $\mathbf{P}^n_\zeta$ is the disjoint union $\mathbf{P}^n_\zeta = \mathbf{C}^n_w \cup \mathcal{H}_{n-1}$. Let $\Omega = D \times \mathbf{P}^n_\zeta$ and let $\widehat{\Sigma}$ be the analytic set in Ω defined by $\widehat{P}_j(u,\zeta) = 0$ $(j = 1, \dots, \nu)$. Then $\Sigma = \widehat{\Sigma} \cap \Omega'$. We let $\mathcal{E}$ denote the irreducible component of $\widehat{\Sigma}$ which contains σ (precisely, $\mathcal{E} \cap \Omega' \supset \sigma$). We set $\mathcal{F} = \mathcal{E} \cap (D \times \mathcal{H}_{n-1})$, which we consider as an analytic set in $\Omega^{n-1} := D \times \mathbf{P}^{n-1}_{\zeta'}$. We use the notation from the beginning of 2.5.3 for $\mathcal{E}$ in Ω. In this case, our assumption on Σ implies that $r_0 = 0$, and hence from Lemma 2.12 we conclude that there exists an analytic set E_{r_1} in D of dimension at most $m-1$ such that $\dim \mathcal{E}(u) = 0$ over each $u \in D \setminus E_{r_1}$, i.e., the section $\mathcal{E}(u)$ is non-empty and consists of a finite number of distinct points $\{p_j(u)\}_{j=1,\dots,l(u)}$ in $\mathbf{P}^n_\zeta$. Note that $l(u)$ is bounded in $D \setminus E_{r_1}$ since $\mathcal{E}$ is an irreducible analytic set of dimension m in Ω. Furthermore, if we set $l := \max \{l(u) \mid u \in D \setminus E_{r_1}\}$ and $e_1 := \{u \in D \setminus E_{r_1} \mid l(u) \leq l - 1\}$, then $\overline{e_1}$ is an analytic set of dimension at most $m - 1$ in D. We again use the notation from 2.5.3 for the analytic set $\mathcal{F}$ in Ω^{n-1}. Since $\mathcal{F} \subset \mathcal{E}$, $\mathcal{F} \neq \mathcal{E}$, and $\mathcal{E}$ is irreducible in Ω, it follows that $r_0 = -1$ for $\mathcal{F}$ in Ω^{n-1}; hence there exists an analytic set F_{r_1} in D of dimension at most $m - 1$ such that the section $\mathcal{F}(u) = \emptyset$ over each $u \in D \setminus F_{r_1}$. Thus, if we set $e := E_{r_1} \cup e_1 \cup F_{r_1}$, then e is an analytic set in D of dimension at most $m - 1$ and $\mathcal{E}(u) = \Sigma_0(u)$ for $u \in D \setminus e$; moreover, $\mathcal{E}(u)$ consists of l distinct points in $\mathbf{C}^n_w$, as desired. □

CHAPTER 3

The Poincaré, Cousin, and Runge Problems

3.1. Meromorphic Functions

3.1.1. Poincaré Problem. Let D be a domain in $\mathbf{C}^n$. If a function $g(z)$ in D can be locally represented as a quotient of two holomorphic functions, then $g(z)$ is called a **meromorphic function** in D. To be precise, g is meromorphic in D if for each point $p \in D$, we can find a neighborhood δ_p of p in D and functions $h_p(z)$, $k_p(z)$ holomorphic in δ_p such that for any p, $q \in D$ with $\delta_p \cap \delta_q \neq \emptyset$, we have

$$k_p(z)h_q(z) = k_q(z)h_p(z) \quad \text{in } \delta_p \cap \delta_q, \tag{3.1}$$

and $g(z) = h_p(z)/k_p(z)$ in δ_p.

From the Weierstrass preparation theorem, by choosing a smaller neighborhood δ_p if necessary, we may assume that $h_p(z)$ and $k_p(z)$ are relatively prime at p; i.e., if we choose the coordinates $(z_1, \dots, z_n)$ satisfying the Weierstrass condition for the analytic hypersurfaces $h_p(z) = 0$ and $k_p(z) = 0$ at p, then $h_p(z)$ and $k_p(z)$ have no common factor which is an irreducible distinguished pseudopolynomial in z_n at p of positive degree. If we let σ_p denote the zero set of $k_p(z)$ in δ_p, then the union of the sets σ_p defines an analytic set Σ in D. Note that Σ does not depend on the choice of $h_p(z)$ and $k_p(z)$. We call Σ the **set of singularities** or **pole set** of g; the function $g(z)$ is holomorphic in $D \setminus \Sigma$.

Let p be a pole of $g(z)$. If $h_p(p) \neq 0$, then clearly $g(p) = \infty$. On the other hand, even though $h_p(z)$ and $k_p(z)$ are assumed to be relatively prime at p, they may simultaneously vanish at p (e.g., take $h_p(z_1, z_2) = z_1$ and $k_p(z_1, z_2) = z_2$ at $p = (0,0)$ in $\mathbf{C}^2$). Then, given *any* number $c \in \mathbf{C}$, the analytic hypersurface in D defined by

$$h_p(z) - ck_p(z) = 0 \qquad \text{in } \delta_p$$

passes through the point p. Thus the value $g(p)$ is not uniquely determined. Such a pole p is called a **point of indeterminacy** of $g(z)$. The set of all indeterminacy points of $g(z)$ in D is a pure $(n-2)$-dimensional analytic set in D. This follows since the non-empty intersection of two distinct irreducible analytic hypersurfaces Σ_1, Σ_2 in a domain $G \subset \mathbf{C}^n$ is a pure $(n-2)$-dimensional analytic set in G.

Since the definition of meromorphic function is local, the problem arises as to when we can write a meromorphic function in D as a quotient of two global holomorphic functions.

Poincaré Problem Let $g(z)$ be a meromorphic function in D. Find two holomorphic functions $h(z)$ and $k(z)$ in D such that $h(z)$ and $k(z)$ are relatively prime at each point $p \in D$ and satisfy $g(z) = h(z)/k(z)$ in D.

This problem in the case of $D = \mathbf{C}^2$ was solved in the affirmative by Poincaré [**59**]. An example of a product domain D where the Poincaré problem is not solvable for g in D will be given in Remark 3.5. We mention that even though the Poincaré problem as stated is not always solvable for g in D, there always exist holomorphic functions $h(z)$ and $k(z)$ in D satisfying $g(z) = h(z)/k(z)$ in any domain of holomorphy D but where $h(z)$ and $k(z)$ are **not** necessarily relatively prime at each $p \in D$. This will be shown in Theorem 8.19 in Chapter 8.

3.1.2. Cousin Problems. The Poincaré problem for a general domain is related to the following problems, known as problems I and II of Cousin.

Let D be a domain in $\mathbf{C}^n$. For each $p \in D$, we assume the pairs (g_p, δ_p) are given, where δ_p is a neighborhood of p in D and $g_p(z)$ is a meromorphic function in δ_p; furthermore, we assume that for any points $p, q \in D$ with $\delta_p \cap \delta_q \neq \emptyset$, the function $g_p(z) - g_q(z)$ is holomorphic in $\delta_p \cap \delta_q$. We call the collection $\{(g_p, \delta_p)\}_p$ for $p \in D$ **Cousin I data** in D or a **Cousin I distribution** in D. In other words, Cousin I data simply gives the analogue in several variables of the principal parts at the poles of a meromorphic function. For a closed set E in $\mathbf{C}^n$, we say that Cousin I data is given in E if Cousin I data is given in a neighborhood D of E.

Cousin I Problem Given Cousin I data $\{(g_p, \delta_p)\}_p$ in D, find a meromorphic function $g(z)$ in D such that $g(z) - g_p(z)$ is holomorphic in δ_p, $p \in D$.

In brief, this is the problem of finding a meromorphic function with a prescribed pole set and prescribed principal parts. If such a $g(z)$ exists, we call $g(z)$ a **solution of the Cousin I problem** for the given data $\{(g_p, \delta_p)\}_p$. In the case of one complex variable, a solution always exists; this is the content of the classical Mittag-Leffler theorem.

Let D be a domain in $\mathbf{C}^n$. For each $p \in D$, let (f_p, δ_p) be given, where δ_p is a neighborhood of p in D and $f_p(z)$ is a holomorphic function in δ_p. Moreover, we assume that if $\delta_p \cap \delta_q \neq \emptyset$, then $f_p(z)/f_q(z)$ is a nonvanishing holomorphic function in $\delta_p \cap \delta_q$. We call the collection $\{(f_p, \delta_p)\}_p$ for $p \in D$ **Cousin II data** in D or a **Cousin II distribution** in D. In other words, we are specifying the zero set and order of vanishing of a family of holomorphic functions. For a closed set E in $\mathbf{C}^n$, we say that Cousin II data in E is given if Cousin II data is given in a neighborhood D of E.

Cousin II Problem Given Cousin II data $\{(f_p, \delta_p)\}_p$ in a domain D, find a holomorphic function $f(z)$ in D such that $f(z)/f_p(z)$ is a nonvanishing holomorphic function in δ_p, $p \in D$.

In short, this is the problem of finding a holomorphic function with a prescribed zero set. If such an $f(z)$ exists, we say that $f(z)$ is a **solution of the Cousin II problem** for the given data $\{(f_p, \delta_p)\}_p$. In the case of one complex variable, a solution always exists; this is the content of the classical Weierstrass theorem.

We remark that for both Cousin I and Cousin II, if we replace each set δ_p by a finite union of subsets δ'_p which cover δ_p, then the collection of pairs $\{(f'_p, \delta'_p)\}_p$, where $f'_p = f_p|_{\delta'_p}$, again forms Cousin data for D. This fact will be used many times.

In $\mathbf{C}$, the Poincaré problem is always solvable in any domain D; a standard proof uses the classical Weierstrass theorem. Similarly, in $\mathbf{C}^n$ we have the following relation between the Poincaré problem and the Cousin II problem.

PROPOSITION 3.1. *Let D be a domain in $\mathbf{C}^n$. If the Cousin II problem is solvable for any Cousin II data in D, then the Poincaré problem is always solvable in D.*

PROOF. Let $g(z)$ be a meromorphic function in D. By definition, for each $p \in D$, there exist a neighborhood δ_p of p in D and holomorphic functions $h_p(z)$, $k_p(z)$ in δ_p which satisfy equation (3.1) and are relatively prime at p. It is easy to see that the collection $\{(k_p, \delta_p)\}_p$ defines Cousin II data in D. Since we are assuming that the Cousin II problem is always solvable in D, we can find a holomorphic function $k(z)$ in D such that $k(z)/k_p(z)$ is a nonvanishing holomorphic function in δ_p. Therefore, if we set $h(z) := g(z)k(z)$, then $h(z)$ becomes a holomorphic function in D. Furthermore, since $h_p(z)$ and $k_p(z)$ are relatively prime at p, it follows that $h(z)$ and $k(z)$ are also relatively prime at each point $p \in D$. Thus $g(z) = h(z)/k(z)$ is a solution of the Poincaré problem for $g(z)$. □

Later on we shall give an example of a Cousin II distribution in a Reinhardt product domain D in $\mathbf{C}^n$ for which no solutions of a Cousin II problem exist. From this example we shall also obtain a meromorphic function in D which cannot be represented as a quotient of two holomorphic functions in D which are relatively prime at each point in D (see Remark 3.5 in section 3.5.3). Thus, even for Reinhardt product domains, the Poincaré problem is not always solvable.

3.1.3. Runge Problem.

Often in attempting to construct holomorphic functions possessing a certain property, as is required in solving Cousin problems, questions on approximation of holomorphic functions arise.

Runge problem Let K_1, K_2 be subsets of $\mathbf{C}^n$ with $K_1 \subset\subset K_2^o$, where K_2^o denotes the interior of K_2. Given a holomorphic function $f(z)$ on K_1, for each $E \subset\subset K_1$ and each $\varepsilon > 0$, find a holomorphic function $F(z)$ on K_2 such that $|F(z) - f(z)| < \varepsilon$ on E.

If this problem is solvable for (K_1, K_2) for any holomorphic function $f(z)$ on K_1, we say that the **Runge theorem** holds for the pair (K_1, K_2). In the case $K_2 = \mathbf{C}^n$, this is the classical Runge problem.

We have the following relation between the Runge problem and the Cousin I problem.

PROPOSITION 3.2. *Let D be a domain in $\mathbf{C}^n$ and let K_j $(j = 1, 2, \dots)$ be a sequence of subsets in D such that each K_j is compact or open and*

$$K_j \subset\subset K_{j+1}^o, \qquad D = \bigcup_{j=1}^{\infty} K_j.$$

If we assume that

1. *the Cousin I problem is solvable on each K_j $(j = 1, 2, \dots)$, and*
2. *the Runge theorem holds for each pair (K_j, K_{j+1}) $(j = 1, 2, \dots)$,*

then the Cousin I problem is solvable for D.

PROOF. We give the proof in the case where each K_j is compact; the general case follows with minor modifications. Let a Cousin I distribution $\mathcal{C}_1 = \{(f_p, \delta_p)\}_p$ be given in D. From 1 let $g_j(z)$ and $g_{j+1}(z)$ be any solutions of the Cousin I problem for the restrictions of $\mathcal{C}_1$ in K_j and K_{j+1}, and let $\varepsilon > 0$. Since $f_j(z) = g_{j+1}(z) - g_j(z)$ is holomorphic on K_j, it follows from 2 that we can find a holomorphic function

$F_{j+1}(z)$ on K_{j+1} such that $|F_{j+1}(z) - f_j(z)| < \varepsilon$ on K_{j-1}. Hence, $G_{j+1}(z) := g_{j+1}(z) - F_{j+1}(z)$ is a solution of the Cousin I problem for $\mathcal{C}_1$ in K_{j+1} satisfying $|G_{j+1}(z) - g_j(z)| < \varepsilon$ on K_{j-1}.

Now let $\varepsilon_j > 0$ $(j = 1, 2, \dots)$ with $\sum_{j=1}^{\infty} \varepsilon_j < \infty$. By induction, we construct a solution $G_j(z)$ $(j = 1, 2, \dots)$ of the Cousin I problem for $\mathcal{C}_1$ in K_j with the property that $|G_{j+1}(z) - G_j(z)| < \varepsilon_j$ on K_{j-1}. Hence, the limit

$$G(z) := \lim_{j \to \infty} G_j(z)$$

converges uniformly on any compact set in D. Thus, $G(z)$ is a solution of the Cousin I problem for $\mathcal{C}_1$ in D. □

3.1.4. Cousin Problems and Domains of Holomorphy.

Cousin problems are not always solvable.

EXAMPLE 3.1. In $\mathbf{C}^2 = \mathbf{C}_z \times \mathbf{C}_w$, consider the following three Reinhardt product domains:

$$\begin{array}{lcrr} \Delta_1 & : & |z| < 2, & 2 < |w| < 3, \\ \Delta_2 & : & |z| < 2, & |w| < 1, \\ \Delta_3 & : & 1 < |z| < 2, & |w| < 3, \end{array}$$

and set $\Delta := \Delta_1 \cup \Delta_2 \cup \Delta_3$. In the domain Δ we define a Cousin I distribution

$$\mathcal{C}_1 : \quad (1, \Delta_1), \qquad (1/z, \Delta_2), \qquad (1, \Delta_3)$$

and a Cousin II distribution

$$\mathcal{C}_2 : \quad (1, \Delta_1), \qquad (z, \Delta_2), \qquad (1, \Delta_3).$$

From Osgood's theorem, it follows that neither $\mathcal{C}_1$ for Cousin I nor $\mathcal{C}_2$ for Cousin II is solvable in Δ.

Related to the Cousin I problem, we have the following result of H. Cartan [**9**].

PROPOSITION 3.3. *Let D be a domain in $\mathbf{C}^n$ satisfying:*

1. *for any $(n-1)$-dimensional complex hyperplane L in $\mathbf{C}^n$, the domain $D \cap L$ is a domain of holomorphy in L; and*
2. *the Cousin I problem for any Cousin I distribution in D is solvable in D.*

Then D is a domain of holomorphy.

PROOF. The proof is by contradiction. Assume that D is not a domain of holomorphy. Then there exist at least one boundary point Q of D and a neighborhood V of Q in $\mathbf{C}^n$ such that each holomorphic function in D has a holomorphic extension to V. We take an $(n-1)$-dimensional complex hyperplane L passing through Q such that Q is a boundary point of $D^0 := D \cap L$. To simplify the notation, we assume that L is the hyperplane given by $z_n = 0$, so that D^0 is an open set in $\mathbf{C}^{n-1}$ with variables $z' := (z_1, \dots, z_{n-1})$. From 1, there exists a holomorphic function $f(z')$ in D^0 whose domain of holomorphy is D^0 itself. By regarding $f(z')$ as a function of all n variables $(z_1, \dots, z_n)$ which is independent of z_n, we see that $f(z')$ is holomorphic in a neighborhood U of $L \cap D$ in D.

We consider the following Cousin I distribution $\mathcal{C}_1 = \{(g_p, \delta_p)\}_p$ in D:

1. If $p \in L$, we take a neighborhood δ_p of p in U and the meromorphic function $g_p(z) = f(z)/z_n$ in δ_p.

2. If $p \notin L$, we take a neighborhood δ_p of p in D such that $\delta_p \cap L = \emptyset$ and set $g_p(z) \equiv 1$ in δ_p.

Then the data $\mathcal{C}_1 = \{(g_p, \delta_p)\}_p$ defines a Cousin I distribution in D. From 2 we have a solution $g(z)$ of the Cousin I problem for $\mathcal{C}_1$ in D. We define $F(z) := z_n\, g(z)$ in D. Then $F(z)$ is a holomorphic function in D, and we claim that $F(z)|_{D^0} = f(z')$. To verify this claim, take $p \in L \cap D$. Then

$$g(z) = \frac{f(z)}{z_n} + h_p(z) \qquad \text{in } \delta_p,$$

where $h_p(z)$ is a holomorphic function in δ_p. Therefore,

$$F(z) = f(z) + z_n\, h_p(z) \qquad \text{in } \delta_p,$$

and hence $F(z', 0) = f(z')$. Since $F(z)$ is holomorphic in D, it has a holomorphic extension to the neighborhood V of Q in $\mathbf{C}^n$. Thus it follows that $f(z')$ has a holomorphic extension to $L \cap V$. This is a contradiction to D^0 being the domain of holomorphy of $f(z')$; thus D is a domain of holomorphy. □

REMARK 3.1. In the case of one complex variable, every domain is a domain of holomorphy; thus condition 1 in the proposition in the case $n = 2$ is always satisfied. Hence we have shown that *any domain D in $\mathbf{C}^2$ such that the Cousin I problem is always solvable in D is necessarily a domain of holomorphy.*

This result suggests that the Cousin I problem should be studied in domains of holomorphy.

3.2. Cousin Problems in Polydisks

3.2.1. Cousin Integral. P. Cousin [**15**] solved in 1895 both of the Cousin problems in polydisks in $\mathbf{C}^n$. In this section, we introduce the notion of a Cousin integral, which will be used in the following section to solve the Cousin I problem in a closed polydisk in $\mathbf{C}^n$.

Let a and b be distinct points in the complex plane $\mathbf{C}$ and let l be a simple smooth arc with initial point a and terminal point b. Take a neighborhood V of l in $\mathbf{C}$ and a holomorphic function $f(z)$ in V, and form the integral

$$F(z) = \frac{1}{2\pi i}\int_l \frac{f(\zeta)}{\zeta - z}\, d\zeta \tag{3.2}$$

for $z \in \mathbf{C}^n \setminus l$. We study the behavior of $F(z)$ near l.

Clearly $F(z)$ is a holomorphic function in $\mathbf{C} \setminus l$ satisfying $\lim_{z\to\infty} F(z) = 0$. Next, we note that

$$\frac{f(\zeta)}{\zeta - z} = \frac{f(\zeta) - f(z)}{\zeta - z} + \frac{f(z)}{\zeta - z}.$$

Since the first term on the right-hand side is a holomorphic function of the two complex variables z and ζ, it follows that

$$F(z) + \frac{f(z)}{2\pi i}\log(a - z) \qquad \text{and} \qquad F(z) - \frac{f(z)}{2\pi i}\log(b - z)$$

are single-valued holomorphic functions in neighborhoods of a and b, respectively.

We next describe the behavior of $F(z)$ near $z' \in l \setminus \{a, b\}$. Note that at such a point $F(z)$ can be analytically continued across the arc l from each side, and the difference between these extensions is equal to $f(z)$ in a neighborhood of z'. To be precise, let $z' \in l \setminus \{a, b\}$ and let $\delta : |z - z'| < \rho$ be a sufficiently small disk contained in $V \setminus \{a, b\}$ so that $\gamma = \partial\delta$ intersects l at exactly two points a' and b'. We let δ_1 and γ_1 denote the portions of δ and γ situated on one side of the oriented arc l, and by δ_2 and γ_2 the portions of δ and γ on the other side of l. Set $\beta := l \cap \delta$. We then define

$$\varphi_j(z) := F(z), \quad z \in \delta_j \ (j = 1, 2).$$

Then $\varphi_1(z)$ ($\varphi_2(z)$) can be analytically continued across the arc β to δ_2 (δ_1), so that $\varphi_1(z)$ and $\varphi_2(z)$ become holomorphic functions on δ and we have

$$\varphi_2(z) - \varphi_1(z) = f(z), \qquad z \in \delta.$$

To verify this last statement, we let l' denote the subarc of l connecting a with a', while l'' denotes the subarc connecting b' and b. Then $l = l' + \beta + l''$. Let $l_1 := l' - \gamma_2 + l''$ and $l_2 := l' + \gamma_1 + l''$. Using the Cauchy integral formula, we obtain, for $z \in \delta_j$ $(j = 1, 2)$,

$$\varphi_j(z) = \frac{1}{2\pi i} \int_l \frac{f(\zeta)}{\zeta - z}\, d\zeta = \frac{1}{2\pi i} \int_{l_j} \frac{f(\zeta)}{\zeta - z}\, d\zeta.$$

Since the function defined by the integral on the right-hand side is a holomorphic function for z on δ, it follows that $\varphi_j(z)$ $(j = 1, 2)$ can be analytically continued to δ. Again using the Cauchy integral formula, we obtain

$$\varphi_2(z) - \varphi_1(z) = \frac{1}{2\pi i} \int_{l_2 - l_1} \frac{f(\zeta)}{\zeta - z}\, d\zeta = \frac{1}{2\pi i} \int_{\gamma} \frac{f(\zeta)}{\zeta - z}\, d\zeta = f(z)$$

for any $z \in \delta$, as claimed. □

This is the idea of Cousin. We call the integral in (3.2) a **Cousin integral** of $f(z)$ along l, and we say that the two holomorphic functions $\varphi_j(z)$ $(j = 1, 2)$ have a **jump** of $f(z)$ along l.

We note from the construction that if $f(z)$ is of the form $f(z, w)$, where w is a complex parameter such that $f(z, w)$ is holomorphic with respect to w, then the Cousin integral $F(z, w)$ of $f(z, w)$ along l as well as the functions $\varphi_j(z, w)$ $(j = 1, 2)$ are also holomorphic in w.

3.2.2. Cousin I Problem in Polydisks. Let $\mathbf{C}^n = \mathbf{C}_{z_1} \times \cdots \times \mathbf{C}_{z_n}$ and for each $\mathbf{C}_{z_j}$ $(j = 1, \ldots, n)$ consider two concentric disks centered at the origin:

$$\Delta_j : |z_j| < r_j \qquad \text{and} \qquad \Delta'_j : |z_j| \le r'_j \qquad (0 < r'_j < r_j).$$

We set

$$\Delta = \Delta_1 \times \cdots \times \Delta_n \qquad \text{and} \qquad \Delta' = \Delta'_1 \times \cdots \times \Delta'_n.$$

Directly from the Taylor series expansion of holomorphic functions, we obtain the following simple lemma.

LEMMA 3.1. *The Runge theorem holds for the pair* (Δ', Δ).

In order to show that the Cousin I problem is always solvable in *open* polydisks, from the lemma and Proposition 3.2 it suffices to show that the Cousin I problem is always solvable in any *closed* polydisk. Using the Cousin integral, we proceed to show that the Cousin I problem can always be solved in a closed polydisk. We remark that the Cousin I problem being solvable on a closed polydisk $\overline{\Delta}$ in $\mathbf{C}^n$ means that if $\mathcal{C} := \{(g_p, \delta_p)\}_p$ is a Cousin I distribution on an open polydisk U containing $\overline{\Delta}$, then the corresponding Cousin I problem is solvable for $\mathcal{C}$ in an open

U' such that $\overline{\Delta} \subset U' \subset U$. The open sets U and U' may depend on the data $\mathcal{C}$; i.e., if we have another set of Cousin I data $\mathcal{C}_1$, then this data may be defined and solvable on a (perhaps) smaller open polydisk U_1 containing $\overline{\Delta}$.

Thus we begin with $\Delta = \Delta_1 \times \cdots \times \Delta_n$ and we let $\mathcal{C}_1 = \{(g_p, \delta_p)\}_p$ be a Cousin I distribution on $\overline{\Delta}$. Setting $z_j = x_j + iy_j$ $(j = 1, \ldots, n)$, we let

$$\Omega_j \ : \ |x_j| \leq 2r_j, \qquad |y_j| \leq 2r_j$$

be a rectangle on $\mathbf{C}_{z_j}$ so that $\Omega_j \supset \overline{\Delta}_j$. We subdivide Ω_j into N^2 rectangles using N lines parallel to the x_j-axis and N lines parallel to the y_j-axis. Let ω_j denote the intersection of Δ_j and one of these rectangles, and define $\omega := \omega_1 \times \cdots \times \omega_n \ \subset \Delta$. We assume that N is chosen sufficiently large so that each cube ω is contained in δ_p for some p in $\overline{\Delta}$; this is where we are using the fact that we have Cousin data on the *closed* polydisk.

Our goal is to replace the meromorphic Cousin I data g_p on a cube $\omega \subset \delta_p$ by a holomorphic function. To this end, let Λ_j $(j = 1, \ldots, n)$ be a closed convex domain in $\mathbf{C}_{z_j}$ bounded by a simple smooth closed curve, and let $\Lambda = \Lambda_1 \times \cdots \times \Lambda_n \ \subset \mathbf{C}^n$. For $\rho > 0$, we define

$$\begin{aligned} \Lambda^1 &:= \{(z_1, \ldots, z_n) \in \Lambda \mid x_1 \leq \rho\}, \\ \Lambda^2 &:= \{(z_1, \ldots, z_n) \in \Lambda \mid x_1 \geq -\rho\}, \end{aligned}$$

and we set $\Lambda^0 := \Lambda^1 \cap \Lambda^2$, which we assume to be nonempty. Then we have the following lemma.[1]

LEMMA 3.2. *Let $g_1(z)$ and $g_2(z)$ be meromorphic functions in Λ^1 and Λ^2 such that $g_1(z) - g_2(z)$ is holomorphic in Λ^0. Then there exist holomorphic functions $h_1(z)$ and $h_2(z)$ in Λ^1 and Λ^2 such that the function*

$$g(z) := \begin{cases} g_1(z) - h_1(z), & z \in \Lambda^1, \\ g_2(z) - h_2(z), & z \in \Lambda^2, \end{cases}$$

defines a single-valued meromorphic function in Λ.

PROOF. Fixing the complex parameters $z_2, \ldots, z_n$ in $\Lambda_2 \times \cdots \times \Lambda_n$, we consider the holomorphic function $f(z) = g_1(z) - g_2(z)$ in Λ^0 as a holomorphic function of z_1 in $\lambda_1^0 = \{z_1 \in \Lambda_1 \mid -\rho \leq x_1 \leq \rho\}$. We let ia, ib $(a < b)$ in $\mathbf{C}_{z_1}$ be the points where $\partial\Lambda_1$ intersects the y_1-axis and we fix a segment $l = [ia', ib']$ $(a' < a, b < b')$ on which $f(z)$ is holomorphic. Using the Cousin integral of $f(z)$ along l, we construct holomorphic functions $h_1(z)$ and $h_2(z)$ in Λ^1 and Λ^2 such that $f(z) = h_1(z) - h_2(z)$ in Λ^0. This gives the desired result. □

Using Lemma 3.2 repeatedly for the meromorphic functions $g_p(z)$ on the cubes ω, we construct a meromorphic function $g(z)$ on $\overline{\Delta}$ which is a solution of the original Cousin I problem for the distribution $\mathcal{C}_1$. Thus the Cousin I problem is solvable on closed polydisks; hence we have proved the following result.

THEOREM 3.1. *The Cousin I problem in an open polydisk in $\mathbf{C}^n$ is always solvable.*

---

[1]The lemma is valid without the convexity assumption on Λ_j; we impose this condition in order to simplify the notation and to clarify the idea of the proof.

3.3. Cousin I Problem in Polynomially Convex Domains

3.3.1. Lifting Principle. Kiyoshi Oka [**53**] proved in 1937 that the Cousin I problem in an arbitrary domain of holomorphy in $\mathbf{C}^n$ is always solvable. Here we introduce his theory in its simplest form. First we will show that the Cousin I problem in polynomially convex domains is always solvable. The key idea is the **lifting principle.** [2]

In $\mathbf{C}^n$ with variables $z = (z_1, \dots, z_n)$, fix m polynomials $P_k(z)$ $(k = 1, \dots, m)$ and define a closed domain $\mathcal{P}$ in $\mathbf{C}^n$,

$$\mathcal{P} \,:\, |z_j| \leq r_j \;\; (j = 1, \dots, n), \qquad |P_k(z)| \leq 1 \;\; (k = 1, \dots, m). \tag{3.3}$$

We call $\mathcal{P}$ a **polynomial polyhedron** in $\mathbf{C}^n$. We will always assume that the collection of polynomials is minimal in the sense that deleting any one of the sets $\{|P_k(z)| \leq 1\}$ from this intersection defines a strictly larger polynomial polyhedron $\widetilde{\mathcal{P}}$, and we call this minimal number m the **rank** of $\mathcal{P}$. Note that the dimension n of $\mathbf{C}^n$ and the rank m of a polynomial polyhedron are independent quantities. For example, a polydisk in $\mathbf{C}^n$ is a polynomial polyhedron of rank 0, regardless of n.

We introduce $\mathbf{C}^m$ with variables $w = (w_1, \dots, w_m)$, and consider the closed polydisk $\overline{\Delta}$ in $\mathbf{C}^{n+m} = \mathbf{C}^n_z \times \mathbf{C}^m_w$ defined by

$$\overline{\Delta} \,:\, |z_j| \leq r_j \;\; (j = 1, \dots, n), \qquad |w_k| \leq 1 \;\; (k = 1, \dots, m). \tag{3.4}$$

Define

$$\Sigma = \{(z, w) \in \overline{\Delta} \mid w_k = P_k(z) \; (k = 1, \dots, m)\}, \tag{3.5}$$

which is a pure n-dimensional analytic set in $\overline{\Delta}$. Using the mapping

$$z \in \mathcal{P} \to M = (z, P_1(z), \dots, P_m(z)) \in \Sigma,$$

we see that $\mathcal{P}$ is homeomorphically equivalent to Σ and $\partial\mathcal{P}$ corresponds to $(\partial\Delta) \cap \Sigma$ (see Figure 1).

In this setting, we consider the following problem.

Lifting Problem. Let $f(z)$ be a holomorphic function on $\mathcal{P}$. Find a holomorphic function $F(z, w)$ on $\overline{\Delta}$ such that

$$f(z) = F(z, P_1(z), \dots, P_m(z)) \qquad \text{for } \; p \in \mathcal{P}.$$

If this problem can be solved for an arbitrary holomorphic function $f(z)$ on $\mathcal{P}$, we say that the **lifting principle** holds for $\mathcal{P}$; and we call $F(z, w)$ an **extension of** $f(z)$ on $\overline{\Delta}$.[3]

[2] The lifting principle is a central idea throughout all of Oka's work. In the footnote of his paper I (p. 249), he has written " Je dois l'idée à M. H. Cartan pour ce mode d'application de théorème de M. Cousin, voir : [**7**]."

[3] Oka's terminology in Japanese for the lifting principle translates literally into *the principle of going up to the sky* in English.

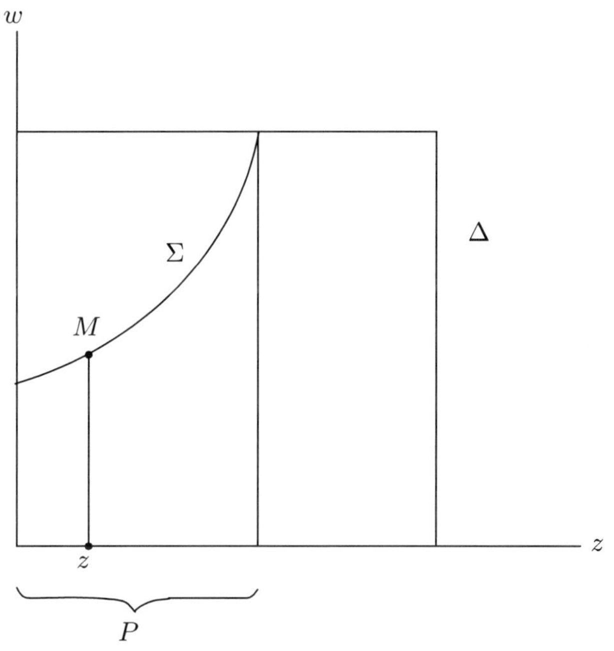

FIGURE 1. Representation of a polynomial polyhedron

3.3.2. Polynomial Polyhedra. The lifting principle is closely related to the Cousin I problem. In fact, on a polynomial polyhedron $\mathcal{P}$ in $\mathbf{C}^n$, the solvability of the Cousin I problem and the solvability of the lifting problem can be proved simultaneously by use of a double induction on the rank m of $\mathcal{P}$. We have already seen that the Cousin I problem in polynomial polyhedra $\mathcal{P}$ of rank 0 (i.e., polydisks) in $\mathbf{C}^n$ is always solvable. Moreover, the lifting principle is trivially true for polydisks. We next prove two lemmas which comprise the double induction proof of solvability of Cousin I and of the lifting problem on polynomial polyhedra.

LEMMA 3.3. *Let $m \geq 1$. Assume that both the Cousin I problem and the lifting problem in any polynomial polyhedron of rank $m-1$ are solvable. Then the lifting problem in any polynomial polyhedron of rank m is solvable.*

PROOF. Let $\mathcal{P}$ be a polynomial polyhedron of rank m in $\mathbf{C}^n$ given by (3.3) and use notation Σ in (3.5). Let $f(z)$ be a holomorphic function on $\mathcal{P}$.

We introduce the $(n+1)$-dimensional Euclidean space $\mathbf{C}^{n+1} = \mathbf{C}^n_z \times \mathbf{C}_{w_1}$ and define

$$\mathcal{P}^* \;:\; |z_j| \leq r_j \quad (j = 1, \dots, n), \quad |w_1| \leq 1, \quad |P_k(z)| \leq 1 \quad (k = 2, \dots, m).$$

Thus $\mathcal{P}^*$ is a polynomial polyhedron of rank $m-1$ in $\mathbf{C}^{n+1}$. We put

$$\Sigma^* \;:\; w_k = P_k(z) \quad (k = 2, \dots, m), \quad (z, w_1) \in \mathcal{P}^*, \tag{3.6}$$

which is an $(n+1)$-dimensional analytic set in $\overline{\Delta}^{n+1} \times \overline{\Delta}^{m-1}$ with $\Sigma \subset \Sigma^*$; moreover, Σ^* is bijective to $\mathcal{P}^*$: $(z, w_1) \in \mathcal{P}^* \to (z, w_1, P_2(z), \dots, P_m(z)) \in \Sigma^*$.

In $\mathcal{P}^*$ we consider the set Σ_1^* defined by

$$\Sigma_1^* : \; w_1 = P_1(z), \quad z \in \mathcal{P}.$$

Note that Σ_1^* is a pure n-dimensional analytic set in $\mathcal{P}^*$ which is homeomorphically equivalent to $\mathcal{P}$ via the mapping $z \in \mathcal{P} \to (z, P_1(z)) \in \Sigma_1^*$, and $\partial\mathcal{P}$ corresponds to $\Sigma_1^* \cap (\partial\mathcal{P}^*)$.

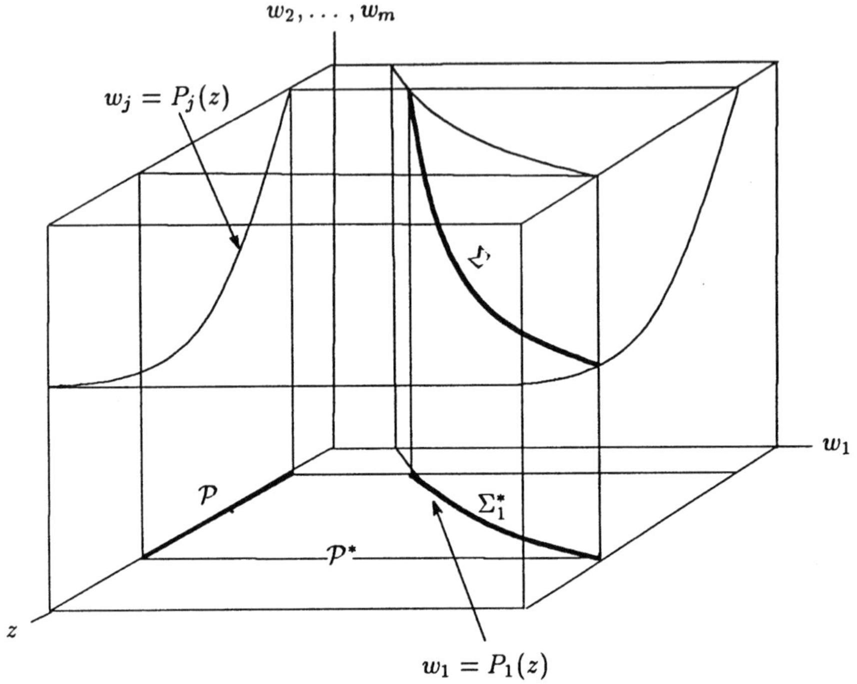

FIGURE 2. Relation between $\mathcal{P}^*$ and Σ_1^*

We can find a neighborhood V of Σ_1^* in $\mathcal{P}^*$ in which $f(z)$ is holomorphic (here we regard f as a function which is independent of w_1).

For each point $p \in \mathcal{P}^*$, we choose a neighborhood δ_p of p in $\mathbf{C}^{n+1}$ and a meromorphic function $\varphi_p(z, w)$ in δ_p such that the following conditions are satisfied:

1. If $p \in \Sigma_1^*$, then we choose δ_p to be contained in V and

$$\varphi_p(z, w_1) = f(z)/(w_1 - P_1(z)) \qquad \text{in } \delta_p.$$

2. If $p \notin \Sigma_1^*$, then we choose δ_p such that $\delta_p \cap \Sigma_1^* = \emptyset$ and $\varphi_p(z, w_1) \equiv 1$ in δ_p.

Then $\mathcal{C}_1 = \{(\varphi_p, \delta_p)\}_{p \in \mathcal{P}^*}$ defines a Cousin I distribution in $\mathcal{P}^*$. Since $\mathcal{P}^*$ is a polynomial polyhedron of rank $m - 1$, it follows from the inductive hypothesis on solvability of the Cousin I problem that we can find a meromorphic function $\Phi(z, w_1)$ in $\mathcal{P}^*$ such that

$$\Phi(z, w_1) - \frac{f(z)}{w_1 - P_1(z)}$$

is holomorphic in each $\delta_p \subset V$ (case 1 above). Thus, if we define

$$f^*(z, w_1) := (w_1 - P_1(z))\Phi(z, w_1) \qquad \text{in } \mathcal{P}^*,$$

then $f^*(z, w_1)$ defines a holomorphic function in $\mathcal{P}^*$ such that

$$f(z) = f^*(z, P_1(z)) \qquad \text{in } \mathcal{P}.$$

Since $\mathcal{P}^*$ is of rank $m-1$ and $f^*(z, w_1)$ is holomorphic in $\mathcal{P}^*$, it follows from the inductive hypothesis on the validity of the lifting principle from (3.6) that we can find an extension $F(z, w)$ of $f^*(z, w_1)$ in $\overline{\Delta}^{n+m}$, i.e.,

$$f^*(z, w_1) = F(z, w_1, P_2(z), \dots, P_m(z)) \qquad \text{in } \mathcal{P}^*,$$

so that $F(z, P_1(z), \dots, P_m(z)) = f(z)$ in $\mathcal{P}$. Therefore, $F(z, w)$ is an extension of $f(z)$ in $\overline{\Delta}^{n+m}$. □

LEMMA 3.4. *If the lifting principle holds for each polynomial polyhedron of rank $m \geq 1$, then every Cousin I problem in each polynomial polyhedron of rank m is solvable.*

PROOF. We proceed as follows. Let $\mathcal{P}$ be a polynomial polyhedron of rank m in $\mathbf{C}^n$ given by (3.3). Let $z_1 = x_1 + iy_1$ and let $\rho > 0$. We consider the intersections

$$\mathcal{P}^1 := \mathcal{P} \cap \{x_1 \leq \rho\} \qquad \text{and} \qquad \mathcal{P}^2 := \mathcal{P} \cap \{x_1 \geq -\rho\},$$

and we set $\mathcal{P}^0 := \mathcal{P}^1 \cap \mathcal{P}^2$, which we assume is non-empty. Let $g_1(z)$ and $g_2(z)$ be meromorphic functions in $\mathcal{P}^1$ and $\mathcal{P}^2$ chosen so that $f(z) = g_1(z) - g_2(z)$ is a holomorphic function in $\mathcal{P}^0$. We claim that we can find holomorphic functions $h_1(z)$ and $h_2(z)$ in $\mathcal{P}^1$ and $\mathcal{P}^2$ such that

$$g(z) := \begin{cases} g_1(z) - h_1(z), & z \in \mathcal{P}^1, \\ g_2(z) - h_2(z), & z \in \mathcal{P}^2, \end{cases} \tag{3.7}$$

is a single-valued meromorphic function in $\mathcal{P}$.

To verify this, we consider the n-dimensional analytic set Σ in the $(n+m)$-dimensional polydisk $\overline{\Delta}$ of (3.4) defined by

$$w_k = P_k(z) \quad (k = 1, \dots, m), \qquad z \in \mathcal{P},$$

where $P_k(z)$ are polynomials defining $\mathcal{P}$. We consider the intersections

$$\Delta^1 = \overline{\Delta} \cap \{x_1 \leq \rho\} \qquad \text{and} \qquad \Delta^2 = \overline{\Delta} \cap \{x_1 \geq -\rho\},$$

and define $\Delta^0 := \Delta^1 \cap \Delta^2$. The n-dimensional analytic set Σ^0 in Δ^0 defined by

$$\Sigma^0 : \ w_k = P_k(z) \ (k = 1, \dots, m), \qquad z \in \mathcal{P}^0$$

is the restriction of Σ to Δ^0. Using the mapping

$$z \in \mathcal{P}^0 \ \rightarrow (z, P_1(z), \dots, P_m(z)) \in \Sigma^0,$$

we see that the domain $\mathcal{P}^0$ in $\mathbf{C}^n$ is homeomorphically equivalent to Σ^0 and the boundary $\partial \mathcal{P}^0$ corresponds to $\partial \Sigma^0$. By definition, $\mathcal{P}^0$ is a polynomial polyhedron in $\mathbf{C}^n$ of rank m.[4] It follows from the assumption of validity of the lifting problem

[4]We note that our previous arguments about polynomial polyhedra of the form

$$|z_j| \leq r_j \ (j = 1, \dots, m), \quad |P_k(z)| \leq 1 \ (k = 1, \dots, m)$$

are valid without change for a set in $\mathbf{C}^n$ given by

$$z_j \in A_j \ (j = 1, \dots, n), \quad |P_k(z)| \leq 1 \ (k = 1, \dots, m),$$

where A_j is a domain in the complex plane $\mathbf{C}_{z_j}$ bounded by a piecewise-smooth closed curve. Thus we also call such a domain a polynomial polyhedron of rank m. In the present situation, $A_1 = \{|z_1| \leq r_1, \ |x_1| \leq \rho\}$ and $A_j = \{|z_j| \leq r_j\}$ $(j = 2, \dots, n)$.

of rank m that we can find a holomorphic function $F(z, w)$ in Δ^0 such that

$$F(z, P_1(z), \dots, P_m(z)) = f(z) \qquad \text{on } \mathcal{P}^0.$$

For simplicity we write $z' := (z_2, \dots, z_m)$. We consider $F(z, w)$ in Δ^0 as a holomorphic function $F(z_1, z', w)$ of the variable z_1 in the domain A_1 (here, A_1 is defined in the footnote); i.e., with $(n + m - 1)$ complex parameters $(z', w) \in \Delta'_z \times \Delta_w$, where

$$\Delta'_z : \ |z_j| \leq r_j, \ j = 2, \dots, n$$

and

$$\Delta_w : \ |w_j| \leq 1, \ j = 1, \dots, m.$$

We form the Cousin integral of $F(z_1, z', w)$ along the segment $l = [-ir'_1, ir'_1]$, where $r'_1 > r_1$ is chosen sufficiently close to r_1 to insure that $F(z_1, z', w)$ is holomorphic near l, and we obtain holomorphic functions $\Phi_1(z, w)$ and $\Phi_2(z, w)$ in Δ^1 and Δ^2 such that

$$F(z, w) = \Phi_1(z, w) - \Phi_2(z, w) \qquad \text{in } \Delta^0.$$

If we define

$$h_i(z) := \Phi_i(z, P_1(z), \dots, P_m(z)) \qquad \text{in } \mathcal{P}^i \ (i = 1, 2),$$

then $h_i(z)$ is a holomorphic function in $\mathcal{P}^i$ and satisfies

$$h_1(z) - h_2(z) = f(z) \qquad \text{in } \mathcal{P}^0.$$

Thus, $h_1(z)$ and $h_2(z)$ satisfy the requirements for (3.7).

Let $\mathcal{C}_1 = \{(g_p(z), \delta_p)\}_p$ be a Cousin I distribution in $\mathcal{P}$. We apply the same method as in the proof of Lemma 3.2, replacing $\overline{\Delta}$ by $\mathcal{P}$. We then construct sufficiently small sets $\omega' := (\omega_1 \times \cdots \times \omega_n) \cap \mathcal{P}$ so that each set ω' is contained in some δ_p. We remarked in the footnote that each ω', as well as $\mathcal{P}$ itself, is a polynomial polyhedron of rank m. Hence, using the above procedure, we obtain Lemma 3.4. □

We have now established the following proposition.

PROPOSITION 3.4 ([**44**]). *For polynomial polyhedra in* $\mathbf{C}^n$, *the Cousin I problem is always solvable and the lifting principle holds.*

3.3.3. Cousin I Problem in Polynomially Convex Domains. In this section, we show that the Cousin I problem is always solvable on a polynomially convex domain. Let $\mathcal{P}$ be a polynomial polyhedron in $\mathbf{C}^n$,

$$\mathcal{P} : \ |z_j| \leq r_j \ (j = 1, \dots, n), \qquad |P_k(z)| \leq 1 \ (k = 1, \dots, m).$$

We first show the following.

THEOREM 3.2 ([**44**]). *The Runge theorem holds for* $(\mathcal{P}, \mathbf{C}^n)$.

PROOF. Define the polydisk

$$\overline{\Delta}^{n+m} : \ |z_j| \leq r_j \ (j = 1, \dots, n), \qquad |w_k| \leq 1 \ (k = 1, \dots, m).$$

Let $f(z)$ be a holomorphic function in $\mathcal{P}$. From Proposition 3.4, we can find a holomorphic function $F(z, w)$ in $\overline{\Delta}^{n+m}$ such that

$$F(z, P_1(z), \dots, P_m(z)) = f(z) \qquad \text{in } \mathcal{P}.$$

Let $\epsilon > 0$ be given. From the Taylor expansion of $F(z,w)$ in $\overline{\Delta}^{n+m}$, we can find a polynomial $\Phi(z,w)$ in $\mathbf{C}^{n+m}$ such that

$$|F(z,w) - \Phi(z,w)| < \epsilon \qquad \text{in } \overline{\Delta}^{n+m}.$$

If we set

$$\varphi(z) := \Phi(z, P_1(z), \ldots, P_m(z)), \qquad z \in \mathbf{C}^n,$$

then φ is a polynomial in $\mathbf{C}^n$ which satisfies $|f(z) - \varphi(z)| < \epsilon$ in $\mathcal{P}$. Thus the theorem is proved. □

Let G be a polynomially convex domain in $\mathbf{C}^n$. Following the argument in Proposition 1.5 in Chapter 1, given any $E \subset\subset G$, we can find a polynomial polyhedron in $\mathbf{C}^n$ such that $E \subset\subset \mathcal{P} \subset\subset G$. In particular, if K is a polynomially convex compact subset of $\mathbf{C}^n$, i.e., the polynomial hull of K in $\mathbf{C}^n$ is identical with K, then any function $f(z)$ which is holomorphic on K is holomorphic on a sufficiently small polynomial polyhedron containing K. Thus, as a corollary to the proof of Theorem 3.2, we have the following approximation result.

COROLLARY 3.1 (Oka-Weil theorem). *Let K be a polynomially convex compact subset in $\mathbf{C}^n$. Then for any function $f(z)$ which is holomorphic on K and any $\varepsilon > 0$, there exists a polynomial $p(z)$ with $|f(z) - p(z)| < \varepsilon$ on K.*

Note also from Theorem 3.2 that the Runge theorem holds for any pair of polynomial polyhedra $(\mathcal{P}_1, \mathcal{P}_2)$ with $\mathcal{P}_1 \subset\subset \mathcal{P}_2$. Thus Proposition 3.4, Theorem 3.2 and Proposition 3.2 imply the following.

THEOREM 3.3 ([**44**]). *The Cousin I problem in polynomially convex domains in $\mathbf{C}^n$ is always solvable.*

3.4. Cousin I Problem in Domains of Holomorphy

3.4.1. Polynomial Hulls. In this section we study the Cousin I problem in a general domain of holomorphy in $\mathbf{C}^n$.[5] The key to its solution is a result about polynomial hulls (see (1.14)) of analytic sets of a special form in polydisks.

We first discuss Oka's lemma. Let E be a compact set in $\mathbf{C}^n$, and let A be a closed set in $\mathbf{C}^n$ such that $E \subset A$. Let $p \in A$, and let δ be a neighborhood of p in $\mathbf{C}^n$. Let $T = [0,1]$ be the unit interval on the real axis of the complex plane $\mathbf{C}_t$ and let V be a neighborhood of T in $\mathbf{C}_t$. Let $f(z,t)$ be a holomorphic function in $\delta \times V$, and define

$$\sigma_t := \{z \in \delta \mid f(z,t) = 0\}$$

for each $t \in T$. If the family of analytic sets $\{\sigma_t\}_{t \in T}$ in δ satisfies

1. $\sigma_t \cap E = \emptyset$ for any $t \in T$;
2. $\sigma_0 \cap A \neq \emptyset$ and $\sigma_1 \cap A = \emptyset$; and
3. $(\partial \sigma_t) \cap A = \emptyset$ for all $t \in T$,

[5]Once the lifting principle for *analytic* polyhedra in $\mathbf{C}^n$ has been established, one can verify, using the same method as in the previous section, that the Cousin I problem in domains of holomorphy is always solvable. However, we cannot establish the lifting principle for analytic polyhedra using the ideas of Part I. We shall establish it in Part II by using the new notion "ideal with indeterminate domain" introduced by Oka [**50**]. In fact, we will show that the lifting principle for analytic polyhedra in a *ramified* domain over $\mathbf{C}^n$ (see Theorem 8.2 and Remark 8.4).

then we say that the family $\{\sigma_t\}_{t\in T}$ satisfies **Oka's condition at p for the pair (E, A)**. Note we require that $\sigma_0 \cap A \neq \emptyset$, but we need not have $p \in \sigma_0$. We emphasize that the analytic sets $\{\sigma_t\}_{t\in T}$ are of codimension one; i.e., each σ_t is an analytic *hypersurface*. Using this notation, we state and prove Oka's lemma.

LEMMA 3.5 (Oka's lemma). *Let E be a compact set in $\mathbf{C}^n$ and let A denote the polynomial hull of E in $\mathbf{C}^n$. Then for each $p \in A$, there does* **not** *exist a family of analytic hypersurfaces $\{\sigma_t\}_{t\in T}$ which satisfies Oka's condition at p for the pair (E, A).*

PROOF. The proof is by contradiction. Assume that for some point $p \in A$ we can find a neighborhood δ of p in $\mathbf{C}^n$ and a neighborhood V of $T = [0, 1]$ in $\mathbf{C}_t$ such that there exists a family of analytic hypersurfaces

$$\sigma_t : \ f(z,t) = 0, \qquad (z,t) \in \delta \times V,$$

which satisfies Oka's condition at p for the pair (E, A).

Let G be a neighborhood of A in $\mathbf{C}^n$ such that $G \cap \sigma_1 = \emptyset$ and $G \cap (\partial\sigma_t) = \emptyset$ for each $t \in T$. Then, since $A \times T$ is the polynomial hull of $E \times T$ in $\mathbf{C}^n \times \mathbf{C}_t$, there exists a polynomial polyhedron $\mathcal{P}$ in $\mathbf{C}^n \times \mathbf{C}_t$ such that

$$A \times T \subset\subset \mathcal{P} \subset\subset G \times V.$$

We define a Cousin I distribution in $\mathcal{P}$ as follows: given any $q = (z', t') \in \mathcal{P}$, we take a neighborhood δ_q in $\mathbf{C}^n \times \mathbf{C}_t$ and a meromorphic function $g_q(z,t)$ in δ_q in such a way that

1. if $(z', t') \in \delta \times V$ and $f(z', t') = 0$, then we take $\delta_q \subset \delta \times V$ and set $g_q(z,t) = 1/f(z,t)$;
2. if $(z', t') \in \delta \times V$ and $f(z', t') \neq 0$ or if $(z', t') \notin \delta \times V$, then we take δ_q so that $f(z,t) \neq 0$ on δ_q and set $g_q(z,t) \equiv 1$.

It is clear from Oka's condition for $\{\sigma_t\}_{t\in T}$ that the collection $\mathcal{C}_1 = \{(g_q, \delta_q)\}_{q\in\mathcal{P}}$ forms a Cousin I distribution in $\mathcal{P}$. From Theorem 3.3 we can find a solution $g(z,t)$ of the Cousin I problem for $\mathcal{C}_1$ in $\mathcal{P}$. Thus, $g(z,t)$ can have poles only on $[\bigcup_{t\in V} \sigma_t] \cap \mathcal{P}$.

Let $t' = \max\{t \in T \mid \sigma_t \cap A \neq \emptyset\} < 1$ and set $T' := [t', 1]$. Then $g(z,t)$ is holomorphic in $E \times T'$, so that

$$M := \max\{|g(z,t)| \mid (z,t) \in E \times T'\} < +\infty.$$

On the other hand, $g(z,t')$ has a pole at some point $z_0 \in A$. Therefore, if we fix t_0 with $t' < t_0 < 1$ chosen sufficiently close to t', then $g(z, t_0) \equiv g^0(z)$ is holomorphic on A and satisfies

$$|g^0(z_0)| > M + 1 \geq \max\{|g^0(z)| \mid z \in E\}.$$

Let $\varepsilon = (|g^0(z_0)| - M - 1)/3 > 0$. Since A is a polynomial hull in $\mathbf{C}^n$, from Corollary 3.1 we can find a polynomial $P(z)$ in $\mathbf{C}^n$ such that $|g^0(z) - P(z)| < \varepsilon$ on A. It follows that

$$|P(z_0)| > \frac{2|g^0(z_0)| + M + 1}{3} > \frac{|g^0(z_0)| + 2(M+1)}{3} > |P(z)| \qquad \text{for } z \in E.$$

This contradicts the fact that $z_0 \in A$. □

In Chapter 9, we will see that the family of analytic hypersurfaces $\{\sigma_t\}_{t\in T}$ need only vary continuously; i.e., only continuity of $f(z,t)$ in $t \in T$ is needed.

3.4.2. Preparation Theorem. Let G be a domain in $\mathbf{C}^n$ with variables $z = (z_1, \dots, z_n)$. Let

$$\mathcal{P} \;:\; |z_j| \le r_j \;\; (j = 1, \dots, n), \qquad |f_k(z)| \le 1 \;\; (k = 1, \dots, m)$$

be an analytic polyhedron such that $\mathcal{P} \subset\subset G$, where $f_k(z)$ $(k = 1, \dots, m)$ is a holomorphic function in G. We introduce $\mathbf{C}^m$ with variables $w = (w_1, \dots, w_m)$; then in the polydisk $\overline{\Delta}$ in $\mathbf{C}^{n+m} = \mathbf{C}^n \times \mathbf{C}^m$,

$$\overline{\Delta} \;:\; |z_j| \le r_j \;\; (j = 1, \dots, n), \qquad |w_k| \le 1 \;\; (k = 1, \dots, m),$$

we consider the pure n-dimensional analytic set

$$\Sigma : w_k = f_k(z) \; (k = 1, \dots, m), \qquad z \in \mathcal{P}. \tag{3.8}$$

THEOREM 3.4 ([**45**]). *Σ is a polynomially convex compact set in $\mathbf{C}^{n+m}$.*[6]

PROOF. Let A denote the polynomial hull of Σ. For $z' = (z'_1, \dots, z'_n) \in \mathbf{C}^n$, we set

$$\begin{aligned} \Sigma(z') &:= \{w \in \mathbf{C}^m \mid (z', w) \in \Sigma\}, \\ A(z') &:= \{w \in \mathbf{C}^m \mid (z', w) \in A\}, \end{aligned}$$

the sections of Σ and A over $z_j = z'_j$ $(j = 1, \dots, n)$. Thus $\Sigma(z') \subset A(z')$; and $A(z')$ may be empty for some $z' \in \mathbf{C}^n$. To prove the theorem it suffices to show that

$$\Sigma(z') = A(z') \qquad \text{for each } z' \in \mathbf{C}^n.$$

Without loss of generality, we may assume that the origin 0 of $\mathbf{C}^n$ is not contained in $\overline{G}$. Given $R > 0$, we define the closed ball $\mathcal{Q}(R)$ in $\mathbf{C}^n$,

$$\mathcal{Q}(R) \;:\; \sum_{j=1}^{n} |z_j|^2 \le R^2.$$

If R is sufficiently large so that $\mathcal{Q}(R) \supset \mathcal{P}$, then it is clear that $\Sigma(z') = A(z') = \emptyset$ for each $z' \notin \mathcal{Q}(R)$.

Fix $R > 0$ such that

$$\Sigma(z') = A(z') \qquad \text{for } z' \notin \mathcal{Q}(R).$$

We will show that for any $p \in \partial\mathcal{Q}(R)$, there exists a neighborhood δ_p^* of p in $\mathbf{C}^n$ such that

$$\Sigma(z') = A(z') \qquad \text{for all } z' \in \delta_p^*.$$

For simplicity, we assume $p = (0, \dots, 0, R)$.

We first assume that $p \notin \mathcal{P}$. Then we can find a ball δ_p centered at p with radius $r > 0$ in $\mathbf{C}^n$ such that $\overline{\delta_p} \cap \mathcal{P} = \emptyset$. Thus, $\Sigma(z') = \emptyset$ for any $z' \in \delta_p$, and it suffices to show that $A(p) = \emptyset$. For, since A is closed in $\mathbf{C}^{n+m}$, we can find a neighborhood $\delta_p^* \subset \delta_p$ of p in $\mathbf{C}^n$ such that $A(z') = \emptyset$ for any $z' \in \delta_p^*$. Thus, if $A(p) = \emptyset$, then $\Sigma(z') = A(z') = \emptyset$ for $z' \in \delta_p^*$. We prove $A(p) = \emptyset$ by contradiction. Assume that $A(p) \ne \emptyset$; suppose $w^0 \in A(p)$. We consider the following family of analytic hypersurfaces $\{\sigma_t\}_t$ in $\delta_p \times \mathbf{C}^m$:

$$\sigma_t \;:\; z_n = R + t, \qquad t \in T = [0, 1].$$

[6]This theorem is the main theorem in Oka's paper II. The proof given here is due to A. Takeuchi [**71**].

Then, $\{\sigma_t\}_{t\in T}$ satisfies Oka's condition for the pair (Σ, A) at the point (p, w^0). Indeed, $\sigma_t \cap \Sigma = \emptyset$ for any $t \in T$ from the condition $\overline{\delta_p} \cap \mathcal{P} = \emptyset$; furthermore, $(p, w^0) \in \sigma_0 \cap A$ and $\sigma_t \cap A = \emptyset$ for each $t \in T \setminus \{0\}$, since $\sigma_t \subset (\mathbf{C}^n \setminus \mathcal{Q}(R)) \times \mathbf{C}^m$ and $A(z') = \Sigma(z') = \emptyset$ for $z' \notin \mathcal{Q}(R)$. Finally, $(\partial\sigma_t) \cap A = \emptyset$ for each $t \in T$, since A is compact in $\mathbf{C}^{n+m}$ and σ_t has empty boundary relative to $\mathbf{C}^{n+m}$. By Lemma 3.5 this contradicts the fact that A is the polynomial hull of Σ in $\mathbf{C}^{n+m}$.

We next consider the case when $p = (0, \dots, 0, R) \in \mathcal{P}$. Let δ_p be a ball centered at p in G. Let $z_n = x_n + iy_n$ and consider the real $(2n-1)$-dimensional hyperplane H in $\mathbf{C}^n$ of the form

$$H = \{z \in \mathbf{C}^n \mid x_n = R - \rho_0\},$$

where ρ_0 is the unique positive number so that $(\partial\mathcal{Q}(R)) \cap \delta_p \subset H$. Fix $\rho > 0$ with $0 < \rho < \rho_0$, and define $\delta_p^* := \delta_p \cap \{x_n > R - \rho\}$, which is a neighborhood of p in $\mathbf{C}^n$. Our claim is that

$$\Sigma(z') = A(z') \qquad \text{for all } z' \in \delta_p^*. \tag{3.9}$$

We prove this by contradiction. Assume that there exists a point $z^* = (z_1^*, \dots, z_n^*) \in \delta_p^*$ such that

$$\Sigma(z^*) \neq A(z^*). \tag{3.10}$$

We set $z_n^* = x_n^* + iy_n^*$, so that $R - \rho < x_n^* \leq R$ since $A(z') = \Sigma(z')$ for $z' \notin \mathcal{Q}(R)$. By (3.10), there exists a point $w^* = (w_1^*, \dots, w_n^*) \in A(z^*)$ such that

$$w_k^* \neq f_k(z^*) \qquad \text{for some } k \ (1 \leq k \leq m).$$

We fix this k and set $c_0 := w_k^* - f_k(z^*) \neq 0$. Consider the family of analytic hypersurfaces $\{\sigma_t\}_t$ in $\delta_p^* \times \mathbf{C}^m$ defined by the equations

$$\sigma_t \ : \ w_k - f_k(z) = c_0(1+t)e^{-\lambda(z_n - z_n^*)}, \quad t \in T = [0, M],$$

where λ, $M > 0$ are chosen large enough so that

$$\text{(i)} \qquad |c_0| e^{-\lambda(R - \rho - x_n^*)} > \max_{(z,w)\in A\cap(\delta_p^*\times\mathbf{C}^m)} \{|w_k - f_k(z)|\};$$

$$\text{(ii)} \qquad |c_0|(1+M) > \max_{(z,w)\in A\cap(\delta_p^*\times\mathbf{C}^m)} \{|w_k - f_k(z)| e^{\lambda(x_n - x_n^*)}\}.$$

We claim that the family $\{\sigma_t\}_{t\in T}$ satisfies Oka's condition for the pair (Σ, A) at the point (z^*, w^*).

Clearly $\sigma_t \cap \Sigma = \emptyset$ for $t \in T$, since $w_k - f_k(z) \neq 0$ on σ_t; also, $(z^*, w^*) \in \sigma_0 \cap A$ and $\sigma_M \cap A = \emptyset$ from (ii). Finally, to prove that $(\partial\sigma_t) \cap A = \emptyset$ for all $t \in T$, we divide $\partial\delta_p^*$ in $\mathbf{C}^n$ into two parts:

$$l_1 = (\partial\delta_p^*) \cap \{x = R - \rho\} \quad \text{and} \quad l_2 = (\partial\delta_p^*) \setminus l_1.$$

Since $f_k(z)$ is defined and holomorphic on $\overline{\delta_p^*}$, we note that $\partial\sigma_t \cap [\delta_p^* \times \mathbf{C}^m] = \emptyset$ $(t \in T)$. Thus, each boundary $\partial\sigma_t$ in $\delta_p^* \times \mathbf{C}^m$ $(t \in T)$ consists of two parts:

$$(\partial\sigma_t)_i := (\partial\sigma_t) \cap (l_i \times \mathbf{C}^m), \quad i = 1, 2.$$

We set

$$A(l_i) := \{(z, w) \in A \mid z \in l_i\}, \quad i = 1, 2.$$

Then (i) implies that

$$(\partial\sigma_t)_1 \cap A(l_1) = \emptyset \quad \text{for all } t \in T.$$

Furthermore, since $l_2 \subset\subset \mathbf{C}^n \setminus \mathcal{Q}(R)$ and $A(z) = \Sigma(z)$ for $z \notin \mathcal{Q}(R)$ by our choice of $R > 0$, it follows that $w_k - f_k(z) = 0$ for all $z \in l_2$. Hence $\overline{\sigma_t} \cap A(l_2) = \emptyset$ for all $t \in T$, by the defining equation for σ_t. Consequently, $(\partial\sigma_t) \cap A = \emptyset$ for all $t \in T$. We conclude that $\{\sigma_t\}_{t\in T}$ satisfies Oka's condition for the pair (Σ, A) at (z^*, w^*). From Lemma 3.5, this contradicts the fact that A is the polynomial hull of Σ in $\mathbf{C}^{n+m}$. Hence, we must have $\Sigma(z') = A(z')$ for all $z' \in \delta_p^*$, and our claim (3.9) is true.

Since $\partial\mathcal{Q}(R)$ is compact, it follows from the Heine-Borel theorem that the infimum of the set of all $R > 0$ such that $\Sigma(z') = A(z')$ for all $z' \notin \mathcal{Q}(R)$ must be 0. This fact, together with the information that $0 \notin \overline{G}$, implies that $\Sigma(z') = A(z')$ for all $z \in \mathbf{C}^n$. □

As will be shown in Remark 7.12 in Chapter 7, this theorem has another quite different proof.

3.4.3. Cousin I Problem in Domains of Holomorphy. Assume that G is a domain of holomorphy in $\mathbf{C}^n$. We use the same notation $\mathcal{P} \subset\subset G$, $\Delta \subset\subset \mathbf{C}^{n+m}$, and $\Sigma : w_k = f_k(z)$ $(k = 1, \ldots, m)$, $z \in \mathcal{P}$, in Δ from the previous section.

THEOREM 3.5 ([**45**]). *The Runge theorem holds for the pair* $(\mathcal{P}, G)$.

PROOF. Let $\varphi(z)$ be a holomorphic function in a neighborhood v of $\mathcal{P}$ in $\mathbf{C}^n$. If we regard $\varphi(z)$ as being independent of $w \in \mathbf{C}^m$, then $\varphi(z)$ is a holomorphic function in a neighborhood V of Σ in $\mathbf{C}^{n+m}$, where $V = v \times \mathbf{C}^m$. From Theorem 3.4 there exists a polynomial polyhedron $\mathcal{P}^*$ in $\mathbf{C}^{n+m}$ such that

$$\Sigma \subset\subset \mathcal{P}^* \subset\subset V.$$

Now Theorem 3.2 implies that the Runge theorem holds for $(\mathcal{P}^*, \mathbf{C}^{n+m})$. Hence, given $\varepsilon > 0$ and an open set V_0 in $\mathbf{C}^{n+m}$ such that $\Sigma \subset\subset V_0 \subset\subset \mathcal{P}^*$, we can find a polynomial $P(z, w)$ in $\mathbf{C}^{n+m}$ with

$$|\varphi(z) - P(z, w)| < \varepsilon \qquad \text{in } V_0.$$

If we set

$$\Phi(z) := P(z, f_1(z), \ldots, f_m(z)), \quad z \in G,$$

then $\Phi(z)$ defines a holomorphic function in G such that

$$|\varphi(z) - \Phi(z)| < \varepsilon, \quad z \in \mathcal{P}.$$

Thus, the Runge theorem holds for $(\mathcal{P}, G)$. □

In order to solve the Cousin I problem in a domain of holomorphy G, the above theorem, combined with Proposition 3.2, shows that it suffices to solve the Cousin I problem in an analytic polyhedron contained in G. We now show this is always the case.

LEMMA 3.6. *Let G be a domain of holomorphy, and let $\mathcal{P} \subset G$ be an analytic polyhedron. Then the Cousin I problem in $\mathcal{P}$ is always solvable.*

PROOF. We use the same notation Σ in Δ for $\mathcal{P}$ defined in (3.8). Let $\mathcal{C}_1 = \{(g_p, \delta_p)\}_{p\in v}$ be a Cousin I distribution defined in a neighborhood v of $\mathcal{P}$ in $\mathbf{C}^n$. If we regard $g_p(z)$ as independent of $w \in \mathbf{C}^m$, then $\mathcal{C}_1$ may be regarded as a Cousin I distribution $\widehat{\mathcal{C}_1}$ in a neighborhood V of Σ in $\mathbf{C}^{n+m}$, where $V = v \times \mathbf{C}^m$. That is, let $p' \in V$ and denote by $p \in v$ the projection of p' into $\mathbf{C}^n$. Then set $\delta_{p'} := \delta_p \times \mathbf{C}^m$, a neighborhood of p' in $\mathbf{C}^{n+m}$, and define $g_{p'}(z,w) := g_p(z)$, which is a meromorphic function in $\delta_{p'}$; then $\widehat{\mathcal{C}_1} := \{(g_{p'}(z,w), \delta_{p'})\}_{p'\in V}$ is a Cousin I distribution in V. Once again using Theorem 3.4, we obtain a polynomial polyhedron $\mathcal{P}^*$ in $\mathbf{C}^{n+m}$ such that

$$\Sigma \subset\subset \mathcal{P}^* \subset\subset V.$$

From Proposition 3.4, there exists a solution $G(z,w)$ of the Cousin I problem for $\widehat{\mathcal{C}_1}$ in $\mathcal{P}^*$. If we set

$$g(z) := G(z, f_1(z), \dots, f_m(z)) \quad \text{in } \mathcal{P},$$

then $g(z)$ is a solution of the Cousin I problem for the original Cousin I data $\mathcal{C}_1$ in $\mathcal{P}$. □

Summarizing the results above, we have proved the main theorem of this section.

THEOREM 3.6 ([**45**]). *The Cousin I problem in domains of holomorphy is always solvable.*

3.4.4. Example. We noted in Remark 3.1 (Cartan) that if D is a domain in $\mathbf{C}^2$ in which the Cousin I problem is always solvable, then D must be a domain of holomorphy. Cartan [**10**] showed that this is not necessarily true for a domain in $\mathbf{C}^3$. We present his example in this section.

First we need a preliminary result. We let $0 < r_1 < r_2$, and we consider the following three product domains in $\mathbf{C}^3 = \mathbf{C}_{z_1} \times \mathbf{C}_{z_2} \times \mathbf{C}_{z_3}$:

$$\begin{array}{llll} \Delta_1 & : & r_1 < |z_1| < r_2, & |z_2| < r_2, & |z_3| < r_2, \\ \Delta_2 & : & |z_1| < r_2, & r_1 < |z_2| < r_2, & |z_3| < r_2, \\ \Delta_3 & : & |z_1| < r_2, & |z_2| < r_2, & r_1 < |z_3| < r_2. \end{array}$$

Set

$$\Delta^1 = \Delta_2 \cap \Delta_3, \qquad \Delta^2 = \Delta_3 \cap \Delta_1, \qquad \Delta^3 = \Delta_1 \cap \Delta_2,$$

and

$$\Delta^0 = \Delta_1 \cap \Delta_2 \cap \Delta_3, \qquad \Delta = \Delta_1 \cup \Delta_2 \cup \Delta_3.$$

Note that Δ is homeomorphic to a punctured ball $\{0 < |z_1|^2 + |z_2|^2 + |z_3|^2 < 1\}$ in $\mathbf{C}^3$. It follows from Osgood's theorem (Theorem 1.10) that Δ is not a domain of holomorphy in $\mathbf{C}^3$.

THEOREM 3.7 (Three ring theorem). *For $j = 1, 2, 3$, let $g^j(z)$ be a holomorphic function on Δ^j satisfying*

$$g^1(z) + g^2(z) + g^3(z) = 0 \quad \text{on } \Delta^0. \tag{3.11}$$

Then there exist holomorphic functions $f_j(z)$ $(j = 1, 2, 3)$ on Δ_j such that

$$g^1(z) = f_2(z) - f_3(z), \;\; g^2(z) = f_3(z) - f_1(z), \;\; g^3(z) = f_1(z) - f_2(z)$$

on Δ^1, Δ^2 and Δ^3, respectively.

PROOF. We expand $g^1(z)$ in a Laurent series with respect to z_1, z_2, z_3 about the origin $0 \in \mathbf{C}^3$. Clearly, the coefficients of $z_1^k z_2^m z_3^l$ with $m < 0$ vanish for all $m, l = 0, \pm 1, \dots$. By (3.11) and the uniqueness of the Laurent expansion, the coefficients of $z_1^k z_2^m z_3^l$ where both $m < 0$ and $l < 0$ vanish. Hence, we can write a unique decomposition of g^1 as

$$g^1(z) = G_1^1(z) + G_2^1(z) + G_3^1(z),$$

where $G_1^1(z)$ is holomorphic on Δ, and $G_j^1(z)$ $(j = 2, 3)$ is holomorphic on Δ_j but not necessarily in Δ. For example, $G_2^1(z)$ is the sum of all terms of the expansion in powers $z_1^k z_2^m z_3^l$ with $m < 0$ and $k, l \geq 0$. In a similar fashion, we have

$$g^i(z) = G_1^i(z) + G_2^i(z) + G_3^i(z), \qquad i = 2, 3,$$

where $G_i^i(z)$ is holomorphic in Δ and $G_j^i(z)$ $(j \neq i)$ is holomorphic in Δ_j but not necessarily in Δ. Once again using (3.11) and the uniqueness of the Laurent expansion, we have

$$G_1^1(z) + G_2^2(z) + G_3^3(z) = 0,$$

$$G_1^2(z) + G_1^3(z) = 0, \quad G_2^1(z) + G_2^3(z) = 0, \quad G_3^1(z) + G_3^2(z) = 0$$

on $\Delta, \Delta_1, \Delta_2$ and Δ_3. Therefore, if we define

$$f_1(z) := \frac{-G_2^2(z) + G_3^3(z)}{3} + G_1^3(z) \quad \text{on } \Delta_1$$

$$f_2(z) := \frac{-G_3^3(z) + G_1^1(z)}{3} + G_2^1(z) \quad \text{on } \Delta_2$$

$$f_3(z) := \frac{-G_1^1(z) + G_2^2(z)}{3} + G_3^2(z) \quad \text{on } \Delta_3,$$

then $f_j(z)$ $(j = 1, 2, 3)$ are the desired functions. □

From this theorem we obtain the following result.

PROPOSITION 3.5. *The Cousin I problem in the above domain Δ in $\mathbf{C}^3$ is always solvable.*

PROOF. Let $\mathcal{C}_1 = \{(g_p, \delta_p)\}_{p \in \Delta}$ be a Cousin I distribution on Δ. Since Δ_j $(j = 1, 2, 3)$ is a product domain, the Cousin I problem is always solvable in Δ_j. Thus we can find a solution $\varphi_j(z)$ of the Cousin I problem for $\mathcal{C}_1$ in Δ_j. If we set

$$g^1(z) = \varphi_2(z) - \varphi_3(z), \quad g^2(z) = \varphi_3(z) - \varphi_1(z), \quad g^3(z) = \varphi_1(z) - \varphi_2(z)$$

on Δ^1, Δ^2 and Δ^3, then each $g^j(z)$ $(j = 1, 2, 3)$ is a holomorphic function on Δ^j and

$$g^1(z) + g^2(z) + g^3(z) = 0 \quad \text{on } \Delta^0.$$

By Theorem 3.7, we can find holomorphic functions $f_j(z)$ $(j = 1, 2, 3)$ on Δ_j such that

$$g^1(z) = f_2(z) - f_3(z), \quad g^2(z) = f_3(z) - f_1(z), \quad g^3(z) = f_1(z) - f_2(z)$$

on Δ^1, Δ^2 and Δ^3. It follows that

$$G(z) := \varphi_j(z) - f_j(z) \quad \text{on } \Delta_j \ (j = 1, 2, 3)$$

defines a single-valued meromorphic function on all of Δ. Hence, $G(z)$ is a solution of the Cousin I problem for $\mathcal{C}_1$ on Δ. □

3.5. Cousin II Problem

3.5.1. Oka's Counterexample. The Cousin II problem in product domains in $\mathbf{C}^n$ is not always solvable; we give a counterexample due to Oka in this section. To illustrate the key idea, we first give an example of Oka [**46**] which indicates a difference between zero sets of real-valued and complex-valued continuous functions.

EXAMPLE 3.2. We consider the domain D in $\mathbf{R}^3$ defined by

$$D := \{(x, y, z) \in \mathbf{R}^3 \mid x^2 + y^2 < 4, \ -2 < z < 2\}$$

and let

$$L := \{(0, 0, z) \in \mathbf{R}^3 \mid -1 \leq z \leq 1\} \subset\subset D.$$

There are many real-valued continuous functions $F(x, y, z)$ in D such that $F(x, y, z) = 0$ if and only if $(x, y, z) \in L$. However, suppose we look for a complex-valued continuous function $F(x, y, z)$ in D satisfying the following conditions:

(i) $F(x, y, z) = 0$ if and only if $(x, y, z) \in L$;

(ii) in the disk δ: $x^2 + y^2 \leq \rho^2 < 4$ on the (x, y)-plane, we require that

$$F(x, y, 0) = (x + iy)\lambda(x, y) \qquad (i^2 = -1),$$

where $\lambda(x, y) \neq 0$ for $(x, y) \in \delta$.

We claim that there does not exist such a function $F(x, y, z)$ in D.

For if $F(x, y, z)$ exists satisfying (i) and (ii), we consider

$$V(z) := \int_{\partial\delta} d(\arg F(x, y, z)) \qquad \text{for } z \in (-2, 2).$$

Then (i) implies that $V(z)$ does not depend on $z \in (-2, 2)$, and also implies $V(3/2) = 0$. However, (ii) implies that $V(0) = 2\pi$, which is a contradiction.

T. Gronwall [**27**] was the first to give an example of a product domain in $\mathbf{C}^n$ in which the Cousin II problem is not always solvable. Below we will give Oka's example, which more clearly indicates the essence of the Cousin II problem and is based on the idea of the example described above.

Oka's counterexample for the Cousin II problem. In $\mathbf{C}^2$ with variables z and w, we consider the product domain

$$\Delta \; : \; 2/3 < |z| < 1, \qquad 2/3 < |w| < 1.$$

We write $z = x + iy$, and denote by Δ' and Δ'' the points of Δ such that $y \geq 0$ and $y \leq 0$. Let

$$\Sigma \; : \; w - z + 1 = 0 \qquad \text{in } \Delta.$$

Note that Σ consists of two connected components $\Sigma' \subset \Delta'$ and $\Sigma'' \subset \Delta''$, since $\Sigma \cap \{y = 0\} = \emptyset$. We take open neighborhoods G' of Δ' and G'' of Δ'' with $\Delta' \subset\subset G'$ and $\Delta'' \subset\subset G''$ such that $G' \cap \Sigma'' = \emptyset$ and $G'' \cap \Sigma' = \emptyset$. If we set

$$\begin{aligned} f_1 &:= w - z + 1, \qquad && \text{in } G', \\ f_2 &:= 1, && \text{in } G'', \end{aligned}$$

then $\mathcal{C}_2 = \{(f_1, G'), (f_2, G'')\}$ defines a Cousin II distribution in Δ. Then there is no solution of the Cousin II problem for $\mathcal{C}_2$ in Δ.

Indeed, assume that there does exist a solution $F(z, w)$ for $\mathcal{C}_2$ in Δ; then $F(z, w)$ vanishes in Δ only at points of Σ', and we can find a nonvanishing holomorphic function $\omega(z, w)$ on Δ' such that

$$F(z, w) = (w - z + 1)\omega(z, w) \quad \text{on } \Delta'. \tag{3.12}$$

We define the circles

$$\gamma_1 : |z| = 5/6 \quad \text{in } \mathbf{C}_z \qquad \text{and} \qquad \gamma_2 : |w| = 5/6 \quad \text{in } \mathbf{C}_w;$$

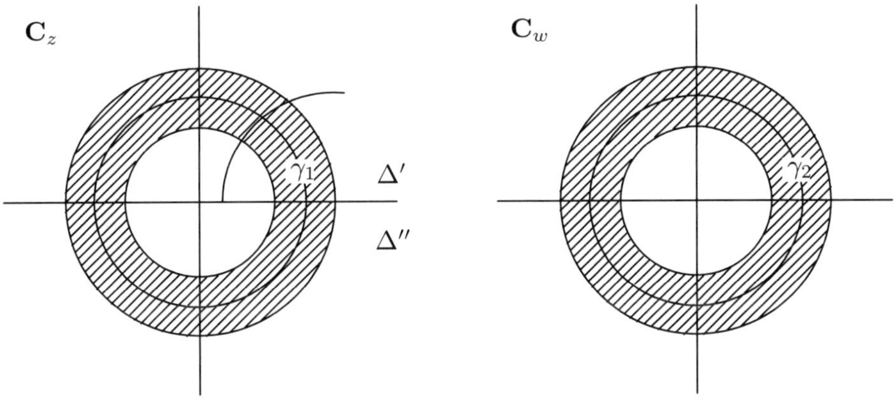

FIGURE 3. Oka's counterexample for the Cousin II problem

and we set $\gamma_1' := \gamma_1 \cap \{y \geq 0\}$ and $\gamma_1'' := \gamma_1 \cap \{y \leq 0\}$. Now we vary z from $5/6$ to $-5/6$ in a continuous fashion along $\gamma_1' \subset \Delta'$. Since $\omega(z,w) \neq 0$ in Δ', the total variation of the argument of $\omega(z,w)$ along γ_2,

$$\int_{\gamma_2} d \arg \omega(z,w),$$

varies continuously with $z \in \gamma_1'$, and, being an integer multiple of 2π, does not depend on $z \in \gamma_1'$. Since

$$\int_{\gamma_2} d \arg(w - 5/6 + 1) = 2\pi \quad \text{and} \quad \int_{\gamma_2} d \arg(w + 5/6 + 1) = 0,$$

it follows from (3.12) that

$$\int_{\gamma_2} d \arg F(5/6, w) = \int_{\gamma_2} d \arg F(-5/6, w) + 2\pi.$$

On the other hand, we note that $F(z,w) \neq 0$ in G''. Therefore, varying z from $5/6$ to $-5/6$ continuously along $\gamma_1'' \subset G''$ and arguing as before, we have

$$\int_{\gamma_2} d \arg F(5/6, w) = \int_{\gamma_2} d \arg F(-5/6, w).$$

This is a contradiction. □

This example will be used again in section 3.6.2.

3.5.2. Oka's Principle. The counterexample in the previous section shows that one of the obstructions to solving a Cousin II problem is topological. Thus we now generalize the holomorphic Cousin II problem to the continuous case. For this purpose, we introduce the following terminology.

Let D be a domain in $\mathbf{C}^n$, and let $\mathcal{C}_2 = \{(f_p, \delta_p)\}_p$ be a Cousin II distribution in D. If we can find a complex-valued *continuous* function $F(z)$ in D such that, at each point $p \in D$,

$$\lambda_p(z) := F(z)/f_p(z)$$

is a nonvanishing *continuous* function in δ_p, then we say that $F(z)$ is a **continuous solution** of the Cousin II problem for $\mathcal{C}_2$ in D. In this section, the "usual" solutions will be called holomorphic solutions to distinguish them from the continuous ones.

THEOREM 3.8 (Oka's principle). *Let G be a domain of holomorphy in $\mathbf{C}^n$, and let $\mathcal{C}_2$ be a Cousin II distribution in G. If G admits a continuous solution of the Cousin II problem for $\mathcal{C}_2$, then G admits a holomorphic solution of the Cousin II problem for $\mathcal{C}_2$.*

PROOF. By taking a refinement, we may assume that each set δ_p, $p \in G$, is a polydisk in $\mathbf{C}^n$. Let $\Phi(z)$ be a continuous solution of the Cousin II problem for $\mathcal{C}_2$ in G. Then for each $p \in G$, $\Phi(z)/f_p(z)$ is a nonvanishing continuous function in δ_p. Hence the function

$$\zeta_p(z) := \log\left(\Phi(z)/f_p(z)\right),$$

where we take an appropriate branch of the logarithm, defines a single-valued continuous function in δ_p with the following property: for any δ_p, δ_q such that $\delta_p \cap \delta_q \neq \emptyset$, the function $\zeta_p(z) - \zeta_q(z)$ is holomorphic in $\delta_p \cap \delta_q$.

We recall the method used to solve the Cousin I problem in polydisks in section 3.2.2, in particular, Lemma 3.2: given a Cousin I distribution $\{(g_p, \delta_p)\}_p$ in G, we constructed a holomorphic function $h_p(z)$ on each set δ_p, $p \in G$, such that $g_p(z) - h_p(z)$ defined a single-valued meromorphic function in all of G. This was achieved by utilizing the relation $g_p(z) - g_q(z) = h_p(z) - h_q(z)$ in $\delta_p \cap \delta_q$. In the present situation, we replace the meromorphic function $g_p(z)$ by the continuous function $\zeta_p(z)$. Then, following the same procedure under the condition that D is a domain of holomorphy, we can find a holomorphic function $h_p(z)$ on each δ_p, $p \in G$, such that the collection of functions $\zeta_p(z) - h_p(z)$ on δ_p defines a single-valued continuous function $\Xi(z)$ on all of G. If we define

$$F(z) := \Phi(z)e^{-\Xi(z)} \quad \text{on } G,$$

then $F(z)$ is a well-defined single-valued function on G. Moreover, since $F(z) = f_p(z)e^{h_p(z)}$ on each δ_p, $F(z)$ is holomorphic on G and yields a holomorphic solution of the Cousin II problem for $\mathcal{C}_2$ in G. □

3.5.3. Generalized Cousin II. A holomorphic Cousin II distribution on a domain D in $\mathbf{C}^n$ can also be generalized to a continuous Cousin II distribution. At each point $p \in D$, let the data (h_p, δ_p) be given, where δ_p is a neighborhood of p in D and $h_p(z)$ is a complex-valued continuous function in δ_p. We require this data to satisfy the condition that for any $p, q \in D$ with $\delta_p \cap \delta_q \neq \emptyset$, we can find a nonvanishing continuous function $\lambda_{pq}(z)$ in $\delta_p \cap \delta_q$ such that $h_p(z) = \lambda_{pq}(z)h_q(z)$ in $\delta_p \cap \delta_q$. We call the collection of pairs $\mathcal{C}_2 = \{(h_p, \delta_p)\}_{p\in D}$ a **generalized Cousin II distribution** in D. Thus, locally, we are given the zero sets of continuous functions in D; we want to find a globally defined continuous function with this zero set.

Generalized Cousin II Problem Given a generalized Cousin II distribution $\mathcal{C}_2 = \{(h_p, \delta_p)\}_{p\in D}$, find a complex-valued continuous function $h(z)$ in D with the property that for each $p \in D$ there exists a nonvanishing continuous function $\lambda_p(z)$ in δ_p such that $h(z) = \lambda_p(z)h_p(z)$ in δ_p.

If such a function $h(z)$ exists, we say that the generalized Cousin II problem for $\mathcal{C}_2$ is solvable in D, and we call the continuous function $h(z)$ in D a solution for $\mathcal{C}_2$ of this generalized Cousin II problem.

REMARK 3.2. In the following example we show that, in general, a generalized Cousin II problem in a polydisk D need not have a solution for a generalized Cousin II distribution $\mathcal{C}_2 = \{(h_p, \delta_p)\}_{p\in D}$ if $\{z \in \delta_p : h_p(z) = 0\}$ has an interior point in δ_p.

Example: We begin with an example in $\mathbf{R}^3$ with variables x, y, z. Consider the real-valued continuous function

$$\psi(x, y, z) := \max\{0, x^2 + y^2 - 1\}.$$

We consider two half-spaces $\Delta^{\pm} := \{(x, y, z) \in \mathbf{R}^3 \mid \pm z > -1\}$ and two cylinders $\delta' := \{x^2 + y^2 < 1/3\} \times \{|z| < 1\}$ and $\delta'' := \{x^2 + y^2 < 1/2\} \times \{|z| < 2\}$. We put

$$\begin{aligned} h_1(x, y, z) &= (x + iy)\psi(x, y, z) && \text{on } \Delta_1 := \Delta^+ \setminus \delta', \\ h_2(x, y, z) &= \psi(x, y, z) && \text{on } \Delta_2 := \Delta^- \setminus \delta', \\ h_3(x, y, z) &= 0 && \text{on } \Delta_3 := \delta''. \end{aligned}$$

Then the collection of pairs $\mathcal{C}_2 = \{(h_i, \Delta_i)\}_{i=1,2,3}$ defines a generalized Cousin II distribution in $\mathbf{R}^3$. Note that the zero sets of the functions h_i in $\mathcal{C}_2$ comprise the set $B \times \mathbf{R}_z$, where $B = \{x^2 + y^2 \le 1\} \subset \mathbf{R}_x \times \mathbf{R}_y$. Following the reasoning in Example 3.2, it is clear that there is no solution of the generalized Cousin II problem for $\mathcal{C}_2$ in $\mathbf{R}^3$.

Now to get a similar example in $\mathbf{C}^2 = \mathbf{R}^4$ with variables $z_1 := x + iy$, $z_2 := z + iv$, we simply take ψ, h_p as above, making them independent of the variable v.

Thus from now on we assume that the zero sets of any generalized Cousin II distribution $\mathcal{C}_2 = \{(h_p, \delta_p)\}_{p \in D}$ have empty interior. This implies that for any $p, q \in D$ such that $\delta_p \cap \delta_q \ne \emptyset$, if a nonvanishing continuous function $\lambda_{pq}(z)$ in $\delta_p \cap \delta_q$ satisfying $h_p(z) = \lambda_{pq}(z) h_q(z)$ in $\delta_p \cap \delta_q$ exists, then it is uniquely determined.

Since the usual (holomorphic) Cousin II distribution satisfies this condition, Theorem 3.8 implies the following: Let D^* be a domain of holomorphy in $\mathbf{C}^n$. Assume that D^* is homeomorphic to a domain D in $\mathbf{C}^n$ which has the property that the generalized Cousin II problem is always solvable in D. Then the Cousin II problem is always solvable in D^*.

LEMMA 3.7. *In the polydisk $\Delta : |z_j| < 1$ $(j = 1, \dots, n)$ in $\mathbf{C}^n$, the generalized Cousin II problem is always solvable.*

PROOF. Let $\mathcal{C}_2 = \{(h_p, \delta_p)\}_{p \in \Delta}$ be a generalized Cousin II distribution in Δ. For any $0 < r < 1$, we define $\overline{\Delta}_r : |z_j| \le r$ $(j = 1, \dots, n)$, and we will find a solution of the generalized Cousin II problem for $\mathcal{C}_2$ in $\overline{\Delta}_r$. Using the same arguments as in solving the Cousin I problem (stated in Lemma 3.2), we see that it suffices to prove the following.

Let Λ_j be a closed convex domain in the unit disk $\{|z_j| < 1\}$ in $\mathbf{C}_{z_j}$ $(j = 1, \dots, n)$ and define $\Lambda := \Lambda_1 \times \cdots \times \Lambda_n \subset \mathbf{C}^n$. Let $z_1 = x + iy$ and fix $\rho > 0$. We set

$$\Lambda^1 := \{z \in \Lambda \mid x \le \rho\}, \qquad \Lambda^2 := \{z \in \Lambda \mid x \ge -\rho\},$$

and $\Lambda^0 := \Lambda^1 \cap \Lambda^2$, which we assume is non-empty. If a generalized Cousin II distribution $\mathcal{C}_2$ defined in Λ has solutions $h_1(z)$ and $h_2(z)$ in Λ^1 and Λ^2, then $\mathcal{C}_2$ has solutions in all of Λ.

To verify this, note that $h^0(z) := h_1(z)/h_2(z)$ defines a nonvanishing continuous function in the simply connected domain Λ^0. Thus, a branch of $\log h^0(z)$ defines a single-valued continuous function in Λ^0. We define the following real-valued continuous function in $\mathbf{C}_{z_1}$:

$$\alpha(z_1) := \begin{cases} 0, & x \le -\rho, \\ (x + \rho)/2\rho, & -\rho \le x \le \rho, \\ 1, & x \ge \rho. \end{cases}$$

If we set

$$\begin{aligned} k_1(z) &:= \alpha(z_1)\log h^0(z), & z\in\Lambda^1,\\ k_2(z) &:= (\alpha(z_1)-1)\log h^0(z), & z\in\Lambda^2, \end{aligned}$$

then $k_i(z)$ $(i=1,2)$ defines a continuous function in Λ^i such that $\log h^0(z) = k_1(z)-k_2(z)$ in Λ^0; equivalently, $h^0(z)=e^{k_1(z)}/e^{k_2(z)}$ in Λ^0. Thus the function

$$h(z) := \begin{cases} h_1(z)e^{-k_1(z)}, & z\in\Lambda^1,\\ h_2(z)e^{-k_2(z)}, & z\in\Lambda^2, \end{cases}$$

defines a single-valued continuous function in Λ, and hence a solution for $\mathcal{C}_2$ in Λ.

Now we let r_k $(k=1,2,\dots)$ be a sequence of positive numbers such that

$$r_k<r_{k+1}, \qquad \lim_{k\to\infty} r_k = 1,$$

and we let $\Delta_k := \{|z_1|\le r_k\}\times\cdots\times\{|z_n|\le r_k\}\subset\subset\Delta$. For each $k=1,2,\dots$, the above argument yields a solution $h_k(z)$ for $\mathcal{C}_2$ of the generalized Cousin II problem in Δ_k. For $k=2,3,\dots$ we consider the following continuous function in the disk $\{|z_j|\le r_{k+1}\}$ $(j=1,\dots,n)$:

$$\beta_k^0(z_j) = \begin{cases} 1, & |z_j|\le r_k-1,\\ 1-\frac{|z_j|-r_{k-1}}{r_k-r_{k-1}}, & r_{k-1}\le|z_j|\le r_k,\\ 0, & |z_j|\ge r_k, \end{cases}$$

and we set $\beta_k(z) := \beta_k^0(z_1)\cdots\beta_k^0(z_n)$ in Δ_{k+1}. We inductively define continuous functions $h_k^0(z)$ $(k=1,2,\dots)$ in Δ_k in such a way that

$$\begin{aligned} h_1^0(z) &= h_1(z) & &\text{in } \Delta_1,\\ h_{k+1}^0(z) &= h_{k+1}(z)e^{\beta_k(z)q_k(z)} & &\text{in } \Delta_{k+1}, \end{aligned}$$

where $q_k(z)$ is a branch of the continuous function $\log\{h_k^0(z)/h_{k+1}(z)\}$ in the polydisk Δ_k. On each Δ_k $(k=2,3,\dots)$, $h_k^0(z)$ is a solution for $\mathcal{C}_2$, and $h_{k+1}^0(z)=h_k^0(z)$ in Δ_{k-1}. Thus, $h(z) := \lim_{k\to\infty} h_k^0(z)$ is a solution for $\mathcal{C}_2$ of the generalized Cousin II problem in all of Δ. □

From Lemma 3.7 and Theorem 3.8 we have the following theorem.

THEOREM 3.9 (Oka [**46**]). *Let D be a domain of holomorphy in $\mathbf{C}^n$ such that D is homeomorphic to the unit polydisk Δ in $\mathbf{C}^n$. Then the Cousin II problem is always solvable.*

REMARK 3.3. Theorem 3.9 is also true in the case when the unit polydisk Δ is replaced by $\widetilde{\Delta}=\widetilde{\Delta}_1\times\Delta_2\times\cdots\times\Delta_n$, where $\widetilde{\Delta}_1$ is any domain in $\mathbf{C}_{z_1}$.

To verify the remark, it suffices to show that Lemma 3.7 is valid if Δ is replaced by $\widetilde{\Delta}$. Let $\mathcal{C}_2=\{(h_p,\delta_p)\}_{p\in\tilde{\Delta}}$ be a generalized Cousin II distribution in $\widetilde{\Delta}$. We take an increasing sequence of domains $\widetilde{\Delta}_{1k}\subset\subset\widetilde{\Delta}_1$ $(k=1,2,\dots)$ such that each $\widetilde{\Delta}_{1k}$ is bounded by a finite number of closed curves $\gamma_{k,l}$ $(1\le l\le m(k))$, and such that $\bigcup_{k=1}^{\infty}\widetilde{\Delta}_{1k}=\widetilde{\Delta}_1$. Let $\gamma_{k,1}$ be the outer boundary component of $\widetilde{\Delta}_{1k}$. For each k,l $(k=1,2,\dots;\ 2\le l\le m(k))$, we choose a point $a_{k,l}$ in the domain in $\mathbf{C}_{z_1}$ bounded by the closed curve $\gamma_{k,l}$ such that $a_{k,l}\notin\widetilde{\Delta}_{1,k+1}$. We set $\widetilde{\Delta}_k = \widetilde{\Delta}_{1k}\times\Delta_{2k}\times\cdots\times\Delta_{nk}\subset\subset\widetilde{\Delta}$ $(k=1,2,\dots)$, where $\Delta_{ik}=\{|z_i|\le r_k\}$ $(i=2,\dots,n)$, and the radii r_k increase up to 1.

First of all, since, for each $c \in \mathbf{R}$, the set $\{z_1 \in \widetilde{\Delta}_{1k} \mid x = c\}$ consists of simply connected sets in $\mathbf{C}_{z_1}$, the same is true of $\{z \in \widetilde{\Delta}_k \mid x = c\}$ in $\mathbf{C}^n$; thus it follows from the same argument as in the proof of Lemma 3.7 that we have a solution $h_k(z)$ of the generalized Cousin II problem with data $\mathcal{C}_2$ in $\widetilde{\Delta}_k$.

Next, let $\widetilde{\beta}_k(z)$ $(k = 2, 3, \dots)$ be a continuous function in $\widetilde{\Delta}_{k+1}$ such that $\widetilde{\beta}_k(z) = 1$ in $\widetilde{\Delta}_{k-1}$ and $\widetilde{\beta}_{k-1}(z) = 0$ in $\widetilde{\Delta}_{k+1}/\widetilde{\Delta}_k$. We inductively define continuous functions $\widetilde{h}_k(z)$ $(k = 1, 2, \dots)$ in $\widetilde{\Delta}_k$ in such a way that

$$
\begin{aligned}
&\widetilde{h}_1(z) = h_1(z) && \text{in } \widetilde{\Delta}_1, \\
&\widetilde{h}_{k+1}(z) = h_{k+1}(z) \left[\prod_{l=2}^{m(k)} (z_1 - a_{k,l})^{N_{k,l}} \right] e^{\widetilde{\beta}_k(z) q_k(z)} && \text{in } \widetilde{\Delta}_{k+1},
\end{aligned}
$$

where

$$
N_{k,l} = \frac{1}{2\pi i} \int_{\gamma_{k,l}} d \log [\widetilde{h}_k(z)/h_{k+1}(z)]
$$

and $q_k(z)$ is one of the single-valued branches of the continuous function

$$
\log [\widetilde{h}_k(z)/h_{k+1}(z)] - \sum_{l=2}^{m(k)} N_{k,l} \log (z_1 - a_{k,l}) \quad \text{in } \widetilde{\Delta}_k.
$$

Then $\widetilde{h}_k(z)$ $(k = 1, 2, \dots)$ is a solution for $\mathcal{C}_2$ of the generalized Cousin II problem in $\widetilde{\Delta}_k$, and $\widetilde{h}_{k+1}(z) = \widetilde{h}_k(z)$ in $\widetilde{\Delta}_{k-1}$. Hence, $\widetilde{h}(z) := \lim_{k \to \infty} \widetilde{h}_k(z)$ defines a solution for $\mathcal{C}_2$ in all $\widetilde{\Delta}$. □

REMARK 3.4. There have been studies related to the Cousin II problem using methods other than Oka's principle. In [**46**], Oka introduced the notion of a *balayable Cousin distribution*; this was the context of the original Oka principle. K. Stein [**66**][7] gave an interesting topological condition for the solvability of the Cousin II problem.

REMARK 3.5. The Poincaré problem does not always admit a solution in product domains in $\mathbf{C}^n$.

PROOF. To see this, we recall Oka's counterexample to Cousin II in $\Delta = \Delta_1 \times \Delta_2$ in $\mathbf{C}^2$ with variables (z, w), where $\Delta_1 = \{2/3 < |z| < 1\}$ and $\Delta_2 = \{2/3 < |w| < 1\}$ (section 3.5.1). Writing $z = x + iy$ and denoting by Δ' and Δ'' the points of Δ such that $y \geq 0$ and $y \leq 0$, we set

$$
\Sigma \; : \; w - z + 1 = 0 \qquad \text{in } \Delta;
$$

then Σ consists of two connected components $\Sigma' \subset \Delta'$ and $\Sigma'' \subset \Delta''$. We take open neighborhoods $G' \supset\supset \Delta'$ and $G'' \supset\supset \Delta''$ such that $G' \cap \Sigma'' = \emptyset$ and $G'' \cap \Sigma' = \emptyset$. If we set

$$
\begin{aligned}
F_1 &:= 1/(w - z + 1) && \text{in } G', \\
F_2 &:= 1 && \text{in } G'',
\end{aligned}
$$

then $\mathcal{C}_1 = \{(F_1, G'), (F_2, G'')\}$ defines a Cousin I distribution in Δ. We can solve the Cousin I problem for $\mathcal{C}_1$ in the product domain Δ, and we obtain a meromorphic function $g(z, w)$ in Δ such that the pole set of $g(z, w)$ is defined by $1/(w - z + 1)$

[7] K. Stein [**67**] found an example of a domain of holomorphy D which admits a Cousin II distribution $\mathcal{C}_2$ such that $\mathcal{C}_2$ has a solution in any subdomain $D_0 \subset\subset D$ but which does not have a solution in all of D.

in Δ'. We claim that this function $g(z,w)$ cannot be written as a quotient of holomorphic functions $h(z,w)$ and $f(z,w)$ in Δ which are relatively prime at each point in Δ. For if we could write $g(z,w) = h(z,w)/f(z,w)$ in Δ, with $h(z,w)$ and $f(z,w)$ relatively prime at each point, then $f(z,w)$ would be a solution of the Cousin II problem for the Cousin II distribution $\mathcal{C}_2$ in Δ given in the Oka counterexample in 3.5.1; i.e., recall that if

$$\begin{aligned} f_1 &:= w - z + 1 && \text{in } G', \\ f_2 &:= 1 && \text{in } G'', \end{aligned}$$

then $\mathcal{C}_2 = \{(f_1, G'), (f_2, G'')\}$ defines a Cousin II distribution in Δ. Thus the Poincaré problem cannot always be solved in Δ.

Moreover, if we set $h(z,w) := g(z,w)(w-z+1)$ in Δ, then h is holomorphic in Δ. Letting $k(z,w) := w - z + 1$, we have $g(z,w) = h(z,w)/k(z,w)$ in Δ. Thus g is a quotient of holomorphic functions in Δ, but $h(z,w)$ and $k(z,w)$ are not relatively prime at any point in Δ'' at which $w - z + 1 = 0$. □

3.6. Runge Problem

3.6.1. General Expansion Theorem. Let G be a domain of holomorphy in $\mathbf{C}^n$ with variables $z_1, \ldots, z_n$. If $\mathcal{K}$ is a class of holomorphic functions in G satisfying the conditions

1. $\mathcal{K}$ contains the coordinates functions z_k $(k = 1, \ldots, n)$, and
2. given a polynomial $P(w_1, \ldots, w_m)$ in $\mathbf{C}^m$ with $\mathbf{C}$-coefficients, and given $f_1(z), \ldots, f_m(z)$ in $\mathcal{K}$, we have $P(f_1(z), \ldots, f_m(z)) \in \mathcal{K}$,

then $\mathcal{K}$ is called a **normal class**. As examples of normal classes $\mathcal{K}$, we have the class of all polynomials; the class of all holomorphic functions in G; and, for a given $G^* \supset G$, the class of all holomorphic functions in G^* (compare with a **regular** class of holomorphic functions in G from section 1.5.3 in Chapter 1).

THEOREM 3.10. *Let G be a domain of holomorphy in $\mathbf{C}^n$ and let $\mathcal{K}$ be a normal class of holomorphic functions in G. Then any holomorphic function in G can be developed into a locally uniformly convergent series of holomorphic functions belonging to $\mathcal{K}$ if and only if G is a $\mathcal{K}$-convex domain.*

PROOF. Assume that G is convex with respect to $\mathcal{K}$. Then we can find a sequence of analytic polyhedra $\mathcal{P}_j$ $(j = 1, 2, \ldots)$ of the form

$$\mathcal{P}_j : \ |z_i| \le r_i \ \ (i = 1, \ldots, n), \qquad |f_{jk}(z)| \le 1 \ \ (k = 1, \ldots, n_j),$$

where each $f_{jk} \in \mathcal{K}$, and these analytic polyhedra satisfy

$$\mathcal{P}_j \subset\subset \mathcal{P}^o_{j+1} \subset\subset G \ \ (j = 1, 2, \ldots), \qquad G = \lim_{j\to\infty} \mathcal{P}_j.$$

Let $f(z)$ be a holomorphic function in G and let $\epsilon_j > 0$ $(j = 1, 2, \ldots)$ satisfy $\lim_{j\to\infty} \epsilon_j = 0$. From the proof of Theorem 3.5, for each $j = 1, 2, \ldots$, we can find a polynomial $P_j(z,w)$ in $\mathbf{C}^{n+n_j}$ such that

$$\Phi_j(z) := P_j(z, f_{j1}(z), \ldots, f_{jn_j}(z)) \quad \text{in } G$$

satisfies

$$|\Phi_j(z) - f(z)| < \epsilon_j \quad \text{in } \mathcal{P}_{j-1}.$$

Consequently, $\lim_{j\to\infty} \Phi_j(z) = f(z)$ uniformly on each compact set K in G. Thus, the necessity of the convexity of G with respect to $\mathcal{K}$ is proved. The sufficiency is clear from the fact (Theorem 1.11) that a domain of holomorphy G is convex with respect to the class of all holomorphic functions in G. □

Oka's lemma (Lemma 3.5) can be generalized to a normal class $\mathcal{K}$.

LEMMA 3.8. *Let G be a domain of holomorphy. Assume that G is convex with respect to a normal class $\mathcal{K}$ of holomorphic functions in G. Let E be a compact set in G and let A denote the $\mathcal{K}$-convex hull of E. For any $p \in A$, there does* **not** *exist a family of analytic hypersurfaces $\{\sigma_t\}_{t\in T}$ which satisfies Oka's condition at p for the pair (E, A).*

PROOF. Since the Cousin I problem in an analytic polyhedron in G is always solvable, the proof given for polynomial hulls in Lemma 3.5 is valid here; together with Theorem 3.10, this yields the lemma. □

3.6.2. Rationally Convex Domains. In 1.5.3 of Chapter 1, we defined the notion of convexity of a domain D in $\mathbf{C}^n$ with respect to a regular class $\mathcal{K}$ of holomorphic functions in D; in particular, D is said to be convex with respect to rational functions if D is convex with respect to rational functions which are holomorphic in D. However, this definition has some drawbacks. For example, using this definition, the unbounded domain $D_f = \mathbf{C}^n \setminus S_f$, where S_f is the zero set of an entire transcendental function $f(z)$ in $\mathbf{C}^n$, is not convex with respect to rational functions. Thus we must extend the definition of rational convexity.

A domain D in $\mathbf{C}^n$ is said to be **rationally convex** if there exists a sequence of relatively compact subdomains $D_n \subset\subset D$ $(n = 1, 2, \dots)$ such that each D_n is convex with respect to rational functions which are holomorphic in D_n, $D_n \subset D_{n+1}$ $(n = 1, 2, \dots)$, and $\bigcup_{n=1}^{\infty} D_n = D$. The unbounded domain D_f in the previous paragraph is thus rationally convex in $\mathbf{C}^n$.

In the case of one complex variable, every domain is convex with respect to rational functions. However, in the case of several complex variables, even a domain of holomorphy need not be rationally convex. The following example is due to Oka [**47**].

Oka's counterexample for rational convexity. In $\mathbf{C}_z \times \mathbf{C}_w$ we consider $\Delta = \Delta_1 \times \Delta_2$, where

$$\Delta_1 \,:\, 2/3 < |z| < 1, \qquad \Delta_2 \,:\, 2/3 < |w| < 1.$$

Then for $z = x + iy$, we denote by Δ' the subset of Δ with $y \geq 0$. We consider the analytic set $\Sigma : w - z + 1 = 0$ in $\mathbf{C}^2$, and set $\Sigma' := \Sigma \cap \Delta'$. Thus Σ' is also an analytic set in Δ, and $G := \Delta \setminus \Sigma'$ is a domain of holomorphy in $\mathbf{C}^2$ which is *not* rationally convex in $\mathbf{C}^2$.

PROOF. Let $g(z, w)$ be a meromorphic function in Δ whose pole set is given by $1/(w-z+1)$ on Σ' (such a function exists by solvability of Cousin I in Δ; cf., Remark 3.5 in 3.5.3). By considering the holomorphic functions $\{z, w, 1/z, 1/w, g(z, w)\}$ in G, we see that G is holomorphically convex; hence G is a domain of holomorphy in $\mathbf{C}^2$.

We also use the meromorphic function $g(z, w)$ in Δ to prove that G is not rationally convex in $\mathbf{C}^2$. For suppose that G is rationally convex. Fix $0 < d < 1/12$, and set $\Delta^0 := \Delta_1^0 \times \Delta_2^0 \subset\subset \Delta$, where

$$\begin{aligned} \Delta_1^0 &= \{2/3 + d \leq |z| \leq 1 - d\} \subset\subset \Delta_1, \\ \Delta_2^0 &= \{2/3 + d \leq |w| \leq 1 - d\} \subset\subset \Delta_2. \end{aligned}$$

Let

$$\alpha := \{|z| = 5/6\} \subset \Delta_1^0, \qquad \partial\Delta_2^0 = \beta_2 - \beta_1 \subset \Delta_2^0,$$

where $\beta_2 = \{|w| = 1-d\}$ and $\beta_1 = \{|w| = 2/3+d\}$. We let σ denote the projection of $\Sigma' \cap \Delta^0$ onto Δ_1^0. Given $0 < \epsilon < d$, we define, for $z \in \sigma$,

$$\gamma_\epsilon(z) := \{w \in \mathbf{C}_w \mid |w - (z-1)| < \epsilon\} \subset\subset \Delta_2.$$

Then we have

$$K_\epsilon := \Delta^0 \setminus \left[\bigcup_{z\in\sigma} (z, \gamma_\epsilon(z))\right] \subset\subset G. \tag{3.13}$$

Let z_0 be the point of intersection of the circles α and $\{|z-1| = 5/6\}$ in $\mathbf{C}_z$ with $\operatorname{Im} z_0 > 0$. In particular, $z_0 \in \sigma$ and $\gamma_\epsilon(z_0) \subset \Delta_2^0$. Finally, let α' and α'' be the subarcs of α connecting z_0 and $-5/6$ in the counterclockwise and clockwise directions, respectively. Then we have

$$(\alpha' \times \beta_1) \cup (\alpha'' \times \beta_2) \cup (\{-5/6\} \times \Delta_2^0) \subset K_\epsilon.$$

We fix ϵ sufficiently small with $0 < \epsilon < d$ so that

$$\min\{|g(z_0, w)| \mid w \in \partial\gamma_\epsilon(z_0)\} > \max\{|g(z_0, w)| \mid w \in \partial\Delta_2^0\}. \tag{3.14}$$

This is possible because $g(z_0, z_0 - 1) = \infty$. Since we are assuming G is rationally convex in $\mathbf{C}^2$, it follows from (3.13) that given any $\eta > 0$, we can find a rational function $R(z,w) = P(z,w)/Q(z,w)$ in $\mathbf{C}^2$, where $P(z,w)$ and $Q(z,w)$ are relatively prime polynomials in $\mathbf{C}^2$, such that $R(z,w)$ is holomorphic in a neighborhood V of K_ϵ in G and satisfies

$$|g(z,w) - R(z,w)| < \eta \quad \text{on } K_\epsilon.$$

Hence, if $\eta > 0$ is sufficiently small, we see from (3.14) that $R(z_0, w)$ is holomorphic as a function of w in $\Delta_2^0 \setminus \gamma_\epsilon(z_0)$ and satisfies

$$\min\{|R(z_0, w)| \mid w \in \partial\gamma_\epsilon(z_0)\} > \max\{|R(z_0, w)| \mid w \in \partial\Delta_2^0\}.$$

The maximum modulus principle for holomorphic functions implies that $R(z_0, w)$ cannot be holomorphic in all Δ_2^0; thus the denominator $Q(z_0, w)$ has at least one zero in $\gamma_\epsilon(z_0)$ and hence in Δ_2^0. Therefore,

$$2\pi \le \int_{\partial\Delta_2^0} d\arg Q(z_0, w) = \int_{\beta_2} d\arg Q(z_0, w) - \int_{\beta_1} d\arg Q(z_0, w). \tag{3.15}$$

On the other hand, since $P(z,w)/Q(z,w)$ is holomorphic on the neighborhood V of K_ϵ, we have $Q(z,w) \neq 0$ on V. In fact, if not, we have a point (z_0, w_0) in V such that $Q(z_0, w_0) = 0$. Thus, $P(z_0, w_0) = 0$. Since $P(z,w)$ and $Q(z,w)$ are relatively prime, it follows that (z_0, w_0) is a point of indeterminacy of $P(z,w)/Q(z,w)$. This is a contradiction. In particular, $Q(z,w) \neq 0$ on $\alpha' \times \beta_1$, $\alpha'' \times \beta_2$, and $\{-5/6\} \times \Delta_2^0$. These statements imply that

$$\begin{aligned} \int_{\beta_1} d\arg Q(z_0, w) &= \int_{\beta_1} d\arg Q(-5/6, w), \\ \int_{\beta_2} d\arg Q(z_0, w) &= \int_{\beta_2} d\arg Q(-5/6, w), \end{aligned}$$

and

$$\int_{\beta_1} d\arg Q(-5/6, w) = \int_{\beta_2} d\arg Q(-5/6, w).$$

Putting these together gives a contradiction to (3.15). □

3.6.3. Approximation by Algebraic Functions. In this section, we prove the following theorem concerning approximation of a holomorphic function by algebraic functions.

THEOREM 3.11 (Oka [**48**]). *Let G be a domain of holomorphy in $\mathbf{C}^n$. Let E be a compact set in G, let $\varepsilon > 0$, and let $f(z)$ be a holomorphic function in G. Then we can find a single-valued branch $u = \varphi(z)$ of an algebraic function over a neighborhood $\mathcal{P}$ of E in G such that*

$$A_0(z)u^l + A_1(z)u^{l-1} + \ldots + A_l(z) = 0 \quad \text{for } z \in \mathcal{P}, \tag{3.16}$$

where $A_i(z)$ $(i = 0, \ldots, l)$ is a polynomial of z in $\mathbf{C}^n$, and

$$|f(z) - \varphi(z)| < \varepsilon \quad \text{on } E.$$

PROOF. It suffices to show that we can find single-valued holomorphic functions $\zeta_i(z)$ $(i = 1, \ldots, m)$ in a neighborhood $\mathcal{P}$ of E in G such that

(i) $w_k = \zeta_k(z)$ $(k = 1, \ldots, m)$, $z \in \mathcal{P}$, satisfy the m algebraic equations

$$P_k(z, w_1, \ldots, w_m) = 0 \qquad (k = 1, \ldots, m),$$

where $P_k(z, w_1, \ldots, w_m)$ $(k = 1, \ldots, m)$ is a polynomial in $\mathbf{C}^{n+m}$ and

$$\frac{\partial(P_1, \ldots, P_m)}{\partial(w_1, \ldots, w_m)} \neq 0 \quad \text{at } w_k = \zeta_k(z) \ (k = 1, \ldots, m), \ z \in \mathcal{P};$$

(ii) we can find a polynomial $\varphi(z)$ in $z, \zeta_1(z), \ldots, \zeta_m(z)$ such that

$$|f(z) - \varphi(z)| < \epsilon \quad \text{on } E.$$

For it follows by standard techniques (using symmetric functions) that $u = \varphi(z)$ is a single-valued branch of an algebraic function of the form (3.16).

To construct $\zeta_i(z)$ $(i = 1, \ldots, m)$ satisfying (i) and (ii), we take an analytic polyhedron $\mathcal{P}$ in G,

$$\mathcal{P}: \quad |z_j| \leq r_j \ \ (j = 1, \ldots, n), \quad |f_k(z)| \leq 1 \ \ (k = 1, \ldots, m),$$

where $f_k(z)$ $(k = 1, \ldots, m)$ is a holomorphic function in G, and

$$E \subset\subset \mathcal{P} \subset\subset G.$$

As usual, we introduce the space $\mathbf{C}^m$ of the variables $w_1, \ldots, w_m$, and we consider the polydisk in $\mathbf{C}^{n+m}$,

$$\overline{\Delta}: \quad |z_j| \leq r_j \ (j = 1, \ldots, n), \quad |w_k| \leq 1 \ (k = 1, \ldots, m).$$

Furthermore, we define

$$\Sigma: \quad w_k = f_k(z) \ (k = 1, \ldots, m), \qquad z \in \mathcal{P},$$

which is a pure n-dimensional analytic set in $\overline{\Delta}$.

As the first step we approximate $f_k(z)$ $(k = 1, \ldots, m)$ by an algebraic function $\zeta_k(z)$ with conditions (i) and (ii). Regarding $f_k(z)$ as independent of $w_1, \ldots, w_m$, we have that

$$w_k - f_k(z)$$

is a holomorphic function in a neighborhood V_k of Σ in $\mathbf{C}^{n+m}$. By Theorem 3.4, $\Sigma \subset \mathbf{C}^{n+m}$ is polynomially convex. Thus there exists a polynomial polyhedron $\mathcal{P}^*$ in $\mathbf{C}^{n+m}$ such that

$$\Sigma \subset\subset \mathcal{P}^* \subset\subset V_k \qquad (k = 1 \ldots, m).$$

We can choose $\rho > 0$ sufficiently small so that

$$Q := \bigcup_{z \in \mathcal{P}} (z, \gamma_1(z), \dots, \gamma_m(z)) \subset\subset \mathcal{P}^*, \tag{3.17}$$

where $\gamma_k(z) = \{w_k \in \mathbf{C}_{z_k} \mid |w_k - f_k(z)| < \rho\}$ $(k = 1, \dots, m)$.

Fix η with $0 < \eta < \rho$. By applying Corollary 3.1 for $(\mathcal{P}^*, \mathbf{C}^{n+m})$ and the function $w_k - f_k(z)$ in $\mathcal{P}^*$, we can find a polynomial $P_k(z, w)$ in $\mathbf{C}^{n+m}$ such that

$$|P_k(z, w) - (w_k - f_k(z))| < \eta/2 \quad \text{on } \mathcal{P}^* \ (k = 1, \dots m). \tag{3.18}$$

We consider the following holomorphic transformation

$$T \ : \ (z, w) \in \mathcal{P} \times \mathbf{C}^m_w \to (z, W) = (z, w_1 - f_1(z), \dots, w_m - f_m(z)) \in \mathcal{P} \times \mathbf{C}^m_W.$$

Then T is one-to-one and maps Q onto the product domain $\mathcal{P} \times B_\rho$, where B_r $(r > 0)$ denotes the polydisk in $\mathbf{C}^m_W$ of center 0 and radius r. We set

$$H_k(z, W) := P_k(z, W_1 + f_1(z), \dots, W_m + f_m(z)) \qquad (k = 1, \dots, m),$$

which is a holomorphic function in $\mathcal{P} \times B_\rho$ and which, for each fixed $z \in \mathcal{P}$, is a polynomial of W in $\mathbf{C}^m$. Since $B_\eta \subset B_\rho$, we obtain from (3.17) and (3.18) that

$$|H_k(z, W) - W_k| < \eta/2 \quad \text{on } \mathcal{P} \times B_\eta \qquad (k = 1, \dots, m).$$

Fix $z \in \mathcal{P}$. Using Rouché's theorem from one complex variable, for each $k = 1, \dots, m$, the above inequalities imply that there exists a *unique* complex number $W_k = h_k(z)$ $(k = 1, \dots, m)$ with

$$H_k(z, h_1(z), \dots, h_m(z)) = 0, \quad \text{and} \quad |h_k(z)| < \eta \ (k = 1, \dots, m).$$

Furthermore, each $h_k(z)$ $(k = 1, \dots, m)$ depends holomorphically on $z \in \mathcal{P}$. Thus, if we define

$$\zeta_k(z) := h_k(z) + f_k(z) \qquad (k = 1, \dots, m), \quad z \in \mathcal{P},$$

then $w_k = \zeta_k(z)$ $(k = 1, \dots, m)$ is a single-valued function in $\mathcal{P}$ which satisfies

$$P_k(z, w_1, \dots, w_m) \ = \ 0 \ \ (k = 1, \dots, m), \quad z \in \mathcal{P}, \tag{3.19}$$

$$|\zeta_k(z) - f_k(z)| \ \le \ \eta \ \ (k = 1, \dots, m), \quad z \in \mathcal{P}. \tag{3.20}$$

The uniqueness of $W_k = h_k(z)$ $(k = 1, \dots, m)$ also implies that any solution $w_k(z)$ $(k = 1, \dots, m)$ of the simultaneous algebraic equations (3.19) for $w_1, \dots, w_m$ (we regard $z \in \mathcal{P}$ as parameters) such that $|w_k(z) - f_k(z)| < \eta$ $(k = 1, \dots, m)$ coincides with $\zeta_k(z)$ $(k = 1, \dots, m)$. Thus

$$\frac{\partial(P_1, \dots, P_m)}{\partial(w_1, \dots, w_m)} \neq 0 \quad \text{at} \quad w_k = \zeta_k(z) \ (k = 1, \dots, m), \ z \in \mathcal{P},$$

which proves the first step.

To prove the second step, and hence the theorem, let $\epsilon > 0$ and let $f(z)$ be a holomorphic function in G. From the proof of Theorem 3.5, there exists a polynomial

$$\Phi(z) = P(z, f_1(z), \dots, f_m(z))$$

of $z, f_1(z), \dots, f_m(z)$ such that

$$|\Phi(z) - f(z)| < \epsilon/2 \quad \text{on } E.$$

Therefore, if we take $\eta > 0$ sufficiently small, and use the functions $w_k = \zeta_k(z)$ ($k = 1, \dots, m$), $z \in \mathcal{P}$, which were constructed above to satisfy $|\zeta_k(z) - f_k(z)| < \eta$ ($k = 1, \dots, m$) on $\mathcal{P}$ (and condition (i)), to define the polynomial

$$\varphi(z) := P(z, \zeta_1(z), \dots, \zeta_m(z))$$

in $z, \zeta_1(z), \dots, \zeta_m(z)$, then we have

$$|\Phi(z) - \varphi(z)| < \epsilon/2 \quad \text{on } E.$$

Thus

$$|\varphi(z) - f(z)| < \epsilon \quad \text{on } E,$$

and the theorem is proved. □

3.6.4. Polynomially Convex Domains. In the case of one complex variable, a domain D in $\mathbf{C}$ is polynomially convex if and only if D is simply connected. In the case of several complex variables, there is no topological characterization of polynomial convexity. Indeed, a polynomially convex domain D in $\mathbf{C}^n$ is not necessarily simply connected.

EXAMPLE 3.3. In $\mathbf{C}^2$ with variables z, w, consider the domain

$$D: \quad |z| < 2, \qquad |w| < 2, \qquad |zw - 1| < 1/2.$$

Then D is polynomially convex but not simply connected. To see that D is not simply connected, assume the contrary. Since $D \bigcap (\{0\} \times \mathbf{C}_w) = \emptyset$, the function $\log z$ has a single-valued branch in D. On the other hand, the closed curve $\gamma := \{(z, w) = (e^{i\theta}, e^{-i\theta}) \mid 0 \le \theta \le 2\pi\}$ in $\mathbf{C}^2$ is contained in D; hence $\log z$ has no single-valued branch in D, a contradiction.

Conversely, a simply connected domain of holomorphy D in $\mathbf{C}^n$ ($n \ge 2$) is not necessarily polynomially convex.

Wermer's Example [78]. In $\mathbf{C}^3$ with variables z, w, t, consider the compact set

$$K: \quad |z| \le 1, \qquad |w| \le 1, \qquad t = 0.$$

For $0 < a < 1$, define

$$\Delta_a: \quad |z| < 1 + a, \qquad |w| < 1 + a, \qquad |t| < a$$

so that $K \subset\subset \Delta_a$. Let $T : (z, w, t) \in \mathbf{C}^3 \to (z_1, z_2, z_3) \in \mathbf{C}^3_z$ be the holomorphic mapping defined as

$$T \ : \ z_1 = z, \quad z_2 = zw + t, \quad z_3 = zw^2 - w + 2tw.$$

Define

$$E := T(K), \qquad G_a := T(\Delta_a).$$

Then T is a one-to-one mapping from K onto E, and $(0, 1, 0) \notin E$. Since the determinant of the Jacobian matrix of T is

$$\left| \frac{\partial(z_1, z_2, z_3)}{\partial(\, z, \, w, \, t\,)} \right| = 1 - 2t,$$

we can choose a sufficiently small so that the domain G_a in $\mathbf{C}^3_z$ is biholomorphically equivalent to the polydisk Δ_a in $\mathbf{C}^3$ and also to insure that $(0, 1, 0) \notin G_a$. We show that, with such a choice of a, G_a is not polynomially convex.

PROOF. We consider the closed curve $\gamma = \{(e^{i\theta}, e^{-i\theta}, 0) \mid 0 \leq \theta \leq 2\pi\}$ in K. Then $T(\gamma)$ is the unit circle lying in the complex plane $L : z_2 = 1,\ z_3 = 0$; i.e., $T(\gamma)$ is given by $|z_1| = 1,\ z_2 = 1,\ z_3 = 0$.

However, in general, any polynomially convex domain D in $\mathbf{C}^n$ satisfies the property that for any m-dimensional complex plane L in $\mathbf{C}^n$ $(0 < m < n)$, $D \cap L$ is also polynomially convex in $\mathbf{C}^m$. In particular, for $m = 1$, $D \cap L$ must be a disjoint union of simply connected domains.

Taking D to be G_a and L to be the 1-plane defined by $z_2 = 1,\ z_3 = 0$ in $\mathbf{C}^3_z$, we see that $G_a \cap L$ contains the circle $|z_1| = 1$ in L but does not contain the center $z_1 = 0$ in $\mathbf{C}_{z_1}$; hence $G_a \cap L$ is not simply connected. Thus G_a is not polynomially convex in $\mathbf{C}^3_z$. □

REMARK 3.6. Let D be a rationally convex domain in $\mathbf{C}^n$. Then for any m-dimensional complex plane L $(0 < m < n)$, the intersection $L \cap D$ is rationally convex in $\mathbf{C}^m$. For the case $m = 1$, this imposes no restriction on $D \cap L$, as every planar domain is rationally convex. However, if we assume, in addition, that D is simply connected in $\mathbf{C}^n$, then for L with $\dim L = 1$, $D \cap L$ must be simply connected in $\mathbf{C}$ (G. Stolzenberg [**70**]).

To verify this last statement, assume, for the sake of obtaining a contradiction, that for some L with $\dim L = 1$, $D_1 := D \cap L$ is not simply connected. For simplicity, we take $L = \mathbf{C}_{z_1}$; i.e., L is defined by $z_2 = \cdots = z_n = 0$. Since D_1 is not simply connected, we can take a point $a \in \mathbf{C}_{z_1} \setminus \overline{D_1}$ and a closed curve γ in D_1 such that $\int_\gamma d\arg(z_1 - a) = 2\pi$. If we let L_2 be the 2-plane $z_3 = \cdots = z_n = 0$, then we can construct a holomorphic function $f_2(z_1, z_2)$ in $D \cap L_2$ such that $f_2(z_1, 0) = 1/(z_1 - a)$ in D_1. To do this, we proceed by using the solution of a Cousin I problem similar to that used in Lemma 3.3. We repeat this procedure to obtain a holomorphic function $f(z)$ in D satisfying $f(z_1, 0, \ldots, 0) = 1/(z_1 - a)$ in D_1.

Now let $\epsilon > 0$ and let G be a simply connected domain such that $\gamma \subset\subset G \subset\subset D$. Since D is rationally convex in $\mathbf{C}^n$, we can find a rational function $R(z) = P(z)/Q(z)$, where $P(z)$ and $Q(z)$ are relatively prime polynomials in $\mathbf{C}^n$, such that $R(z)$ is holomorphic in G and satisfies $|R(z) - f(z)| < \epsilon$ in G. In particular, the denominator $Q(z)$ of $R(z)$ cannot vanish at any point in the simply connected domain G; hence

$$\int_\gamma d\arg Q(z) = 0.$$

On the other hand, if $0 < \epsilon < \min\{\frac{1}{|z_1 - a|} : z_1 \in \gamma\}$, then Rouché's theorem implies

$$\int_\gamma d\arg \frac{P(z)}{Q(z)} = \int_\gamma d\arg \frac{1}{z_1 - a} = -2\pi.$$

Hence,

$$\int_\gamma d\arg Q(z) = \int_\gamma d\arg P(z) + 2\pi \geq 2\pi,$$

which is a contradiction. □

It follows that the domain G_a in Wermer's example is not rationally convex, but it is a simply connected domain in $\mathbf{C}^3$. We mention that there is an example due to J. Duval [**16**] of a domain in $\mathbf{C}^2$ which is both rationally convex and simply connected but which is not polynomially convex.

CHAPTER 4

Pseudoconvex Domains and Pseudoconcave Sets

4.1. Pseudoconvex Domains

4.1.1. Domains of Holomorphy, Domains of Meromorphy, and Domains of Normality. The notion of a **pseudoconvex domain** was developed by K. Oka. This will give a geometric characterization of the boundary of a domain of holomorphy (section 4.2). Preliminary results motivating this concept were obtained by F. Hartogs, E. E. Levi and G. Julia, and we begin our discussion of this topic with some classical results based on their work.

1. Hartogs' Theorem

In $\mathbf{C}^{n+1}$ with variables $z = (z_1, \dots, z_n)$ and w, we consider a polydisk $\Lambda = \Delta \times \Gamma$ where

$$\Delta \ : \ |z_j| < r_j \quad (j = 1, \dots, n), \qquad \Gamma \ : \ |\, w \,| < \rho,$$

with $r_j > 0$ $(j = 1, \dots, n)$ and $\rho > 0$. For r'_j and ρ' with $0 < r'_j < r_j$ $(j = 1, \dots, n)$ and $0 < \rho' < \rho$, we set

$$\Delta' \ : \ |z_j| < r'_j \quad (j = 1, \dots, n), \qquad \Gamma^* \ : \ \rho' < |\, w \,| < \rho,$$

and define

$$E_1 := \Delta' \times \Gamma, \quad E_2 := \Delta \times \Gamma^*, \quad \text{and} \quad E := E_1 \cup E_2$$

(see Figure 1).

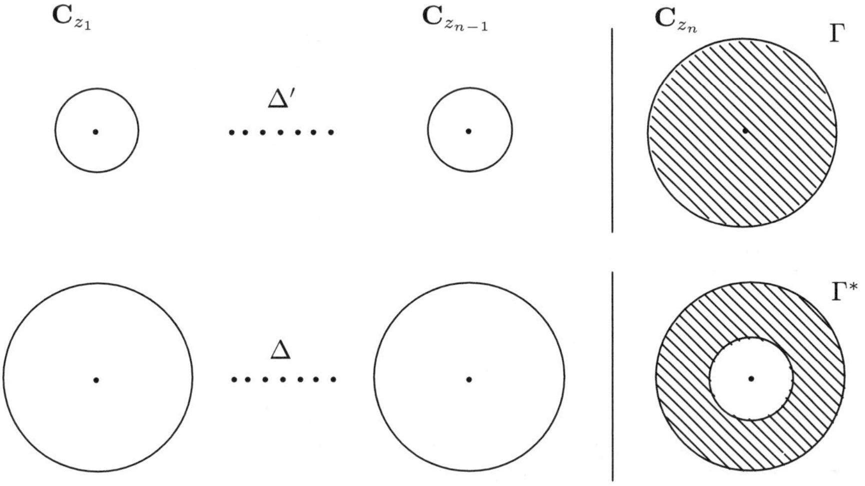

FIGURE 1. Hartogs' Theorem

We have the following theorem.

THEOREM 4.1 (Hartogs [**29**]). [1] *Each holomorphic function $f(z,w)$ on E extends holomorphically to the polydisk Λ.*

PROOF. Since $f(z,w)$ is holomorphic in E_2, we can consider the Hartogs-Laurent series

$$f(z,w) = \sum_{j=-\infty}^{\infty} a_j(z)w^j \qquad \text{in } E_2,$$

where $a_j(z)$ $(j = 0, \pm 1, \dots)$ are holomorphic functions in Δ. Since $f(z,w)$ is holomorphic on E_1, from the uniqueness of the Hartogs-Laurent expansion it follows that $a_j(z) \equiv 0$ on Δ' for $j = -1, -2, \dots$ and hence $a_j(z) \equiv 0$ on Δ for these values of j. Thus $f(z,w)$ is the restriction to E of the holomorphic function $\sum_{j=0}^{\infty} a_j(z)w^j$ in Λ. □

In particular, let D be a domain in $\mathbf{C}^n$ and let σ be an analytic set of dimension at most $n-2$. Then each holomorphic function $f(z)$ on $D \setminus \sigma$ extends holomorphically to the domain D.

2. E. E. Levi's Theorem

Using the same notation as above, we prove the following lemmas.

LEMMA 4.1. *Let $g(z,w)$ be a holomorphic function in E_2 with Hartogs-Laurent series*

$$g(z,w) = \sum_{j=-\infty}^{\infty} a_j(z)w^j \qquad \textit{in } E_2,$$

where $a_j(z)$ $(j = 0, \pm 1, \dots)$ is holomorphic in Δ. Then $g(z,w)$ can be extended to a meromorphic function in the polydisk Λ if and only if there exist a finite number of holomorphic functions $\{b_1(z), \dots, b_l(z)\}$ on Δ which satisfy the following infinite set of simultaneous equations for z in Δ:

$$a_{-(\nu+l)}(z) + a_{-(\nu+l-1)}(z) \cdot b_1(z) + \cdots + a_{-\nu}(z) \cdot b_l(z) \equiv 0 \quad (\nu \geq 1). \tag{4.1}$$

PROOF. Suppose first of all that $g(z,w)$ can be extended to a meromorphic function $f(z,w)/h(z,w)$ on Λ where $f(z,w)$ and $h(z,w)$ are holomorphic on Λ and relatively prime at each point $(z,w) \in \Lambda$. Since $g(z,w)$ is holomorphic on E_2, it follows that $h(z,w) \neq 0$ on E_2. Using Remark 2.3 in section 2.1.3, the zero set of $h(z,w)$ in Λ (which may be empty) coincides with that of a distinguished pseudopolynomial $P(z,w)$ in Λ,

$$P(z,w) = w^l + b_1(z)w^{l-1} + \cdots + b_l(z), \tag{4.2}$$

where each $b_k(z)$ $(k = 1, \dots, l)$ is holomorphic in Δ. Since $g(z,w)P(z,w)$ can be holomorphically extended to Λ, the coefficient of $w^{-\nu}$ $(\nu = 1, 2, \dots)$ of the Hartogs-Laurent series of $g(z,w)P(z,w)$ vanishes; this coefficient equals the left-hand side of equation (4.1).

For the converse, assume that there exist a finite number of holomorphic functions $\{b_1(z), \dots, b_l(z)\}$ on Δ which satisfy the equations in (4.1). We then define $P(z,w)$ by (4.2) so that $P(z,w)$ is holomorphic on Λ and by (4.1) each coefficient of $w^{-\nu}$ $(\nu = 1, 2, \dots)$ of the Hartogs-Laurent series of $g(z,w)P(z,w)$ vanishes on Δ. Thus $g(z,w)P(z,w)$ can be considered as a holomorphic function in Λ, so that $g(z,w)$ extends to a meromorphic function on Λ. □

---

[1] In this paper, Hartogs proved Theorem 4.1 using Cauchy's integral formula.

LEMMA 4.2. *Let $g(z,w)$ be a meromorphic function on E. If $g(z,w)$ is holomorphic on E_2, then $g(z,w)$ can be extended to a meromorphic function on the polydisk Λ.*

PROOF. We first develop $g(z,w)$ into the Hartogs-Laurent series

$$g(z,w) = \sum_{j=-\infty}^{\infty} a_j(z)w^j \quad \text{on } E_2,$$

where $a_j(z)$ $(j = 0, \pm 1, \dots)$ are holomorphic on Δ. We would like to use Lemma 4.1 to construct a finite number of holomorphic functions $\{b_1(z), \dots, b_l(z)\}$ on Δ satisfying (4.1). Since $g(z,w)$ is meromorphic in E and hence in the domain $E_1 = \Delta' \times \Gamma$, and the functions $a_j(z)$ are holomorphic on Δ and hence in Δ', we can appeal to Lemma 4.1 to find a finite number of holomorphic functions $\{b_1(z), \dots, b_l(z)\}$ on the smaller polydisk Δ' which satisfy

$$a_{-(\nu+l)}(z) + a_{-(\nu+l-1)}(z)b_1(z) + \cdots + a_{-\nu}(z)b_l(z) \equiv 0 \quad (\nu \geq 1) \tag{4.3}$$

on Δ'. Consider the matrix $A(z)$, $z \in \Delta$, with infinitely many rows, each of length $l+1$, defined by

$$A(z) := \begin{pmatrix} a_{-l-1}(z) & a_{-l}(z) & \cdots & a_{-1}(z) \\ a_{-l-2}(z) & a_{-l-1}(z) & \cdots & a_{-2}(z) \\ \vdots & \vdots & & \vdots \end{pmatrix},$$

and the corresponding infinite set of homogeneous linear equations for $z \in \Delta$:

$$A(z) \begin{pmatrix} X_0(z) \\ X_1(z) \\ \vdots \\ X_l(z) \end{pmatrix} = \begin{pmatrix} 0 \\ 0 \\ \vdots \\ 0 \end{pmatrix}. \tag{4.4}$$

We seek non-trivial holomorphic solutions $\{X_0(z), \dots, X_l(z)\}$ of these equations in Δ. Let $r := \max\{\text{rank } A(z) \mid z \in \Delta'\}$. From (4.3), we know that $0 \leq r \leq l$. If $r = 0$, then $a_j(z) \equiv 0$ on Δ' and hence on Δ for $j = -1, -2, \dots$, so that $g(z,w)$ is holomorphic in Λ and there is nothing to prove. We may therefore assume that $1 \leq r \leq l$ and we fix a point $z_0 \in \Delta'$ such that rank $A(z_0) = r$. Next we fix a neighborhood V of z_0 in Δ and an $r \times r$ minor matrix of $A(z)$ whose determinant $D(z)$ does not vanish at any point $z \in V$; say, e.g.,

$$D(z) = \begin{vmatrix} a_{-l-1}(z) & \cdots & a_{-l-2+r}(z) \\ \vdots & \ddots & \vdots \\ a_{-l-2-r}(z) & \cdots & a_{-l-3}(z) \end{vmatrix} \neq 0 \quad \text{on } V,$$

while the determinant of any $s \times s$ minor matrix of $A(z)$ with $s \geq r+1$ necessarily vanishes identically in V. Using Cramer's rule, we get polynomials $A_{\mu k}(z)$ $(0 \leq \mu \leq r-1\,; r \leq k \leq l)$ in the coefficient functions $a_j(z)$ $(j = 0, \pm 1, \dots)$ such that, if we define

$$B_\mu(z) := \frac{A_{\mu r}(z)}{D(z)} B_r(z) + \cdots + \frac{A_{\mu l}(z)}{D(z)} B_l(z),$$

where $B_r(z), \dots, B_l(z)$ are *arbitrary* holomorphic functions in Δ, then the functions $\{B_0(z), \dots, B_l(z)\}$ on Δ satisfy the equations (4.4) for $z \in V$. We now set

$$X_k(z) := D(z)B_k(z) \qquad (0 \leq k \leq l) \quad \text{on } \Delta.$$

Then $X_k(z)$ $(0 \le k \le l)$ are holomorphic on Δ and satisfy the equations (4.4) for z in V and hence for all z in Δ. From $1 \le r \le l$, we have thus constructed non-trivial holomorphic solutions $\{X_0(z), \dots, X_l(z)\}$ of the system of equations (4.4) on all of Δ. Defining

$$Q(z,w) := X_0(z)w^l + X_1(z)w^{l-1} + \cdots + X_l(z) \ (\not\equiv 0) \quad \text{on } \Lambda,$$

for each $\nu \ge 1$ the coefficient of $w^{-\nu}$ of the Hartogs-Laurent series of $h(z,w) \equiv g(z,w)Q(z,w)$ on E_2 is equal to zero on Δ; i.e., $h(z,w)$ is holomorphic in Λ. Hence, $g(z,w)$ extends to a meromorphic function on all of Λ. □

We now consider $\mathbf{C}^{n+2}$ with variables $z = (z_1, \dots, z_n)$ and $w = (w_1, w_2)$. Fix $r > \rho > 0$ and define two balls in $\mathbf{C}^2_w$:

$$Q(r) \ : \ |w_1 + r|^2 + |w_2|^2 \le r^2,$$
$$q(\rho) \ : \ |w_1|^2 + |w_2|^2 < \rho^2.$$

Finally, set

$$\begin{aligned} \beta &:= q(\rho) \setminus Q(r) && \text{in } \mathbf{C}^2_w, \\ B &:= \{0\} \times \beta && \text{in } \mathbf{C}^{n+2}. \end{aligned} \tag{4.5}$$

We have the following theorem.

THEOREM 4.2 (E. E. Levi [**17**]). *Any meromorphic function $g(z,w)$ on B has a meromorphic extension to the origin $(z,w) = (0,0)$; precisely, $g(z,w)$ has a meromorphic extension to the set $\widehat{B}$ in $\mathbf{C}^{n+2}$ defined by*

$$\widehat{B} \ : \ z = 0, \quad (w_1, w_2) \in q(\rho) \cap \{\text{Re } w_1 > -\rho^2/2r\}.$$

PROOF. Let D be a neighborhood of $\{0\} \times \beta$ in $\mathbf{C}^{n+2}$ on which $g(z,w)$ is meromorphic; we assume the origin $(0,0) \in \mathbf{C}^{n+2}$ is not contained in D (otherwise there is nothing to prove). Given $w' \in \beta$, we let $D(w') \subset \mathbf{C}^n_z$ denote the section of D over $w = w'$.

We let σ denote the set of poles of $g(z,w)$ in D, and we let $\sigma(0) \subset \beta \subset \mathbf{C}^2_w$ denote the section of σ over $z = 0$. We have two cases to consider:

$$\text{Case I:} \quad \dim \sigma(0) \le 1; \text{ i.e., } \sigma(0) \ne \beta;$$

$$\text{Case II:} \quad \dim \sigma(0) = 2; \text{ i.e., } \sigma(0) = \beta.$$

We first prove the theorem for Case I; we proceed in several steps.

First step. $g(z,w)$ has a meromorphic extension to $\{0\} \times (q(\rho) \cap \partial Q(r))$ in $\mathbf{C}^{n+2}$.

It suffices to prove that $g(z,w)$ has a meromorphic extension to the origin $(z,w) = (0,0)$ in $\mathbf{C}^{n+2}$.

We first consider the case when

$$\{(w_1, w_2) \in \beta \mid w_1 = 0\} \not\subset \sigma(0). \tag{4.6}$$

Since $\{(w_1, w_2) \in \beta \mid w_1 = 0\} \cap \sigma(0)$ consists of isolated points in the punctured disk $0 < |w_2| < \rho$, we can find a circle $|w_2| = \rho_2$ in $\mathbf{C}_{w_2}$ with $0 < \rho_2 < \rho/2$ such that $g(z,w)$ is holomorphic on $\{z = 0\} \times \{w_1 = 0\} \times \{|w_2| = \rho_2\}$. Thus $g(z,w)$ is holomorphic on $\{z = 0\} \times \{|w_1| \le \rho_1\} \times \{|w_2| = \rho_2\}$ if $0 < \rho_1 < \rho/2$ is sufficiently small. Fix a point $w_1^0 \in \{|w_1| < \rho_1\} \cap \{\text{Re } w_1 > 0\}$ sufficiently close to the origin $w_1 = 0$ so that $K := \{(w_1^0, w_2) \in \mathbf{C}^2_w \mid |w_2| \le \rho/2\} \subset\subset \beta$. It follows from Lemma

4.2 that $g(z,w)$ has a meromorphic extension to $\{0\} \times \{|w_1| \le \rho_1\} \times \{|w_2| \le \rho_2\}$ in $\mathbf{C}^{n+2}$, and, in particular, to the origin $(z,w) = (0,0)$.

Now we consider the case when

$$\{(w_1, w_2) \in \beta \mid w_1 = 0\} \subset \sigma(0).$$

For $\varepsilon > 0$, the holomorphic mapping T of $\mathbf{C}^2$,

$$T \;:\; w_1' = w_1 - \varepsilon w_2^2, \quad w_2' = w_2$$

is one-to-one and maps onto $\mathbf{C}^2$ fixing the origin $(0,0)$. We choose $\varepsilon > 0$ sufficiently small so that

$$\{w_1 = \varepsilon w_2^2\} \cap \beta \not\subset \sigma(0) \tag{4.7}$$

(since $\dim \sigma(0) \le 1$). An elementary calculation shows that

$$\{w_1 = \varepsilon w_2^2\} \cap (q(r) \setminus \{(0,0)\}) \subset \beta$$

provided ε is sufficiently small. Given $r' > r$ and $0 < \rho' < \rho$, we set

$$\widetilde{Q}(r') \;:\; |w_1' + r'|^2 + |w_2'|^2 \le r'^2, \qquad \widetilde{q}(\rho') \;:\; |w_1'|^2 + |w_2'|^2 < \rho'^2$$

and $\widetilde{\beta} := \widetilde{q}(\rho') \setminus \widetilde{Q}(r')$. If r' is sufficiently large and $\rho' > 0$ is sufficiently small, it can be shown that

$$\widetilde{\beta} \subset T(\beta).$$

We set $\tilde{g}(z,w') := g(z,w)$, where $w' = Tw$. Then $\tilde{g}(z,w')$ is meromorphic in a neighborhood of $\{0\} \times T(\beta)$, and, in particular, $\tilde{g}(z,w')$ is meromorphic on $\{0\} \times \widetilde{\beta}$. If we let $\tilde{\sigma}$ denote the set of poles of $\tilde{g}(z,w')$, then $\{(w_1', w_2') \in \widetilde{\beta} \mid w_1' = 0\} \not\subset \tilde{\sigma}(0)$ from (4.7). It now follows from the previous case (4.6) that $\tilde{g}(z,w')$ has a meromorphic extension to the origin $(z,w') = (0,0)$, and hence $g(z,w)$ has a meromorphic extension to the origin $(z,w) = (0,0)$. Thus, the first step is proved.

Second step. $g(z,w)$ has a meromorphic extension to $\widehat{B}$.

We note that $(\partial Q(r)) \cap (\partial q(\rho)) \cap \{w_2 = 0\}$ lies on $\Re w_1 = -\rho^2/2r$. From the first step, it suffices to prove the second step under the condition that $g(z,w)$ is meromorphic on $(\partial Q(r)) \cap (\partial q(\rho))$ (for we can take a smaller $q(\rho)$ sufficient close to the original $q(\rho)$, if necessary). Given $0 \le a < \rho^2/2r$, we consider the ball $B(a)$ in $\mathbf{C}^2_w$ centered at $(-R-a, 0)$ with radius R, where R is chosen so that the sphere $\partial B(a)$ intersects $(\partial Q(r)) \cap (\partial q(\rho))$. Precisely, we take

$$B(a): \quad |w_1 + R + a|^2 + |w_2|^2 < R^2,$$

where

$$R^2 = \left(R + a - \frac{\rho^2}{2r}\right)^2 + \rho^2 - \left(\frac{\rho^2}{2r}\right)^2.$$

Then we have $B(0) = Q(r)$; $q(\rho) \setminus B(a') \subset q(\rho) \setminus B(a'')$ for all a', a'' with $0 \le a' < a'' < \rho^2/2r$; $\lim_{a \to \rho^2/2r} B(a) = \{(w_1, w_2) \in \mathbf{C}^2 \mid \Re w_1 < -\rho^2/2r\}$; and

$$\widehat{B} = \bigcup_{0 \le a < \rho^2/2r} \{0\} \times [q(\rho) \setminus \overline{B(a)}]. \tag{4.8}$$

We set

$$a^* := \sup\{a \mid g(z,w) \text{ has a meromorphic extension to } \{0\} \times [q(\rho) \setminus \overline{B(a)}]\}.$$

Using (4.8), we see that our goal is to show that $a^* = \rho^2/2r$. We prove this by contradiction; hence, we assume $a^* < \rho^2/2r$. From the first step, $g(z,w)$ has a meromorphic extension to each point of $\{0\} \times [\overline{q(\rho)} \cap (\partial B(0))]$, so that we certainly have $a^* > 0$. Since $g(z,w)$ has a meromorphic extension to each point of $\{0\} \times [\overline{q(\rho)} \cap (\partial B(a))]$ for $a < a^*$, $g(z,w)$ has a meromorphic extension to each point of $\{0\} \times [\overline{q(\rho)} \cap (\partial B(a^*))]$ by the first step; thus, we again get $a > a^*$ sufficiently close to a^* such that $g(z,w)$ has a meromorphic extension to each point of $\{0\} \times [\overline{q(\rho)} \cap (\partial B(a))]$, contradicting the definition of a^*. Thus the second step is proved, and hence the theorem is valid in Case I.

We now turn to Case II. Fix $\mathbf{a} \in \mathbf{C}^n \setminus \{0\}$ and the one-dimensional complex line $L = L_{\mathbf{a}} := \{t\mathbf{a} \in \mathbf{C}^n \mid t \in \mathbf{C}\}$ such that

$$D \cap (L \times \mathbf{C}^2_w) \not\subset \sigma. \tag{4.9}$$

For $0 < \varepsilon < 1$, consider the following linear bijection T_ε of $\mathbf{C}^{n+2}$:

$$T_\varepsilon : \quad z' = z + (\varepsilon w_1)\mathbf{a}, \quad w_1' = w_1, \quad w_2' = w_2,$$

which fixes the origin $(0,0)$. We set

$$\widehat{g}_\varepsilon(z', w') := g(z,w), \quad \text{where } (z',w') \in T_\varepsilon(D).$$

Then $\widehat{g}_\varepsilon(z',w')$ is a meromorphic function in $\widehat{D}_\varepsilon := T_\varepsilon(D)$ whose set of poles $\widehat{\sigma}_\varepsilon$ in $\widehat{D}_\varepsilon$ satisfies $\dim \widehat{\sigma}_\varepsilon(0) \le 1$ (here $\widehat{\sigma}_\varepsilon(0)$ denotes the section of $\widehat{\sigma}_\varepsilon$ over $z' = 0$). This follows from (4.9). Note that the section $\widehat{D}_\varepsilon(0)$ of $\widehat{D}_\varepsilon$ over $z' = 0$ does not contain a set of the form β (defined by (4.5)) at $w' = 0$.

For η satisfying $0 < \eta < \rho$, we define the following subsets of $\mathbf{C}^2_w$:

$$\begin{aligned} \widehat{Q}(r+\eta) \quad &: \quad |w_1 + r|^2 + |w_2|^2 \le (r+\eta)^2, \\ \widehat{q}(\rho - \eta) \quad &: \quad |w_1 - \eta|^2 + |w_2|^2 < (\rho - \eta)^2, \end{aligned}$$

and we let $\widehat{\beta}(\eta) = \widehat{q}(\rho-\eta) \setminus \widehat{Q}(r+\eta)$. Note that $\widehat{\beta}(\eta) \subset\subset \beta$. There exists a unique real number α_η satisfying

$$(\partial \widehat{Q}(r+\eta)) \cap (\partial \widehat{q}(\rho - \eta)) = \{\mathrm{Re}\, w_1 = -\alpha_\eta\} \cap (\partial \widehat{q}(\rho-\eta)).$$

We have $\widehat{\beta}(\eta) \to \beta$ and $\alpha_\eta \to \rho^2/2r$ as $\eta \to 0$; hence we can choose $\eta > 0$ with $\alpha_\eta > 0$ sufficiently small so that

$$(0,0) \in \widehat{q}(\rho - \eta) \setminus \{\mathrm{Re}\, w_1 > -\alpha_\eta\}.$$

Since $\widehat{\beta}(\eta) \subset\subset \beta$, we can find a polydisk $\delta : |z_j| < s_j$ $(j = 1, \dots, n)$ in $\mathbf{C}^n_z$ such that $\delta \times \widehat{\beta}(\eta) \subset\subset D$. We set $\delta' : |z_j'| < s_j/2$ $(j = 1, \dots, n)$ in $\mathbf{C}^n_{z'}$ and fix $\varepsilon > 0$ sufficiently small so that

$$\{z' - (\varepsilon w_1)\mathbf{a} \in \mathbf{C}^n_z \mid z' \in \delta', \ |w_1| \le \rho\} \subset\subset \delta.$$

It follows that $\delta' \times \widehat{\beta}(\eta) \subset \widehat{D}_\varepsilon$. Since $\widehat{\beta}(\eta)$ in $\mathbf{C}^2_{w'}$ is of the same form as β, but with $w = (0,0)$ replaced by $w' = (\eta, 0)$, and since $\dim \widehat{\sigma}(0) \le 1$, it follows from Case I that $\widehat{g}_\varepsilon(z',w')$ has a meromorphic extension to

$$\{0\} \times \{w' \in \widehat{q}(\rho - \eta) \mid \mathrm{Re}\, w_1' > -\alpha_\eta\} \quad \text{in } \mathbf{C}^{n+2},$$

and, in particular, to the origin $(z',w') = (0,0)$. Thus, $g(z,w)$ has a meromorphic extension to the origin $(z,w) = (0,0)$. Using the same method as in the proof of the second step in Case I, we get the proof of the second step in Case II; hence the theorem is true in Case II. □

3. Julia's Theorem

Let D be a domain in $\mathbf{C}^{n+1}$ containing the origin $(z,w) := (z_1,\ldots,z_n,\ w) = (0,0)$. Let $\mathcal{F}$ be a family of holomorphic functions in D. Consider the set

$$L_0 \ :\ z_j = 0 \ \ (j=1,\ldots,n), \qquad 0 < |w| < r$$

in D and assume that $\mathcal{F}$ is a normal family at each point of L_0; i.e., for any $p \in L_0$, there exists a connected neighborhood V of p in D such that $\mathcal{F}$ is normal on V. The definition of normality means that for any sequence $\{f_n\}_n \subset \mathcal{F}$, we can find a subsequence $\{f_{n_j}\}_j$ of $\{f_n\}_n$ such that $f_{n_j} \to f$ $(j \to \infty)$ uniformly on compact subsets of V where f is either a holomorphic function in V or $f \equiv \infty$ in V.

Under this assumption we have the following theorem.

THEOREM 4.3 (Julia [**32**]). *Suppose $\mathcal{F}$ is not normal at the origin $(z,w) = (0,0)$. Then, given any r' with $0 < r' < r$, there exists $\rho > 0$ such that, for any $z' = (z'_1,\ldots,z'_n) \in \mathbf{C}^n_z$ with $|z'_j| < \rho$ $(j=1,\ldots,n)$, there must be at least one point q on the set*

$$L_{z'} \ :\ z_j = z'_j \ \ (j=1,\ldots,n), \qquad |w| \le r'$$

such that $\mathcal{F}$ is not normal at q.

PROOF. Since $\{0\} \times \{|w| = r'\} \subset L_0$, it follows from our assumptions that there exist a $\rho > 0$ sufficiently small and $0 < \varepsilon < r'$ such that, setting $E_2 := \overline{\Delta} \times \Gamma^*$ where

$$\overline{\Delta} \ :\ |z_j| \le \rho \quad (j=1,\ldots,n), \qquad \Gamma^* \ :\ r' - \varepsilon \le |\,w\,| \le r' + \varepsilon,$$

we have $E_2 \subset D$ and $\mathcal{F}$ is normal on E_2. We prove that this $\rho > 0$ yields the conclusion of the theorem. For suppose not. Then there exists some $z'_0 = (z'_{01},\ldots,z'_{0n})$ with $|z'_{0j}| < \rho$ $(j=1,\ldots,n)$ such that $\mathcal{F}$ is normal on $L_{z'_0} = \{z'_0\} \times \{|w| \le r'\}$. We can thus find a neighborhood E_1 of $L_{z'_0}$ in D such that $\mathcal{F}$ is normal on E_1; indeed, we can take E_1 of the form $E_1 = \overline{\Delta}' \times \Gamma$, where

$$\overline{\Delta}' :\ |z_j - z'_{0j}| \le \rho'_0 \ (j=1,\ldots,n)\ , \qquad \Gamma :\ |w| \le r' + \varepsilon$$

with $\overline{\Delta}' \subset \Delta$. Now let $\{f_n\}_n$ be any sequence in $\mathcal{F}$. We can find a subsequence $\{f_{n_j}\}_j$ of $\{f_n\}_n$ such that $f_{n_j} \to f$ as $j \to \infty$ uniformly on $E_1 \cup E_2$. If f is holomorphic on E_2, then from Cauchy's integral formula applied to f_{n_j} in the polydisk $\Delta \times \Gamma \subset D$, we conclude that $f_{n_j} \to f$ as $j \to \infty$ uniformly on $\Delta \times \Gamma$. This contradicts our assumption that $\mathcal{F}$ is not normal at the origin $(0,0)$. Thus we may assume that the limiting function $f \equiv \infty$ on E_2 and hence on $E_1 \cup E_2$ (since $E_1 \cup E_2$ is connected). If infinitely many of the f_{n_j} are non-vanishing on $\Delta \times \Gamma$, it follows that $f_{n_j} \to \infty$ uniformly on $\Delta \times \Gamma$, which also contradicts our assumption at $(0,0)$. Thus there exist a subsequence $\{g_l\}_l$ of $\{f_{n_j}\}_j$ and points (a_l, b_l) with $a_l \in \Delta \backslash \overline{\Delta}'$ and $|b_l| < r' - \varepsilon$ such that $g_l(a_l,b_l) = 0$. On the other hand, since $g_l \to \infty$ as $l \to \infty$ on E_2, say $|g_l| \ge 1$ on E_2 for all $l \ge l_0$, it follows that for any $z' \in \Delta$, each g_l for $l \ge l_0$ vanishes at some point $(z', w_l(z')) \in \mathbf{C}^{n+1}$ with $|w_l(z')| < r' - \varepsilon$ by the Weierstrass preparation theorem. In particular, if we fix $z' = z'_0$ in $\overline{\Delta}'$ and consider a limit point w^* of $\{w_l(z'_0)\}_{l \ge l_0}$ on $L_{z'_0}$, then $|w^*| \le r' - \varepsilon$. We conclude that $\{g_l\}$ cannot converge uniformly to ∞ on any neighborhood of (z'_0, w^*). This contradicts our assumption that $g_l \to \infty$ as $l \to \infty$ uniformly on E_1, and proves the theorem. □

4.1.2. Definition of Pseudoconvex Domains. Motivated by the three theorems in the previous section, we give three equivalent definitions of pseudoconvexity, following the ideas of Oka in **[52]**.[2] We recall from 1.3.5 that a one-to-one holomorphic mapping T of a domain $D \subset \mathbf{C}^n$ onto a domain $D' \subset \mathbf{C}^n$ with a holomorphic inverse T^{-1} is called a biholomorphic mapping; if $D = D'$, we call T an automorphism of D.

Definition A. Let D be a domain in $\mathbf{C}^n$ $(n \geq 2)$ and let $P = (a_1, \ldots, a_n) \in \partial D$. We say that D **satisfies the continuity theorem of type A at the boundary point** P if the following holds: under the assumption that there exists $r > 0$ such that the punctured disk

$$L_a \ : \ z_j = a_j \ (j = 1, \ldots, n-1) \ , \quad 0 < |z_n - a_n| < r$$

is contained in D, we have, for any r' satisfying $0 < r' < r$, that there exists $\rho > 0$ such that for each $(z'_1, \ldots, z'_{n-1}) \in \mathbf{C}^{n-1}$ with $|z'_j - a_j| < \rho$ $(j = 1, \ldots, n-1)$, the disk

$$L_{z'} \ : \ z_j = z'_j \ (j = 1, \ldots, n-1) \ , \quad |z_n - a_n| < r'$$

intersects ∂D.

Now if $P \in \partial D$ satisfies the continuity theorem of type A and if this theorem remains valid under any biholomorphic transformation T of a neighborhood V of P in $\mathbf{C}^n$ (i.e., if $T(D \cap V)$ satisfies the continity theorem of type A at the point $T(P)$), then we say that D is **pseudoconvex of type A at** P.

Finally, if D is pseudoconvex of type A at all boundary points of D, then we say that D is a **pseudoconvex domain of type A**.

Definition B. Let D be a domain in $\mathbf{C}^n$ $(n \geq 2)$ and let $P = (a_1, \ldots, a_n)$ be a point in ∂D. In $\mathbf{C}^2$ with variables z_{n-1} and z_n, we fix a point $Q = (b_{n-1}, b_n)$ such that $Q \neq (a_{n-1}, a_n)$, and we let $r > 0$ be the Euclidean distance between (a_{n-1}, a_n) and Q. For $0 < \rho < r$, we define the set β_ρ in $\mathbf{C}^2$ by the following inequalities:

$$\begin{array}{rcl} |z_{n-1} - b_{n-1}|^2 + |z_n - b_n|^2 & > & r^2, \\ |z_{n-1} - a_{n-1}|^2 + |z_n - a_n|^2 & < & \rho^2. \end{array} \tag{4.10}$$

If for each $Q \in \mathbf{C}^2$ and each $0 < \rho < r$, the set B in $\mathbf{C}^2$ defined by

$$B \ : \ z_j = a_j \ (j = 1, \ldots, n-2) \ , \ (z_{n-1}, z_n) \in \beta_\rho$$

is not contained in D, we say that D **satisfies the continuity theorem of type B at the boundary point** P **of** D.

If $P \in \partial D$ satisfies the continuity theorem of type B and if this theorem remains valid under any biholomorphic transformation of a neighborhood of P in $\mathbf{C}^n$, then we say that D **is pseudoconvex of type B at** P.

Finally, if D is pseudoconvex of type B at every boundary point of D, then we say that D is a **pseudoconvex domain of type B**.

Definition C. Let D be a domain in $\mathbf{C}^n$ $(n \geq 2)$. Let $P = (a_1, \ldots, a_n)$ be a point in $\mathbf{C}^n$ and let Δ be a polydisk centered at P with polyradius r_j $(j = 1, \ldots, n)$:

$$\Delta \ : \ |z_j - a_j| < r_j \ (j = 1, \ldots, n).$$

[2] In part I we are restricted to univalent domains in $\mathbf{C}^n$. The definition of pseudoconvexity stated here will later be extended to *unramified* covering domains over $\mathbf{C}^n$ without any change.

For $0 < r_j' < r_j$, $j = 1, \dots, n$, we consider the following two sets in Δ:

$$\begin{aligned} E_1 &: |z_j - a_j| < r_j' \ (j = 1, \dots, n-1), \quad |z_n - a_n| < r_n, \\ E_2 &: |z_j - a_j| < r_j \ (j = 1, \dots, n-1), \quad r_n' < |z_n - a_n| < r_n, \end{aligned} \tag{4.11}$$

and we set $E := E_1 \cup E_2$. If for any such Δ and E in $\mathbf{C}^n$, $E \subset D$ implies that $\Delta \subset D$, then we say that D **satisfies the continuity theorem of type C**.

Next, if for any polydisk K, $K \cap D$ satisfies the continuity theorem of type C and if the image of $K \cap D$ under any biholomorphic transformation from $K \cap D$ into $\mathbf{C}^n$ also satisfies the continuity theorem of type C, then we say that D **is a pseudoconvex domain of type C**.

Note that we assume $n \geq 2$. In the case $n = 1$, we say that any domain in $\mathbf{C}$ is a pseudoconvex domain.[3]

At first glance, these definitions of pseudoconvexity may seem rather difficult to understand. We remark that the notion of a pseudoconvex domain of type A or B is a *local* notion dependent on the boundary; the definition gives a property which is to be satisfied at each boundary point P of the domain. On the other hand, pseudoconvexity of type C is a global property of the domain.

4.1.3. Equivalence of Definitions. In this section we show that the three definitions of pseudoconvex domains in $\mathbf{C}^n$, $n \geq 2$, are equivalent.

1. Pseudoconvex Domains of Type A are of Type B.

PROOF. Let D be a domain in $\mathbf{C}^n$ of type A. Let P be any point in ∂D; we show that D satisfies the continuity theorem of type B at P. For simplicity we assume that P is the origin $z = 0$ in $\mathbf{C}^n$ and the set β_ρ defined by (4.10) is of the form

$$\beta_\rho \ : \ |z_{n-1} + r|^2 + |z_n|^2 > r^2, \qquad |z_{n-1}|^2 + |z_n|^2 < \rho^2.$$

Our claim is thus to show that the set B in $\mathbf{C}^n$ defined by

$$B \ : \ z_j = 0 \ (j = 1, \dots, n-2), \qquad (z_{n-1}, z_n) \in \beta_\rho$$

is not contained in D. We prove this by contradiction; hence we assume $B \subset D$. Then the subset L_0 of B defined by

$$L_0 \ : \ z_j = 0 \ (j = 1, \dots, n-1), \qquad 0 < |z_n| < \rho$$

is contained in D and $0 \in \partial D$. Further, for any $0 < s < \rho/2$ the set

$$z_j = 0 \ (j = 1, \dots, n-2), \quad z_{n-1} = s, \quad |z_n| < \rho/2$$

is contained in D. It follows that D does not satisfy the continuity theorem of type A at the origin. This is a contradiction. □

2. Pseudoconvex Domains of Type B are of Type C.

PROOF. Let D be a pseudoconvex domain in $\mathbf{C}^n$ of type B. We prove the assertion by contradiction; i.e., we assume that D does not satisfy the continuity theorem of type C. Thus there exist a polydisk Δ centered at a point P in $\mathbf{C}^n$ and a set $E_1 \cup E_2$ defined by (4.11) in Δ such that $E_1 \cup E_2 \subset D$ but $\Delta \not\subset D$. We fix a

[3]Any function of one complex varable z may be considered as a function of two complex variables z and w which is independent of w. Thus, it is natural to consider any domain D in the complex plane $\mathbf{C}_z$ as a domain $D \times \mathbf{C}_w$ in $\mathbf{C}^2 = \mathbf{C}_z \times \mathbf{C}_w$. Since $D \times \mathbf{C}_w$ is proved to be a pseudoconvex domain of type C in $\mathbf{C}^2$, the case $n = 1$ need not be treated as an exceptional case.

point $z' = (z_1', \ldots, z_n')$ in $\Delta \setminus D$ through this proof. Again for simplicity we may assume that P is the origin $z = 0$ in $\mathbf{C}^n$, so that

$$\begin{aligned} \Delta &: |z_j| < r_j \ (j = 1, \ldots, n), \\ E_1 &: |z_j| < r_j' \ (j = 1, \ldots, n-1), \ |z_n| < r_n, \\ E_2 &: |z_j| < r_j \ (j = 1, \ldots, n-1), \ r_n' < |z_n| < r_n. \end{aligned}$$

Thus the point $z' \in \Delta \setminus D$ satisfies

$$r_j' \leq |z_j'| < r_j \quad \text{for some } j = 1, \ldots, n-1, \qquad |z_n'| < r_n'.$$

For simplicity we assume that $r_{n-1}' \leq |z_{n-1}'| < r_{n-1}$. We fix $c > 0$ sufficiently large so that

$$\left(\frac{c}{r_{n-1}}\right)^2 + r_n^2 < \left(\frac{c}{|z_{n-1}'|}\right)^2 + |z_n'|^2. \tag{4.12}$$

We then consider the following automorphism T of $\mathbf{C}_n' := \mathbf{C}^n \setminus \{z_{n-1} = 0\}$:

$$T : w_j = z_j \ (j = 1, \ldots, n \ ; \ j \neq n-1), \qquad w_{n-1} = \frac{c}{z_{n-1}}.$$

We let G and S denote the sections of the image $T(\Delta \cap D \cap \mathbf{C}_n')$ and $T(\Delta \cap \partial D \cap \mathbf{C}_n')$ over $w_j = z_j'$ $(j = 1, \ldots, n-2)$. These sets are subsets of $\mathbf{C}^2$ in the variables w_{n-1} and w_n. We set

$$\eta_0 := \max \{|w_{n-1}|^2 + |w_n|^2 \mid (w_{n-1}, w_n) \in \overline{S}\} < \infty$$

and take a point $(w_{n-1}^0, w_n^0) \in \partial S$ such that $|w_{n-1}^0|^2 + |w_n^0|^2 = \eta_0$. Then $Q := (z_1', \ldots, z_{n-2}', w_{n-1}^0, w_n^0) \in \partial T(\Delta \cap D \cap \mathbf{C}_n')$. We see from (4.12) that, if we take $\rho > 0$ sufficiently small and define the set β_ρ in $\mathbf{C}^2$ by

$$\beta_\rho : |w_{n-1}|^2 + |w_n|^2 > \eta_0^2, \quad |w_{n-1} - w_{n-1}^0|^2 + |w_n - w_n^0|^2 \leq \rho^2,$$

then β_ρ is contained in G. Consequently, the set

$$B : w_j = z_j' \ (j = 1, \ldots, n-2), \qquad (w_{n-1}, w_n) \in \beta_\rho$$

is contained in $T(\Delta \cap D \cap \mathbf{C}_n')$, so that $T(\Delta \cap D \cap \mathbf{C}_n')$ does not satisfy the continuity theorem of type B at the boundary point Q. This contradicts our assumption. □

3. Pseudoconvex Domains of Type C are of Type A.

PROOF. Let D be a pseudoconvex domain in $\mathbf{C}^n$ of type C. Let $P = (a_1, \ldots, a_n)$ be a point in ∂D; we prove that D satisfies the continuity theorem of type A at P. Suppose not. Then we can find a set

$$L_a : z_j = a_j \ (j = 1, \ldots, n-1), \quad 0 < |z_n - a_n| < r$$

contained in D and a number r' with $0 < r' < r$ such that, for *any* given $0 < \rho \ll 1$, we can find a point $(z_1', \ldots, z_{n-1}') \in \mathbf{C}^{n-1}$ with $|z_j' - a_j| < \rho$ $(j = 1, \ldots, n-1)$ such that

$$L_{z'} : z_j = z_j' \ (j = 1, \ldots, n-1), \qquad |z_n - a_n| \leq r'$$

is contained in D.

On the other hand, since $\{(a_1, \ldots, a_{n-1}, z_n) | \ |z_n - a_n| = r'\} \subset\subset D$, we can find $\rho > 0$ sufficiently small so that the subset

$$\sigma : |z_j - a_j| < \rho \ (j = 1, \ldots, n-1), \qquad |z_n - a_n| = r'$$

of $\mathbf{C}^n$ is contained in D. Fixing this $\rho > 0$, we can find a point $(z'_1 \dots, z'_{n-1}) \in \mathbf{C}^{n-1}$ with $|z'_j - a_j| < \rho$ $(j = 1, \dots, n-1)$ and with the property that $L_{z'} \subset D$. Since $L_{z'}$ and σ are compact in D, our assumption that D satisfies the continuity theorem of type C implies that the polydisk Δ centered at P defined by

$$\Delta : \ |z_j - a_j| < \rho \ (j = 1, \dots, n-1), \qquad |z_n - a_n| < r'$$

is contained in D, which contradicts $P \in \partial D$. □

Using these three equivalent conditions, we call a domain in $\mathbf{C}^n$ of type A, B, or C a **pseudoconvex domain**. Theorem 4.1 (Hartogs) states that any domain of holomorphy is a pseudoconvex domain.

As with the definition of a domain of holomorphy in 1.5.2, we can define a domain of meromorphy: if a domain D in $\mathbf{C}^n$ admits at least one meromorphic function $f(z)$ which cannot extend meromorphically across any point of ∂D, then D is called a **domain of meromorphy**. Theorem 4.2 (Levi) states that any domain of meromorphy is a pseudoconvex domain.

Let $\mathcal{F}$ be a family of holomorphic functions in $D \subset \mathbf{C}^n$. The set D' consisting of all points z in D such that $\mathcal{F}$ is normal in a neighborhood of z is called the **domain of normality** of $\mathcal{F}$. Theorem 4.3 (Julia) states that, if D is a pseudoconvex domain, so is D'.[4]

Let $D \subset \mathbf{C}^n$ and let $\{f_j\}_{j=1,2,\dots}$ be a sequence of holomorphic functions in D. The set D' of points z in D such that $\{f_j\}_j$ converges uniformly in a neighborhood of z is called a **domain of uniform convergence** of $\{f_j\}_j$. Clearly such a domain D' satisfies the continuity theorem of type C. Therefore, if D is a pseudoconvex domain, so is D'.

4.1.4. Properties of Pseudoconvex Domains. We list some elementary properties of pseudoconvex domains which follow from the definition.

1. If D_1 and D_2 are pseudoconvex domains in $\mathbf{C}^n$, then $D_1 \cap D_2$ is a pseudoconvex domain.

2. Let $I = \{\iota\}$ be any index set and let D_ι $(\iota \in I)$ be a family of pseudoconvex domains. Then the interior of $\bigcap_{\iota \in I} D_\iota$ is a pseudoconvex domain.

3. Let D_j $(j = 1, 2, \dots)$ be a sequence of pseudoconvex domains in $\mathbf{C}^n$ such that $D_j \subset D_{j+1}$. Then $D_0 := \bigcup_{j=1}^{\infty} D_j$ is a pseudoconvex domain.

4. Let D be a pseudoconvex domain in $\mathbf{C}^n$ and let L be any r-dimensional $(0 < r < n)$ complex hyperplane. Then each connected component of $D \cap L$ is a pseudoconvex domain in $\mathbf{C}^r = L$.

5. (Invariance under holomorphic mappings) Let T be a biholomorphic mapping from a domain D in $\mathbf{C}^n$ onto a domain D' in $\mathbf{C}^n$. Then D is pseudoconvex if and only if D' is pseudoconvex.

[4] In Oka [**43**] the definition of a normal family of analytic hypersurfaces in a domain in $\mathbf{C}^n$ was given, and it was proved that the domain of normality of such a family in a pseudoconvex domain is also a pseudoconvex domain. This study was developed in his last paper [**54**].

4.2. Pseudoconvex Domains with Smooth Boundary

4.2.1. Levi Problem. E. E. Levi [**35**][5] was the first to pose the problem of determining whether a pseudoconvex domain is necessarily a domain of holomorphy. He restricted his study of this problem to the consideration of a neighborhood of a boundary point of a domain in $\mathbf{C}^2$ with smooth boundary. In this section we extend his results to $\mathbf{C}^n$.

In $\mathbf{C}^n$ with variables $z_1, \dots, z_n$, we let $z_j = x_j + iy_j$ $(j = 1, \dots, n)$. Let $D \subset \mathbf{C}^n$ be a domain and let $p \in \partial D$. If there exists a C^2 function $\varphi(z)$ in a neighborhood δ of p such that

$$\begin{aligned} \delta \cap D &= \{z \in \delta \mid \varphi(z) < 0\}, \\ \delta \cap \partial D &= \{z \in \delta \mid \varphi(z) = 0\}, \end{aligned}$$

and $\nabla\varphi := (\partial\varphi/\partial z_1, \dots, \partial\varphi/\partial z_n) \neq 0$ on $\delta \cap \partial D$, then we say that D has **smooth boundary** at p. We call $\varphi(z)$ a **defining function** for $\delta \cap D$ at p.

We have the following proposition.

PROPOSITION 4.1. *Let $D \subset \mathbf{C}^n$ and let $p \in \partial D$. Assume that D has smooth boundary at p. Then:*

1. *If D is pseudoconvex at p, then there does not exist a nonsingular, one-dimensional analytic curve C in a neighborhood δ of p such that C contains p and $C \setminus \{p\} \subset D$.*
2. *If there exists an $(n-1)$-dimensional analytic hypersurface S in a neighborhood δ of p such that S contains p and $S \subset \delta \setminus D$, then D is pseudoconvex at p.*

PROOF. Let $D \subset \mathbf{C}^n$ be a pseudoconvex domain and let $p \in \partial D$. We prove 1 by contradiction; i.e., we assume that there exists a nonsingular analytic curve C in a neighborhood δ of p such that $p \in C$ and $C \setminus \{p\} \subset D$. By shrinking δ, if necessary, we can find a one-to-one holomorphic mapping T from δ onto a neighborhood δ^0 of the origin $z = 0$ in $\mathbf{C}^n$ such that $T(p) = 0$ and $C^0 := T(C) = \{z \in \delta^0 \mid z_j = 0,\ j = 1, \dots, n-1\}$. We set $\varphi^0 := \varphi \circ T^{-1}$ in δ^0. By hypothesis, $\varphi^0(0, \dots, 0, z_n)$ attains a local maximum at $z_n = 0$; thus it follows that $\partial\varphi^0/\partial z_n = 0$ at $z = 0$. Since $\nabla\varphi \neq 0$ on $\delta \cap \partial D$, we may assume that $\partial\varphi^0/\partial y_1 \neq 0$ at $z = 0$, so that $v^0 := T(D \cap \delta)$ can be written in the form

$$v^0 = \{z \in \delta^0 \mid y_1 < \xi(x_1, z_2, \dots, z_n)\},$$

where ξ is a C^2 function defined in a neighborhood $\delta' \subset \mathbf{R}^{2n-1}$ of the point $(x_1, z_2, \dots, z_n) = (0, 0, \dots, 0)$. From the assumption that $C^0 \setminus \{0\} \subset v^0$, we can find $r > 0$ sufficiently small so that the set

$$z_j = 0 \quad (j = 1, \dots, n-1), \qquad 0 < |z_n| \leq r$$

is contained in v^0. Hence

$$0 < \xi(0, \dots, 0, z_n), \qquad 0 < |z_n| \leq r.$$

It follows that for any $\varepsilon > 0$, all points of the form

$$(-i\varepsilon, 0, \dots, 0, z_n) \in \mathbf{C}^n$$

with $|z_n| < r$ lie in v^0. This contradicts the assumption that D is pseudoconvex of type A at $z = 0$, and 1 is proved.

[5] E. E. Levi died in 1917 at the age of 34.

To prove 2, let $D \subset \mathbf{C}^n$ and $p \in \partial D$. Suppose that we can find an analytic hypersurface S in a neighborhood δ of p such that $p \in S \subset \delta \setminus D$. We show that D satisfies the continuity theorem of type A at p. For simplicity we take $p = 0$. Assume that the set

$$L_0 \ : \ z_j = 0 \ \ (j = 1, \ldots, n-1), \quad 0 < |z_n| \leq r$$

is contained in $v := \delta \cap D$. Fix r' with $0 < r' < r$. We then take $\rho > 0$ and $0 < \varepsilon < r'$ such that

$$\Gamma^* : \ |z_j| \leq \rho \ (j = 1, \ldots, n-1), \qquad r' - \varepsilon \leq |z_n| \leq r' + \varepsilon$$

is contained in v. Since S is an analytic hypersurface in the polydisk

$$\Lambda \ : \ |z_j| < \rho \ (j = 1, \ldots, n-1), \qquad |z_n| < r' + \varepsilon$$

with $0 \in S$ and $S \cap \Gamma^* = \emptyset$, it follows from Remark 2.3 that for any z_j' ($j = 1, \ldots, n-1$) with $|z_j'| < \rho$, we have

$$S \cap \{(z_1', \ldots, z_{n-1}', z_n) : |z_n| < r' - \varepsilon\} \neq \emptyset.$$

Since $S \subset \delta \setminus D$, we see that D satisfies the continuity theorem of type A at p, and 2 is proved. □

REMARK 4.1. Let $D \subset \mathbf{C}^n$ be a domain and let $p \in \partial D$. In the proof of 2, the condition that ∂D is smooth at p was not used. Thus we have the following result. *Let $D \subset \mathbf{C}^n$ be a domain. If for each $p \in \partial D$ there exists an analytic hypersurface S in a neighborhood δ of p which contains p and lies entirely outside of D, i.e., $S \subset \delta \setminus D$, then D is a pseudoconvex domain.*

We now write down the conditions at p described in Proposition 4.1 in terms of a defining function $\varphi(z)$ of $\delta \cap D$ at p. For simplicity, we take $p = 0$ and write $\varphi(z) = \varphi(z_1, \ldots, z_n, \overline{z}_1, \ldots, \overline{z}_n)$. Since $\varphi(z)$ is of class C^2, we consider the following expansion at $z = 0$:

$$\begin{aligned} \varphi(z) \ &= \ 2\Re\left(\sum_{j=1}^n \frac{\partial \varphi}{\partial z_j}(0) z_j\right) + 2\Re\left(\sum_{j \leq k} \frac{\partial^2 \varphi}{\partial z_j \partial z_k}(0) z_j z_k\right) \\ &\quad + \sum_{j,k=1}^n \frac{\partial^2 \varphi}{\partial z_j \partial \overline{z}_k}(0) z_j \overline{z}_k + o(\|z\|^2), \end{aligned} \tag{4.13}$$

where $\|z\|^2 = \sum_{j=1}^n |z_j|^2$, $\Re(\alpha)$ is the real part of the complex number α, and $\lim_{r \to 0} o(r^2)/r^2 = 0$.

We have the following proposition.

PROPOSITION 4.2. *Let $D \subset \mathbf{C}^n$ and let $p \in \partial D$. Assume that D has smooth boundary at p. Let $\varphi(z)$ be a defining function of $\delta \cap D$ at p, where δ is a neighborhood of p.*

1. *If D is pseudoconvex at p, then for any $\mathbf{a} = (a_1, \ldots, a_n) \in \mathbf{C}^n$ satisfying*

$$\sum_{j=1}^n \frac{\partial \varphi}{\partial z_j}(p) a_j = 0, \tag{4.14}$$

we have

$$\sum_{j,k=1}^{n} \frac{\partial^2 \varphi}{\partial z_j \partial \bar{z}_k}(p) a_j \bar{a}_k \geq 0. \tag{4.15}$$

2. *If we have, for any* $\mathbf{a} \neq 0$ *satisfying (4.14),*

$$\sum_{j,k=1}^{n} \frac{\partial^2 \varphi}{\partial z_j \partial \bar{z}_k}(p) a_j \bar{a}_k > 0, \tag{4.16}$$

then D *is pseudoconvex at* p.

By a simple calculation we see that conditions 1 and 2 depend on neither the choice of defining function φ nor on a biholomorphic transformation of a neighborhood of p. Equation (4.14) states that $\mathbf{a}$ belongs to the **complex tangent space of** ∂D **at** p.

PROOF. For simplicity we take $p = 0$. To prove 1, we assume that D is pseudoconvex at 0. If assertion 1 is not true, then there exists an $\mathbf{a} = (a_1, \dots, a_n) \neq 0$ which satisfies (4.14) and

$$A := \sum_{j,k=1}^{n} \frac{\partial^2 \varphi}{\partial z_j \partial \bar{z}_k}(0) a_j \bar{a}_k < 0.$$

Given $\mathbf{b} = (b_1, \dots, b_n)$ (which will be determined later), we consider the analytic curve C in $\mathbf{C}^n$ passing through 0 defined by

$$C \; : \; z_j = a_j t + b_j t^2 \qquad (j = 1, \dots, n), \qquad t \in \mathbf{C}.$$

From (4.13), for $z = t\mathbf{a} + t^2\mathbf{b}$ with t sufficiently small, we have

$$\begin{aligned} \varphi(z) &= 2\Re\left\{\left(\sum_{j=1}^{n} \frac{\partial \varphi}{\partial z_j}(0) b_j + \sum_{j \leq k}^{n} \frac{\partial^2 \varphi}{\partial z_j \partial z_k}(0) a_j a_k\right) t^2\right\} \\ &\quad + \left(\sum_{j,k=1}^{n} \frac{\partial^2 \varphi}{\partial z_j \partial \bar{z}_k}(0) a_j \bar{a}_k\right) |\,t\,|^2 + o(|\,t\,|^2). \end{aligned}$$

Since $\nabla\varphi(0) \neq 0$, we can choose $\mathbf{b}$ to make the coefficient of t^2 on the right-hand side vanish. Then we have

$$\varphi(t\mathbf{a} + t^2\mathbf{b}) = A|\,t\,|^2 + o(|\,t\,|^2) < 0, \qquad 0 < |t| << 1.$$

Thus we can find a neighborhood δ of 0 in $\mathbf{C}^n$ such that $(C \cap \delta) \setminus \{0\} \subset D$. This contradicts 1 of Proposition 4.1, and assertion 1 is proved.

To prove 2, we apply 2 of Proposition 4.1. We consider the following algebraic hypersurface of degree 2:

$$S : \sum_{j=1}^{n} \frac{\partial \varphi}{\partial z_j}(0) z_j + \sum_{j \leq k}^{n} \frac{\partial^2 \varphi}{\partial z_j \partial z_k}(0) z_j z_k = 0,$$

which contains 0. It suffices to show that there exists a neighborhood δ of 0 such that $S \cap \delta \subset \delta \setminus D$. To prove this we consider the complex tangent space L of ∂D at 0 defined by

$$L : \sum_{j=1}^{n} \frac{\partial \varphi}{\partial z_j}(0) z_j = 0.$$

By assumption there exist a neighborhood δ_1 of 0 in $\mathbf{C}^n$ and $\varepsilon > 0$ such that

$$\sum_{j,k=1}^{n} \frac{\partial^2 \varphi}{\partial z_j \partial \overline{z}_k}(0) z_j \overline{z}_k \geq \varepsilon \|z\|^2, \qquad z \in L \cap \delta_1.$$

For each $z' \in S$, we consider the nearest point $Z = Z(z')$ to z' on L; then $z' = Z + c(z')$, where $c(z') = o(\|z'\|)$ at $z' = 0$. It follows that

$$\begin{aligned} \varphi(z') &= \sum_{j,k=1}^{n} \frac{\partial^2 \varphi}{\partial z_j \partial \overline{z}_k}(0) z'_j \overline{z}'_k + o(\|z'\|^2) \\ &\geq \varepsilon \|Z\|^2 + o(\|Z\|^2). \end{aligned}$$

We can thus find a neighborhood $\delta \subset \delta_1$ of 0 in $\mathbf{C}^n$ such that $\varphi(z') > 0$ on $\delta \cap S$ except for $z' = 0$, and assertion 2 is proved. □

REMARK 4.2. The proof of assertion 2 implies the following fact: Let $D \subset \mathbf{C}^n$ be a domain with smooth boundary at $p \in \partial D$. Assume that for any $\mathbf{a} = (a_1, \ldots, a_n) \neq 0$ satisfying (4.14), we have (4.16). By continuity there exists a neighborhood δ of p such that for each $q \in \delta \cap (\partial D)$ there exists an analytic hypersurface S_q which passes through q and lies completely outside of D, i.e., $S_q \setminus \{q\} \subset \delta \setminus \overline{D}$. Furthermore, we can assume that δ is a ball centered at p and that S_q is an analytic hypersurface in δ. It follows that $D \cap \delta$ is a domain of holomorphy; i.e., *D is locally a domain of holomorphy at p.*

The conditions in Proposition 4.2 are called **Levi's conditions**. We note that Levi's condition is not linear. In $\mathbf{C}^2$ with variables z and w, the vector $\mathbf{a} = (a_1, a_2)$ in (4.14) is uniquely determined (up to multiplicative constants) by the equation

$$a_1 \frac{\partial \varphi}{\partial z}(p) + a_2 \frac{\partial \varphi}{\partial w}(p) = 0.$$

Therefore, if we set

$$\begin{aligned} L(\varphi) &= \begin{vmatrix} 0 & \partial\varphi/\partial z & \partial\varphi/\partial w \\ \partial\varphi/\partial\overline{z} & \partial^2\varphi/\partial z\partial\overline{z} & \partial^2\varphi/\partial w\partial\overline{z} \\ \partial\varphi/\partial\overline{w} & \partial^2\varphi/\partial z\partial\overline{w} & \partial^2\varphi/\partial w\partial\overline{w} \end{vmatrix} \\ &= \frac{\partial^2\varphi}{\partial w\partial\overline{w}} \left|\frac{\partial\varphi}{\partial z}\right|^2 - 2\Re\left\{\frac{\partial^2\varphi}{\partial\overline{w}\partial z}\frac{\partial\varphi}{\partial w}\frac{\partial\varphi}{\partial\overline{z}}\right\} + \left|\frac{\partial\varphi}{\partial w}\right|^2 \frac{\partial^2\varphi}{\partial z\partial\overline{z}}, \end{aligned}$$

then 1 and 2 of Proposition 4.2 may be restated as follows:

1. If D is pseudoconvex at p, then $L(\varphi) \geq 0$ at p.
2. If $L(\varphi) > 0$ at p, then D is pseudoconvex at p.

Originally, E. E. Levi wrote down the operator $L(\varphi)$ using the coordinates x_1, x_2, y_1, y_2, where $z = x_1 + ix_2$, $w = y_1 + iy_2$, via:

$$\begin{aligned} &\left(\frac{\partial^2\varphi}{\partial x_1^2} + \frac{\partial^2\varphi}{\partial x_2^2}\right)\left[\left(\frac{\partial\varphi}{\partial y_1}\right)^2 + \left(\frac{\partial\varphi}{\partial y_2}\right)^2\right] + \left(\frac{\partial^2\varphi}{\partial y_1^2} + \frac{\partial^2\varphi}{\partial y_2^2}\right)\left[\left(\frac{\partial\varphi}{\partial x_1}\right)^2 + \left(\frac{\partial\varphi}{\partial x_2}\right)^2\right] \\ &\quad -2\left(\frac{\partial\varphi}{\partial x_1}\frac{\partial\varphi}{\partial y_1} + \frac{\partial\varphi}{\partial x_2}\frac{\partial\varphi}{\partial y_2}\right)\left(\frac{\partial^2\varphi}{\partial x_1\partial y_1} + \frac{\partial^2\varphi}{\partial x_2\partial y_2}\right) \\ &\quad -2\left(\frac{\partial\varphi}{\partial x_1}\frac{\partial\varphi}{\partial y_2} - \frac{\partial\varphi}{\partial x_2}\frac{\partial\varphi}{\partial y_1}\right)\left(\frac{\partial^2\varphi}{\partial x_1\partial y_2} - \frac{\partial^2\varphi}{\partial x_2\partial y_1}\right). \end{aligned}$$

We call $L(\varphi)$ the **Levi form** of $\varphi(z,w)$. We note that whether $L(\varphi)$ is positive, negative or 0 at $p \in \partial D$ does not depend on the choice of defining function φ nor on a biholomorphic mapping of a neighborhood of p.

4.2.2. Levi Flat Surfaces. Let $D \subset \mathbf{C}^n$ be a domain and let φ be a real-valued C^2 function in D. We set $\Sigma := \{z \in D \,|\varphi(z) = 0\}$. We assume that $\nabla\varphi \neq 0$ at any point of Σ, so that Σ is a real $(2n-1)$-dimensional smooth hypersurface in D. If both of the domains $\{\varphi(z) < 0\}$ and $\{\varphi(z) > 0\}$ are pseudoconvex at each point of Σ, then the hypersurface Σ is called **Levi flat**. E. E. Levi [**35**] was the first one who studied such hypersurfaces in $\mathbf{C}^2$. In this section, we follow the ideas of E. Cartan [**6**] and study Levi flat hypersurfaces in the case where $\varphi(z)$ is real analytic in $D \subset \mathbf{C}^n$ $(n \geq 2)$.[6]

Let D be a domain in $\mathbf{C}^n$ with variables $z_j = x_j + iy_j$ $(j = 1, \dots, n)$. Let φ be a real-valued real analytic function in D; to be precise, we write

$$\varphi(z, \overline{z}) := \varphi(z_1, \dots, z_n, \overline{z}_1, \dots, \overline{z}_n).$$

We set $\Sigma := \{z \in D \mid \varphi(z,\overline{z}) = 0\}$, $D^+ := \{z \in D \mid \varphi < 0\}$ and $D^- := \{z \in D \mid \varphi > 0\}$. We assume that $\nabla\varphi \neq 0$ at all points of Σ.

Then we have the following lemma.

LEMMA 4.3. *Let $z^0 \in \Sigma$. Then we have:*

1. *Assume that there exists an analytic hypersurface S in a neighborhood of z^0 such that $z^0 \in S \subset \Sigma$. Then S is unique and is given by the equation*
$$\varphi(z, \overline{z}^0) = 0$$
in a neighborhood of z^0.
2. *Assume that $\partial\varphi/\partial z_n \neq 0$ at z^0. If both D^+ and D^- are pseudoconvex at z^0, then, at z^0,*

$$\begin{aligned} L_{jk}(z^0, \overline{z}^0) :=& \frac{\partial^2\varphi}{\partial z_j \partial \overline{z}_k}\frac{\partial\varphi}{\partial z_n}\frac{\partial\varphi}{\partial \overline{z}_n} - \frac{\partial^2\varphi}{\partial z_j \partial \overline{z}_n}\frac{\partial\varphi}{\partial z_n}\frac{\partial\varphi}{\partial \overline{z}_k} \\ &- \frac{\partial^2\varphi}{\partial z_n \partial \overline{z}_k}\frac{\partial\varphi}{\partial z_j}\frac{\partial\varphi}{\partial \overline{z}_n} + \frac{\partial^2\varphi}{\partial z_n \partial \overline{z}_n}\frac{\partial\varphi}{\partial z_j}\frac{\partial\varphi}{\partial \overline{z}_k} = 0 \quad (j,k = 1, \dots, n-1). \end{aligned} \tag{4.17}$$

PROOF. We use the following elementary fact about real-analytic functions based on the Taylor expansion: *Let $h(z,\overline{z})$ be a real analytic function in a neighborhood δ of a point a in $\mathbf{C}^n$ $(n \geq 1)$. If $h(z, \overline{z}) = 0$ for $z \in \delta$, then $h(z, \overline{w}) = 0$ for $(z,w) \in \delta \times \delta$. In particular, $h(z, \overline{a}) = 0$ for $z \in \delta$.* Equivalently, *if $f(z,w)$ is holomorphic in $\delta \times \overline{\delta}$ in $\mathbf{C}^{2n}$ with $f(z,\overline{z}) = 0$ for $z \in \delta$ in $\mathbf{C}^n$, then $f(z,w) \equiv 0$ in $\delta \times \overline{\delta}$.*

For the proof of 1, let $z^0 = (z_1^0, \dots, z_n^0) \in \Sigma$ and let S be an analytic hypersurface in a neighborhood δ centered at z^0 such that $z^0 \in S \subset \Sigma$. We may assume that $\partial\varphi/\partial z_n \neq 0$ at any $z \in S$, and that S can be described by

$$S \; : \; z_n = \xi(z_1, \dots, z_{n-1}),$$

where ξ is a holomorphic function in a neighborhood $\underline{\delta}$ of $(z_1^0, \dots, z_{n-1}^0)$ in $\mathbf{C}^{n-1}$. Since $S \subset \Sigma$, we have

$$\varphi(z_1, \dots, z_{n-1}, \xi(z_1, \dots, z_{n-1}), \overline{z}_1, \dots, \overline{z}_{n-1}, \overline{\xi(z_1, \dots, z_{n-1})}) = 0$$

[6] The description of Levi flat hypersurfaces in the case where $\varphi(z)$ is of class C^2 may be found, e.g., in the textbook by V. S. Vladimirov [**76**].

for any $(z_1, \dots, z_{n-1}) \in \underline{\delta}$. It follows from the fact stated above that

$$\varphi(z_1, \dots, z_{n-1}, \xi(z_1, \dots, z_{n-1}), \overline{z}_1^0, \dots, \overline{z}_{n-1}^0, \overline{\xi(z_1^0, \dots, z_{n-1}^0)}) = 0$$

for any $(z_1, \dots, z_{n-1}) \in \underline{\delta}$. Since $z_n^0 = \xi(z_1^0, \dots, z_{n-1}^0)$, this impies that S coincides with the analytic hypersurface given by the equation $\varphi(z, \overline{z^0}) = 0$ in a neighborhood of z^0 in $\mathbf{C}^n$. Thus 1 is proved.

To prove 2, assume that D^+ and D^- are pseudoconvex at $z^0 \in \Sigma$ and that $\partial\varphi/\partial z_n \neq 0$ at z^0. Given any $\mathbf{a}' = (a_1, \dots, a_{n-1}) \in \mathbf{C}^{n-1}$, we set

$$a_n := -\sum_{j=1}^{n-1} \left[\frac{\partial \varphi}{\partial z_j} / \frac{\partial \varphi}{\partial z_n}\right]_{z=z^0} a_j.$$

Then $\mathbf{a} := (\mathbf{a}', a_n)$ satisfies (4.14) at z^0. Since D^+ and D^- are pseudoconvex at z^0, it follows from 1 of Proposition 4.2 that

$$\mathcal{L}\varphi(\mathbf{a}, z^0) := \sum_{j,k=1}^{n} \frac{\partial^2 \varphi}{\partial z_j \partial \overline{z}_k}(z^0) a_j \overline{a}_k = 0.$$

We substitute a_n given above into this equation and obtain

$$-\left|\frac{\partial \varphi}{\partial z_n}(z^0, \overline{z}^0)\right|^{-2} \sum_{j,k=1}^{n-1} L_{jk}(z^0, \overline{z}^0) a_j \overline{a}_k = 0.$$

Since $\mathbf{a}'$ is an arbitrary point in $\mathbf{C}^{n-1}$, we have $L_{jk}(z^0, \overline{z}^0) = 0$ $(j, k = 1, \dots, n-1)$. Thus 2 is proved. □

Let $z^0 \in \Sigma$. If there is a holomorphic mapping T defined in a neighborhood δ of z^0 mapping onto a neighborhood δ^0 of $w = 0$ in $\mathbf{C}^n$ with

$$T(\delta \cap \Sigma) = \{w \in \delta^0 \mid \Re\, w_n = 0\},$$

then Σ is called a **hypersurface of planar type at** z^0.

We make the following remark.

REMARK 4.3. Let $\Sigma : \varphi(z, \overline{z}) = 0$ be a real $(2n-1)$-dimensional real-analytic hypersurface in a neighborhood δ of a point z_0 in $\mathbf{C}^n$ with $\nabla\varphi(z_0) \neq 0$. Assume that for any fixed $\zeta \in \Sigma$, there exists a complex-analytic hypersurface S_ζ in δ such that $\zeta \in S_\zeta \subset \Sigma$. Then Σ is a hypersurface of planar type at z_0.

PROOF. We may assume that $z_0 = 0$ and $\partial\varphi/\partial z_n(0) \neq 0$. We set

$$\ell = \Sigma \cap (0, \dots, 0, \mathbf{C}_{z_n}),$$

which is a real 1-dimensional real-analytic nonsingular curve in the complex plane $\mathbf{C}_{z_n}$. There exists a conformal mapping $w_n = h(z_n)$ from a neighborhood δ_n of $z_n = 0$ onto a neighborhood $\tilde{\delta}_n$ of $w_n = 0$ such that $\ell \cap \delta_n$ gets mapped to $\{\Re\, w_n = 0\} \cap \tilde{\delta}_n$. Thus we may assume from the beginning that ℓ is given by $\{\Re z_n = 0\}$ in δ_n. Fix a point $iy_n \in \ell$. By hypothesis there exists an analytic hypersurface $S_{i\,y_n}$ in a neighborhood of $z = 0$ such that $(0, \dots, 0, i\,y_n) \in S_{i\,y_n} \subset \Sigma$. By 1 of Lemma 4.3 we have

$$S_{i\,y_n} : \varphi(z_1, \dots, z_n, 0, \dots, 0, \overline{i\,y_n}) = 0. \tag{4.18}$$

We solve the equation $\varphi(z_1, \dots, z_n, 0, \dots, 0, w_n) = 0$ for w_n; i.e.,

$$w_n = \eta(z_1, \dots, z_n),$$

where η is a holomorphic function in a neighborhood of $z = 0$ in $\mathbf{C}^n$. We note from (4.18) that $\eta\,|_{S_{i\,y_n}} = \overline{i\,y_n}$. We then consider the analytic transformation

$$T:\ w_i = z_i \quad (i = 1, \dots, n-1), \quad w_n = \eta(z_1, \dots, z_n)$$

from a neighborhood of $z = 0$ onto a neighborhood of $w = 0$. Note that, for each fixed y_n in ℓ, $T(S_{i\,y_n})$ is a domain in the $(n-1)$-dimensional plane $\mathbf{C}^{n-1} \times \{\overline{i\,y_n}\}$ which contains $(0, \overline{i\,y_n})$. It follows that $T(\Sigma) \subset \{\Re w_n = 0\}$, so that Σ is of planar type at the origin 0. □

We shall prove the following theorem.

THEOREM 4.4. *Let Σ be a real $(2n-1)$-dimensional real analytic smooth hypersurface in $D \subset \mathbf{C}^n$. Then Σ is Levi flat if and only if Σ is of planar type at each point of Σ.*

PROOF. Assume that Σ is of planar type at each point of Σ. Fix $z^0 \in \Sigma$. Then we can find an $(n-1)$-dimensional analytic hypersurface S in a neighborhood of z^0 in $\mathbf{C}^n$ such that $z^0 \in S \subset \Sigma$; indeed, after applying a holomorphic change of coordinates, we can take $S = \{w_n = 0\}$. It follows from 2 of Proposition 4.1 that both D^+ and D^- are pseudoconvex at z^0. Thus Σ is Levi flat in D.

To prove the converse, we assume that Σ is Levi flat in D. Fix $z^0 \in \Sigma$. Since the result is local, we can assume that $z^0 = 0$ and that $\varphi(z, w)$ is convergent in a polydisk $\Delta \times \Delta$ centered at $(0,0)$ in $\mathbf{C}^{2n}$, and that $\partial\varphi/\partial z_n \neq 0$ in $\Delta \times \Delta$. Since $\varphi(z, \overline{z})$ is real-valued, we note that

$$\overline{\varphi(z, \overline{w})} = \varphi(w, \overline{z}). \tag{4.19}$$

For a fixed $\overline{\zeta} \in \Delta$, we consider the analytic hypersurface in Δ defined by

$$S_\zeta := \{z \in \Delta \mid \varphi(z, \overline{\zeta}) = 0\}.$$

We can write this hypersurface as

$$S_\zeta\ :\ z_n = \xi_n(z_1, \dots, z_{n-1}, \overline{\zeta}), \tag{4.20}$$

where ξ_n is a holomorphic function of $(z_1, \dots, z_{n-1})$ in a neighborhood Δ' of 0 in $\mathbf{C}^{n-1}$. We want to construct a biholomorphic mapping $\mathbf{S} : w = \mathbf{S}(z)$ from a neighborhood $\delta' \subset \Delta$ of $z = 0$ in $\mathbf{C}^n$ onto a neighborhood δ'' of $w = 0$ in $\mathbf{C}^n$ such that, for each fixed $\zeta \in \delta'$, we have

$$\mathbf{S}(S_\zeta) = \{w = (w_1, \dots, w_n) \in \delta'' \mid w_n = c(\zeta)\}, \tag{4.21}$$

where $c(\zeta)$ is a constant. For this purpose, we prove the following.

First claim: There exists a completely integrable system of partial differential equations in Δ,

$$\frac{\partial z_n}{\partial z_j} = F_j(z_1, \dots, z_n) \qquad (j = 1, \dots, n-1), \tag{4.22}$$

such that each function $z_n = \xi_n(z_1, \dots, z_{n-1}, \overline{\zeta})$, $\overline{\zeta} \in \Delta$, satisfies (4.22). In particular, $F_j(z_1, \dots, z_n)$ does not depend on $\overline{\zeta} \in \Delta$.

Indeed, set $z' := (z_1, \dots, z_{n-1})$, $\overline{\zeta}' = (\overline{\zeta}_1, \dots, \overline{\zeta}_{n-1})$, $z = (z_1, \dots, z_n) = (z', z_n)$ and $\overline{\zeta} = (\overline{\zeta}_1, \dots, \overline{\zeta}_n) = (\overline{\zeta}', \overline{\zeta}_n)$. Since $\varphi(z', \xi_n(z', \overline{\zeta}), \overline{\zeta}) = 0$ in $\Delta' := \{z' \in \mathbf{C}^{n-1} \mid (z', z_n) \in \Delta\}$, we have, for $j = 1, \dots, n-1$,

$$\frac{\partial \xi_n}{\partial z_j}(z', \overline{\zeta}) = -\left\{\left(\frac{\partial\varphi}{\partial z_j}\right) / \left(\frac{\partial\varphi}{\partial z_n}\right)\right\}(z', \xi_n(z', \overline{\zeta}), \overline{\zeta}). \tag{4.23}$$

Fix $\overline{\zeta}' \in \Delta'$ and consider $\overline{\zeta}_n$ as a parameter in $\varphi(z, \overline{\zeta}', \overline{\zeta}_n) = 0$ for $z \in \Delta$. We solve this equation for $\overline{\zeta}_n$ and write $\overline{\zeta}_n = h_n(z, \overline{\zeta}')$, so that

$$\varphi(z, \overline{\zeta}', h_n(z, \overline{\zeta}')) = 0, \qquad (z, \overline{\zeta}') \in \Delta \times \Delta'. \tag{4.24}$$

From (4.23) we see that $z_n = \xi_n(z', \overline{\zeta})$ satisfies the following system of partial differential equations:

$$\begin{aligned} \frac{\partial z_n}{\partial z_j} &= -\left\{ \frac{\partial \varphi}{\partial z_j} \Big/ \frac{\partial \varphi}{\partial z_n} \right\} (z, \overline{\zeta}', h_n(z, \overline{\zeta}')) \\ &\equiv F_j(z, \overline{\zeta}') \qquad (j = 1, \ldots, n-1). \end{aligned} \tag{4.25}$$

To prove the first claim, it suffices to show that

(1) $F_j(z, \overline{\zeta}')$ $(j = 1, \ldots, n-1)$ does not depend on $\overline{\zeta}' \in \Delta'$; and
(2) the system (4.25) is completely integrable;

To prove (1), it suffices to show that

$$\frac{\partial F_j}{\partial \overline{\zeta}_k}(z, \overline{\zeta}') = 0 \qquad (j, k = 1, \ldots, n-1) \quad \text{in } \Delta \times \Delta'.$$

From (4.24) we have

$$\frac{\partial h_n}{\partial \overline{z}_j}(z, \overline{\zeta}') = -\left\{ \frac{\partial \varphi}{\partial \overline{z}_j} \Big/ \frac{\partial \varphi}{\partial \overline{z}_n} \right\} (z, \overline{\zeta}', h_n(z, \overline{\zeta}')) \qquad (j = 1, \ldots, n-1).$$

Using these equations, we compute that

$$\begin{aligned} \frac{\partial F_j}{\partial \overline{\zeta}_k}(z, \overline{\zeta}') &= \frac{1}{(\frac{\partial \varphi}{\partial z_n})^2 (\frac{\partial \varphi}{\partial \overline{z}_n})^2} \cdot \left\{ \frac{\partial^2 \varphi}{\partial z_j \partial \overline{z}_n} \frac{\partial \varphi}{\partial z_n} \frac{\partial \varphi}{\partial \overline{z}_k} - \frac{\partial^2 \varphi}{\partial z_j \partial \overline{z}_k} \frac{\partial \varphi}{\partial z_n} \frac{\partial \varphi}{\partial \overline{z}_n} \right. \\ &\qquad \left. + \frac{\partial^2 \varphi}{\partial z_n \partial \overline{z}_k} \frac{\partial \varphi}{\partial z_j} \frac{\partial \varphi}{\partial \overline{z}_n} - \frac{\partial^2 \varphi}{\partial z_n \partial \overline{z}_n} \frac{\partial \varphi}{\partial z_j} \frac{\partial \varphi}{\partial \overline{z}_k} \right\} \\ &\equiv \frac{1}{(\frac{\partial \varphi}{\partial z_n})^2 (\frac{\partial \varphi}{\partial \overline{z}_n})^2} \cdot \widetilde{\mathcal{L}}_{jk}, \end{aligned}$$

where the right-hand side is evaluated at $(z, \overline{z}) = (z, \overline{\zeta}', h_n(z, \overline{\zeta}'))$. Since Σ is Levi flat, it follows from 1 of Proposition 4.2 that $L_{jk}(z, \overline{z}) = 0$ on Σ; hence

$$L_{jk}(z, \overline{z}) = A_{jk}(z, \overline{z}) \cdot \varphi(z, \overline{z}), \qquad z \in \Delta,$$

where $A_{jk}(z, \overline{z})$ is real analytic for $z \in \Delta$. Since both sides of the above equation are real analytic in Δ, it follows that

$$L_{jk}(z, w) = A_{jk}(z, w) \cdot \varphi(z, w), \qquad (z, w) \in \Delta \times \Delta.$$

Observing that $L_{jk}(z, \overline{z}) = \widetilde{\mathcal{L}}_{jk}(z, \overline{z})$ in $\Delta \times \Delta$, we see from (4.24) that for any $(z, \overline{\zeta}') \in \Delta \times \Delta'$,

$$\widetilde{\mathcal{L}}_{jk}(z, \overline{\zeta}', h_n(z, \overline{\zeta}')) = A_{jk}(z, \overline{\zeta}', h_n(z, \overline{\zeta}')) \cdot \varphi(z, \overline{\zeta}', h_n(z, \overline{\zeta}')) = 0.$$

It follows that $(\partial F_j / \partial \overline{\zeta}_k)(z, \overline{\zeta}') = 0$ on $\Delta \times \Delta'$, which proves (1).

To prove (2), it suffices to show that

$$\frac{\partial F_j}{\partial z_k} + \frac{\partial F_j}{\partial z_n} F_k = \frac{\partial F_k}{\partial z_j} + \frac{\partial F_k}{\partial z_n} F_j \qquad (j, k = 1, \ldots, n-1)$$

for $(z,\overline{\zeta}') \in \Delta \times \Delta'$. This can be verified by direct calculation, using the explicit form of F_j and F_k in (4.25) and the following equalities from (4.24):

$$\frac{\partial h_n}{\partial z_j}(z,\overline{\zeta}') = -\left\{\frac{\partial \varphi}{\partial z_j} \Big/ \frac{\partial \varphi}{\partial \overline{\zeta}_n}\right\}(z,\overline{\zeta}', h_n(z,\overline{\zeta}')) \qquad (j=1,\dots,n-1).$$

This proves (2) and thus the first claim.

Second claim: Assertion (4.21) holds.

If necessary, we may take a smaller polydisk Δ centered at $z=0$ in $\mathbf{C}^n$, so that the solutions of the system of differential equations (4.22) give a foliation of complex $(n-1)$-dimensional analytic hypersurfaces

$$\mathcal{S}_c : \ G(z_1,\dots,z_n) = c, \qquad c \in \delta_1,$$

where δ_1 is a neighborhood of the origin 0 in the complex plane $\mathbf{C}$. We consider the analytic mapping $\mathbf{S}$ from a neighborhood of $z=0$ in $\mathbf{C}^n_z$ onto a neighborhood of $w=0$ in $\mathbf{C}^n_w$ defined by

$$\mathbf{S} : \ w_j = z_j \ (j=1,\dots,n-1), \quad w_n = G(z_1,\dots,z_n).$$

From the first claim it follows that each analytic hypersurface S_{ζ}, $\overline{\zeta} \in \Delta$, defined by (4.20) is mapped to a complex hyperplane of the form $w_n = c(\overline{\zeta})$ =const. in a neighborhood of $w=0$ in $\mathbf{C}^n_w$. Thus, the second step is proved.

Finally we shall show that Σ is of planar type. Since all arguments are invariant under analytic mappings of a neighborhood of the origin, we may assume from the beginning that for each $\overline{\zeta} \in \Delta$, the analytic hypersurface $S_\zeta : \varphi(z,\overline{\zeta}) = 0$ can be written in the form $z_n = \xi_n(z',\overline{\zeta}) = c(\overline{\zeta})$. Therefore,

$$\varphi(z,\overline{\zeta}) = (z_n - c(\overline{\zeta}))\, H(z,\overline{\zeta}),$$

where $H(z,\overline{\zeta}) \neq 0$ in a neighborhood of $(0,0)$ in $\mathbf{C}^{2n}$. Formula (4.19) implies that $c(\overline{\zeta})$ is independent of $(\overline{\zeta_1},\dots,\overline{\zeta_{n-1}})$, i.e., $c(\overline{\zeta}) = c(\overline{\zeta_n})$. Thus $S_\zeta : \varphi(z,\overline{\zeta}) = 0$ is of the form $z_n = c(\overline{\zeta_n})$. In particular, $\Sigma : \varphi(z,\overline{z}) = 0$ is of the form $z_n = c(\overline{z_n})$, and hence $c(\overline{z}_n) = \overline{z}_n$. It follows that $\Sigma = \{z \in \Delta \mid y_n = 0\}$, where $z_n = x_n + iy_n$. Consequently, Σ is of planar type, and Theorem 4.4 is completely proved. □

We see from the above theorem that if $\Sigma := \{z \in D \mid \varphi(z) = 0\}$ is Levi-flat in a domain D, then both domains D^+ and D^- are locally domains of holomorphy at each point z of Σ.

4.3. Boundary Problem

The Levi conditions for C^2 functions $\varphi(z)$ look very similar to the condition of plurisubharmonicity of $\varphi(z)$. Plurisubharmonic functions are considered today as the natural extension to several complex variables of subharmonic functions in one complex variable. However, plurisubharmonic functions were first introduced by K. Oka [**49**] to investigate pseudoconvex domains.[7] He wanted to find a *linear* condition on $\varphi(z)$ which would imply the (nonlinear) Levi conditions on $\varphi(z)$. The reason he wanted to do this is the following: an arbitrary pseudoconvex domain can have a non-smooth and complicated boundary; to approximate such a domain

[7] In his paper Oka called plurisubharmonic functions *pseudoconvex functions.* The name *plurisubharmonic functions* was given by P. Lelong. In Part II in this book the author will use the terminology "pseudoconvex functions" in the setting of analytic spaces.

by pseudoconvex domains with smooth boundary, one could begin with the non-smooth function $\varphi(z) = -\log d_D(z)$ (see section 4.3.2) and take integral averages to get smoother approximations. A linear condition on $\varphi(z)$ will be preserved under this averaging process. In this section we study the relationship between plurisubharmonic functions and pseudoconvex domains.

4.3.1. Strictly Pseudoconvex Domains and Strictly Plurisubharmonic Functions. Let D be a domain in $\mathbf{C}^n$ and let $p \in \partial D$. Let δ be an open neighborhood of p in $\mathbf{C}^n$ and let $I = [0,1]$ be a closed interval in the real axis of the complex t-plane $\mathbf{C}_t$. Let $f(z,t)$ be a holomorphic function of (z,t) on $\delta \times I$. This means there exists a neighborhood G of I in $\mathbf{C}_t$ such that $f(z,t)$ is holomorphic in $\delta \times G$. We set

$$\sigma_t := \{ z \in \delta \mid f(z,t) = 0 \} \qquad (t \in I),$$

so that $\{\sigma_t\}_{t \in I}$ is a family of analytic hypersurfaces in δ. If $\{\sigma_t\}_{t \in I}$ satisfies the conditions

1. $p \in \sigma_0$ and $\sigma_0 \setminus \{p\} \subset \delta \setminus \overline{D}$,
2. $\sigma_t \subset \delta \setminus \overline{D}$ for each $0 < t \leq 1$,

then we call $\{\sigma_t\}_{t \in I}$ a **family of analytic hypersurfaces touching p from the complement of D.**

If a boundary point p of D admits such a family of analytic hypersurfaces, we say that D is **strictly pseudoconvex at p.** Furthermore, if D is strictly pseudoconvex at each boundary point of D, then we say that D is a **strictly pseudoconvex domain.**

We see from 2 of Proposition 4.1 that if D is strictly pseudoconvex at a point $p \in \partial D$, then D is pseudoconvex at p.

REMARK 4.4. Assume that D is strictly pseudoconvex at a point $p \in \partial D$. Given a neighborhood $\delta' \subset \delta$ of p, we can choose $\varepsilon > 0$ sufficiently small so that

(1) $\beta := \{ z \in \delta \cap D \mid |f(z,0)| < \varepsilon \} \subset\subset \delta'$; and
(2) each branch of $\log f(z,0)$ is single-valued on β.

For (1) is clear from condition 1 on $\{\sigma_t\}_{t \in I}$. To prove (2), let γ be a 1-cycle in β. If $\varepsilon > 0$ is sufficiently small, we can find a ball B such that $\beta \subset B \subset \delta'$ by 1, and $B \cap \{ z \in \delta \mid f(z,1) = 0 \} = \emptyset$ by 2. Hence, $\int_\gamma d\arg\, f(z,1) = 0$. Since $f(z,t) \neq 0$ on γ for all $t \in I$, it follows from the continuity of $f(z,t)$ on $\gamma \times I$ that $\int_\gamma d\arg\, f(z,0) = 0$, and (2) is verified.

REMARK 4.5. Let $D = \{ (z,w) \in \mathbf{C}^2 \mid |w| < |z| \}$. Then D is a pseudoconvex domain whose boundary ∂D contains the origin 0. The complex line $L : z = 0$ in $\mathbf{C}^2$ passes through 0, and $L \setminus \{0\} \subset \mathbf{C}^2 \setminus \overline{D}$. However, D is not strictly pseudoconvex at 0.

Let $\phi(z)$ be plurisubharmonic in a domain $D \subset \mathbf{C}^n$. For a real number c, we set

$$D_c := \{ z \in D \mid \phi(z) < c \}.$$

If $D_c \neq \emptyset$, then D_c is pseudoconvex at each point of ∂D_c in D. This follows from the definitions of plurisubharmonicity and of pseudoconvexity of type C. In addition, we have the following result.

PROPOSITION 4.3. *Let $\phi(z)$ be a C^2 function in a domain $D \subset \mathbf{C}^n$. If $\phi(z)$ is strictly plurisubharmonic at a point z^0 in D, then, setting $c = \phi(z^0)$, we have that D_c is strictly pseudoconvex at z^0.*

PROOF. We may assume $z^0 = 0$ and $c = \phi(z^0) = 0$. Since $\phi(z)$ is of class C^2 at 0, we can write

$$\begin{aligned}\phi(z) &= 2\Re\left\{\sum_{j=1}^{n} \frac{\partial \phi}{\partial z_j}(0) z_j + \sum_{j\le k}^{n} \frac{\partial^2 \varphi}{\partial z_j \partial z_k}(0) z_j z_k\right\} \\ &\quad + \sum_{j,k=1}^{n} \frac{\partial^2 \varphi}{\partial z_j \partial \overline{z}_k}(0) z_j \overline{z}_k + o(\|\, z\,\|^2)\end{aligned}$$

near $z = 0$. Since $\phi(z)$ is strictly plurisubharmonic at 0, we can find a neighborhood δ of 0 in D such that

$$\sum_{j,k=1}^{n} \frac{\partial^2 \varphi}{\partial z_j \partial \overline{z}_k}(0) z_j \overline{z}_k \;>\; o(\|\, z\,\|^2) \quad \text{in } \delta \setminus \{0\}.$$

For each $0 \le t \le 1$, we define

$$S_t := \left\{ z \in \delta \mid \sum_{j=1}^{n} \frac{\partial \phi}{\partial z_j}(0) z_j + \sum_{j\le k}^{n} \frac{\partial^2 \varphi}{\partial z_j \partial z_k}(0) z_j z_k \;= t \right\}.$$

Then $S_0 \cap \overline{D}_0 = \{0\}$, and $S_t \cap \overline{D}_0 = \emptyset$ for $0 < t \le 1$. Thus, $\{S_t\}_t$ is a family of analytic hypersurfaces touching $0 \in \partial D_0$ from the complement of D_0. Hence, D_0 is strictly pseudoconvex at 0. □

Note that in the proof we did not require $\nabla\phi(z^0) \neq 0$.

4.3.2. Boundary Distance Function on a Pseudoconvex Domain. One of the most significant properties of a pseudoconvex domain in $\mathbf{C}^n$ can be described in terms of the boundary distance function. Given a domain D in $\mathbf{C}^n$ and a point $z \in D$, recall from section 1.1.5 that

$$d_D(z) := \inf\{\|z - \zeta\| \mid \zeta \in \partial D\ \},$$

the boundary distance function on D. This is a continuous, positive-valued function on D satisfying $\lim_{z\to\partial D}\ d_D(z) = 0$.

We have the following theorem.

THEOREM 4.5. *If D is a pseudoconvex domain in $\mathbf{C}^n$, then $-\log d_D(z)$ is a continuous plurisubharmonic function on D.*

To prove this theorem, we begin with three lemmas.

LEMMA 4.4. *Let D be a domain in the complex z-plane $\mathbf{C}_z$. Then $-\log d_D(z)$ is a continuous subharmonic function on D.*

PROOF. If $D = \mathbf{C}_z$, then $-\log d_D(z) \equiv -\infty$. Thus we assume $D \neq \mathbf{C}_z$. It is clear that $-\log d_D(z)$ is continuous at points z where it is finite-valued; i.e., on D. Fix $\zeta \in \partial D$. Then $z \to -\log|z - \zeta|$ is a harmonic function on D. Therefore,

$$-\log d_D(z) = \sup\ \{-\log|z-\zeta| \mid \zeta \in \partial D\ \}, \quad z \in D$$

is a subharmonic function on D. □

Let D be a domain in the complex z-plane $\mathbf{C}_z$ and let $\mathcal{G}$ be a subdomain in the product space $D \times \mathbf{C}_w \subset \mathbf{C}^2$. Given $z \in D$, we set

$$\mathcal{G}(z) := \{w \in \mathbf{C}_w \mid (z,w) \in \mathcal{G}\},$$

the section of $\mathcal{G}$ over z. We assume $D \times \{0\} \subset \mathcal{G}$, i.e., $0 \in \mathcal{G}(z)$ for each $z \in D$. We let $\mathcal{R}_{\mathcal{G}}(z) > 0$ denote the boundary distance from $w = 0$ to $\partial\mathcal{G}(z)$,

$$\mathcal{R}_{\mathcal{G}}(z) = \inf\{|w - \xi| \mid \xi \in \partial\mathcal{G}(z)\},$$

and we call this the **Hartogs radius** of $\mathcal{G}$ for $z \in D$.

We have the following lemma.

LEMMA 4.5. *Let $\mathcal{G}$ be a domain in $D \times \mathbf{C}_w \subset \mathbf{C}^2$, where D is a domain in $\mathbf{C}_z$ with $D \times \{0\} \subset \mathcal{G}$. If $\mathcal{G}$ is a pseudoconvex domain in $\mathbf{C}^2$, then $-\log \mathcal{R}_{\mathcal{G}}(z)$ is a subharmonic function on D.*

PROOF. For simplicity we write $\mathcal{R}(z) = \mathcal{R}_{\mathcal{G}}(z)$ and $\phi(z) = -\log \mathcal{R}_{\mathcal{G}}(z)$ for $z \in D$. Since $\mathcal{G}$ is an open set in $\mathbf{C}^2$, $\phi(z)$ is uppersemicontinuous on D. It suffices to verify the subaveraging property. We proceed by contradiction: suppose there exist a point $a \in D$ and a positive radius ρ so that ϕ does not satisfy the subaveraging property on $\overline{\delta} := \{z : |z-a| \leq \rho\} \subset D$, i.e.,

$$\phi(a) > \frac{1}{2\pi}\int_0^{2\pi} \phi(a + \rho e^{i\theta})\, d\theta.$$

Since $\phi(z)$ is uppersemicontinuous on $\partial\delta = \{|z-a| = \rho\}$, we can find a real analytic function $u(z)$ such that

$$u(a+\rho e^{i\theta}) > \phi(a+\rho e^{i\theta}), \qquad 0 \leq \theta \leq 2\pi,$$
$$\phi(a) > \frac{1}{2\pi}\int_0^{2\pi} u(a+\rho e^{i\theta})\, d\theta.$$

By the Poisson formula we construct the harmonic function $h(z)$ on δ with $h(z) = u(z)$ on $\partial\delta$. Then $\phi(a) > h(a)$. We take a harmonic conjugate $k(z)$ of $h(z)$ so that $\xi(z) := h(z) + ik(z)$ is holomorphic on δ. We then consider the automorphism

$$T \;:\; z = z, \qquad w' = e^{\xi(z)} w$$

of the product domain $\Omega = \delta \times \mathbf{C}_w$, and set $\mathcal{G}^* := T(\Omega \cap \mathcal{G})$. Then $\Omega \cap \mathcal{G}$ and $\mathcal{G}^*$ are pseudoconvex domains in $\mathbf{C}^2$ and $\delta \times \{w' = 0\} \subset \mathcal{G}^*$. The Hartogs radius $\mathcal{R}^*(z)$ about $w' = 0$ for $z \in \delta$ is equal to

$$\mathcal{R}^*(z) = e^{h(z)}\, \mathcal{R}(z).$$

From the relations between $u(z)$ and $\phi(z)$, we have

$$\mathcal{R}^*(a) < 1 < \mathcal{R}^*(a+\rho e^{i\theta}), \qquad 0 \leq \theta < 2\pi.$$

Thus we can find a point $w'^0 \in \partial\mathcal{G}^*(a)$ with $|w'^0| = \mathcal{R}^*(a) < 1$, while $\{|w'| \leq 1\} \subset\subset \mathcal{G}^*(a+\rho e^{i\theta})$ for $0 \leq \theta < 2\pi$. Since $\mathcal{R}^*(z) > 0$ on $\{|z-a| \leq \rho\}$ (for $D \times \{0\} \subset G$), $\mathcal{G}^*$ does not satisfy the continuity theorem of type C, contradicting the pseudoconvexity of $\mathcal{G}^*$. □

Let $D \subset \mathbf{C}^n$ be a domain and let $z' = (z_1', \dots, z_n') \in D$. We let $D(z') \subset \mathbf{C}_{z_n}$ denote the section of D over the complex line $z_j = z_j'$ $(j = 1, \dots, n-1)$ in $\mathbf{C}^n$, i.e.,

$$D(z') = \{z_n \in \mathbf{C}_{z_n} \mid (z_1', \dots, z_{n-1}', z_n) \in D\}.$$

We let $\mathcal{R}_n(z')$ denote the boundary distance from z'_n to $\partial D(z')$ in $\mathbf{C}_{z_n}$. Then $\mathcal{R}_n(z)$ is a positive-valued function on D, which we call the **Hartogs radius of D with respect to z_n**.

We have the following lemma.

LEMMA 4.6. *Let D be a pseudoconvex domain in $\mathbf{C}^n$ and let $\mathcal{R}_n(z)$ be the Hartogs radius of D with respect to z_n. Then $-\log \mathcal{R}_n(z)$ is a plurisubharmonic function on D.*

PROOF. Since D is an open set in $\mathbf{C}^n$, $-\log \mathcal{R}_n(z)$ is uppersemicontinuous on D. Thus we must show that the restriction of $-\log \mathcal{R}_n(z)$ to any complex line L in D is subharmonic. Let $a = (a_1, \dots, a_n) \in D$ and fix a complex line L passing through a. If L is of the form $z_j = a_j\ (j = 1, \dots, n-1)$, then from Lemma 4.4 it follows that the restriction of $-\log \mathcal{R}_n(z)$ to L is subharmonic. Thus we may assume that L is of the form

$$L: \quad z_j = L_j(z_1) = c_j(z_1 - a_1) + a_j \qquad (j = 2, \dots, n),$$

where $c_j \neq 0$ for some $j = 2, \dots, n$. Fix a disk $\delta := \{|z_1 - a_1| < \rho\}$ in $\mathbf{C}_{z_1}$ such that $(\overline{\delta} \times \mathbf{C}^{n-1}) \cap L \subset D$. We show that $s(z_1) := -\log d_D(z_1, L_2(z_1), \dots, L_n(z_1))$ is subharmonic for $z_1 \in \delta$.

For each $z_1 \in \delta$ we consider the subset of $\mathbf{C}_{z_n}$ given by

$$D_n(z_1) := \{z_n \in \mathbf{C}_{z_n} \mid (z_1, L_2(z_1), \dots, L_{n-1}(z_1), z_n) \in D\}.$$

Let

$$G := \{(z_1, z_n) \in \mathbf{C}^2 \mid z_1 \in \delta, \quad z_n \in D_n(z_1)\ \}.$$

Then $G = D \cap (L' \times \mathbf{C}_{z_n})$ where L' denotes the projection of L onto $\mathbf{C}_{z_1}$; thus G is a pseudoconvex domain in $\mathbf{C}^2$. We consider the automorphism

$$T\ :\quad z_1 = z_1, \qquad w = z_n - c_n(z_1 - a_1) - a_n$$

of $\delta \times \mathbf{C}_w$ and set $\mathcal{G} := T(G)$. Then $\mathcal{G}$ is a pseudoconvex domain in $\mathbf{C}_{z_1} \times \mathbf{C}_w$ with $\delta \times \{0\} \subset \mathcal{G}$. Since $\mathcal{R}_{\mathcal{G}}(z_1) = d_D(z_1, L_2(z_1), \dots, L_n(z_1))$ for $z_1 \in \delta$, it follows from Lemma 4.5 that $s(z_1)$ is subharmonic on δ. □

PROOF OF THEOREM 4.5. Letting $z = (z_1, \dots, z_n)$ denote the usual coordinates in $\mathbf{C}^n$, we fix a unitary matrix U and form the coordinate transformation $z' := (z'_1, \dots, z'_n) = (z_1, \dots, z_n) \cdot U$ of $\mathbf{C}^n$. Consider the Hartogs function $\mathcal{R}_n^U(z')$ of D with respect to z'_n. We have

$$-\log d_D(z) = \sup_U \{-\log \mathcal{R}_n^U(z')\},$$

where the supremum is taken over all unitary matrices U. From Lemma 4.6 we conclude that $-\log \mathcal{R}_n^U(z')$ is a plurisubharmonic function on D for each such U; since $-\log d_D(z)$ is continuous in D we conclude that $-\log d_D(z)$ is a plurisubharmonic function on D. □

4.3.3. Approximating the Boundary. The boundary of an arbitrary pseudoconvex domain D may be rather complicated, and thus we would like to be able to approximate D from inside by pseudoconvex domains with simpler boundaries. Indeed, this procedure is indispensable in order to verify that any pseudoconvex domain is a domain of holomorphy (which will be discussed in Chapter 9).

We note that a pseudoconvex domain D in $\mathbf{C}^n$ admits a continuous plurisubharmonic **exhaustion function** $\xi(z)$. This means that for any real number a,

$D_a := \{z \in D \mid \xi(z) < a\} \subset\subset D$. To see this, in the case when D is bounded, Theorem 4.5 implies that

$$\xi(z) := -\log\, d_D(z)$$

is a continuous plurisubharmonic exhaustion function for D. If D is unbounded,

$$\xi(z) := -\log\, d_D(z) + \|z\|^2$$

satisfies this property (here, $\|z\|^2 = |z_1|^2 + \ldots + |z_n|^2$).

Therefore any pseudoconvex domain D in $\mathbf{C}^n$ can be approximated from inside by an increasing sequence of relatively compact pseudoconvex domains $\{D_a\}_a$ with continuous boundaries. We present a method which modifies D_a to a pseudoconvex domain with smoother boundary.

Let $D \subset \mathbf{C}^n$ be a domain (not necessarily pseudoconvex) and let $\phi(z)$ be a plurisubharmonic function in D. If for each $p \in D$ we can find a neighborhood δ of p and a finite number of plurisubharmonic functions $\psi_k(z)$ $(k = 1, \ldots, l)$ of class C^2 in δ such that

$$\phi(z) = \max_k \{\psi_k(z)\} \qquad \text{in } \delta,$$

then we say that $\phi(z)$ is a piecewise smooth plurisubharmonic function on D. In addition, if each $\psi_k(z)$ is a strictly plurisubharmonic function on δ, we say that $\phi(z)$ is a **piecewise smooth strictly plurisubharmonic function** on D.

Let D be a domain in $\mathbf{C}^n$. For $c > 0$ we let D^c denote the set of all points z in D such that the polydisk distance from z to ∂D is greater than c.

Let $f(z)$ be a locally (Lebesgue) integrable function on D. Given $0 < \eta < c$, we let Δ_η : $|\zeta_j| < \eta$ $(j = 1, \ldots, n)$ be a polydisk in $\mathbf{C}^n$. Then we can define, for $z \in D^c$, the average value of f,

$$A_\eta[f](z) := \frac{1}{V} \int_{\Delta_\eta} f(z_1 + \zeta_1, \ldots, z_n + \zeta_n) dv_\zeta,$$

where dv_ζ is the volume element of $\mathbf{C}^n$ at ζ and $V = (\pi\eta^2)^n$.

LEMMA 4.7. *If $f(z)$ is a locally integrable function on D, then $A_\eta[f](z)$ is a continuous function on D^c. If $f(z)$ is continuous (resp. of class C^1), then $A_\eta[f](z)$ is of class C^1 (resp. of class C^2) in D^c.*

The proof is standard, and is omitted.

LEMMA 4.8. *Assume that $f(z)$ is plurisubharmonic on D. Then:*

1. *$A_\eta[f](z)$ is a continuous plurisubharmonic function on D^{2c}.*
2. *$A_{\eta_1}[f](z) \le A_{\eta_2}[f](z)$ if $0 < \eta_1 < \eta_2 < c$ and $f(z) = \lim_{\eta \to 0} A_\eta[f](z)$ pointwise on D.*
3. *If $f(z)$ is continuous on D, then $f(z) = \lim_{\eta \to 0} A_\eta[f](z)$ uniformly on each compact set $K \subset D$.*

PROOF. Since $f(z)$ is plurisubharmonic on D, it is clear that $f(z)$ is locally integrable on D. Fix $z' = (z_1', \ldots, z_n') \in D^c$ and let

$$l : z_j = a_j t + z_j' \quad (j = 1, \ldots, n), \qquad t \in \mathbf{C}_t,$$

be a complex line in $\mathbf{C}^n$ passing through z'. Fix $\varepsilon > 0$ such that the restriction l_ε of l for $|t| \le \varepsilon$ is contained in D^c, and consider the boundary of $l_\varepsilon : = \{z(\theta) = (z_1(\theta), \ldots, z_n(\theta))\}$, where

$$z_j(\theta) = a_j \varepsilon e^{i\theta} + z_j' \qquad (j = 1, \ldots, n).$$

Since $f(z)$ is plurisubharmonic on D, we have

$$\frac{1}{2\pi}\int_0^{2\pi} f(z+z(\theta))d\theta \geq f(z+z'), \qquad z,\ z+z' \in D^c.$$

Since $A_\eta[f](z)$ is continuous on D^c, to verify assertion 1 it suffices to show that

$$m := \frac{1}{2\pi}\int_0^{2\pi} A_\eta[f](z(\theta))d\theta \geq A_\eta[f](z'), \qquad z' \in D^{2c}.$$

We have

$$m = \frac{1}{2\pi}\int_0^{2\pi}\left\{\frac{1}{V}\int_{\Delta_\eta} f(z(\theta)+\zeta)dv_\zeta\right\}d\theta.$$

Since $f(z)$ is uppersemicontinuous on D, it is bounded above on any compact set in D. Thus $f(z(\theta)+\zeta)$ is bounded above on $(\theta,\zeta) \in [0,2\pi]\times\Delta_\eta$, and we can interchange the order of integration to obtain

$$\begin{aligned} m &= \frac{1}{V}\int_{\Delta_\eta}\left\{\frac{1}{2\pi}\int_0^{2\pi} f(z(\theta)+\zeta)d\theta\right\}dv_\zeta \\ &\geq \frac{1}{V}\int_{\Delta_\eta} f(z'+\zeta)dv_\zeta = A_\eta[\,f\,](z'). \end{aligned}$$

This proves 1.

To prove 2, we note that for any subharmonic function $s(z)$ on a disk $|z-a|\leq\rho$ in the complex plane $\mathbf{C}_z$, we have, for $0<\rho_1<\rho_2<\rho$,

$$s(a) \leq \frac{1}{2\pi}\int_0^{2\pi} s(a+\rho_1 e^{i\theta})d\theta \leq \frac{1}{2\pi}\int_0^{2\pi} s(a+\rho_2 e^{i\theta})d\theta. \tag{4.26}$$

Set $\zeta_j := r_j e^{i\theta_j}$ $(j=1,\dots,n)$. Using the change of variables $r_j = \eta s_j$ $(0\leq s_j\leq 1;\ j=1,\dots,n)$, for any $z'\in D^c$ and $0<\eta<c$ we obtain

$$\begin{aligned} &A_\eta[f](z') \\ &= \frac{1}{\pi^n}\int_{[0,2\pi]^n\times[0,1]^n} f(z_1'+\eta s_1 e^{i\theta_1},\dots,z_n'+\eta s_n e^{i\theta_n})\, s_1\cdots s_n\, d\Theta\, dS, \end{aligned}$$

where $d\Theta = d\theta_1\cdots d\theta_n$ and $dS = ds_1\cdots ds_n$. Together with (4.26), this implies the first part of statement 2. The second part follows from the uppersemicontinuity of $f(z)$. Assertion 3 follows from 2 and Dini's theorem.[8] □

Thus, for any continuous plurisubharmonic function $\psi(z)$ on D and for any $\varepsilon>0$ and $K \subset\subset D^c$, we can find a plurisubharmonic function $\phi^*(z)$ of class C^2 on D^c such that

$$|\phi(z)-\phi^*(z)|<\varepsilon, \qquad z\in K.$$

Furthermore, if D is bounded, we can assume that $\phi^*(z)$ is strictly plurisubharmonic on D^c. Indeed, it suffices to take $\lambda>0$ sufficiently small and replace $\phi^*(z)$ by

$$\phi^0(z) := \phi^*(z) + \lambda\,\|z\|^2.$$

We now prove the following theorem.

THEOREM 4.6 (Oka [**52**]). *A pseudoconvex domain D in $\mathbf{C}^n$ admits a piecewise smooth, strictly plurisubharmonic exhaustion function.*

[8]The idea of smoothing plurisubharmonic functions by using integral averages is due to F. Riesz [**62**].

PROOF. We take a continuous plurisubharmonic exhaustion function $\xi(z)$ for D. Let α_j $(j = 1, 2, \dots)$ be a sequence of real numbers with $\alpha_{j+1} - \alpha_j > 2$ $(j = 1, 2, \dots)$ such that if we set $D_j := \{z \in D \mid \xi(z) < \alpha_j\}$ $(\subset\subset D)$, then $D_1 \neq \emptyset$. On each domain D_{2l+3} $(l = 1, 2, \dots)$, we can construct a strictly plurisubharmonic function $\xi_l(z)$ of class C^2 such that $|\xi(z) - \xi_l(z)| < 1$ on D_{2l+3}. We have

$$\xi_l(z) - \alpha_{2l} < -1 \qquad \text{for } z \in \partial D_{2l-1};$$
$$\xi_l(z) - \alpha_{2l} > 1 \qquad \text{for } z \in D_{2l+3} \setminus D_{2l+1}.$$

We shall construct, by induction, a piecewise smooth strictly plurisubharmonic function $\xi_l^*(z)$ on D_{2l+1} $(l = 1, 2, \dots)$.

We first set $\xi_1^*(z) := \xi_1(z)$ on D_3 so that $\xi_1^*(z) \geq 1$ on $D_3 \setminus D_1$. Next, having constructed a piecewise smooth, strictly plurisubharmonic function $\xi_l^*(z)$ on D_{2l+1} $(l \geq 1)$ such that $\xi_l^*(z) \geq l$ on $D_{2l+1} \setminus D_{2l-1}$, we take $c_{l+1} > 0$ sufficiently large so that, if we set $\eta_{l+1}(z) := c_{l+1}(\xi_l(z) - \alpha_{2l})$ on D_{2l+3}, then $\eta_{l+1}(z) \geq l + 1$ on $D_{2l+3} \setminus D_{2l+1}$ and

$$\xi_l^*(z) - \eta_{l+1}(z) > 0 \qquad \text{for } z \in \partial D_{2l-1},$$
$$\xi_l^*(z) - \eta_{l+1}(z) < 0 \qquad \text{for } z \in \partial D_{2l+1}.$$

Now for $z \in D_{2l+3}$ we define

$$\xi_{l+1}^*(z) := \begin{cases} \xi_l^*(z) & \text{for } z \in D_{2l-1}, \\ \max(\xi_l^*(z), \eta_{l+1}(z)) & \text{for } z \in D_{2l+1} - D_{2l-1}, \\ \eta_{l+1}(z) & \text{for } z \in D_{2l+3} - D_{2l+1}. \end{cases}$$

Then $\xi_{l+1}^*(z)$ is a piecewise smooth, strictly plurisubharmonic function in D_{2l+3} satisfying $\xi_{l+1}^*(z) = \xi_l^*(z)$ on D_{2l-1}, $\xi_{l+1}^*(z) \geq \xi_l^*(z)$ on D_{2l+1}, and $\xi_{l+1}^*(z) \geq l + 1$ on $D_{2l+3} \setminus D_{2l+1}$. It follows that the limit $\xi^*(z) := \lim_{l \to \infty} \xi_l^*(z)$ exists on D and defines a piecewise smooth, strictly plurisubharmonic exhaustion function on D. □

4.4. Pseudoconcave Sets

4.4.1. Definition of Pseudoconcave Sets. The complement of a pseudoconvex domain has a certain analytic property, if it is "small" as a set. This fact was first discovered by F. Hartogs [**30**] and was carefully studied by K. Oka. We follow the ideas of Oka [**43**] and extend their study from the two-dimensional case to the general n-dimensional case, $n \geq 2$.[9]

Let D be a domain in $\mathbf{C}^n$ and let $\mathcal{E}$ be a closed set in D. For each point $z' = (z'_1, \dots, z'_n)$ of $\mathcal{E}$ and each polydisk δ : $|z_j - z'_j| < r$ $(j = 1, \dots, n)$ in D centered at z', if each connected component of $\delta \setminus (\delta \cap \mathcal{E})$ is a pseudoconvex domain in $\mathbf{C}^n$, then $\mathcal{E}$ is called a **pseudoconcave set in** D. As an example, an analytic hypersurface in D is a pseudoconcave set in D.

REMARK 4.6. We will simply say that $\delta \setminus (\delta \cap \mathcal{E})$ is a pseudoconvex domain if each connected component of $\delta \setminus (\delta \cap \mathcal{E})$ is an open, connected pseudoconvex set; i.e., we take a domain to be an open (but not necessarily connected) set.

The following properties of pseudoconcave sets are a consequence of the elementary properties of pseudoconvex domains.

[9]See Oka's posthumous work No. 7 in [**55**], in which he called a pseudoconcave set an *(H)-set*. See also T. Nishino [**40**].

1. If $\mathcal{E}_1$ and $\mathcal{E}_2$ are pseudoconcave sets in D, then so is $\mathcal{E}_1 \bigcup \mathcal{E}_2$.

2. Let $I = \{\iota\}$ be an index set. If $\mathcal{E}_\iota$ $(\iota \in I)$ is a family of pseudoconcave sets in D, then the closure of $\bigcup_{\iota \in I} \mathcal{E}_\iota$ in D is a pseudoconcave set in D.

3. Let $\mathcal{E}_j$ $(j = 1, 2, \dots)$ be a decreasing sequence of pseudoconcave sets in D; i.e., $\mathcal{E}_{j+1} \subset \mathcal{E}_j$ for all j. Then $\mathcal{E}_0 := \bigcap_{j=1}^\infty \mathcal{E}_j$ is a pseudoconcave set in D.

4. Let $\mathcal{E}$ be a pseudoconcave set in D. For any r-dimensional complex analytic plane L with $0 < r < n$, $\mathcal{E} \cap L$ is a pseudoconcave set in $D \cap L$ (which we identify with $\mathbf{C}^r$).

5. Let $\mathcal{E}$ be a closed set in D. Suppose that for each $p \in \mathcal{E} \cap D$, there exist a neighborhood δ of p in D and an analytic hypersurface σ_p in δ such that $p \in \sigma_p \subset \mathcal{E}$. Then $\mathcal{E}$ is a pseudoconcave set in D.

6. Let S be an irreducible analytic hypersurface in a domain $D \subset \mathbf{C}^n$ and let $\mathcal{E}$ be a nonempty pseudoconcave set in D. If $\mathcal{E} \subset S$, then $\mathcal{E} = S$. This fact can be proved by use of the continuity theorem of type A.

7. If $\mathcal{E}$ is a pseudoconcave set in D, and T is a biholomorphic mapping of D onto $T(D)$, then $T(\mathcal{E})$ is a pseudoconcave set in $T(D)$.

Let $\mathcal{E}$ be a pseudoconcave set in a domain $D \subset \mathbf{C}^n$ and let $p \in \partial\mathcal{E}$. If there exists a neighborhood δ of p in D such that the domain $\delta \setminus (\delta \cap \mathcal{E})$ is strictly pseudoconvex at p, then we say that $\mathcal{E}$ is **strictly pseudoconcave** at p. If $\delta \setminus (\delta \cap \mathcal{E})$ is a piecewise smooth strictly pseudoconvex domain at p, then we say that $\mathcal{E}$ is a **piecewise smooth pseudoconcave set at** p. If $\mathcal{E}$ is strictly pseudoconcave (resp., piecewise smooth) at each boundary point of $\mathcal{E}$ in D, then we say that $\mathcal{E}$ is a strictly pseudoconcave (resp., piecewise smooth) set in D. Theorem 4.6 implies that any pseudoconcave set $\mathcal{E}$ in a pseudoconvex domain D can be approximated by a decreasing sequence of piecewise smooth strictly pseudoconcave sets in D.

4.4.2. Hartogs' Theorem. Consider $\mathbf{C}^{n+1}$ as the product of $\mathbf{C}^n$ with variables $z = (z_1, \dots, z_n)$ and $\mathbf{C}_w$ with variable w. Let D be a domain in $\mathbf{C}^n$ and set $G := D \times \mathbf{C}_w$. Let $\mathcal{E}$ be a pseudoconcave set in G. For each $z' \in D$, the fiber of $\mathcal{E}$ over z' is defined by

$$\mathcal{E}(z') := \{w \in \mathbf{C}_w \mid (z', w) \in \mathcal{E}\}.$$

We make the assumption that each $\mathcal{E}(z')$, $z' \in D$, is bounded in $\mathbf{C}_w$.

We prove the following theorem.

THEOREM 4.7 (Hartogs). *Let $\mathcal{E}$ be a pseudoconcave set in $G = D \times \mathbf{C}_w$, where D is a domain in $\mathbf{C}^n$. If each fiber $\mathcal{E}(z)$, $z \in D$, consists of exactly one point $\zeta(z)$ in $\mathbf{C}_w$, then $z \to \zeta(z)$ is a holomorphic function of z in D.*

PROOF. Let $z_0 \in D$ and let $p_0 = (z_0, \zeta(z_0)) \in \mathcal{E}$. Since $G \setminus \mathcal{E}$ satisfies the continuity theorem of type A at p_0, it follows that $\zeta(z)$ is continuous at z_0 in D. We now show that $\zeta(z)$ is holomorphic at z_0 in D.

Fix a point $w_0 \in \mathbf{C}_w$ such that $w_0 \neq \zeta(z_0)$. We take a ball δ in D centered at z_0 such that $\zeta(z) \neq w_0$ for $z \in \delta$. We set

$$h(z) := \log|\zeta(z) - w_0|, \qquad z \in \delta.$$

Since $|\zeta(z)-w_0|$ is the Hartogs radius of $D\backslash\mathcal{E}$ with respect to w, Lemma 4.6 implies that $-h(z)$ is a plurisubharmonic function on δ. Consider the following analytic transformation T of $\delta\times(\mathbf{C}_w\setminus\{w_0\})$ onto $\delta\times(\mathbf{C}_{w'}\setminus\{0\})$:

$$T_1\ :\ z_j=z_j\ \ (j=1,\dots,n),\qquad w'=\frac{1}{w-w_0}.$$

We set $\omega:=\delta\times\mathbf{C}_w$ and $\mathcal{E}^0:=T_1(\mathcal{E}\cap\omega)$. Then $\mathcal{E}^0$ is a pseudoconcave set in ω with the property that each fiber $\mathcal{E}^0(z)$, $z\in\delta$, consists of one point $\zeta^0(z)$ in $\mathbf{C}_{w'}$ with

$$w'=\zeta^0(z)=\frac{1}{\zeta(z)-w_0}.$$

Since $\zeta^0(z)\neq 0$, $z\in\delta$, it follows by the same reasoning that $-\log(1/|\zeta(z)-w_0|)$ is a plurisubharmonic function on δ. Consequently, $h(z)$ is a pluriharmonic function on δ.

We now take a conjugate pluriharmonic function $k(z)$ of $h(z)$ on δ so that

$$\xi(z):=h(z)+ik(z),\qquad z\in\delta,$$

is a holomorphic function on δ. Then we form the following automorphism of ω:

$$T_2\ :\ z_j=z_j\ \ (j=1,\dots,n),\qquad w''=(w-w_0)e^{-\xi(z)}.$$

The image $\mathcal{E}^*:=T_2(\mathcal{E})$ is thus a pseudoconcave set in ω with the property that each fiber $\mathcal{E}^*(z)$ consists of one point $\zeta^*(z)$ with

$$w''=\zeta^*(z)=(\zeta(z)-w_0)e^{-\xi(z)}.$$

Note that $|\zeta^*(z)|\equiv 1$ on δ.

We next fix a point $w^*\in\mathbf{C}_{w''}$ such that

$$|w^*|<1,\qquad \arg w^*=\arg\zeta^*(z_0).$$

Since $|\zeta^*(z)|\equiv 1$ on δ, $|\zeta^*(z)-w^*|$ for $z\in\delta$ attains its minimum at the center z_0 of δ. On the other hand, Lemma 4.6 again implies that $h^*(z):=-\log\ |\zeta^*(z)-w^*|$ is a plurisubharmonic function on δ. It follows that $h^*(z)$ is constant on δ. This implies, together with the fact that $|\zeta^*(z)|\equiv 1$ on δ, that $\zeta^*(z)$ is constant on δ, say $\zeta^*(z)\equiv\alpha$. Hence,

$$\zeta(z)=\alpha e^{\xi(z)}+w_0,\qquad z\in\delta,$$

so that $\zeta(z)$ is a holomorphic function on δ. □

4.4.3. Preparation Theorem. Let E be a compact set in $\mathbf{C}_w$. We fix an integer $m\geq 2$ and take m points w_j $(j=1,\dots,m)$ in E. We set

$$V_m(w_1,\dots,w_m):=\sqrt[\frac{m(m-1)}{2}]{\prod_{\nu<\mu}|w_\nu-w_\mu|}\ ,$$

and define

$$D_m(E):=\max\{V(w_1,\dots,w_m)\mid w_1,\dots,w_m\in E\},$$

the m**-th diameter of** E. It is easy to verify that

$$D_m(E)\geq D_{m+1}(E)\qquad (m=2,3,\dots).$$

Thus the limit

$$D_\infty(E):=\lim_{n\to\infty}D_m(E)$$

exists and is called the **transfinite diameter of** E.[10]

Now let $G = D \times \mathbf{C}_w$ where D is a domain in $\mathbf{C}^n$, and let $\mathcal{E}$ be a closed set in G such that each fiber $\mathcal{E}(z)$ over $z \in D$ is a bounded set in $\mathbf{C}_w$. Let $m \geq 2$ be an integer or let $m = \infty$. For each $z \in D$, we consider the m-th diameter $D_m(\mathcal{E}(z))$ of the fiber $\mathcal{E}(z)$. We set $D_m(z) := D_m(\mathcal{E}(z))$ for $z \in D$.

We have the following theorem.

THEOREM 4.8. [11] *If $\mathcal{E}$ is a pseudoconcave set in G, then* $\log D_m(z)$ $(m = 2, 3, \dots, \infty)$ *is a plurisubharmonic function on D.*

PROOF. We know that the decreasing limit of a sequence of plurisubharmonic functions on D is a plurisubharmonic function on D and that any pseudoconcave set in G is a decreasing limit of a sequence of piecewise smooth, strictly pseudoconcave sets in G. Thus it suffices to prove the theorem for each fixed finite integer $m \geq 2$ under the assumption that $\mathcal{E}$ is a piecewise smooth strictly pseudoconcave set in G.

Since $\mathcal{E}$ is closed in G, we first note that $\log\ D_m(z)$ is uppersemicontinuous on D. Fix $z^0 = (z_1^0, \dots, z_n^0) \in D$. It suffices to show that for any $(a_1, \dots, a_n) \in \mathbf{C}^n$ and any sufficiently small $\epsilon_j, 0 < \varepsilon_j < 1$ $(j = 1, \dots, n)$,

$$\log D_m(z^0) \leq \frac{1}{2\pi} \int_0^{2\pi} \log D_m(z_1^0 + a_1\varepsilon_1 e^{i\theta}, \dots, z_n^0 + a_n\varepsilon_n e^{i\theta})\, d\theta. \tag{4.27}$$

We can find m points w_ν^0 $(\nu = 1, \dots, m)$ in $\mathcal{E}(z^0)$ such that

$$D_m(z_0) = \sqrt[\frac{m(m-1)}{2}]{\prod_{\nu<\mu} |w_\nu^0 - w_\mu^0|}.$$

By the maximum principle we observe that $w_\nu^0 \in \partial\mathcal{E}(z^0)$ $(\nu = 1, \dots, m)$. We set $p_\nu = (z^0, w_\nu^0)$ $(\nu = 1, \dots, m)$. Since $\mathcal{E}$ is a piecewise smooth, strictly pseudoconcave set at p_ν, we can find an analytic hypersurface S_ν in a neighborhood δ_ν of p_ν in G such that $p_\nu \in S_\nu \subset \mathcal{E} \cap \delta_\nu$. Since $w_\nu^0 \in \partial\mathcal{E}(z^0)$ and $\mathcal{E}(z^0)$ is bounded in $\mathbf{C}_w$, it follows from Lemma 2.1 that if we choose a suitably small polydisk $\delta_0 \times \Delta_\nu$ in G centered at p_ν where $\delta_0 = \{|z_j - z_j^0| < r\}$ $(j = 1, \dots, n)$ and $\Delta_\nu = \{|w - w_\nu| < \rho_\nu\}$ $(\nu = 1, \dots, m)$, then $S_\nu \cap [\delta_0 \times (\partial\Delta_\nu)] = \emptyset$ and S_ν in $\delta_0 \times \Delta_\nu$ can be written in the form

$$P_\nu(z, w - w_\nu^0) = (w - w_\nu^0)^{k_\nu} + a_1^{(\nu)}(z)(w - w_\nu^0)^{k_\nu - 1} + \dots + a_{k_\nu}^{(\nu)}(z) = 0 \tag{4.28}$$

where P_ν has no multiple factors. In this equation, each coefficient $a_i^{(\nu)}(z)$ is a holomorphic function on δ_0 with the property that $a_i^{(\nu)}(z^0) = 0$. We consider the discriminant $d_\nu(z)$ of $P_\nu(z, w - w_\nu^0)$ with respect to $w - w_\nu^0$, and set

$$\sigma_\nu := \{z \in \delta_0 \mid d_\nu(z) = 0\} \qquad \text{and} \qquad \delta_0' = \delta_0 \setminus \left(\bigcup_{\nu=1}^m \sigma_\nu\right).$$

Fix $z^* \in \delta_0'$. We take a single-valued branch $\eta_\nu(z)$ $(\nu = 1, \dots, m)$ of the algebraic function given by the solution of equation (4.28) on a neighborhood δ^* of z^* in δ_0', and we consider the following vector-valued holomorphic function on δ^*:

$$\eta(z) := (\eta_1(z), \dots, \eta_m(z)).$$

---

[10] The notion of transfinite diameter was introduced by M. Fékete [**19**]. He also proved that $D_\infty(E)$ is equal to the logarithmic capacity of E.

[11] This theorem was first proved by K. Oka [**43**] for the case $m = 2$. The proof given here for the general case is due to H. Yamaguchi [**79**].

Let γ be any arc in δ_0' with initial point z^* and terminal point $\tilde{z}$. Then $\eta(z)$ can be analytically continued along γ. If we denote the resulting function by $\eta(z) = (\tilde{\eta}_1(z), \dots, \tilde{\eta}_m(z))$ near $\tilde{z}$, then each $\tilde{\eta}_\nu(z)$ is a branch of the algebraic function given by equation (4.28) in a neighborhood of $\tilde{z}$ in δ_0'. We form the analytic continuation of $\eta(z)$ over all arcs in δ_0' with initial point z^*, and the resulting function is a bounded, vector-valued function on δ_0'. We use the same notation

$$\eta(z) = (\eta_1(z), \dots, \eta_m(z)) \quad \text{on} \quad \delta_0';$$

then the function

$$f(z) = \prod_{1 \le \nu < \mu \le m} (\eta_\nu(z) - \eta_\mu(z)), \qquad z \in \delta_0',$$

becomes a bounded, single-valued holomorphic function on δ_0'. From Riemann's removable singularity theorem, $f(z)$ can be holomorphically extended to δ_0. Since $\eta_\nu(z) \in \mathcal{E}(z)$, $z \in \delta$ $(\nu = 1, \dots, m)$, and $\eta_\nu(z^0) = w_\nu^0$, it follows that

$$\sqrt[\frac{m(m-1)}{2}]{|f(z)|} \le D_m(z), \qquad \sqrt[\frac{m(m-1)}{2}]{|f(z^0)|} = D_m(z^0).$$

Since $\frac{2}{m(m-1)}$ log $|f(z)|$ is a plurisubharmonic function on δ_0, these two formulas imply the desired inequality (4.27). □

4.4.4. Pluripolar Sets. For a set E in the complex plane **C** we can canonically define its potential theoretic size, called the **logarithmic capacity** of E. We summarize the well-known results for the logarithmic capacity in **C**. If E is compact, this coincides with the transfinite diameter of E. Sets of logarithmic capacity zero coincide with **polar sets**: a set $E \subset \mathbf{C}$ is polar if for each point $z_0 \in E$ there exists a subharmonic function $u(z) \not\equiv -\infty$ defined on a neighborhood δ of z_0 with

$$E \cap \delta \subset \{z \in \delta : u(z) = -\infty\}.$$

This local notion is actually a global one: if E is polar, then one can find $u(z)$ subharmonic in a neighborhood D of E, $u(z) \not\equiv -\infty$, with

$$E \subset \{z \in D : u(z) = -\infty\}.$$

Indeed, D can be taken to be all of **C**. Thus if $\phi(z)$ is a subharmonic function on a domain D in **C** and the set $E_\phi := \{z \in D \mid \phi(z) = -\infty\}$ is of positive logarithmic capacity, then $\phi(z) \equiv -\infty$ on D.

For a set E in $\mathbf{C}^n$ for $n \ge 2$, we have an analogous notion of **pluripolar sets**: a set $E \subset \mathbf{C}^n$ is pluripolar if for each point $z_0 \in E$ there exists a plurisubharmonic function $u(z) \not\equiv -\infty$ defined on a neighborhood δ of z_0 with

$$E \cap \delta \subset \{z \in \delta : u(z) = -\infty\}.$$

Again, this local notion is a global one: if E is pluripolar, then one can find $u(z)$ plurisubharmonic in a neighborhood D of E with

$$E \subset \{z \in D : u(z) = -\infty\}$$

(cf. M. Klimek [**34**], Theorem 4.7.4). Indeed, D can be taken to be all of $\mathbf{C}^n$.

Let $e \subset \mathbf{C}^n$ and $p \in e$. If for any neighborhood δ of p in $\mathbf{C}^n$ and any plurisubharmonic function $\phi(z)$ on δ with $\phi(z) = -\infty$ on $e \cap \delta$ we have $\phi(z) \equiv -\infty$ on δ, then p is called a **point of type** (β) in e. If p is not of type (β) in e, p is called a

point of type (α) in e. Thus if e consists entirely of points of type (α) in e, then e is a pluripolar set in $\mathbf{C}^n$. In general, we define

$$e_\beta := \{p \in e \mid p \text{ is of type } (\beta) \text{ in } e\}. \tag{4.29}$$

If e is contained in a domain $D \subset \mathbf{C}^n$, then e_β is closed in D, and clearly e_β is pluripolar – and hence empty! – if and only if e is pluripolar.

We easily have the following:

1. A countable union of pluripolar sets in $\mathbf{C}^n$ is pluripolar.
2. A non-empty open set G in $\mathbf{C}^n$ is not pluripolar.
3. Let e be a pluripolar set in D, where D is a domain in $\mathbf{C}^n$. If $f(z)$ is a bounded holomorphic function in $D \setminus e$, then $f(z)$ has a holomorphic extension to all of D.
4. The pluripolarity or non-pluripolarity of a set $e \subset \mathbf{C}^n$ is not a metric property of e and depends on the complex structure of $\mathbf{C}^n$. For example, any analytic hypersurface S in a domain D in $\mathbf{C}^n$ (hence S is real $(2n-2)$-dimensional) is pluripolar in $\mathbf{C}^n$. On the other hand, the set
$$e = \{z = (z_1, \dots, z_n) \mid \Re z_j = 0 \, (j = 1, \dots, n)\ \}$$
(which is real n-dimensional) is not pluripolar in $\mathbf{C}^n$. Similarly, the distinguished boundary $e = \{|z_j| = 1 \ (j = 1, \dots, n)\}$ of the unit polydisk in $\mathbf{C}^n$, which is also real n-dimensional, is not pluripolar in $\mathbf{C}^n$.

4.4.5. Oka's First Theorem. We utilize the notion of pluripolar sets to prove the following theorem.

THEOREM 4.9 (Oka). *Let $G = D \times \mathbf{C}_w$, where D is a domain in $\mathbf{C}^n$. Let $\mathcal{E}$ be a pseudoconcave set in G such that each fiber $\mathcal{E}(z)$ for $z \in D$ is bounded in $\mathbf{C}_w$. Define*

$$e := \{z \in D \mid \mathcal{E}(z) \text{ consists of a finite number of points in } \mathbf{C}_w\}.$$

If e is not pluripolar, then $\mathcal{E}$ is an analytic hypersurface in G.

PROOF. From the continuity theorem of type A it follows that $\mathcal{E}(z) \neq \emptyset$ for each $z \in D$. For an integer $\nu \geq 1$, we let e_ν denote the set of points z in e such that the fiber $\mathcal{E}(z)$ consists of at most ν distinct points in $\mathbf{C}_w$ so that

$$e_1 \subset e_2 \subset \cdots, \qquad e = \bigcup_{\nu=1}^{\infty} e_\nu.$$

Since e is not pluripolar, it follows that some e_ν is not pluripolar. We fix e_{ν_0}, where $\nu_0 \geq 1$ is the smallest such integer; thus $e_{\nu_0 - 1}$ is pluripolar (in the case $\nu_0 = 1$, we set $e_0 := \emptyset$). We consider the (ν_0+1)-st diameter $D_{\nu_0+1}(z)$ of $\mathcal{E}(z)$, $z \in D$. From Theorem 4.8, $\log\ D_{\nu_0+1}(z)$ is a plurisubharmonic function on D. Since $D_{\nu_0+1}(z) = 0$ for $z \in e_{\nu_0}$ and e_{ν_0} is not pluripolar, it follows that $\log D_{\nu_0+1} \equiv -\infty$ on D, i.e., $e_{\nu_0} = D$. We set $D' = D \setminus e_{\nu_0 - 1}$ and write $\mathcal{E}(z) = \{\xi_1(z), \dots, \xi_{\nu_0}(z)\}$ for $z \in D'$, where $\xi_i(z) \neq \xi_j(z)$ $(i \neq j)$. Define

$$\begin{aligned} P(z,w) &:= \prod_{j=1}^{\nu_0} [w - \xi_j(z)] \\ &= w^{\nu_0} + a_1(z) w^{\nu_0 - 1} + \cdots + a_{\nu_0}(z). \end{aligned}$$

We claim that each $a_j(z)$ $(j = 1, \dots, \nu_0)$ is a holomorphic function on D'.

To verify this, fix $z_0 \in D'$. Using Theorem 4.7, we can find a polydisk δ in D' centered at z_0 such that if we set $\omega := \delta \times \mathbf{C}_w$, then $\mathcal{E} \cap \omega$ can be described by the equations

$$w = \xi_j(z) \quad (j = 1, \ldots, \nu_0),$$

where each $\xi_j(z)$ is a single-valued holomorphic function on δ. Since the $a_j(z)$ $(j = 1, \ldots, \nu_0)$ are symmetric functions of $\{\xi_1(z), \ldots, \xi_{\nu_0}(z)\}$, it follows that the $a_j(z)$ are holomorphic functions on δ and hence on D'.

Furthermore, at each point $z^* \in e_{\nu_0 - 1}$, we can find a neighborhood δ^* of z^* in D such that each $a_j(z)$ $(j = 1, \ldots, \nu_0)$ is bounded on $\delta^* \cap D'$. Since e_{ν_0-1} is pluripolar, it follows that each $a_j(z)$ has a holomorphic extension to δ^*, and hence to all of D. Then $P(z, w)$ is a polynomial in w with coefficients that are holomorphic in D; thus $P(z, w)$ is holomorphic in G, and it is easy to see that

$$\mathcal{E} = \{(z, w) \in G \mid P(z, w) = 0\}.$$

Thus $\mathcal{E}$ is an analytic hypersurface in G. □

This theorem gives us a generalization of Theorem 4.7.

COROLLARY 4.1. *Let $\mathcal{E}$ be a pseudoconcave set in $G = D \times \mathbf{C}_w$ such that each $\mathcal{E}(z)$, $z \in D$, is bounded in $\mathbf{C}_w$. Assume that the set of points $z \in D$ such that $\mathcal{E}(z)$ consists of exactly one point in $\mathbf{C}_w$ is a non-pluripolar set. Then $\mathcal{E}$ can be described as the set of points*

$$w = \xi(z), \qquad z \in D,$$

where $\xi(z)$ is a single-valued holomorphic function in D.

Furthermore, using Theorem 4.8 in the case $m = \infty$, we obtain the following theorem.

THEOREM 4.10 (Yamaguchi). *Let $\mathcal{E}$ be a pseudoconcave set in $G = D \times \mathbf{C}_w$ such that each $\mathcal{E}(z)$, $z \in D$, is bounded in $\mathbf{C}_w$. Assume that the set of points $z \in D$ such that $\mathcal{E}(z)$ is of logarithmic capacity zero is a non-pluripolar set. Then each $\mathcal{E}(z)$, $z \in D$, is of logarithmic capacity zero.*

4.5. Analytic Derived Sets

4.5.1. Definition of Analytic Derived Sets. Let D be a domain in $\mathbf{C}^n$ and let $\mathcal{E}$ be a pseudoconcave set in D. Fix $p \in \mathcal{E}$. If there exists a neighborhood δ of p in D such that $\mathcal{E} \cap \delta$ is an analytic hypersurface in δ, then we say that p is a point of $\mathcal{E}$ of the first kind. If $p \in \mathcal{E}$ is not of the first kind, we say that p is of the second kind. We call the set $\mathcal{E}'$ of all points $z \in \mathcal{E}$ which are of the second kind the **analytic derived set** of $\mathcal{E}$.

REMARK 4.7. In standard set-theoretic topology, given a closed set E in $\mathbf{C}^n$, one considers the subset E' of E, called the **derived set** of E, which is obtained by excluding from E all isolated points of E. Thus the analytic derived set $\mathcal{E}'$ of a pseudoconcave set $\mathcal{E}$ may be regarded as a type of analytic modification of the usual derived set E' of a closed set E, where we consider "analytic hypersurface points" of $\mathcal{E}$ as isolated points of E.

The following theorem concerning analytic derived sets will be essential in the following sections.

THEOREM 4.11 (Oka [**43**], Nishino [**40**]). *Let $\mathcal{E}$ be a pseudoconcave set in a domain D in $\mathbf{C}^n$. Then the analytic derived set $\mathcal{E}'$ is also a pseudoconcave set in D.*

PROOF. From the definition of analytic derived sets, $\mathcal{E}'$ is a closed set in D. Fix $z^0 \in \mathcal{E}'$. It suffices to prove that $D \setminus \mathcal{E}'$ satisfies the continuity theorem of type B at z^0. For simplicity we may assume that z^0 is the origin 0 in $\mathbf{C}^n$. We fix a set $\beta \subset \mathbf{C}^2 = \mathbf{C}_{z_{n-1}} \times \mathbf{C}_{z_n}$ of the form

$$\beta \,:\, |z_{n-1}+r|^2 + |z_n|^2 > r^2, \qquad |z_{n-1}|^2 + |z_n|^2 < \rho$$

and we consider the set

$$B \,:\, z_j = 0 \ \ (j = 1, \dots, n-2), \qquad (z_{n-1}, z_n) \in \beta$$

in $\mathbf{C}^n$. Our goal is to show that $B \not\subset D \setminus \mathcal{E}'$.

We remark that the content of the theorem is similar in spirit to that of Theorem 4.2 (Levi's theorem). Indeed, the method of proof will be similar to that of Theorem 4.2.

For the sake of obtaining a contradiction, we assume that $B \subset D \setminus \mathcal{E}'$. Recalling the proof of Theorem 4.2, we see that it suffices to deduce a contradiction under the assumption that, if we let l denote the complex line

$$l \,:\, z_j = 0 \qquad (j = 1, \dots, n-1)$$

in $\mathbf{C}^n$, then the restriction of l to any fixed neighborhood of the origin 0 in D is not contained in the original pseudoconcave set $\mathcal{E}$. Since $B \subset D \setminus \mathcal{E}'$, for any point $p \in B$ we can find a neighborhood δ_p of p in $\mathbf{C}^n$ such that $\mathcal{E} \cap \delta_p$ is an analytic hypersurface in δ_p (possibly empty). We thus see that under our assumption about l, for any ρ_1, ρ_2 with $0 < \rho_2 < \rho_1 < \rho$, the set

$$\mathcal{E} \cap \{(0, \dots, 0, z_n) : \rho_2 \le |z_n| \le \rho_1\}$$

consists of a finite number of points in D. Thus we can choose η with $\rho_2 < \eta < \rho_1$ and $\delta > 0$ such that

$$\mathcal{E} \cap \{(z_1, \dots, z_n) : |z_j| \le \delta \ \ (j = 1, \dots, n-1), \ \ |z_n| = \eta\} = \emptyset. \tag{4.30}$$

We consider the open polydisk $\Lambda = \Delta \times \Gamma$ centered at 0 in D, where

$$\Delta \,:\, |z_j| < \delta \ \ (j = 1, \dots, n-1), \qquad \Gamma \,:\, |z_n| < \eta.$$

By choosing smaller values of δ and η, if necessary, we may assume that $\Lambda \subset D$. Set $\mathcal{E}_0 := \Lambda \cap \mathcal{E}$. It follows from (4.30) that $\mathcal{E}_0$ is a pseudoconcave set in $\omega := \Delta \times \mathbf{C}_{z_n}$. Fix a point $a > 0$ sufficiently close to $z_{n-1} = 0$ in $\mathbf{C}_{z_{n-1}}$ so that the set

$$z_j = 0 \ (j = 1, \dots, n-2), \quad z_{n-1} = a, \quad |z_n| \le \eta$$

is contained in $B \cap \Lambda$. Since $B \subset D \setminus \mathcal{E}'$, we can choose δ' with $0 < \delta' < a$ such that, setting

$$\Delta' :\, |z_j'| \le \delta' \ \ (j = 1, \dots, n-2), \ \ |z_{n-1}' - a| \le \delta',$$

each fiber $\mathcal{E}'(z_1', \dots, z_{n-1}')$ of $\mathcal{E}_0$ over $(z_1', \dots, z_{n-1}') \in \Delta'$ consists of a finite number of points in $\mathbf{C}_{z_n}$. Since Δ' is not pluripolar, it follows from Theorem 4.9 that $\mathcal{E}_0$ is an analytic hypersurface in ω. Hence $\mathcal{E}_0' = \emptyset$, which contradicts the fact that $0 \in \Lambda \cap \mathcal{E}' = \mathcal{E}_0'$. □

We will use the following lemma in the next section. Recall that for a subset e in $\mathbf{C}^n$, $e_\beta := \{p \in e \mid p \text{ is of type } (\beta) \text{ in } e\}$. Let D be a domain in $\mathbf{C}^n$. For $a = (a_1, \dots, a_n) \in D$, $r > 0$ sufficiently small, $b \in \mathbf{C}_w$, and $\rho > 0$, we let $\Delta_r(a)$ denote the polydisk centered at a with radius r in $D \subset \mathbf{C}^n$ and we let $\gamma_\rho(b)$ be the disk centered at b with radius ρ in $\mathbf{C}_w$.

LEMMA 4.9. *Let D be a domain in $\mathbf{C}^n$ and let $\mathcal{F}$ be a pseudoconcave set in $G := D \times \mathbf{C}_w$ such that each fiber $\mathcal{F}(z)$, $z \in D$, is bounded in $\mathbf{C}_w$. Let e be a non-pluripolar set in D. Suppose there exists a point $(a,b) \in \mathcal{F}'$ (the analytic derived set of $\mathcal{F}$) such that $a \in e_\beta$ and such that there exists a sequence of circles $c_\nu = \{|w - b| = \rho_\nu\}$ $(\nu = 1, 2, \dots)$ in $\mathbf{C}_w$ with $\rho_\nu \to 0$ $(\nu \to \infty)$ such that $c_\nu \cap \mathcal{F}(a) = \emptyset$. Then for each $r > 0$ and $\rho > 0$, there exists at least one point $z' \in \Delta_r(a) \cap e_\beta$ such that $\mathcal{F}(z') \cap \gamma_\rho(b)$ contains infinitely many distinct points in $\mathbf{C}_w$.*

PROOF. The proof is by contradiction. Thus we assume that there exist $r > 0$ and $\rho > 0$ such that for each $z \in \Delta_r(a) \cap e_\beta$, the set $\mathcal{F}(z) \cap \gamma_\rho(b)$ contains at most finitely many distinct points in $\mathbf{C}_w$. We take a sufficiently large integer ν such that the radius $\rho_\nu > 0$ of the circle c_ν is smaller than ρ. We let γ_ν denote the disk bounded by c_ν; then by hypothesis $(\partial\gamma_\nu) \cap \mathcal{F}(a) = \emptyset$. We can find $r_0 > 0$ with $r_0 < r$ such that $(\partial\gamma_\nu) \cap \mathcal{F}(z) = \emptyset$ for all $z \in \Delta_{r_0}(a)$. Let $\omega := \Delta_{r_0}(a) \times \gamma_\nu$, a polydisk centered at (a,b) in G. Since $e_\beta \cap \Delta_{r_0}(a)$ is not pluripolar, it follows from Theorem 4.9 that $\mathcal{F} \cap \omega$ is an analytic hypersurface in ω. Thus, $(a,b) \notin \mathcal{F}'$, which is a contradiction. □

The hypothesis in Lemma 4.9 does not imply that $\mathcal{F}'(a)$ contains infinitely many distinct points in $\mathbf{C}_w$. For example, let D be a domain in $\mathbf{C}_z$ and consider the pseudoconcave set $\mathcal{F}$ in $D \times \mathbf{C}_w \subset \mathbf{C}^2$ defined as

$$\mathcal{F} := \left[\bigcup_{j=1}^{\infty} \{(z,w) \mid z \in D,\ w = 1/j\}\right] \cup \{(z,w) \mid z \in D,\ w = 0\}. \tag{4.31}$$

Then $(0,0) \in \mathcal{F}'$ and $\mathcal{F}'(0) = \{0\}$, although each $\mathcal{F}'(z)$, $z \in D$ with $z \neq 0$ contains infinitely many points in $\mathbf{C}_w$.

4.5.2. Kernel of a Pseudoconcave Set. We now define higher order derived sets of pseudoconcave sets in $\mathbf{C}^n$ in order to generalize Theorem 4.9 as Theorem 4.12 below.

We let $\mathcal{N}$ denote the set of all ordinal numbers up until the first uncountable ordinal Ω.; i.e., $\mathcal{N}$ is the set of all so-called **countable ordinals**. We will only need the following properties of $\mathcal{N}$:

1. $\mathcal{N}$ is a well-ordered set; i.e.,
   (i) there is a total order relation on $\mathcal{N}$, which we denote by $\leq$; i.e., $\leq$ is transitive, anti-symmetric, and, for any $\alpha, \beta \in \mathcal{N}$, either $\alpha \leq \beta$ or $\beta \leq \alpha$;
   (ii) every non-empty subset S of $\mathcal{N}$ contains a minimal element; i.e., there exists an element $\alpha \in S$ such that $\alpha \leq \beta$ for all $\beta \in S$.
2. Each $\alpha \in \mathcal{N}$ has a successor, which we denote by $\alpha + 1$, in $\mathcal{N}$, i.e., $\alpha < \alpha + 1$ and $\alpha + 1 \leq \beta$ for all $\alpha < \beta$. In particular, $\{0, 1, 2 \dots\} \subset \mathcal{N}$ since $0 < 1 < 2 < \dots$.

3. For any increasing sequence $\{\alpha_n\}_n$ in $\mathcal{N}$, i.e.,
$$\alpha_1 < \alpha_2 < \dots ,$$
$\alpha := \sup\{\alpha_n : n = 1, 2, \dots\}$ exists and is a member of $\mathcal{N}$.
4. For each $\alpha \in \mathcal{N}$, define
$$I(\alpha) := \{\gamma \in \mathcal{N} \mid \gamma < \alpha\},$$
which is called the initial interval determined by α. Then $I(\alpha)$ is at most countable.

We note that $\mathcal{N}$ is not countable. We may divide $\mathcal{N}$ into two distinct parts $\mathcal{N}'$ and $\mathcal{N}''$, where
$$\begin{aligned} \mathcal{N}' &= \{\alpha \in \mathcal{N} \mid \text{there exists } \beta \in \mathcal{N} \text{ such that } \alpha = \beta + 1\}, \\ \mathcal{N}'' &= \mathcal{N} \setminus \mathcal{N}'. \end{aligned}$$
The elements belonging to $\mathcal{N}'$ are said to be **successor ordinals**, while the elements of $\mathcal{N}''$ are said to be **limit ordinals**.

Let E be a compact set in $\mathbf{C}^n$. We let E' denote the *usual* derived set of E; i.e., E' is the subset of E consisting of all non-isolated points of E. We define $E^{(\alpha)}$ for each $\alpha \in \mathcal{N}$ by transfinite induction as follows. First define $E^{(0)} := E$. If $0 < \alpha$ and $E^{(\gamma)}$ has been defined for each $\gamma \in I(\alpha)$, then we define $E^{(\alpha)}$ by:

(i) $E^{(\alpha)} := [E^{(\beta)}]'$ if $\alpha = \beta + 1$ for some $\beta < \alpha$, i.e., if α is a successor ordinal;
(ii) $E^{(\alpha)} := \bigcap\{E^{(\gamma)} : \gamma < \alpha\} = \bigcap_{I(\alpha)} E^{(\gamma)}$ if α is a limit ordinal.

It now follows that $E^{(\alpha)}$ is well-defined for each $\alpha \in \mathcal{N}$. Each $E^{(\alpha)}, \alpha \in \mathcal{N}$, is a compact subset of E with $E^{(\alpha)} \subset E^{(\beta)}$ for $\beta < \alpha$. We call
$$E^{(\Omega)} := \bigcap_{\alpha \in \mathcal{N}} E^{(\alpha)} \qquad \text{in } \mathbf{C}^n$$
the **kernel** of the compact set E.

PROPOSITION 4.4. *Let E be a compact set in $\mathbf{C}^n$. Then:*

1. *There exists a unique $\alpha_0 \in \mathcal{N}$ such that*
   (1) *$E^{(\alpha_0)} = E^{(\Omega)}$, and hence $E^{(\gamma)} = E^{(\alpha_0)}$ for all $\gamma \in \mathcal{N}$ with $\alpha_0 < \gamma$;*
   (2) *$E^{(\gamma+1)}$ is a proper subset of $E^{(\gamma)}$ for each $\gamma \in I(\alpha_0)$;*
   (3) *$E = E^{(\Omega)} \cup \left(\bigcup_{\gamma \in I(\alpha_0)} [E^{(\gamma)} - E^{(\gamma+1)}]\right)$, and this is a disjoint union.*
2. *E is countable if and only if $E^{(\Omega)} = \emptyset$.*

PROOF. The proof of (1), (2) and (3) in 1 follows from the fact that α_0 is the smallest (i.e., first) element in $\mathcal{N}$ such that $E^{(\alpha_0)} = E^{(\alpha_0+1)}$, which is easily proved by the above properties about $\mathcal{N}$.

To prove assertion 2, we first assume that $E^{(\Omega)} = \emptyset$. We have from (3) in 1,
$$E = \bigcup_{\gamma \in I(\alpha_0)} [E^{(\gamma)} \setminus E^{(\gamma+1)}].$$
Now $I(\alpha_0)$ is countable, and since $E^{(\gamma)} \setminus E^{(\gamma+1)}$ consists of the isolated points of $E^{(\gamma)}$, each $E^{(\gamma)} \setminus E^{(\gamma+1)}$ is at most a countable set. Therefore E is countable.

To prove the converse, we assume that $E^{(\Omega)} = E^{(\alpha_0)} \neq \emptyset$. Then $E^{(\Omega)}$ is a perfect subset of E; i.e., $E^{(\Omega)}$ has no isolated points. It follows from the Baire category theorem in $\mathbf{C}^n$ that $E^{(\Omega)}$ must be uncountable and hence E is uncountable. $\square$

We return to the case of pseudoconcave sets. Let D be a domain in $\mathbf{C}^n$ and let $\mathcal{E}$ be a pseudoconcave set in $G := D \times \mathbf{C}_w$ with the property that each fiber $\mathcal{E}(z)$, $z \in D$, is bounded in $\mathbf{C}_w$. In order to define the **analytic kernel** $\mathcal{E}^{(\Omega)}$ of $\mathcal{E}$ we first define $\mathcal{E}^{(\alpha)}$ for each $\alpha \in \mathcal{N}$.

We first set $\mathcal{E}^{(0)} := \mathcal{E}$. Given $\alpha \in \mathcal{N}$ with $0 < \alpha$, we assume that $\mathcal{E}^{(\gamma)}$ has been defined as a pseudoconcave set in G for each $\gamma \in I(\alpha)$. Then if α is a successor ordinal, i.e., if there exists $\beta \in \mathcal{N}$ with $\alpha = \beta + 1$, we define

$$\mathcal{E}^{(\alpha)} := [\mathcal{E}^{(\beta)}]'$$

(here A' denotes the analytic derived set of a pseudoconcave set A in G). If α is a limit ordinal, we define

$$\mathcal{E}^{(\alpha)} := \bigcap\{\mathcal{E}^{(\gamma)} : \gamma < \alpha\} = \bigcap_{\gamma \in I(\alpha)} \mathcal{E}^{(\gamma)}.$$

Using Theorem 4.11 and property 3 in 4.4.1, we see that $\mathcal{E}^{(\alpha)}$ is a pseudoconcave set in G for each $\alpha \in \mathcal{N}$. Note that $\mathcal{E}^{(\alpha)} \subset \mathcal{E}^{(\beta)}$ for $\beta < \alpha$. Finally, we call

$$\mathcal{E}^{(\Omega)} := \bigcap_{\alpha \in \mathcal{N}} \mathcal{E}^{(\alpha)}$$

the **analytic kernel** of the pseudoconcave set $\mathcal{E}$.

PROPOSITION 4.5 (cf. Baire [**1**]). *Let D be a domain in $\mathbf{C}^n$ and let $\mathcal{E}$ be a pseudoconcave set in $G := D \times \mathbf{C}_w$ with the property that each fiber $\mathcal{E}(z)$, $z \in D$, is bounded in $\mathbf{C}_w$. Then there exists a unique $\alpha_0 \in \mathcal{N}$ such that*

(1) *$\mathcal{E}^{(\alpha_0)} = \mathcal{E}^{(\Omega)}$, and hence $\mathcal{E}^{(\gamma)} = \mathcal{E}^{(\alpha_0)}$ for all $\gamma \in \mathcal{N}$ with $\alpha_0 < \gamma$;*
(2) *$\mathcal{E}^{(\gamma+1)}$ is a proper subset of $\mathcal{E}^{(\gamma)}$ for each $\gamma \in I(\alpha_0)$; and*
(3) *$\mathcal{E} = \mathcal{E}^{(\Omega)} \cup \left(\bigcup_{\gamma \in I(\alpha_0)} [\mathcal{E}^{(\gamma)} - \mathcal{E}^{(\gamma+1)}]\right)$, and this is a disjoint union.*

PROOF. The proof is similar to that of the preceding proposition. □

In particular, from (1) it follows that $\mathcal{E}^{(\Omega)}$ *is a pseudoconcave set in G which satisfies $[\mathcal{E}^{(\Omega)}]' = \mathcal{E}^{(\Omega)}$.*

4.5.3. Oka's Second Theorem. We now state and prove a result for a pseudoconcave set $\mathcal{E}$ analogous to the second part of Proposition 4.4 for a compact set E in $\mathbf{C}^n$.

THEOREM 4.12 (Oka). *Let D be a domain in $\mathbf{C}^n$ and let $\mathcal{E}$ be a pseudoconcave set in $G := D \times \mathbf{C}_w$ with the property that each fiber $\mathcal{E}(z)$, $z \in D$, is bounded in $\mathbf{C}_w$.*

1. *Suppose that $\mathcal{E}^{(\Omega)} = \emptyset$. Then each fiber $\mathcal{E}(z)$, $z \in D$, is a countable set in $\mathbf{C}_w$. Furthermore, for any point $p \in \mathcal{E}$, there exists an analytic hypersurface σ defined in a neighborhood of p which is contained in $\mathcal{E}$ and which contains the point p.*
2. *If the subset e of D defined by*

$$e = \{z \in D \mid \mathcal{E}(z) \text{ is a countable set in } \mathbf{C}_w\}$$

is not pluripolar, then $\mathcal{E}^{(\Omega)} = \emptyset$.

PROOF. To prove 1, we assume that $\mathcal{E}^{(\Omega)} = \emptyset$ and we fix $z_0 \in D$. Note that for any pseudoconcave set $\mathcal{A}$ in G, we have $[\mathcal{A}(z_0)]' \subset \mathcal{A}'(z_0)$ (here, on the left-hand side we are taking the set-theoretic derived set of the fiber $\mathcal{A}(z_0)$; on the right-hand side we are taking the fiber over z_0 of the analytic derived set of $\mathcal{A}$). Hence

$[\mathcal{E}(z_0)]^{(\alpha)} \subset \mathcal{E}^{(\alpha)}(z_0)$ for each $\alpha \in \mathcal{N}$; thus $\mathcal{E}^{(\Omega)}(z_0) = [\mathcal{E}(z_0)]^{(\Omega)} = \emptyset$. It follows from the second part of Proposition 4.4 that $\mathcal{E}(z_0)$ is countable.

Fix $p \in \mathcal{E}$. From (3) of Proposition 4.5 we can find a $\gamma \in I(\alpha_0)$ such that $p \in \mathcal{E}^{(\gamma)} \setminus \mathcal{E}^{(\gamma+1)}$. Therefore there exists an analytic hypersurface σ defined in a neighborhood δ in G such that $p \in \sigma \subset \mathcal{E}^{(\gamma)} \subset \mathcal{E}$; hence 1 is proved.

We prove 2 by contradiction. Thus we assume that $\mathcal{E}^{(\Omega)} \neq \emptyset$. Let e_β be the set of all points $z \in e$ of type (β). Thus e_β is a closed non-pluripolar set in D.

Fix $z^{(0)} \in e_\beta$ and $w^{(0)} \in \mathcal{E}^{(\Omega)}(z^{(0)})$. Since $\mathcal{E}^{(\Omega)} = [\mathcal{E}^{(\Omega)}]'$ and since the fiber $\mathcal{E}^{(\Omega)}(z^{(0)}) \subset \mathcal{E}(z^{(0)})$ is a closed countable set in $\mathbf{C}_w$, we can apply Lemma 4.9 with $\mathcal{F} = \mathcal{E}^{(\Omega)}$, $a = z^{(0)}, b = w^{(0)}, r = r_0 = 1$, and $\rho = \rho_0 = 1$ to obtain $z^{(1)} \in e_\beta \cap \Delta_{r_0}(z^{(0)})$ such that $\mathcal{E}^{(\Omega)}(z^{(1)}) \cap \gamma_{\rho_0}(w^{(0)})$ contains infinitely many distinct points in $\mathbf{C}_w$ (recall that $\gamma_{\rho_0}(w^{(0)})$ denotes the disk of radius ρ_0 centered at $w^{(0)}$). We choose two of these points $w^{(1)}_{\mu_1}$ $(\mu_1 = 0, 1)$, and we take disjoint disks $\gamma_{\rho_1}(w^{(1)}_{\mu_1})$ centered at $w^{(1)}_{\mu_1}$ with radius ρ_1 which are contained in our original disk $\gamma_{\rho_0}(w^{(0)})$; i.e.,

$$\gamma_{\rho_1}(w^{(1)}_0) \cap \gamma_{\rho_1}(w^{(1)}_1) = \emptyset, \qquad \gamma_{\rho_1}(w^{(1)}_0) \cup \gamma_{\rho_1}(w^{(1)}_1) \subset\subset \gamma_{\rho_0}(w^{(0)}).$$

For each $\mu_1 = 0, 1$, we can again apply Lemma 4.9 with $\mathcal{F} = \mathcal{E}^{(\Omega)}$, $a = z^{(1)}, b = w^{(1)}_{\mu_1}$, $r = r_1 = 1/2$, and $\rho = \rho_1 < 1/2$. We obtain $z^{(2)} \in e_\beta \cap \Delta_{r_1}(z^{(1)})$ and two distinct points $w^{(2)}_{\mu_1,\mu_2}$ $(\mu_2 = 0, 1)$. For each $\mu_2 = 0, 1$, we again take a disk $\gamma_{\rho_2}(w^{(2)}_{\mu_1,\mu_2})$ centered at $w^{(2)}_{\mu_1,\mu_2}$ such that

$$\gamma_{\rho_2}(w^{(2)}_{\mu_1,0}) \cap \gamma_{\rho_2}(w^{(2)}_{\mu_1,1}) = \emptyset, \qquad \gamma_{\rho_2}(w^{(2)}_{\mu_1,0}) \cup \gamma_{\rho_2}(w^{(2)}_{\mu_1,1}) \subset\subset \gamma_{\rho_1}(w^{(1)}_{\mu_1}).$$

We inductively repeat this procedure to obtain the countable subset

$$\mathcal{K} := \{(z^{(l)}, w^{(l)}_{\mu_1,\dots,\mu_l}) \in \mathcal{E}^{(\Omega)} \mid l = 1, 2\dots;\ \mu_h = 0, 1;\ h = 1, \dots, l\}$$

of $\mathcal{E}^{(\Omega)}$ which satisfies the following conditions:

(i) Each $z^{(l)}$ $(l = 1, 2, \dots)$ belongs to e_β and the limit $z^{(*)} := \lim_{l\to\infty} z^{(l)}$ exists; hence $z^{(*)} \in e_\beta$.

(ii) For each $z^{(l)}$ $(l = 1, 2, \dots)$, we can find 2^l distinct points $w^{(l)}_{\mu_1,\dots,\mu_l}$ $(\mu_h = 0, 1;\ h = 1, \dots, l)$ which belong to $\mathcal{E}^{(\Omega)}(z^{(l)})$.

(iii) For each $l = 1, 2, \dots$, we can find 2^l disjoint disks $\gamma^{(l)}_{\rho_l}(w^{(l)}_{\mu_1,\dots,\mu_l})$ centered at $w^{(l)}_{\mu_1,\dots,\mu_l}$ with radius ρ_l $(0 < \rho_l < 1/2^l)$ in $\mathbf{C}_w$ such that

$$\gamma^{(l)}_{\rho_l}(w^{(l)}_{\mu_1,\dots,\mu_l}) \subset\subset \gamma^{(l-1)}_{\rho_{l-1}}(w^{(l-1)}_{\mu_1,\dots,\mu_{l-1}}) \qquad (\mu_l = 0, 1).$$

Since $\mathcal{E}^{(\Omega)}$ is closed in G, the set $\mathcal{K}_1$ of all accumulation points of $\mathcal{K}$ is contained in $\mathcal{E}^{(\Omega)}$. By condition (i), $\mathcal{K}_1$ lies over $z^{(*)}$, and (iii) implies that the fiber $\mathcal{K}_1(z^{(*)})$ is uncountable (in fact, its cardinality is equal to that of the real number system $\mathbf{R}$). This contradicts the fact that $z^{(*)} \in e_\beta \subset e$, since $\mathcal{K}_1(z^{(*)}) \subset \mathcal{E}(z^{(*)})$ and $\mathcal{E}(z^{(*)})$ is countable. Consequently $\mathcal{E}^{(\Omega)} = \emptyset$, which proves Theorem 4.12. □

We make a remark on 1 of Theorem 4.12. Part of the conclusion is that there exists an analytic hypersurface σ defined in a neighborhood δ of p which is contained in $\mathcal{E}$ and which contains the point p; i.e.,

$$\sigma \subset \mathcal{E} \cap \delta.$$

If we assume the fibers $\mathcal{E}(z)$ are *discrete*, then we can get

$$\sigma = \mathcal{E} \cap \delta.$$

Precisely, let $\mathcal{E}$ be a pseudoconcave set in $D \times \mathbf{C}_w$. If each fiber $\mathcal{E}(z)$, $z \in D$, is a discrete subset of $\mathbf{C}_w$, then for any point $p = (z_0, w_0) \in \mathcal{E}$, there exists an analytic hypersurface σ defined in a neighborhood of p such that $\sigma = \mathcal{E} \cap \delta$.

To verify this, fix $p = (z_0, w_0) \in \mathcal{E}$. Since $\mathcal{E}(z_0)$ is discrete, we can choose $r > 0$ sufficiently small so that the circle $z = z_0$, $|w - w_0| = r$ does not intersect $\mathcal{E}$. Since $\mathcal{E}$ is closed, we can choose $\eta > 0$ sufficiently small so that the intersection of $\mathcal{E}$ with the polydisk $\delta : \ |z - z_0| < \eta, \ |w - w_0| < r$ has the properties that

1. $\mathcal{E} \cap \{|z - z_0| < \eta, \ |w - w_0| = r\} = \emptyset$, and
2. for $|z - z_0| < \eta$, the fiber $\mathcal{E}(z)$ is finite.

Applying Theorem 4.9 to $\mathcal{E} \cap \delta$, we conclude that $\mathcal{E} \cap \delta$ coincides with an analytic hypersurface σ in δ.

If we only assume the fibers $\mathcal{E}(z)$ are *countable* but not necessarily discrete, the conclusion is not true. For example, consider the pseudoconcave set $\mathcal{F}$ in $D \times \mathbf{C}_w$ from (4.31).

4.5.4. Thullen's Theorem. Using analytic derived sets, we shall prove the following theorem on analytic continuation of analytic sets.[12]

THEOREM 4.13 (Thullen). *Let $\mathcal{E}$ be a pure r-dimensional analytic set in a domain D in $\mathbf{C}^n$ $(n \geq 2)$. Let $\mathcal{F}$ be an analytic set in $D' = D \setminus \mathcal{E}$. If each irreducible component of $\mathcal{F}$ has dimension greater than or equal to r and if $\mathcal{F}$ can be analytically continued to at least one point of each irreducible component of $\mathcal{E}$, then $\mathcal{F}$ can be analytically continued to all points of $\mathcal{E}$, and the closure $\overline{\mathcal{F}}$ of $\mathcal{F}$ in D is an analytic set in D.*

PROOF. From Theorem 2.5 we may assume that $\mathcal{F}$ is pure r-dimensional and that $\mathcal{E}$ is irreducible in D (but $\mathcal{F}$ need not be irreducible in D).

We first consider the case when $r = n - 1$; thus $\mathcal{E}$ is an analytic hypersurface in D and $\mathcal{F}$ is an analytic hypersurface in D'. Then $S := \mathcal{E} \cup \mathcal{F}$ is a closed, pseudoconcave set in D (using 5 in section 4.4.1). Thus the analytic derived set S' of S in D is contained in $\mathcal{E}$. Since $\mathcal{F}$ can be analytically continued to at least one point p of $\mathcal{E}$, we can find a neighborhood δ of p such that $S' \cap \delta = \emptyset$. It follows from 6 in section 4.4.1 that $S' = \emptyset$. This means that $\mathcal{F}$ can be analytically continued to all points of $\mathcal{E}$ and implies that the closure $\overline{\mathcal{F}}$ of $\mathcal{F}$ in D is an analytic set in D.

In the case when $1 \leq r < n - 1$, we choose complex coordinates $(z_1, \ldots, z_n)$ which satisfy the Weierstrass condition at each point of $\mathcal{E}$ and $\mathcal{F}$ (Theorem 2.3). Fix $a = (a_1, \ldots, a_n) \in \mathcal{E}$. We let $D(a)$, $\mathcal{E}(a)$, and $\mathcal{F}(a)$ denote the sections of D, $\mathcal{E}$, and $\mathcal{F}$ over the $(n - r)$-dimensional plane $z_j = a_j$ $(j = 1, \ldots, r)$. Then $\mathcal{D}(a)$ is a domain in $\mathbf{C}^{n-r} = \mathbf{C}_{z_{r+1}} \times \ldots \times \mathbf{C}_{z_n}$, $\mathcal{E}(a)$ consists of isolated points in $D(a)$, and $\mathcal{F}(a)$ consists of isolated points in $D(a) \setminus \mathcal{E}(a)$. More precisely, $\mathcal{F}(a)$ may have accumulation points in D, but these points will lie in $\mathcal{E}(a)$. We choose $\eta > 0$ and then $\rho > 0$ sufficiently small so that, if we let $\Lambda_{(a)} = \Delta_{(a)} \times \Gamma_{(a)}$ denote the polydisk centered at a in D given by

$$\Delta_{(a)} : \ |z_j - a_j| \leq \rho \qquad (j = 1, \ldots, r),$$
$$\Gamma_{(a)} : \ |z_k - a_k| \leq \eta \qquad (k = r + 1, \ldots, n),$$

[12]This theorem was first discovered in the case of analytic hypersurfaces by P. Thullen [**74**]. In the case of analytic sets it was proved by R. Remmert and K. Stein [**69**]. The proof given here is due to K. Kato [**33**]. See also W. Rothstein [**64**].

then $\mathcal{E}(a) \cap \Gamma_{(a)}$ consists of only one point $(a_{r+1}, \dots, a_n)$ and

$$[\Delta_{(a)} \times (\partial\Gamma_{(a)})] \cap (\mathcal{E} \cup \mathcal{F}) = \emptyset.$$

We set

$$\mathcal{E}_{(a)} := \mathcal{E} \cap \Lambda_{(a)}, \qquad \mathcal{F}_{(a)} := \mathcal{F} \cap \Lambda_{(a)}.$$

In each complex plane $\mathbf{C}_{z_k}$ $(k = r+1, \dots, n)$, we let $\Gamma^k_{(a)}$ denote the disk $|z_k - a_k| \leq \eta$ centered at a_k, i.e., $\Gamma_{(a)} = \Gamma^{(1)}_{(a)} \times \cdots \times \Gamma^{(n)}_{(a)}$, and we define the polydisk

$$\Lambda^k_{(a)} := \Delta_{(a)} \times \Gamma^k_{(a)}$$

centered at $(a_1, \dots, a_r, a_k)$ in $\mathbf{C}^{r+1} = \mathbf{C}_{z_1} \times \cdots \times \mathbf{C}_{z_r} \times \mathbf{C}_{z_k}$. For each $k = r+1, \dots, n$, we have

$$[\Delta^k_{(a)} \cap \partial(\Gamma^{r+1}_{(a)} \times \cdots \times \widehat{\Gamma^k_{(a)}} \times \cdots \times \Gamma^n_{(a)})] \cap (\mathcal{E} \cup \mathcal{F}) = \emptyset$$

(where $\widehat{A}$ means that we omit A). From Proposition 2.3 it follows that the projection $\mathcal{E}^k_{(a)}$ of the analytic set $\mathcal{E}_{(a)}$ onto $\Lambda^k_{(a)}$ is an analytic hypersurface in $\Lambda^k_{(a)}$. By taking a linear coordinate transformation which is sufficiently close to the identity, if necessary, we may impose the assumption $(*)$ that there exists at least one point $z^j = (z^j_1, \dots, z^j_n)$ on each irreducible component $\mathcal{F}^j_{(a)}$ $(j = 1, 2, \dots)$ of $\mathcal{F}_{(a)}$ such that $(z^j_1, \dots, z^j_r, z^j_k) \notin \mathcal{E}^k_{(a)}$ for each $k = r+1, \dots, n$. The projection $\mathcal{F}^k_{(a)}$ of the analytic set $\mathcal{F}_{(a)}$ in $\Lambda_{(a)} \backslash \mathcal{E}_{(a)}$ onto $\Lambda^k_{(a)}$ is an analytic hypersurface in $\Lambda^k_{(a)} \backslash \mathcal{E}^k_{(a)}$ (to be precise, the non-empty set $\mathcal{F}^k_{(a)} \setminus \mathcal{E}_{(a)}$ is analytic in $\Lambda_{(a)} \setminus \mathcal{E}_{(a)}$). We note that if $\mathcal{F}_{(a)}$ can be analytically continued to all points of $\mathcal{E}_{(a)}$, then each $\mathcal{F}^k_{(a)}$ $(k = r+1, \dots, n)$ can be analytically continued to all points of $\mathcal{E}^k_{(a)}$. The converse is also true under the assumption $(*)$ from the definition of an analytic set (cf. Theorem 2.2). Hence, using the case when $r = n-1$, we see that if $\mathcal{F}_{(a)}$ can be analytically continued to at least one point of $\mathcal{E}_{(a)}$, then $\mathcal{F}_{(a)}$ can be analytically continued to all points of $\mathcal{E}_{(a)}$ in $\Lambda_{(a)}$. Thus if we set

$$\mathcal{E}_0 := \{z \in \mathcal{E} \mid \mathcal{F} \text{ can be analytically continued to the point } z\},$$

then $\mathcal{E}_0$ is a non-empty open subset without relative boundary in $\mathcal{E}$. Since $\mathcal{E}$ is irreducible, we have $\mathcal{E}_0 = \mathcal{E}$. Thus the theorem is proved in the case when $1 \leq r < n-1$. □

Isolated essential singular points Let D be a domain in $\mathbf{C}^n$. Let $\mathcal{E}$ be an analytic hypersurface in D, and let $f(z)$ be a holomorphic function in $D \setminus \mathcal{E}$. Fix $p \in \mathcal{E}$. If $f(z)$ cannot be extended holomorphically or meromorphically to p, then we say that p is an **essential singular point** of $f(z)$. If $\mathcal{E}$ is irreducible and if at least one point p of $\mathcal{E}$ is an essential singular point of $f(z)$, then all points of $\mathcal{E}$ are essential singular points of $f(z)$. This fact follows immediately from Theorems 4.1 and 4.2. Moreover, we have the following result.

COROLLARY 4.2. *Let $\mathcal{E}$ be an irreducible analytic hypersurface in a domain D in $\mathbf{C}^n$ and let $f(z)$ be a holomorphic function in $D \setminus \mathcal{E}$. Assume that $f(z)$ has at least one essential singular point p on $\mathcal{E}$. Then for any complex number $a \in \mathbf{C}$, with at most one exception, the analytic hypersurface defined by*

$$S_a := \{z \in D \setminus \mathcal{E} : f(z) = a\}$$

cannot be analytically continued to any point of $\mathcal{E}$.

PROOF. We prove this by contradiction. Hence assume that there exist at least two distinct complex numbers a_i $(i = 1, 2)$ such that there exists at least one point b_i on $\mathcal{E}$ to which S_{a_i} can be analytically continued. Thullen's theorem implies that the closure Σ_i of S_{a_i} in D is an analytic hypersurface in D. Therefore, $f(z)$ is a holomorphic function in $D \setminus [\mathcal{E} \cup \Sigma_1 \cup \Sigma_2]$ which does not attain the values a_1 or a_2. By Picard's theorem for one complex variable, $f(z)$ can be holomorphically extended to $D \setminus \sigma$, where σ is the analytic set of non-regular points of $\mathcal{E} \cup \Sigma_1 \cup \Sigma_2$ in D. Since dim $\sigma \leq n - 2$, it follows that $f(z)$ is holomorphic on all of D, contradicting our assumption. □

CHAPTER 5

Holomorphic Mappings

5.1. Holomorphic Mappings of Elementary Domains

In the theory of functions of one complex variable, conformal mappings and conformal equivalence play an important role. We will analyze the analogous notions in several complex variables.

Let D_1 and D_2 be domains in $\mathbf{C}^n$. If there exists a one-to-one holomorphic mapping from D_1 onto D_2, then we say that D_1 and D_2 are **biholomorphically equivalent**. In one complex variable, the Riemann mapping theorem states that all simply connected proper subdomains of $\mathbf{C}$ are biholomorphically equivalent to the unit disk. However, in $\mathbf{C}^n$ for $n \geq 2$, the unit polydisk is not biholomorphically equivalent to the unit ball. This was discovered by H. Poincaré [**60**]. We give a proof of this fact by elementary methods in the next section.

5.1.1. Schwarz Lemma. We first extend the Schwarz lemma of one complex variable to the case of several complex variables. Let D be a domain in $\mathbf{C}^n$ $(n \geq 2)$ which contains the origin 0. If the intersection $D \cap l$ of D with a complex line l passing through 0 is always a disk in l centered at the origin, we say that D is of **disk type with respect to** 0 or that D is **completely circled with respect to** 0. Equivalently, this means that whenever $(z_1, \ldots, z_n) \in D$, then $\{(tz_1, \ldots, tz_n) \mid |t| \leq 1,\ t \in \mathbf{C}\} \subset D$. For example, balls, polydisks, and, more generally, complete Reinhardt domains are of disk type about their centers.

Let D be a domain in $\mathbf{C}^n$. Fix $r > 0$ and consider the homothetic transformation T_r of $\mathbf{C}^n$ given by

$$T_r \; : \; z'_j = r z_j \qquad (j = 1, \ldots, n).$$

Setting

$$D^{(r)} = T_r(D),$$

we have that $D^{(r)}$ is a domain homothetic to D with ratio r. Let $A = (a_{jk})_{j,k=1,\ldots,n}$ be a non-singular matrix and define

$$S_A \; : \; z = (z_1, \ldots, z_n) \in \mathbf{C}^n \to z' = (z'_1, \ldots, z'_n) = Az \in \mathbf{C}^n.$$

If D is a domain in $\mathbf{C}^n$ of disk type about the origin 0, then $S_A(D)$ is also of disk type about 0, and clearly $S_A(D^{(r)}) = S_A(D)^{(r)}$ for any $r > 0$. If A is a unitary matrix, we call S_A a **unitary transformation** of $\mathbf{C}^n$.

We prove the following generalization of the Schwarz lemma.

LEMMA 5.1 (Schwarz lemma). *Let D be a domain in $\mathbf{C}^n$ which is of disk type about the origin 0. Let $f(z)$ be a holomorphic function in D with $f(0) = 0$. If $|f(z)| \leq M$ on D, then for any $0 < r < 1$,*

$$|f(z)| \leq Mr \quad \textit{for } z \in D^{(r)}.$$

PROOF. Let $0 < r < 1$ and let $z^0 \in D^{(r)}$. We want to show that $|f(z^0)| \le Mr$. By use of a unitary transformation of $\mathbf{C}^n$ we may assume $z^0 = (z_1^0, 0, \ldots, 0)$.

Let l be the complex line $l : z_2 = \cdots = z_n = 0$ and set $D_1 := D \cap l$. We regard D_1 as a disk centered at $z_1 = 0$ with radius $R > 0$ in $\mathbf{C}_{z_1}$. We restrict $f(z)$ to D_1, and set

$$\phi(z_1) := f(z_1, 0, \ldots, 0), \quad z_1 \in D_1.$$

Then $|\phi(z_1)| \le M$ on D_1 and $\phi(0) = 0$. Since $|z_1^0| \le rR$, it follows from the Schwarz lemma in one complex variable that $|\phi(z_1^0)| \le M|z_1^0| \le Mr$. Thus $|f(z^0)| \le Mr$. □

Using this lemma, we will deduce the following result.

THEOREM 5.1 (Poincaré). *In $\mathbf{C}^n$ for $n \ge 2$, the ball*

$$\mathcal{Q} \,:\, ||z||^2 := |z_1|^2 + \cdots + |z_n|^2 < 1$$

and the polydisk

$$\Delta \,:\, |z_j| < 1 \qquad (j = 1, \ldots, n)$$

are not biholomorphically equivalent.

PROOF. For the sake of obtaining a contradiction, we assume that there exists a one-to-one holomorphic mapping T from $\mathcal{Q}$ onto Δ. Composing with a linear transformation

$$z_j' = \frac{z_j - a_j}{1 - \bar{a}_j z_j} \quad (j = 1, \ldots, n)$$

which maps the unit disk Δ_j in $\mathbf{C}_{z_j}$ onto itself, we may assume that $T(0) = 0$. We shall prove that

$$T(\mathcal{Q}^{(r)}) = \Delta^{(r)} \quad \text{for any } 0 < r < 1. \tag{5.1}$$

Set

$$T: \quad z \in \mathcal{Q} \to w = (f_1(z), \ldots, f_n(z)) \in \Delta,$$

$$T^{-1} \quad : w \in \Delta \to z = (g_1(w), \ldots, g_n(w)) \in \mathcal{Q}.$$

Fix $0 < r < 1$. Since $|f_j(z)| \le 1$ $(j = 1, \ldots, n)$ in $\mathcal{Q}$ and $f_j(0) = 0$, from the Schwarz lemma we obtain that $|f_j(z)| \le r$ in $\mathcal{Q}^{(r)}$, and hence $T(\mathcal{Q}^{(r)}) \subset \Delta^{(r)}$. Conversely, let $w^0 \in \Delta^{(r)}$ and let $z^0 = T^{-1}(w^0)$. We take a unitary transformation $\zeta = S_0(z)$ of $\mathbf{C}^n$ with $S_0(z^0) = (\zeta_1^0, 0, \ldots, 0)$; thus $\|z^0\| = |\zeta_1^0|$. We consider the holomorphic mapping

$$\zeta = S_0 \circ T^{-1}(w) = (\phi_1(w), \ldots, \phi_n(w))$$

from Δ onto $\mathcal{Q}$. Note that $\phi_1(w^0) = \zeta_1^0$. Since $|\phi_1(w)| \le 1$ on Δ and $\phi_1(0) = 0$, again using the Schwarz lemma we see that $|\phi_1(w^0)| \le r$, and hence $\|z^0\| \le r$. Thus, $z^0 \in \mathcal{Q}^{(r)}$. We obtain (5.1). The boundary of $\mathcal{Q}^{(r)}$ is smooth everywhere; this is not the case for the boundary of $\Delta^{(r)}$. This contradicts the fact that $T(\partial\mathcal{Q}^{(r)}) = \partial\Delta^{(r)}$, which follows from (5.1). □

REMARK 5.1. Using a similar proof, we obtain the following fact. Let D be a ball or a polydisk centered at the origin 0 in $\mathbf{C}^n$. Let Φ be a holomorphic mapping from D into D such that $\Phi(0) = 0$. Then $\Phi(D^{(r)}) \subset D^{(r)}$ for each $0 < r < 1$. Thus if Φ is a one-to-one holomorphic mapping from D onto D such that $\Phi(0) = 0$, then $\Phi(D^{(r)}) = D^{(r)}$ for each $0 < r < 1$.

5.1.2. Automorphisms of the Polydisk. Let D be a domain in $\mathbf{C}^n$. A one-to-one holomorphic mapping from D onto itself is called an **automorphism** of D. The set of all automorphisms of D forms a group under composition, which we call the **automorphism group** $\mathcal{A}(D)$ of D. In this section we determine the automorphism group $\mathcal{A}(\Delta)$ of the unit polydisk $\Delta : |z_j| < 1 \quad (j = 1, \ldots, n)$. Given $a = (a_1, \ldots, a_n) \in \Delta$, we define the component-wise linear fractional transformation

$$\mathcal{T}_{(a)} : z'_j = \frac{z_j - a_j}{1 - \bar{a}_j z_j} \qquad (j = 1, \ldots, n).$$

Given $\theta = (\theta_1, \ldots, \theta_n) \in \mathbf{R}^n$, we have the rotation

$$\mathcal{R}_{(\theta)} : z'_j = e^{i\theta_j} z_j \qquad (i^2 = -1;\ j = 1, \ldots, n).$$

Given a permutation $(k) = \left(\begin{smallmatrix} 1, \ldots, n \\ k_1, \ldots, k_n \end{smallmatrix} \right)$ of $(1, \ldots, n)$, we define

$$\mathcal{P}_{(k)} : z'_j = z_{k_j} \qquad (j = 1, \ldots, n).$$

Clearly these linear fractional transformations, rotations and permutations are elements of $\mathcal{A}(\Delta)$, and we have $\mathcal{T}_{(a)}^{-1} = \mathcal{T}_{(-a)}$, $\mathcal{R}_{(\theta)}^{-1} = \mathcal{R}_{(-\theta)}$ and $\mathcal{P}_{(k)}^{-1} = \mathcal{P}_{(k^{-1})}$, where $(k^{-1}) = \left(\begin{smallmatrix} k_1, \ldots, k_n \\ 1, \ldots, n \end{smallmatrix} \right)$.

We have the following theorem.

THEOREM 5.2. *$\mathcal{A}(\Delta)$ is generated by the elements $\mathcal{T}_{(a)}$, $\mathcal{R}_{(\theta)}$ and $\mathcal{P}_{(k)}$.*

PROOF. Let $T \in \mathcal{A}(\Delta)$ and let $T(0) = a$. Setting $T_1 := \mathcal{T}_{(a)} \circ T \in \mathcal{A}(\Delta)$, we have $T_1(0) = 0$. We write $T_1 : w_j = f_j(z)$ $(j = 1, \ldots, n)$. By 1 of Remark 5.1, $T_1(\Delta^{(r)}) = \Delta^{(r)}$ for each $0 < r < 1$. Hence

(1) $T_1(\partial\Delta^{(r)}) = \partial\Delta^{(r)}$, and

(2) $|f_j(z)| \leq r$ $(j = 1, \ldots, n)$ in $\Delta^{(r)}$.

We set $\Delta_j := \{|z_j| < 1\}$ and $\Delta_j^{(r)} := \{|z_j| < r\}$ $(j = 1, \ldots, n)$. Fix $0 < r < 1$. Since $(r, 0, \ldots, 0) \in \partial\Delta^{(r)}$, (1) implies that $T_1(r, 0, \ldots, 0) \in \partial\Delta^{(r)}$. In addition, since

$$\partial\Delta^{(r)} = \bigcup_{i=1}^{n} [\Delta_1^{(r)} \times \cdots \times (\partial\Delta_i^{(r)}) \times \cdots \times \Delta_n^{(r)}],$$

there exists k_1 $(1 \leq k_1 \leq n)$ with $|f_{k_1}(r, 0, \ldots, 0)| = r$. Since $|f_{k_1}(z_1, 0, \ldots, 0)| \leq 1$ for z_1 in Δ_1 and $f_{k_1}(0) = 0$, from the one-variable Schwarz lemma we conclude that

(3) $f_{k_1}(z_1, 0, \ldots, 0) = e^{i\theta_1} z_1$ for z_1 in Δ_1,

where θ_1 is a constant with $0 \leq \theta_1 < 2\pi$. Fix $z_1^0 \in \Delta_1$ and set

$$\Delta_{n-1}^0 := \{z' = (z_2, \ldots, z_n) \mid |z_j| \leq |z_1^0| \ (j = 2, \ldots, n)\}.$$

We set

$$\phi(z') := f_{k_1}(z_1^0, z_2, \ldots, z_n) \quad \text{in } \Delta_{n-1}^0,$$

so that $|\phi(z')| \leq |z_1^0|$ (from (2)). Using (3) we conclude that $\phi(z')$ attains its maximum modulus $|z_1^0|$ at $z' = 0$. Thus $\phi(z')$ is constant on Δ_{n-1}^0, and hence $f_{k_1}(z_1^0, z_2, \ldots, z_n) \equiv e^{i\theta_1} z_1^0$ in Δ_{n-1}^0. Since $z_1^0 \in \Delta_1$ was arbitrary, $f_{k_1}(z) \equiv e^{i\theta_1} z_1$ in Δ.

Similarly, for each $j = 1, \ldots, n$, we can find an integer $1 \leq k_j \leq n$ and a constant θ_j with $0 \leq \theta_j < 2\pi$ such that

$$w_{k_j} = f_{k_j}(z) \equiv e^{i\theta_j} z_j \quad (j = 1, \ldots, n).$$

Set $(\theta) := (\theta_1, \dots, \theta_n)$ and $(k) := \left(\begin{smallmatrix} 1, \dots, n \\ k_1, \dots, k_n \end{smallmatrix}\right)$. Then $T_1 = \mathcal{P}_{(k)} \circ \mathcal{R}_{(\theta)}$, so that $T = \mathcal{T}_{(-a)} \circ \mathcal{P}_{(k)} \circ \mathcal{R}_{(\theta)}$. □

5.1.3. Uniqueness Theorem. In this section we prove a uniqueness theorem of H. Cartan [**8**] for holomorphic mappings of bounded domains in $\mathbf{C}^n$. We then give some applications of this result.

THEOREM 5.3 (Cartan). *Let D be a bounded domain in $\mathbf{C}^n$ $(n \geq 1)$ containing the origin 0. Let*

$$T \; : \; z = (z_1, \dots, z_n) \to w = (f_1(z), \dots, f_n(z))$$

be a holomorphic mapping (not necessarily one-to-one) from D into D with $T(0) = 0$. Assume that

$$f_j(z) = z_j + \sum_{\nu=2}^{\infty} f_{j,\nu}(z) \quad (j = 1, \dots, n) \tag{5.2}$$

near $z = 0$, where $f_{j,\nu}$ is a homogeneous polynomial of degree $\nu \geq 2$. Then T is the identity mapping.

PROOF. For the sake of obtaining a contradiction, we assume that T is not the identity mapping, i.e., $f_{j,\nu}(z) \not\equiv 0$ for some $1 \leq j \leq n$ and some $\nu \geq 2$. For $j = 1, \dots, n$, let ν_j be the smallest integer greater than or equal to 2 such that $f_{j,\nu_j}(z) \not\equiv 0$. Since $T(D) \subset D$, we can consider the iterates $T^{(l)} = T \circ \cdots \circ T$ of T for $l = 1, 2, \dots$. These all map D into D with $T^{(l)}(0) = 0$. We write

$$T^{(l)} \; : \; z = (z_1, \dots, z_n) \to w = (f_1^{(l)}(z), \dots, f_n^{(l)}(z)).$$

Since D is bounded, for each $j = 1, \dots, n$ the family of holomorphic functions

$$\mathcal{F}_j = \{f_j^{(l)}(z) \mid l = 1, 2, \dots\}$$

in D forms a normal family.

However, a simple calculation using (5.2) yields

$$f_j^{(l)}(z) = z_j + l f_{j,\nu_j}^{(l)}(z) + F_{\nu_j+1}(z) \quad (j = 1, \dots, n)$$

near $z = 0$, where $F_{\nu_j+1}(z)$ consists of sums of homogeneous polynomials of degree $\geq \nu_j + 1$. Since $f_{j,\nu_j}^{(l)}(z) \not\equiv 0$ for some $1 \leq j \leq n$ and some $\nu_j \geq 2$, this contradicts the normality of the corresponding family $\mathcal{F}_j$ in a neighborhood of $z = 0$. □

REMARK 5.2. Let D be a bounded domain in $\mathbf{C}^n$ and fix $z^0 \in D$. We consider the **isotropy subgroup** $\mathcal{A}_0(D)$ of the automorphism group $\mathcal{A}(D)$ of D consisting of the elements $T \in \mathcal{A}(D)$ which fix z_0; i.e., $T(z^0) = z^0$. Cartan's theorem implies that each $T \in \mathcal{A}_0(D)$ is uniquely determined by its Jacobian matrix at z^0.

As an application of Cartan's theorem, we prove the following.

COROLLARY 5.1. *Let D_1, D_2 be bounded domains of disk type with respect to the origin 0 in $\mathbf{C}^n$. Let ζ be a biholomorphic mapping of D_1 onto D_2 with $\zeta(0) = 0$. Then ζ is the restriction to D_1 of a linear transformation of $\mathbf{C}^n$.*

PROOF. Given $0 \leq \theta < 2\pi$, we consider the rotation

$$\mathcal{R}_{(\theta)} \; : \; z_j' = e^{i\theta} z_j \qquad (i^2 = -1; \; j = 1, \dots, n).$$

Since D_2 is of disk type with respect to the origin, $\mathcal{R}_{(\theta)}$ is an automorphism of D_2. The same is true for D_1 and $\mathcal{R}_{(-\theta)}$. Therefore,

$$\zeta^* := \mathcal{R}_{(-\theta)} \circ \zeta^{-1} \circ \mathcal{R}_{(\theta)} \circ \zeta$$

is an automorphism of D_1 with $\zeta^*(0) = 0$; moreover, if we set $\zeta^*(z) := (f_1(z), \dots, f_n(z))$, then each $f_j(z)$ $(j = 1, \dots, n)$ is of the form

$$f_j(z) = z_j + \sum_{\nu=2}^{\infty} f_{j,\nu}(z) \quad (j = 1, \dots, n),$$

where $f_{j,\nu}$ is a homogeneous polynomial of degree $\nu \geq 2$. Since D_1 is bounded in $\mathbf{C}^n$, it follows from Cartan's theorem that ζ^* is the identity mapping on D_1. We thus have

$$\zeta \circ \mathcal{R}_{(\theta)} = \mathcal{R}_{(\theta)} \circ \zeta \quad \text{for all } 0 \leq \theta < 2\pi.$$

It is easy to see that this implies that ζ must be a linear mapping of $\mathbf{C}^n$. □

REMARK 5.3. The proof shows that the same conclusion is valid for any bounded domains D_i $(i = 1, 2)$ in $\mathbf{C}^n$ such that $\mathcal{A}(D_i)$ contains all rotations $\mathcal{R}_{(\theta)}$ $(0 \leq \theta < 2\pi)$. As a simple application we see that a complex ellipsoid

$$E : \sum_{j=1}^{n} a_j |z_j|^{\nu_j} < 1,$$

where a_j, $\nu_j > 0$ $(j = 1, \dots, n)$, is biholomorphically equivalent to the unit ball $\mathcal{Q}$ if and only if $\nu_j = 2$ $(j = 1, \dots, n)$. In fact, assume that there exists a biholomorphic mapping T of $\mathcal{E}$ onto $\mathcal{Q}$. We may assume $T(0) = 0$ by Remark 5.4 (which will appear after Theorem 5.4). Thus, T is a linear transformation of $\mathbf{C}^n$, and hence $T(\partial\mathcal{E}) = \partial\mathcal{Q}$. By a simple calculation, this implies $\nu_j = 2$ for each $j = 1, \dots, n$.

5.1.4. Automorphisms of the Ball. In this section we determine the automorphism group $\mathcal{A}(\mathcal{Q})$ of the unit ball $\mathcal{Q} : \sum_{j=1}^{n} |z_j|^2 < 1$. First of all, we let $A = (a_{j,k})_{j,k=1,\dots,n}$ be an $n \times n$ unitary matrix. Then

$$S_A : z = (z_1, \dots, z_n) \to z' = (z'_1, \dots, z'_n) = Az$$

is a unitary transformation of $\mathbf{C}^n$. Clearly $S_A \in \mathcal{A}(\mathcal{Q})$ and $(S_A)^{-1} = S_{A^{-1}}$. Next we let a be a complex number with $|a| < 1$. For each $1 \leq j \leq n$, define

$$T_a^i \; : \; z'_i = \frac{z_i - a}{1 - \bar{a} z_i}, \qquad z'_j = \frac{\sqrt{1 - |a|^2}}{1 - \bar{a} z_i} z_j \quad (j \neq i).$$

We note that $T_a^i(0, \dots, 0, a, z_{i+1}, \dots, z_n) = (0, \dots, 0, 0, z'_{i+1}, \dots, z'_n)$. In particular, $T_a^1(a, 0, \dots, 0) = (0, \dots, 0)$. Clearly $T_a^i \in \mathcal{A}(\mathcal{Q})$ and $(T_a^i)^{-1} = T_{-a}^i$.

THEOREM 5.4. *$\mathcal{A}(\mathcal{Q})$ is generated by S_A, A unitary, and T_a^i, $|a| < 1$, $i = 1, \dots, n$.*

PROOF. Let $T \in \mathcal{A}(\mathcal{Q})$. Composing T with a finite number (at most n) of the mappings $T_{a_i}^i$ $(i = 1, \dots, \nu;\ \nu \leq n)$ (or composing T with an S_A and a T_a), which we denote by F, we can form an automorphism $\widetilde{T} := F \circ T \in \mathcal{A}(\mathcal{Q})$ with $\widetilde{T}(0) = 0$. Applying Corollary 5.1, we conclude that $\widetilde{T}$ is a linear mapping of $\mathbf{C}^n$. Since $\widetilde{T}(\mathcal{Q}) = \mathcal{Q}$, it follows that $\widetilde{T}$ is a unitary transformation S_A of $\mathbf{C}^n$. Hence $T = F^{-1} \circ S_{A^{-1}}$. □

REMARK 5.4. Let D be a domain in $\mathbf{C}^n$ $(n \geq 1)$. If for any two distinct points p, q in D there exists an automorphism T of D such that $T(p) = q$, then D is called a **homogeneous domain** and $\mathcal{A}(D)$ is said to act **transitively** on D. As shown above, the ball and the polydisk in $\mathbf{C}^n$ are homogeneous domains. In $\mathbf{C}$, every simply connected domain is homogeneous. This is no longer true in $\mathbf{C}^n$ if $n \geq 2$.

For example, in $\mathbf{C}^2$ with variables z and w, we consider the domain

$$D \ : \ |zw| < 1.$$

Then D is simply connected, and every automorphism T of D satisfies $T(0,0) = (0,0)$. For let $T : (z,w) \to (z',w') := (f(z,w), g(z,w))$. If we set $F(z) := f(z,0)g(z,0)$ for $z \in \mathbf{C}_z$, then F is an entire function in $\mathbf{C}_z$ with $|F(z)| < 1$. Thus, $F(z) \equiv a$ (constant) in $\mathbf{C}_z$. We claim that $a = 0$. For if not, under T the z-axis $w = 0$ is mapped into the set $z'w' = a \neq 0$ in a one-to-one fashion. This is impossible, since the set $z'w' = a$ is not simply connected. Therefore $a = 0$, which means that $f(z,0) \equiv 0$ or $g(z,0) \equiv 0$ in $\mathbf{C}_z$ (note that we can't have both $f(z,0) \equiv 0$ and $g(z,0) \equiv 0$, since if we did we could have $T(z,0) \equiv (0,0)$, contradicting the fact that T is an automorphism). Thus the z-axis $w = 0$ is mapped by T onto either the z'-axis or the w'-axis. In a similar manner, either $f(0,w) \equiv 0$ or $g(0,w) \equiv 0$ in $\mathbf{C}_w$, and it follows that the w-axis $z = 0$ is mapped by T onto the w'-axis (if $f(z,0) \equiv 0$) or the z'-axis (if $g(z,0) \equiv 0$). In either case, since if we did we could have $T(0,0) = (0,0)$ and D is not homogeneous.

Indeed, if $D \subset \mathbf{C}^n$, $n \geq 2$, is a bounded domain with smooth boundary and if D has a transitive automorphism group $\mathcal{A}(D)$, then D is biholomorphically equivalent to the unit ball $\mathcal{Q}$. This is a result of J.-P. Rosay [**63**].

5.2. Holomorphic Mappings of $\mathbf{C}^n$

There are many interesting phenomena concerning holomorphic mappings of $\mathbf{C}^n$ for $n > 1$ which do not occur in one complex variable. In this section we discuss certain holomorphic mappings of $\mathbf{C}^n$ which were studied by Poincaré and Picard (see Picard [**57**]).

5.2.1. Transcendental Entire Mappings of Poincaré –Picard. In this section, we consider polynomial mappings of $\mathbf{C}^n$ into $\mathbf{C}^n$, i.e.,

$$T_P \ : \ z'_j = P_j(z_1, \dots, z_n) \qquad (j = 1, \dots, n),$$

where each $P_j(z_1, \dots, z_n)$ $(j = 1, \dots, n)$ is a polynomial in $z_1, \dots, z_n$. We assume that $P_j(z_1, \dots, z_n)$ is of the form

$$P_j(z) = a_j z_j + p_j(z_1, \dots, z_n) \qquad (j = 1, \dots, n),$$

where

(1) $|a_j| > 1$ $(j = 1, \dots, n)$;
(2) $p_j(z_1, \dots, z_n) = \sum_{\nu=2}^{m_j} f_{j,\nu}(z_1, \dots, z_n)$ $(j = 1, \dots, n)$, where $f_{j,\nu}$ is a homogeneous polynomial of degree $\nu \geq 2$; and
(3) for each $j = 1, \dots, n$ and nonnegative integers $k_1, \dots, k_n$ with $k_1 + \cdots + k_n \geq 2$, we have
$$a_1^{k_1} \cdots a_n^{k_n} \neq a_j.$$

We write $a := (a_1, \dots, a_n)$ and $az := (a_1 z_1, \dots, a_n z_n)$.

We have the following proposition.

PROPOSITION 5.1. *Given a polynomial mapping $T_P : z'_j = P_j(z_1, \dots, z_n)$ $(j = 1, \dots, n)$ satisfying (1), (2) and (3), there exist n entire functions $F_j(z)$ $(j = 1, \dots n)$ in $\mathbf{C}^n$ which satisfy the simultaneous functional equations*

$$F_j(az) = P_j(F_1(z), \dots, F_n(z)) \qquad (j = 1, \dots, n) \tag{5.3}$$

and are of the form

$$F_j(z) = z_j + \sum_{\nu=2}^{\infty} F_{j,\nu}(z_1, \dots, z_n) \quad (j = 1, \dots, n), \tag{5.4}$$

where $F_{j,\nu}$ is a homogeneous polynomial of degree $\nu \geq 2$. Furthermore, the F_j are unique.

Using the entire functions $F_j(z)$ $(j = 1, \dots, n)$, we form the holomorphic mapping

$$S_F \; : \; z'_j = F_j(z_1, \dots, z_n) \qquad (j = 1, \dots, n)$$

of $\mathbf{C}^n$. This is called the **Poincaré–Picard entire mapping** of $\mathbf{C}^n$ associated to the polynomial mapping T_P of $\mathbf{C}^n$. It satisfies

$$S_F(az) = T_P \circ S_F(z) \qquad \text{in } \mathbf{C}^n. \tag{5.5}$$

We note that if $P(z)$ is of degree at least 2, i.e., the degree of at least one $P_j(z)$ $(j = 1, \dots, n)$ is greater than or equal to 2, then S_F is a transcendental mapping of $\mathbf{C}^n$.

REMARK 5.5.

1. Let $T_Q : z'_j = Q_j(z_1, \dots, z_n)$ $(j = 1, \dots, n)$ be a polynomial mapping with $T_Q(0) = 0$ and let $J_{T_Q}(z) = \partial(Q_1, \dots, Q_n)/\partial(z_1, \dots, z_n)$ be the Jacobian matrix of Q at $z \in \mathbf{C}^n$. If $J_{T_Q}(0)$ is diagonalizable and has eigenvalues λ_j $(j = 1, \dots, n)$ with $|\lambda_j| > 1$ $(j = 1, \dots, n)$, then the polynomial mapping T_Q satisfies (1), (2) and (3) at $z = 0$ (after a coordinate change to diagonalize $J_{T_Q}(0)$).
2. Equation (5.5) for S_F may be regarded as a generalization of the type of relation certain transcendental entire functions satisfy in the complex plane. For example, if we set $z' = P(z) = -4z^3 + 3z$ in $\mathbf{C}$ and take $a = 3$, then the unique solution $F(z)$ of the equation $F(az) = P \circ F(z)$ with $F(z) = z + o(z^2)$ is $F(z) = \sin z$.

In this section we will use the notation

$$\Delta_\rho \; : \; |z_j| < \rho \qquad (j = 1, \dots, n)$$

for the polydisk centered at the origin $z = 0$ with radius $\rho > 0$, and for $a = (a_1, \dots, a_n) \in \mathbf{C}^n$ with $a_j \neq 0$ we write

$$\Delta_\rho^{(-1)} \; : \; |z_j| < \rho/|a_j| \qquad (j = 1, \dots, n).$$

Finally, if g is a holomorphic function in a neighborhood of the origin, we write

$$g(z) = g_2(z) + o(|z|^2)$$

to signify that the Taylor series expansion of g about the origin contains neither constant nor linear terms; $g_2(z)$ denotes the quadratic terms. To prove Proposition 5.1 we need two lemmas.

LEMMA 5.2. *Let $\varphi(z)$ be a holomorphic function in Δ_ρ with $\varphi(z) = \varphi_2(z) + o(|z|^2)$. Let $|a_j| > 1$ $(j = 0, 1, \ldots, n)$ and let $a = (a_1, \ldots, a_n) \in \mathbf{C}^n$ satisfy*

$$a_1^{k_1} \cdots a_n^{k_n} \neq a_0 \tag{5.6}$$

for all integers $k_j \geq 0$ $(j = 1, \ldots, n)$ such that $\sum_{j=1}^n k_j \geq 2$. Then there exists a holomorphic function $g(z)$ in the polydisk Δ_ρ with $g(z) = g_2(z) + o(|z|^2)$ which satisfies the functional equation

$$g(az) = a_0 g(z) + \varphi(z) \quad \text{in } \Delta_\rho^{(-1)}. \tag{5.7}$$

Furthermore, there exists a number $k > 0$, depending only on a_j $(j = 0, 1, \ldots, n)$, such that if $|\varphi(z)| \leq M$ in Δ_ρ, then

$$|g(z)| \leq kM \quad \text{in } \Delta_\rho.$$

The function g with these properties is unique.

REMARK 5.6. We will see from the proof that we can take, for example,

$$k = \sum_{k_1 + \cdots + k_n \geq 2} \left| \frac{1}{a_1^{k_1} \cdots a_n^{k_n} - a_0} \right| < \infty.$$

PROOF. Let $g(z)$ be a holomorphic function in the polydisk Δ_ρ with $g(z) = g_2(z) + o(|z|^2)$, and consider the Taylor series expansions of $\varphi(z)$ and $g(z)$ about $z = 0$:

$$\begin{aligned} \varphi(z) &= \sum_{k_1 + \cdots + k_n \geq 2} c_{k_1, \ldots, k_n} z_1^{k_1} \cdots z_n^{k_n}, \\ g(z) &= \sum_{k_1 + \cdots + k_n \geq 2} u_{k_1, \ldots, k_n} z_1^{k_1} \cdots z_n^{k_n}. \end{aligned}$$

Assume that $g(z)$ satisfies the functional equation (5.7). From condition (5.6) we obtain

$$u_{k_1, \ldots, k_n} = \frac{c_{k_1, \ldots, k_n}}{a_1^{k_1} \cdots a_n^{k_n} - a_0} \tag{5.8}$$

provided $\sum_{j=1}^n k_j \geq 2$. Since $|a_j| > 1$ $(j = 1, \ldots, n)$, the infinite series

$$k := \sum_{k_1 + \cdots + k_n \geq 2} \left| \frac{1}{a_1^{k_1} \cdots a_n^{k_n} - a_0} \right|$$

is convergent. Also, since $|\varphi(z)| \leq M$ on Δ_ρ, the Cauchy estimates give

$$|c_{k_1, \ldots, k_n}| \leq \frac{M}{\rho^{k_1 + \cdots + k_n}}.$$

Thus for any $z \in \Delta_\rho$,

$$\begin{aligned} |g(z)| &\leq \sum_{k_1 + \cdots + k_n \geq 2} |u_{k_1, \ldots, k_n} z_1^{k_1} \cdots z_n^{k_n}| \\ &= \sum_{k_1 + \cdots + k_n \geq 2} \left| \frac{c_{k_1, \ldots, k_n}}{a_1^{k_1} \cdots a_n^{k_n} - a_0} \right| |z_1^{k_1} \cdots z_n^{k_n}| \\ &\leq M \sum_{k_1 + \cdots + k_n \geq 2} \left| \frac{1}{a_1^{k_1} \cdots a_n^{k_n} - a_0} \right| = kM. \end{aligned}$$

We conclude that if we now *define*

$$g(z) := \sum_{k_1+\cdots+k_n\geq 2} \frac{c_{k_1,\ldots,k_n}}{a_1^{k_1}\cdots a_n^{k_n} - a_0} z_1^{k_1}\cdots z_n^{k_n},$$

then $g(z)$ is a holomorphic function in Δ_ρ; $|g(z)| \leq kM$ in Δ_ρ; and $g(z)$ satisfies the functional equation (5.7) in the polydisk $\Delta_\rho^{(-1)}$ centered at 0. The uniqueness of $g(z)$ follows from the uniqueness of the Taylor series coefficients (5.8). □

LEMMA 5.3. *Let $\psi(z, Z)$ be a holomorphic function in a polydisk $\Delta \times \Delta$ centered at $(0,0)$ in $\mathbf{C}^n \times \mathbf{C}^n$ such that the Taylor series expansion of $\psi(z, Z) = \psi(z_1, \ldots, z_n, Z_1, \ldots, Z_n)$ about the origin contains neither constant nor linear terms. Then, for each $\rho > 0$ with $\Delta_\rho \subset\subset \Delta$, there exists a number $\lambda_\rho > 0$ such that*

$$|\psi(z, Z) - \psi(z, W)| \leq \lambda_\rho \sum_{j=1}^n |Z_j - W_j| \quad \textit{for } (z, Z, W) \textit{ in } \Delta_\rho \times \Delta_\rho \times \Delta_\rho.$$

Furthermore,

$$\lim_{\rho\to 0} \lambda_\rho = 0.$$

PROOF. Fix $\rho > 0$ with $\Delta_\rho \subset\subset \Delta$. We observe that

$$Z_1^{k_1}\cdots Z_n^{k_n} - W_1^{k_1}\cdots W_n^{k_n} = \sum_{j=1}^n (Z_j^{k_j} - W_j^{k_j}) Z_{j+1}^{k_{j+1}}\cdots Z_n^{k_n} W_1^{k_1}\cdots W_{j-1}^{k_{j-1}};$$

thus writing out the Taylor series expansion of $\psi(z, Z) - \psi(z, W)$ about the origin in $\mathbf{C}^n \times \mathbf{C}^n \times \mathbf{C}^n$, we obtain n holomorphic functions $H_j(z, Z, W)$ $(j = 1, \ldots, n)$ in $\Delta_\rho \times \Delta_\rho \times \Delta_\rho$ such that

$$\psi(z, Z) - \psi(z, W) = \sum_{j=1}^n (Z_j - W_j) H_j(z, Z, W)$$

in $\Delta_\rho \times \Delta_\rho \times \Delta_\rho$. Thus if we set

$$\lambda_\rho := \max_{(z,Z,W)\in\Delta_\rho\times\Delta_\rho\times\Delta_\rho} \{|H_1(z, Z, w)|, \ldots, |H_n(z, Z, w)|\} < \infty,$$

then we have

$$|\psi(z, Z) - \psi(z, W)| \leq \lambda_\rho \sum_{j=1}^n |Z_j - W_j|$$

in $\Delta_\rho \times \Delta_\rho \times \Delta_\rho$. Since $\psi(z, Z)$ contains neither constant nor linear terms, we have that $H_j(0, 0, 0) = 0$ $(j = 1, \ldots, n)$. Consequently, $\lim_{\rho\to 0} \lambda_\rho = 0$. □

PROOF OF PROPOSITION 5.1. The first step is to find a solution $F_j(z)$ $(j = 1, \ldots, n)$ of the functional equation (5.3) valid in a certain polydisk centered at 0. We write

$$P_j(z) = a_j z_j + p_j(z), \quad F_j(z) = z_j + f_j(z) \qquad (j = 1, \ldots, n),$$

where neither $p_j(z)$ nor $f_j(z)$ contains constant or linear terms. We also define

$$p_j^*(z, Z) := p_j(z_1 + Z_1, \ldots, z_n + Z_n) \quad (j = 1, \ldots, n),$$

which is a polynomial in $\mathbf{C}^{2n}$ with neither constant nor linear terms in the variables $z_1, \ldots, z_n, Z_1, \ldots, Z_n$. From a direct calculation we see that the functions $F_j(z)$ $(j = 1, \ldots, n)$ defined in a polydisk Δ_ρ satisfy the functional equation (5.3) in

$\Delta_\rho^{(-1)}$: $|z_j| < \rho/|a_j|$ $(j = 1, \dots, n)$ if and only if the functions $f_j(z)$ $(j = 1, \dots, n)$ satisfy the functional equation

$$f_j(az) = a_j f_j(z) + p_j^*(z, f(z)) \quad (j = 1, \dots, n) \ \text{ in } \Delta_\rho^{(-1)}, \tag{5.9}$$

where $f(z) := (f_1(z), \dots, f_n(z))$.

Thus we look for functions $f_j(z)$ $(j = 1, \dots, n)$ defined in a certain polydisk Δ_ρ which satisfy (5.9) in the polydisk $\Delta_\rho^{(-1)}$. To do this, we use the *method of alternation.* We set

$$k := \max_{j=1,\dots,n} \left\{ \sum_{k_1+\cdots+k_n \geq 2} \left| \frac{1}{a_1^{k_1} \cdots a_n^{k_n} - a_j} \right| \right\} < \infty. \tag{5.10}$$

From Lemma 5.3 we can find a sufficiently small polydisk Δ_ρ centered at 0 in $\mathbf{C}^n$ and a constant $\lambda_\rho > 0$ such that

$$|p_j^*(z, Z) - p_j^*(z, W)| \leq \lambda_\rho \sum_{j=1}^{n} |Z_j - W_j| \quad (j = 1, \dots, n) \quad \text{in } \Delta_\rho \times \Delta_\rho \times \Delta_\rho$$

and such that

$$kn\lambda_\rho < 1/2.$$

Furthermore, we may also assume that

$$|p_j^*(z, 0)| < \rho/(2kn) \qquad (j = 1, \dots, n) \ \text{ in } \Delta_\rho,$$

since $p_j^*(z, 0)$ contains neither constant nor linear terms.

We begin by setting

$$f_j^0(z) \equiv 0 \qquad (j = 1, \dots, n) \ \text{ in } \Delta_\rho.$$

Then for $\nu \geq 1$, having determined holomorphic functions $f_j^{\nu-1}(z)$ $(j = 1, \dots, n)$ in Δ_ρ such that each $f_j^{\nu-1}$ contains neither constant nor linear terms, we define holomorphic functions $f_j^\nu(z)$ $(j = 1, \dots, n)$ containing neither constant nor linear terms in Δ_ρ in the following manner. We let $f^{\nu-1}(z) := (f_1^{\nu-1}(z), \dots, f_n^{\nu-1}(z))$, and for each $j = 1, \dots, n$, we apply Lemma 5.2 to $\varphi(z) := p_j^*(z, f^{\nu-1}(z))$ and $a_0 = a_j$ to find a unique holomorphic function $f_j^\nu(z)$ in Δ_ρ such that

$$f_j^\nu(az) = a_j f_j^\nu(z) + p_j^*(z, f^{\nu-1}(z)) \quad (j = 1, \dots, n) \ \text{ in } \Delta_\rho^{(-1)}. \tag{5.11}$$

It follows that we have now inductively defined a sequence of analytic mappings $f^\nu(z) = (f_1^\nu(z), \dots, f_n^\nu(z))$ $(\nu = 0, 1, \dots)$ on Δ_ρ. To verify the first step of our proof, using (5.9) and (5.11) it suffices to prove that the sequence of holomorphic functions $\{f_j^\nu\}_{\nu=1,2,\dots}$ converges uniformly in Δ_ρ for each $j = 1, \dots, n$. To do this, we shall show that

$$|f_j^\nu(z) - f_j^{\nu-1}(z)| < \rho/2^\nu \quad (\nu = 1, 2, \dots;\ j = 1, \dots, n) \quad \text{in } \Delta_\rho. \tag{5.12}$$

Indeed, using Lemma 5.2, for $z \in \Delta_\rho$,

$$|f_j^1(z) - f_j^0(z)| = |f_j^1(z)| \leq k \max_{z \in \Delta_\rho} |p_j^*(z, 0)| < k \cdot \rho/(2kn) \leq \rho/2,$$

which proves the case $\nu = 1$ of (5.12). Now we assume that

$$|f_j^\mu(z) - f_j^{\mu-1}(z)| < \rho/2^\mu \quad (1 \leq \mu \leq \nu;\ j = 1, \dots, n) \quad \text{in } \Delta_\rho.$$

In particular, we have $|f_j^\mu(z)| \leq \rho$ $(j = 1, \dots, n)$ in Δ_ρ; i.e., $f^\mu(z) \in \Delta_\rho$ $(1 \leq \mu \leq \nu)$.

From (5.9), for $z \in \Delta_\rho^{(-1)}$ we have

$$f_j^{\nu+1}(az) - f_j^\nu(az) = a_j(f_j^{\nu+1}(z) - f_j^\nu(z)) + \{p_j^*(z, f^\nu(z)) - p_j^*(z, f^{\nu-1}(z))\}.$$

Thus if we set

$$\begin{aligned} g_j(z) &:= f_j^{\nu+1}(z) - f_j^\nu(z) \quad \text{in } \Delta_\rho, \\ \varphi_j(z) &:= p_j^*(z, f^\nu(z)) - p_j^*(z, f^{\nu-1}(z)) \quad \text{in } \Delta_\rho, \end{aligned}$$

then $g_j(z)$ $(j = 1, \ldots, n)$ satisfies the functional equation

$$g_j(az) = a_j g_j(z) + \varphi_j(z) \quad \text{in } \Delta_\rho^{(-1)},$$

where $|a_j| > 1$ and $\varphi_j(z)$ is a holomorphic function in Δ_ρ with neither constant nor linear terms. It follows from Lemma 5.2 and (5.10) that

$$|g_j(z)| \leq k \left(\max_{z \in \Delta_\rho} |\varphi_j(z)| \right) \quad \text{in } \Delta_\rho.$$

Therefore, for any $z \in \Delta_\rho$

$$\begin{aligned} |f_j^{\nu+1}(z) - f_j^\nu(z)| &\leq k \left(\max_{z \in \Delta_\rho} |p_j^*(z, f^\nu(z)) - p_j^*(z, f^{\nu-1}(z))| \right) \\ &\leq k\lambda_\rho \max_{z \in \Delta_\rho} \left(\sum_{i=1}^n |f_i^\nu(z) - f_i^{\nu-1}(z)| \right) \\ &\leq k\lambda_\rho \sum_{i=1}^n \rho/2^\nu \\ &= k\lambda_\rho n\rho/2^\nu < \rho/2^{\nu+1}. \end{aligned}$$

Thus (5.12) is verified and $\{f_j^\nu(z)\}_{\nu=1,2,\ldots}$ $(j = 1, \ldots, n)$ converges uniformly to a holomorphic function $f_j(z)$ in Δ_ρ, which satisfies equation (5.9). Our first step is proved.

We now set

$$F(z) := (F_1(z), \ldots, F_n(z)),$$

where

$$F_j(z) = z_j + f_j(z) \quad (j = 1, \ldots, n) \text{ in } \Delta_\rho$$

and

$$S_F : \ z \in \Delta_\rho \to w = F(z) \in \mathbf{C}^n.$$

From the first step, we have

$$F_j(az) = P_j(F_1(z), \ldots, F_n(z)) \quad (j = 1, \ldots, n) \text{ in } \Delta_\rho^{(-1)}. \tag{5.13}$$

For the second step we now want to show that $F(z)$ has a holomorphic extension to all of $\mathbf{C}^n$. To this end, we consider the linear automorphism T_a of $\mathbf{C}^n$ defined by

$$T_a : \ z'_j = a_j z_j \qquad (j = 1, \ldots, n).$$

The functional equations (5.13) in $\Delta_\rho^{(-1)}$ are equivalent to

$$S_F \circ T_a = T_P \circ S_F \quad \text{in } \Delta_\rho^{(-1)}. \tag{5.14}$$

Now for $l = 1, 2, \ldots$ we set

$$\begin{aligned} \Delta_\rho^{(-l)} &: \quad |z_j| < \rho/|a_j|^l \quad (j = 1, \ldots, n), \\ \Delta_\rho^{(l)} &: \quad |z_j| < |a_j|^l \rho \quad (j = 1, \ldots, n), \end{aligned}$$

which are polydisks centered at 0; note that as $l \to \infty$, the polydisks $\Delta_\rho^{(-l)}$ shrink to the origin while the polydisks $\Delta_\rho^{(l)}$ increase to all of $\mathbf{C}^n$. Noting that $(T_a^{-1})^l(\Delta_\rho) = \Delta_\rho^{(-l)}$, we iterate (5.14) l times to obtain

$$S_F \circ T_a^l = T_P^l \circ S_F \quad \text{in } \Delta_\rho^{(-l)} \quad (l = 1, 2, \dots). \tag{5.15}$$

On the other hand, since the right-hand side in this equation is defined in Δ_ρ and since $(T_a^{-1})^l(\Delta_\rho^{(l)}) = \Delta_\rho$, we can extend the domain of definition of S_F from Δ_ρ to $\Delta_\rho^{(l)}$ by use of the equation

$$S_F(z) = T_P^l \circ S_F \circ (T_a^{-1})^l(z), \quad z \in \Delta_\rho^{(l)}.$$

From (5.15), it follows that this extension of S_F is independent of $l = 1, 2, \dots$. Since $\lim_{l\to\infty} \Delta_\rho^{(l)} = \mathbf{C}^n$, $S_F(z)$ is thus a holomorphic mapping on all of $\mathbf{C}^n$. Furthermore, S_F satisfies the equation $S_F \circ T_a = T_P \circ S_F$ in $\mathbf{C}^n$ (this may be proved directly or by using analytic continuation). This finishes our second step of the proof.

Finally, we must verify the uniqueness of $F_j(z)$ $(j = 1, \dots, n)$ satisfying the conditions stated in Proposition 5.1. To do this, it suffices to show the following. Let $|a_j| > 1$ $(j = 1, \dots, n)$ satisfy condition (3), i.e., $\sum a_1^{k_1} \cdots a_n^{k_n} \neq a_j$ $(j = 1, \dots, n)$, and let $p_j^*(z, Z)$ be a polynomial in $\mathbf{C}^n \times \mathbf{C}^n$ with neither constant nor linear terms in $z_1, \dots, z_n, Z_1, \dots, Z_n$. Then the holomorphic functions $f_j(z)$ $(j = 1, \dots, n)$ defined in a polydisk Δ_ρ centered at 0 which satisfy the functional equations (5.9) and which contain neither constant nor linear terms are uniquely determined by a_j $(j = 1, \dots, n)$ and $p_j^*(z, Z)$ $(j = 1, \dots, n)$. Indeed, for $j = 1, \dots, n$, writing the Taylor series development in Δ_ρ, we have

$$\begin{aligned} f_j(z) &= \sum_{k_1+\cdots+k_n\geq 2} v_{k_1,\dots,k_n}^{(j)} z_1^{k_1} \cdots z_n^{k_n}, \\ p_j^*(z, Z) &= \sum_{\substack{l_1+\cdots+l_n \\ +m_1+\cdots+m_n \geq 2}} C_{l_1,\dots,l_n,m_1,\dots,m_n}^{(j)} z_1^{l_1} \cdots z_n^{l_n} Z_1^{m_1} \cdots Z_n^{m_n}. \end{aligned}$$

Using (5.9), we have

$$\begin{aligned} &\sum_{k_1+\cdots+k_n\geq 2} (a_1^{k_1} \cdots a_n^{k_n} - a_j) v_{k_1,\dots,k_n}^{(j)} z_1^{k_1} \cdots z_n^{k_n} \\ &= \sum_{\substack{l_1+\cdots+l_n \\ +m_1+\cdots+m_n \geq 2}} C_{l_1,\dots,l_n,m_1,\dots,m_n}^{(j)} z_1^{l_1} \cdots z_n^{l_n} \\ &\qquad \times \left(\textstyle\sum_{k_1+\cdots+k_n\geq 2} v_{k_1,\dots,k_n}^{(1)} z_1^{k_1} \cdots z_n^{k_n}\right)^{m_1} \\ &\qquad \times \cdots \times \left(\sum_{k_1+\cdots+k_n\geq 2} v_{k_1,\dots,k_n}^{(n)} z_1^{k_1} \cdots z_n^{k_n} \right)^{m_n}. \end{aligned} \tag{5.16}$$

It suffices to prove that each $v_{k_1,\dots,k_n}^{(j)}$ $(j = 1, \dots, n;\ s := k_1 + \ldots + k_n \geq 2)$ is determined uniquely by $\mathbf{a} = \{a_j \mid j = 1, \dots, n\}$ and $\mathcal{C}_s$, where

$$\mathcal{C}_s := \{C_{l_1,\dots,l_n,m_1,\dots,m_n}^{(i)} \mid l_1 + \cdots + l_n + m_1 + \cdots + m_n \leq s;\ i = 1, \dots, n\}.$$

We verify this by induction on $s = k_1 + \ldots + k_n \geq 2$. First of all, let $k_1, \dots, k_n \geq 0$ be integers such that $k_1 + \cdots + k_n = 2$. Then, by comparing the expressions for the coefficient of $z_1^{k_1}\cdots z_n^{k_n}$, we have

$$v^{(j)}_{k_1,\dots,k_n} = C^{(j)}_{k_1,\dots,k_n,0,\dots,0}/(a_1^{k_1}\cdots a_n^{k_n} - a_j) \quad (j = 1,\dots,n),$$

so that the case for $s = 2$ is true. Next we let $s \geq 3$ and assume that each $v^{(j)}_{k_1,\dots,k_n}$ $(k_1 + \cdots + k_n \leq s-1; j = 1,\dots,n)$ is determined uniquely by $\mathbf{a}$ and $\mathcal{C}_{s-1}$. Then for each $j = 1,\dots,n$ we compare the expressions for the coefficient of $z_1^{k_1}\cdots z_n^{k_n}$, where $k_1 + \cdots + k_n = s$, in equation (5.16). On the left-hand side this coefficient is $(a_1^{k_1}\cdots a_n^{k_n} - a_j)v^{(j)}_{k_1,\dots,k_n}$; on the right-hand side we obtain a polynomial in

$$C^{(j)}_{l_1,\dots,l_n,m_1,\dots,m_n} \;:\; l_1 + \cdots + l_n + m_1 + \cdots + m_n \leq s$$

and

$$v^{(i)}_{k_1,\dots,k_n} \;:\; k_1 + \cdots + k_n \leq s-1, \quad i = 1,\dots,n.$$

This can be seen by noting that if one of the $v^{(i)}_{k_1,\dots,k_n}$ for some $i = 1,\dots,n$ and $k_1 + \cdots + k_n \geq s$ occurs, then $m_i = 1$, $l_k = 0$ $(k = 1,\dots,n)$, and $m_k = 0$ $(k \neq i)$. This contradicts $\sum_{j=1}^n (l_j + m_j) \geq 2$. Therefore, $v^{(j)}_{k_1,\dots,k_n}$ is uniquely determined by $\mathbf{a}$ and $\mathcal{C}_s$, and the uniqueness of $f_j(z)$ $(j = 1,\dots,n)$ is proved. □

5.2.2. Bieberbach's Example. Let $T_P : z_j = P_j(z)$ $(j = 1,\dots,n)$ be a polynomial mapping of $\mathbf{C}^n$ $(n \geq 1)$. If a point $z \in \mathbf{C}^n$ satisfies $T_P(z) = z$, then z is called a **fixed point** of T_P. Let $z^0 \in \mathbf{C}^n$ be a fixed point of T_P and let

$$J_{T_P}(z) = \frac{\partial(P_1,\dots,P_n)}{\partial(z_1,\dots,z_n)}$$

be the Jacobian matrix of T_P in $\mathbf{C}^n$. We let λ_j $(j = 1,\dots,n)$ denote the eigenvalues of $J_{T_P}(z^0)$. If $|\lambda_j| > 1$ for $j = 1,\dots,n$, we call z^0 a **repelling fixed point** of T_P. If $|\lambda_j| < 1$ for $j = 1,\dots,n$, we call z^0 an **attracting fixed point** of T_P. In all other cases we call z^0 a **loxodromic fixed point** of T_P.

If z^0 is a repelling fixed point of T_P, then we can find a neighborhood γ of z^0 in $\mathbf{C}^n$ such that $\gamma \subset\subset T_P(\gamma)$. If z_0 is an attracting fixed point of T_P, then we can find a neighborhood γ of z_0 in $\mathbf{C}^n$ such that $T_P(\gamma) \subset\subset \gamma$. For the polynomial mapping T_P studied in section 5.2.1, the origin $z = 0$ is a repelling fixed point of T_P.

If a polynomial mapping T_P of $\mathbf{C}^n$ is one-to-one from $\mathbf{C}^n$ onto $\mathbf{C}^n$, then we say that T_P is a **polynomial automorphism** of $\mathbf{C}^n$. In the case $n = 1$, any automorphism of $\mathbf{C}$ is linear. However, for $n \geq 2$ there are many polynomial automorphisms of degree at least two.

Let

$$T_P \;:\; z'_j = P_j(z_1,\dots,z_n) \qquad (j = 1,\dots,n)$$

be a polynomial automorphism of $\mathbf{C}^n$ such that $P_j(z)$ $(j = 1,\dots,n)$ satisfies conditions (1), (2), and (3) stated at the begining of section 5.2.1 (a specific example will be given at the end of this section). We fix a polydisk $\gamma^0 : |z_j| < \rho$ $(j = 1,\dots,n)$ such that $\gamma^0 \subset\subset T_P(\gamma^0)$. We recursively define

$$\gamma^{l+1} := T_P(\gamma^l) \qquad (l = 0,1,2,\dots).$$

Since $\gamma^l \subset\subset \gamma^{l+1}$ $(l = 0,1,2,\dots)$ and T_P is an automorphism of $\mathbf{C}^n$,

$$\Gamma_{T_P} := \lim_{l=\infty} \gamma^l$$

is a domain in $\mathbf{C}^n$. We note that Γ_{T_P} does not depend on the choice of the initial polydisk γ^0 as long as $\gamma^0 \subset\subset T_P(\gamma^0)$.

We consider the Poincaré-Picard entire mapping S_F with respect to the above polynomial mapping T_P; this mapping is defined via the equation

$$S_F \circ T_a = T_P \circ S_F \qquad \text{in } \mathbf{C}^n.$$

We show the following.

PROPOSITION 5.2. *The Poincaré-Picard entire mapping S_F maps $\mathbf{C}^n$ onto the domain Γ_{T_P} in a one-to-one manner.*

PROOF. By (5.4) we fix a neighborhood δ^0 of the origin $z = 0$ such that S_F is one-to-one on δ^0 and such that $S_F(\delta^0) = \gamma^0$ (recall that we can start with any polydisk γ^0 such that $\gamma^0 \subset\subset T_P(\gamma^0)$). Since

$$S_F \circ T_a^l = T_P^l \circ S_F \qquad \text{in } \mathbf{C}^n \ (l = 1, 2, \dots), \tag{5.17}$$

we have

$$S_F(T_a^l(\delta^0)) = \gamma^l \qquad (l = 1, 2, \dots).$$

Since $T_a^l : z \in \mathbf{C}^n \to w = (a_1^l z_1, \dots, a_n^l z_n)$ and $|a_j| > 1$ $(j = 1, \dots, n)$, we see that the domains $T_a^l(\delta^0)$ increase to $\mathbf{C}^n$. Thus S_F maps $\mathbf{C}^n$ onto the domain Γ_{T_P}.

We show that S_F is one-to-one. For if not, there exist $z_1, z_2 \in \mathbf{C}^n$ with $z_1 \neq z_2$ such that $S_F(z_1) = S_F(z_2)$. We fix an integer $l_0 \geq 1$ sufficiently large so that if we let $\zeta_i := T_a^{-l_0}(z_i)$ $(i = 1, 2)$, then $\zeta_1, \zeta_2 \in \delta^0$. Then $\zeta_1 \neq \zeta_2$, and hence $S_F(\zeta_1) \neq S_F(\zeta_2)$. Since $T_a^{l_0}(\zeta_i) = z_i$ $(i = 1, 2)$, it follows from (5.17) that

$$S_F(z_i) = T_P^{l_0} \circ S_F(\zeta_i).$$

Therefore, $T_P^{l_0} \circ S_F(\zeta_1) = T_P^{l_0} \circ S_F(\zeta_2)$, which contradicts the condition that T_P (and hence $T_P^{l_0}$) is an automorphism of $\mathbf{C}^n$. □

We now impose the following additional condition on the above algebraic automorphism T_P of $\mathbf{C}^n$: *there exists another repelling fixed point $z^* \neq 0$ in $\mathbf{C}^n$*. Thus we can find a polydisk γ^* centered at z^* in $\mathbf{C}^n$ such that $\gamma^* \subset\subset T_P(\gamma^*)$. We define

$$\Gamma_{T_P}^* := \lim_{l \to \infty} T_P^l(\gamma^*),$$

which is a domain in $\mathbf{C}^n$. Furthermore, since T_P is an automorphism of $\mathbf{C}^n$, we have $\Gamma_{T_P}^* \cap \Gamma_{T_P} = \emptyset$. From Proposition 5.2, we see that the domain Γ_{T_P} in $\mathbf{C}^n$ is biholomorphically equivalent to $\mathbf{C}^n$. We will give an example of a polynomial automorphism T_P of $\mathbf{C}^n$ which satisfies conditions (1), (2), and (3) (at $z = 0$) stated at the beginning of section 5.2.1 and which has another repelling fixed point $z^* \neq 0$ in $\mathbf{C}^n$. Thus we have the following proposition, which indicates another major difference between $\mathbf{C}^n$ for $n \geq 2$ and $\mathbf{C}$.

PROPOSITION 5.3. *There exists a domain D in $\mathbf{C}^n$ $(n \geq 2)$ such that D is biholomorphically equivalent to $\mathbf{C}^n$ and such that $\mathbf{C}^n \setminus D$ has non-empty interior.*

EXAMPLE 5.1. In $\mathbf{C}^2$ with variables z and w, we set

$$T_P : \begin{cases} z' &= w, \\ w' &= 2z + w(w-1)(2w-1) - w. \end{cases}$$

Then T_P is a polynomial automorphism of $\mathbf{C}^2$ such that both $(0, 0)$ and $(1, 1)$ are repelling fixed points of T_P with eigenvalues $\pm\sqrt{2}$ whose Jacobian matrix is diagonizable at both points.

5.2.3. Picard's Theorem. We consider a holomorphic mapping

$$T \ : \ w_j = f_j(z) \qquad (j = 1, \ldots, m)$$

from $\mathbf{C}^n$ to $\mathbf{C}^m$ with $m, n \geq 1$. We call $T : \mathbf{C}^n \to \mathbf{C}^m$ an **entire mapping**. We call

$$\mathcal{E}_T := \mathbf{C}^m \setminus T(\mathbf{C}^n)$$

the set of **exceptional values of** T. As a particular example, if S is an algebraic hypersurface in $\mathbf{C}^m$, i.e., $S = \{P(w) = 0\}$ where $P(w)$ is a non-zero polynomial in $w = (w_1, \ldots, w_m)$, and if S satisfies $S \subset \mathcal{E}_T$, we call S an **algebraic exceptional set of** T.

In the case when $n = m = 1$, from Picard's theorem in one complex variable, it follows that $\mathcal{E}_T$ consists of at most one point for any non-constant entire function T. In Proposition 5.3 we observed that in the case when $n = m = 2$, there are examples of entire mappings T such that $\mathcal{E}_T$ contains interior points.

Let $T : \mathbf{C}^n \to \mathbf{C}^m$ be an entire mapping. If there exists an algebraic hypersurface Σ in $\mathbf{C}^n$ such that $T(\mathbf{C}^n) \subset \Sigma$, then we say that T is **degenerate**. In this case we may assume that Σ is irreducible in $\mathbf{C}^n$. We shall show in Theorem 5.6 that if T is non-degenerate, then the number of irreducible algebraic exceptional sets of T is limited. This fact may be regarded as a generalization of Picard's theorem in one complex variable. To state the theorem, we first discuss the following generalization to several variables of Borel's theorem from the theory of functions of one complex variable.

THEOREM 5.5 (Borel). *Let $\nu \geq 1$ and let $f_j(z)$ $(j = 1, \ldots, \nu)$ be entire functions in $\mathbf{C}^n$ $(n \geq 1)$ such that $f_j(z) \neq 0$ on $\mathbf{C}^n$. If there exist non-zero complex numbers a_j $(j = 1, \ldots, \nu)$ such that*

$$a_1 f_1(z) + \cdots + a_\nu f_\nu(z) \equiv 0 \quad \textit{in } \mathbf{C}^n, \tag{5.18}$$

then there exists at least one pair h, k $(h \neq k;\ 1 \leq h, k \leq \nu)$ such that the ratio $f_k(z)/f_h(z)$ is constant in $\mathbf{C}^n$.

PROOF. We prove this fact by induction on the dimension $n \geq 1$. In the case $n = 1$, the result is Borel's theorem for one complex variable.[1] We use this fact without proof. Assume the result is true in $\mathbf{C}^n$ for $n \geq 1$ fixed, and we shall prove that the result is true in $\mathbf{C}^{n+1}$.

Let H be a complex hyperplane in $\mathbf{C}^{n+1}$ which passes through the origin 0. We restrict each $f_j(z)$ $(j = 1, \ldots, \nu)$ and the relation (5.18) in $\mathbf{C}^{n+1}$ to H. Since H can be regarded as $\mathbf{C}^n$, it follows from the inductive hypothesis that there exists at least one pair h, k $(h \neq k,\ 1 \leq h, k \leq \nu)$ depending on H such that $f_k(z)/f_h(z) = C_H$ (constant) on H. Since there exist infinitely many complex hyperplanes H passing through 0 in $\mathbf{C}^{n+1}$ and since there exist at most a finite number of pairs h, k $(h \neq k,\ 1 \leq h, k \leq \nu)$, it follows that there exists at least one pair h, k $(h \neq k,\ 1 \leq h, k \leq \nu)$ such that

$$f_h(z)/f_k(z) = C_H$$

on infinitely many distinct hyperplanes H. Since $f_h(z)/f_k(z)$ is holomorphic at $z = 0$, this implies that the complex numbers C_H coincide with $c := f_h(0)/f_k(0)$. Hence $f_h(z)/f_k(z) \equiv c$ on the union of the hyperplanes H. It follows that $f_h(z)/f_k(z) \equiv c$ in all of $\mathbf{C}^{n+1}$. For, if $F(z) := f_h(z)/f_k(z) \not\equiv c$ in a neighborhood V of the origin,

[1] See, for example, the classic textbook [**38**].

then, since $F(z)$ is a non-constant holomorphic function in V, the set Σ defined by $F(z) = c$ in V consists of a finite number of irreducible analytic hypersurfaces in V. This contradicts the fact that Σ contains infinitely many distinct irreducible hyperplanes H passing through 0. □

From Borel's theorem we obtain the following.

THEOREM 5.6 ([**42**]). *Let* $T : z \in \mathbf{C}^n \to w = (f_1(z), \ldots , f_m(z)) \in \mathbf{C}^m$ $(n, m \geq 1)$ *be an entire mapping. If the set* $\mathcal{E}_T$ *of exceptional values of* T *contains at least* $m+1$ *irreducible algebraic hypersurfaces* S_k $(k = 1, \ldots , m+1)$, *then* T *is degenerate.*

PROOF. Let S_k $(k = 1, \ldots , m + 1)$ be given in the form

$$S_k : P_k(w_1, \ldots , w_m) = 0 \quad \text{in } \mathbf{C}^m,$$

where $P_k(w)$ is a polynomial in $w \in \mathbf{C}^m$. We set

$$W_k := P_k(w_1, \ldots , w_m) \qquad (k = 1, \ldots , m + 1) \tag{5.19}$$

and we use these $m + 1$ equations to get an algebraic relation between $W_1, \ldots ,$ W_{m+1}:

$$Q(W_1, \ldots , W_{m+1}) := \sum_{(j_1,\ldots,j_{m+1})\in J} a_{j_1,\ldots,j_{m+1}} W_1^{j_1} \cdots W_{m+1}^{j_{m+1}} = 0 \text{ in } \mathbf{C}_W^{m+1}; \tag{5.20}$$

here, we have only a finite number of indices in J. Thus

$$\begin{aligned} 0 &= \sum_{(j_1,\ldots,j_{m+1})\in J} a_{j_1,\ldots,j_{m+1}} P_1(w)^{j_1} \cdots P_{m+1}(w)^{j_{m+1}} \qquad (5.21)\\ &=: \sum_{(j_1,\ldots,j_{m+1})\in J} a_{j_1,\ldots,j_{m+1}} \wp_{j_1,\ldots,j_{m+1}}(w) \quad \text{for } w \in \mathbf{C}^m, \end{aligned}$$

where $\wp_{j_1,\ldots,j_{m+1}}(w) := P_1(w)^{j_1} \cdots P_{m+1}(w)^{j_{m+1}}$ is a polynomial in $w \in \mathbf{C}^m$. We set

$$\begin{aligned} w_{j_1,\ldots,j_{m+1}}(z) &:= \wp_{j_1,\ldots,j_{m+1}}(T(z)) \\ &= \wp_{j_1,\ldots,j_{m+1}}(f_1(z), \ldots , f_m(z)) \quad \text{for } z \in \mathbf{C}^n, \end{aligned}$$

which is an entire function on $\mathbf{C}^n$. Since $S_j \subset \mathcal{E}_T$ $(j = 1, \ldots , m + 1)$, we have $P_j(f_1(z), \ldots , f_m(z)) \neq 0$ for all $z \in \mathbf{C}^n$, and hence $w_{j_1,\ldots,j_{m+1}}(z) \neq 0$ for all $z \in \mathbf{C}^n$. Furthermore, from (5.21) we have

$$\sum_{(j_1,\ldots,j_{m+1})\in J} a_{j_1,\ldots,j_{m+1}} w_{j_1,\ldots,j_{m+1}}(z) \equiv 0 \quad \text{on } \mathbf{C}^n.$$

It follows from Borel's theorem that there exists at least one pair $(j_1, \ldots , j_{m+1}) \neq (k_1, \ldots , k_{m+1})$ in J such that

$$w_{j_1,\ldots,j_{m+1}}(z)/w_{k_1,\ldots,k_{m+1}}(z) \equiv c = \text{ const.} \quad \text{in } \mathbf{C}^n.$$

Thus if we set

$$P^*(w) := \wp_{j_1,\ldots,j_{m+1}}(w) - c\, \wp_{k_1,\ldots,k_{m+1}}(w) \quad \text{for } w \in \mathbf{C}^m, \tag{5.22}$$

which is a polynomial in $w \in \mathbf{C}^m$, then

$$T(\mathbf{C}^n) \subset \{w \in \mathbf{C}^n \mid P^*(w) = 0\},$$

so that T is degenerate. □

REMARK 5.7. In the proof, if $P_k(w)$ $(k = 1, \ldots, m+1)$ in (5.19) are homogeneous polynomials in $w = (w_1, \ldots, w_m)$, then $P^*(w)$ in (5.22) is also a homogeneous polynomial in w.

To prove this, we let $q_k \geq 1$ denote the degree of the homogeneous polynomial $P_k(w)$ $(k = 1, \ldots, m+1)$. For $(j_1, \ldots, j_{m+1}) \in J$, $\wp_{j_1,\ldots,j_{m+1}}(w)$ is a homogeneous polynomial in w of degree

$$d := j_1 q_1 + \cdots + j_{m+1} q_{m+1};$$

moreover, we claim that this degree may be assumed to be independent of $(j_1, \ldots, j_{m+1}) \in J$. To see this, fix $\lambda \neq 0$. Then $P_k(\lambda w_1, \ldots, \lambda w_m) = \lambda^{q_k} P_k(w_1, \ldots, w_m)$, so that, setting $W_i = P_i(w)$ $(i = 1, \ldots, m+1)$, $w \in \mathbf{C}^m_w$, (5.20) implies that

$$Q(W_1/\lambda^{q_1}, \ldots, W_{m+1}/\lambda^{q_{m+1}}) = 0,$$

and hence

$$\sum_{(j_1,\ldots,j_{m+1}) \in J} a_{j_1,\ldots,j_{m+1}} \lambda^{-(q_1 j_1 + \cdots + q_{m+1} j_{m+1})} P_1(w)^{j_1} \cdots P_{m+1}(w)^{j_{m+1}} = 0$$

for $w \in \mathbf{C}^m_w$. Since this equation is valid for all $\lambda \neq 0$, it follows that d may be assumed to be independent of $(j_1, \ldots, j_{m+1}) \in J$ (for if not, collect all terms of the highest degree of λ). Thus $P^*(w) = \wp_{j_1,\ldots,j_{m+1}}(w) - c\ \wp_{k_1,\ldots,k_{m+1}}(w)$ is a homogeneous polynomial in w of degree d.

In a similar manner, we can treat the case of complex projective space $\mathbf{P}^m$; i.e., we consider a holomorphic mapping

$$T^0 : \mathbf{C}^n \to \mathbf{P}^m$$

and the set $\mathcal{E}_{T^0}$ of exceptional values of T^0: $\mathcal{E}_{T^0} := \mathbf{P}^m \setminus T^0(\mathbf{C}^n)$. In the case $n = m = 1$, T^0 is a meromorphic function on $\mathbf{C}$ and it follows from Picard's theorem in one complex variable that $\mathcal{E}_{T^0}$ consists of at most two points. In the general case, we have the following theorem.

THEOREM 5.7. *Let $T^0 : \mathbf{C}^n \to \mathbf{P}^m$ $(n, m \geq 1)$ be a holomorphic mapping. If $\mathcal{E}_{T^0}$ contains at leasts $m+2$ irreducible algebraic hypersurfaces S_i $(i = 1, \ldots, m+2)$ in $\mathbf{P}^m$, then T^0 is degenerate; i.e., there exists an algebraic hypersurface Σ in $\mathbf{P}^m$ such that $T^0(\mathbf{C}^n) \subset \Sigma$.*

PROOF. We consider the canonical mapping $\pi : \mathbf{C}^{m+1} \setminus \{0\} \to \mathbf{P}^m$ given by $\pi(w_0, \ldots, w_m) = [w_0 : \ldots : w_m]$. Since $T^0 : \mathbf{C}^n \to \mathbf{P}^m$ is a holomorphic mapping, by solving the Cousin II problem we can find a holomorphic mapping

$$\widetilde{T^0} : z \in \mathbf{C}^n \to (f_0(z), \ldots, f_m(z)) \in \mathbf{C}^{m+1} \setminus \{0\},$$

where each $f_j(z)$ $(j = 0, \ldots, m)$ is an entire function on $\mathbf{C}^n$, such that $\pi(\widetilde{T^0}(z)) = T^0(z)$ for $z \in \mathbf{C}^n$. Since $S_i \subset \mathcal{E}_{T^0}$ $(i = 1, \ldots, m+2)$, we can find a homogeneous polynomial $P_i(w)$ in $\mathbf{C}^{m+1}$ such that if we set

$$\widetilde{S}_i := \{w \in \mathbf{C}^{m+1} \mid P_i(w) = 0\} \quad (i = 1, \ldots, m+2),$$

then $\pi(\widetilde{S}_i \setminus \{0\}) = S_i$ and $\widetilde{S}_i \subset \mathcal{E}_{\widetilde{T}^0}$. By applying Theorem 5.6 and Remark 5.7 to $T = \widetilde{T^0} : \mathbf{C}^n \to \mathbf{C}^{m+1}$, we can find a homogeneous polynomial $P^*(w) (\not\equiv 0)$ in $\mathbf{C}^{m+1}$ such that $\widetilde{T^0}(\mathbf{C}^n) \subset \Sigma^* := \{w \in \mathbf{C}^{m+1} \mid P^*(w) = 0\}$. Hence $T^0(\mathbf{C}^n)$ is contained in the algebraic hypersurface $\Sigma := \pi(\Sigma^* \setminus \{0\})$ in $\mathbf{P}^m$. □

Part 2

Theory of Analytic Spaces

CHAPTER 6

Ramified Domains

6.1. Ramified Domains

Let f be a holomorphic function at a point p in $\mathbf{C}^n$. In general, when f is analytically continued to a domain in $\mathbf{C}^n$, we obtain a multiple-valued function $\tilde{f}$. In the theory of several complex variables, as in the theory of one complex variable, we consider a multiply sheeted domain $\mathcal{D}$ over $\mathbf{C}^n$ on which $\tilde{f}$ becomes a single-valued function. If we do not consider a branch point as an interior point of $\mathcal{D}$, so that $\mathcal{D}$ is *unramified*, then the study of holomorphic functions in $\mathcal{D}$ is very similar to that in the case of one complex variable. However, if we consider a branch point as an interior point of $\mathcal{D}$ so that $\mathcal{D}$ is *ramified*, then we encounter interesting new phenomena in the study of holomorphic functions on the ramified domain $\mathcal{D}$ over $\mathbf{C}^n$ for $n \geq 2$ which do not occur in the case of $n = 1$. The major portion of this chapter is devoted to a proof of the local existence of a so-called *simple function* f on a ramified domain $\mathcal{D}$ over $\mathbf{C}^n$ (Theorem 6.1). Such an f provides a local m-fold cover $\tilde{\Delta}$ of a polydisk Δ in $\mathbf{C}^n$; the essential point is that, if we consider the graph $\mathcal{C} : X = f(p),\ p \in \tilde{\Delta}$ in the $(n+1)$-dimensional product space $\Delta \times \mathbf{C}_X$, then $\tilde{\Delta}$ and $\mathcal{C}$ are one-to-one except for an analytic set of dimension at most $n-1$; moreover, each branch point p of $\tilde{\Delta}$ corresponds to a nonsingular point of $\mathcal{C}$ except for an analytic set of dimension at most $n-2$. A corollary of this result, Theorem 6.4, which establishes local existence of a fundamental system for $\mathcal{D}$, will be used in Chapter 7.

6.1.1. Unramified Domains. Let $\mathcal{G}$ be a connected Hausdorff space. Assume that there exists a continuous mapping π from $\mathcal{G}$ into $\mathbf{C}^n$ which is locally one-to-one; i.e., for any point $p \in \mathcal{G}$ we can find a neighborhood V of p in $\mathcal{G}$ and a ball B centered at $\pi(p)$ in $\mathbf{C}^n$ such that π is continuous and bijective from V onto B. We say that $\mathcal{G}$ is an **unramified domain over $\mathbf{C}^n$** (or a **Riemann domain over $\mathbf{C}^n$**) and π is the **canonical projection** of $\mathcal{G}$ into $\mathbf{C}^n$. Given a point $p \in \mathcal{G}$, $\pi(p)$, denoted by $\underline{p}$, is called the **projection** of p or the **base point** of p. We say that p **lies over** $\underline{p}$. Given any set $E \subset \mathcal{G}$, we call $\pi(E)$ the projection of E into $\mathbf{C}^n$ and we write $\underline{E} := \pi(E)$. Given any $\underline{p} \in \mathbf{C}^n$, we consider the number $m(\underline{p})$ of points in the pre-image $\pi^{-1}(\underline{p})$ in $\mathcal{G}$; this number is at most countable. We call $m(\mathcal{G}) := \max \{m(\underline{p}) \mid \underline{p} \in \mathbf{C}^n\}$ the **number of sheets** of $\mathcal{G}$, which may be infinite. If there exists a domain D such that $\pi(\mathcal{G}) \subset D$, then $\mathcal{G}$ is called an unramified domain over D.

A connected and open subset in $\mathcal{G}$ is called a domain in $\mathcal{G}$, although we will have occasion to drop the connectivity assumption as in the case of $\mathbf{C}^n$. Let v be a domain in $\mathcal{G}$. If $\pi\,|_v$ is bijective from v onto $\pi(v)$, then v is called a **univalent domain** in $\mathcal{G}$ over $\mathbf{C}^n$. Given $p \in \mathcal{G}$, there exists a univalent neighborhood δ of p in $\mathcal{G}$, which will be called a coordinate neighborhood of p. Then by letting $q \in \delta$

correspond to $\underline{q} = \pi(q) \in \mathbf{C}^n$, we consider $\pi(\delta)$ as giving local coordinates of δ at p. We introduce analytic structure into $\mathcal{G}$ and define holomorphic functions in the following manner. Let $f(p)$ be a single-valued, complex-valued function defined on a domain v in $\mathcal{G}$. For $p \in v$, let $\delta \subset v$ be a coordinate neighborhood of p. If $f(\pi^{-1}(\underline{q}))$ is a holomorphic function for $\underline{q}$ in $\pi(\delta) \subset \mathbf{C}^n$, then we say that $f(p)$ is a **holomorphic function** in v.

Boundary points. Let $\mathcal{G}$ be an unramified domain over $\mathbf{C}^n$ with variables $z_1, \dots, z_n$. Let π be the canonical projection of $\mathcal{G}$ into $\mathbf{C}^n$. We want to define a boundary point of $\mathcal{G}$. Let $\underline{p} = (z_1', \dots, z_n') \in \mathbf{C}^n$. Let $r_{j,k} > 0$ $(j = 1, \dots, n\ ;\ \ k = 1, 2, \dots)$ be n sequences of positive numbers such that $r_{j,k} > r_{j,k+1}$ and $\lim_{k\to\infty} r_{j,k} = 0$ $(j = 1, \dots, n)$. We consider the nested sequence of polydisks $\underline{\delta}_k$ centered at $\underline{p}$ in $\mathbf{C}^n$ defined by

$$\underline{\delta}_k \ :\ |z_j - z_j'| < r_{j,k} \qquad (j = 1, \dots, n;\ k = 1, 2, \dots).$$

Set $\delta_k := \pi^{-1}(\underline{\delta}_k) \subset \mathcal{G}$. Then each $\pi^{-1}(\underline{\delta}_k)$ $(k = 1, 2, \dots)$ can be decomposed into at most a countable number of connected components. If there exists a connected component $\delta_k^0 \neq \emptyset$ in δ_k for each $k = 1, 2, \dots$ such that

$$\delta_{k+1}^0 \subset \delta_k^0 \ \ (k = 1, 2, \dots), \qquad \bigcap_{k=1}^{\infty} \delta_k^0 = \emptyset,$$

then the sequence $\{\delta_k^0\}_k$ defines a boundary point $\tilde{p}$ of $\mathcal{G}$ over $\underline{p}$. We say that each component δ_k^0 is a fundamental neighborhood of $\tilde{p}$ in $\mathcal{G}$ and the sequence $\{\delta_k^0\}_k$ is a **fundamental neighborhood system** of $\tilde{p}$ in $\mathcal{G}$. Let $\{\delta_k^0\}_k$ and $\{\eta_k^0\}_k$ be two fundamental neighborhood systems of the boundary points $\tilde{p}_1$ and $\tilde{p}_2$ of $\mathcal{G}$ over the same base point $\underline{p} \in \mathbf{C}^n$. If, for any δ_k^0 (resp. η_k^0), we can find $\eta_{k'}^0$ (resp. $\delta_{k'}^0$) such that $\eta_{k'}^0 \subset \delta_k^0$ (resp. $\delta_{k'}^0 \subset \eta_k^0$), then $\tilde{p}_1$ and $\tilde{p}_2$ define the same boundary point of $\mathcal{G}$ over $\underline{p}$.

Let $\mathcal{G}$ be an unramified domain over a domain $D \subset \mathbf{C}^n$ and let π denote the canonical projection of $\mathcal{G}$ into D. The boundary point $\tilde{p}$ of $\mathcal{G}$ over a point $\underline{p}$ in D is called a **relative boundary point** of $\mathcal{G}$ with respect to D. We say that an unramified domain $\mathcal{G}$ over $D \subset \mathbf{C}^n$ which has no relative boundary points with respect to D is an unramified domain over D without relative boundary. Let $\mathcal{G}$ be an unramified domain over D without relative boundary. From standard covering space theory, given any point $p \in \mathcal{G}$ and any continuous arc $\underline{\ell}$ in D starting from $\underline{p}$, we can find a unique continuous arc l in $\mathcal{G}$ with initial point at p such that $\pi(\ell) = \underline{\ell}$. This shows that the number of sheets $n(\underline{p})$ of $\mathcal{G}$ over $\underline{p} \in D$ is independent of $\underline{p}$. In particular, if D is simply connected, then any unramified domain $\mathcal{G}$ over D without relative boundary is a univalent domain.

Intersection of domains. Let $\{\mathcal{G}_j\}_{j=1}^l$ be a finite number of unramified domains over $\mathbf{C}^n$ and let π_j denote the canonical projection of $\mathcal{G}_j$ into $\mathbf{C}^n$. Let $\underline{p}$ be a point in $\mathbf{C}^n$ such that there is at least one point p_j in $\mathcal{G}_j$ $(j = 1, \dots, l)$ with $\pi(p_j) = \underline{p}$. We set $p := (p_1, \dots, p_l)$ and consider the set $\widetilde{\mathcal{G}}$ of all such p for each $\underline{p} \in \mathbf{C}^n$. Define $\widetilde{\pi}$ to be the projection from $\widetilde{\mathcal{G}}$ into $\mathbf{C}^n$ via $\widetilde{\pi}(p) = \underline{p}$. We next introduce a topology on $\widetilde{\mathcal{G}}$ as follows: let $p = (p_1, \dots, p_l) \in \widetilde{\mathcal{G}}$ and let $\widetilde{\pi}(p) = \underline{p} \in \mathbf{C}^n$. We can find a polydisk $\underline{\delta}$ containing $\underline{p}$ in $\mathbf{C}^n$ such that there exists a univalent neighborhood δ_j of p_j in $\mathcal{G}_j$ with $\pi_j(\delta_j) = \underline{\delta}$. The set $\widetilde{\delta}$ of all points $q = (q_1, \dots, q_l)$ of $\widetilde{\mathcal{G}}$, where $q_j \in \delta_j$ $(j = 1, \dots, l)$ are associated to some $\underline{q} \in \underline{\delta}$, constitutes a neighborhood of

$p \in \widetilde{\mathcal{G}}$. These form a neighborhood basis at p for the topology on $\widetilde{\mathcal{G}}$. The space $\widetilde{\mathcal{G}}$ equipped with this topology becomes a Hausdorff space. Since $\widetilde{\pi} \mid_{\tilde{\delta}}$ is bijective from $\widetilde{\delta}$ onto δ, it follows that $\widetilde{\mathcal{G}}$ is an unramified domain over $\mathbf{C}^n$, and $\widetilde{\pi}$ is the canonical projection from $\widetilde{\mathcal{G}}$ into $\mathbf{C}^n$. We say that $\widetilde{\mathcal{G}}$ is the **intersection of the unramified domains** $\{\mathcal{G}_j\}_{j=1}^l$ over $\mathbf{C}^n$, and we write $\widetilde{\mathcal{G}} = \mathcal{G}_1 \cap \ldots \cap \mathcal{G}_l$.

In general, $\widetilde{\mathcal{G}}$ is not connected even if each $\mathcal{G}_j$ $(j = 1, \ldots, l)$ is connected; cf. Remark 6.2. Let $p = (p_1, \ldots, p_l) \in \widetilde{\mathcal{G}}$ lie over $\underline{p} \in \mathbf{C}^n$ and let $\widetilde{\mathcal{G}}_p$ be the connected component of $\widetilde{\mathcal{G}}$ which contains p. Let $q = (q_1, \ldots, q_l) \in \widetilde{\mathcal{G}}_p$ lie over $\underline{q} \in \widetilde{\mathcal{G}}$. We can find an arc $\underline{\ell}$ which connects $\underline{p}$ and $\underline{q}$ such that there exists an arc L_j in $\mathcal{G}_j$ $(j = 1, \ldots, l)$ which connects p_j and q_j with $\pi_j(L_j) = \underline{\ell}$; then $L = (L_1, \ldots, L_l)$ is an arc connecting p and q in $\widetilde{\mathcal{G}}_p$. By convention we say that $\widetilde{\mathcal{G}}_p$ is the intersection of $\{\mathcal{G}_j\}_{j=1}^l$ determined by the initial point p. We can also consider the intersection of infinitely many (countable or uncountable) unramified domains in a similar fashion to that of finite intersections of unramified domains.

REMARK 6.1. Let $f_j(p)$ be a holomorphic function on an unramified domain $\mathcal{G}_j$ $(j = 1, \ldots, l)$, where $l < \infty$. Then $\sum_{j=1}^l f_j^{a_j}(p)$ and $\prod_{j=1}^l f_j^{a_j}(p)$ ($a_j \geq 0$ is an integer) define holomorphic functions on the intersection $\mathcal{G}_1 \cap \cdots \cap \mathcal{G}_l$.

REMARK 6.2. Even when $\mathcal{G}_1 = \mathcal{G}_2$ and $\mathcal{G}_1$ is connected, the intersection $\mathcal{G}_1 \cap \mathcal{G}_2$ need not be equal to $\mathcal{G}_1 = \mathcal{G}_2$; nor must the intersection be connected. For example, let $\mathcal{G}_1 = \mathcal{G}_2$ be the unramified domain over $D = \{0 < |z| < \infty\}$ in the complex plane $\mathbf{C}_z$ determined by the function $\sqrt{z}$, i.e., the Riemann surface of $\sqrt{z}$ over D. Let $z = 1 \in D$. Then each $\mathcal{G}_j$ $(j = 1, 2)$ contains two different points p_j and q_j lying over $z = 1$. In $\mathcal{G}_1 \cap \mathcal{G}_2$ we set $p = (p_1, p_2)$, $q = (q_1, q_2)$, and $r = (p_1, q_2)$. The connected component $\widetilde{\mathcal{G}}_p$ of $\mathcal{G}_1 \cap \mathcal{G}_2$ determined by the initial point p coincides with the connected component $\widetilde{\mathcal{G}}_q$ determined by the initial point q. However, the connected component $\widetilde{\mathcal{G}}_r$ of $\mathcal{G}_1 \cap \mathcal{G}_2$ determined by the initial point r is not the same as $\widetilde{\mathcal{G}}_p$. Thus, $\mathcal{G}_1 \cap \mathcal{G}_2$ consists of two connected components, $\widetilde{\mathcal{G}}_p$ and $\widetilde{\mathcal{G}}_r$.

6.1.2. Locally Ramified Domains. We next define ramified domains over $\mathbf{C}^n$; here, the branch points will be regarded as interior points of the domain. First we consider the local case. Fix $a = (a_1, \ldots, a_n) \in \mathbf{C}^n$ and let

$$\Delta \ : \ |z_j - a_j| < r_j \qquad (j = 1, \ldots, n)$$

be a polydisk in $\mathbf{C}^n$. Let Σ be an analytic hypersurface in $\overline{\Delta}$ and set $\sigma := \Delta \cap \Sigma$ and $\Delta' := \Delta \setminus \sigma$. We note that for any $\underline{p} \in \sigma$ and any sufficiently small polydisk δ centered at $\underline{p}$ in $\mathbf{C}^n$ the intersection $\delta \cap \Delta'$ is connected. Let $\mathcal{D}'$ be an unramified domain over Δ' without relative boundary; let π be its canonical projection; and let m be the number of sheets of $\mathcal{D}'$ over Δ', which is assumed to be *finite*. Fix $\underline{p} \in \sigma$. There exist boundary points $\widetilde{p}$ of $\mathcal{D}'$ over $\underline{p}$. The number of such points $\{\widetilde{p}\}$ is at most m. We form the union of all $\widetilde{p}$ over each $p \in \sigma$ with $\mathcal{D}'$, and we denote the resulting set by $\mathcal{D}$. For each $\widetilde{p} \in \mathcal{D} \setminus \mathcal{D}'$, we introduce a fundamental neighborhood basis as follows. Let δ^0 be any fundamental neighborhood of the boundary point $\widetilde{p}$ of $\mathcal{D}'$ and set $\underline{\delta}^0 := \pi(\delta^0)$. Let $\underline{q} \in \sigma \cap \underline{\delta}^0$ and let $\widetilde{q}$ be one of the boundary points of $\mathcal{D}'$ over $\underline{q}$. If there exists a fundamental neighborhood η of $\widetilde{q}$ which is contained in δ^0, then we say that $\widetilde{q}$ touches δ^0. We let $\widetilde{\delta^0}$ denote the union of δ^0 and all points $\widetilde{q}$ for each $\underline{q} \in \sigma \cap \underline{\delta}^0$ which touches δ^0. We then define $\widetilde{\delta^0}$ to be a fundamental neighborhood of $\widetilde{p}$ of $\mathcal{D}$. For a fundamental neighborhood basis δ_k^0 $(k = 1, 2, \dots)$ of a boundary point $\widetilde{p}$ of the unramified domain $\mathcal{D}'$, we construct $\widetilde{\delta}_k^0$ $(k = 1, 2, \dots)$ in $\mathcal{D}$ by the above procedure and we let $\widetilde{\delta}_k^0$ $(k = 1, \dots)$ be a fundamental neighborhood basis of $\widetilde{p}$ in $\mathcal{D}$. Using this neighborhood basis, $\mathcal{D}$ becomes a Hausdorff space which contains $\mathcal{D}'$. Note that since the topology of $\mathcal{D}$ does not depend on the choice of a fundamental neighborhood system $\widetilde{\delta}_k^0$ $(k = 1, 2, \dots)$ of $\widetilde{p}$ over $\underline{p} \in \sigma$, $\mathcal{D}$ is uniquely determined by $\mathcal{D}'$. We call $\mathcal{D}$ the **ramified domain associated to** $\mathcal{D}'$. In general, such a domain $\mathcal{D}$ is called a **locally ramified domain over** Δ, precisely, a locally ramified domain over Δ without relative boundary (or a **branched cover of** Δ). The canonical projection π defined in $\mathcal{D}'$ extends continuously to $\mathcal{D}$ in a unique fashion. We call this the **canonical projection** of $\mathcal{D}$ onto Δ, and we use the same notation π. We also call m (the number of sheets of $\mathcal{D}'$ over Δ') the **number of sheets** of $\mathcal{D}$ over Δ. For any $p \in \mathcal{D}$, we call $\underline{p} := \pi(p)$ the projection of p, or the base point of p.

Using the same notation, if Σ (the analytic hypersurface in the polydisk $\overline{\Delta}$) does not intersect

$$E \; : \; |z_j - a_j| \le r_j \quad (j = 1, \dots, n-1), \qquad |z_n - a_n| = r_n,$$

then $\mathcal{D}$ is said to be **standard with respect to** z_n. Let $\mathcal{D}$ be any locally ramified domain over the polydisk Δ and let $p \in \mathcal{D}$. Then we can choose a coordinate system $(z_1, \dots, z_n)$ of $\mathbf{C}^n$ such that there exists a polydisk $\Delta_0 \subset \Delta$ centered at $\underline{p}$ with the property that the portion $\mathcal{D}_0$ of $\mathcal{D}$ lying over Δ_0 contains p and is standard with respect to z_n.

Branch sets. Let $\mathcal{D}$ be a ramified domain over a polydisk Δ in $\mathbf{C}^n$ and let Σ be an analytic hypersurface in $\overline{\Delta}$. Let $\sigma := \Delta \cap \Sigma$; let $\widetilde{p} \in \mathcal{D}$ lie over $\underline{p} \in \sigma$ and let $\widetilde{\delta}_k^0$ $(k = 1, 2, \dots)$ be a fundamental neighborhood basis of $\widetilde{p}$ in $\mathcal{D}$. Then each $\widetilde{\delta}_k^0$ becomes a ramified domain over $\pi(\widetilde{\delta}_k^0) \subset \Delta$. Furthermore, the number of sheets m_k of $\widetilde{\delta}_k^0$ is independent of k provided k is sufficiently large. We denote this number by $\nu \ge 1$, and we call $\nu - 1$ the **ramification number of $\mathcal{D}$ at** $\widetilde{p}$. If $\nu \ge 2$, we say that $\widetilde{p}$ is a **branch point** of $\mathcal{D}$. If $\nu = 1$, we say that $\widetilde{p}$ (as well as each point $p \in \mathcal{D}'$) is a **regular point** of $\mathcal{D}$. The set $\mathcal{D}^0$ of all regular points of $\mathcal{D}$ is called the **regular part** of $\mathcal{D}$. Clearly $\mathcal{D}^0$ is a connected subdomain of $\mathcal{D}$ and may be considered as an unramified domain over Δ. We let S denote the set of all branch points of $\mathcal{D}$ and we call S the **branch set** of $\mathcal{D}$. Note that if $S \ne \emptyset$, then the projection $\underline{S}$ of S into Δ consists of some irreducible components of σ in Δ. Furthermore, suppose $p \in S$ is chosen so that $\underline{p}$ is a non-singular point of $\underline{S}$. Let l be the ramification number of $\mathcal{D}$ at p. For simplicity, suppose $\underline{p} = 0 \in \Delta$ and $\underline{S} : \; z_n = 0$ near $\underline{p} = 0$. Then, near p, $\mathcal{D}$ has a representation over a neighborhood of the origin in $\mathbf{C}^n$ as the product of $\mathbf{C}^{n-1}$ with variables $z_1, \dots, z_{n-1}$ and the Riemann surface of $\sqrt[l]{z_n}$ over $\mathbf{C}_{z_n}$. We call such a branch point p a **regular branch point** of $\mathcal{D}$. If $\sigma \ne \emptyset$ and the number of sheets m of $\mathcal{D}'$ is at least 2, then $\mathcal{D}$ always contains branch points. This follows since Δ is simply connected.

Let $\mathcal{D}$ be a ramified domain over Δ. If a *continuous*, complex-valued function $f(p)$ on $\mathcal{D}$ is holomorphic in $\mathcal{D}'$, then we say that $f(p)$ is a **holomorphic function** on $\mathcal{D}$. We now give the prototypical example of such a function.

EXAMPLE 6.1. Let

$$P(z, w) = w^m + a_1(z) w^{m-1} + \cdots + a_m(z)$$

be an irreducible polynomial which is monic in w, where $a_i(z)$ $(i = 1, \dots, m)$ are holomorphic functions in a polydisk Δ in $\mathbf{C}^n$, and let

$$\Sigma: \ P(z, w) = 0 \qquad \text{in } \Delta \times \mathbf{C}_w.$$

We let $d(z)$ denote the discriminant of $P(z, w)$ with respect to w; thus $d(z)$ is not identically 0 in Δ. Let $\sigma := \{z \in \Delta \mid d(z) = 0\}$ and $\Delta' := \Delta \setminus \sigma$. Then we have the algebraic (single-valued) function $w = \eta(p)$ defined implicitly by $P(z, w) = 0$ on the unramified m-sheeted domain $\mathcal{D}'$ over Δ' without relative boundary. Furthermore, if we consider the ramified domain $\mathcal{D}$ associated to $\mathcal{D}'$, then $\eta(p)$ becomes a holomorphic function on $\mathcal{D}$. We call the locally ramified domain $\mathcal{D}$ over Δ the **Riemann domain determined by the algebraic function** $w = \eta(p)$, and we call $\mathcal{D}$ the **projection of the analytic hypersurface** Σ **over** Δ.

Let $(z_0, w_0) \in \Sigma$ with $z_0 \in \sigma$. If (z_0, w_0) is a non-singular point of the analytic hypersurface Σ in $\Delta \times \mathbf{C}_w$, then we can find a unique point $p \in \mathcal{D} \setminus \mathcal{D}'$ such that p is a regular branch point of $\mathcal{D}$; and there exist neighborhoods δ of (z_0, w_0) in $\Delta \times \mathbf{C}_w$ and V of p in $\mathcal{D}$ with $\delta \cap \Sigma$ bijective to V. In general, there exist a finite number of points p_i $(i = 1, \dots, \nu)$ in $\mathcal{D}$ which correspond to (z_0, w_0), i.e., $\underline{p_i} = z_0$ and $\eta(p_i) = w_0$. For example, let $P(z_1, z_2, w) = w^6 - z_1^2 z_2^3$. Then $\sigma = \{z_1 = 0\} \cup \{z_2 = 0\}$ and $\{z_1 = w = 0\} \cup \{z_2 = w = 0\} \subset \Sigma$. The origin $(0, 0, 0)$ in Σ corresponds to the single point p_1 of $\mathcal{D}$ lying over the origin $(0, 0)$. However, to any point $(z_1, 0, 0)$ or $(0, z_2, 0)$ of Σ other than the origin there correspond three or two points of $\mathcal{D}$, respectively.

Conversely, let $\mathcal{D}$ be any m-sheeted locally ramified domain over a polydisk Δ in $\mathbf{C}^n$ with branch set S. We let $\sigma := \underline{S}$. Let $\eta(p)$ be any holomorphic function on $\mathcal{D}$ such that $\eta(p)$ has m different values over some point $\underline{p} \in \Delta' := \Delta \setminus \sigma$. Then we can construct a monic polynomial $P(z, w)$ in w such that the coefficients are holomorphic functions in Δ and such that the algebraic function determined by $P(z, w) = 0$ coincides with $w = \eta(z)$. Indeed, it suffices to define $P(z, w) := \prod_{j=1}^{m} (w - \eta_j(z))$, where $\eta_j(z)$ $(j = 1, \dots, m)$ are the values which $\eta(p)$ assumes at p over z. In particular, if we set $A := \{p \in \mathcal{D} | \ \eta(p) = 0\}$, then the projection $\underline{A}$ is the analytic hypersurface in Δ defined by $\{z \in \Delta \mid P(z, 0) = 0\}$. Thus, the zeros of the holomorphic function $\eta(p)$ on the ramified domain $\mathcal{D}$ are not isolated, as in the case of a univalent domain in $\mathbf{C}^n$ $(n \geq 2)$.

6.1.3. Ramified Domains. Next we define a (globally) ramified domain over $\mathbf{C}^n$. Let $\mathcal{G}$ be a connected Hausdorff space and let π be a continuous mapping from $\mathcal{G}$ into $\mathbf{C}^n$ with the following property: for any point $p \in \mathcal{G}$, there exists a polydisk $\underline{\delta}$ centered at $\pi(p)$ in $\mathbf{C}^n$ such that the connected component δ of $\pi^{-1}(\underline{\delta})$ which contains p is a locally ramified domain over $\underline{\delta}$ without relative boundary; and the canonical projection from δ to $\underline{\delta}$ is the restriction of π onto δ. In this case, we call $\mathcal{G}$ a **ramified domain over** $\mathbf{C}^n$. We call δ a fundamental neighborhood of p, and a branch point of δ is called a branch point of $\mathcal{G}$. The set of all branch points of $\mathcal{G}$ is called the **branch set** of $\mathcal{G}$. A point of $\mathcal{G}$ which is not a branch point of $\mathcal{G}$ is said to be regular. The set of all regular points of $\mathcal{G}$ is called the **regular part** of $\mathcal{G}$. For a point $\underline{p} \in \mathbf{C}^n$ such that $\pi^{-1}(\underline{p})$ contains no branch points of $\mathcal{G}$, the cardinality of $\pi^{-1}(\underline{p})$ is called the **number of sheets** of $\mathcal{G}$ at $\underline{p}$. The maximum $m(\mathcal{G})$ of such $m(\underline{p})$, $\underline{p} \in \mathbf{C}^n$, is called the number of sheets of $\mathcal{G}$. This number may be $+\infty$. A boundary point of $\mathcal{G}$ over $\mathbf{C}^n$ is defined in a manner similar to the case of a boundary point in an unramifed domain over $\mathbf{C}^n$. If $\pi(\mathcal{G})$ is contained in a domain

$D \subset \mathbf{C}^n$, then we say that $\mathcal{G}$ is a ramified domain over D. A boundary point p of the ramified domain $\mathcal{G}$ over D such that $\pi(p) \in D$ is called a relative boundary point of $\mathcal{G}$ with respect to D. A ramified domain $\mathcal{G}$ over D which has no relative boundary points with respect to D will be called a ramified domain over D without relative boundary. An open and connected set $\mathcal{D}$ in $\mathcal{G}$ is called a domain in $\mathcal{G}$, although we often drop the connectivity assumption. Let $f(p)$ be a complex-valued function on a domain v in $\mathcal{G}$. If $f(p)$ is holomorphic on a fundamental neighborhood δ_p of each point $p \in v$, then we say that $f(p)$ is a **holomorphic function** on v.

Let $\mathcal{G}_1$ and $\mathcal{G}_2$ be ramified domains over $\mathbf{C}^n$ and $\mathbf{C}^m$. Let $\varphi(p)$ be a mapping from $\mathcal{G}_1$ into $\mathcal{G}_2$. If for any domain v in $\mathcal{G}_2$ and any holomorphic function $f(q)$ on v the composite function $\widetilde{f}(p) := f(\varphi(p))$ is a holomorphic function on the domain $\varphi^{-1}(v) \subset \mathcal{G}_1$, then we say that $\varphi(p)$ is an **analytic** (or **holomorphic**) **mapping** from $\mathcal{G}_1$ to $\mathcal{G}_2$. Moreover, if $m = n$ and if there exists a bijective analytic mapping from $\mathcal{G}_1$ to $\mathcal{G}_2$, then $\mathcal{G}_1$ and $\mathcal{G}_2$ are **analytically** (or **biholomorphically**) **equivalent**. Note that an analytic mapping does not always map branch points to branch points.

Let $\mathcal{G}'$ be an unramified domain over $\mathbf{C}^n$ and let π be the canonical projection of $\mathcal{G}'$ into $\mathbf{C}^n$. We will construct in a canonical way a ramified domain $\mathcal{G}$ associated to $\mathcal{G}'$. First take $a \in \mathbf{C}^n$ and a polydisk Δ centered at a. Let Σ be an analytic hypersurface in $\overline{\Delta}$, and set $\sigma := \Sigma \cap \Delta$ and $\Delta' := \Delta \setminus \sigma$. If there exists a connected component $\mathcal{D}'$ of $\pi^{-1}(\Delta') \subset \mathcal{G}'$ which is a finitely sheeted unramified domain over Δ' without relative boundary, then we can construct the ramified domain $\mathcal{D}$ associated to $\mathcal{D}'$ as defined in 6.1.2. We replace each such component $\mathcal{D}'$ by the corresponding domain $\mathcal{D}$. Then we have constructed a ramified domain $\mathcal{G}$ over $\mathbf{C}^n$, which we call the **ramified domain associated to** $\mathcal{G}'$.

We now define the intersection of a finite number of ramified domains $\mathcal{G}_j$ ($j = 1, \ldots, l$) over $\mathbf{C}^n$. Let $\mathcal{G}_j'$ be the regular part of $\mathcal{G}_j$. Since $\mathcal{G}_j'$ is an unramified domain over $\mathbf{C}^n$, we can construct $\mathcal{G}_1' \cap \ldots \cap \mathcal{G}_l'$, which consists of a finite number of unramified domains H_k' ($k = 1, \ldots, L$). We form the ramified domain H_k over $\mathbf{C}^n$ associated to H_k', and the totality of these domains $H_1, \ldots, H_L$ is called the **intersection of ramified domains** $\{\mathcal{G}_j\}_{j=1}^{l}$ and is denoted by $\mathcal{G}_1 \cap \ldots \cap \mathcal{G}_l$. Given $p_j \in \mathcal{G}_j$ ($j = 1, \ldots, l$) such that $\pi_j(p_j)$ is the same base point $\underline{p} \in \mathbf{C}^n$, we can define the connected component of $\mathcal{G}_1 \cap \ldots \cap \mathcal{G}_l$ determined by the initial point $p = (p_1, \ldots, p_l)$ in a similar fashion as in the case of unramified domains. Similarly, we can define the intersection of infinitely many (not necessarily countable) ramified domains over $\mathbf{C}^n$.

The following example gives the relationship between an analytic set of pure dimension $r < n$ in a domain $D \subset \mathbf{C}^n$ and an associated ramified domain over $\mathbf{C}^r$.

EXAMPLE 6.2. Let $\mathcal{E}$ be an irreducible r-dimensional analytic set in a domain $D \subset \mathbf{C}^n$. We take Euclidean coordinates $(z_1, \ldots, z_n)$ of $\mathbf{C}^n$ which satisfy the Weierstrass condition for Σ at each point of Σ. Then $\mathcal{E}$ can be represented in the form $(z_{r+1}, \ldots, z_n) = \eta(z_1, \ldots, z_r)$; i.e.,

$$\begin{array}{rcl} z_{r+1} & = & \eta_{r+1}(z_1, \ldots, z_r), \\ \vdots & & \vdots \\ z_n & = & \eta_n(z_1, \ldots, z_r), \end{array}$$

where $\eta_j(z_1, \ldots, z_r)$ ($j = r + 1, \ldots, n$) is a holomorphic function in a ramified domain $\mathcal{D}_j$ over $\mathbf{C}^r$ with variables $z_1, \ldots, z_r$ and canonical projection π_j. Fix a non-singular point $p' = (z'_1, \ldots, z'_n)$ of $\mathcal{E}$. Then, over the point $(z'_1, \ldots, z'_r)$, there exists a unique point $p'_j \in \mathcal{D}_j$ $(j = r+1, \ldots, n)$ such that $z'_j = \eta_j(p'_j)$. If we construct the intersection $\widetilde{\mathcal{D}}$ of $\mathcal{D}_j$ $(j = r+1, \ldots, n)$ determined by the initial point $(p'_{r+1}, \ldots, p'_n)$, then $\eta(z)$ is a single-valued holomorphic vector-valued function on $\widetilde{\mathcal{D}}$. Since $\mathcal{E}$ is irreducible, the graph $(z_{r+1}, \ldots, z_n) = \eta(z')$, $z' = (z_1, \ldots, z_r) \in \widetilde{\mathcal{D}}$ in $\mathbf{C}^n$ coincides with $\mathcal{E}$. Thus we call $\widetilde{\mathcal{D}}$ the **ramified domain over $\mathbf{C}^r$ determined by $\eta(z')$**, or the **projection of $\mathcal{E}$ over $\mathbf{C}^r$**.

6.1.4. Properties of Locally Ramified Domains. We now exhibit some interesting phenomena for ramified domains over $\mathbf{C}^n$ $(n \geq 2)$, noted by K. Oka, which do not occur in the case $n = 1$.

1. Non-uniformizable branch points. Let $\mathcal{G}$ be a ramified domain over $\mathbf{C}^n$ and let π be the canonical projection of $\mathcal{G}$ into $\mathbf{C}^n$. Let p be a branch point of $\mathcal{G}$. If there exists a neighborhood v of p in $\mathcal{G}$ such that v is biholomorphically equivalent to a polydisk, then p is called a **uniformizable branch point** of $\mathcal{G}$. In the case of $\mathbf{C}$, any branch point of any Riemann surface is uniformizable.

Let p be a branch point of $\mathcal{G}$. Let δ be a fundamental neighborhood of p in $\mathcal{G}$ and let σ be the branch set of $\mathcal{G}$ in δ. We let $\underline{\delta} = \pi(\delta)$ and $\underline{\sigma} = \pi(\sigma)$, so that $\underline{p} \in \underline{\sigma}$. It is clear that if $\underline{p}$ is a non-singular point of $\underline{\sigma}$, then p is a uniformizable branch point of $\mathcal{G}$. We now give an example of a branch point which is not uniformizable.

EXAMPLE 6.3. In $\mathbf{C}^2$ with variables z_1 and z_2, we consider the ramified domain $\mathcal{G}$ over $\mathbf{C}^2$ determined by the algebraic function

$$w^2 - z_1^2 + z_2^2 = 0, \quad \text{i.e.,} \ w = \sqrt{z_1^2 - z_2^2}.$$

Then $\mathcal{G}$ is two-sheeted over $\mathbf{C}^2$ with branch set $z_1^2 - z_2^2 = 0$. The point O of $\mathcal{G}$ over the origin $(0,0) \in \mathbf{C}^2$ is not a uniformizable branch point of $\mathcal{G}$.

We prove this by contradiction. Assume that O is uniformizable. In particular, there exists a simply connected neighborhood δ of O in $\mathcal{G}$ such that $\delta \setminus \{O\}$ is also simply connected. We may assume that δ has no relative boundary points with respect to $\underline{\delta}$. We consider the function

$$g(z_1, z_2) = \sqrt{z_1 - z_2},$$

which is *locally* holomorphic on $\delta \setminus \{O\}$. Since $\delta \setminus \{O\}$ is simply connected, $g(z_1, z_2)$ must be single-valued on $\delta \setminus \{O\}$. On the other hand, fix an ϵ with $0 < \epsilon < 1$ such that $\{|z_1| \leq 2\epsilon\} \times \{|z_2| \leq \epsilon\} \subset \underline{\delta}$. We can choose two distinct points P_ϵ^+, P_ϵ^- in δ over the point $(\epsilon, 0) \in \underline{\delta}$. Consider the closed circle $\ell : (z_1, z_2) = (\epsilon e^{i\theta}, 0)$, where $0 \leq \theta \leq 2\pi$, in $\underline{\delta}$. If we traverse ℓ in $\mathcal{G}$ starting at P_ϵ^+, then we return to P_ϵ^+. However, $g(P_\epsilon^+) = \sqrt{\epsilon} \neq 0$ will vary through the values $\sqrt{\epsilon\, e^{i\theta}}$ and has final value $-\sqrt{\epsilon}$. This contradicts the single-valuedness of $g(z_1, z_2)$ on $\delta \setminus \{O\}$.

2. Analytic sets in a ramified domain. As in the case of univalent domains in $\mathbf{C}^n$, we shall define analytic sets A in a ramified domain $\mathcal{G}$ over $\mathbf{C}^n$. Let A be a closed set in $\mathcal{G}$. If, for each point $a \in A$, there exist a neighborhood δ of a in $\mathcal{G}$ and a finite number of holomorphic functions $f_j(p)$ $(j = 1, \ldots, \mu)$ in δ such that $\delta \cap A = \{f_j(p) = 0 \ (j = 1, \ldots, \mu)\}$, then we say that A is an **analytic set** in $\mathcal{G}$. We note that the zero set Σ of a non-constant holomorphic function $f(p)$ in $\mathcal{G}$ is called an **analytic hypersurface** in $\mathcal{G}$ (provided $\Sigma \neq \emptyset$). Such a set Σ contains no isolated points in $\mathcal{G}$ (see 2 of Remark 6.1).

In the case of an analytic hypersurface Σ in a univalent domain D in $\mathbf{C}^n$, for any point $z_0 \in \Sigma$, we can find a holomorphic function $f(z)$ defined in a neighborhood δ of z_0 in D such that $f(z) = 0$ precisely on $\delta \cap \Sigma$ with order 1. This result is no longer true in the case of analytic hypersurfaces in ramified domains over $\mathbf{C}^n$. We give an example of an analytic hypersurface Σ in a ramified domain $\mathcal{G}$ over $\mathbf{C}^n$ (which passes through a non-uniformizable branch point p_0 of $\mathcal{G}$) such that there does not exist a neighborhood δ of p_0 in $\mathcal{G}$ on which $\Sigma \cap \delta$ may be written as the zero set of a holomorphic function $f(p)$ with order 1.

EXAMPLE 6.4. Let $\mathcal{G}$ be the same ramified domain over $\mathbf{C}^2$ as in Example 6.3. Take two (one-dimensional) analytic hypersurfaces Σ^+ and Σ^- over $z_2 = 0$ in $\mathcal{G}$, and consider the holomorphic function $f = f(p)$ in $\mathcal{G}$, defined as

$$f(p) := \sqrt{z_1^2 - z_2^2} - z_1. \tag{6.1}$$

If we choose a suitable branch of the function $\sqrt{z_1^2 - z_2^2}$, then $f(p) = 0$ on Σ^+ and $f(p) \neq 0$ in $\mathcal{G} \setminus \Sigma^+$. Note that the order of the zero of $f(p)$ at each point on Σ^+ is two. This follows because if we fix $(\epsilon, 0) \in \Sigma^+$ with $\epsilon > 0$, then in a neighborhood of $(\epsilon, 0)$ we can write

$$f(\epsilon, z_2) = \sqrt{\epsilon^2 - z_2^2} - \epsilon = -\frac{z_2^2}{2\epsilon} + O(z_2^4).$$

On the other hand, we now proceed to show that there does not exist a holomorphic function $F(p)$ defined in a neighborhood δ of the point O in $\mathcal{G}$ over the origin $(0,0) \in \mathbf{C}^2$ such that $F(p)$ vanishes to order one at each point of $\Sigma^+ \cap \delta$ and $F(p) \neq 0$ on $\delta \setminus \Sigma^+$.[1] We prove this by contradiction. Assume that there exists such a holomorphic function $F(p)$ defined in a neighborhood δ of O in $\mathcal{G}$. We may assume that δ has no relative boundary on $\underline{\delta}$. We consider the holomorphic mapping

$$T: \quad w_1 = z_1, \quad w_2 = F(p)$$

from δ into $\mathbf{C}^2_w$. Define $\kappa := T(\delta)$, which is a ramified domain over $\mathbf{C}^2_w$ such that the point $\widetilde{O} = T(O)$ is the only point of κ lying over $(w_1, w_2) = (0,0)$. We fix a bidisk $B := B_1 \times B_2 \subset \mathbf{C}^2_w$, where $B_1 = \{|w_1| < \rho_1\}$ and $B_2 = \{|w_2| < \rho_2\}$, such that there exists a subset κ_0 of κ over B which has no relative boundary points on B. Thus the number of sheets $m \geq 1$ of κ_0 is determined. If we show that, in fact, $m = 1$, it follows that $\kappa_0 = B$. In this case, the point O is thus a uniformizable point of $\mathcal{G}$, which contradicts the fact stated in Example 6.3. Hence it suffices to verify that $m = 1$.

We begin by choosing a bidisk $\Delta := \Delta_1 \times \Delta_2 \subset\subset \underline{\delta}$, where

$$\Delta_1 = \{|z_1| < \rho_1^*\} \quad \text{and} \quad \Delta_2 = \{|z_2| < \rho_2^*\},$$

and with $\rho_2 > 0$ chosen so small that $[(\partial \Delta_1) \cap \Delta_2] \cap \{z_1^2 = z_2^2\} = \emptyset$. We let $\widetilde{\Delta}$ denote the subset of δ over Δ. We fix an annulus $\Gamma_1 := \{\rho_1'' < |z_1| < \rho_1'\}$ containing $\partial \Delta_1$ such that if we let $\Lambda := \Gamma_1 \times \Delta_2$, then $\Lambda \cap \{z_1^2 = z_2^2\} = \emptyset$. We thus have two regular parts $\Lambda^{\pm} = \Gamma_1^{\pm} \times \Delta_2$ of $\widetilde{\Delta}$ over Λ. We assume $\Lambda^+ \cap \Sigma^+ \neq \emptyset$ and $\Lambda^- \cap \Sigma^+ = \emptyset$. Over each point $(z_1, z_2) \in \Lambda$, there exists a point $(z_1^{\pm}, z_2) \in \Lambda^{\pm}$. We fix $z_1 \in \Gamma_1$. By assumption, as a holomorphic function of the complex variable z_2 in Δ_2, $F(z_1^+, z_2)$ vanishes if and only if $z_2 = 0$ (with order 1); also $F(z_1^-, z_2) \neq 0$ at any point $z_2 \in \Delta_2$. Thus there exist small disks $\Delta_2^0 := \{|z_2| < \alpha_2\} \subset \Delta_2$ and $B_2^0 := \{|w_2| < \beta_2\} \subset B_2$

[1] This proof is due to T. Kizuka.

such that $F(z_1^+, z_2)$ is univalent on Δ_2^0 with $B_2^0 \subset F(z_1^+, \Delta_2^0)$; in addition, we can choose $\eta_2 > 0$ sufficiently small so that $|F(z_1^-, z_2)| \geq \eta_2$ for any $z_2 \in \Delta_2$. Letting z_1 vary over Γ_1, we may assume that α_2, β_2, $\eta_2 > 0$ are independent of $z_1 \in \Gamma_1$. Furthermore, since $F(p)$ vanishes only on Σ^+ in $\mathcal{G}$, it follows that we can find $\xi_2 > 0$ such that $|F(p)| \geq \xi_2$ for any $p \in \mathcal{G}$ with $\underline{p} \in \underline{\delta} \cap (\Gamma_1 \times \{|z_2| \geq \alpha_2\})$. Thus, if we set $\epsilon_2 := \min\{\beta_2, \eta_2, \xi_2\} > 0$ and $\lambda := \{\rho_1'' < |w_1| < \rho_1'\} \times \{|w_2| < \epsilon_2\} \subset \mathbf{C}^2_w$, then the subset of κ_0 lying over λ consists of a single univalent part. Thus $m = 1$, as desired.

Note that if we set $g(p) := \sqrt{z_1^2 - z_2^2} - z_1(1 + z_2)$, then $g(p) = 0$ is of order 1 along Σ^+ but $g(p)$ has additional zeros near Σ^+.

3. Intersection of two analytic hypersurfaces in a ramified domain.
Let D be a (univalent) domain in $\mathbf{C}^n$ and let S_1 and S_2 be two distinct irreducible analytic hypersurfaces in D. If the intersection $S_1 \cap S_2$ is nonempty, it is a pure $(n-2)$-dimensional analytic set in D. This result no longer holds in the case of ramified domains over $\mathbf{C}^n$.

EXAMPLE 6.5. [2] We consider $\mathbf{C}^{2n}$ with variables $z = (z_1, \ldots, z_n)$ and $w = (w_1, \ldots, w_n)$ (here $n \geq 2$). Let

$$\Sigma := \{(z, w) \in \mathbf{C}^{2n} \mid z_i w_j = z_j w_i \ (1 \leq i, j \leq n)\},$$

or, as is usually written,

$$\Sigma : \ \frac{w_1}{z_1} = \cdots = \frac{w_n}{z_n}.$$

Thus, Σ is an irreducible $(n+1)$-dimensional analytic set in $\mathbf{C}^{2n}$ passing through the origin $(0, 0)$. Choose coordinates $(u_1, \ldots, u_{2n})$ of $\mathbf{C}^{2n}$ which satisfy the Weierstrass condition for Σ at each point of Σ. If $\mathcal{D}$ denotes the projection of Σ over the space $\mathbf{C}^{n+1}$ generated by the first $n+1$ variables $u_1, \ldots, u_{n+1}$, it follows that $\mathcal{D}$ is a ramified domain over $\mathbf{C}^{n+1}$. We usually identify $\mathcal{D}$ with Σ, and we let O denote the point of $\mathcal{D}$ which corresponds to the origin $(0, 0)$ in Σ. For any complex number $c \in \mathbf{C}$, we define the n-dimensional analytic plane

$$L_c : \ w_i = c z_i \qquad (i = 1, \ldots, n)$$

in $\mathbf{C}^{2n}$. Then $L_c \subset \Sigma$ for each $c \in \mathbf{C}$, and $L_{c_1} \cap L_{c_2} = \{(0, 0)\}$ in $\mathbf{C}^{2n}$ if $c_1 \neq c_2$.

Let $\mathcal{L}_c$ denote the set in $\mathcal{D}$ corresponding to L_c in $\mathbf{C}^{2n}$. Then $\mathcal{L}_c$ is an irreducible analytic hypersurface in $\mathcal{D}$ with $\mathcal{L}_{c_1} \cap \mathcal{L}_{c_2} = \{O\}$ $(c_1 \neq c_2)$. This is 0-dimensional, which yields the example since $n + 1 \geq 3$.

Furthermore, we note that for each $c \in \mathbf{C}$, there does not exist a holomorphic function $f(p)$ defined in a neighborhood V of O in $\mathcal{D}$ which vanishes precisely on $V \cap \mathcal{L}_c$ (regardless of the order of vanishing of $f(p)$ along $\mathcal{L}_c$). We prove this statement by contradiction. Thus we assume that there exists such a function $f(p)$ in a neighborhood V of O in $\mathcal{D}$. We fix $c' \in \mathbf{C}$ with $c' \neq c$. Denote the restriction of $f(p)$ to $V \cap \mathcal{L}_{c'}$ (which is an n-dimensional ramified domain) by $f_0(p)$. Then $f_0(p)$ vanishes only at the origin O in $V \cap \mathcal{L}_{c'}$; in particular, the zeros of $f_0(p)$ are isolated. This is impossible since $n \geq 2$. □

4. Meromorphic functions in a ramified domain.
Let $\mathcal{D}$ be a ramified domain over $\mathbf{C}^n$ $(n \geq 2)$ and let $A \subset \mathcal{D}$. We say that A has dimension at most k if, for each point $a \in A$, there exists a neighborhood δ of

[2] This example is due to H. Grauert [**23**].

a in $\mathcal{D}$ lying over a polydisk $\underline{\delta}$ centered at $\underline{a}$ in $\mathbf{C}^n$ without relative boundary such that $\underline{A \cap \delta}$ is contained in a k-dimensional analytic set in $\underline{\delta}$.

Let $g(p)$ be a function defined in $\mathcal{D}$. If $g(p)$ can be represented locally as the quotient of two holomorphic functions, then we say that $g(p)$ is a meromorphic function in $\mathcal{D}$. More precisely, $g(p)$ is a single-valued function on $\mathcal{D}$ (taking values in $\mathbf{C} \cup \{\infty\}$) except for an at most $(n-1)$-dimensional set A in $\mathcal{D}$; moreover, at each point $p \in \mathcal{D}$, there exist a neighborhood δ of p in $\mathcal{D}$ and holomorphic functions $h_\delta(p)$, $k_\delta(p)$ such that $\{h_\delta(p) = k_\delta(p) = 0\} \subset A$ and $g(p) = h_\delta(p)/k_\delta(p)$ in $\delta \setminus A$. A point p at which $h_\delta(p) \neq 0$ and $k_\delta(p) = 0$ is called a **pole** of $g(p)$. The points p at which $h_\delta(p) = k_\delta(p) = 0$ are called the **points of indeterminacy** of $g(p)$. Thus, the set A is considered as the set of all indeterminacy points of $g(p)$ in $\mathcal{D}$.

Assume now that $\mathcal{D}$ is a ramified domain over a polydisk $\Delta \subset \mathbf{C}^n$ without relative boundary such that the number of sheets m is finite. Let $g(p)$ be a meromorphic function in $\mathcal{D}$. Then there exists a polynomial $Q(z, w)$ of one complex variable w of degree m,

$$Q(z,w) = a_0(z)w^m + a_1(z)w^{m-1} + \cdots + a_m(z),$$

where each $a_i(z)$ $(i = 0, 1, \ldots, m)$ is a holomorphic function in $\mathcal{D}$, such that for each fixed $z = \underline{p}$, the set of points w satisfying $Q(z, w) = 0$ coincides with $w = g(p)$.

To see this, fix $z \in \Delta \setminus \underline{A}$, where A is the set of indeterminacy of $g(p)$ in $\mathcal{D}$. Then there exist m points $p_1, \ldots, p_m$ of $\mathcal{D}$ such that $\underline{p_i} = z$, and we denote by $g_1(z), \ldots, g_m(z)$ the values of $g(p)$ at $p_1, \ldots, p_m$. We form the product

$$R(z,w) = \prod_{i=1}^{m}(w - g_i(z)) = w^m + b_1(z)w^{m-1} + \cdots + b_m(z),$$

where each $b_i(z)$ $(i = 1, \ldots, m)$ is a single-valued meromorphic function on $\Delta \setminus \underline{A}$. Since $g(p)$ can be locally represented as the quotient of two holomorphic functions, $b_i(z)$ is a meromorphic function on Δ. Thus, $b_i(z) = a_{1i}(z)/a_{2i}(z)$, where $a_{1i}(z)$ and $a_{2i}(z)$ are holomorphic functions in Δ, and setting $Q(z, w) := R(z, w)a_{21}(z) \cdots a_{2m}(z)$ gives the desired representation.

An indeterminacy point $p \in \mathcal{D}$ of $g(p)$ satisfies $a_0(\underline{p}) = \cdots = a_m(\underline{p}) = 0$. In general, the set of indeterminacy points of a meromorphic function in a ramified domain over $\mathbf{C}^n$ is no longer of dimension $n-2$, in contrast to the case of univalent domains in $\mathbf{C}^n$.

EXAMPLE 6.6. We recall the ramified domain $\mathcal{D}$ over $\mathbf{C}^{n+1}$ $(n \geq 2)$ from Example 6.5, and we use the same notation $\Sigma, \mathcal{D}, L_c, \mathcal{L}_c$. We note that

$$\Sigma = \bigcup_{c \in \mathbf{P}} L_c, \qquad \text{where } L_\infty = \{(0, w) \in \mathbf{C}^{2n} \mid w \in \mathbf{C}^n\}.$$

We set

$$\Phi(z, w) := \frac{w_1}{z_1} \qquad \text{in } \mathbf{C}^{2n},$$

and consider the restriction of Φ to Σ, which we denote by $\varphi(z, w)$. Thus, $\varphi|_{L_c} = c$. If we let $\widetilde{\varphi}(p)$ denote the function on $\mathcal{D}$ which corresponds to $\varphi(z, w)$ on Σ, then $\widetilde{\varphi}$ is a meromorphic function on $\mathcal{D}$ with pole set $\mathcal{L}_\infty \setminus \{O\}$, zero set $\mathcal{L}_0 \setminus \{O\}$, and only one indeterminacy point, namely $\{O\}$, which is not of codimension 2.

EXAMPLE 6.7. In Example 6.4, consider the meromorphic function $g(p) := f(p)/z_2$, where $\underline{p} = (z_1, z_2)$, in the two-dimensional ramified domain $\mathcal{G}$, and $f(p)$ is defined by (6.1). Then the set of indeterminacy points of $g(p)$ is one-dimensional.

6.1.5. Ramified Domains of Holomorphy. Let $\mathcal{D}$ be a ramified domain over $\mathbf{C}^n$. Let $f(p)$ be a holomorphic function in $\mathcal{D}$. If $f(p)$ satisfies the following two conditions:

1. $f(p)$ has different function elements at any two distinct points of $\mathcal{D}$; i.e., for any two distinct regular points $p_1, p_2 \in \mathcal{D}$ such that $\underline{p}_1 = \underline{p}_2 = z_0$ in $\mathbf{C}^n$, $f(p)$ has different Taylor expansions in powers of $z - z_0$ in neighborhoods of p_1 and p_2, and
2. there is no ramified domain $\widetilde{\mathcal{D}}$ over $\mathbf{C}^n$ with $\mathcal{D} \subset \widetilde{\mathcal{D}}$ and $\widetilde{\mathcal{D}} \neq \mathcal{D}$ such that $f(p)$ can be holomorphically extended to $\widetilde{\mathcal{D}}$,

then we say that $\mathcal{D}$ is a **ramified domain of holomorphy of** $f(p)$. Furthermore, a ramified domain $\mathcal{D}$ over $\mathbf{C}^n$ is called a **ramified domain of holomorphy** if there exists at least one holomorphic function $f(p)$ such that $\mathcal{D}$ is a domain of holomorphy of $f(p)$. Given a ramified domain $\mathcal{D}$ over $\mathbf{C}^n$, there exists a smallest ramified domain of holomorphy $\widehat{\mathcal{D}}$ which contains $\mathcal{D}$. For this purpose, it suffices to consider the intersection $\bigcap \widetilde{\mathcal{D}}$ of all ramified domains of holomorphy $\widetilde{\mathcal{D}}$ such that $\mathcal{D} \subset \widetilde{\mathcal{D}}$, since, in particular, $\mathbf{C}^n$ is one such $\widetilde{\mathcal{D}}$.

Now let $\mathcal{D}$ be a ramified domain over $\mathbf{C}^n$ and let $K \subset\subset \mathcal{D}$. We define

$$\widehat{K}_{\mathcal{D}} := \{q \in \mathcal{D} \mid |f(q)| \leq \max_{p \in K} |f(p)| \text{ for all } f \text{ holomorphic in } \mathcal{D}\},$$

which is called the **holomorphic hull** of K in $\mathcal{D}$.

If a ramified domain $\mathcal{D}$ over $\mathbf{C}^n$ satisfies the two conditions:

1. there exists a holomorphic function $f(p)$ in $\mathcal{D}$ such that $f(p)$ has different function elements at any two distinct points of $\mathcal{D}$, and
2. for any $K \subset\subset \mathcal{D}$ we have $\widehat{K}_{\mathcal{D}} \subset\subset \mathcal{D}$,

then we say that $\mathcal{D}$ is **holomorphically convex**. In Theorem 1.13 in Part I we showed that a (univalent) domain D in $\mathbf{C}^n$ is a domain of holomorphy if and only if D is holomorphically convex. In the case of ramified domains $\mathcal{D}$ over $\mathbf{C}^n$ for $n \geq 2$, $\mathcal{D}$ holomorphically convex implies $\mathcal{D}$ is a domain of holomorphy, but the converse is no longer true in general. This is shown by the following example of H. Grauert and R. Remmert [**23**].

EXAMPLE 6.8. We consider the ramified domain $\mathcal{D}$ over $\mathbf{C}^{n+1}$ $(n \geq 2)$ in Example 6.5 and use the same notation. We set $\mathcal{D}' := \mathcal{D} \setminus \mathcal{L}_0$, which is also a ramified domain over $\mathbf{C}^{n+1}$. Then $\mathcal{D}'$ is clearly a domain of holomorphy for the function $1/\widetilde{\varphi}(p)$ (where $\widetilde{\varphi}(p)$ is defined in Example 6.6). However, $\mathcal{D}'$ is not holomorphically convex.

To prove this, we fix a non-zero complex number c and consider the following subsets in $\Sigma \subset \mathbf{C}^{2n}$:

$$\begin{aligned} K &= L_c \cap \{\|z\|^2 + \|w\|^2 = 1\}, \\ I &= L_c \cap \{0 < \|z\|^2 + \|w\|^2 \leq 1\}. \end{aligned}$$

We let $\mathcal{K}$ and $\mathcal{I}$ denote the sets in $\mathcal{D}$ which correspond to K and I. Then $\mathcal{K} \subset\subset \mathcal{D}'$ and $K \subset\subset \Sigma \setminus L_0$. Furthermore, $\mathcal{I} \subset \mathcal{D}'$ and $I \subset \Sigma \setminus L_0$, while $\mathcal{I}$ is not compactly contained in $\mathcal{D}'$ nor is I in $\Sigma \setminus L_0$. Let $f(p)$ be a holomorphic function on $\mathcal{D}'$. We restrict $f(p)$ to $\mathcal{L}_c \setminus \{O\}$ and denote this restriction by $f_c(p)$, and we let $\widetilde{f_c}(z, w)$

denote the corresponding function on $L_c \setminus \{(0,0)\}$. Then $\widetilde{f}_c(z, cz)$ is holomorphic for $z \in \mathbf{C}^n \setminus \{0\}$, and hence in all of $\mathbf{C}^n$. It follows that

$$|\widetilde{f}_c(z, cz)| \leq \max_{\|z\|=1/(1+|c|^2)} |\widetilde{f}_c(z, cz)| \qquad \text{in} \quad 0 < \|z\| \leq (1+|c|^2)^{-1}.$$

Hence $\mathcal{I} \subset \widehat{\mathcal{K}}_{\mathcal{D}'}$, so that $\mathcal{D}'$ is not holomorphically convex.

6.1.6. Ramified Pseudoconvex Domains. The notion of pseudoconvexity of a domain is extracted from some geometric properties which a domain of holomorphy satisfies, and it was conjectured that, conversely, a pseudoconvex domain is a domain of holomorphy. As will be shown in Chapter 9, Oka proved that this is true in the case of a univalent domain in $\mathbf{C}^n$ and even in the case of an unramified domain over $\mathbf{C}^n$. However, in the case of a ramified domain over $\mathbf{C}^n$, the problem of finding necessary and sufficient geometric conditions for the domain to be a ramified domain of holomorphy over $\mathbf{C}^n$ is not yet completely solved. Thus, there is no precise notion of pseudoconvexity for a ramified domain over $\mathbf{C}^n$.

We first give the definition of pseudoconvexity of an unramified domain over $\mathbf{C}^n$, even though it is very similar to the case of a univalent domain in $\mathbf{C}^n$. Let $\mathcal{D}$ be an unramified domain over $\mathbf{C}^n$ with variables $z_1, \dots, z_n$ and let π be the canonical projection from $\mathcal{D}$ to $\mathbf{C}^n$. Let $z^0 = (z_1^0, \dots, z_n^0) \in \mathbf{C}^n$. For positive numbers $r' < r$ and $\rho' < \rho$ we consider two open sets in $\mathbf{C}^n$ defined by

$$|z_j - z_j^0| < r \quad (j = 1, \dots, n-1), \qquad \rho' < |z_n - z_n^0| < \rho;$$
$$|z_j - z_j^0| < r' \quad (j = 1, \dots, n-1), \qquad |z_n - z_n^0| < \rho.$$

We let E denote the union of these two open sets, and we let C be the open polydisk in $\mathbf{C}^n$ given by

$$C \ : \ |z_j - z_j^0| < r \ \ (j = 1, \dots, n-1), \qquad |z_n - z_n^0| < \rho.$$

If there exist univalent parts v and V of $\mathcal{D}$ such that $\pi(v) = E$ and $\pi(V) = C$, then we denote these sets by $v = \widetilde{E}$ and $V = \widetilde{C}$.

We say that the unramified domain $\mathcal{D}$ satisfies the **continuity theorem** if for any $z^0 \in \mathbf{C}^n$ and any r, r', ρ, ρ', whenever the set $\widetilde{E}$ exists as described, then a corresponding set $\widetilde{C}$ exists with $\widetilde{E} \subset \widetilde{C}$.

Now let Δ be a polydisk in $\mathbf{C}^n$. If the subdomain $\pi^{-1}(\Delta)$ of $\mathcal{D}$ satisfies the continuity theorem and if this property remains invariant under an analytic mapping of Δ,[3] then we say that $\mathcal{D}$ is **pseudoconvex**.

This definition of pseudoconvexity of an unramified domain over $\mathbf{C}^n$ corresponds to the pseudoconvexity of type C for a univalent domain in $\mathbf{C}^n$. One may also define the pseudoconvexity of an unramified domain over $\mathbf{C}^n$ which corresponds to that of type A or of type B for a univalent domain in $\mathbf{C}^n$; we will not state these definitions here.

We will temporarily define a pseudoconvex ramified domain over $\mathbf{C}^n$ as follows. Let $\mathcal{D}$ be a ramified domain over $\mathbf{C}^n$ with branch set $\mathcal{S}$ and let $\mathcal{D}^0 = \mathcal{D} \setminus \mathcal{S}$. Since $\mathcal{D}^0$ is an unramified domain over $\mathbf{C}^n$, we have the ramified domain $\mathcal{D}^*$ over $\mathbf{C}^n$ associated to $\mathcal{D}^0$. We let $\mathcal{S}^*$ denote the branch set of $\mathcal{D}^*$. In general, $\mathcal{S} \subset \mathcal{S}^*$. If $\mathcal{D}$ satisfies the following three conditions, then we say that $\mathcal{D}$ is **pseudoconvex**:

[3]Precisely, let $w = \phi(z)$ be an analytic mapping from Δ onto a univalent domain Δ^* in $\mathbf{C}_w^n$, so that $\phi(z)$ maps $\pi^{-1}(\Delta)$ to an unramified domain $\tilde{d}$ over Δ^*. Then $\tilde{d}$ satisfies the continuity theorem.

(i) $\mathcal{D}^0$ is an unramified pseudoconvex domain.
(ii) Let σ^* be the set of all regular points of the branch set $\mathcal{S}^*$. If there exists at least one point of σ^* contained in $\mathcal{S}$, then $\sigma^* \cap \mathcal{S}$ is pseudoconvex in σ^* (as an $(n-1)$-dimensional domain).
(iii) Let p be a branch point of $\mathcal{D}^*$. If there exists a neighborhood v of p in $\mathcal{D}^*$ such that $v \cap \mathcal{S}^* \subset \mathcal{S}$ except for at most an $(n-2)$-dimensional analytic set, then $p \in \mathcal{S}$.

According to this definition, a ramified domain of holomorphy over $\mathbf{C}^n$ is pseudoconvex, but, as stated earlier, the converse problem remains open.

6.2. Fundamental Theorem for Locally Ramified Domains

6.2.1. Characteristic Functions in Ramified Domains. Let $\Delta : |z_j| < r_j$ $(j = 1, \ldots, n)$ be a polydisk in $\mathbf{C}^n$ with variables $z = (z_1, \ldots, z_n)$. Let $\mathcal{R}^1$ be a ramified domain over a neighborhood of $\overline{\Delta}$ such that the part $\mathcal{R}^0$ of $\mathcal{R}^1$ over Δ has no relative boundary. We let m denote the number of sheets of $\mathcal{R}^0$, which we assume is finite. We also let σ denote the branch set of $\mathcal{R}^0$ and $\underline{\sigma} = \pi(\sigma)$ the projection of σ onto Δ, so that $\underline{\sigma}$ is an analytic hypersurface in Δ; finally, we set $\Delta' := \Delta - \underline{\sigma}$.

Let $f(p)$ be a holomorphic function on $\mathcal{R}^0$. If $f(p)$ has m different function elements at the m distinct points of $\mathcal{R}^0$ lying over a base point $z_0' \in \Delta'$, then $f(p)$ is called a **characteristic function** on $\mathcal{R}^0$. In this case, $f(p)$ has m different function elements at each of the m distinct points of $\mathcal{R}^0$ lying over *any* base point $z' \in \Delta'$.

We introduce an additional complex plane $\mathbf{C}_X$ and consider the product space $\Lambda = \Delta \times \mathbf{C}_X \subset \mathbf{C}^{n+1}$. Given a holomorphic function $f(p)$ on $\mathcal{R}^0$, we consider the set $\mathcal{C}$ in Λ defined as

$$\mathcal{C} : \; X = f(p) \qquad \text{for } p \in \mathcal{R}^0,$$

which defines an analytic hypersurface in Λ. If $f(p)$ is a characteristic function on $\mathcal{R}^0$, then we say that $\mathcal{C}$ is the **graph of $f(p)$ on $\mathcal{R}^0$**. There is a bijection between $\mathcal{R}^0$ and $\mathcal{C}$ except on at most an analytic hypersurface in $\mathcal{R}^0$. We call the set of points $p \in \mathcal{R}^0$ such that there exists a point $q \in \mathcal{R}^0$, $q \neq p$, with $f(p) = f(q)$ the set of **multiple points** or simply **double points** of $f(p)$.

Let $f(p)$ be a characteristic function on $\mathcal{R}^0$ and let $\mathcal{C}$ be the graph of $f(p)$ on $\mathcal{R}^0$ in Λ. We let S denote the $(n-1)$-dimensional analytic subset of $\mathcal{C}$ which corresponds to the branch set σ of $\mathcal{R}^0$. If each point of S except for at most an $(n-2)$-dimensional analytic set is a non-singular point of $\mathcal{C}$, then we say that $f(p)$ is a **simple function** on $\mathcal{R}^0$. This does not necessarily mean that the singular set of $\mathcal{C}$ has dimension at most $n-2$. Let $f(p)$ be a characteristic function on $\mathcal{R}^0$. If there exists at least one non-singular point of $\mathcal{C}$ on each irreducible component of S, then $f(p)$ is a simple function on $\mathcal{R}^0$.

THEOREM 6.1 (Fundamental Theorem). *A ramified domain $\mathcal{D}$ over $\mathbf{C}^n$ locally carries a simple function.*

Unlike the case of one complex variable, the local existence of simple functions on $\mathcal{D}$ in several complex variables is non-trivial. This theorem was first proved by H. Grauert and R. Remmert [**24**]. In this chapter we shall give an elementary proof of the theorem.

We begin with some preliminaries. Let $\mathcal{D}$ be a ramified domain over $\mathbf{C}^n$ $(n \geq 2)$ and let $p \in \mathbf{C}^n$. We may assume $\underline{p} = 0 \in \mathbf{C}^n$. We can always choose Euclidean coordinates $(z_1, \dots, z_n)$ and a neighborhood $\mathcal{R}^0$ of p in $\mathcal{D}$ lying over a polydisk $\Delta: \; |z_j| < r_j \; (j = 1, \dots, n)$ such that $\mathcal{R}^0$ is standard with respect to z_n. Using the above notation $\sigma, \Delta', \underline{\sigma}$, etc., this means that $\underline{\sigma}$ does not intersect $\Delta^{n-1} \times \partial\Delta_n$, where

$$\Delta^{n-1}: \; |z_j| < r_j \; (j = 1, \dots, n-1) \qquad \text{and} \qquad \Delta_n: \; |z_n| < r_n.$$

To prove the fundamental theorem, it suffices to verify the existence of a simple function on the standard ramified domain $\mathcal{R}^0$. Then, for any fixed $z' \in \Delta^{n-1}$, the fiber

$$\mathcal{R}^0(z') = \{z_n \mid (z', z_n) \in \mathcal{R}^0\}$$

is an m-sheeted Riemann surface over the disk Δ_n in $\mathbf{C}_{z_n}$ (with m finite), which may have branch points. Thus $\mathcal{R}^0$ can be considered as a variation of Riemann surfaces over the disk Δ_n without relative boundary with variation parameter $z' \in \Delta^{n-1}$,

$$\mathcal{R}^0 : z' \to \mathcal{R}^0(z'), \quad z' \in \Delta^{n-1}.$$

Let $\mathbf{P}_w$ denote the Riemann sphere $\{|w| \le \infty\}$ and let $\Delta : |z_j| < r_j$ $(j = 1, \dots, n)$ be a polydisk in $\mathbf{C}^n$ $(n \ge 1)$. In the product space $\mathbf{C}^n \times \mathbf{P}_w$, we consider the product domain

$$\Lambda = \Delta \times \mathbf{P}_w.$$

Let $\mathcal{R}$ be an m-sheeted ramified domain over Λ without relative boundary (m finite) and let

$$\pi: \; \mathcal{R} \to \Lambda$$

be the canonical projection. We let π_n denote the projection from Λ to Δ. We assume that the projection $\underline{\sigma}$ of the branch set σ of $\mathcal{R}$ does not contain any line of the form $\{z_0\} \times \mathbf{P}_w$. For any subset e of Δ, we define

$$\mathcal{R}(e) := \pi^{-1}(\pi_n^{-1}(e)).$$

If e is a domain δ in Δ, then $\mathcal{R}(\delta)$ is a ramified domain over $\delta \times \mathbf{P}_w$ without relative boundary. If e is a point $z \in \Delta$, then $\mathcal{R}(z)$ is an m-sheeted compact Riemann surface over $\mathbf{P}_w$. We let π_z denote the restriction of π to $\mathcal{R}(z)$.

Finally, for ρ^0 with $0 < \rho^0 < \infty$, we let

$$\begin{array}{lll} \Delta & : \; |z_j| < r_j \; (j = 1, \dots, n) & \text{in } \mathbf{C}^n \; (n \ge 1), \\ \Gamma^0 & : \; |w| < \rho^0 & \text{in } \mathbf{P}_w, \\ \Lambda^0 & : \; \Delta \times \Gamma^0 & \text{in } \Lambda = \Delta \times \mathbf{P}_w. \end{array}$$

Let $\mathcal{R}^0$ be a finitely sheeted ramified domain over Λ^0 without relative boundary. If there exists a finitely sheeted ramified domain $\mathcal{R}$ over Λ without relative boundary such that $\mathcal{R}|_{\Lambda^0} = \mathcal{R}^0$, then $\mathcal{R}$ is called an **algebraic extension** of $\mathcal{R}^0$.

With this terminology, we now state the following result.

PROPOSITION 6.1. *Let $\mathcal{R}^0$ be a finitely sheeted ramified domain over $\overline{\Lambda^0} := \overline{\Delta} \times \overline{\Gamma^0}$ without relative boundary such that the projection $\underline{\sigma}^0$ of the branch set σ_0 of $\mathcal{R}^0$ onto $\overline{\Lambda^0}$ does not intersect $\overline{\Delta} \times \partial\Gamma^0$, i.e., $\mathcal{R}^0$ is standard with respect to the coordinate w. Then there exists an algebraic extension $\mathcal{R}$ of $\mathcal{R}^0$ which satisfies the following conditions:*

1. *$\mathcal{R}$ has no branch set lying over $\Delta \times \{w = \infty\}$.*
2. *For any $z^0 \in \Delta$, $\mathcal{R}(z^0)$ is a connected, compact Riemann surface over $\mathbf{P}_w$.*

PROOF. Since $\underline{\sigma} \cap (\underline{\Delta} \times \partial\Gamma^0) = \emptyset$, we can find a sufficiently thin annulus $A := \{\rho' < |w| < \rho''\}$ in $\mathbf{P}_w$ which contains $\partial\Gamma^0$ and is such that the part $\mathcal{A}$ of $\mathcal{R}^0$ over $\Delta \times A$ consists of a finite number of connected unramified domains without relative boundary, i.e., $\mathcal{A}$ is a finite number of disjoint unions of product sets of the form $\Delta \times S_j$, where S_j $(j = 1, \dots, j_0)$ is a finitely sheeted Riemann surface over A without relative boundary. We construct an m-sheeted connected Riemann surface $\widetilde{B}$ over $\{\rho' < |w| \le \infty\}$ without relative boundary such that the part of $\widetilde{B}$ over A coincides with S_j $(j = 1, \dots, j_0)$ and $\tilde{B}$ has no branch points over $w = \infty$. Then we attach $\mathcal{R}^0$ to the ramified domain $\Delta \times \widetilde{B}$ along the common part $\mathcal{A}$, and the resulting ramified domain $\mathcal{R}$ over $\Delta \times \mathbf{P}_w$ satisfies the conclusion of the proposition. □

We set $\mathcal{R}' = \mathcal{R} \setminus \pi^{-1}(\Delta \times \{\infty\})$. From this proposition, we see that to prove Theorem 6.1, it suffices *to construct a simple function on $\mathcal{R}'$ instead of on $\mathcal{R}^0$.*[4]

6.2.2. Algebraic Functions of One Complex Variable. We recall a fundamental result about algebraic functions of one complex variable. Let R be a compact Riemann surface of genus g. Let p_j $(j = 1, \dots, \mu)$ be a finite set of points of R and let e_j $(j = 1, \dots, \mu)$ be positive integers. We set

$$e := e_1 + \cdots + e_\mu.$$

We let $\mathcal{M} = \mathcal{M}(R)$ denote the complex-linear space of meromorphic functions $f(z)$ on R such that $f(z)$ is holomorphic in $R \setminus \{p_j\}_{j=1,\dots,\mu}$ and has a pole of order at most e_j at p_j $(j = 1, \dots, \mu)$. We let Ω denote the complex-linear space of holomorphic differentials ω on R such that ω has a zero of order at least e_j at p_j $(j = 1, \dots, \mu)$. We recall the Riemann-Roch theorem.

THEOREM 6.2.

$$\dim\ \mathcal{M} = \dim\ \Omega + e - g + 1. \tag{6.2}$$

In particular,

$$\dim\ \mathcal{M} = e - g + 1 \qquad \textit{if} \quad e \ge 2g - 1. \tag{6.3}$$

The last statement (6.3) folows from the fact that any non-zero holomorphic differential ω on a compact Riemann surface R of genus g has just $2g - 2$ zeros (counted with multiplicity).

Let R be a compact Riemann surface of genus g lying m-sheeted over $\mathbf{P}_w$. We will assume in this section that all points of R over $w = \infty$ are regular points, i.e., non-branch points of R; we list them as L^j_∞ $(j = 1, \dots, m)$. Thus we can choose and fix a large number $\rho_0 > 0$ such that, setting $E = E_{\rho_0} = \{w \in \mathbf{P}_w \mid |w| > \rho_0\}$, there are no branch points over E. Therefore, there are m different copies E_j $(j = 1, \dots, m)$ of E with $L^j_\infty \in E_j$. We set

$$R' := R \setminus \{L^j_\infty\}_{j=1,\dots,m}.$$

Let $\mathcal{L}(R)$ denote the linear space of meromorphic functions $f(p)$ on R such that $f(p)$ is holomorphic in R'; i.e., $f(p)$ may have poles only at the points L^j_∞ $(j = 1, \dots, m)$. If we restrict $f(p)$ to E_j $(j = 1, \dots, m)$ and denote this restriction by $f_j(w)$, then $f_j(w)$ is a single-valued meromorphic function on E which may have poles only at $w = \infty$.

[4]This idea is due to H. Behnke and K. Stein.

Given an integer $\nu \geq 1$, we let $\mathcal{L}_\nu(R)$ denote the set of all $f(p) \in \mathcal{L}(R)$ which have poles of order at most ν at each point L^j_∞ $(j = 1, \dots, m)$. We also define

$$\mathcal{L}^*_\nu(R) = \{f(p) \in \mathcal{L}_\nu(R) \mid f(p) \text{ has poles of order } \nu \text{ at each } L^j_\infty \ (j = 1, \dots, m)\}.$$

Note that $f(p) \in \mathcal{L}_\nu(R)$ belongs to $\mathcal{L}^*_\nu(R)$ if and only if the total order of $f(p)$ is equal to $m\nu$.

Let $f(p) \in \mathcal{L}(R)$. Then by constructing the fundamental symmetric functions in the m branches of $f(p)$, we obtain an irreducible polynomial of a new complex variable X of the form

$$F(w, X) = X^m + \alpha_1(w)X^{m-1} + \cdots + \alpha_m(w) \tag{6.4}$$

such that for each X, the solutions of $F(w, X) = 0$ coincide with $X = f(w)$ and such that each $\alpha_k(w)$ $(k = 1, \dots, m)$ is a *polynomial* in $w \in \mathbf{C}_w$. Thus, $F(w, X)$ is a polynomial in both variables w and X in $\mathbf{C}^2$. We call $F(w, X)$ the **defining polynomial for** $f(p)$.

We have the following lemma.

LEMMA 6.1. *Let $f(p) \in \mathcal{L}(R)$. Then $f(p) \in \mathcal{L}_\nu(R)$ if and only if each coefficient $\alpha_k(w)$ $(k = 1, \dots, m)$ of the defining polynomial $F(w, X)$ for $f(p)$ is a polynomial of degree at most νk.*

PROOF. Let $f(p) \in \mathcal{L}_\nu(R)$. We consider the branch $f_j(w)$ of $f(p)$ on E_j $(j = 1, \dots, m)$. For each $k = 1, \dots, m$ we have

$$\alpha_k(w) = \sum f_{j_1}(w) \cdots f_{j_k}(w) \qquad \text{on } E.$$

It follows that $\alpha_k(w)$ is a polynomial in w of degree at most νk.

We prove the converse by contradiction. Hence we assume that some $f_j(w)$ has a pole of order greater than ν at $w = \infty$. We let $\nu^* \geq \nu + 1$ denote the maximum such order and we suppose $f_{j_1}, \dots, f_{j_l}$ $(1 \leq l \leq m)$ have poles of order ν^* at $w = \infty$. Then $\alpha_l(w)$ is a polynomial in w of degree $\nu^* l$, which is a contradiction. □

REMARK 6.3. Let $f(p) \in \mathcal{L}_\nu(R)$ and let $F(w, X)$ in (6.4) denote the defining polynomial for $f(p)$. Then $f(p) \in \mathcal{L}^*_\nu(R)$ if and only if $\alpha_m(w)$ is of order $m\nu$.

Let $f(p) \in \mathcal{L}(R)$ and let $F(w, X)$ be the defining polynomial for $f(p)$. We set

$$\mathcal{C}_f := \{(w, X) \in \mathbf{C}^2 \mid F(w, X) = 0\},$$

which is called the **graph of** $f(p)$ in $\mathbf{C}^2$. Thus, $\mathcal{C}_f$ is a one-dimensional analytic set in $\mathbf{C}^2$. We consider the points P_h $(h = 1, \dots, h_0)$ of $\mathcal{C}_f$ which correspond to the branch points of R and the points Q_k $(k = 1, \dots, k_0)$ which correspond to the singular points of $\mathcal{C}_f$. We write

$$P_h = (\xi_h, \xi'_h) \ (h = 1, \dots, h_0), \qquad Q_k = (\eta_k, \eta'_k) \ (k = 1, \dots, k_0).$$

It may happen that $P_h = Q_k$ for some h and k. We note that the ξ_h $(h = 1, \dots, h_0)$ are uniquely determined by the Riemann surface R, but of course this is not the case for ξ'_h, η_k and η'_k. If $\mathcal{C}_f$ satisfies the three conditions:

1. each P_h $(h = 1, \dots, h_0)$ is a regular point of $\mathcal{C}_f$ and each Q_k $(k = 1, \dots, k_0)$ is a normal double point of $\mathcal{C}_f$;

2. if $i \neq j$ $(1 \leq i, j \leq m)$, then

$$\lim_{w \to \infty} f_i(w)/f_j(w) \neq 0, 1, \text{ or } \infty; \tag{6.5}$$

3. if $k \neq l$ $(1 \leq k, l \leq k_0)$, then $\eta_k \neq \eta_l$; furthermore, for each $k = 1, \dots, k_0$, we have $\eta_k \neq \xi_h$ $(h = 1, \dots, h_0)$;

then we say that $f(p)$ has a **simple graph** $\mathcal{C}_f$ in $\mathbf{C}^2$. Here we say that $Q = (\eta, \eta')$ is a normal double singular point of $\mathcal{C}_f$ if there exists a bidisk $\lambda = \delta \times \gamma \subset \mathbf{C}^2_{w,X}$ centered at Q such that $\Lambda \cap \mathcal{C}_f$ can be written as

$$\{(w, X) \in \lambda \mid (X - f_1(w))(X - f_2(w)) = 0\},$$

where $f_1(w), f_2(w)$ are holomorphic functions in δ with $f_1(\eta) = f_2(\eta) = \eta'$ and $f_1'(\eta) \neq f_2'(\eta)$. We note that condition 2 implies that the function $f(p)$ is a characteristic function on R, and hence $\mathcal{C}_f$ can be considered as a graph of $f(p)$ on R.

We let $\mathcal{L}^s_\nu(R) \subset \mathcal{L}_\nu(R)$ denote the set of all $f(p) \in \mathcal{L}_\nu(R)$ whose graphs $\mathcal{C}_f$ are simple in $\mathbf{C}^2$.

It is easy to see the following fact. Let $f(p) \in \mathcal{L}^s_\nu(R)$ and let $f_n(p) \in \mathcal{L}_\nu(R)$ $(n = 1, 2, \dots)$ with $\lim_{n \to \infty} f_n(p) = f(p)$ uniformly on R; i.e.,

1. $\lim_{n \to \infty} f_n(p) = f(p)$ uniformly on each subset $K \subset\subset R'$;
2. for suffciently large n, $f_n(p)$ has the same order as $f(p)$ at each L^j_∞ $(j = 1, \dots, m)$.

Then $f_n(p) \in \mathcal{L}^s_\nu(R)$ for sufficiently large n.

We have the following theorem.

THEOREM 6.3. *Let R be an m-sheeted compact Riemann surface over $\mathbf{P}_w$ of genus g. Let h_0 be the number of branch points of R, and set $\nu_0 := (h_0 + 2)m + g$. Then for any function $g(p) \in \mathcal{L}^*_\nu(R)$ with $\nu > m\,\nu_0$ satisfying condition (6.5), there exist a finite number of functions $\phi_i(p) \in \mathcal{L}_{\nu_0}(R)$ $(i = 1, \dots, q)$ such that for sufficiently small $\varepsilon_i \neq 0$ $(i = 1, \dots, q)$, $G_\varepsilon(p) := g(p) + \sum_{i=1}^q \varepsilon_i \phi_i(p)$ is a simple function on R.*

This is a classical result in the theory of algebraic functions of one complex variable. The proof will be given in Appendix 1 to this chapter.

6.2.3. Meromorphic Functions on $\mathcal{R}(z)$. We return to the subject of 6.2.1. Let $\Delta \subset \mathbf{C}^n$ and let $\mathcal{R}$ be a ramified domain over $\Lambda = \Delta \times \mathbf{P}_w$ which has no relative boundary and which satisfies conditions 1 and 2 in Proposition 6.1. Let $\pi : \mathcal{R} \to \Lambda$ be the canonical projection and let m be the number of sheets of $\mathcal{R}$. Then the subset of $\mathcal{R}$ lying over $w = \infty$, i.e., $\pi^{-1}(\Delta \times \{\infty\})$, consists of m different analytic hyperplanes which will be denoted by L^j_∞ $(j = 1, \dots, m)$. We set

$$\mathcal{R}' = \mathcal{R} \setminus \left(\bigcup_{j=1}^m L^j_\infty \right).$$

For $z \in \Delta$, we write $\mathcal{R}(z) = \pi^{-1}(z)$ for the fiber over z, which is a compact Riemann surface lying m-sheeted over $\mathbf{P}_w$. We set

$$\mathcal{R}'(z) := \mathcal{R}' \cap \mathcal{R}(z), \qquad L^j_\infty(z) := L^j_\infty \cap \mathcal{R}(z) \quad (j = 1, \dots, m).$$

To prove the fundamental theorem (Theorem 6.1), it suffices to *construct a meromorphic function $G(z,p)$ of $n+1$ complex variables (z,p) in $\mathcal{R}$ which is holomorphic in $\mathcal{R}'$ and such that for* **some** *point $a \in \Delta$, $X = G(a,p)$ has a simple graph in $\mathbf{C}^2_{w,X}$.*

Let Σ be the branch set of $\mathcal{R}$ and set $\underline{\Sigma} = \pi(\Sigma)$. For $z \in \Delta$, we let $\Sigma(z)$ denote the branch points of $\mathcal{R}(z)$, so that the section of Σ over z' coincides with $\Sigma(z')$ for all but a finite set of points $z' \in \Delta$. From condition 1 in Proposition 6.1, if we fix a sufficiently large number $\rho > 0$ and set

$$\Gamma : \ |w| < \rho \quad \text{in } \mathbf{P}_w, \qquad \Lambda^0 = \Delta \times \Gamma,$$

then $\underline{\Sigma}$ is an analytic set in Λ^0 with $\underline{\Sigma} \cap (\Delta \times \partial\Gamma) = \emptyset$. It follows that the part of $\mathcal{R}$ over $\Delta \times (\rho \le |w| \le \infty)$ consists of m disjoint univalent parts and that $\underline{\Sigma}$ can be written as

$$\underline{\Sigma} = \{(z,w) \in \Delta \times \mathbf{P}_w \mid P(z,w) = 0\},$$

where

$$P(z,w) = w^\kappa + \alpha_1(z) w^{\kappa-1} + \cdots + \alpha_\kappa(z)$$

and each $\alpha_j(z)$ $(j = 1,\ldots,\kappa)$ is a holomorphic function on Δ. We decompose $P(z,w)$ into the prime factorization

$$P(z,w) = \prod_{\chi=1}^{\chi_0} P_\chi(z,w),$$

where each $P_\chi(z,w)$ $(\chi = 1,\ldots,\chi_0)$ has the same form as $P(z,w)$. We note that $P(z,w)$ has no multiple factors. We let $d(z)$ denote the discriminant of $P(z,w)$ with respect to w, and we set

$$\sigma := \{z \in \Delta \mid d(z) = 0\}, \qquad \Delta' = \Delta \setminus \sigma.$$

Then σ is an $(n-1)$-dimensional analytic hypersurface in Δ. We note that for each $z \in \Delta'$, $R(z)$ is a compact Riemann surface of the same genus, say g. However, for $z \in \sigma$, $R(z)$ is of genus g' with $0 \le g' \le g$.

Let $\nu \ge 1$ be an integer such that $m\nu \ge 2g - 1$. For $z \in \Delta$ we write $\mathcal{L}_\nu(z) = \mathcal{L}_\nu(R(z))$, i.e., $\mathcal{L}_\nu(z)$ is the linear space of meromorphic functions on $\mathcal{R}(z)$ which are holomorphic in $\mathcal{R}'(z)$ and which have poles at $L^j_\infty(z)$ $(j = 1,\ldots,m)$ of order at most ν. By the Riemann-Roch theorem, dim $\mathcal{L}_\nu(z) = m\nu - g + 1$. For simplicity, we set

$$l_0 := m\nu - g, \qquad \dim\ \mathcal{L}_\nu(z) = l_0 + 1.$$

We need the following lemma, related to normal families of holomorphic functions of one complex variable; this will form the basis of our proof of the Fundamental Theorem from 6.2.1.

LEMMA 6.2. *Let z^j $(j = 1,2,\ldots)$ be a sequence of points in Δ which converges to a point $z^0 \in \Delta$. Let $f^j(p) \in \mathcal{L}_\nu(z^j)$ $(j = 1,2,\ldots)$ be non-constant on $\mathcal{R}'(z^j)$. Assume that one of the zeros ξ^j of $f^j(p)$ converges to a point ξ^0 in $\mathcal{R}'(z^0)$ as $j \to \infty$. Then there exists a subsequence $f^{j_\nu}(p)$ $(\nu = 1,2,\ldots)$ of $f^j(p)$ $(j = 1,2,\ldots)$ and a sequence b^{j_ν} $(\nu = 1,2,\ldots)$ of complex numbers such that, if we set*

$$g^{j_\nu}(p) := b^{j_\nu} f^{j_\nu}(p) \qquad (\nu = 1,2,\ldots),$$

then the $g^{j_\nu}(p)$ $(\nu = 1,2,\ldots)$ converge locally uniformly to a non-constant holomorphic function $g^0(p)$ on $\mathcal{R}'(z^0)$ with $g^0(p) \in \mathcal{L}_\nu(z^0)$.

We need to explain the terminology of local uniform convergence on $\mathcal{R}'(z^0)$. Fix $p_0 = (z^0, w^0) \in \mathcal{R}'(z^0) \setminus \Sigma(z^0)$, i.e., w^0 is a regular point of $\mathcal{R}'(z^0)$. Take a relatively compact polydisk $\Delta_0 \times \delta \subset \Delta \times \mathbf{C}_w$ centered at (z^0, w^0) in $\mathcal{R}' \setminus \Sigma$. We restrict $g^{j_\nu}(p)$ $(\nu = 1, 2, \dots)$ to $\{z^j\} \times \delta$, which is thus a holomorphic function for $w \in \delta$ which we denote by $g^{j_\nu}(w)$. Similarly, we set $g^0(w) = g^0(p)$ for $w \in \delta$, where $p = (z^0, w)$. The conclusion of Lemma 6.2 is that $\lim_{\nu\to\infty} g^{j_\nu}(w) = g^0(w)$ uniformly on δ.

PROOF. Since there are at most $m\nu$ zeros of $f^j(p)$ in $\mathcal{R}(z^j)$, we can extract from $\{f^j(p)\}_{j=1,2,\dots}$ a subsequence $\{f^{j_k}(p)\}_{k=1,2,\dots}$ in such a way that the zeros of $f^{j_k}(p)$ converge in $\mathcal{R}$ either to the points ξ_ι^0 $(\iota = 1, \dots, m')$ in $\mathcal{R}'(z^0)$ or to points of $L_\infty^j(z^0)$ $(j = 1, \dots, m)$. By assumption one of the points ξ_ι^0 $(\iota = 1, \dots, m')$ coincides with $\xi^0 \in \mathcal{R}'(z^0)$ described in the lemma, say $\xi_1^0 = \xi^0$. We select a regular point η_1^0 of $\mathcal{R}'(z^0)$ such that $\eta_1^0 \neq \xi_\iota^0$ $(\iota = 1, \dots, m')$ and we fix a polydisk $\Delta^1 \times \delta^1$ centered at (z^0, η_1^0) in $\mathcal{R}' \setminus \Sigma$ such that $f^{j_k}(p) \neq 0$ for any $p = (z^{j_k}, w)$ with $w \in \delta^1$. We set $\eta_1^{j_k} := (z^{j_k}, \eta_1^0) \in \mathcal{R}'(z^{j_k})$. Since $f^{j_k}(\eta_1^{j_k}) \neq 0$ for any sufficiently large $k \geq k_0$, we can define

$$b^{j_k} = 1/f^{j_k}(\eta_1^{j_k}), \qquad g^{j_k}(p) = b^{j_k} f^{j_k}(p) \quad \text{in} \quad \mathcal{R}(z^{j_k}).$$

Since there are at most $m\nu$ points p satisfying $g^{j_k}(p) = 1$ in $\mathcal{R}(z^{j_k})$, we can extract from $\{g^{j_k}(p)\}_{k(\geq k_0)}$ a subsequence $\{g^{j_h}(p)\}_{h=1,2,\dots}$ such that all points satisfying $g^{j_h}(p) = 1$ converge in $\mathcal{R}$ either to the points η_ι^0 $(\iota = 1, \dots, m'')$ in $\mathcal{R}'(z^0)$ or to points of $L_\infty^j(z^0)$ $(j = 1, \dots, m)$.

Take any $p_0 = (z^0, w^0) \in \mathcal{R}'(z^0) \setminus \Sigma(z^0)$ such that $w^0 \neq \xi_\iota^0$, η_j^0 $(\iota = 1, \dots, m'$; $j = 1, \dots, m'')$. Choose a polydisk $\Delta^0 \times \delta^0$ centered at (z^0, w^0) in $\mathcal{R}' \setminus \Sigma$ such that $g^{j_h}(p) \neq 0, 1$ for any point p of the form $p = (z^{j_h}, w)$ with $w \in \delta^0$ and $h > 1$ sufficiently large so that $z^{j_h} \in \Delta^0$. If we restrict each $g^{j_h}(p)$ to δ^0 and call this restriction $g^{j_h}(w)$, then $g^{j_h}(w)$ is a holomorphic function on δ^0 which omits the values 0 and 1. Thus, by Picard's theorem, we can extract from $\{g^{j_h}(w)\}_{h=1,2,\dots}$ a subsequence $\{g^{j_\nu}(w)\}_{\nu=1,2,\dots}$ which converges uniformly to $g^0(w)$ on δ^0. We can consider $g^0(w)$ as a holomorphic function $g^0(p)$ on $\mathcal{R}'(z^0) \cap \delta^0$. By the standard diagonal method we may assume that $g^{j_\nu}(p)$ $(\nu = 1, 2, \dots)$ converges locally uniformly to a holomorphic function $g^0(p)$ on

$$\mathcal{R}^*(z^0) := \mathcal{R}'(z^0) \setminus (\Sigma(z^0) \cup \{\xi_\iota^0\}_{\iota=1,\dots,m'} \cup \{\eta_j^0\}_{j=1,\dots,m''}).$$

Note that we have not ruled out the possibility that $g^0(p) \equiv 0, 1$, or ∞ on $\mathcal{R}^*(z^0)$. Since $\eta_1^0 = (z^0, w^1)$ was a regular point of $\mathcal{R}'(z^0)$, there exists a sufficiently small polydisk $\Delta' \times \delta'$ centered at (z^0, w^1) in $\mathcal{R}' \setminus \Sigma$ and such that $\partial\delta'$ contains neither ξ_ι^0 $(\iota = 1, \dots, m')$ nor η_j^0 $(j = 1, \dots, m'')$. It follows from Weierstrass' theorem that the uniform convergence of $g^{j_\nu}(w)$ $(\nu = 1, 2, \dots)$ on $\partial\delta'$ implies the uniform convergence of $g^{j_\nu}(w)$ on δ'. Consequently, $g^0(p)$ is holomorphic at η_1^0 and $g^0(\eta_1^0) = 1$. Similarly, $g^0(p)$ is holomorphic at ξ_1^0 and $g^0(\xi_1^0) = 0$. Thus, $g^0(p) \not\equiv \infty$ on $\mathcal{R}^*(z^0)$. It follows from the Riemann removable singularity theorem that $g^0(p)$ is holomorphic on $\mathcal{R}'(z^0)$. Hence, $g^0(p)$ is a non-constant holomorphic function on $\mathcal{R}'(z^0)$. Since $g^{j_\nu}(p) \in \mathcal{L}_\nu(z^{j_\nu})$ $(\nu = 1, 2, \dots)$, we also have $g^0(p) \in \mathcal{L}_\nu(z^0)$. □

For $h = 1, \dots, l_0 := m\nu - g$, let ζ_h be an irreducible analytic hypersurface in $\mathcal{R}'$ such that the projection $\pi(\zeta_h)$ of ζ_h onto $\Lambda = \Delta \times \mathbf{P}_w$ is an n-dimensional complex hyperplane $w = \omega_h$. In particular, ζ_h lies over $\Delta \times \{w_h\}$ in Λ. We assume that $w_h \neq w_k$ for $h \neq k$ and that $|\omega_h| > \rho$, and we set

$$\zeta_h(z) := \zeta_h \cap \mathcal{R}(z), \qquad z \in \Delta.$$

For $z \in \Delta$, this is a point in $\mathcal{R}(z)$ lying over w_h in $\mathbf{P}_w$, i.e., $\underline{\zeta_h(z)} = w_h$.

For each $z \in \Delta$, we consider the subset $\mathcal{L}^0_\nu(z)$ of $\mathcal{L}_\nu(z)$ defined as

$$\mathcal{L}^0_\nu(z) := \{ f(z,p) \in \mathcal{L}_\nu(z) \mid f(z, \zeta_h(z)) = 0 \ (l = 1, \dots, l_0) \ \}.$$

When we need to emphasize the dependence on ζ_h $(h = 1, \dots, l_0)$, we will write $\mathcal{L}^0_\nu(z) = \mathcal{L}^0_\nu(z, \{\zeta_h\}_{h=1,\dots,l_0})$. We also define, for $z \in \Delta$,

$$\begin{aligned} \mathcal{L}^*_\nu(z) \quad = \quad & \{ f(z,p) \in \mathcal{L}_\nu(z) \mid f(z,p) \text{ has poles} \\ & \qquad \text{of order } \nu \text{ at each } L^j_\infty(z) \ (j = 1, \dots, m) \}. \end{aligned}$$

We prove the following one complex variable lemma.

LEMMA 6.3. *Assume* $\nu > 2g - 1$. *Then:*

1. *Each* $\mathcal{L}^0_\nu(z)$, $z \in \Delta$, *is a complex-linear space with* dim $\mathcal{L}^0_\nu(z) \geq 1$.
2. *Fix* $a \in \Delta' := \Delta \setminus \sigma$. *Then we can choose the points* $\zeta_h(a)$ $(h = 1, \dots, l_0)$ *and* $\zeta_0(a)$ *on* $\mathcal{R}'(a)$ *with* $|w_h| = |\zeta_h(a)| > \rho$ $(h = 0, 1, \dots, l_0)$ *and* $w_h \neq w_k$ *if* $h \neq k$ *such that*
   (a) dim $\mathcal{L}^0_\nu(a, \{\zeta_h\}_{h=1,\dots,l_0}) = 1$, *and*
   (b) *there exists a function* $g(a,p) \in \mathcal{L}^0_\nu(a, \{\zeta_h\}_{h=1,\dots,l_0})$ *such that*
      (i) $g(a,p) \in \mathcal{L}^*_\nu(a)$;
      (ii) $\mathcal{L}^0_\nu(a, \{\zeta_h\}_{h=1,\dots,l_0}) = \{c\, g(a,p)\}_{c \in \mathbf{C}}$;
      (iii) $g(a, \zeta_0(a)) = 1$;
      (iv) $g(a,p)$ *does not vanish at any points of* $\mathcal{R}(a)$ *lying over* w_l $(l = 1, \dots, l_0)$ *except at* $\zeta_l(a)$, *and* $g(a,p)$ *does not assume the value* 1 *at any points of* $\mathcal{R}(a)$ *over* w_0 *except at* $\zeta_0(a)$.

PROOF. Assertion 1 follows Theorem 6.2. To prove assertion 2, since dim $\mathcal{L}_\nu(a)$ $= l_0 + 1$, we can choose a basis $\{f_\alpha(p)\}_{\alpha=1,\dots,l_0+1}$ of $\mathcal{L}_\nu(a)$ such that, on $\mathcal{R}(a)$, $f_1(p)$ has poles of total order $m\nu$ and each $f_\alpha(p)$ for $\alpha = 2, \dots, l_0 + 1$ has poles of total order at most $m\nu - 1$. Thus, by analyticity, we can choose a point $\zeta_1(a)$ in $\mathcal{R}'(a)$ such that $|w_1| = |\zeta_1(a)| > \rho$; $f_{l_0+1}(\zeta_1(a)) \neq 0$; and such that $f_1(p) - [f_1(\zeta_1(a))/f_{l_0+1}(\zeta_1(a))] \cdot f_{l_0+1}(p)$ does not vanish at any point of $\mathcal{R}(a)$ lying over w_1 except at $\zeta_1(a)$. Note there are at most $m - 1$ such points. Then, $\{g_\alpha(p)\}_{\alpha=1,\dots,l_0}$, where

$$g_\alpha(p) := f_\alpha(p) - [f_\alpha(\zeta_1(a))/f_{l_0+1}(\zeta_1(a))] \cdot f_{l_0+1}(p),$$

forms a base of $\mathcal{L}^0_\nu(a, \{\zeta_1(a)\})$; dim $\mathcal{L}^0_\nu(a, \{\zeta_1(a)\} = l_0$; $g_1(p) \in \mathcal{L}^*_\nu(a)$; each $g_\alpha(p)$ for $\alpha = 2, \dots, l_0$ has poles of total order at most $m\nu - 1$; and $g_1(p)$ does not vanish at any point of $\mathcal{R}(a)$ lying over w_1 except at $\zeta_1(a)$. Thus we can recursively find l_0 different points $\zeta_h(a)$ $(h = 1, \dots, l_0)$ in $\mathcal{R}'(a)$ such that $|w_h| = |\zeta_h(a)| > \rho$ and $w_h \neq w_k$ for $h \neq k$, and a function $h_1(p)$ on $\mathcal{R}(a)$ of the form $h_1(p) = f_1(p) - \sum_{\alpha=2}^{l_0+1} c_\alpha f_\alpha(p)$ such that $\{h_1(p)\}$ forms a base of $\mathcal{L}^0_\nu(a, \{\zeta_h(a)\}_{h=1,\dots,l_0})$ and $h_1(p)$ does not vanish at any point of $\mathcal{R}(a)$ lying over w_l $(l = 1, \dots, l_0)$ except at $\zeta_l(a)$. Thus, dim $\mathcal{L}^0_\nu(a, \{\zeta_h(a)\}_{h=1,\dots,l_0}) = 1$ and $h_1(p) \in \mathcal{L}^*_\nu(a)$. We finally choose a point $\zeta_0(a) \in \mathcal{R}'(a)$ such that $|w_0| = |\zeta_0(a)| > \rho$, $w_0 \neq w_l$ $(l = 1, \dots, h_0)$, and $h_1(a, \zeta_0(a)) \neq 0$. If we set $g(a,p) = h_1(p)/h_1(\zeta_0(a))$ on $\mathcal{R}(a)$, then this function $g(a,p)$ satisfies all the conditions in 2. $\square$

We remark that the function $g(a,p)$ is necessarily a characteristic function on $R(a)$. In fact, if not, there exist distinct points p', p'' on $R(a)$ with $\underline{p'} = \underline{p''} \in \mathbf{C}_w$ such that $g(a,p)$ has the same Taylor development about p' and p''. We connect p' and $\zeta_1(a)$ by an arc γ in $R'(a)$ such that $\underline{\gamma}$ does not pass through $\underline{\Sigma(a)}$ in $\mathbf{C}_w$. As we move p'' along $\underline{\gamma}$ in $R'(a)$, we reach a point $\widetilde{\zeta_1} \neq \zeta_1(a)$ over $\underline{\zeta_1(a)}$ in $R'(a)$. Then $g(a, \zeta_1(a)) = g(a, \widetilde{\zeta_1})$ by analytic continuation. This contradicts (b)-(iv) for $g(a,p)$ in the lemma.

Throughout this section, we fix a point $a \in \Delta'$, points ζ_h $(h = 0, 1, \ldots, l_0)$ and a function $g(a,p) \in \mathcal{L}^0_\nu(a) = \mathcal{L}^0_\nu(a, \{\zeta_h\}_{h=1,\ldots,l_0})$ satisfying 2 in Lemma 6.3.

We have the following lemma.

LEMMA 6.4. *(Stability)*

1. *Let* $z^j \in \Delta$ $(j = 1, 2, \ldots)$ *converge to the point* a *and let*
$$f(z^j, p) \in \mathcal{L}^0_\nu(z^j) \qquad (j = 1, 2, \ldots) \qquad \text{with } f(z^j, p) \not\equiv 0.$$
Any limit function $f(a,p)$ *of* $b^{j_\nu} f(z^{j_\nu}, p)$ $(\nu = 1, 2, \ldots)$ *on* $\mathcal{R}'(a)$ *obtained by applying Lemma 6.2 for* $z^0 = a$ *and* $f_j(p) = f(z^j, p)$ $(j = 1, 2, \ldots)$ *must be of the form* $c\, g(a,p)$ *for some nonzero constant* c. *Hence,* $f(z^j, p) \in \mathcal{L}^*_\nu(z^j)$ *for sufficiently large* j.
2. *There exists a neighborhood* V_h *of* $\zeta_h(a)$ $(h = 0, 1, \ldots, l_0)$ *in* $\mathcal{R}'(a)$ *such that*
   (i) $\dim \mathcal{L}^0_\nu(a, \{\xi_h\}_{h=1,\ldots,l_0}) = 1$ *for each* $\xi_h \in V_h$ $(h = 1, \cdots, l_0)$*;*
   (ii) *there exists a function* $f(a,p) \in \mathcal{L}^0_\nu(a, \{\xi_h\}_{h=1,\ldots,l_0}) \cap \mathcal{L}^*_\nu(a)$ *such that* $f(a, \xi_0) \neq 0$.

PROOF. Assertion 1 is clear from Lemma 6.2 and the uniqueness of $g(a,p)$. We prove (i) of 2 by contradiction. Assume that there exist $\{\xi^i_h\}_{i=1,2,\ldots} \subset \mathcal{R}'(a)$ $(h = 1, \ldots, l_0)$ such that $\lim_{i \to \infty} \xi^i_h = \zeta_h$ in $\mathcal{R}'(a)$ and $\dim\ \mathcal{L}^0_\nu(a, \{\xi^i_h\}_{h=1,\ldots,l_0}) \geq 2$ $(i = 1, 2, \ldots)$. Then for each $i = 1, 2, \ldots$, we can find a non-constant function $f_i(p) \in \mathcal{L}^0_\nu(a, \{\xi^i_h\}_{h=1,\ldots,l_0})$ such that $f_i(\zeta_0(a)) = 0$. We can follow the same argument as in the proof of Lemma 6.2 in the case $z^j = a$ $(j = 1, 2, \ldots)$, and we find that $b^{j_\nu} f_{i_\nu}(p)$ $(\nu = 1, 2, \ldots)$ converges locally uniformly to a nonconstant multiple $cg(a,p)$ on $\mathcal{R}'(a)$ by 1. This gives a contradiction at $\zeta_0(a)$, and part (i) of 2 is proved. Part (ii) of 2 can also be proved by the technique in Lemma 6.2. □

From this lemma we deduce the following corollaries with
$$\mathcal{L}^0_\nu(z) = \mathcal{L}^0_\nu(z, \{\zeta_h\}_{h=1,\ldots,l_0})$$
as above.

COROLLARY 6.1. *There exists a neighborhood* δ' *of* a *in* Δ' *such that for* $z \in \delta'$, *each function* $f(z,p) \in \mathcal{L}^0_\nu(z)$ *which is not identically zero does not vanish at* $\zeta_0(z)$ *and belongs to* $\mathcal{L}^*_\nu(z)$.

COROLLARY 6.2. *There exists a neighborhood* δ'' *of* a *in* Δ' *such that* $\dim\ \mathcal{L}^0_\nu(z) = 1$ *for all* $z \in \delta''$.

Let δ be a neighborhood of a in Δ' which satisfies both Corollaries 6.1 and 6.2. Then for each $z \in \delta$ there exists a unique function $g(z,p) \in \mathcal{L}^0_\nu(z) \cap \mathcal{L}^*_\nu(z)$ such that $g(z, \zeta_0(z)) = 1$. Thus $g(z,p)$ can be considered as a function of $n+1$ complex variables (z,p) in $\mathcal{R}(\delta)$.

We have the following proposition.

PROPOSITION 6.2. *$g(z,p)$ is a continuous function of (z,p) in $\mathcal{R}(\delta)$.*

PROOF. Let $z' \in \delta$ and let $z^j \to z'$ as $j \to \infty$ in δ. Using 1 of Lemma 6.4 and the fact that $g(z^j, \zeta_0(z^j)) = 1$ $(j = 1, 2, \dots)$, we can extract a subsequence $g(z^{j_\nu}, p)$ $(\nu = 1, 2, \dots)$ of $\{g(z^j, p)\}_{j=1,2,\dots}$ which converges locally uniformly to $g(a,p)$ on $\mathcal{R}'(a)$. Hence $g(z^j,p) \to g(a,p)$ $(j \to \infty)$ locally uniformly on $\mathcal{R}'(a)$, and $g(z,p)$ is a continuous function on $\mathcal{R}'(\delta)$. Since the $m\nu$ zeros of $g(z^j,p)$ in $\mathcal{R}'(z^j)$ converge to the $m\nu$ zeros $\{p^{(s)}\}_{s=1,\dots,m\nu}$ of $g(a,p)$ in $\mathcal{R}'(a)$ by Hurwitz' theorem, it follows that $1/g(z^j,p)$ converges uniformly to $1/g(a,p)$ in $\mathcal{R}(\delta) \cap [\delta \times (\rho_0 \le |w| \le \infty)]$, where $\rho_0 > |\underline{p^{(s)}}|$ $(s = 1, \dots, m\nu)$. Hence $g(z,p)$ is continuous in $\mathcal{R}(\delta)$. □

This proposition has the following interpretation: for each $z \in \delta$, consider the graph $\mathcal{C}_z : X = g(z,p)$ in $\mathbf{C}^2_{w,X}$. Then $\mathcal{C}_z$ varies continuously with the parameter $z \in \delta$ not only in $\mathbf{C}^2_{w,X}$, but also in $\mathbf{P}_w \times \mathbf{P}_X$. Since the simpleness of the graph is stable, by taking a smaller neighborhood δ of a, if necessary, it follows that each $\mathcal{C}_z$, $z \in \delta$, is a simple graph in $\mathbf{C}^2$. As a modification of Lemma 6.3, we need the following fact.

REMARK 6.4. As in Theorem 6.3, we let h_0 denote the number of branch points of $R(a)$ and set

$$\nu_0 := (h_0 + 2)m + g > 2g + 1. \tag{6.6}$$

If $\nu > m\nu_0$, then we can choose $\zeta_l(a)$ $(l = 0, 1, \dots, l_0 = m\nu - g)$ with $|w_l| = |\underline{\zeta_l(a)}| > \rho$ in $\mathcal{R}'(a)$ such that the function $g(a,p)$ in 2 of Lemma 6.3 satisfies conditions (i)-(iv) in the lemma as well as the following condition:

(v) *the graph $\mathcal{C}_{g(a,p)} :\ X = g(a,p)$ in $\mathbf{C}^2_{w,X}$ is simple.*

In fact, for each $i = 1, \dots, m$, there exists a meromorphic function $h_i(p)$ on $R(a)$ such that $h_i(p)$ has a pole at $L^i_\infty(a)$ of order ν but such that the poles at $L^j_\infty(a)$ $(j \neq i)$ are of order at most $\nu - 1$. We set $G^\varepsilon(p) := g(a,p) + \sum_i^m \varepsilon_i h_i(p)$ on $R(a)$. If ε_i $(i = 1, \dots, m)$ are suitably small, then $G^\varepsilon(p)$ satisfies condition (6.5): if $i \neq j$ $(1 \le i, j \le m)$, then

$$\lim_{z \to \infty} G^\varepsilon_i(z)/G^\varepsilon_j(z) \neq 0, 1, \text{ or } \infty.$$

Thus, l_0 of the zeros $\zeta'_h(a)$ $(h = 1, \dots, l_0)$ of $G^\varepsilon(p)$ and one of the zeros $\zeta'_0(a)$ of $G^\varepsilon(p) - 1$ are arbitrarily close to $\zeta_h(a)$ and $\zeta_0(a)$. From 2 of Lemma 6.4, $G^\varepsilon(p)$ and $\zeta'_h(a)$ $(h = 0, 1, \dots, l_0)$ in $\mathcal{R}'(a)$ satisfy the conditions (i)-(v) as well as (6.5). From Theorem 6.3 we see that there exists a function $G(p) \in \mathcal{L}^*_\nu(a)$ on $R(a)$ arbitrarily close to the function $G^\varepsilon(a,p)$ such that the graph $\mathcal{C}_G$ is simple in $\mathbf{C}^2_{w,X}$. Thus, l_0 of the zeros $\zeta^*_h(a)$ $(h = 1, \dots, l_0)$ of $G(p)$ and one of the zeros $\zeta^*_0(a)$ of $G(p) - 1$ are arbitrarily close to $\zeta'_h(a)$ and $\zeta'_0(a)$. Again using 2 of Lemma 6.4, we see that $G(p)$ and $\zeta^*_h(a)$ $(h = 0, 1, \dots, h_0)$ satisfy (i)-(v).

Throughout this section, we will assume that $\nu > m\nu_0$ and we will work with the function $g(z,p)$ having a pole of order ν along L^j_∞ $(j = 1, \dots, m)$ in $\mathcal{R}(\delta)$ constructed so that $g(a,p)$ satisfies all conditions in Lemma 6.3 as well as (v).

6.2.4. Analyticity of $g(z,p)$. We now prove that $g(z,p)$ is a *holomorphic* function of (z,p) in $\mathcal{R}'(\delta)$. For each $z \in \delta$, we let $G(z,w,X)$ denote the defining polynomial for $X = g(z,p)$:

$$G(z,w,X) = X^m + \alpha_1(z,w)X^{m-1} + \cdots + \alpha_m(z,w), \tag{6.7}$$

where

$$\alpha_i(z,w) = c_{i,0}(z)w^{\nu i} + c_{i,1}(z)w^{\nu i-1} + \cdots + c_{i,\nu i}(z) \quad (i = 1, \ldots, m). \tag{6.8}$$

From Proposition 6.2, $c_{i,j}(z)$ $(i = 1, \ldots, m;\ j = 0, 1, \ldots, \nu i)$ is a continuous function on δ.

We summarize the conditions satisfied by $c_{i,j}(z)$. Since $g(z,p)$ vanishes at $\zeta_l(z)$ $(l = 1, \ldots, l_0)$, where $\underline{\zeta_l(z)} = w_l$, and since $g(z,p)$ assumes the value 1 at $\zeta_0(z)$, where $\underline{\zeta_0(z)} = w_0$, it follows that the $c_{i,j}(z)$ satisfy the following $l_0 + 1$ simultaneous linear equations with constant coefficients:

$$c_{m,0}(z)\omega_l^{\nu m} + c_{m,1}(z)\omega_l^{\nu m-1} + \cdots + c_{m,\nu m}(z) = 0 \quad (l = 1, \ldots, l_0); \tag{6.9}$$

$$1 + \sum_{i=1}^{m} [c_{i,0}(z)\omega_0^{\nu i} + c_{i,1}(z)\omega_0^{\nu i-1} + \cdots + c_{i,\nu i}(z)] = 0. \tag{6.10}$$

We note that the graph $\mathcal{C}_a = \mathcal{C}_{g(a,p)} : X = g(a,p)$ in $\mathbf{C}^2 = \mathbf{C}_{w,X}$ is simple. We let

$$P_h(a) := (\xi_h(a), \xi'_h(a)) \qquad (h = 1, \ldots, h_0)$$

denote the points on the graph $\mathcal{C}_a$ which correspond to the branch points of $\mathcal{R}(a)$. Each $P_h(a)$ $(h = 1, \ldots, h_0)$ is a non-singular point of $\mathcal{C}_a$ in $\mathbf{C}^2$. Let

$$Q_k(a) = (\eta_k(a), \eta'_k(a)) \qquad (k = 1, \ldots, k_0)$$

denote the singular points of $\mathcal{C}_a$ in $\mathbf{C}^2$, which are all normal double points.

We set

$$Z_0(a) = (\omega_0, 1), \qquad Z_l(a) = (\omega_l, 0) \quad (l = 1, \ldots, l_0);$$

these are the points on $\mathcal{C}_a$ which correspond to $\zeta_l(a)$ $(l = 0, 1, \ldots, l_0)$ on $\mathcal{R}(a)$. We fix

$$E_\rho := \{w \in \mathbf{P}_w \mid |w| > \rho\}$$

so that there exist m disjoint univalent parts E_ρ^j $(j = 1, \ldots, m)$ of $\mathcal{R}(a)$ over E_ρ. We let $g_j(a,w)$ denote the branch of $g(z,p)$ on E_ρ^j $(j = 1, \ldots, m)$.

In $\mathbf{C}^2$ with variables w and X, we take a closed bidisk

$$\Gamma_h = \gamma_h \times \gamma'_h \quad (h = 1, \ldots, h_0)$$

containing the point $P_h(a) = (\xi_h(a), \xi'_h(a))$ and such that

(1) $\Gamma_h \cap \Gamma_k = \emptyset$ $(h \neq k)$;
(2) $(\mathcal{C}_a \cap \Gamma_h) \cap \{w = \xi_h(a)\}$ $(h = 1, \ldots, h_0)$ consists of only one point, $P_h(a)$;
(3) $\mathcal{C}_a \cap (\gamma_h \times \partial\gamma'_h) = \emptyset$ $(h = 1, \ldots, h_0)$.

We take a closed bidisk

$$\Lambda_k = \lambda_k \times \lambda'_k \quad (k = 1, \ldots, k_0)$$

containing the point $Q_k(a) = (\eta_k(a), \eta'_k(a))$ and such that

(1) $\lambda_h \cap \lambda_k = \emptyset$ $(h \neq k)$, so that $\Lambda_h \cap \Lambda_k = \emptyset$ $(h \neq k)$;
(2) $(\mathcal{C}_a \cap \Lambda_k) \cap \{w = \eta_k(a)\}$ $(k = 1, \ldots, k_0)$ consists of only one point, $Q_k(a)$;
(3) $\mathcal{C}_a \cap (\lambda_k \times \partial\lambda'_k) = \emptyset$ $(k = 1, \ldots, k_0)$.

On each complex line $w = \omega_l$ $(l = 0, 1, \ldots, l_0)$ in $\mathbf{C}^2_{w,X}$, by (iv) we can take a disk $\beta_0 : |X - 1| < \varepsilon_0$ and $\beta_l : |X| < \varepsilon_l$ $(l = 1, \ldots, l_0)$ such that $\mathcal{C}_a \cap (\{w_l\} \times \beta_l) = \{\zeta_l(a)\}$ $(l = 0, 1, \ldots, l_0)$.

We say that the graph C_a is **standard** with respect to $\{\Gamma_h, \Lambda_k, \beta_l\}_{h,k,l}$.

From Proposition 6.2 we obtain the following lemma.

LEMMA 6.5. *For a sufficiently small neighborhood δ of a in Δ', each graph $\mathcal{C}_z$ of $X = g(z,p)$ in $\mathbf{C}^2_{w,X}$ for $z \in \delta$ is simple and is standard with respect to $\{\Gamma_h, \Lambda_k, \beta_l\}_{h,k,l}$ defined above. Here, condition (2) for Γ_h $(h = 1, \dots, h_0)$ and Λ_k $(k = 1, \dots, k_0)$ above are replaced by the following conditions:*

(2') for Γ_h: $(\mathcal{C}_a \cap \Gamma_h) \cap \{w = \xi_h(z)\}$ consists of exactly one point, denoted $P_h(z)$;

(2') for Λ_k: Λ_k contains exactly one normal double point of $\mathcal{C}_z$, denoted $Q_k(z)$.

Clearly the points $Q_k(z)$ $(k = 1, \dots, k_0)$ coincide with the set of all singular points of the graph $\mathcal{C}_z$ in $\mathbf{C}^2$. For $z \in \delta$, we set

$$\begin{aligned} P_h(z) &= (\xi_h(z), \xi'_h(z)) \qquad (h = 1, \dots, h_0), \\ Q_k(z) &= (\eta_k(z), \eta'_k(z)) \qquad (k = 1, \dots, k_0). \end{aligned}$$

We observe that $\xi_h(z)$ $(h = 1, \dots, h_0)$ are single-valued *holomorphic* functions on δ (determined by the given ramified domain $\mathcal{R}$), and $\eta_k(z)$ $(k = 1, \dots, k_0)$ become single-valued *continuous* functions on δ by Proposition 6.2.

The main result in this section is the following.

Claim *Both $c_{i,j}(z)$ $(i = 1, \dots, m;\ j = 0, 1, \dots, \nu i)$ and $\eta_k(z)$ $(k = 1, \dots, k_0)$ are single-valued holomorphic functions on δ.*

PROOF. Fix $z \in \delta$. We let $D(z,w)$ denote the discriminant of the polynomial $G(z,w,X)$ with respect to the variable X, i.e., $D(z,w)$ is obtained by eliminating the variable X from the equations

$$G(z,w,X) = 0, \qquad \frac{\partial}{\partial X} G(z,w,X) = 0.$$

Thus $D(z,w)$ is a polynomial in w whose coefficients are polynomials in $c_{i,j}(z)$; hence $D(z,w)$ is of the form

$$D(z,w) = A_0(c_{i,j}(z)) w^N + A_1(c_{i,j}(z)) w^{N-1} + \cdots + A_N(c_{i,j}(z)), \tag{6.11}$$

where N is a positive integer.

On the other hand, it is known that $D(z,w)$ coincides with the product of the square of the differences of any two solutions of $G(z,w,X) = 0$ with respect to X, i.e.,

$$D(z,w) = \prod_{i \neq j} (g_i(z,w) - g_j(z,w))^2, \quad w \in \mathbf{C}_w,$$

where $g_i(z,w)$ $(i = 1, \dots, m)$ is a branch of $g(z,p)$ lying over a neighborhood of w. Since the graph $\mathcal{C}_z$, $z \in \delta$, is simple, the zeros of $D(z,w) = 0$ coincide with

$$w = \xi_h(z) \quad (h = 1, \dots, h_0), \qquad w = \eta_k(z) \quad (k = 1, \dots, k_0),$$

and the order of $\xi_h(z)$ is equal to the order d_h of ramification of $\mathcal{R}(z)$ at $w = \xi_h(z)$; note the order of $\eta_k(z)$ is always equal to 2. It follows that

$$D(z,w) = A_0(c_{i,j}(z)) \prod_{h=1}^{h_0} (w - \xi_h(z))^{d_h} \prod_{k=1}^{k_0} (w - \eta_k(z))^2. \tag{6.12}$$

We formally develop the right-hand side into a polynomial in w whose coefficients are polynomials in $c_{i,j}(z)$, $\xi_h(z)$, and $\eta_k(z)$:

$$\begin{aligned} D(z,w) &= A_0(c_{i,j}(z)) w^N + B_1(c_{i,j}(z), \xi_h(z), \eta_k(z)) w^{N-1} + \\ &\quad \cdots + B_N(c_{i,j}(z), \xi_h(z), \eta_k(z)). \end{aligned} \tag{6.13}$$

We compare the coefficients of w^s $(s = 0, 1, \dots, N)$ in (6.11) and (6.13), and obtain N equations which $c_{i,j}(z)$ and $\eta_k(z)$ must satisfy:

$$A_\lambda(c_{i,j}(z)) - B_\lambda(c_{i,j}(z), \xi_h(z), \eta_k(z)) = 0 \qquad (\lambda = 1, \dots, N). \tag{6.14}$$

Now we introduce new complex variables

$$u_{i,j} \ (i = 1, \dots, m; j = 0, 1, \dots, \nu i), \qquad v_k \ (k = 1, \dots, k_0)$$

and consider $l_0 + 1 + N$ equations obtained by replacing $c_{i,j}(z)$ and $\eta_k(z)$ by $u_{i,j}$ and v_k in (6.9), (6.10), and (6.14), i.e.,

$$u_{m,0}\omega_l^{\nu m} + u_{m,1}\omega_l^{\nu m-1} + \cdots + u_{m,\nu m} = 0 \qquad (l = 1, \dots, l_0), \tag{6.15}$$

$$1 + \sum_{i=1}^{m}[u_{i,0}\omega_0^{\nu i} + u_{i,1}\omega_0^{\nu i-1} + \cdots + u_{i,\nu i}] = 0, \tag{6.16}$$

$$A_\lambda(u_{i,j}) - B_\lambda(u_{i,j}, \xi_h(z), v_k) = 0 \qquad (\lambda = 1, \dots, N). \tag{6.17}$$

We consider the space $\mathbf{C}^M$ with variables $u_{i,j}$ and v_k, and the product space

$$\Pi_\delta = \delta \times \mathbf{C}^M.$$

We let Ω denote the analytic set in Π_δ defined by the analytic equations (6.15), (6.16) and (6.17). These are all algebraic for $u_{i,j}$ and v_k with holomorphic coefficients on δ; i.e., each of them is a polynomial in the variables $u_{i,j}$ and v_k whose coefficients are single-valued holomorphic functions of z in δ. From the construction, Ω contains the set

$$\mathcal{E}^* \ : \ u_{i,j} = c_{i,j}(z), \quad v_k = \eta_k(z), \quad z \in \delta.$$

We let Ω^0 denote the irreducible component of Ω which contains $\mathcal{E}^*$. To verify the claim, using (ii) in Remark 2.8 in Chapter 2, it suffices to show that Ω^0 is of dimension n. Therefore, we let $\Omega^0(a)$ denote the section of Ω^0 over $z = a$:

$$\Omega^0(a) = \{(u_{i,j}, v_k) \in \mathbf{C}^M \mid (a, u_{i,j}, v_k) \in \Omega^0\};$$

thus $(c_{i,j}(a), \eta_k(a)) \in \Omega^0(a)$. We want to show that

$$\text{the point } (c_{i,j}(a), \eta_k(a)) \text{ is isolated in } \Omega^0(a). \tag{6.18}$$

We prove this by contradiction. Let $Q = (c_{i,j}(a), \eta_k(a))$, and let $Q^* = (u^*_{i,j}, v^*_k) \neq Q$ be a point in $\Omega^0(a)$ arbitrarily close to Q in $\mathbf{C}^M$. We construct the polynomial $\alpha^*_i(w)$ in w,

$$\alpha^*_i(w) = u^*_{i,0}w^{\nu i} + u^*_{i,1}w^{\nu i-1} + \cdots + u^*_{i,\nu i} \qquad (i = 1, \dots, m), \tag{6.19}$$

and the polynomial $G^*(w, X)$ in X with coefficients $\alpha^*_i(w)$,

$$G^*(w, X) = X^m + \alpha^*_1(w)X^{m-1} + \cdots + \alpha^*_m(w). \tag{6.20}$$

We recall the definition of the analytic set Ω^0; thus, from (6.15) and (6.16),

$$G^*(\omega_j, 0) = 0 \quad (j = 1, \dots, l_0), \qquad G^*(\omega_0, 1) = 0.$$

If we let $D^*(w)$ denote the discriminant of $G^*(w, X)$ with respect to X, then our condition (6.17) implies

$$D^*(w) = A_0(u^*_{i,j}) \prod_{h=1}^{h_0}(w - \xi_h(a))^{d_k} \prod_{k=1}^{k_0}(w - v^*_k)^2.$$

We consider the algebraic function $X = g^*(p)$ defined by $G^*(w, X) = 0$. We let R^* denote the Riemann surface over $\mathbf{P}_w$ determined by $g^*(w)$, and we let $\mathcal{C}^*$ denote the graph of $g^*(p)$ in $\mathbf{C}^2_{w,X}$,

$$\mathcal{C}^* : X = g^*(p) \quad \text{in} \quad \mathbf{C}^2_{w,X}.$$

If Q^* approaches Q on Ω^0, then the graph $\mathcal{C}^*$ approaches the graph $\mathcal{C}_a$, not only in $\mathbf{C}^2_{w,X}$, but also in $\mathbf{P}_w \times \mathbf{P}_X$ by Remark 6.3. We thus assume that the graph $\mathcal{C}^*$ is simple and standard with respect to $\{\Gamma_h, \Lambda_k, \beta_l\}$, where condition (2) for Γ_h $(h = 1, \dots, h_0)$ and Λ_k $(k = 1, \dots, k_0)$ is replaced by the following condtions: $(\mathcal{C}^* \cap \Gamma_h) \cap \{w = \xi_h(z)\}$ (resp., $(\mathcal{C}^* \cap \Lambda_k) \cap \{w = v_k^*\}$) consists of only one point; we call this point $P_h^* = (\xi_h(a), \xi_h^{*\prime})$ (resp., $Q_k^* = (v_k^*, v_k^{*\prime})$).

With this notation we state the following lemma; this will be needed to reach a contradiction to finally verify (6.18).

LEMMA 6.6. *If $Q^* = (u_{i,j}^*, v_k^*) \in \Omega^0(a)$ is close enough to $Q = (c_{i,j}(a), \eta_k(a))$ but $Q^* \neq Q$, then*

(1) *$R^* = \mathcal{R}(a)$, and*
(2) *$g^*(p) = g(a, p)$ on $\mathcal{R}(a)$.*

PROOF. We observe that the graph $\mathcal{C}^*$ approaches $\mathcal{C}_a$ in $\mathbf{P}_w \times \mathbf{P}_X$, and $\mathcal{C}^* \neq \mathcal{C}_a$. Thus, the number of sheets of R^* over $\mathbf{P}_w$ is equal to the number of sheets m of $\mathcal{R}(a)$ over $\mathbf{P}_w$. Also, the solutions of the discriminant equation $D^*(w) = 0$ coincide with $\xi_h(a)$ $(h = 1, \dots, h_0)$ and v_k^* $(k = 1, \dots, k_0)$. Since $P_h(a)$ $(h = 1, \dots, h_0)$ is a non-singular point of $\mathcal{C}_a$ in $\mathbf{C}^2_{w,X}$, P_h^* is a non-singular point of $\mathcal{C}^*$. Furthermore, the order of ramification of R^* at $w = \xi_h(a)$ is d_h, the same as that of $\mathcal{R}(a)$ at $w = \xi_h(a)$. Since $Q_k(a)$ $(k = 1, \dots, k_0)$ is a normal double point of $\mathcal{C}_a$, Q_k^* is a normal double point of $\mathcal{C}^*$. Thus Q_k^* does not correspond to a branch point of R^*. It turns out that R^* is an m-sheeted Riemann surface over $\mathbf{P}_w$ with branch points only at $\xi_h(a)$ $(h = 1, \dots, h_0)$ over the same projection $\underline{\xi_h^*(a) = \xi_h(a)}$ with the same order d_h of ramification as $\xi_h(a)$ has for $\mathcal{R}(a)$. Since the graph $\mathcal{C}^*$ approaches $\mathcal{C}_a$, we see that $R^* = \mathcal{R}(a)$, which proves (1).

To prove (2), we note from (1) that $g^*(p)$ is a meromorphic function on $\mathcal{R}(a)$. Since each coefficient $\alpha_i^*(w)$ of the polynomial $G^*(w, X)$ is of order at most νi, it follows from Lemma 6.1 that $g^*(p) \in \mathcal{L}_\nu(a)$. From (6.15) (resp., (6.16)) $g^*(p)$ (resp., $g^*(p) - 1$) vanishes for at least one point of $R(a)$ over w_l $(l = 1, \dots, l_0)$ (resp., w_0). Since $\mathcal{C}^*$ approaches C_a, it follows from condition (iv) of 2 in Lemma 6.3 that there is only one such point, namely $\zeta_l(a)$ (resp., $\zeta_0(a)$). Equivalently, $g^*(p) \in \mathcal{L}_\nu^0(a)$ with $g^*(\zeta_0(a)) = 1$. From the uniqueness of $g(a, p)$ on $R(a)$ (using the fact that dim $\mathcal{L}_\nu^0(a) = 1$), we have $g^*(p) = g(a, p)$. Hence (2) is proved. □

We return to the proof of (6.18). Note that, for a given $Q^* = (u_{i,j}^*, v_k^*) \in \Omega^0(a)$, the construction of the meromorphic function $g^*(p)$ on R^* (using $\alpha_k^*(w)$ and $G^*(w, X)$) yields a one-to-one function. This contradicts 2 of Lemma 6.6. Hence, (6.18) is true. Thus, the claim is proved. □

In the proof, since Ω^0 is irreducible in Π_δ and $\mathcal{E}^* \subset \Omega^0$, it follows that

$$\Omega^0 = \mathcal{E}^*. \tag{6.21}$$

Since $c_{i,j}(z)$ $(i = 1, \dots, m; j = 1, \dots, \nu i)$ were shown to be single-valued holomorphic functions on δ, we thus obtain from (6.7) and (6.8) the following result.

PROPOSITION 6.3. *$g(z,p)$ is a meromorphic function of (z,p) in $\mathcal{R}(\delta)$ which is holomorphic in $\mathcal{R}'(\delta)$.*

6.2.5. Analytic Continuation of $g(z,p)$. We next prove that $g(z,p)$ in $\mathcal{R}(\delta)$ can be meromorphically continued to all of $\mathcal{R}$. We recall the simultaneous equations (6.15), (6.16), and (6.17) which define the analytic set Ω in Π_δ. From our condition for the branch set of $\mathcal{R}$: $w = \xi_h(z)$ $(h = 1, \ldots, h_0)$ with order of ramification d_h, we have

$$\prod_{h=1}^{h_0} (w - \xi_h(z))^{d_h} = \prod_{\chi=1}^{\chi_0} [P_\chi(z,w)]^{e_\chi},$$

where $P_\chi(z,w)$ is of the form

$$P(z,w) = w^\kappa + \beta_1(z) w^{\kappa-1} + \cdots + \beta_\kappa(z)$$

with the $\beta_i(z)$ $(i = 1, \ldots, \kappa)$ being single-valued holomorphic functions on all of Δ. Here, the e_χ $(\chi = 1, \ldots, \chi_0)$ are positive integers which are determined by the order of ramification of the branch set of $\mathcal{R}$ over $P_\chi(z,w) = 0$. Thus, from (6.12) all the equations (6.15), (6.16), and (6.17) are algebraic with respect to $u_{i,j}$ and v_k with coefficients which are single-valued holomorphic functions on Δ. It follows that Ω and Ω^0 defined in Π_δ can be analytically continued and considered as analytic sets in the product space $\Pi_\Delta := \Delta \times \mathbf{C}^M$ or even in $\Pi^*_\Delta := \Delta \times \mathbf{P}^M$. We use the same notation Ω^0 for the analytic set considered in Π_Δ or in Π^*_Δ obtained by this analytic continuation of Ω^0 in Π_δ. We can apply the results from § 2.5 in Chapter 2 to Ω^0. From (6.21) we see that $\dim \Omega^0 = n$ and the projection of Ω^0 onto Δ contains δ (and hence the point a). Given $z' \in \Delta$, we let $\Omega^0(z')$ denote the section of Ω^0 over $z = z'$. It follows from Corollary 2.8 that there exists an analytic set e in Δ of dimension at most $n-1$ such that for $z \in \Delta \setminus e$, $\Omega^0(z)$ consists of m^* distinct points in $\mathbf{C}^M$, where $m^* \geq 1$ is an integer independent of $z \in \Delta \setminus e$.

Therefore, if we set $\Delta_1 := \Delta' \setminus e$ (where $\Delta' = \Delta \setminus \sigma$ and σ is the zero set of the discriminant $d(z)$ of $P(z,w)$), and we set $\Omega^0_1 := \Omega^0 \cap (\Delta_1 \times \mathbf{C}^M)$, then Ω^0_1 can be written in the form

$$\Omega^0_1 : \quad u_{i,j} = c^*_{i,j}(\widetilde{z}), \quad v_k = \eta^*_k(\widetilde{z}), \quad z \in \widetilde{\Delta}_1,$$

where $\widetilde{\Delta}_1$ is an unramified, finitely sheeted domain over Δ_1 without relative boundary, and $c_{i,j}(\widetilde{z})$ and $\eta^*_k(\widetilde{z})$ are holomorphic functions on $\widetilde{\Delta}_1$. Thus, m^* is the number of sheets of $\widetilde{\Delta}_1$ over Δ_1. Our next claim is that $\widetilde{\Delta}_1$ is univalent over Δ_1, i.e.,

Claim $\quad m^* = 1$.

Indeed, from (6.21), there exists an open, univalent part δ_0 of $\widetilde{\Delta}_1$ over $\delta \cap \Delta_1$ such that $c^*_{i,j}(z) = c_{i,j}(z)$ and $\eta^*_k(z) = \eta_k(z)$ for $z \in \delta_0$. Take $z \in \Delta'_1$ and let $(c^*_{i,j}(\widetilde{z}), \eta^*_k(\widetilde{z})) \in \Omega^0_1(z)$. As in (6.19) and (6.20), we construct, for $i = 1, \ldots, m$,

$$\alpha^*_i(\widetilde{z}, w) = c^*_{i,0}(\widetilde{z}) w^{\nu i} + c^*_{i,1}(\widetilde{z}) w^{\nu i - 1} + \cdots + c^*_{i,\nu i}(\widetilde{z}), \tag{6.22}$$

$$G^*(\widetilde{z}, w, X) = X^m + \alpha^*_1(\widetilde{z}, w) X^{m-1} + \cdots + \alpha^*_m(\widetilde{z}, w). \tag{6.23}$$

These functions are holomorphic for $\widetilde{z} \in \widetilde{\Delta}$. We let $X = g^*(\widetilde{z}, w)$ denote the algebraic function determined by $G^*(\widetilde{z}, w, X) = 0$ and we write $R^*(\widetilde{z})$ for the Riemann surface of $g^*(\widetilde{z}, w)$. We also set

$$\mathcal{C}^*(\widetilde{z}) : \quad X = g^*(\widetilde{z}, w), \quad \widetilde{z} \in \widetilde{\Delta}_1, \tag{6.24}$$

the graph of $g^*(\widetilde{z}, w)$ in $\mathbf{C}^2_{w,X}$.

Fix $a_0 \in \delta_0$ and let γ be any *closed* curve in Δ_1 starting at a_0 and terminating at $a_1 = a_0$. We obtain a variation of graphs

$$\widetilde{z} \in \gamma \to \mathcal{C}^*(\widetilde{z})$$

starting from the graph $\mathcal{C}_{a_0}$ of $X = g(a_0, w)$. If $\widetilde{z}$ lies in a sufficiently small neighborhood $\delta_0' \subset \delta_0$ of the starting point a_0, then $\mathcal{C}^*(\widetilde{z}) = \mathcal{C}_z$, where $\underline{\widetilde{z}} = z$ and $\mathcal{C}_z$ is the graph of $g(z,p)$. This does not necessarily hold for $\widetilde{z}$ in a neighborhood of the terminal point a_1. In any case, we have $R^*(\widetilde{z}) = R(z)$ for $\widetilde{z} \in \delta_0'$. For, since the Riemann surface $R(z)$ ($R^*(\widetilde{z})$) varies holomorphically with respect to $z \in \Delta'$ ($\widetilde{z} \in \Delta_1$), and since $\Delta_1 \subset \Delta'$, it follows that $R^*(\widetilde{z}) = R(z)$ for $\widetilde{z}$ lying in a neighborhood of the terminal point a_1 with $\underline{\widetilde{z}} = z$. For such points $\widetilde{z}$ close to the terminal point a_1, $g^*(\widetilde{z}, p)$ is thus a meromorphic function on the Riemann surface $R(z)$ with $\underline{\widetilde{z}} = z$. By construction of $X = g^*(\widetilde{z}, p)$, we have $g^*(\widetilde{z}, p) \in \mathcal{L}_\nu^0(z)$ and $g^*(\widetilde{z}, \zeta_0(z)) = 1$ since $Q^* = (u_{i,j}^*, v_k^*) \in \Omega^0$ and ω^0 satisfies (6.16). Since $\mathcal{L}_\nu^0(z) = \{c\, g(z,p)\}_{c \in \mathbf{C}}$ for $z \in \delta$, it follows that $g^*(\widetilde{z}, p) = g(z,p)$ for all $\widetilde{z}$ sufficiently close to the terminal point a_1. Since $c_{i,j}^*(\widetilde{z})$, $\eta_k^*(\widetilde{z})$ can be constructed from $g^*(\widetilde{z}, p)$, this means that $c_{i,j}^*(\widetilde{z})$, $\eta_k^*(\widetilde{z})$ vary holomorphically with $\widetilde{z} \in \gamma$, starting with the values $c_{i,j}(z)$, $\eta_k(z)$ in δ_0 and returning to the same values $c_{i,j}(z)$, $\eta_k(z)$. We thus have $m^* = 1$. □

In particular, we have $a \in \Delta_1$. We finally arrive at the last step of the proof of the fundamental theorem (Theorem 6.1). Since $m^* = 1$, we can write $c_{i,j}^*(z) = c_{i,j}(z)$, $\eta_k^*(z) = \eta_k(z)$ for $z \in \Delta_1$; these are single-valued holomorphic functions on Δ_1. As in (6.22) and (6.23), we write $\alpha_i^*(z) = \alpha_i(z)$ on Δ_1 and $G(z, w, X) = G^*(z, w, X)$ for $z \in \Delta_1 \times \mathbf{C}_{w,X}$. We also write $g^*(z,p) = g(z,p)$; this is a meromorphic function of (z,p) in $\mathcal{R}(\Delta_1)$ which is holomorphic in $\mathcal{R}'(\Delta_1)$. Since $(c_{i,j}(z), \eta_k(z))$, $z \in \Delta_1$ is a subset of the n-dimensional analytic set Ω^0 in $\Pi_\Delta^* = \Delta \times \mathbf{P}^M$ and since e_1 and σ are analytic sets in Δ of dimension at most $n-1$ (where $\Delta_1 = \Delta \backslash (e_1 \cup \sigma)$), it follows that $c_{i,j}(z)$ and $\eta_k(z)$ are meromorpohic function on all of Δ whose poles $\wp$ are contained in $e \cup \sigma$. Using the solvability of the Poincaré problem in Δ (by Theorem 3.9), we can construct a holomorphic function $\varphi(z)$ on Δ such that $\varphi(z) = 0$ on $\wp$; $\varphi(a) \neq 0$; and each $\varphi(z) c_{i,j}(z)$ $(i = 1, \dots m; j = 1, \dots \nu i)$ can be holomorphically extended to all of Δ. Therefore, equations (6.23) and (6.24) imply that the holomorphic function $X = \widehat{g}(z,p) := \varphi(z) g(z,p)$ on $\mathcal{R}$ satisfies the following equation:

$$X^m + \alpha_1(z,w)\varphi(z)X^{m-1} + \cdots + \alpha_m(z,w)(\varphi(z))^m = 0,$$

where each coefficient function $\alpha_i(z,w)(\varphi(z))^i$ $(i = 1, \dots, m)$ is a holomorphic function on $\mathbf{C}_{z,w}^{n+1}$. Since $\widehat{g}(a,p)$ as well as $g(a,p)$ have simple graphs in $\mathbf{C}_{w,X}^2$, it follows that $\widehat{g}(z,p)$ is a simple function on $\mathcal{R}'$. The fundamental theorem is completely proved.

□

6.2.6. Fundamental System of Locally Ramified Domains. Let Δ : $|z_j| < r_j$ $(j = 1, \dots, n)$ be a polydisk in $\mathbf{C}^n$. Let $\mathcal{R}$ be a ramified domain over Δ without relative boundary which is standard with respect to the variable z_n, and let $\pi : \mathcal{R} \to \Delta$ be the canonical projection. We let m be the number of sheets of $\mathcal{R}$ over Δ; we let Σ be the branch set of $\mathcal{R}$; and we set $\underline{\Sigma} = \pi(\Sigma)$. Then $\underline{\Sigma}$ can be written as the zero set of a pseudopolynomial $P(z)$ with respect to z_n:

$$P(z) = z_n^\kappa + \alpha_1(z') z_n^{\kappa-1} + \cdots + \alpha_\kappa(z'),$$

where $\alpha_i(z')$ $(i = 1, \dots, \kappa)$ is a holomorphic function for $z' = (z_1, \dots, z_{n-1})$ in $\Delta' : |z_j| < r_j$ $(j = 1, \dots, n-1)$.

Let $\Phi_i(p)$ $(i = 1, \dots, N)$ be a holomorphic function on $\mathcal{R}$. We introduce $\mathbf{C}^N$ with variables w_i $(i = 1, \dots, N)$ and consider the following analytic set $\widetilde{\mathcal{C}}$ in the $(n + N)$-dimensional product space $\Lambda = \Delta \times \mathbf{C}^N$:

$$\widetilde{\mathcal{C}} \ : \ w_i = \Phi_i(p) \qquad (p \in \mathcal{R};\ i = 1, \dots, N).$$

We let $\mathcal{S}$ denote the analytic set of singular points of $\widetilde{\mathcal{C}}$ in Λ. If $\mathcal{S}$ is at most $(n-2)$-dimensional as an analytic set in Λ, we say that $\{\Phi_i(p)\}_{i=1,\dots,N}$ is a **fundamental system** for $\mathcal{R}$.

We have the following theorem.

THEOREM 6.4. *There exists a fundamental system for* $\mathcal{R}$.

PROOF. By the fundamental theorem (Theorem 6.1), there exists a characteristic and simple function $\Phi_1(p)$ on $\mathcal{R}$. We consider the product space $\Lambda_1 = \Delta \times \mathbf{C}_{w_1}$ and the analytic hypersurface

$$\mathcal{C}_1 : \quad w_1 = \Phi_1(p), \quad p \in \mathcal{R}.$$

We let Σ_1 denote the set of points of $\mathcal{C}_1$ which correspond to the branch points p of $\mathcal{R}$, i.e., to the points $p \in \Sigma$. Since $\Phi_1(p)$ is a simple function for $\mathcal{R}$, Σ_1 consists of regular points of $\mathcal{C}_1$ except for an analytic set in Λ of dimension at most $n-2$.

We let $\mathcal{S}$ denote the set of singular points of $\mathcal{C}_1$ in Λ; this consists of a finite number of irreducible components S_j $(j = 1, \dots, \mu)$ with each S_j being an analytic set of dimension at most $n-1$. We fix a component S_1 of $\mathcal{S}$ which is $(n-1)$-dimensional (if such a component exists). Since $\dim(S_1 \cap \Sigma_1) \le n-2$, we can find a point $z^* \in \underline{S}_1 \setminus \Sigma$ such that there exist distinct regular points p_i $(i = 1, \dots, m_1; m_1 \le m)$ on $\mathcal{R}$ over z^* with $\Phi_1(p)$ being a single-valued holomorphic function on a (univalent) neighborhood of each p_i in $\mathcal{R}$ and with $\Phi_1(p_i) = \Phi_1(p_j)$ if $i \ne j$. Take any two distinct points p_i and p_j among the points $\{p_i\}_{i=1,\dots,m_1}$. Since $\Phi_1(p)$ is characteristic on $\mathcal{R}$, it has different function elements at p_i and p_j. Thus, among the partial differentiations

$$\frac{\partial^{j_1+\cdots+j_n}}{\partial z_1^{j_1} \cdots \partial z_n^{j_n}} \Phi_1(p)$$

of $\Phi_1(p)$ with respect to z_j $(j = 1, \dots, n)$, there exists at least one, say $\Psi(p)$, which attains different values at p_i and p_j. We note that $\Psi(p)$ is a meromorphic function on $\mathcal{R}$ whose poles are contained in the branch surface Σ of $\mathcal{R}$. Thus, if we take a sufficiently large integer λ, the function $\Psi_1(p) := (P(z))^\lambda \Psi(p)$ is holomorphic on $\mathcal{R}$ and satisfies $\Psi_1(p_i) \ne \Psi_1(p_j)$. Repeating this method for any pair (p_i, p_j) $(i \ne j)$, we obtain holomorphic functions

$$\Psi_1^{(1)}(p), \dots, \Psi_{k_1}^{(1)}(p) \quad \text{ on } \mathcal{R}$$

with the following properties: for distinct i and j $(1 \le i, j \le m_1)$, there exists some $\Psi_k^{(1)}(p)$ $(1 \le k \le k_1)$ such that $\Psi_k^{(1)}(p_i) \ne \Psi_k^{(1)}(p_j)$. We thus consider $\mathbf{C}^{1+k_1} := \mathbf{C}_{w_1} \times \mathbf{C}^{k_1}$ with variables $w_1, w_{1,k}$ $(k = 1, \dots, k_1)$ and form the graph

$$\mathcal{C}_2 : w_1 = \Phi_1(p),\ w_{1,k} = \Psi_k^{(1)}(p) \ \ (k = 1, \dots, k_1), \quad p \in \mathcal{R}$$

in the product space $\Lambda_2 := \Delta \times \mathbf{C}^{1+k_1}$. Then $\mathcal{C}_2$ is an n-dimensional analytic set in Λ_2 which is non-singular at all points corresponding to a point of $S_1 \setminus \Sigma$ except perhaps for an analytic set in Λ_2 of dimension at most $n-2$. Repeating this procedure for S_l $(l = 2, \dots \mu)$, we obtain a holomorphic function $\Psi_k^{(l)}(p)$ $(k = 1, \dots, k_l)$ on $\mathcal{R}$. In $\mathbf{C}^M$, where $M = 1 + k_1 + \cdots + k_\mu$ with variables $w_1, w_{l,k}$ $(l = 1, \dots, \mu;\ k = 1, \dots, k_l)$, we form the graph

$$\mathcal{C} : w_1 = \Phi_1(p), \quad w_{l,k} = \Psi_{\iota}^{(l)}(p) \ (l = 1, \dots, \mu;\ k = 1, \dots, k_l), \qquad p \in \mathcal{R}$$

lying in $\Lambda := \Delta \times \mathbf{C}^M$. The singular set of $\mathcal{C}$ consists of analytic sets of dimension at most $n-2$ in Λ. Thus $\{\Phi_1(p), \Psi_k^{(l)}\}_{l,k}$ is a fundamental system for $\mathcal{R}$. □

6.3. Appendix 1

In this section we give a proof of Theorem 6.3. Let R be an m-sheeted compact Riemann surface over $\mathbf{P}_w$ of genus g. Let $\pi : R \to \mathbf{P}_w$ denote the projection and let P_h $(h = 1, \dots, h_0)$ be the set of branch points of R. We assume that $\pi(P_h) \neq \infty$.

We let R' denote the part of R lying over $\mathbf{C}_w$. We set $c_h = \pi(P_h)$, and we let $e_h - 1$ denote the order of ramification of R at P_h. Then we can choose a local parameter t_h at P_h of the form $w = c_h + t_h^{e_h}$. We let L_j $(j = 1, \dots, m)$ denote the m distinct points over $w = \infty$, and we use a local parameter t_j at L_j of the form $t_j = 1/w$. For a regular point p of R' we use a local parameter t_p at P of the form $w = \pi(P) + t_p$.

Given a meromorphic function $f(p)$ on R, we write $f'(p)$ to denote the derivative of $f(p)$ with respect to the local parameter at p.

We let c_h $(h = 1, \dots, h_0')$ denote the distinct points among the points c_h $(h = 1, \dots, h_0)$, and we write

$$\mathbf{C}_w' := \mathbf{C} \setminus \{c_h\}_{h=1,\dots,h_0'}.$$

We write $\mathcal{L}_\nu(R)$ for the complex-linear space of meromorphic functions $f(p)$ on R such that $f(p)$ is holomorphic on R' and the order of the pole of $f(p)$ at L_j $(j = 1, \dots, m)$ is less than or equal to ν. Given a non-constant function $f(p) \in \mathcal{L}_\nu(R)$, we consider the graph

$$\mathcal{C}_f : \quad X = f(p), \quad p \in R,$$

in $\mathbf{C}^2_{w,X}$ with variables w and X; then $\mathcal{C}_f$ is a one-dimensional analytic set in $\mathbf{C}^2_{w,X}$. For $p \in R'$, we call $(\pi(p), f(p)) \in \mathcal{C}_f$ the point corresponding to p. If $f(p)$ is a characteristic function on R, i.e., if there exists a point $w_0 \in \mathbf{C}_w'$ such that $f(p)$ has m different function elements over a neighborhood of w_0, then this correspondence $R \to \mathcal{C}_f$ is one-to-one except at a finite set of points.

We have the following proposition.

PROPOSITION 6.4. *Let q_ι $(\iota = 1, \dots, \kappa)$ be κ distinct points of R' and let $w_\iota = \pi(q_\iota)$ $(\iota = 1, \dots, \kappa)$. Let α_ι, β_ι $(\iota = 1, \dots, \kappa)$ be complex numbers. Then there exists a function $f(p) \in \mathcal{L}_{\mu_0}$ with $\mu_0 := m(\kappa + 2) + g$ and*

$$f(q_\iota) = \alpha_\iota, \quad f'(q_\iota) = \beta_\iota \qquad (\iota = 1, \dots, \kappa).$$

PROOF. We let t_ι be a local parameter at q_ι $(\iota = 1, \dots, \kappa)$. In a neighborhood of each q_ι we prescribe a principal part $\wp_\iota$ of a meromorphic function as follows:

(i) if q_ι is a branch point of order $e_\iota - 1$, then $\wp_\iota = \alpha_\iota / t_\iota^{2e_\iota} + \beta_\iota / t_\iota^{2e_\iota - 1}$;
(ii) if q_ι is an ordinary point of R', then $\wp_\iota = \alpha_\iota / t_\iota^2 + \beta_\iota / t_\iota$.

Since $e_\iota \leq m$, we easily see from the Riemann-Roch theorem that there exists a meromorphic function $\varphi(p)$ on R such that

(1) $\varphi(p)$ has principal part $\wp_\iota$ at each q_ι $(\iota = 1, \dots, \kappa)$;
(2) $\varphi(p)$ has poles at L_j $(j = 1, \dots, m)$ of order at most μ_0;

(3) $\varphi(p)$ is holomorphic on $R' \setminus \{q_\iota\}_{\iota=1,\dots,\kappa}$.

We let w_ι $(\iota = 1,\dots,\kappa_0)$ denote the set of distinct points among all the points w_ι $(\iota = 1,\dots,\kappa)$. If we set

$$f(p) := \varphi(p) \prod_{\iota=1}^{\kappa_0} (w - w_\iota)^2, \quad p \in R,$$

then $f(p)$ satisfies all the conditions in the proposition. □

We put $\nu_0 := m(h_0+2)+g$, which is determined by the given Riemann surface R. From this proposition we obtain the following lemma.

LEMMA 6.7. *There exists a meromorphic function $f(p) \in \mathcal{L}_{\nu_0}(R)$ such that if $\mathcal{C}_f : X = f(p)$, $p \in R$, denotes the graph of $f(p)$ in $\mathbf{C}^2_{w,X}$, then each intersection point of $\mathcal{C}_f$ and the complex line $w = c_h$ $(h = 1,\dots,h_0')$ in $\mathbf{C}^2_{w,X}$ is a non-singular point of $\mathcal{C}_f$ in $\mathbf{C}^2_{w,X}$.*

PROOF. For each $h = 1,\dots,h_0'$, we let $q_{h,\nu}$ $(\nu = 1,\dots,\nu_h)$ denote the points of R' lying over $w = c_h$. From the above proposition we can find a meromorphic function $f(p) \in \mathcal{L}_{\nu_0}(R)$ such that

$$f(q_{h,\nu}) = \nu, \qquad f'(q_{h,\nu}) = 1.$$

Then the graph $\mathcal{C}_f : X = f(p)$, $p \in R'$, in $\mathbf{C}^2_{w,X}$ satisfies the conditions in the lemma. □

Now let $\nu > m\nu_0$ and let $g(p)$ be a characteristic function on R such that $g(p) \in \mathcal{L}^*_\nu(R)$ (i.e., the order of the pole at each L^j_∞ $(j = 1,\dots,m)$ is equal to ν). Using the function $f(p)$ from the above lemma, we set

$$G(p) := g(p) + \varepsilon f(p) \quad \text{on } R \tag{6.25}$$

for $\varepsilon > 0$. If ε is sufficiently small, then the graph $\mathcal{C}_G : X = G(p)$, $p \in R'$, in $\mathbf{C}^2_{w,X}$ satisfies the following:

Condition $(*)$: All points of R' over $w = c_h$ $(h = 1,\dots,h_0')$ correspond to non-singular points of $\mathcal{C}_G$ in $\mathbf{C}^2_{w,X}$.

Since $G(p)$ as well as $g(p)$ is a characteristic function on R, the correspondence $p \in R' \to (\pi(p), G(p)) \in \mathcal{C}_G$ is one-to-one except for a finite point set

$$Q_k = (a_k, b_k) \qquad (k = 1,\dots,k_0)$$

of $\mathcal{C}_G$. Here $a_k \in \mathbf{C}'_w$. We let $\eta_k \geq 2$ be the number of points of R' which correspond to Q_k. In the present case Q_k $(k = 1,\dots,k_0)$ coincides with the set of singular points of $\mathcal{C}_G$ in $\mathbf{C}^2_{w,X}$. We study the behavior of $\mathcal{C}_G$ in a neighborhood of a point Q_k in $\mathbf{C}^2_{w,X}$. For simplicity, we write $Q_k = Q = (a,b)$ and $\eta_k = \eta$. We let $p_1,\dots,p_\eta$ in R' denote the points corresponding to Q through $X = G(p)$. There exists a closed bidisk $\Lambda := \Delta \times \Gamma \subset \mathbf{C}'_w \times \mathbf{C}_X$ centered at Q such that $\mathcal{C}_G \cap \Lambda$ can be written in the form

$$P_Q(w, X) := \prod_{i=1}^{\eta} (X - \psi_i(w)) = 0, \tag{6.26}$$

where $\psi_i(w)$ $(i = 1, \dots, \eta)$ is a single-valued holomorphic function on Δ with $b = \psi_j(a)$ $(i = 1, \dots, \eta)$ and $\psi_i(w) \neq \psi_j(w)$ if $w \neq a$ and $i \neq j$. The discriminant $D_Q(w)$ of the polynomial $P_Q(w, X)$ with respect to X can be written in the form

$$D_Q(w) = A(w)\ (w - a)^{2\rho},$$

where $\rho \geq 1$ and $A(w)$ is a non-vanishing holomorphic function on Δ. We call the integer $\rho \geq 1$ the **order of the singularity** of $\mathcal{C}_G$ at the singular point Q. We observe that Q is a normal double singular point of $\mathcal{C}_G$ if and only if $\rho = 1$.

We have the following reduction for $G(p)$.

LEMMA 6.8. *Let $\nu > m\,\nu_0$ and let $G(p)$ be a characteristic function on R such that $G(p) \in \mathcal{L}_\nu^*(R)$ and the graph $\mathcal{C}_G : X = G(p)$ of $G(p)$ in $\mathbf{C}^2_{w,X}$ satisfies condition $(*)$. Then there exist a finite number of meromorphic functions $\phi_j(p) \in \mathcal{L}_{\nu_0}(R)$ $(j = 1, \dots, M)$ such that for suitably small $\varepsilon_j \neq 0$ $(j = 1, \dots, M)$, if we set*

$$K(p) := G(p) + \sum_{j=1}^{M} \varepsilon_j \phi_j(p) \qquad \text{on } R$$

and if we let $\mathcal{C}_K : X = K(p),\ p \in R'$ be the graph of $K(p)$ in $\mathbf{C}^2_{w,X}$, then:

(1) *the graph $\mathcal{C}_K$ satisfies condition $(*)$;*
(2) *all singular points $Z_\kappa = (x_\kappa, y_\kappa)$ $(\kappa = 1, \dots, \kappa_0)$ of the graph $\mathcal{C}_K$ in $\mathbf{C}^2_{w,X}$ consist of normal double points;*
(3) *$x_k \neq x_l$ if $k \neq l$; $k, l = 1, \dots, \kappa_0$.*

PROOF. *First step.* In order to modify $G(p)$ to satisfy conditions (1) and (2), we let $\rho_k \geq 1$ $(k = 1, \dots, k_0)$ denote the order of the singularity of $\mathcal{C}_G$ at the singular point $Q_k = (a_k, b_k)$. We consider closed bidisks $\Lambda_k := \overline{\Delta}_k \times \Gamma_k \subset \mathbf{C}_w \times \mathbf{C}_X$ centered at Q_k such that $\Lambda_k \cap \Lambda_l = \emptyset$ if $k \neq l$ and such that $\mathcal{C}_G \cap \Lambda_k$ is of the form

$$P_k(w, X) := \prod_{i=1}^{\eta_k} (X - \psi_{k,i}(w)) = 0,$$

where $b_k = \psi_{k,i}(a_k)$ $(k = 1, \dots, k_0)$ and $\psi_{k,i}(w) \neq \psi_{k,j}(w)$ for $w \neq a_k$; $i \neq j$. The discriminant $D_k(w)$ of $P_k(w, X)$ with respect to X is of the form

$$D_k(w) = A_k(w)\ (w - a_k)^{2\rho_k},$$

where $A_k(w) \neq 0$ for $w \in \overline{\Delta}_k$.

Assume that $\rho_k \geq 2$ for some k, say $k = 1$ for simplicity. Let $p_1, \dots, p_{\eta_1}$ be the points of R' which correspond to Q_1 through $\mathcal{C}_G$. From Proposition 6.4 there exists a meromorphic function $\varphi(p) \in \mathcal{L}_{\nu_0}(R)$ such that

$$\begin{aligned} &\varphi(p_1) = \varphi(p_2) = 0, \quad \varphi'(p_1) = 1, \quad \varphi'(p_2) = 2, \\ &\varphi(p_\mu) = \mu, \quad \varphi'(p_\mu) = 0 \qquad (\mu = 3, \dots, \eta_1). \end{aligned}$$

We set

$$H(p) := G(p) + \varepsilon\ \varphi(p) \qquad \text{on } R$$

and consider the graph $\mathcal{C}_H :\ X = H(p),\ p \in R'$, in $\mathbf{C}^2_{w,X}$. If $\varepsilon \neq 0$ is sufficiently small, then $H(p)$ satisfies condition $(*)$ and, moreover, the singular points of $\mathcal{C}_H$ in $\mathbf{C}^2_{w,X}$ are all located in Λ_k $(k = 1, \dots, k_0)$.

In fact, since $G(p) \in \mathcal{L}_\nu^*(R)$, $G(p)$ has pole of order ν at each L_j^∞ $(j = 1, \dots, m)$. On the other hand, $\varphi(p) \in \mathcal{L}_{\nu_0}(R)$ has pole of order at most $m\nu_0$ at any L_j^∞ $(j = 1, \dots, m)$. It follows from $\nu > m\,\nu_0$ that $|G(p)| > 2|\varphi(p)|$ outside the domains of R

over the large disk $\Delta_r := \{|w| < r\}$ in $\mathbf{C}_w$. Hence $\mathcal{C}_H$ as well as $\mathcal{C}_G$ has no singular points outside $\Delta_r \times \mathbf{C}_X$. Moreover, if ε is sufficiently small, then $\mathcal{C}_H$ as well as $\mathcal{C}_G$ has no singular points in $(\Delta_r \times \mathbf{C}_X) \setminus \bigcup_{k=1}^{k_0} \Lambda_k$.

In the bidisk Λ_1, the graph $\mathcal{C}_H \cap \Lambda_1$ has the form

$$\widetilde{P}_1(w, X) := \prod_{i=1}^{\eta_1} [X - (\psi_{1,i}(w) + \varepsilon\varphi(w))].$$

We see that Q_1 is also a singular point of $\mathcal{C}_H$ (as well as of $\mathcal{C}_G$), but it becomes a normal double singular point of $\mathcal{C}_H$, so that the order of the singularity of $\mathcal{C}_H$ at Q_1 is equal to 1. Besides Q_1, some new singular points of $\mathcal{C}_H$ may be created in Λ_1; these will be denoted by T_j $(j = 1, \ldots, j_0)$. We let $\widetilde{\rho}_j$ be the order of the singularity of $\mathcal{C}_H$ at T_j $(j = 1, \ldots, j_0)$. Let $\widetilde{D}_1(w)$ be the discriminant of $\widetilde{P}_1(w, X)$ with respect to X; then the number of zeros of $\widetilde{D}_1(w)$ in $\overline{\Delta}_1$ (counted with multiplicity) is equal to $2\,\rho_1$ – the same as that of $D_1(w)$ – and it also equals the sum of the order of the singularities of $\mathcal{C}_H$ in Λ_1. Hence

$$\rho_1 = 1 + \widetilde{\rho}_1 + \cdots + \widetilde{\rho}_{j_0}.$$

It follows that $\widetilde{\rho}_j \le \rho_1 - 1$ $(j = 1, \ldots, j_0)$. Thus all singular points of $\mathcal{C}_H$ in the bidisk Λ_1 have order of singularity at most $\rho_1 - 1$.

Similarly, in the other bidisks Λ_k $(k = 2, \ldots, k_0)$ there may be finitely many singular points $\widetilde{T}_{k,j}$ $(j = 1, \ldots, j_k)$ of $\mathcal{C}_H$ even though Λ_k has only one singular point Q_k of $\mathcal{C}_G$ and the order of singularity at Q_k is $\rho_k \ge 1$. Let $\widetilde{\rho}_{k,j}$ be the order of the singularity of $\mathcal{C}_H$ at the singular point $\widetilde{T}_{k,j}$ $(j = 1, \ldots, j_k)$; then we see from the argument above involving the discriminants that $\rho_k = \sum_{j=1}^{j_k} \widetilde{\rho}_{k,j}$, so that each $\widetilde{\rho}_{k,j} \le \rho_k$ $(j = 1, \ldots, j_k)$.

Repeating the same procedure, step by step, we construct a finite number of meromorphic functions $\phi_j(p) \in \mathcal{L}_{\nu_0}(R)$ $(j = 1, \ldots, N)$ such that for sufficiently small $\varepsilon_j \ne 0$ $(j = 1, \ldots, M)$, if we define

$$L(p) := G(p) + \sum_{j=1}^{N} \varepsilon_j \phi_j(p) \qquad \text{on } R$$

and if we let $\mathcal{C}_L :\ X = L(p),\ p \in R'$, denote the graph of $L(p)$ in $\mathbf{C}^2_{w,X}$, then $\mathcal{C}_L$ satisfies condition $(*)$ and the set $\mathcal{A}$ of all singular points A_κ $(\kappa = 1, \ldots, \kappa_0)$ of $\mathcal{C}_L$ in $\mathbf{C}^2_{w,X}$ consists of those whose orders of singularity ρ_κ are all equal to 1, i.e., each A_κ $(\kappa = 1, \ldots, \kappa_0)$ is a normal double singular point of $\mathcal{C}_L$.

Second step. In order to modify $L(p)$ to satisfy condition (3) we set $A_\kappa = (\alpha_\kappa, \beta_\kappa)$, $\kappa = 1, \ldots, \kappa_0$, and let α_κ $(\kappa = 1, \ldots, \kappa_0')$ be the set of distinct points among all the points α_κ $(\kappa = 1, \ldots, \kappa_0)$ in $\mathbf{C}_w$. For each $\kappa = 1, \ldots, \kappa_0'$ we let j_κ be the number of points of $\mathcal{A}$ lying over the complex line $w = \alpha_\kappa$.

Fix $\kappa = 1$. Let $A_{1,j}$ $(j = 1, \ldots, j_1)$ be the points of $\mathcal{A}$ lying over the complex line $w = \alpha_1$, and let $p_j', p_j'' \in R'$ be the points of R' which correspond to the normal double singular point $A_{1,j}$ of $\mathcal{C}_L$ through $X = L(p)$. From Proposition 6.4 there exists a function $\phi_1(p) \in \mathcal{L}_{\nu_0}(R)$ such that

$$\begin{aligned} \phi_1(p_1') &= \phi_1(p_1'') = 0, & \phi_1'(p_1') = \phi_1'(p_1'') = 0,& \\ \phi_1(p_j') &= 1,\ \phi_1(p_j'') = 2, & \phi_1'(p_j') = \phi_1'(p_j'') = 0\ \ (j = 2, \ldots, j_1).& \end{aligned} \tag{6.27}$$

For sufficiently small η_1, set

$$M(p) := L(p) + \eta_1 \phi_1(p) \quad \text{on } R$$

and let $\mathcal{C}_M : X = M(p)$, $p \in R'$, denote the graph of $M(p)$ in $\mathbf{C}^2_{w,X}$. Then the graph $\mathcal{C}_M$ satisfies condition $(*)$, and the set $\mathcal{B}$ of all singular points of $\mathcal{C}_M$ consists of κ_0 normal double points as well as the set $\mathcal{C}_H$. This follows since η_1 is sufficiently small. Furthermore, if we set $\mathcal{B} := \{B_\iota = (\gamma_\iota, \delta_\iota)\}_{\iota=1,\cdots,\kappa_0}$, then (6.27) implies that the number of distinct points among the γ_ι $(\iota = 1, \ldots, \kappa_0)$ in $\mathbf{C}'_w$ is greater than or equal to $\kappa'_0 + 1$.

We continue this procedure and obtain $\phi_s(p) \in \mathcal{L}_{\nu_0}(R)$ $(s = 1, \ldots, s_0)$ with the property that, if we define

$$K(p) := H(p) + \sum_{s=1}^{s_0} \eta_s \, \phi_s(p), \quad p \in R,$$

and we let $\mathcal{C}_K : X = K(p)$, $p \in R'$, denote the graph of $K(p)$ in $\mathbf{C}^2_{w,X}$, then for sufficiently small η_s, the graph $\mathcal{C}_K$ satisfies condition $(*)$ and all singular points $\mathcal{Z}$ of $\mathcal{C}_K$ consist of κ_0 double points. Moreover, if we write $\mathcal{Z} = \{Z_\kappa = (x_\kappa, y_\kappa)\}_{\kappa=1,\ldots,\kappa_0}$, then the points x_κ $(\kappa = 1, \ldots, \kappa_0)$ are distinct. Thus the lemma is proved. □

PROOF OF THEOREM 6.3. Let $\nu > m\nu_0$. Fix any $g(p) \in \mathcal{L}^*_\nu(R)$ satisfying condition (6.5). Then $g(p)$ is a characteristic function on R. Thus, after constructing $G(p) = g(p) + \varepsilon f(p)$ as in (6.25), we can use Lemma 6.8 to obtain $K(p) = G(p) + \sum_{j=1}^{M} \varepsilon_j \phi_j(p)$ on R. This function $K(p)$ for sufficiently small ε and ε_j $(j = 1, \ldots, M)$ satisfies all the conditions of the theorem. □

6.4. Appendix 2

Let $\Lambda = \Delta \times \Gamma \subset \mathbf{C}^n_z \times \mathbf{C}_w$ be a polydisk, where

$$\Delta : \ |z_j| < 1 \ (j = 1, \ldots, n), \qquad \Gamma : \ |w| < 1.$$

We set $z = (z_1, \ldots, z_n) = (z', z_n)$ and $\Delta = \Delta^{(n-1)} \times \Delta_n$, where $\Delta^{(n-1)} = \Delta_1 \times \ldots \times \Delta_{n-1}$ and $\Delta_j := \{z_j : |z_j| < 1\}$. Let Σ be an analytic hypersurface in Λ such that $\Sigma \cap (\Delta \times \partial\Gamma) = \emptyset$. Then there exists a monic pseudopolynomial in w,

$$P(z, w) = w^\nu + a_1(z) w^{\nu-1} + \cdots + a_\nu(z),$$

where $a_j(z)$ $(j = 1, \ldots, \nu)$ is a holomorphic function on Δ, with

$$\Sigma = \{(z, w) \in \Delta \times \mathbf{C}_w \mid P(z, w) = 0\} \tag{6.28}$$

and such that $P(z, w)$ has no multiple factors. We let $d(z) \not\equiv 0$ denote the discriminant of $P(z, w)$ with respect to w, and we set

$$\sigma = \{z \in \Delta \mid d(z) = 0\},$$

which is an analytic hypersurface in Δ. We set $\Delta' = \Delta \setminus \sigma$ and $\Lambda' = \Lambda \setminus \Sigma$. For $z_0 \in \Delta$, we let $\Lambda(z_0), \Lambda'(z_0)$ and $\Sigma(z_0)$ denote the sections of Λ, Λ', and Σ over z_0. We usually identify these sets in $\{z_0\} \times \Gamma$ with sets in the disk Γ; $\Sigma(z_0)$ consists of at most ν distinct points and $\Lambda'(z_0) = \Lambda(z_0) \setminus \Sigma(z_0)$ is a punctured disk with at most ν punctures.

We have the following lemma, which is stated on pp. 68-69 in Picard and Simard [**58**].

LEMMA 6.9. [5] *In the above setting, let $z^* \in \Delta\setminus\sigma$. Then any real 1-dimensional closed curve γ in $\Lambda\setminus\Sigma$ can be continuously (i.e., homotopically) deformed in $\Lambda\setminus\Sigma$ to a closed curve γ^* in $\Lambda'(z^*)$.*

The conclusion is not necessarily true for $z^* \in \sigma$. For example, let $n = 1$; $P(z,w) = w(w - z/2)$; $\gamma : \theta \in [0, 2\pi] \to (z,w) = (1/2, 1/5e^{i\theta})$; and $z^* = 0$. Then γ cannot be continuously deformed in $\Lambda\setminus\Sigma$ to a closed curve in $\Lambda'(0)$.

PROOF OF THE LEMMA. We may assume the following:

(1) $\Delta = \Delta_1 \times \cdots \times \Delta_n$, where Δ_j $(j = 1, \ldots, n)$ is a rectangle $(|x_j| < 1) \times (|y_j| < 1)$ in the complex plane $\mathbf{C}_{z_j}$; $z_j = x_j + iy_j$.

(2) The hypersurface Σ in Λ contains no complex lines of the form $z' = c', w = d$, where $c' = (c_1, \ldots, c_{n-1})$ and d are constant, i.e., the coordinates (z', w, z_n) of $\mathbf{C}^{n+1}$ as well as the coordinates (z', z_n, w) satisfy the Weierstrass condition for Σ.

(3) If $n \geq 2$, we may assume that the hypersurface σ in the rectangle Δ contains no irreducible component of the form $z_n = c$ where c is a constant.

(4) The closed curve γ is a real analytic, one-dimensional closed curve in $\Delta\setminus\Sigma$ of the form $\gamma: \ z' = \phi(s), \ z_n = \chi(s), \ w = \psi(s)$, where $\phi(s), \chi(s), \psi(s)$ are real analytic functions on $(-\infty, +\infty)$ with period 2π and the projection of γ to each axis $x_1, y_1, \ldots, x_n, y_n, u, v$ (where $w = u + iv$) does not reduce to a point. To emphasize the z_n-component, we set $z_n = x + iy$ and $\chi(s) = \chi'(s) + i\chi''(s)$, so that

$$\begin{aligned} \gamma : s \in [0, 2\pi] \to M &= \gamma(s) \\ &= (\phi(s), \chi'(s), \chi''(s), \psi(s)) \in \Lambda\setminus\Sigma. \end{aligned} \tag{6.29}$$

(5) For $M = \gamma(s) \in \gamma$, consider the real one-dimensional segment X_M and the real analytic 2-dimensional set $\mathbf{X}_\gamma$ in the rectangle Δ defined via

$$X_M := \{(\phi(s), x, \chi''(s)) \in \Delta \mid -1 \leq x \leq 1\},$$
$$\mathbf{X}_\gamma := \bigcup_{M \in \gamma} X_M.$$

The set $\mathbf{X}_\gamma$ intersects the complex $(n-1)$-dimensional analytic set σ in Δ in at most a finite number of points; i.e.,

$$\mathbf{X}_\gamma \cap \sigma = \{A_1, \ldots, A_{l_0}\}. \tag{6.30}$$

Thus we can find $a^{(k)} \in [0, 2\pi]$ and $x_h^{(k)} \in (-1, 1)$ $\ (k = 1, \ldots, l_0; h = 1, \ldots, h(k))$ such that $A_k = (\phi(a^{(k)}), x_h^{(k)}, \chi''(a^{(k)}))$ and such that $X_M \cap \sigma = \emptyset$ for each $s \neq a^{(k)}$ $(k = 1, \ldots, l_0)$, where $M = \gamma(s)$.

(6) If $n \geq 2$, let $z^* = ((z^*)', z_n^*) = ((z^*)', x_n^* + iy_n^*)$ be the point in Δ in the lemma. For any fixed $M = \gamma(s) \in \gamma$, we consider the real one-dimensional segment $Y_M(z^*)$ and the real analytic 2-dimensional set $\mathbf{Y}_\gamma$ in Δ defined via

$$Y_M(z^*) := \{(\phi(s), x_n^*, y) \in \Delta \mid -1 \leq y \leq 1\},$$
$$\mathbf{Y}_\gamma := \bigcup_{M \in \gamma} Y_M(z^*).$$

Then $\mathbf{Y}_\gamma$ intersects σ in at most a finite number of points, say

$$\mathbf{Y}_\gamma \cap \sigma = \{B_1, \ldots, B_{p_0}\}. \tag{6.31}$$

[5] This proof follows Oka in his posthumous work No. 2, pp.113-123 in [**55**].

Thus we can find $b^{(k)} \in [0, 2\pi]$ and $y_l^{(k)} \in (-1, 1)$ $(k = 1, \ldots, p_0; l = 1, \ldots, l(k))$ such that $B_{l^k} = (\phi(b^{(k)}), x_n^*, y_l^{(k)})$ and such that $Y_M(z^*) \cap \sigma = \emptyset$ for each $s \neq b^{(k)}$ $(k = 1, \ldots, p_0)$, where $M = \gamma(s)$.

Condition (1) follows from the Riemann mapping theorem. Conditions (2) and (3) are obtained by taking a linear transformation of the coordinates (z, w) of $\mathbf{C}^{n+1}$ as close as we want to the original coordinates according to Lemma 2.9. Conditions (4), (5), and (6) are obtained by taking a small perturbation (under condition (3) in case $n \geq 2$) of the given closed curve γ in Δ, if necessary, since the proof of the lemma remains valid after such a small continuous deformation. Thus it suffices to prove the lemma under the conditions (1)-(6).

In addition, we will use the following facts:

(I) In $\mathbf{R}^3$ with variables (t, u, v), let $D = I \times \Gamma$ be a solid cylinder, where $I = \{|t| < 1\}$ and $\Gamma = \{u^2 + v^2 < 1\}$. Let $\mathcal{L}_j$ $(j = 1, \ldots, \nu)$ be a smooth arc in $\overline{D}$ of the form

$$\mathcal{L}_j: \quad (t, u, v) = (t, u_j(t), v_j(t)), \qquad \text{where } t \in \overline{I} = [-1, 1],$$

and $u_j(t), v_j(t)$ are continuous functions on $[-1, 1]$ with $\mathcal{L}_j \cap (\overline{I} \times \partial\Gamma) = \emptyset$ $(j = 1, \ldots, \nu)$ and $\mathcal{L}_j \cap \mathcal{L}_k = \emptyset$ $(j \neq k)$. We set $\mathcal{L} := \bigcup_{j=1}^{m} \mathcal{L}_j$ and $D' := D \setminus \mathcal{L}$. For $t \in I$, we set $D(t) = \{t\} \times \Gamma$ and $D'(t) = D(t) \setminus \mathcal{L}(t)$; the latter is an m-punctured disk.

In this setting, fix $t_0 \in I$ and let γ be an arc or a closed curve in D' of the form

$$\gamma: \ (t, u, v) = (t(s), u(s), v(s)), \qquad s \in [a, b],$$

such that $t(a) = t(b) = t_0$, i.e., the initial point $\gamma(a)$ and the terminal point $\gamma(b)$ both lie on $D'(t_0)$. Then γ can be continuously deformed in D' to an arc or a closed curve $\tilde{\gamma}$ on $D'(t_0)$ with the same initial and terminal points $\gamma(a)$ and $\gamma(b)$.

In fact, since $\mathcal{L}_i \cap \mathcal{L}_j = \emptyset$ $(i \neq j)$, there exists a homeomorphism $\Phi : D \to \tilde{D} = I \times \tilde{\Gamma}$, where $\tilde{\Gamma} = \{\tilde{u}^2 + \tilde{v}^2 < 1\}$, such that $\Phi(D(t)) = \tilde{D}(t)$ for each $t \in I$, $\Phi(\mathcal{L}_j) = I \times \{(a_j, b_j)\}$ $(j = 1, \ldots, \nu)$, where $(a_i, b_i) \neq (a_j, b_j)$ $(i \neq j)$, and such that $\Phi|_{D(t_0)}$ is the identity mapping. This yields fact (I).

(II) The analytic hypersurface Σ in Λ defined by (6.28) can be written in the form

$$w = \xi(\tilde{z}), \quad \tilde{z} \in \tilde{\Delta},$$

where $\tilde{\Delta}$ is a ν-sheeted ramified domain over the rectangle Δ without relative boundary and $\xi(\tilde{z})$ is a single-valued holomorphic function on $\tilde{\Delta}$. We let π denote the projection from $\tilde{\Delta}$ onto Δ and we let $\mathcal{S}$ denote the branch set of $\tilde{\Delta}$, so that its projection $\underline{\mathcal{S}}$ onto Δ coincides with σ.

We fix $M \in \gamma$ and we choose $s \in [0, 2\pi]$ such that $M = \gamma(s) = (\phi(s), \chi'(s), \chi''(s), \psi(s))$. Using the set $X_M \subset \Delta$, we define the set $\mathcal{X}_M$ in Λ:

$$\mathcal{X}_M := X_M \times \Gamma = \{(\phi(s), x, \chi''(s)) \in \Delta \mid -1 \leq x \leq 1\} \times \Gamma;$$

this is a real three-dimensional solid cylinder in Λ with “left-hand” cover given by $K_M^- := \{(\phi(s), -1, \chi''(s)\} \times \Gamma$ and “right-hand” cover given by $K_M^+ := \{(\phi(s), +1, \chi''(s)\} \times \Gamma$. Then:

(a) The set $\Sigma \cap \mathcal{X}_M$ consists of ν distinct (but not necessarily disjoint) piecewise real analytic arcs $\mathcal{L}_j$ $(j = 1, \dots, \nu)$ in the solid cylinder $\mathcal{X}_M$. The arcs $\mathcal{L}_i$ and $\mathcal{L}_j$ $(i \neq j)$ may intersect at finitely many points. Moreover, each $\mathcal{L}_j$ starts at a point on the left-hand cover K_M^- of the solid cylinder $\mathcal{X}_M$ and terminates at a point on the right-hand cover K_M^+.

(b) For all $M \in \gamma$ except perhaps for a finite number of points, $\mathcal{L}_j$ $(j = 1, \dots, \nu)$ in (a) is a real analytic arc, and $\mathcal{L}_i \cap \mathcal{L}_j = \emptyset$ $(i \neq j)$.

(c) If we let $\underline{\mathcal{L}_j}$ denote the projection of $\mathcal{L}_j$ onto Γ, then $\underline{\mathcal{L}_j}$ does not reduce to a point.

In fact, we have

$$\begin{aligned} \Sigma \cap \mathcal{X}_M &= \{(\phi(s), x, \chi''(s), w) \mid P(\phi(s), x + i\chi''(s), w) = 0,\ |x| < 1\} \\ &= \{w = \xi(\phi(s), x + i\chi''(s)) \mid |x| < 1\}. \end{aligned}$$

We have two cases to consider: either $s \neq a^{(k)}$ for any $k = 1, \dots, l_0$, or $s = a^{(k)}$ for some $k = 1, \dots, l_0$ (where $a^{(k)}$ is defined in condition (5)). In the first case, since $\sigma \cap X_M = \emptyset$, the part of $\tilde{\Delta}$ over X_M consists of ν disjoint segments $\tilde{L}_j$ $(j = 1, \dots, \nu)$. On each $\tilde{L}_j$ $(j = 1, \dots, n)$, if we set $\xi(\tilde{z}) = \xi_j(\phi(s), x, \chi''(s)) \equiv f_j(x)$ for $x \in (-1, 1)$, then $f_i(x) \cap f_j(x) = \emptyset$ $(i \neq j)$ for $x \in (-1, 1)$. Therefore, $\Sigma \cap \mathcal{X}_M$ consists of ν different arcs $\mathcal{L}_j$ $(j = 1, \dots, \nu)$ in the solid cylinder $\mathcal{X}_M$ with $\mathcal{L}_i \cap \mathcal{L}_j = \emptyset$ $(i \neq j)$ and such that each $\mathcal{L}_j$ starts at a point on the left-hand cover K_M^- of the solid cylinder $\mathcal{X}_M$ and terminates at a point on the right-hand cover K_M^+. Thus (b) is proved. In the second case, suppose for simplicity that $s = a^{(1)}$. Then, by an argument similar to the first case, we see that $\Sigma \cap \mathcal{X}_M$ consists of ν different piecewise real analytic arcs $\mathcal{L}_j$ $(j = 1, \dots, \nu)$ in the solid cylinder $\mathcal{X}_M$, where $\mathcal{L}_i$ and $\mathcal{L}_j$ $(i \neq j)$ may intersect each other at finitely many points (corresponding to the points $x_h(1)$ $(h = 1, \dots, h(1))$ which are defined in condition (5)) and each $\mathcal{L}_j$ starts at a point on the left-hand cover K_M^- of $\mathcal{X}_M$ and terminates at a point on the right-hand cover K_M^+. Thus (a) is proved; (c) is clear from condition (2).

Therefore, if we set

$$\widetilde{\mathcal{X}_\gamma} = \bigcup_{M \in \gamma} \mathcal{X}_M$$

and we consider $\widetilde{\mathcal{X}_\gamma}$ as a variation of the solid cylinder $\mathcal{X}_M$ with parameter $M \in \gamma$, then each solid cylinder $\mathcal{X}_M$ with corresponding arcs $\mathcal{L}_j$ $(j = 1, \dots, \nu)$ satisfies the condition in (I) except for at most a finite number of parameter values M.

(III) If $n \geq 2$, using the notation in (6): $z^* = ((z^*)', x_n^* + iy_n^*) \in \Delta'$, $Y_M(z^*) \subset \Delta$ for $M \in \gamma$, we define

$$\begin{aligned} \mathcal{Y}_M(z^*) &= Y_M(z^*) \times \Gamma \\ &= \{(\phi(s), x_n^*, y) \in \Delta \mid -1 \leq y \leq 1\} \times \Gamma, \end{aligned}$$

which is a real, three-dimensional solid cylinder in Λ with "bottom" $H_M^- := \{(\phi(s), x_n^*, -1)\} \times \Gamma$ and "top" $H_M^+ := \{(\phi(s), x_n^*, +1)\} \times \Gamma$. Then:

(a) The set $\Sigma \cap \mathcal{Y}_M(z^*)$ consists of ν different piecewise real analytic arcs $\mathcal{L}_j$ $(j = 1, \dots, \nu)$ in the solid cylinder $\mathcal{Y}_M(z^*)$, where $\mathcal{L}_i$ and $\mathcal{L}_j$ $(i \neq j)$ may intersect each other at finitely many points and each $\mathcal{L}_j$ starts at a point on the bottom H_M^- of the solid cylinder $\mathcal{Y}_M(z^*)$ and terminates at a point of the top H_M^+.

(b) For all $M \in \gamma$ except for a finite number of points, $\mathcal{L}_j$ $(j = 1, \dots, \nu)$ in (a) is a real analytic arc and $\mathcal{L}_i \cap \mathcal{L}_j = \emptyset$ $(i \neq j)$.

(c) The projection $\underline{\mathcal{L}_j}$ of $\mathcal{L}_j$ onto Γ does not reduce to a point.
Therefore, if we set

$$\widetilde{\mathcal{Y}_\gamma} = \bigcup_{M\in\gamma} \mathcal{Y}_M(z^*) \tag{6.32}$$

and we consider $\widetilde{\mathcal{Y}_\gamma}$ as a variation of the solid cylinder $\mathcal{Y}_M(z^*)$ with parameter $M \in \gamma$, then each solid cylinder $\mathcal{Y}_M(z^*)$ with corresponding arcs $\mathcal{L}_j$ $(j = 1, \ldots, \nu)$ satisfies the condition in (I) except for at most a finite number of parameter values M.

This is proved as in (II) using conditions (6) and (2).

Proof of the lemma: Case $n = 1$. In this case we note that $\sigma : d(z) = 0$ consists of a finite number of points $A_k = A_k' + iA_k''$ $(k = 1, \ldots, k_0)$, where A_k', A_k'' are real. We claim that it suffices to prove the lemma in this case under the following assumption:

(#) The point $z^* = x^* + iy^* \in \Delta \setminus \sigma$ in the lemma satisfies $y^* \neq A_k''$ $(k = 1, \ldots, k_0)$.

To prove this claim, we take a point $\widehat{z} = \widehat{x} + i\widehat{y} \in \Delta \setminus \sigma$ such that $\widehat{y} \neq A_k''$ $(k = 1, \ldots, k_0)$. If the lemma were true under assumption (#), then γ could be continuously deformed in $\Lambda \setminus \Sigma$ to a closed curve $\widehat{\gamma}$ in $\Lambda'(\widehat{z})$. We connect $\widehat{z}$ and z^* by an arc ℓ in $\Delta \setminus \sigma$. Since $\bigcup_{z\in\ell} \Lambda'(z)$ is homeomorphic to the product set $\ell \times \Lambda'(\widehat{z})$ with the fibers being preserved, it follows that $\widehat{\gamma}$ (and hence γ) can be continuously deformed in $\Lambda \setminus \Sigma$ to a closed curve γ^* in $\Lambda'(z^*)$. Thus we may proceed under assumption (#).

We divide the proof of case $n = 1$ into three steps.

First step. Let $M = (z_M, w_M) \in \gamma$, where $z_M = x_M + iy_M$ and $w_M = u_M + iv_M$, and set

$$X_M = \{|x| < 1\} \times \{y_M\} \subset \Delta, \quad \mathcal{X}_M = X_M \times \Gamma \subset \Lambda.$$

We can find a real 1-dimensional line segment $L(M)$ in the solid cylinder $\mathcal{X}_M$ passing through the point M such that

(i) $L(M) \cap \Sigma = \emptyset$,
(ii) $L(M) \cap [\{|x| \leq 1\} \times \{y_M\} \times \partial\Gamma] = \emptyset$.

To see this, we consider the line segment $L(M)$ in $\mathcal{X}_M$ passing through M given by

$$L(M) : (x, y, u, v) = (x, y_M, u_M + \alpha(x - x_M), v_M + \beta(x - x_M)) \tag{6.33}$$

where $x \in \overline{I} = [-1, 1]$ and α, β are real numbers. If we let $\underline{L(M)}$ denote the projection of $L(M)$ onto Γ, then condition (ii) means that $\underline{L(M)} \subset\subset \Gamma$. Thus (ii) is satisfied for sufficiently small $|\alpha|, |\beta|$. To choose α, β in order that $L(M)$ satisfies (i), we consider the set $\Sigma \cap \mathcal{X}_M$. As shown in (II), this set consists of ν different piecewise real analytic arcs $\mathcal{L}_j$ $(j = 1, \ldots, \nu)$ in the solid cylinder $\mathcal{X}_M$, where $\mathcal{L}_i$ and $\mathcal{L}_j$ $(i \neq j)$ may intersect and where each $\mathcal{L}_j$ starts at a point of the left-hand cover K_M^- of the solid cylinder $\mathcal{X}_M$ and terminates at a point of the right-hand cover K_M^+; finally, the projection $\underline{\mathcal{L}_j}$ of $\mathcal{L}_j$ onto Δ does not reduce to a point. We set $\mathcal{L} = \bigcup_{j=1}^{\nu} \mathcal{L}_j = \Sigma \cap \mathcal{X}_M$ and $\underline{\mathcal{L}} = \bigcup_{j=1}^{\nu} \underline{\mathcal{L}_j}$.

If $w_M \notin \underline{\mathcal{L}}$, the segment $L(M)$ satisfies condition (i) for sufficiently small α, β. For the second step we exclude the case $\alpha = \beta = 0$.

If $w_M \in \underline{\mathcal{L}}$, there exist points $(x_M, y^{(j)}, w_M) \in \mathcal{L}_j$ for certain j, say $j = 1, \dots, \nu' \leq \nu$. We set $\mathcal{L}' = \bigcup_{j=1}^{\nu'} \mathcal{L}_j$ and $\mathcal{L}'' = \mathcal{L} \setminus \mathcal{L}'$. Since $M \in \gamma$ and $\gamma \cap \Sigma = \emptyset$, we have $y^{(j)} \neq y_M$ $(j = 1, \dots, \nu')$. We choose two real numbers α, β with $(\alpha, \beta) \neq (0,0)$ and such that the slope β/α of the line segment $\underline{L(M)}$ in Γ is not equal to the slope of the tangent line to any $\underline{\mathcal{L}_j}$ $(j = 1, \dots, \nu')$ at the point w_M. Furthermore, if α and β are sufficiently small, then we have $\underline{L(M)} \cap \underline{\mathcal{L}'} = \{w_M\}$ and $\underline{L(M)} \cap \underline{\mathcal{L}''} = \emptyset$. Since there is only one point M of $L(M)$ over w_M and since $M \notin \mathcal{L}_j$ $(j = 1, \dots, \nu')$, it follows that $L(M) \cap \mathcal{L}' = \emptyset$ and hence $L(M) \cap \mathcal{L} = \emptyset$.

We make the following essential step in the proof of the lemma.

Second step. We set $z^* = x^* + iy^* \in \Delta'$ in the lemma under condition (#), and we set

$$Y(z^*) = \{(x^*, y) \in \Delta \mid -1 \leq y \leq 1\}, \qquad \mathcal{Y}(z^*) = Y(z^*) \times \Gamma.$$

Then $\mathcal{Y}(z^*)$ is a three-dimensional solid cylinder in Λ with bottom $H^- := \{(x^*, -1)\} \times \Gamma$ and top $H^+ := \{(x^*, +1)\} \times \Gamma$. We claim that we can continuously deform the curve γ in $\Lambda \setminus \Sigma$ to a closed curve $\tilde{\gamma}$ in $\mathcal{Y}(z^*) \setminus \Sigma$.

To verify this claim, let $M_0 \in \gamma$ and let α_0, β_0 be the constants corresponding to the line segment $L(M_0)$ in (6.33). From the first step, there exists a subarc $[M_0' M_0'']$ of γ which contains M as an interior point such that $L(M)$ satisfies conditions (i) and (ii) for any point $M \in [M_0' M_0'']$ using the same constants α_0, β_0. Since γ is compact in $\Lambda \setminus \Sigma$, it follows that we can find a finite number of points $M_1, \dots, M_q$ on γ such that each subarc $[M_i' M_i'']$ $(i = 1, \dots, q)$ of γ satisfies the above conditions, i.e., for any point $M \in [M_i M_{i+1}]$, the line segment $L(M)$ in (6.33) satisfies (i) and (ii) with the same α_i, β_i, and the union of the subarcs $[M_i M_{i+1}]$ covers γ. We set $M_i : (z_i, w_i) = (x_i, y_i, u_i, v_i)$ $(i = 1, \dots, q)$. Note that $M_{q+1} = M_1$. We may assume

$$y_i \neq A_k'' \qquad (i = 1, \dots, q;\ k = 1, \dots, k_0), \tag{6.34}$$

for if $y_i = A_k''$ for some i and k, we perturb M_i on γ. From condition (4), a slightly modified M_i will satisfy $y_i \neq A_k''$. Since a small deformation will not affect the above situation, we can assume $y_i \neq A_k''$ for each i and k.

Fix $i \in \{1, \dots, q\}$. To each $M = (z_M, w_M) = (x_M, y_M, u_M, v_M) \in [M_i M_{i+1}]$ there corresponds a point $p_i(M)$ on the solid cylinder $\mathcal{Y}(z^*)$ such that

$$p_i(M) = L_i(M)|_{x=x^*} = (x^*, y_M, u_M + \alpha_i(x^* - x_M), v_M + \beta_i(x^* - x_M)).$$

For simplicity we set $p_i(M_i) = p_i'$ and $p_i(M_{i+1}) = p_i''$, and we consider the following arc on the solid cylinder $\mathcal{Y}(z^*)$:

$$[p_i' p_i''] = \{p_i(M) \mid M \in [M_i M_{i+1}]\}.$$

The arc $[M_i M_{i+1}]$ can be continuously deformed in Λ to the arc $[p_i' p_i'']$ in such a manner that $M \in [M_i' M_i'']$ moves to $p_i(M)$ along the line segment $L_i(M)$. Since $L_i(M) \cap \Sigma = \emptyset$, this continuous deformation takes place entirely in $\Lambda \setminus \Sigma$. We note that each point M_i $(i = 1, \dots, q)$ corresponds to two points $p_i' = L_i(M_i)$ and $p_{i-1}'' = L_{i-1}(M_i)$, which both lie in the solid cylinder $\mathcal{X}_{M_i}$. Note that $L_q(M_{q+1}) = L_q(M_1) = p_1''$. In the solid cylinder $\mathcal{X}_{M_i}$ we form an arc λ_i such that λ_i consists of two line segments λ_i' and λ_i'', where λ_i' joins p_{i-1}'' and M_i on the segment $L_{i-1}(M_i)$ and λ_i'' joins M_i and p_i' on the segment $L_i(M_i)$. Note that $p_0'' = p_q''$. By condition

(i) for the segment $L(M)$, we have $\lambda_i \subset \mathcal{X}_{M_i} \setminus \Sigma$. Thus, if we form

$$\widehat{\gamma} := \lambda_1 + [p_1' \, p_1''] + \lambda_2 + \cdots + [p_q' \, p_q''],$$

it follows that $\widehat{\gamma}$ is a closed curve in $\Lambda \setminus \Sigma$ such that γ can be continuously deformed to $\widehat{\gamma}$ in $\Lambda \setminus \Sigma$.

Again fix $i \in \{1, \ldots, q\}$. The initial and terminal points p_{i-1}'' and p_i'' of $\lambda_i = \lambda_i' + \lambda_i''$ lie on $x = x^*$ in Λ. Using (6.34), we have $X_{M_i} \cap \sigma = \emptyset$, so that $\mathcal{L}_i \cap \mathcal{L}_j = \emptyset$ $(i \neq j)$, where $\Sigma \cap \mathcal{X}_{M_i} = \bigcup_{j=1}^{\nu} \mathcal{L}_j$. It follows from (I) that the arc λ_i can be continuously deformed in $\mathcal{X}_{M_i} \setminus \Sigma$ to an arc $\tilde{\lambda}_i$ with the same initial (resp., terminal) point as λ_i in the ν-punctured disk $\Lambda'(x^* + iy_i)$. Thus $\widehat{\gamma}$ can be deformed in $\Lambda \setminus \Sigma$ to a closed curve

$$\tilde{\gamma} := \tilde{\lambda}_1 + [p_1' \, p_1''] + \tilde{\lambda}_2 + \cdots + [p_q' \, p_q''],$$

which lies in $\mathcal{Y}(z^*) \setminus \Sigma$; this proves the second step.

Third step. The closed curve $\tilde{\gamma}$ in $\mathcal{Y}(z^*) \setminus \Sigma$ in the second step can be continuously deformed in $\mathcal{Y}(z^*) \setminus \Sigma$ to a closed curve γ^* in $\Lambda'(z^*)$.

Indeed, since we imposed assumption (#) for the point z^*, the set $\Sigma \cap \mathcal{Y}(z^*)$ consists of ν different arcs $\mathcal{L}_j$ $(j = 1, \ldots, \nu)$ such that $\mathcal{L}_i \cap \mathcal{L}_j = \emptyset$ $(i \neq j)$ and such that each $\mathcal{L}_j$ starts from a point on the bottom H^- of the solid cylinder $\mathcal{Y}(z^*)$ and terminates at a point on the top H^+. Thus, again using (I) for the closed curve $\tilde{\gamma}$, we obtain the third step. Hence the lemma in the case $n = 1$ is completely proved.

Proof of the lemma: Case $n \geq 2$. By taking a linear transformation of $\mathbf{C}_z^n$, if necessary, we we may assume that the point $z^* = (z_1^*, \ldots, z_n^*) = ((z')^*, z_n^*) \in \Delta \setminus \sigma$ in the lemma satisfies the following condition: if we set $\sigma^{(n-1)} = \{z' = (z_1, \ldots, z_{n-1}) \in \Delta^{(n-1)} \mid (z', z_n^*) \in \sigma\}$, then

$$(z')^* \in \Delta^{(n-1)} \setminus \sigma^{(n-1)}. \tag{6.35}$$

First step. Let $M \in \gamma$ and choose $s \in [0, 2\pi]$ such that

$$\begin{aligned} M = \gamma(s) &= (\phi(s), \chi'(s), \chi''(s), \psi(s)) \\ &=: (z_M', x_M, y_M, u_M + iv_M). \end{aligned} \tag{6.36}$$

Using the set $X_M \subset \Delta$ in (5), we set

$$\mathcal{X}_M = X_M \times \Gamma = \{(\phi(s), x, \chi''(s)) \in \Delta \mid -1 \leq x \leq 1\} \times \Gamma,$$

which is a three-dimensional solid cylinder in Λ. We take a line segment $L(M)$ in $\mathcal{X}_M$ passing through the point M of the form

$$L(M) : (z', x, y, u, v) = (z_M', x, y_M, u_M + \alpha(x - x_M), v_M + \beta(x - x_M)),$$

where $-1 \leq x \leq 1$ and α, β are constants, which satisfies
(i) $L(M) \cap \Sigma = \emptyset$,
(ii) $L(M) \cap [\{z_M'\} \times \{|x| \leq 1\} \times \{y_M\} \times \partial\Gamma] = \emptyset$.

This is proved as in the first step of the case $n = 1$, from conditions (2) and (4).

Second step. Let $z^* = ((z')^*, z_n^*) \in \Delta'$, where $z_n^* = x_n^* + iy_n^*$, be the point in the lemma and let

$$\Lambda_{x_n^*} := \Delta^{(n-1)} \times \{x_n^*\} \times \{|y| < 1\} \times \Gamma.$$

Using the notation $\mathcal{Y}_M(z^*)$ and $\widetilde{\mathcal{Y}_\gamma} \subset \Lambda_{x_n^*}$ in (6.32), we claim that we can continuously deform the curve γ in $\Lambda \setminus \Sigma$ to a closed curve $\tilde{\gamma}$ in the set $\widetilde{\mathcal{Y}_\gamma} \setminus \Sigma$, where $\tilde{\gamma} : s \in [0, 2\pi] \to (\phi(s), x_n^*, y(s), u(s), v(s)) \in \Lambda \setminus \Sigma$ and $y(s), u(s), v(s)$ are continuous functions of $s \in [0, 2\pi]$.

This is proved as in the second step of the case $n = 1$, using (II).

Third step. The closed curve $\tilde{\gamma}$ in $\widetilde{\mathcal{Y}_\gamma} \setminus \Sigma$ in the second step can be continuously deformed in $\Lambda_{x_n^*} \setminus \Sigma$ to a closed curve $\widehat{\gamma}$ in

$$\mathbf{Z} := \left(\bigcup_{M \in \gamma} \{(\phi(s), x_n^*, y_n^*)\} \times \Gamma \right) \setminus \Sigma.$$

This is proved by repeating the method used in the first and second steps, using (III) instead of (II).

Forth step. The lemma is true in case $n \geq 2$.

For the curve γ can be continuously deformed in $\Lambda \setminus \Sigma$ to the closed curve $\widehat{\gamma}$ in $\mathbf{Z}$ from the third step. We put

$$\Lambda_{z_n^*} := \Delta^{(n-1)} \times \{z_n^*\} \times \Gamma,$$

so that $\mathbf{Z} \subset \Lambda_{z_n^*} \setminus \Sigma$. Thus, if we set $\Sigma^{(n-1)} = \Sigma \cap \Lambda_{z_n^*}$ and $\Lambda_{z_n^*}$ is identified with $\Delta^{(n-1)} \times \Gamma =: \Lambda^{(n-1)}$, then we have $\widehat{\gamma} \subset \Lambda^{(n-1)} \setminus \Sigma^{(n-1)}$. Therefore, under condition (6.35), the case n reduces to the case $n - 1$. Since the case $n = 1$ was proved, the fourth step follows from induction. □

Now let $\Lambda = \Delta \times \Gamma \subset \mathbf{C}_z^n \times \mathbf{C}_w$, $\Sigma = \{P(z, w) = 0\}$ satisfying condition (6.28), $\sigma = \{d(z) = 0\}$, $\Lambda' = \Lambda \setminus \Sigma$, and $\Delta' = \Delta \setminus \sigma$ be as defined in the beginning of this section. Let $\mathcal{D}$ be a ramified domain over the polydisk Λ without relative boundary such that the projection of the branch set $\mathcal{S}$ of $\mathcal{D}$ onto Δ coincides with Σ. For $z_0 \in \Delta$, we let $D(z_0)$ denote the fiber of $\mathcal{D}$ over $z = z_0$; this is a finitely sheeted Riemann surface over the disk Γ without relative boundary. Let $D'(z_0) := D(z_0) \setminus \Sigma(z_0)$, which is equal to $D(z_0)$ punctured in at most a finite number of points.

We have the following.

COROLLARY 6.3. *Let $z_0 \in \Delta'$. Any closed curve γ in $\mathcal{D} \setminus \mathcal{S}$ can be continuously deformed in $\mathcal{D} \setminus \mathcal{S}$ to a closed curve $\tilde{\gamma}$ in the fiber $D'(z_0)$.*

PROOF. By taking a small continuous deformation of γ in $\mathcal{D} \setminus \mathcal{S}$ we may assume that the projection $\underline{\gamma}$ of γ onto Δ satisfies $\underline{\gamma} \cap \sigma = \emptyset$. Since $z_0 \in \Delta'$, it follows from the above lemma that $\underline{\gamma}$ can be continuously deformed in $\Lambda \setminus \Sigma$ to a closed curve τ in $\Lambda'(z_0)$. We write this deformation as $t \in [0, 1] \to \tau(t)$ so that $\tau(0) = \underline{\gamma}$ and $\tau(1) = \tau$. This deformation uniquely induces a continuous deformation of the closed curves $\gamma(t)$ in the unramified domain $\mathcal{D} \setminus \mathcal{S}$ over Δ, where $\gamma(0) = \gamma$ and $\underline{\gamma(t)} = \tau(t)$ for all $t \in [0, 1]$. Since $\gamma(1)$ lies on the fiber $D'(z_0)$, we obtain the corollary. □

The following corollary will be used in Chapter 9.

COROLLARY 6.4. *Using the same notation as in the above corollary, assume that $\Sigma(0)$ consists of a single point, which we take to be the origin 0 in Γ, i.e., the equation $P(0, w) = 0$ has the unique solution $w = 0$ of order ν. Let $\zeta = f(z, w)$ be* *a non-vanishing holomorphic function on $\mathcal{D}$. Then there is a single-valued branch of the function* $\log f(z, w)$ *defined on $\mathcal{D}$.*

PROOF. Fix $z_0 \in \Delta'$. Let γ be a closed curve in $\mathcal{D} \setminus \mathcal{S}$ and let $C = f(\gamma)$. This is a closed curve in the complex plane $\mathbf{C}_\zeta$ such that the origin 0 is not in C. We let N denote the winding number of C about 0, and our first claim is that $N = 0$. By the lemma, γ can be continuously deformed in $\mathcal{D} \setminus \mathcal{S}$ to a closed curve $\gamma(z_0)$ on the Riemann surface $D'(z_0)$. This Riemann surface is finitely sheeted and is punctured in at most a finite number of points p_j $(j = 1, \dots, \mu)$; moreover the points $\{p_j\}_j$ lie over $\Sigma(z_0)$. We set $C(z_0) = f(\gamma(z_0))$; this is a closed curve in $\mathbf{C}_\zeta \setminus \{0\}$ whose winding number $N(z_0)$ about 0 is equal to N (independent of $z_0 \in \Delta'$). We may assume that the projection $\underline{\gamma(z_0)}$ of $\gamma(z_0)$ onto Γ lies over the disk $\delta_\varepsilon(z_0)$ centered at 0 and of radius $\varepsilon = \varepsilon(z_0)$ in Γ, where $\max_{j=1,\dots,\mu}\{|\underline{p_j}|\} < \varepsilon < 1$ and ε is as close to this maximum as we like. Under the assumption that $P(0, w) = 0$ has only the solution $w = 0$ of order ν, we have $\Sigma(z_0) \to \{0\}$ in Λ as $z_0 \to 0$ in Δ'. Hence we can take $\varepsilon = \varepsilon(z_0)$ such that $\delta_\varepsilon(z_0) \to 0$ as $z_0 \to \{0\}$. Moreover, if we let $\tilde{\delta}_\varepsilon(z_0)$ denote the connected component of the part of $D(z_0)$ over $\delta_\varepsilon(z_0)$ which contains $\gamma(z_0)$, then $\tilde{\delta}_\varepsilon(z_0)$ converges in $\mathcal{D}$ to a point $q_0 \in D(0)$ over $w = 0$. Thus, $C(z_0) \to f(q_0) \neq 0$ as $z_0 \to 0$, and hence $N(z_0) \to 0$ as $z_0 \to 0$, so that $N = 0$. From this claim, it follows that there exists a single-valued branch of the holomorphic function $\log f(z, w)$ on $\mathcal{D} \setminus \mathcal{S}$. Since this function is bounded there, it follows from Riemann's theorem on removable singularities that $\log f(z, w)$ extends to a holomorphic function on all of $\mathcal{D}$. □

CHAPTER 7

Analytic Sets and Holomorphic Functions

7.1. Holomorphic Functions on Analytic Sets

7.1.1. Holomorphic Functions on Analytic Sets. Chapters 7 and 8 will be devoted to establishing the lifting principle for analytic polyhedra in an analytic space.[1] Let D be a domain in $\mathbf{C}^n$ with variables $z_1, \dots, z_n$. Let Σ be an analytic set in D and let $v \subset \Sigma$. We say that v is an open set in Σ if there exists an open set δ in $\mathbf{C}^n$ such that $v = \delta \cap \Sigma$. Let $p \in \Sigma$. An open set v in Σ containing the point p is called a neighborhood of p in Σ. Let $\phi(z)$ be a function defined on $\delta \subset \mathbf{C}^n$ and let $v = \delta \cap \Sigma$. We let $\phi(z)|_v$ denote the restriction of $\phi(z)$ to v.

We shall define holomorphic functions on the analytic set Σ as follows. First, let v be an open set in Σ and let $f(p)$ be a complex-valued function on v. Fix $q \in v$. If we can find an open neighborhood δ_q of q in $\mathbf{C}^n$ and a holomorphic function $\phi(z)$ in δ_q such that the restriction $\phi(z)|_{v_q}$, where $v_q := \delta_q \cap \Sigma \subset v$, coincides with $f(p)$ on v_q, then we say that $f(p)$ is **holomorphic at** q on v. If $f(p)$ is holomorphic at each point $q \in v$, then we say that $f(p)$ is holomorphic on v.

Let Σ be a pure r-dimensional analytic set in $D \subset \mathbf{C}^n$ and let $(z_1, \dots, z_n)$ be coordinates of $\mathbf{C}^n$ which satisfy the Weierstrass condition at each point of Σ. Recall this means that if we project Σ over the space $\mathbf{C}^r$ comprised of the first r complex variables $(z_1, \dots, z_r)$ and we denote the image of Σ by $\mathcal{D}$, then $\mathcal{D}$ is a ramified domain over $\mathbf{C}^r$ and Σ can be described as

$$z_j = \xi_j(z_1, \dots, z_r) \qquad (j = r+1, \dots, n),$$

where $(z_1, \dots, z_r)$ lie in the ramified domain $\mathcal{D}$ and each $\xi_j(z_1, \dots, z_r)$ $(j = r + 1, \dots, n)$ is a single-valued holomorphic function on $\mathcal{D}$. Note that Σ and $\mathcal{D}$ are one-to-one except for an analytic set of dimension at most $r - 1$. If Σ has no singular points in D, then the projection $\pi : \Sigma \to \mathcal{D}$ gives a bijection between Σ and $\mathcal{D}$. In this case, any open set V in $\mathcal{D}$ corresponds to an open set v in Σ where $\pi(v) = V$, and conversely. Furthermore, for any holomorphic function $F(q)$ on V, the function $f(p) := F(\pi(p))$ is holomorphic on v; and for any holomorphic function $g(p)$ on v, the function $G(q) := g(\pi^{-1}(q))$ is holomorphic on V. Thus, in the case when Σ has no singular points in D, the holomorphic functions $F(q)$ on $V \subset \Sigma$ and the holomorphic functions $f(p)$ on $v \subset \mathcal{D}$ are in one-to-one correspondence through the projection π. However, if Σ has singular points in D, this is not necessarily the case.

---

[1]This problem was first solved by K. Oka [**50**], [**51**]. After that, H. Cartan created sheaf theory from Oka's method. In this book we develop the lifting principle using Oka's original ideas. In the textbooks [**25**], [**26**] H. Grauert and R. Remmert explain the lifting principle by means of sheaf theory.

EXAMPLE 7.1. Consider $\mathbf{C}^2$ with variables z and w, and the analytic hypersurface Σ in $\mathbf{C}^2$ defined by the equation

$$w^2 - z(z-1)^2 = 0.$$

We project Σ over the complex plane $\mathbf{C}_z$. This projection of Σ can be identified with the Riemann surface $\mathcal{R}$ determined by $\sqrt{z}$; thus we write $\pi : \Sigma \to \mathcal{R}$. Note Σ has a singularity at $(1,0)$, and Σ and $\mathcal{R}$ are not bijective. Furthermore, $\sqrt{z}$ is a (single-valued) holomorphic function on $\mathcal{R}$, but the corresponding function $f(p) := \sqrt{\pi(p)}$ is not continuous at the point $(1,0)$, so it is not holomorphic at $(1,0)$ on Σ.

EXAMPLE 7.2. We consider the analytic hypersurface Σ in $\mathbf{C}^2$ defined by

$$w^2 - z^3 = 0.$$

We project Σ over the complex plane $\mathbf{C}_z$; again, this projection of Σ can be identified with the Riemann surface $\mathcal{R}$ determined by $\sqrt{z}$, and we write $\pi : \Sigma \to \mathcal{R}$. The hypersurface Σ has a singularity at $(0,0)$, and Σ and $\mathcal{R}$ *are* bijective in this case. Again, $\sqrt{z}$ is a (single-valued) holomorphic function on $\mathcal{R}$, but the corresponding function $f(p) := \sqrt{\pi(p)}$ is not holomorphic at the point $(0,0)$ on Σ.

We prove this by contradiction, thus we assume that $f(z)$ is holomorphic in a neighborhood of $(0,0)$ on Σ. Thus there exists a holomorphic function $\phi(z,w)$ defined on a neighborhood of $(0,0)$ in $\mathbf{C}^2$ such that $\phi(z, z^{3/2}) = \sqrt{z}$ for z in a neighborhood of $z = 0$ in $\mathbf{C}_z$. This is impossible as can be seen from the Taylor expansion of $\phi(z,w)$ about $(0,0)$ in $\mathbf{C}^2$ and from the uniqueness of the Puiseux series.

7.1.2. Weakly Holomorphic Functions on Analytic Sets. We next introduce another notion of holomorphy of functions defined on an analytic set Σ in a domain D in $\mathbf{C}^n$. Let S be the set of singular points of Σ and set $\Sigma' := \Sigma \setminus S$. Let v be an open set in Σ and set $v' := v \cap \Sigma'$. Let $f(p)$ be a complex-valued function on v'. If $f(p)$ is holomorphic on v' and if $f(p)$ is bounded in a neighborhood of each point $q \in S \cap v$ in Σ, then we say that $f(p)$ is a **weakly holomorphic** function on $v \subset \Sigma$. The condition that $f(p)$ be bounded in a neighborhood of $q \in S \cap v$ means that there exists a neighborhood u of q in v such that $f(p)$ is bounded in $v' \cap u$. We also say that a function $f(p)$ is weakly holomorphic at a point $q \in \Sigma$ if there exists a neighborhood $v \subset \Sigma$ of q on which $f(p)$ is weakly holomorphic.

Let Σ be a pure r-dimensional analytic set in D and, as in the previous section, consider the ramified domain $\mathcal{D}$ over $\mathbf{C}^r$ given by the image of Σ by the projection π to $\mathbf{C}^r$. Let S be the set of singular points of Σ; thus S is an analytic set in D of dimension at most $r-1$. We set $\Sigma' := \Sigma \setminus S$ and $\mathcal{D}' := \pi(\Sigma') \subset \mathcal{D}$. Thus $\mathcal{D}'$ and Σ' are bijective via π. For any open set v in Σ, we set $V := \pi(v) \subset \mathcal{D}$, which is a ramified domain over $\mathbf{C}^r$. We set $v' := v \setminus S$ and $V' := \pi(v')$. Then for any holomorphic function $F(q)$ on V, the function $f(p) := F(\pi(p))$ for $p \in v'$ clearly defines a weakly holomorphic function on v. Conversely, let $f(p)$ be a weakly holomorphic function on v. We claim that the function $F(q) := f(\pi^{-1}(q))$ for $q \in V'$ can be uniquely extended as a holomorphic function on V.

To verify this last statement, let $p_0 \in V \setminus V'$. We take a singular point $z^0 = (z_1^0, \ldots, z_n^0)$ of v such that $\pi(z_0) = p_0$, and a polydisk $\delta = \delta^r \times \delta^{n-r}$ (where $\delta^r \subset \mathbf{C}^r$) centered at z_0 such that each irreducible component v_0^j $(j = 1, \ldots, l)$ of $v \cap \delta$ passing through z_0 is bijective to $V_0^j := \pi(v_0^j) \subset V$ and $v_0^j \cap (\delta^r \cap \partial\delta^{n-r}) = \emptyset$. Thus, if we put $v_0 = \bigcup_{j=1}^l v_0^j$ and $V_0 := \bigcup_{j=1}^l V_0^j$, then V_0 is a finitely-sheeted ramified domain over the polydisk δ^r centered at $(z_1^0, \dots, z_r^0)$ without relative boundary, and $F(q)$ is a bounded holomorphic function on V_0 except for an at most $(r-1)$-dimensional set $\pi(v_0 \cap S)$. We let m denote the number of sheets of V_0 over δ^r. By the standard method of taking symmetric functions of branches of $F(q)$, we see that $F(q)$ coincides with the solution $w = \xi(z') = \xi(z_1, \dots, z_r)$ of the equation of a monic polynomial in w,

$$P(z', w) = w^m + a_1(z')w^{m-1} + \cdots + a_m(z') = 0,$$

where each $a_j(z')$ is a holomorphic function in δ^r. Here we use the boundedness of $F(q)$. Since $\xi(z')$ is a holomorphic function on V_0, we have the result.

Thus we obtain the following

REMARK 7.1. For an open set $v \subset \Sigma$ and the projection $V = \pi(v) \subset \mathcal{D}$, we get a one-to-one correspondence between the family of all weakly holomorphic functions $f(p)$ on v and the family of all holomorphic functions $F(q)$ on V with $F(\pi(p)) = f(p)$.

This remark, in conjuction with Hartogs' theorem (Theorem 4.1), implies the following result, which will be useful later when combined with Theorem 6.4 of Chapter 6.

REMARK 7.2. Let D be a domain in $\mathbf{C}^n$ and let Σ be a pure r-dimensional analytic set in D. Let σ be an analytic set in D such that $\sigma \subset \Sigma$ and σ is of dimension at most $r-2$. Then any weakly holomorphic function $f(z)$ on $\Sigma \setminus \sigma$ can be extended to a weakly holomorphic function on all of Σ.

PROOF. Let $z_0 \in \sigma$. We may assume the coordinate system $z = (z_1, \dots, z_r, z_{r+1}, \dots, z_n) = (z', z_{r+1}, \dots, z_n)$ satisfies the Weierstrass condition for Σ at z_0, so that there is a polydisk $\Lambda := \Delta \times \Gamma \subset D$ centered at $z_0 = (z_1^0, \dots, z_r^0, z_{r+1}^0, \dots, z_n^0)$ such that $\Delta \subset \mathbf{C}^r_{z_1, \dots, z_r}$, $\Gamma \subset \mathbf{C}^{n-r}_{z_{r+1}, \dots, z_n}$, and $\Sigma \cap (\Delta \times \partial\Gamma) = \emptyset$. Thus $\Sigma \cap \Lambda$ may be written in the form

$$z_j = \xi_j(z_1, \dots, z_r) \quad (j = r+1, \dots, n),$$

where $\tilde{z}' = (z_1, \dots, z_r)$ varies over a finitely sheeted ramified domain $\tilde{\Delta}$ over Δ without relative boundary.

We let m denote the number of sheets of $\tilde{\Delta}$ over Δ and we let S denote the branch set of $\tilde{\Delta}$. We also write $\underline{S}$ for the projection of S onto Δ, and we let $\underline{\sigma}$ denote the projection of $\sigma \cap \Lambda$ onto Δ. Thus $\underline{S}$ is of dimension $r-1$ and $\underline{\sigma}$ is of dimension at most $r-2$. We set $\Delta_1 = \Delta \setminus (\underline{S} \bigcup \underline{\sigma})$ and $\tilde{\Delta}_1 = \tilde{\Delta}$ over Δ_1. Therefore, the weakly holomorphic function $f(z)$ on $\Sigma \setminus \sigma$ gives rise to a holomorphic function $F(\tilde{z}')$ on $\tilde{\Delta}_1$ by the relation

$$f(z', \xi_{r+1}(z'), \dots, \xi_n(z')) = F(\tilde{z}').$$

Let $\zeta' \in \Delta_1$ and fix a ball δ centered at ζ' in Δ_1. Then the function $F(\tilde{z}')$ for $z' \in \delta$ defines m holomorphic functions $F_j(z')$ $(j = 1, \dots, m)$ on δ. If we construct the function

$$\begin{aligned} P(z', X) &:= (X - F_1(z')) \cdots (X - F_m(z')) \\ &= X^m + a_1(z')X^{m-1} + \cdots + a_m(z') \quad \text{on } \delta \times \mathbf{C}_X, \end{aligned}$$

then each $a_j(z')$ $(j = 1, \dots, m)$ can be extended to a single-valued holomorphic function on Δ_1. Let $\xi' \in \underline{S} \setminus \underline{\sigma}$. Since $f(z)$ is weakly holomorphic on $\Sigma \setminus \sigma$, it follows that $a_j(z')$ $(j = 1, \dots, m)$ is bounded in a neighborhood δ' of $\xi' \in \Delta \setminus \underline{\sigma}$, so that $a_j(z')$ has an extension as a holomorphic function on $\Delta \setminus \underline{\sigma}$. Since $\underline{\sigma}$ is of dimension at most $r - 2$ in the r-dimensional polydisk Δ, it follows from Hartogs' theorem that $a_j(z')$ $(j = 1, \dots, m)$ has an extension as a holomorphic function on Δ, so that $P(z', X)$ is a monic pseudopolynomial in X whose coefficients are holomorphic functions on all of Δ. Thus $F(z')$ can be extended to a holomorphic function $\tilde{F}(\tilde{z})$ on the ramified domain $\tilde{\Delta}$ by use of the relation $P(z', X) = 0$ (cf., Example 6.1). This means that $f(z)$ can be extended to a weakly holomorphic function $\tilde{f}(z)$ on $\Sigma \cap \Lambda$ by means of the relation $\tilde{f}(z', \xi_{r+1}(z'), \dots, \xi_n(z')) = \tilde{F}(z')$ for $\tilde{z}' \in \tilde{\Delta}$. □

Let Σ be an analytic set in a domain D in $\mathbf{C}^n$ and let $q \in \Sigma$. If every function f which is weakly holomorphic at the point q is holomorphic at q, i.e., there exists a holomorphic function $F(z)$ defined in a neighborhood U of q in $\mathbf{C}^n$ such that $F|_{U \cap \Sigma} = f(p)$, then we say that Σ is **normal** at q, or the point q is a **normal point** of Σ. If Σ is normal at each point of Σ, then we say that Σ is a **normal analytic set** in D. Clearly any non-singular point of Σ is a normal point of Σ; however, it may also happen that a singular point of Σ is a normal point of Σ.

EXAMPLE 7.3. In $\mathbf{C}^3$ with variables z_1, z_2 and w, we consider the analytic hypersurface Σ defined by the equation

$$w^2 - z_1 z_2 = 0.$$

Then the origin O in $\mathbf{C}^3$ is the only singular point of Σ, and it is a normal point of Σ.

To prove this, let $\mathcal{D}$ denote the projection of Σ onto $\mathbf{C}^2$ with variables z_1, z_2: $\mathcal{D} = \pi(\Sigma)$. Thus $\mathcal{D}$ is a two-sheeted ramified domain over $\mathbf{C}^2$ determined by $\sqrt{z_1 z_2}$ and the branch set $\mathcal{L}$ of $\mathcal{D}$ lies over $L := \pi(\mathcal{L}) = \Delta \cap (\{z_1 = 0\} \bigcup \{z_2 = 0\})$. Let $f(p)$ be a weakly holomorphic function defined on an open neighborhood v of the origin O in Σ. We let $V \subset \mathcal{D}$ denote the open set which corresponds to v. By taking a smaller set V if necessary, we can assume that V is a two-sheeted ramified domain over a polydisk Δ centered at $(0,0)$ in $\mathbf{C}^2$ without relative boundary. Thus over each $(z_1, z_2) \in \Delta \setminus L$, we can find two points $p_1(z_1, z_2)$ and $p_2(z_1, z_2)$ in V. If we set

$$f_j(z_1, z_2) := f(z_1, z_2, p_j(z_1, z_2)) \qquad (j = 1, 2),$$

then $f_j(z_1, z_2)$ becomes a (single-valued) holomorphic function on the ramified domain V with the property that if (z_1, z_2) traces a closed curve in $\Delta \setminus L$ in such a manner that $p_1(z_1, z_2)$ moves in a continuous fashion to $p_2(z_1, z_2)$, then $f_1(z_1, z_2)$ continuously varies to $f_2(z_1, z_2)$. Thus the pair of functions $f_j(z_1, z_2)$ $(j = 1, 2)$ has the same behavior as the pair $\pm\sqrt{z_1 z_2}$. It follows that if we define

$$a(z_1, z_2) := \frac{f_1(z_1, z_2) - f_2(z_1, z_2)}{2\sqrt{z_1 z_2}},$$

$$b(z_1, z_2) := \frac{f_1(z_1, z_2) + f_2(z_1, z_2)}{2},$$

then $a(z_1, z_2)$ and $b(z_1, z_2)$ are single-valued on all of Δ except perhaps for the complex lines $z_1 = 0$ and $z_2 = 0$. However, it is clear that $b(z_1, z_2)$ is holomorphic on Δ. Moreover, since $f_1 = f_2$ on L except for the origin $(0,0)$, we see that $a(z_1, z_2)$

is holomorphic in $\Delta \setminus \{(0,0)\}$, and hence on all of Δ from Osgood's theorem. Thus we can define the holomorphic function

$$F(z_1, z_2, w) := a(z_1, z_2)\, w + b(z_1, z_2)$$

in $\Delta \times \mathbf{C}_w$. We have $f(p) = F(z_1, z_2, w)|_v$ and hence the origin O is a normal point of Σ.

In the case when Σ is an analytic hypersurface in $D \subset \mathbf{C}^n$ we have the following fact.

REMARK 7.3. Let Σ be an analytic *hypersurface* in a domain $D \subset \mathbf{C}^n$. If each point of Σ except for an analytic set σ of dimension at most $n-3$ is a normal point of Σ, then Σ is a normal analytic set in D.

To prove this, fix $p_0 \in \sigma$ and let $f(p)$ be a weakly holomorphic function on an open neighborhood $v \subset \Sigma$ of p_0. For simplicity, we set $p_0 = 0$ in $\mathbf{C}^n$ and we choose Euclidean coordinates $(z_1, \ldots, z_n)$ which satisfy the Weierstrass condition for σ at 0. We take a polydisk $\Delta = \Delta_1 \times \cdots \times \Delta_n$ centered at 0 in $\mathbf{C}^n$ and a holomorphic function $\phi(z)$ in Δ such that $\Sigma \cap \Delta = \{\phi(z) = 0\}$. Since $\dim \sigma \leq n-3$, we can find a polydisk of the form $\Delta' = (\Delta_1' \times \Delta_2' \times \Delta_3') \times (\Delta_4 \times \cdots \times \Delta_n)$ centered at 0, such that $\Delta_i' \subset\subset \Delta_i$ $(i = 1,2,3)$ and such that $\Delta^0 := \Delta \setminus \Delta'$ does not intersect σ. By assumption, for each point $p \in \Sigma \cap \Delta^0$ we can find a neighborhood $\delta_p \subset \Delta^0$ of p and a holomorphic function $F_p(z)$ in δ_p such that $F_p(z)|_{\delta_p \cap \sigma} = f(p)$ on $\delta_p \cap \sigma$. We define a Cousin I distribution $\mathcal{C} = \{(g_p, \delta_p)\}$ on Δ^0 as follows: for $p \in \Sigma \cap \Delta^0$, we take the above neighborhood δ_p of p and the meromorphic function $g_p(z) = F_p(z)/\phi(z)$ in δ_p, and, for $p \in \Delta^0 \setminus \Sigma$, we take a neighborhood δ_p of p with $\delta_p \cap \Sigma = \emptyset$ and set $g_p(z) \equiv 1$. From Lemma 3.5 (Cartan) there is a solution $G(z)$ of the Cousin I problem for $\mathcal{C}$ in Δ^0. If we define $F(z) := G(z)\phi(z)$, then $F(z)$ is a holomorphic function on Δ^0 and $F(z)|_{\Sigma \cap \Delta^0} = f(p)$ on $\Sigma \cap \Delta^0$. Since $F(z)$ can be holomorphically extended to Δ by Osgood's theorem, we see that $F(z)|_{\Delta \cap \Sigma} = f(p)$ on $\Delta \cap \Sigma$.

7.1.3. Liftings of Analytic Sets. To treat analytic sets in a simpler fashion, we introduce two types of liftings of such sets. Let D be a domain in $\mathbf{C}^n$ with variables $z_1, \ldots, z_n$, and let Σ be an analytic set in D.

Lifting of the first kind Let $\varphi_j(p)$ $(j = 1, \ldots, m)$ be weakly holomorphic functions on Σ. Using the variables $w_1, \ldots, w_m$ for $\mathbf{C}^m$, we consider the product domain $\Lambda = D \times \mathbf{C}^m \subset \mathbf{C}^{n+m}$. In the domain Λ we consider the set

$$E: \; w_j = \varphi_j(p) \qquad (p \in \Sigma, \; j = 1, \ldots, m).$$

The closure $\Sigma^0 := \overline{E}$ in Λ is an analytic set in Λ. We call the analytic set Σ^0 in Λ a **lifting of the first kind** of the analytic set Σ in D through $\varphi_j(p)$ $(j = 1, \ldots, m)$.

The projection $\widetilde{\pi}$ from $\mathbf{C}^{n+m}$ to $\mathbf{C}^n$ induces a projection from Σ^0 onto Σ, which we denote by π_0,

$$\pi_0: \; \Sigma^0 \subset \mathbf{C}^{n+m} \to \Sigma \subset \mathbf{C}^n.$$

If p is a non-singular point of Σ in D, then each q in $\pi_0^{-1}(p)$ is also a non-singular point of Σ^0 in Λ. We let σ be the set of singular points of Σ in D, and we set $\sigma^0 := \pi_0^{-1}(\sigma)$. Note it may occur that each point of $\pi_0^{-1}(p_0)$ is a non-singular point of Σ^0 for some $p_0 \in \sigma$.

The projection π_0 gives a bijection between $\Sigma^0 \setminus \sigma^0$ and $\Sigma \setminus \sigma$. Thus, from the definition of a weakly holomorphic function on an analytic set, for each weakly holomorphic function $f(p)$ at p_0 on Σ we get a weakly holomorphic function $\widetilde{f}(q)$ at a point q_0 in $\pi_0^{-1}(p_0) \subset \Sigma^0$ by setting $\widetilde{f}(q) := f(\pi_0(q))$. The converse is also true; i.e., the family $\mathcal{W}_\Sigma$ of all weakly holomorphic functions on Σ coincides with the family $\mathcal{W}_{\Sigma^0}$ of all weakly holomorphic functions on Σ^0 via the projection π_0. Under this correspondence $f(p) \to \tilde{f}(q)$, if f is holomorphic at a point p_0 in Σ, then $\tilde{f}$ is holomorphic at each point of $\pi_0^{-1}(p_0)$ in Σ^0. Moreover, it may occur that every function f which is weakly holomorphic at p_0 in Σ corresponds to the holomorphic function $\tilde{f}$ at each point of $\pi_0^{-1}(p_0)$ in Σ^0; i.e., even if p_0 is not a normal point of Σ, each point in $\pi_0^{-1}(p_0)$ may be a normal point of Σ^0.

If Σ is pure r-dimensional, then the irreducible components of Σ in D correspond in a one-to-one manner to the irreducible components of Σ^0 in Λ. Moreover, if we choose coordinates $z = (z_1, \dots, z_r, z_{r+1}, \dots, z_n)$ which satisfy the Weierstrass condition for the analytic set Σ, then the coordinates (z, w) satisfy the Weierstrass condition for Σ^0. The projection $\mathcal{D}$ of Σ over $\mathbf{C}^r_{z_1,\dots,z_r}$ coincides with that of Σ^0 as a ramified domain over $\mathbf{C}^r_{z_1,\dots,z_r}$, and $\mathcal{W}_\Sigma$ $(\mathcal{W}_{\Sigma^0})$ *can be identified with the family of all holomorphic functions on* $\mathcal{D}$.

From Example 6.3 we see that there exists an analytic set σ in a domain $D \subset \mathbf{C}^n$ such that, for some point $q \in \sigma$, there do not exist a neighborhood δ of q in D and a lifting of the first kind $\tilde{\sigma}$ of $\sigma \cap \delta$ in a domain $\tilde{\delta} \subset \mathbf{C}^{n+\nu}$ with the property that if we set $\pi_0 : \tilde{\sigma} \to \sigma$, then $\tilde{\sigma}$ is non-singular at any point of $\pi_0^{-1}(q)$. However, we shall show in section 8.2 that for any analytic set σ in D and any point $q \in \sigma$, we can construct a lifting of the first kind $\tilde{\sigma}$ in $\tilde{\delta}$ of $\sigma \cap \delta$ such that $\tilde{\sigma}$ is normal at each point $\pi_0^{-1}(q)$.

Lifting of the second kind We decompose Σ into $\Sigma := \Sigma_0 \cup \cdots \cup \Sigma_r$ $(r < n)$, where each Σ_k $(k = 0, 1, \dots, r)$ is a pure k-dimensional analytic set in D. We introduce $\mathbf{C}^r$ with variables $u_1, \dots, u_r$ and the product space $\Lambda = D \times \mathbf{C}^r$. For $k = 0, 1, \dots, r$ we define the k-dimensional hyperplane

$$H_k : \ u_j = 0 \quad (j = k+1, \dots, r) \qquad \text{in } \mathbf{C}^r,$$

where by convention we set $H_r := \mathbf{C}^r$. In Λ we define

$$\Sigma_k^* = \Sigma_k \times H_{r-k} \ \ (k = 0, 1, \dots, r), \qquad \Sigma^* = \Sigma_0^* \cup \cdots \cup \Sigma_r^*.$$

In case $\Sigma_k = \emptyset$, we set $\Sigma_k \times H_{r-k} = \emptyset$. Then Σ^* is a pure r-dimensional analytic set in Λ, and $\Sigma^* \cap (D \times \{(0, \dots, 0)\}$ in Λ can be identified with Σ in D. We call the analytic set Σ^* in Λ a **lifting of the second kind** of the analytic set Σ in D.

The projection $\widetilde{\pi}$ from $\mathbf{C}^{n+r}$ to $\mathbf{C}^n$ induces a projection from Σ^* onto Σ, which will be denoted by π_*,

$$\pi_* : \ \Sigma^* \subset \mathbf{C}^{n+r} \to \Sigma \subset \mathbf{C}^n.$$

If $p \in \Sigma$ is a non-singular point of Σ in D, then each point q in $\pi_*^{-1}(p)$ is also a non-singular point of Σ^* in Λ. Conversely, if $q \in \Sigma^*$ is a non-singular point of Σ^*, then $\pi_*(q)$ is a non-singular point of Σ. We note that for any weakly holomorphic function $f(p)$ at a point p_0 on Σ, the function $\widetilde{f}(q) := f(\pi_*(q))$ is weakly holomorphic at each point $q_0 \in \pi_*^{-1}(p_0)$ on Σ^*. Conversely, if $\widetilde{f}(q)$ is a weakly holomorphic function on Σ^*, then $\widetilde{f}|_\Sigma$ becomes a weakly holomorphic function on Σ.

7.2. Universal Denominators

7.2.1. Weierstrass Theorem. Let $\mathbf{C}^{n+1} = \mathbf{C}^n \times \mathbf{C}_w$ with variables $z_1, \dots,$ z_n and w. Let $D \subset \mathbf{C}^n$ be a domain and let

$$F(z,w) = w^l + a_1(z)w^{l-1} + \cdots + a_l(z)$$

be a monic polynomial in w such that $a_i(z)$ $(i = 1, \dots, l)$ is a holomorphic function in D. We do not assume that $F(z,w)$ is irreducible, but we do assume that $F(z,w)$ has no multiple factors. We set $\Lambda := D \times \mathbf{C}_w \subset \mathbf{C}^{n+1}$ and consider the analytic hypersurface

$$\Sigma : \ F(z,w) = 0 \qquad \text{in } \Lambda.$$

We note that $(\partial F/\partial w)(z,w) \not\equiv 0$ on each irreducible component of Σ.

Then we have the following proposition concerning the representation of weakly holomorphic functions on Σ.

PROPOSITION 7.1. *Let $\phi(p)$ be a weakly holomorphic function on the analytic hypersurface Σ. Then there exists a unique pseudopolynomial $\Phi(z,w)$ in w of degree at most $l-1$,*

$$\Phi(z,w) = A_0(z)w^{l-1} + \cdots + A_{l-1}(z),$$

where each $A_i(z)$ $(i = 0, 1, \dots, l-1)$ is a holomorphic function on D, such that

$$\phi(p) = \frac{\Phi(z,w)}{(\partial F/\partial w)(z,w)} \qquad \text{on } \Sigma. \tag{7.1}$$

PROOF. Let $d(z)$ be the discriminant of $F(z,w)$ with respect to w. Thus, $d(z)$ is a holomorphic function in D with $d(z) \not\equiv 0$ in D. We set $\sigma := \{z \in D \mid d(z) = 0\}$ and $D' = D \setminus \sigma$. Fix $z \in D'$ and let δ be a simply connected neighborhood of z in D'. Then the equation $F(z,w) = 0$ has l distinct solutions $w = \eta_j(z)$ $(j = 1, \dots, l)$, so that each $\eta_j(z)$ is a holomorphic function on δ and $F(z,w) = \prod_{j=1}^{l}(w - \eta_j(z))$ in $\delta \times \mathbf{C}_w$. We let v_j $(j = 1, \dots, l)$ denote the portion of Σ defined by $w = \eta_j(z)$ for $z \in \delta$. We write $\phi_j(p) = \phi(p)|_{v_j}$ and regard $\phi_j(p)$ as a holomorphic function on δ; thus we denote it by $\phi_j(z)$. Next we consider the following function on $\delta \times \mathbf{C}_w$:

$$\begin{aligned}\Phi(z,w) \quad &:= \quad F(z,w)\left\{\frac{\phi_1(z)}{w - \eta_1(z)} + \cdots + \frac{\phi_l(z)}{w - \eta_l(z)}\right\} \\ &= \quad \sum_{j=1}^{l} \phi_j(z)(w - \eta_1(z)) \cdots \widehat{(w - \eta_j(z))} \cdots (w - \eta_l(z)),\end{aligned}$$

where $\widehat{\circ}$ denotes the omission of $\circ$. Note that $\Phi(z,w)$ is of the form

$$\Phi(z,w) = A_0(z)w^{l-1} + \cdots + A_{l-1}(z) \qquad \text{for } z \in \delta,$$

where each $A_i(z)$ is a holomorphic function on δ, and $\Phi(z,w)$ satisfies the relation

$$\Phi(z, \eta_j(z)) = \phi_j(z)\frac{\partial F}{\partial w}(z, \eta_j(z)) \qquad \text{for } z \in \delta.$$

By analytic continuation and the expression of $A_i(z)$ $(i = 0, 1, \dots, l-1)$ as a symmetric function of $\phi_1(z), \dots, \phi_l(z)$, each $A_i(z)$ becomes a single-valued, bounded holomorphic function in D'. It follows from Riemann's removable singularity theorem that each $A_i(z)$ is holomorphic in all of D. Thus, $\Phi(z,w)$ is a pseudopolynomial in w of degree at most $l-1$ whose coefficients $A_i(z)$ are holomorphic functions on D, and

$$\Phi(z,w) = \phi(p)\,\frac{\partial F}{\partial w}(z,w) \qquad \text{for } (z,w) = p \in \Sigma.$$

Now take any $p = (z,w) \in \Sigma$ such that $z \notin \sigma$. Since $\frac{\partial F}{\partial w}(p) \neq 0$, $\phi(p)$ is of the form (7.1). Furthermore, by analytic continuation (7.1) holds at any non-singular point of Σ. The uniqueness is clear from the Weierstrass preparation theorem (Remark 2.3). □

Proposition 7.1 says that any weakly holomorphic function $\phi(p)$ on an analytic hypersurface Σ in Λ is the restriction of the meromorphic function $G(z,w) := \frac{\Phi(z,w)}{(\partial F/\partial w)(z,w)}$ in Λ to Σ. Note the denominator $\partial F/\partial w$ does not depend on $\phi(p)$. Let $p_0 = (z_0, w_0)$ be a singular point of Σ. If $\phi(p)$ is a weakly holomorphic function but is not a holomorphic function at p_0, then p_0 is a point of indeterminacy of the function $G(z,w)$ associated to $\phi(p)$.

A non-singular point p_0 of Σ such that $z_0 \in \sigma$ may be a point of indeterminacy of $G(z,w)$. For example, take the non-singular hypersurface $\Sigma : \ F(z,w) = w^2 - z = 0$ in $\mathbf{C}^2$ and let $\phi(p) = \sqrt{z}$. Then we have $\partial F/\partial w = 2w$, $\Phi(z,w) = 2z$ and $G(z,w) = z/w$. For another example, take $\Sigma : \ F(z,w) = w(w^2 - z) = 0$, which is singular at $(0,0)$. Let $\phi(p) = 1$ on $w^2 = z$, while $\phi(p) = -1$ on $w = 0$. Then $\phi(p)$ is discontinuous at $(z,w) = (0,0)$. Here, $G(z,w) = (w^2 + z)/(3w^2 - z)$.

Remark 7.4. Using the same notation $\Sigma : F(z,w) = 0$ in $\Lambda = D \times \mathbf{C}_w$ as in Proposition 7.1, the proposition implies the following fact: Let $\tilde{\Sigma}$ be a lifting of Σ of the first kind,

$$\tilde{\Sigma} : w_j = \varphi_j(p) \quad (p \in \Sigma,\ j = 1, \ldots, m),$$

which lies in $\Lambda \times \mathbf{C}^m$. Here each $\varphi_j(p)\,(j = 1, \ldots, m)$ is a weakly holomorphic function on Σ. Then, for each $j = 1, \ldots, m$, there exists a pseudopolynomial $\Phi_j(z,w)$ in w of degree at most $l-1$,

$$\Phi_j(z,w) = A_0^{(j)}(z) w^{l-1} + \cdots + A_{l-1}^{(j)}(z),$$

where each $A_i^{(j)}(z)$ $(i = 0, 1, \ldots, l-1)$ is a holomorphic function on D, such that the *linear* polynomial $\Phi_j^*(z,w,w_j)$ in w_j defined by

$$\Phi_j^*(z,w,w_j) := w_j\,\frac{\partial F}{\partial w}(z,w) - \Phi_j(z,w) \quad \text{in } \Lambda \times \mathbf{C}^m$$

vanishes on $\tilde{\Sigma}$.

7.2.2. Universal Denominators. Let Σ be an analytic set in a domain D in $\mathbf{C}^n$. Let δ be an open set in D and let $W(z)$ be a holomorphic function in δ. We set $v := \delta \cap \Sigma$. Suppose $W(z)$ satisfies the following condition: *for any $q \in v$ and any weakly holomorphic function $f(p)$ at q, the weakly holomorphic function $f(p)W(p)$ at q is a holomorphic function at q.* This means that there exist a neighborhood δ_q of q in δ and a holomorphic function $F(z)$ on δ_q such that $F(z) = f(p)W(p)$ on $v \cap \delta_q$. Then we say that $W(z)$ is a **universal denominator**[2] for Σ in δ. Fix $p \in \Sigma$ and let $W(z)$ be a holomorphic function at p in $\mathbf{C}^n$. If there exists a neighborhood δ of p in $\mathbf{C}^n$ such that $W(z)$ is a universal denominator for Σ in δ, then we say that $W(z)$ is a universal denominator for Σ at p. Clearly if $W(z) \equiv 0$ on Σ, then $W(z)$ is a universal denominator for Σ in D.

[2]In [**51**], Oka calls a universal denominator a *W-function*.

Proposition 7.1 yields the following result.

COROLLARY 7.1. *Using the same notation as in Proposition 7.1, the function* $\partial F/\partial w$ *is a universal denominator at each point of* Σ *in* Λ.

PROOF. Let $q \in \Sigma$ and let $f(p)$ be a weakly holomorphic function at q. We may assume that $f(p)$ is weakly holomorphic on $v := \Sigma \cap \lambda$, where $\lambda := \delta \times \gamma \subset D \times \mathbf{C}_w$ is a polydisk centered at q with $\overline{v} \cap [\delta \times \partial\gamma] = \emptyset$. Then there exists a monic polynomial $F_1(z, w)$ in w,

$$F_1(z, w) = w^k + b_1(z)w^{k-1} + \cdots + b_k(z),$$

whose coefficients are holomorphic functions on δ, with $1 \leq k \leq l$ and $v = \{(z, w) \in \delta \times \mathbf{C}_w \mid F_1(z, w) = 0\}$. By applying Proposition 7.1 to F_1 in $\delta \times \mathbf{C}_w$, we see that $(\partial F_1/\partial w) \cdot f|_v$ has a holomorphic extension $P_1(z, w)$ in $\delta \times \mathbf{C}_w$ of the form

$$P_1(z, w) = B_0(z)w^{k-1} + \cdots + B_{k-1}(z).$$

On the other hand, $F(z, w)$ can be written as

$$F(z, w) = F_1(z, w)F_2(z, w) \qquad \text{in } \delta \times \mathbf{C}_w,$$

where $F_2(z, w)$ is also a monic polynomial in w of degree $l - k \geq 0$ whose coefficients are holomorphic functions in δ with $F_2(z, w) \neq 0$ at each point $(z, w) \in \lambda$. Since

$$\frac{\partial F}{\partial w} = \left(\frac{\partial F_1}{\partial w}\right) / F_2 \qquad \text{on } v,$$

it follows that $\frac{\partial F}{\partial w} \cdot f|_v$ has a holomorphic extension $P_1(z, w)/F_2(z, w)$ in λ. Hence, the function $\partial F/\partial w$ in Λ is a universal denominator at each point of Σ. □

The following two propositions indicate the relation between our two kinds of liftings of analytic sets and universal denominators.

PROPOSITION 7.2. *Let* Σ *be an analytic set in a domain* D *in* $\mathbf{C}^n$. *Let* $\varphi_j(p)$ $(j = 1, \ldots, m)$ *be weakly holomorphic functions on* Σ. *We consider a lifting* Σ^0 *of the first kind of* Σ,

$$\Sigma^0 : \ w_j = \varphi_j(p) \quad (p \in \Sigma, \ j = 1, \ldots, m),$$

in $\Lambda := D \times \mathbf{C}_w^m$. *If a holomorphic function* $W(z)$ *in* $\delta \subset D$ *is a universal denominator for* Σ *in* δ, *then* $W(z)$, *considered as a holomorphic function on* $\delta \times \mathbf{C}_w^m$ *which is independent of* w, *is a universal denominator for* Σ^0 *in* $\delta \times \mathbf{C}_w^m$.

PROOF. The proposition follows from the fact that the weakly holomorphic functions on $\Sigma^0 \cap (\delta \times \mathbf{C}_w^m)$ and the weakly holomorphic functions on $\Sigma \cap \delta$ are in one-to-one correspondence via the standard projection mapping. □

PROPOSITION 7.3. *Let* Σ *be an analytic set in a domain* D *in* $\mathbf{C}^n$ *and let* $\Sigma = \Sigma_0 \cup \cdots \cup \Sigma_r$, *where* Σ_i $(i = 1, \ldots, r)$ *is a pure i-dimensional analytic set in* D. *Suppose* Σ^* *is a lifting of the second kind of* Σ *in* $\Lambda := D \times \mathbf{C}_u^r$. *Let* δ *be an open set in* D *and let* $W(z, u)$ *be a holomorphic function in a neighborhood* λ *of* $\delta \times \{0\}$ *in* Λ. *If* $W(z, u)$ *is a universal denominator for* Σ^* *in* λ, *then* $W(z, 0)$ *is a universal denominator for* Σ *in* δ.

This easily follows from the definition of universal denominators.

Using these propositions, we have the following result.

PROPOSITION 7.4. *Let Σ be an analytic set in a domain D in $\mathbf{C}^n$ and fix $p_0 \in \Sigma$. Then there exists a neighborhood Λ of p_0 in D such that for any given neighborhood Λ_0 of p_0 with $\Lambda_0 \subset\subset \Lambda$ and any non-singular point q_0 of Σ in Λ_0, there exists a universal denominator $W_0(z)$ of Σ in Λ^*, where $\Lambda_0 \subset\subset \Lambda^* \subset\subset \Lambda$, with $W_0(z) \not\equiv 0$ on any irreducible component of Σ in Λ_0, and $W_0(q_0) \neq 0$.*

PROOF. Since the singular points of Σ correspond to the singular points of the lifting Σ^* of the second kind of Σ, it follows from Proposition 7.3 that Σ may be assumed to be a pure r-dimensional analytic set in D. Fix $p_0 = (z_1^0, \dots, z_n^0) \in \Sigma$ and choose coordinates $z = (z_1, \dots, z_n)$ of $\mathbf{C}^n$ which satisfy the Weierstrass condition for Σ at p_0. Then we can take

$$\begin{aligned} \Delta' \quad &: \quad |z_j - z_j^0| \leq \rho_j \qquad (j = 1, \dots, r), \\ \Delta_k \quad &: \quad |z_k - z_k^0| \leq \rho_k \qquad (k = r+1, \dots, n), \end{aligned}$$

so that if we set $\Gamma := \Delta_{r+1} \times \cdots \times \Delta_n$ and $\Lambda = \Delta' \times \Gamma$, then

$$\Lambda \subset\subset D, \qquad \Sigma \cap [\Delta' \times \partial\Gamma] = \emptyset.$$

We let $\mathcal{D}$ denote the projection of $\Sigma \cap \Lambda$ over Δ', so that $\mathcal{D}$ is a ramified domain over Δ' without relative boundary. For simplicity, we write $z' := (z_1, \dots, z_r)$. Then $\Sigma \cap \Lambda$ can be described as

$$(z_{r+1}, \dots, z_n) = (\eta_{r+1}(z'), \dots, \eta_n(z')), \qquad z' \in \mathcal{D},$$

where each $\eta_j(z')$ $(j = r+1, \dots, n)$ is a holomorphic function on $\mathcal{D}$. We put

$$\Delta = \Delta' \times \Delta_{r+1}, \quad \Gamma' = \Delta_{r+2} \times \cdots \times \Delta_n.$$

Since the condition $\Sigma \cap [\Delta' \times \partial\Gamma] = \emptyset$ implies that $\Sigma \cap [\Delta \times \partial\Gamma'] = \emptyset$, it follows from Proposition 2.3 that the projection $\underline{\Sigma}$ of Σ onto $\Delta \subset \mathbf{C}^{r+1}$ is an analytic set in Δ. We note that $\underline{\Sigma}$ is an analytic hypersurface in Δ. We let $\mathcal{D}'$ denote the projection of $\underline{\Sigma}$ over Δ'. Then $\underline{\Sigma}$ can be described as

$$\underline{\Sigma} : \; F(z', z_{r+1}) \equiv z_{r+1}^l + a_1(z') z_{r+1}^{l-1} + \cdots + a_l(z') = 0 \quad \text{in } \Delta, \tag{7.2}$$

where each $a_i(z')$ $(i = 1, \dots, l)$ is a holomorphic function in Δ'; i.e., $z_{r+1} = \eta_{r+1}(z')$, $z' \in \mathcal{D}'$, coincides with the solution set of $F(z', z_{r+1}) = 0$. Furthermore, we may assume that $F(z', z_{r+1})$ has no multiple factors. It follows from Corollary 7.1 that

$$W(z', z_{r+1}) = \frac{\partial F}{\partial z_{r+1}}(z', z_{r+1})$$

is a universal denominator for $\underline{\Sigma}$ in Δ such that $W(z', z_{r+1}) \not\equiv 0$ on each irreducible component of $\underline{\Sigma}$. On the other hand, Σ in Λ is a lifting of the first kind of $\underline{\Sigma}$ through $\eta_k(p)$ $(k = r+2, \dots, n)$. From Proposition 7.2, it follows that $W(z) := W(z', z_{r+1})$, considered as independent of $z_{r+2}, \dots, z_n$, is a universal denominator for Σ in Λ which is not identically zero on each irreducible component of Σ.

Using the variable z_k $(k = r+2, \dots, n)$ instead of z_{r+1} we obtain a universal denominator $W_k(z', z_k)$ for Σ in Λ which is not identically zero on each irreducible component of Σ.

Let Λ_0 be a neighborhood of p_0 with $\Delta_0 \subset\subset \Lambda$ and let $q_0 \in \Lambda_0$ be a non-singular point of Σ. For simplicity we take $q_0 = 0$. If $W_k(0,0) \neq 0$ for some k $(r+1 \leq k \leq n)$, then we can take $W_0(z)$ to be $W_k(z', z_k)$ in Λ and set $\Lambda^* = \Lambda$, and we are done. Thus we assume that $W_k(0,0)=0$ for each $k=r+1,\dots,n$. As in (7.2) we set, for each $k=r+1,\dots,n$,

$$\Sigma_k:\ F_k(z',z_k)\equiv z_k^m+a_{k,1}(z')z_k^{m-1}+\cdots+a_{k,m}(z')=0 \quad \text{in } \Delta'\times\Delta_k,$$

in order that $W_k(z',z_k)=(\partial F_k/\partial z_k)(z',z_k)$ in Λ. Here m depends on k. Since $(0,0)$ is a non-singular point of Σ, it follows that

$$\frac{\partial a_{k,m}}{\partial z_i}(0)\neq 0 \quad \text{for some } k\ (r+1\le k\le n) \text{ and } i\ (1\le i\le r);$$

we take $k=r+1$ and $i=1$ for simplicity. For small $\varepsilon\neq 0$, we consider the following coordinate transformation of $\mathbf{C}^n$:

$$\widetilde{z}_1=z_1+\varepsilon z_{r+1}, \qquad \widetilde{z}_j=z_j \quad (j=2,\dots,n).$$

If we again construct a universal denominator $\widetilde{W}_{r+1}(\widetilde{z}',\widetilde{z}_{r+1}):=\partial\widetilde{F}_{r+1}/\partial\widetilde{z}_{r+1}$ in $\Lambda^*=\widetilde{\Delta}\times\widetilde{\Gamma}$ of the same type as $W_{r+1}(z',z_{r+1})$ in Λ, then $\widetilde{W}_{r+1}(0,0)=-(\frac{\partial a_{r+1,m}}{\partial z_1}(0))\,\varepsilon\neq 0$. Choose $\varepsilon\neq 0$ sufficiently small so that $\Lambda^*\subset\Lambda$ is close enough to Λ to ensure that $\Lambda_0\subset\subset\Lambda^*$. Then, taking $W_0(z)$ to be $\widetilde{W}_{r+1}(\widetilde{z}_{r+1},\widetilde{z}')$, the conclusion of the proposition is satisfied. □

From this proposition we easily deduce the following corollary.

COROLLARY 7.2. *Let Σ be an analytic set in a domain D in $\mathbf{C}^n$ and let S be the set of singular points of Σ in D. Let $p\in\Sigma$. Then there exists a neighborhood δ of p in D such that the common zeros in δ of all universal denominators of Σ on δ are contained in $S\cap\delta$.*

This corollary, combined with the fundamental theorem in Chapter 6, yields the following result, which will play an important role in the next chapter.

COROLLARY 7.3. *Let Σ be a pure r-dimensional analytic set in a domain D in $\mathbf{C}^n$ and let $p\in\Sigma$. There exists a neighborhood δ of p in D such that, if we write $\Sigma_\delta:=\Sigma\cap\delta$, we can find a lifting of Σ_δ of the first kind,*

$$(\Sigma_\delta)^0:\ w_j=\eta_j(p) \qquad (p\in\Sigma_\delta,\ j=1,\dots,m),$$

in $\delta^0:=\delta\times\mathbf{C}^m$, such that the common zeros in δ^0 of all universal denominators for $(\Sigma_\delta)^0$ in δ^0 are contained in an analytic set of dimension at most $r-2$ in δ^0.

PROOF. Fix $p\in\Sigma$ and a closed polydisk $\delta:=\Delta'\times\Delta''\subset\mathbf{C}^r\times\mathbf{C}^{n-r}$ centered at $p=(p',p'')$ which satisfies the Weierstrass condition for Σ, and such that, if we set $\Sigma_\delta:=\Sigma\cap\delta$, then $\Sigma_\delta\cap(\Delta'\times\partial\Delta'')=\emptyset$. We let $\mathcal{D}$ denote the projection of Σ_δ over Δ'. Thus $\mathcal{D}$ is a ramified domain over Δ' without relative boundary. By taking a smaller δ centered at p if necessary, from Theorem 6.4 we can find a fundamental system $\{\Phi_i(p)\}_{i=1,\cdots,m}$ for $\mathcal{D}$; i.e., each $\Phi_i(p)$ is a holomorphic function on $\mathcal{D}$ such that the set S of singular points of the graph

$$\mathcal{C}:\ w_i=\Phi_i(p) \qquad (p\in\mathcal{D},\ i=1,\dots,m)$$

in $\Delta'\times\mathbf{C}^m$ is of dimension at most $r-2$.

Since $\Phi_i(p)$ becomes a weakly holomorphic function on Σ_δ, we have a lifting of the first kind of Σ_δ,

$$(\Sigma_\delta)^0:\ w_i=\Phi_i(p) \qquad (p\in\Sigma_\delta,\ i=1,\dots,m)$$

in $\delta^0:=\delta\times\mathbf{C}^m\subset\mathbf{C}^{n+m}$ such that the set of singular points of $(\Sigma_\delta)^0$ is of dimension at most $(r-2)$. Thus, the corollary follows from Corollary 7.2. □

7.3. $\mathcal{O}$-Modules

7.3.1. Definition of $\mathcal{O}$-Modules. In $\mathbf{C}^n$ with variables $z_1, \dots, z_n$, let $\delta \subset \mathbf{C}^n$ be an open set and let $\lambda \geq 1$ be an integer. Take λ single-valued holomorphic functions $f_j(z)$ $(j = 1, \dots, \lambda)$ on δ and set

$$f(z) := (f_1(z), \dots, f_\lambda(z)).$$

We call $f(z)$ a **holomorphic vector-valued function** on δ of **rank** λ, and $f_j(z)$ $(j = 1, \dots, \lambda)$ is the j-th **component** of $f(z)$. We let $\mathcal{O}^\lambda$ denote the set of all pairs $(f(z), \delta)$, where δ is an open set in $\mathbf{C}^n$ and $f(z)$ is a holomorphic vector-valued function on δ of rank λ. In case $\lambda = 1$ we use the notation $\mathcal{O}$.

Let $\mathcal{J}^\lambda$ be a subset of $\mathcal{O}^\lambda$. Suppose $\mathcal{J}^\lambda$ satisfies the following two conditions:

(1) If $(f_1(z), \delta_1), (f_2(z), \delta_2) \in \mathcal{J}^\lambda$ and $\delta_1 \cap \delta_2 \neq \emptyset$, then $(f_1(z) + f_2(z), \delta_1 \cap \delta_2) \in \mathcal{J}^\lambda$.

(2) Let $\delta' \subset \mathbf{C}^n$ be an open set and let $\alpha(z)$ be a holomorphic function on δ'. If $(f(z), \delta) \in \mathcal{J}^\lambda$ and $\delta \cap \delta' \neq \emptyset$, then $(\alpha(z) f(z), \delta \cap \delta') \in \mathcal{J}^\lambda$.

Then we say that $\mathcal{J}^\lambda$ is an **$\mathcal{O}$-module of rank** λ, or simply, an **$\mathcal{O}$-module**. In case $\lambda = 1$, we call $\mathcal{J}^\lambda = \mathcal{J}$ an **$\mathcal{O}$-ideal.**[3]

Let $\mathcal{J}^\lambda$ be an $\mathcal{O}$-module. If $(f(z), \delta) \in \mathcal{J}^\lambda$, then we say that $f(z)$ **belongs to** $\mathcal{J}^\lambda$ on δ. From condition (2), $(f(z), \delta) \in \mathcal{J}^\lambda$ and $\delta' \subset \delta$ imply that $(f(z), \delta') \in \mathcal{J}^\lambda$. Let $p \in \mathbf{C}^n$ and let $f(z)$ be a holomorphic vector-valued function of rank λ at a point p. If there exists a neighborhood δ of p in $\mathbf{C}^n$ such that $f(z)$ belongs to $\mathcal{J}^\lambda$ on δ, then we say that $f(z)$ belongs to $\mathcal{J}^\lambda$ at the point p.

Let $D \subset \mathbf{C}^n$ be a domain and let $\mathcal{J}^\lambda$ be an $\mathcal{O}$-module. If the open set $\delta \subset \mathbf{C}^n$ is contained in D for each $(f(z), \delta) \in \mathcal{J}^\lambda$, then we say that $\mathcal{J}^\lambda$ is an $\mathcal{O}$-module on D. To emphasize this, we write $\mathcal{J}^\lambda(D)$.

Let $\mathcal{J}^\lambda$ be an $\mathcal{O}$-module and let $\mathcal{I}^\lambda \subset \mathcal{J}^\lambda$. If $\mathcal{I}^\lambda$ itself is an $\mathcal{O}$-module, then we say that $\mathcal{I}^\lambda$ is an **$\mathcal{O}$-submodule** of $\mathcal{J}^\lambda$.

Let $D \subset \mathbf{C}^n$ be a domain and let $\mathcal{J}^\lambda$ be an $\mathcal{O}$-module. Then it is clear that $\mathcal{I}^\lambda := \{(f(z), \delta) \in \mathcal{J}^\lambda \mid \delta \subset D\}$ is an $\mathcal{O}$-submodule of $\mathcal{J}^\lambda$. We say that $\mathcal{I}^\lambda$ is the **restriction of $\mathcal{J}^\lambda$ to D.**

Let $\mathcal{J}_1^\lambda$ and $\mathcal{J}_2^\lambda$ be $\mathcal{O}$-modules. Then $\mathcal{J}^\lambda := \mathcal{J}_1^\lambda \cap \mathcal{J}_2^\lambda$ is also an $\mathcal{O}$-module, which is called the **intersection** of $\mathcal{J}_1^\lambda$ and $\mathcal{J}_2^\lambda$.

Let $\mathcal{J}_1^\lambda$ and $\mathcal{J}_2^\lambda$ be $\mathcal{O}$-modules and let $D \subset \mathbf{C}^n$. Fix a point $p \in D$. Assume that if $f(z)$ belongs to $\mathcal{J}_1^\lambda$ (resp. $\mathcal{J}_2^\lambda$) at p, then $f(z)$ belongs to $\mathcal{J}_2^\lambda$ (resp. $\mathcal{J}_1^\lambda$) at p. If this occurs for each $p \in D$, then we say that $\mathcal{J}_1^\lambda$ and $\mathcal{J}_2^\lambda$ are **equivalent on** D. Furthermore, let $\mathcal{J}_1^\lambda$ and $\mathcal{J}_2^\lambda$ be $\mathcal{O}$-modules and fix $p \in \mathbf{C}^n$. If there exists a neighborhood δ of p in $\mathbf{C}^n$ such that $\mathcal{J}_1^\lambda$ and $\mathcal{J}_2^\lambda$ are equivalent on δ, then we say that $\mathcal{J}_1^\lambda$ and $\mathcal{J}_2^\lambda$ are **equivalent at the point** p.

Let $D \subset \mathbf{C}^n$ and λ, $\nu \geq 1$ be integers. Let

$$\Phi_j(z) = (\Phi_{1,j}(z), \dots, \Phi_{\lambda,j}(z)) \qquad (j = 1, \dots, \nu)$$

be ν holomorphic vector-valued functions of rank λ on D. For an open set $\delta \subset D$ and ν holomorphic functions $\alpha_j(z)$ $(j = 1, \dots, \nu)$ on δ, we form the holomorphic vector-valued function of rank λ

$$f(z) = \alpha_1(z)\Phi_1(z) + \cdots + \alpha_\nu(z)\Phi_\nu(z) \quad \text{on } \delta.$$

[3]The notion of $\mathcal{O}$-ideal was first introduced by Oka [**50**] under the name of *ideal with indeterminate domain.* The $\mathcal{O}$-module defined here is an example of a *presheaf* in sheaf theory.

The totality of such pairs $(f(z), \delta)$ becomes an $\mathcal{O}$-module of rank λ. We call it the **$\mathcal{O}$-module generated by** $\{\Phi\} := \{\Phi_j(z)\}_{j=1,\dots,\nu}$ and denote it by $\mathcal{J}^\lambda\{\Phi\}$.

Let $\mathcal{J}^\lambda$ be an $\mathcal{O}$-module. If there exist a finite number of holomorphic vector-valued functions $\Phi_j(z)$ $(j = 1, \dots, \nu)$ of rank λ on a domain $D \subset \mathbf{C}^n$ such that the restriction of $\mathcal{J}^\lambda$ to D is the $\mathcal{O}$-module $\mathcal{J}^\lambda\{\Phi\}$ generated by $\{\Phi_j(z)\}_{j=1,\cdots,\nu}$, then we say that $\mathcal{J}^\lambda$ is a **finitely generated $\mathcal{O}$-module** on D, and we call $\{\Phi_j(z)\}_{j=1,\dots,\nu}$ a **pseudobase** for $\mathcal{J}^\lambda$ on D.

Let $\mathcal{J}^\lambda$ be an $\mathcal{O}$-module and let $p \in \mathbf{C}^n$. If there exists a neighborhood δ of p in $\mathbf{C}^n$ such that $\mathcal{J}^\lambda$ is equivalent to a finitely generated $\mathcal{O}$-module $\mathcal{I}^\lambda\{\Phi\}$ on δ, where $\{\Phi\} = \{\Phi_j(z)\}_{j=1,\dots,\nu}$, then we say that $\mathcal{J}^\lambda$ is a **locally finitely generated $\mathcal{O}$-module at the point** p and $\{\Phi_j(z)\}_{j=1,\dots,\nu}$ is a **local pseudobase** at the point p; equivalently, we say that $\mathcal{J}^\lambda$ admits a **locally finite pseudobase** at p.

REMARK 7.5. Let $D \subset \mathbf{C}^n$ be a domain and let $\mathcal{R}$ be the ring of all holomorphic functions on D. Then one ordinarily defines an $\mathcal{R}$-ideal $\mathcal{F}$ on D as a set $\mathcal{F}$ of holomorphic functions on D satisfying (1) if $f_1(z), f_2(z) \in \mathcal{F}$, then $f_1(z) + f_2(z) \in \mathcal{F}$; (2) if $\alpha(z) \in \mathcal{R}$ and $f(z) \in \mathcal{F}$, then $\alpha(z)f(z) \in \mathcal{F}$. We will call such an ideal $\mathcal{F}$ an *ideal with determined domain* D, while the $\mathcal{O}$-ideal defined above is an *ideal with indeterminate domain.* These two types of ideals have different properties arising from the structure of the zero sets of holomorphic functions.

For example, in $\mathbf{C}^2$ with variables $z = (z_1, z_2)$, we define $\Delta : |z_1| < 2$, $|z_2| < 2$ and $E : z_1 = 0$, $|z_2| \leq 1$ so that $E \subset\subset \Delta$. We define J and $\mathcal{J}$ as follows:

(1) J is the set of all holomorphic functions $f(z)$ on Δ such that $f(z) = 0$ on E.
(2) $\mathcal{J}$ is the set of all pairs $(f(z), \delta)$ such that $f(z)$ is a holomorphic function on $\delta \subset D$ with $f(z) = 0$ on $\delta \cap E$.

Then J is an ideal with determined domain Δ and $\mathcal{J}$ is an ideal with indeterminate domain in Δ. The common zero set of all of the functions $f(z) \in J$ is the disk $|z_2| < 2$ in the complex line $z_1 = 0$ (which contains E), while the the zero set of any holomorphic function $f(z)$ in δ such that $(f(z), \delta) \in \mathcal{J}$ is necessarily contained in E.

REMARK 7.6. An $\mathcal{O}$-ideal does not always admit a locally finite pseudobase at a given point.

For example, let $\gamma \subset\subset \Gamma$ be concentric open balls centered at the origin in $\mathbf{C}^2$ with variables x and y. Let Σ be the hyperplane $x = y$ in $\mathbf{C}^2$ and let σ denote the portion of Σ in $\Gamma \setminus \gamma$. Consider the set $\mathcal{J}$ of all pairs $\{(f(x, y), \delta)\}$ with $\delta \subset \Gamma$ and $f(x, y)$ holomorphic on δ with $f(x, y) = 0$ on $\sigma \cap \delta$. Then $\mathcal{J}$ is an $\mathcal{O}$-module in Γ which does not admit a locally finite pseudobase at each point of $\Sigma \cap (\partial\gamma)$ in Γ.

As another example, let $\Delta = (|x| < 1) \times (|y| < 1)$ and $\Delta' = (|x| < 1) \times (0 < |y| < 1)$ in $\mathbf{C}^2$ and let $\mathcal{I}$ be the $\mathcal{O}$-ideal in Δ generated by (xy, Δ) and $(1, \Delta')$. To be precise, $(f, \delta) \in \mathcal{I}$ if and only if, in case $\delta \subset \Delta'$, f is an arbitrary holomorphic function in δ, while in case $\delta \subset \Delta$ but $\delta \not\subset \Delta'$, f is of the form $h(x, y)xy$ where $h(x, y)$ is a holomorphic function on δ. Then $\mathcal{I}$ is an $\mathcal{O}$-ideal in Δ which does not admit a locally finite pseudobase at the origin $(0, 0)$.

7.3.2. Main Theorem. Let $D \subset \mathbf{C}^n$ be a domain and let $\lambda, \nu \geq 1$ be integers. Let

$$F_j(z) = (F_{1,j}(z), \dots, F_{\lambda,j}(z)) \qquad (j = 1, \dots, \nu)$$

be ν holomorphic vector-valued functions of rank λ on D. We consider the following system of λ homogenous linear equations involving ν unknown holomorphic functions $f_j(z)$ $(j = 1, \ldots, \nu)$:

$$(\Omega) \qquad f_1(z)F_1(z) + \cdots + f_\nu(z)F_\nu(z) = 0,$$

or equivalently,

$$\begin{cases} F_{1,1}(z)f_1(z) & + \cdots + \quad F_{1,\nu}(z)f_\nu(z) = 0, \\ \quad \vdots & \qquad\qquad \vdots \\ F_{\lambda,1}(z)f_1(z) & + \cdots + \quad F_{\lambda,\nu}(z)f_\nu(z) = 0. \end{cases}$$

If a holomorphic vector-valued function

$$f(z) = (f_1(z), \ldots, f_\nu(z))$$

of rank ν on an open set $\delta \subset D$ satisfies these linear homogeneous equations on δ, then we say that $(f(z), \delta)$ is a solution of equation (Ω). The set of all solutions $(f(z), \delta)$ of equation (Ω) where $\delta \subset D$ clearly becomes an $\mathcal{O}$-module of rank ν. We call it the **$\mathcal{O}$-module with respect to the linear relation** (Ω) and denote it by $\mathcal{L}\{\Omega\}$.

With this terminology we have the following theorem.

THEOREM 7.1 (Oka). *For any given system of homogeneous linear equations* (Ω) *on* $D \subset \mathbf{C}^n$, *the* $\mathcal{O}$*-module* $\mathcal{L}(\Omega)$ *with respect to the linear relation* (Ω) *has a locally finite pseudobase at each point of* D.

This theorem is the main theorem in the theory of $\mathcal{O}$-modules. It was first proved by Oka in 1948 (cf. [**50**]); the proof below is a modification of Oka's proof due to H. Cartan [**12**].

7.3.3. Two Preparation Theorems. Let D be a domain in $\mathbf{C}^n$ with variables $z_1, \ldots, z_n$ and let $\mathbf{C}_w$ be the complex plane with variable w. Let $l \geq 1$ be an integer, and consider a monic pseudopolynomial $P(z, w)$ in w of degree l,

$$P(z, w) = w^l + a_1(z)w^{l-1} + \cdots + a_l(z),$$

where each $a_i(z)$ $(i = 1, \ldots, l)$ is a holomorphic function on $\overline{D}$. ($P(z, w)$ may have multiple factors.) Fix $r > 0$ and define $\Gamma := \{|w| \leq r\} \subset \mathbf{C}_w$ and $\Lambda := D \times \Gamma \subset \mathbf{C}^{n+1}$. We assume $r > 0$ is sufficiently large so that for each $z' \in D$, the l solutions of $P(z', w) = 0$ with respect to w are contained in the interior of Γ, i.e.,

$$\{(z, w) \in D \times \mathbf{C}_w \mid P(z, w) = 0\} \subset\subset \Lambda. \tag{7.3}$$

Then we have the following two theorems.

THEOREM 7.2 (Remainder theorem). *Let* $f(z, w)$ *be a holomorphic function in* Λ.

1. *There exist a holomorphic function* $q(z, w)$ *in* Λ *and* l *holomorphic functions* $c_k(z)$ $(k = 0, 1, \ldots, l-1)$ *in* D *such that*

$$f(z, w) = q(z, w) \cdot P(z, w) + r(z, w) \qquad \text{on } \Lambda, \tag{7.4}$$

where

$$r(z, w) = c_0(z)w^{l-1} + \cdots + c_{l-1}(z) \qquad \text{on } D \times \mathbf{C}_w. \tag{7.5}$$

2. (a) *The holomorphic functions* $q(z, w)$ *and* $r(z, w)$ *which satisfy (7.4) are uniquely determined by* $f(z, w)$.

(b) *Let* $0 < r_0 < r$. *Define* $\Gamma_0 : |w| \le r_0$ *and* $\Lambda_0 = D \times \Gamma_0$. *Then there exists a constant* $K > 0$ *such that if* $|f(z,w)| \le M$ *on* Λ, *then*

$$\begin{aligned} &|q(z,w)|,\ |r(z,w)| \ \le KM \quad on\ \Lambda_0, \\ &|c_k(z)| \le KM \ (k = 0, 1, \dots, l-1) \quad on\ D_0. \end{aligned}$$

PROOF. Following H. Cartan, we consider the following integral for $(z,w) \in \Lambda \setminus (D \times \partial\Gamma)$:

$$I(z,w) := \frac{1}{2\pi i}\int_{|\zeta|=r} \frac{f(z,\zeta)}{P(z,\zeta)} \cdot \frac{P(z,\zeta) - P(z,w)}{\zeta - w} d\zeta.$$

From Cauchy's theorem we have

$$I(z,w) = f(z,w) - P(z,w) \cdot \left(\frac{1}{2\pi i}\int_{|\zeta|=r} \frac{f(z,\zeta)}{(\zeta - w)P(z,\zeta)} d\zeta \right).$$

Since $P(z,\zeta) \ne 0$ on $D \times \partial\Gamma$ by our assumption on Γ, it follows that the integral in the second term of the right-hand side is a holomorphic function $q(z,w)$ for $(z,w) \in \Lambda$.

On the other hand, the integral $I(z,w)$ may be written as

$$I(z,w) = \sum_{j=0}^{l-1} \left(\frac{1}{2\pi i}\int_{|\zeta|=r} \frac{f(z,\zeta)}{P(z,\zeta)} \cdot Q_j(z,\zeta) d\zeta \right) w^j,$$

where each $Q_j(z,\zeta)$ $(j = 0, \dots, l-1)$ is a pseudopolynomial in ζ of degree at most $l-1$ whose coefficients are holomorphic functions for z in D. Thus, the coefficient of w^j $(j = 0, \dots, l-1)$ on the right-hand side is a holomorphic function $c_j(z)$ in D. Hence,

$$f(z,w) - q(z,w)P(z,w) = \sum_{j=0}^{l-1} c_j(z) w^j \quad \text{in } \Lambda,$$

which proves 1.

To prove 2 (a), let $q(z,w)$ and $r(z,w)$ satisfy conditions (7.4) and (7.5). Then Cauchy's theorem applied to $q(z,w)$ yields, for $(z,w) \in \Lambda \setminus (D \times \partial\Gamma)$,

$$\begin{aligned} q(z,w) &= \frac{1}{2\pi i}\int_{|\zeta|=r} \frac{q(z,\zeta)}{\zeta - w} d\zeta \\ &= \frac{1}{2\pi i}\int_{|\zeta|=r} \frac{1}{\zeta - w}\frac{f(z,\zeta)}{P(z,\zeta)} d\zeta - \frac{1}{2\pi i}\int_{|\zeta|=r} \frac{1}{\zeta - w}\frac{r(z,\zeta)}{P(z,\zeta)} d\zeta \\ &=: S(z,w) - T(z,w). \end{aligned}$$

Condition (7.3) implies

$$T(z,w) = \frac{1}{2\pi i}\int_{|\zeta|=R} \frac{1}{\zeta - w}\frac{r(z,\zeta)}{P(z,\zeta)} d\zeta \quad \text{for any } R \ge r.$$

By letting $R \to \infty$, we see from (7.5) that $T(z,w) = 0$. We thus have $q(z,w) = S(z,w)$, which is uniquely determined by $f(z,w)$; hence so is $r(z,w)$.

To prove 2 (b), fix $(z,w) \in \Lambda_0$ so that $|w| \le r_0 < r$. We set

$$\begin{aligned} &A := \max\{|P(z,w)| \mid (z,w) \in \Lambda\}, \\ &a := \min\ \{|P(z,\zeta)| \mid (z,\zeta) \in D \times \partial\Gamma\} > 0. \end{aligned}$$

Suppose $|f(z,w)| \le M$ on Λ. Then the above expression for $q(z,w) = S(z,w)$ for $(z,w) \in \Lambda_0$ implies

$$|q(z,w)| \le \frac{1}{2\pi} \int_{|\zeta|=r} \frac{1}{|\zeta - w|} \frac{|f(z,\zeta)|}{|P(z,\zeta)|} |d\zeta| \le \frac{rM}{(r-r_0)a} =: K_1 M,$$

so that $|r(z,w)| \le M(1 + K_1 A)$. It follows that $|c_j(z)| \le M(1+K_1A)/r^j$ for $j = 0, 1, \dots, l-1$. Thus, $K := \max_{j=0,1,\dots,l-1} \{(1+K_1A)/r^j\} > 0$, and this proves 2 (b). □

THEOREM 7.3 (Division theorem). *Let $\Phi(z,w)$ be a pseudopolynomial in w of degree at most λ whose coefficients are holomorphic functions on D. If there exists a holomorphic function $q(z,w)$ in Λ such that*

$$\Phi(z,w) = q(z,w) \cdot P(z,w) \qquad \text{on } \Lambda,$$

then $q(z,w)$ is also a pseudopolynomial in w of degree at most $\lambda - l$ whose coefficients are holomorphic functions on D.

PROOF. Noting that $q(z,w)$ is holomorphic in $D \times \mathbf{C}_w$, we set

$$q(z,w) := \sum_{n=0}^{\infty} a_n(z) w^n \qquad \text{on } D \times \mathbf{C}_w,$$

where each $a_n(z)$ $(n = 0, 1, \dots)$ is a holomorphic function on D. Fix $z \in \Delta$ and $R > r$. Then we have from condition (7.3) that

$$\begin{aligned} a_n(z) &= \frac{1}{2\pi i} \int_{|w|=r} \frac{q(z,w)}{w^{n+1}} dw \\ &= \frac{1}{2\pi i} \int_{|w|=R} \frac{1}{w^{n+1}} \frac{\Phi(z,w)}{P(z,w)} dw. \end{aligned}$$

Let $n \ge \lambda - l + 1$ and let $R \to \infty$. Since $\deg_w \Phi(z,w) \le \lambda$ and $\deg_w P(z,w) = l$, we have $a_n(z) = 0$, so that $q(z,w) = \sum_{n=0}^{\lambda - l} a_n(z) w^n$, as desired. □

7.3.4. Proof of the Main Theorem. Let $D \subset \mathbf{C}^n$ be a domain and let

$$F_j(z) = (F_{1,j}(z), \dots, F_{\lambda,j}(z)) \quad (j = 1, \dots, \nu)$$

be a given set of ν holomorphic vector-valued functions of rank $\lambda \ge 1$ in D. Let

$$(\Omega) \qquad f_1(z)F_1(z) + \cdots + f_\nu(z)F_\nu(z) = 0$$

be a system of λ simultaneous linear equations for the ν unknown functions $f_j(z)$ $(j = 1, \dots, \nu)$. We form the $\mathcal{O}$-module $\mathcal{L}\{\Omega\}$ which consists of all solutions $(f(z), \delta)$ of (Ω); i.e.,

$$f(z) = (f_1(z), \dots, f_\nu(z))$$

is a holomorphic vector-valued function of rank ν in a domain $\delta \subset D$ which satisfies equation (Ω) in δ.

When we need to emphasize the dimension n, the rank λ, and the domain D, we write $\Omega(n, \lambda, D)$ and $\mathcal{L}\{\Omega(n,\lambda,D)\}$ instead of Ω and $\mathcal{L}\{\Omega\}$. We prove Theorem 7.1 by double induction with respect to the dimension $n \ge 1$ and the rank $\lambda \ge 1$. It suffices to prove the following three steps.

First step. Each $\mathcal{O}$-module $\mathcal{L}\{\Omega(1,1,D)\}$ has a locally finite pseudobase at every point of D.

Second step. If each $\mathcal{O}$-module $\mathcal{L}\{\Omega(n,k,D)\}$ $(k = 1,\dots,\lambda)$ has a locally finite pseudobase at every point in D, then the same is true for each $\mathcal{O}$-module $\mathcal{L}\{\Omega(n,\lambda+1,D)\}$.

Third step. If each $\mathcal{O}$-module $\mathcal{L}\{\Omega(n,\lambda,D)\}$ $(\lambda = 1,2,\dots)$ has a locally finite pseudobase at every point in D, then the same is true for each $\mathcal{O}$-module $\mathcal{L}\{\Omega(n+1,1,D)\}$.

Proof of the first step. Fix $z_0 \in D$. We set $F_j(z) = h_j(z)(z-z_0)^{k_j}$ $(j = 1,\dots,\nu)$, where $h_j(z)$ is a holomorphic function in a neighborhood v of z_0 in D with $h_j(z) \neq 0$ in v. Let $k := \min\{k_1,\dots,k_\nu\}$; for simplicity, assume $k_1 = k$. Then

$$G_j(z) := (-F_j(z)/F_1(z), 0,\dots,0,1,0,\dots,0) \qquad (j = 2,\dots,\nu)$$

(where the "1" occurs in the j-th slot) is a local pseudobase in v. Indeed, we note first that $(G_j(z), v) \in \mathcal{L}\{\Omega(1,1,v)\}$ for $j = 2,\dots,\nu$. Next, fix any $(f,\delta) \in \mathcal{L}\{\Omega(1,1,v)\}$, where $f = (f_1,\dots,f_\nu)$. Then we have $f = f_2G_2 + \cdots + f_\nu G_\nu$ in δ, which concludes the proof of the first step. □

Proof of the second step. Assume that each $\mathcal{O}$-module $\mathcal{L}\{\Omega(n,k,D)\}$ $(k = 1,\dots,\lambda)$ has a local pseudobase at each point in D. Let $\mathcal{L}\{\Omega(n,\lambda+1,D)\}$ be an $\mathcal{O}$-module for a set of linear relations $\Omega(n,\lambda+1,D)$. Precisely, let $D \subset \mathbf{C}^n$ and let

$$F_j(z) = (F_{0,j}(z), F_{1,j}(z),\dots,F_{\lambda,j}(z)) \quad (j = 1,\dots,\nu)$$

be ν given a holomorphic vector-valued functions of rank $\lambda+1$ in D, and let

$$(\Omega) \qquad f_1(z)F_1(z) + \cdots + f_\nu(z)F_\nu(z) = 0$$

be a set of simultaneous linear equations for the unknown holomorphic vector-valued function $f(z) = (f_1(z),\dots,f_\nu(z))$ of rank ν. Then $\mathcal{L}\{\Omega(n,\lambda+1,D)\}$ is the set of all pairs (f,δ) where $f(z)$ is a holomorphic vector-valued function in δ satisfying (Ω) in δ.

Fix $z_0 \in D$. Our goal is to find a neighborhood δ_0 of z_0 in D and a finite number, say κ, of holomorphic vector-valued functions of rank ν

$$K_l(z) = (K_{1,l}(z),\dots,K_{\nu,l}(z)) \qquad (l = 1,\dots,\kappa)$$

in δ_0 such that at any point $z^* \in \delta_0$, any $f(z)$ belonging to $\mathcal{L}\{\Omega(n,\lambda+1,D)\}$ at z^* can be written in the form

$$f(z) = h_1(z)K_1(z) + \cdots + h_\kappa(z)K_\kappa(z) \quad \text{in } \delta^*,$$

where δ^* is a neighborhood of z^* in δ_0 and $h_l(z)$ $(l = 1,\dots,\kappa)$ is a holomorphic function in δ^*.

Set

$$F_j^0(z) := (F_{1,j}(z),\dots,F_{\lambda,j}(z)) \qquad (j = 1,\dots,\nu)$$

and consider the simultaneous linear equations

$$(\Omega^0) \qquad f_1(z)F_1^0(z) + \cdots + f_\nu(z)F_\nu^0(z) = 0$$

involving the unknown holomorphic vector-valued function $f(z) = (f_1(z),\dots,f_\nu(z))$ of rank ν, so that (Ω^0) is of type $\Omega(n,\lambda,D)$.

We also consider the single linear equation

$$(\Omega_1) \qquad F_{0,1}(z)f_1(z) + \cdots + F_{0,\nu}(z)f_\nu(z) = 0$$

involving the unknown holomorphic vector-valued function $f(z) = (f_1(z), \dots, f_\nu(z))$ of rank ν, so that (Ω_1) is of type $\Omega(n, 1, D)$. Note that

$$\mathcal{L}\{\Omega\} = \mathcal{L}\{\Omega^0\} \cap \mathcal{L}\{\Omega_1\}.$$

By the inductive hypothesis, $\mathcal{L}\{\Omega^0\}$ has a local pseudobase at z^0 in D, i.e., there exist a neighborhood δ' of z_0 in D and a finite number, say μ, of holomorphic vector-valued functions of rank ν

$$\Phi_k(z) = (\Phi_{1,k}(z), \dots, \Phi_{\nu,k}(z)) \qquad (k = 1, \dots, \mu)$$

in δ' such that at any $z^* \in \delta'$, any $f(z)$ belonging to $\mathcal{L}\{\Omega^0\}$ at z^* can be written in the form

$$f(z) = g_1(z)\Phi_1(z) + \cdots + g_\mu(z)\Phi_\mu(z) \quad \text{in } e^*, \tag{7.6}$$

where e^* is a neighborhood of z^* in δ' and $g_k(z)$ $(k = 1, \dots, \mu)$ is a holomorphic function in e^*. By substituting this expression for $f(z)$ into (Ω_1), we obtain the following. Let

$$G_k(z) := F_{0,1}(z)\Phi_{1,k}(z) + \cdots + F_{0,\nu}(z)\Phi_{\nu,k}(z) \qquad (k = 1, \dots, \mu)$$

which is a holomorphic function in δ', and consider the single linear equation

$$(\Omega') \qquad g_1(z)G_1(z) + \cdots + g_\mu(z)G_\mu(z) = 0$$

involving the unknown holomorphic function $g(z) = (g_1(z), \dots, g_\mu(z))$ of rank μ, so that (Ω') is of type $\Omega(n, 1, \delta')$. Fix $z^* \in \delta'$. Then $f(z)$ belongs to $\mathcal{L}\{\Omega^0\} \cap \mathcal{L}\{\Omega_1\}$ at z^* if and only if $f(z)$ can be written in the form (7.6) with $g(z)$ belonging to $\mathcal{L}\{\Omega'\}$ at z^*.

Again by the inductive hypothesis, $\mathcal{L}\{\Omega'\}$ has a local pseudobase at z_0. Thus there exist a neighborhood $\delta_0 \subset \delta'$ of z_0 and a finite number, say κ, of holomorphic vector-valued functions of rank μ in δ_0,

$$\Psi_l(z) = (\Psi_{1,l}(z), \dots, \Psi_{\mu,l}(z)) \qquad (l = 1, \dots, \kappa),$$

such that at each $z^* \in \delta_0$, any holomorphic vector-valued function $g(z) = (g_1(z), \dots, g_\mu(z))$ which belongs to $\mathcal{L}\{\Omega'\}$ at z^* can be written in the form

$$g(z) = h_1(z)\Psi_1(z) + \cdots + h_\kappa(z)\Psi_\kappa(z) \quad \text{in } e_0,$$

where e_0 is a neighborhood of z^* in δ_0 and $h_k(z)$ $(k = 1, \dots, \kappa)$ is a holomorphic function on e_0.

We substitute this expression for $g(z)$ into equation (7.6) for $f(z)$. Upon setting

$$K_l(z) := \Psi_{1,l}(z)\Phi_1(z) + \cdots + \Psi_{\mu,l}(z)\Phi_\mu(z) \qquad (l = 1, \dots, \kappa),$$

which is a holomorphic vector-valued function of rank ν in δ_0, we have

$$f(z) = h_1(z)K_1(z) + \cdots + h_\kappa(z)K_\kappa(z)$$

in a neighborhood of z^* in δ_0. We thus conclude that $K_l(z)$ $(k = 1, \dots, \kappa)$ is a local pseudobase of $\mathcal{L}\{\Omega\}$ on δ_0. Thus, the second step is proved. □

Proof of the third step. Let $D \subset \mathbf{C}^{n+1}$ and let F_j $(j = 1, \dots, \nu)$ be a given set of ν holomorphic functions in D. We consider the single linear equation

$$(\Omega) \qquad f_1(z)F_1(z) + \cdots + f_\nu(z)F_\nu(z) = 0$$

involving the unknown vector-valued function $f(z) = (f_1(z), \dots, f_\nu(z))$; i.e., equation (Ω) is the general equation of type $\Omega(n+1, 1, D)$.

Fix $z_0 \in D$. We shall show that $\mathcal{L}\{\Omega\}$ has a local pseudobase at z_0. For simplicity we set $z_0 = 0 \in \mathbf{C}^{n+1}$. If we have $F_j(0) \neq 0$ for some j, say $j = \nu$, then $\mathcal{L}\{\Omega\}$ has a local pseudobase in a neighborhood of $z = 0$. Indeed, fix $\delta_0 := \{|z| < r_0\} \subset\subset \Delta$ so that $F_\nu(z) \neq 0$ on δ_0. Then the following $\nu - 1$ holomorphic vector-valued functions of rank ν on δ_0,

$$G_j(z) := (0, \dots, 1, 0, \dots, -F_j(z)/F_\nu(z)) \quad (j = 1, \dots, \nu - 1),$$

where the "1" occurs in the j-th slot, form a pseudobase of $\mathcal{L}\{\Omega\}$ on δ_0.

Thus, it remains to treat the case when $F_j(0) = 0$ for $j = 1, \dots, \nu$. We may assume that the coordinates $(z_1, \dots, z_n, z_{n+1})$ satisfy the Weierstrass condition for each hypersurface $F_j(z) = 0$ $(j = 1, \dots, \nu)$ at the origin 0. For simplicity, we use the notation $z := (z_1, \dots, z_n)$ and $w := z_{n+1}$. Hence we can find a polydisk Δ centered at $z = 0$ and a disk Γ centered at $w = 0$ such that, upon setting $\Lambda := \Delta \times \Gamma$, we have $\Lambda \subset\subset D$ and $F_j(z, w) \neq 0$ $(j = 1, \dots, \nu)$ on $\Delta \times \partial\Gamma$. Therefore, we can write

$$F_j(z, w) = \omega_j(z, w) \cdot P_j(z, w) \quad \text{in } \Lambda,$$

where

$$P_j(z, w) = w^{l_j} + A_{j, l_j - 1}(z) w^{l_j - 1} + \cdots + A_{j,0}(z)$$

(which may have multiple factors); each $A_{j,k}(z)$ $(0 \le k \le l_j)$ is a holomorphic function in Δ; each $\omega_j(z, w)$ is a non-vanishing holomorphic function in Λ; and

$$\{(z, w) \in \Delta \times \mathbf{C}_w \mid P_j(z, w) = 0\} \subset\subset \Lambda \qquad (j = 1, \dots, \nu). \tag{7.7}$$

We consider the single linear equation

$$(\Omega') \qquad f_1(z, w) P_1(z, w) + \cdots + f_\nu(z, w) P_\nu(z, w) = 0.$$

Since $\omega_j(z, w) \neq 0$ $(j = 1, \dots, \nu)$ on Λ, it thus suffices, to complete the third step, to prove that $\mathcal{L}\{\Omega'\}$ has a local pseudobase at $(z, w) = (0, 0)$. We set $l = \max\{l_1, \dots, l_\nu\}$. We assume, for simplicity, that $l = l_\nu$.

[I] We consider the set of holomorphic vector-valued functions of rank ν,

$$\mathbf{Q}(z, w) = (Q_1(z, w), \dots, Q_\nu(z, w)),$$

such that $Q_j(z, w)$ $(j = 1, \dots, \nu)$ is a pseudopolynomial in w of degree at most $l - 1$:

$$Q_j(z, w) = b_{j,l-1}(z) w^{l-1} + b_{j,l-2}(z) w^{l-2} + \quad \cdots \quad + b_{j,0}(z) \qquad (j = 1, \dots, \nu). \tag{7.8}$$

Here, $b_{j,k}(z)$ $(k = 0, \dots, l - 1)$ is a holomorphic function for z in $\delta \subset \Delta$. If $\mathbf{Q}(z, w)$ is a solution of equation (Ω') on an open set $\delta \times \gamma \subset \Lambda$, then we write $(\mathbf{Q}(z, w), \delta) \in \mathcal{L}^{l-1}\{\Omega'\}$, and we say that $\mathbf{Q}(z, w)$ belongs to $\mathcal{L}^{l-1}\{\Omega'\}$ on δ since this condition does not depend on $\gamma \subset \Gamma$.

We first show that $\mathcal{L}^{l-1}\{\Omega'\}$ has a local pseudobase at $(z, w) = (0, 0)$. Precisely, we will find a neighborhood δ_0 of $z = 0$ in Δ and a finite number μ of vector-valued functions $\Psi_k(z, w)$ $(k = 1, \dots, \mu)$ belonging to $\mathcal{L}^{l-1}\{\Omega'\}$ on δ_0 such that at each point $z^* \in \delta_0$, each vector-valued function $\mathbf{Q}(z, w)$ belonging to $\mathcal{L}^{l-1}\{\Omega'\}$ at z^* may be written in the form

$$\mathbf{Q}(z, w) = q_1(z) \Psi_1(z, w) + \cdots + q_\mu(z) \Psi_\mu(z, w) \tag{7.9}$$

in $\delta^* \times \Gamma$, where δ^* is a neighborhood of z^* in δ_0 and $q_\iota(z)$ $(\iota = 1, \dots, \mu)$ is a holomorphic function in δ^*.

Indeed, assume that $\mathbf{Q}(z,w) = (Q_1(z,w),\dots,Q_\nu(z,w))$ belongs to $\mathcal{L}^{l-1}\{\Omega'\}$ on $\delta \subset \Delta$ and that $Q_j(z,w)$ $(j=1,\dots,\nu)$ is of the form (7.8). Then we have

$$\sum_{j=1}^{\nu} P_j(z,w)Q_j(z,w) \;=\; \sum_{k=0}^{2l-1}\sum_{j=1}^{\nu}\left(\sum_{s+t=k} A_{j,t}(z)b_{j,s}(z)\right) w^k = 0$$

in $\delta \times \Gamma \subset \Lambda$. This is equivalent to

$$(\Omega'') \qquad \sum_{j=1}^{\nu}\sum_{s+t=k} A_{j,s}(z)b_{j,t}(z) = 0 \quad (k=0,\dots,2l-1) \quad \text{in } \delta.$$

We can regard (Ω'') as a set of $2l$ simultaneous linear equations with holomorphic coefficient functions $A_{j,t}(z)$ in Δ involving the unknown vector-valued holomorphic functions of rank $\lambda := \nu\, l$,

$$\mathbf{b}(z) = (b_{j,k}(z)) \qquad (j=1,\dots,\nu;\ k=0,1,\dots,l-1).$$

Thus (Ω'') is of type $\Omega(n,2l,\Delta)$. By the inductive hypothesis, we can find a neighborhood δ_0 of $z=0$ and a finite number μ of holomorphic vector-valued functions of rank λ,

$$\mathbf{c}^\iota(z) := (c^\iota_{j,k}(z)) \qquad (\iota=1,\dots,\mu;\ j=1,\dots,\nu;\ k=0,1,\dots,l-1),$$

such that at any point $z^* \in \delta_0$, each $\mathbf{b}(z)$ belonging to $\mathcal{L}\{\Omega''\}$ at z^* may be written as

$$\mathbf{b}(z) = \beta_1(z)\mathbf{c}^1(z) + \cdots + \beta_\mu(z)\mathbf{c}^\mu(z) \quad \text{in } \delta^*,$$

where δ^* is a neighborhood of z^* in δ_0 and $\beta_\iota(z)$ $(\iota=1,\dots,\mu)$ is a holomorphic function in δ^*.

Fix ι $(\iota=1,\dots,\mu)$. Using a holomorphic vector-valued function $\mathbf{c}^\iota(z)$ of rank λ, we construct ν pseudopolynomials $\Psi^\iota_j(z,w)$ in w of degree at most $l-1$:

$$\Psi^\iota_j(z,w) = c^\iota_{j,l-1}(z)w^{l-1} + c^\iota_{j,l-2}(z)w^{l-2} + \cdots + c^\iota_{j,0}(z) \qquad (j=1,\dots,\nu).$$

Next we set

$$\Psi_\iota(z,w) := (\Psi^\iota_1(z,w),\dots,\Psi^\iota_\nu(z,w)) \qquad (\iota=1,\dots,\mu),$$

which belongs to $\mathcal{L}^{\nu-1}\{\Omega'\}$ on δ_0. We see from the above argument that for any $z^* \in \delta_0$, each $\mathbf{Q}(z,w) = (Q_1(z,w),\dots,Q_\nu(z,w))$ belonging to $\mathcal{L}^{l-1}\{\Omega'\}$ at z^* can be written in the form

$$\mathbf{Q}(z,w) = q_1(z)\Psi_1(z,w) + \cdots + q_\mu(z)\Psi_\mu(z,w)$$

in $\delta^* \times \Gamma$, where δ^* is a neighborhood of z^* in δ_0 and $q_j(z)$ $(j=1,\dots\mu)$ is a holomorphic function in δ^*. We thus have assertion (7.9). □

[II] We set

$$\Phi_j(z,w) = (0,\dots,0,P_\nu(z,w),0,\dots,0,-P_j(z,w)) \qquad (j=1,\dots,\nu-1),$$

where the "1" occurs in the j-th slot, which belongs to $\mathcal{L}\{\Omega'\}$ on Λ. We shall prove that the collection of all

$$\Phi_j(z,w) \quad (j=1,\dots,\nu-1), \qquad \Psi_\iota(z,w) \quad (\iota=1,\dots,\mu)$$

forms a local pseudobase of $\mathcal{L}\{\Omega'\}$ on $\delta_0 \times \Gamma$.

To see this, fix $(z^*, w^*) \in \delta_0 \times \Gamma$ and let

$$f(z,w) = (f_1(z,w), \ldots, f_\nu(z,w))$$

be a holomorphic vector-valued function of rank ν belonging to $\mathcal{L}\{\Omega'\}$ on a neighborhood $\lambda' := \delta' \times \gamma' \subset \delta_0 \times \Gamma$ of (z^*, w^*). If $P_\nu(z^*, w^*) \neq 0$, we have

$$f(z,w) = \frac{f_1(z,w)}{P_\nu(z,w)}\Phi_1(z,w) + \cdots + \frac{f_{\nu-1}(z,w)}{P_\nu(z,w)}\Phi_{\nu-1}(z,w)$$

in a neighborhood of (z^*, w^*) in λ'. Thus it suffices to study the case $P_\nu(z^*, w^*) = 0$. In this case we can find a polydisk $\lambda^* := \delta^* \times \gamma^* \subset \lambda'$ with center (z^*, w^*) such that $P_\nu(z,w) \neq 0$ on $\delta^* \times \partial\gamma^*$. Then we have

$$P_\nu(z,w) = P'(z,w) \cdot P''(z,w) \quad \text{in } \lambda^*, \tag{7.10}$$

where $P''(z,w) \neq 0$ in λ^* and where $P'(z,w)$ is a pseudopolynomial in w,

$$P'(z,w) = w^{l'} + a_1(z)w^{l'-1} + \cdots + a_{l'}(z),$$

where $a_j(z)$ $(j = 1, \ldots, l')$ is a holomorphic function in δ^* such that

$$\{(z,w) \in \delta^* \times \mathbf{C}_w \mid P'(z,w) = 0\} \subset\subset \lambda^*. \tag{7.11}$$

By the division theorem, $P''(z,w)$ is also a monic pseudopolynomial in w of degree $l - l' = l''$.

We can apply the remainder theorem to this $P'(z,w)$ in λ^*, and obtain

$$f_j(z,w) = q_j(z,w) \cdot P'(z,w) + r_j(z,w) \ (j = 1, \ldots, \nu - 1) \quad \text{in } \lambda^*,$$

where $q_j(z,w)$ $(j = 1, \ldots, \nu - 1)$ is a holomorphic function in λ^* and $r_j(z,w)$ $(j = 1, \ldots, \nu - 1)$ is a pseudopolynomial in w of degree at most $l' - 1$ whose coefficients are holomorphic functions in δ^*. Thus, using the fact that $P''(z,w) \neq 0$ on λ^*, we have

$$\begin{aligned} f(z,w) &= (q_1(z,w)P'(z,w), \ldots, q_{\nu-1}(z,w)P'(z,w), 0) \\ &\quad + (r_1(z,w), \ldots, r_{\nu-1}(z,w), f_\nu(z,w)) \\ &= \frac{q_1(z,w)}{P''(z,w)}\Phi_1(z,w) + \cdots + \frac{q_{\nu-1}(z,w)}{P''(z,w)}\Phi_{\nu-1}(z,w) \\ &\quad + (r_1(z,w), \ldots, r_{\nu-1}(z,w), R_\nu(z,w)) \\ &\equiv g(z,w) + r(z,w), \end{aligned}$$

where

$$R_\nu(z,w) = f_\nu(z,w) + \frac{q_1(z,w)}{P''(z,w)}P_1(z,w) + \cdots + \frac{q_{\nu-1}(z,w)}{P''(z,w)}P_{\nu-1}(z,w)$$

(this explicit formula will not be used). To prove [II], since $P''(z,w) \neq 0$ on λ^*, it suffices to show that

$$\begin{aligned} \widetilde{r}(z,w) &:= P''(z,w)r(z,w) \\ &= (P''(z,w)r_1(z,w), \ldots, P''(z,w)r_{\nu-1}(z,w), P''(z,w)R_\nu(z,w)) \end{aligned}$$

belongs to $\mathcal{L}^{l-1}\{\Omega'\}$ on δ^*.

Since $f(z,w)$ and $g(z,w)$ belong to $\mathcal{L}\{\Omega'\}$ in λ^*, so does $r(z,w)$, so that $\widetilde{r}(z,w)$ belongs to $\mathcal{L}\{\Omega'\}$ on λ^*. Next, $P''(z,w)r_j(z,w)$ $(j = 1, \ldots, \nu - 1)$ is clearly a pseudopolynomial in w of degree at most $l-1$. Finally, since $r(z,w)$ belongs to $\mathcal{L}\{\Omega'\}$ on λ^*, we have

$$P_1(z,w)r_1(z,w)+\cdots+P_{\nu-1}(z,w)r_{\nu-1}(z,w)+P_\nu(z,w)R_\nu(z,w)=0$$

in λ^*, so that

$$-(P_1(z,w)r_1(z,w)+\cdots+P_{\nu-1}(z,w)r_{\nu-1}(z,w))=[P''(z,w)R_\nu(z,w)]\cdot P'(z,w)$$

in λ^*. From (7.11), we can apply the division theorem for $P'(z,w)$ in λ^*. Since the left-hand side is a pseudopolynomial in w of degree at most $l+l'-1$, it follows that $P''(z,w)R_\nu(z,w)$ must be a pseudopolynomial in w of degree at most $(l+l'-1)-l'=l-1$. Therefore, $\widetilde{r}(z,w)$ belongs to $\mathcal{L}^{l-1}\{\Omega'\}$ on δ^*, which proves [II]. □

Let $D\subset\mathbf{C}^n$ be a domain. Let $\mathcal{J}_1$ and $\mathcal{J}_2$ be two $\mathcal{O}$-modules of the same rank λ in D. From the main theorem (Theorem 7.1), we obtain the following useful result.

THEOREM 7.4. *If $\mathcal{J}_1$ and $\mathcal{J}_2$ each have a locally finite pseudobase at a point z_0, then $\mathcal{J}_1\cap\mathcal{J}_2$ also has a locally finite pseudobase at z_0.*

PROOF. By assumption we can find a neighborhood δ_0 of z_0 in D and holomorphic vector-valued functions of rank λ on δ_0,

$$\Phi_j(z)\quad(j=1,\dots,\nu),\qquad\Psi_k(z)\quad(k=1,\dots,\mu),$$

which generate $\mathcal{J}_1$ and $\mathcal{J}_2$ on δ_0.

Fix $z'\in\delta_0$. Then $f(z)$ belongs to $\mathcal{J}_1\cap\mathcal{J}_2$ at z' if and only if we have

$$f(z)=\sum_{j=1}^{\nu}a_j(z)\Phi_j(z)=\sum_{k=1}^{\mu}b_k(z)\Psi_k(z)$$

for z in a neighborhood δ' of z' in δ_0, where $a_j(z)$ $(j=1,\dots,\nu)$ and $b_k(z)$ $(k=1,\dots,\mu)$ are holomorphic functions in δ'.

Now we regard $\Phi_j(z)$ $(j=1,\dots,\nu)$ and $\Psi_k(z)$ $(k=1,\dots,\mu)$ as fixed holomorphic functions on δ_0, and we consider the single linear equation

$$(\Omega^0)\qquad\sum_{j=1}^{\nu}a_j(z)\Phi_j(z)-\sum_{k=1}^{\mu}b_k(z)\Psi_k(z)=0$$

involving the unknown holomorphic vector-valued function

$$(a_1(z),\dots,a_\nu(z),-b_1(z),\dots,-b_\mu(z))$$

of rank $\nu+\mu$. By Theorem 7.1 we can find a locally finite pseudobase of the $\mathcal{O}$-module $\mathcal{L}\{\Omega^0\}$ with respect to the linear relation (Ω^0),

$$\mathbf{c}^i(z)=(a_1^i(z),\dots,a_\nu^i(z),-b_1^i(z),\dots,-b_\mu^i(z))\quad(i=1,\dots,\kappa),$$

valid in a neighborhood δ^* of z_0 in δ_0. Then

$$\varphi_i(z):=\sum_{j=1}^{\nu}a_j^i(z)\Phi_j(z)\quad(i=1,\dots,\kappa)\qquad\text{on }\delta^*$$

is a finite pseudobase of $\mathcal{J}_1\cap\mathcal{J}_2$ on δ^*. □

Using the remainder theorem for $P'(z,w)$ (where $P_\nu(z,w)=P'(z,w)P''(z,w)$) in the same manner as it was used in (7.10) with (7.11), we easily obtain the following elementary fact.

REMARK 7.7. Let Σ be a pure r-dimensional analytic set in the polydisk Λ centered at the origin 0 in $\mathbf{C}^n$. Here $\Lambda = \Delta \times \Gamma \subset \mathbf{C}_z^r \times \mathbf{C}_w^s$, where $r + s = n$, and with $\Sigma \cap [\Delta \times \partial\Gamma] = \emptyset$. We set $\Gamma := \Gamma_1 \times \cdots \times \Gamma_s$, where Γ_j $(j = 1, \ldots, s)$ is a disk in $\mathbf{C}_{w_j}$. We let Σ_j denote the projection of Σ onto the polydisk $\Lambda_j := \Delta \times \Gamma_j$; Σ_j is thus an analytic hypersurface in Λ_j. Then we have

$$\Sigma_j = \{(z, w_j) \in \Delta \times \mathbf{C}_{w_j} \mid P_j(z, w_j) = 0\},$$

where

$$P_j(z, w_j) = w_j^{m_j} + a_1^{(j)}(z) w_j^{m_j - 1} + \cdots + a_{m_j}^{(j)}(z)$$

and $a_k^{(j)}(z)$ $(k = 1, \ldots, m_j)$ is a holomorphic function on Δ; moreover, $P_j(z, w_j)$ has no multiple factors. We set $M = (\sum_{j=1}^{s} m_j) - s$. In this setting we let $f(z, w)$ be a holomorphic function near the point (z_0, w_0) in Λ. Then $f(z, w)$ can be written in the following form in a sufficiently small polydisk $\lambda := \delta \times \gamma$ centered at (z_0, w_0) in Λ with $\Sigma \cap (\delta \times \partial\gamma) = \emptyset$:

$$\begin{aligned} f(z, w) &= \varphi_1(z, w) P_1(z, w_1) + \cdots + \varphi_s(z, w) P_s(z, w_s) \\ &\quad + \sum_{|\mathbf{j}|=0}^{M} \beta_{\mathbf{j}}(z) w_1^{j_1} \cdots w_s^{j_s} \\ &\text{for } \mathbf{j} = (j_1, \ldots, j_s);\ |\mathbf{j}| = j_1 + \cdots + j_s;\ 0 \le j_k \le m_j - 1, \end{aligned}$$

where each $\varphi_j(z, w)$ is a holomorphic function of $(z, w) \in \lambda$ and each $\beta_{\mathbf{j}}(z)$ is a holomorphic function of $z \in \delta$.

7.4. Combination Theorems

7.4.1. Combination Problems. Let $D \subset \mathbf{C}^n$ be a domain. Let $F_j(z)$ $(j = 1, \ldots, \nu)$ be ν holomorphic vector-valued functions of rank λ in D,

$$F_j(z) = (F_{1,j}(z), \ldots, F_{\lambda,j}(z)) \quad (j = 1, \ldots, \nu).$$

We let $\mathcal{J}^\lambda\{F\}$ denote the $\mathcal{O}$-module on D generated by $\{F_j(z)\}_{j=1,\ldots,\nu}$. In this setting, we pose the following two problems.

Problem C_1 Let $\Phi(z)$ be a holomorphic vector-valued function of rank λ on D such that $\Phi(z)$ belongs to $\mathcal{J}^\lambda\{F\}$ at each point in D. Find ν holomorphic functions $A_j(z)$ $(j = 1, \ldots, \nu)$ on D such that

$$\Phi(z) = A_1(z) F_1(z) + \cdots + A_\nu(z) F_\nu(z) \quad \text{in } D.$$

Problem C_2 For each $p \in D$, let the pair $(\phi_p(z), \delta_p)$ be given, where δ_p is a neighborhood of p in D and $\phi_p(z)$ is a holomorphic vector-valued function of rank λ in δ_p having the property that for any $p, q \in D$ with $\delta_p \cap \delta_q \neq \emptyset$, the difference $\phi_p(z) - \phi_q(z)$ belongs to $\mathcal{J}^\lambda\{F\}$ at each point of $\delta_p \cap \delta_q$. Find a holomorphic vector-valued function $\Phi(z)$ of rank λ in D such that for each p, $\Phi(z) - \phi_p(z)$ belongs to $\mathcal{J}^\lambda\{F\}$ at each point of δ_p.

We call the collection of pairs $\mathcal{C} := \{(\phi_p(z), \delta_p)\}_{p \in D}$ a **C_2-distribution**, and we call $\Phi(z)$ a **solution of Problem C_2** for the C_2-distribution $\mathcal{C}$.

Let $\mathcal{J}^\lambda$ be an $\mathcal{O}$-module of rank λ in D. We also consider the following problem.

Problem E Assume that $\mathcal{J}^\lambda$ has a locally finite pseudobase at each point of D. Find a finite number of holomorphic vector-valued functions $\Phi_k(z)$ $(k = 1, \ldots, \nu)$

of rank λ on D such that the $\mathcal{O}$-module $\mathcal{J}^\lambda\{\Phi\}$ generated by $\{\Phi_k(z)\}_{k=1,\dots,\nu}$ on D is equivalent to $\mathcal{J}^\lambda$ on D.

We also consider Problems C_1, C_2 and E for a *closed* set $D \subset \mathbf{C}^n$ and their solvability on D. To this end, let $D \subset \mathbf{C}^n$ be a closed set and let $F_j(z)$ $(j = 1, \dots, \nu)$ be ν holomorphic vector-valued functions of rank λ on D,

$$F_j(z) := (F_{1,j}(z), \dots, F_{\lambda,j}(z)) \quad (j = 1, \dots, \nu).$$

Recall that this means there exists an open set G with $D \subset G \subset \mathbf{C}^n$ such that each $F_j(z)$ $(j = 1, \dots, \nu)$ is holomorphic on G. We let $\mathcal{J}^\lambda\{F\}$ denote the $\mathcal{O}$-module on G generated by $\{F_j(z)\}_{j=1,\dots,\nu}$. In this setting we pose the following three problems.

Problem C_1 Let $\Phi(z)$ be a holomorphic vector-valued function of rank λ on D such that $\Phi(z)$ belongs to $\mathcal{J}^\lambda\{F\}$ at each point in D; i.e., there exists an open set G_1 with $D \subset G_1 \subset G$ such that $\Phi(z)$ is holomorphic on G_1 and $\Phi(z)$ belongs to $\mathcal{J}^\lambda\{F\}$ at each point in G_1. Find ν holomorphic functions $A_j(z)$ $(j = 1, \dots, \nu)$ on D such that

$$\Phi(z) = A_1(z)F_1(z) + \cdots + A_\nu(z)F_\nu(z) \quad \text{in } D;$$

i.e., each $A_j(z)$ $(j = 1, \dots, \nu)$ is holomorphic on an open set G_2, where $D \subset G_2 \subset G_1$, and the above relation is satisfied on G_2.

If this holds for any data $\Phi(z)$ on the closed set D in $\mathbf{C}^n$, then we say that Problem C_1 is solvable on the closed set D.

Problem C_2 For each $p \in D$, let the pair $(\phi_p(z), \delta_p)$ be given where δ_p is a neighborhood of p and $\phi_p(z)$ is a holomorphic vector-valued function of rank λ in δ_p having the property that for any $p, q \in D$ with $\delta_p \cap \delta_q \neq \emptyset$, the difference $\phi_p(z) - \phi_q(z)$ belongs to $\mathcal{J}^\lambda\{F\}$ at each point of $\delta_p \cap \delta_q$; i.e., there exists an open set $G_1^{p,q}$ with $\delta_p \cap \delta_q \subset G_1^{p,q} \subset G$ such that $\phi_p(z) - \phi_q(z)$ is holomorphic on $G_1^{p,q}$ and belongs to $\mathcal{J}^\lambda\{F\}$ at each point in $G_1^{p,q}$. Find a holomorphic vector-valued function $\Phi(z)$ of rank λ in D such that for each p,

$$\Phi(z) - \phi_p(z) \text{ belongs to } \mathcal{J}^\lambda\{F\}$$

at each point of δ_p; i.e., $\Phi(z)$ is holomorphic on an open set G_2, where $D \subset G_2 \subset \bigcup_p \delta_p$, and the above relation is satisfied on G_2.

If this holds for any such pair $(\phi_p(z), \delta_p)$, then we say that Problem C_2 is solvable on the closed set D.

Problem E Let $\mathcal{J}^\lambda$ be an $\mathcal{O}$-module of rank λ on D such that $\mathcal{J}^\lambda$ has a locally finite pseudobase at each point of D; i.e., $\mathcal{J}^\lambda$ is an $\mathcal{O}$-module of rank λ on an open set G with $D \subset G \subset \mathbf{C}^n$ and has a locally finite pseudobase at each point of G. Find a finite number of holomorphic vector-valued functions $\Phi_k(z)$ $(k = 1, \dots, \nu)$ of rank λ on D such that the $\mathcal{O}$-module $\mathcal{J}^\lambda\{\Phi\}$ generated by $\{\Phi_k(z)\}_{k=1,\dots,\nu}$ on D is equivalent to $\mathcal{J}^\lambda$ on D; i.e., $\Phi_k(z)$ $(k = 1, \dots, \nu)$ is a holomorphic vector-valued function of rank λ on an open set G_1 with $D \subset G_1 \subset G$ such that the $\mathcal{O}$-module $\mathcal{J}^\lambda\{\Phi\}$ generated by $\{\Phi_k(z)\}_{k=1,\dots,\nu}$ on G_1 is equivalent to $\mathcal{J}^\lambda$ on G_1.

If this holds for any such $\mathcal{O}$-module $\mathcal{J}^\lambda$ of rank λ on D in $\mathbf{C}^n$, then we say that Problem E is solvable on the closed set D.

These three problems were solved in a special case by K. Oka in 1943 in his reports in Japanese.[4] As with the Cousin problems, these problems cannot always

---

[4]See Oka's posthumous work No. 1 in [**55**].

be solved in arbitrary domains D in $\mathbf{C}^n$. In 1948, Oka solved these problems in the polydisk; this was published in French in 1950 (Oka [**50**]). In this section we present his proofs.

REMARK 7.8. This remark is for the reader familiar with sheaf theory; thus we do not explain the (standard) notation and terminology. We state the following two important results in sheaf theory.

1) Let V be an analytic space (to be defined in the next chapter) and let

$$0 \longrightarrow \mathcal{M}_1 \longrightarrow \mathcal{M}_2 \longrightarrow \mathcal{M}_3 \longrightarrow 0$$

be an exact sequence of sheaves on V. Then we have the following exact sequence of cohomology on V:

$$\begin{aligned} 0 &\longrightarrow \Gamma(V, \mathcal{M}_1) \longrightarrow \Gamma(V, \mathcal{M}_2) \longrightarrow \Gamma(V, \mathcal{M}_3) \\ &\longrightarrow H^1(V, \mathcal{M}_1) \longrightarrow H^1(V, \mathcal{M}_2) \longrightarrow H^1(V, \mathcal{M}_3) \longrightarrow \cdots . \end{aligned}$$

Thus if $H^1(V, \mathcal{M}_1) = 0$, then the mapping

$$\Gamma(V, \mathcal{M}_2) \longrightarrow \Gamma(V, \mathcal{M}_3)$$

is surjective.

2) Let $\mathcal{M}$ and $\mathcal{N}$ be two sheaves on V and let $\phi : \mathcal{M} \longrightarrow \mathcal{N}$ be a sheaf homomorphism. We let $\mathcal{K}$ denote the kernel of ϕ, and we let $\mathcal{I}$ denote the image of ϕ. Then

$$0 \longrightarrow \mathcal{K} \longrightarrow \mathcal{M} \longrightarrow \mathcal{M}/\mathcal{K} \longrightarrow 0$$

and

$$0 \longrightarrow \mathcal{I} \longrightarrow \mathcal{N} \longrightarrow \mathcal{N}/\mathcal{I} \longrightarrow 0$$

are exact sequences.

Now let D be a domain in $\mathbf{C}^n$ and let $\phi : \mathcal{O}^q(D) \longrightarrow \mathcal{O}^p(D)$ be a homomorphism with the propery that there exist q holomorphic vector-valued functions F_j ($j = 1, \ldots, q$) on D of rank p such that ϕ maps $\alpha = (\alpha_1, \ldots, \alpha_q) \in \mathcal{O}^q(D)$ to $\alpha_1 F_1 + \cdots + \alpha_q F_q \in \mathcal{O}^p(D)$. In this case we take $\mathcal{I}$ to be the $\mathcal{O}$-module $\mathcal{O}\{F\}$ generated by $\{F_j\}_{j=1,\ldots,q}$ on D, and we take $\mathcal{K}$ to be the $\mathcal{L}$-module $\mathcal{L}(\Omega)$ on D with respect to the linear relation

$$(\Omega) \qquad \alpha_1 F_1 + \cdots + \alpha_q F_q = 0.$$

Problem C_1 is to show that

$$\Gamma(D, \mathcal{O}^p) \longrightarrow \Gamma(D, \mathcal{O}^p/\mathcal{K})$$

is surjective, and Problem C_2 is to show that

$$\Gamma(D, \mathcal{O}^q) \longrightarrow \Gamma(D, \mathcal{O}^q/\mathcal{I})$$

is surjective. Furthermore, it is clear that $\mathcal{I}$ is coherent. The coherence of $\mathcal{K}$ is nothing but Oka's Theorem 7.1 on the existence of a locally finite pseudobase for $\mathcal{L}(\Omega)$ at each point in D.

7.4.2. Two Lemmas. For convenience we consider $\mathbf{C}^{n+1} = \mathbf{C}_z^n \times \mathbf{C}_w$ with variables $z_1, \ldots, z_n$ and w, and we set $w := u + iv$ $(i^2 = -1)$. Let G be a closed region in $\mathbf{C}_z^n$ and consider two closed rectangles K_1, K_2 in $\mathbf{C}_w$ constructed in the following manner: for $a < a' < b' < b$ and c, d, we define

$$\begin{aligned} K_1' &: \quad a \le u \le b', \qquad c \le v \le d, \\ K_2' &: \quad a' \le u \le b, \qquad c \le v \le d. \end{aligned}$$

In $\mathbf{C}_w$, we set

$$D' := K_1' \cap K_2', \qquad K' := K_1' \cup K_2',$$

and finally in $\mathbf{C}^{n+1}$ we define

$$K_1 := G \times K_1', \quad K_2 := G \times K_2', \quad D := G \times D', \quad K := G \times K'. \tag{7.12}$$

We let $l = 2(b' - a' + d - c)$ be the perimeter of D' and set $L = l/\pi$.

Choose $e > 0$ sufficiently small so that

$$e < \min \left\{ \frac{a' - a}{2}, \frac{b' - b}{2}, \frac{d - c}{2} \right\}.$$

In $\mathbf{C}_w$, we define

$$\begin{aligned} K_1'(e) &: \quad a - e \le u \le b' + e, \qquad c - e \le v \le d + e, \\ K_2'(e) &: \quad a' - e \le u \le b + e, \qquad c - e \le v \le d + e. \end{aligned}$$

Note that $K_1' \subset\subset K_1'(e)$ and $K_2' \subset\subset K_2'(e)$. We also set

$$D'(e) := K_1'(e) \cap K_2'(e), \qquad K'(e) := K_1'(e) \cup K_2'(e).$$

Finally, we define the following subsets of $\mathbf{C}^{n+1}$:

$$\begin{aligned} K_1(e) &:= G \times K_1'(e), \qquad & K_2(e) &:= G \times K_2'(e), \\ D(e) &:= G \times D'(e), \qquad & K(e) &:= G \times K'(e). \end{aligned}$$

We can now state the first lemma in this section.

LEMMA 7.1 (Cousin's lemma). [5] *Let $f_0(z, w)$ be a holomorphic function on $D(e)$.*

1. *There exist holomorphic functions $f_1(z, w)$ and $f_2(z, w)$ in K_1 and K_2 such that*
$$f_0(z, w) = f_1(z, w) + f_2(z, w) \quad \text{in } D.$$
2. *If $|f_0(z, w)| \le \rho$ on $D(e)$, then we can find $f_1(z, w)$ and $f_2(z, w)$ as in 1 which satisfy*
$$|f_1(z, w)| \le L\rho/e \text{ in } K_1 \quad \text{and} \quad |f_2(z, w)| \le L\rho/e \text{ in } K_2.$$

PROOF. We let C denote the boundary of the rectangle $D'(e)$ in $\mathbf{C}_w$. Fix two points $p := ((a' + b')/2, c - e)$ and $q := ((a' + b')/2, d + e)$ on C and let C_1 and C_2 denote the right- and left-hand portions of C divided by p and q. We form the Cousin integrals

$$\begin{aligned} f_1(z, w) &:= \frac{1}{2\pi i} \int_{C_1} \frac{f_0(z, \zeta)}{\zeta - w} d\zeta, \quad (z, w) \in K_1, \\ f_2(z, w) &:= \frac{1}{2\pi i} \int_{C_2} \frac{f_0(z, \zeta)}{\zeta - w} d\zeta, \quad (z, w) \in K_2, \end{aligned}$$

[5] This lemma was essentially proved in Part I; we repeat the statement and proof due to its importance.

where C_1 and C_2 are oriented so that $\partial D'(e) = C_1 + C_2$. Thus, in particular, $f_1(z,w)$ and $f_2(z,w)$ are holomorphic functions in K_1 and K_2. Cauchy's theorem implies that

$$f_0(z,w) = f_1(z,w) + f_2(z,w), \quad (z,w) \in D,$$

which proves assertion 1.

To prove 2, assume that $|f_0(z,w)| \leq \rho$ on $D(e)$. Let $(z,w) \in K_i$ $(i = 1,2)$. From the integral formula above for $f_i(z,w)$, since $|\zeta - w| \geq e$ if $\zeta \in C_i$, we obtain the estimate

$$|f_i(z,w)| \leq \frac{1}{2\pi}\int_{C_i} \frac{|f_0(z,w)|}{|\zeta - w|}|d\zeta| \leq \frac{1}{2\pi}\frac{\rho}{e}[(b'-a') + (d-c) + 4e] < \frac{L\rho}{e},$$

which proves 2. □

For each nonnegative integer n, we consider the sets

$$K_1'(e/2^n),\ K_2'(e/2^n), \dots, D(e/2^n),\ K(e/2^n);$$

clearly each corresponding sequence of closed sets is nested; e.g., $K_1'(\frac{e}{2^{n+1}}) \subset K_1'(\frac{e}{2^n})$, and these sequences decrease to

$$K_1',\ K_2', \dots, D \text{ and } K.$$

Hence, in the proof of Lemma 7.1 (replacing $C = \partial D'(e)$, $D(e)$, and D by $C_n = \partial D'(\frac{e}{2^n})$, $D(\frac{e}{2^n})$, and $D(\frac{e}{2^{n+1}})$, and using $|\zeta - w| \geq e/2^{n+1}$ for $\zeta \in \partial D(\frac{e}{2^n})$ and $w \in D(\frac{e}{2^{n+1}})$ in the last estimate), we obtain the following remark.

REMARK 7.9. Let $f_{0,n}(z,w)$ be a holomorphic function on $D(\frac{e}{2^n})$ with inequality $|f_{0,n}(z,w)| \leq \rho_n$ on $D(\frac{e}{2^n})$, where $\rho_n > 0$ is a constant. Then we obtain holomorphic functions $f_{1,n}(z,w)$ and $f_{2,n}(z,w)$ in $K_1(\frac{e}{2^{n+1}})$ and $K_2(\frac{e}{2^{n+1}})$ such that

(1) $f_{1,n}(z,w) + f_{2,n}(z,w) = f_{0,n}(z,w)$ in $D(e/2^{n+1})$;
(2) $|f_{j,n}(z,w)| \leq \frac{L}{e} 2^{n+1}\rho_n$ in $K_j(e/2^{n+1})$ $(j = 1,2)$.

Using 1 of Lemma 7.1 we have the following corollary.

COROLLARY 7.4. *Let $\Phi_j(z,w)$ $(j = 1, \dots, \nu)$ be a holomorphic vector-valued function of rank λ on the set K and let $\mathcal{J}^\lambda\{\Phi\}$ denote the $\mathcal{O}$-module generated by $\{\Phi_j(z,w)\}_{j=1,\dots,\nu}$ on K. Let $f_1(z,w)$ and $f_2(z,w)$ be holomorphic vector-valued functions of rank λ on K_1 and K_2 such that $f_1(z,w) - f_2(z,w)$ belongs to $\mathcal{J}^\lambda\{\Phi\}$ at each point in D. If Problem C_1 is always solvable on D, there exists a holomorphic vector-valued function $F(z,w)$ of rank λ on K such that $F(z,w) - f_1(z,w)$ belongs to $\mathcal{J}^\lambda\{\Phi\}$ on K_1 and $F(z,w) - f_2(z,w)$ belongs to $\mathcal{J}^\lambda\{\Phi\}$ on K_2.*

PROOF. From the hypothesis, for any $p \in D$, we can find ν holomorphic vector-valued functions $\alpha_j(z,w)$ $(j = 1, \dots, \nu)$ in a neighborhood δ_p of p in D such that

$$f_1(z,w) - f_2(z,w) = \alpha_1(z,w)\Phi_1(z,w) + \cdots + \alpha_\nu(z,w)\Phi_\nu(z,w) \qquad \text{in } \delta_p.$$

Since Problem C_1 is assumed to be solvable on D, we can find ν holomorphic vector-valued functions $A_j(z,w)$ $(j = 1, \dots, \nu)$ on D such that

$$f_1(z,w) - f_2(z,w) = A_1(z,w)\Phi_1(z,w) + \cdots + A_\nu(z,w)\Phi_\nu(z,w) \qquad \text{on } D.$$

Using 1 of Lemma 7.1 (note that if we take a sufficiently small $e > 0$, each $A_j(z,w)$ $(j = 1, \dots, \nu)$ is defined and holomorphic on $D(e)$), for each $j = (1, \dots, \nu)$ we can find holomorphic functions $A_j'(z,w)$ and $A_j''(z,w)$ on K_1 and K_2 such that

$$A_j(z,w) = A_1'(z,w) - A_j''(z,w) \quad \text{on } D.$$

Therefore, if we set

$$F(z,w) := \begin{cases} f_1(z,w) - (A_1'(z,w)\Phi_1(z,w) + \cdots + A_\nu'(z,w)\Phi_\nu(z,w)), & (z,w) \in K_1, \\ f_2(z,w) - (A_1''(z,w)\Phi_1(z,w) + \cdots + A_\nu''(z,w)\Phi_\nu(z,w)), & (z,w) \in K_2, \end{cases}$$

then $F(z,w)$ is a single-valued holomorphic vector-valued function of rank λ on K such that $F(z,w) - f_i(z,w)$ $(i = 1,2)$ belongs to $\mathcal{J}^\lambda\{\Phi\}$ on K_i. □

Repeating the same procedure step by step (similar to the solution of the Cousin I problem in 3.2.2), we obtain the following proposition.

PROPOSITION 7.5. *If Problem C_1 is always solvable in any closed polydisk in $\mathbf{C}^n$, then Problem C_2 is always solvable in any closed polydisk in $\mathbf{C}^n$.*

We want to show that under the hypothesis of Proposition 7.5, Problem E is always solvable in any closed polydisk in $\mathbf{C}^n$; then we will show that, indeed, Problem C_1 (and hence Problem C_2 and Problem E) is always solvable in any closed polydisk in $\mathbf{C}^n$. To do this, we will need a lemma of Cartan on holomorphic matrix-valued functions. First we introduce some notation involving these functions.

Let $\mathbf{C}^\nu$ be the space of ν complex variables $u_1, \dots, u_\nu$ and let $V \subset \mathbf{C}^\nu$ be a domain. We call an (m,n)-matrix $A(u) = (a_{j,k}(u))_{j,k}$ whose coefficients $a_{j,k}(u)$ are holomorphic functions in V, a **holomorphic (m,n)-matrix-valued function**, or simply an **(m,n)-holomorphic matrix** in V. We let $\mathcal{M}_{m,n}(V)$ denote the set of all (m,n)-holomorphic matrices in V. In case $m = n$, we call $A(u)$ a square holomorphic matrix of order m in V, and we write $\mathcal{M}_m(V) := \mathcal{M}_{m,m}(V)$. We let E denote the identity matrix of order m.

Given $A(u) \in \mathcal{M}_m(V)$ and an integer $l \geq 1$, for each $u \in V$, we write $A^l(u)$ for the l-th the power of matrix $A(u)$. Thus $A^l(u) \in \mathcal{M}_m(V)$. If $A(u)$ has an inverse matrix for each $u \in V$, we denote it by $A^{-1}(u)$; then $A^{-1}(u) \in \mathcal{M}_m(V)$, and we say that $A(u)$ is invertible in V.

Fix $\xi = (\xi_1, \dots, \xi_m) \in \mathbf{C}^m$ with $\|\xi\| = 1$ and fix $A(u) = (a_{j,k}(u))_{j,k} \in \mathcal{M}_{m,n}(V)$. Given $u \in V$, we define

$$\|A(u)\| := \max_{\|\xi\|=1} \{\|\xi \cdot A(u)\|\},$$

where $\|\xi \cdot A(u)\|$ denotes the Euclidean length in $\mathbf{C}^n$ of the image of ξ under the linear transformation $A(u) : \mathbf{C}^m \to \mathbf{C}^n$, and we define

$$\|A\|_V := \max_{u \in V} \{\|A(u)\|\}.$$

It is clear that $|a_{j,k}(u)| \leq \|A\|_V$ for each $1 \leq j \leq m$, $1 \leq k \leq n$ and $u \in V$; conversely, if $|a_{j,k}(u)| \leq M$ for each j,k and $u \in V$, then $\|A\| \leq \sqrt{mn}M$. Furthermore, for $A(u), B(u) \in \mathcal{M}_m(V)$,

$$\|A + B\|_V \leq \|A\|_V + \|B\|_V, \quad \|A \cdot B\|_V \leq \|A\|_V \cdot \|B\|_V.$$

Therefore, for $A(u) \in \mathcal{M}_m(V)$,

$$e^{A(u)} := E + \frac{A(u)}{1!} + \frac{A^2(u)}{2!} + \cdots$$

is well-defined, belongs to $\mathcal{M}_m(V)$, and is invertible (since $(e^{A(u)})^{-1} = e^{-A(u)}$). We note that the "usual" law of exponents $e^{A(u)+B(u)} = e^{A(u)} \cdot e^{B(u)}$ does not necessarily hold. It is valid, for example, if $A(u) \cdot B(u) = B(u) \cdot A(u)$.

PROPOSITION 7.6. *Let $A_j(u) \in \mathcal{M}_m(V)$ $(j = 1, 2, \dots)$ and let $\varepsilon_j, 0 < \varepsilon_j < 1$ $(j = 1, 2, \dots)$, satisfy $\sum_{j=1}^{\infty} \varepsilon_j < \infty$. If $\|A_j\|_V \le \varepsilon_j$ $(j = 1, 2, \dots)$, then*

$$B(u) := \lim_{n\to\infty} B_n(u) := \lim_{n\to\infty} (E - A_1(u))(E - A_2(u)) \cdots (E - A_n(u)),$$
$$C(u) := \lim_{n\to\infty} C_n(u) := \lim_{n\to\infty} (E - A_n(u))(E - A_{n-1}(u)) \cdots (E - A_1(u))$$

are uniformly convergent in V. Furthermore, $B(u)$ and $C(u)$ belong to $\mathcal{M}_m(V)$ and are invertible in V.

PROOF. We write $B_n(u) = (b_{j,k}^{(n)}(u))_{j,k}$. We note that, for $n = 1, 2, \dots,$

$$\|B_n\|_V \le \prod_{i=1}^{\infty} (1 + \varepsilon_i) =: M \ < \infty.$$

Let $l > k$ and set $\delta_k := \sum_{j=k+1}^{\infty} \varepsilon_j$. Then we have

$$\|B_l - B_k\|_V \le M\|E - (E - A_{k+1}) \cdots (E - A_l)\|_V \le M(\delta_k + \delta_k^2 + \cdots).$$

It follows that for each $j, k = 1, \dots, m$, the sequence of holomorphic functions $\{b_{j,k}^{(n)}(u)\}_n$ in V forms a Cauchy sequence, so that $\lim_{n\to\infty} B_n(u) =: B(u)$ converges uniformly in V and $B(u) \in \mathcal{M}_m(V)$. Moreover, each factor $E - A_j(u)$ $(j = 1, 2, \dots)$ is invertible in V, i.e.,

$$(E - A_j(u))^{-1} = E - A_j(u) + A_j^2(u) + \cdots,$$

which is uniformly convergent in V from the estimate $\|A_j\|_V < \varepsilon_j < 1$. So, $B_n(u)$ is invertible in V. Since $\| - A_j(u) + A_j^2(u) + \cdots \|_V \le K\varepsilon_j$ (where K is independent of $j = 1, 2, \dots$), we can similarly prove that $\lim_{n\to\infty} B_n^{-1}(u) =: B^*(u)$ converges uniformly in V and belongs to $\mathcal{M}_m(V)$. Since $B_n(u) \cdot B_n^{-1}(u) = E$ for $u \in V$, it follows that $B(u) \cdot B^*(u) = E$ for $u \in V$, so that $B(u)$ is invertible in V. Similarly, $C(u)$ belongs to $\mathcal{M}_m(V)$ and is invertible in V. □

We fix an integer $m \ge 1$, and use the notation $D(e), K_1, K_2, D = K_1 \cap K_2$, and $K = K_1 \cup K_2$ defined at the beginning of this section. We fix a small $e > 0$ such that $L/e > 1$. Recall that we consider $\mathbf{C}^{n+1} = \mathbf{C}_z^n \times \mathbf{C}_w$ with variables $z_1, \dots, z_n$ and w. Let $A(z, w) = (a_{j,k}(z, w))_{j,k} \in \mathcal{M}_m(D(e)) =: \mathcal{M}(D(e))$ and define $B(z, w) = (b_{j,k}(z, w))_{j,k}$ via

$$A(z, w) = E + B(z, w).$$

Set $\rho = \|B\|_{D(e)} \ge 0$. Applying Remark 7.9 to each $b_{j,k}(z, w)$ $(j, k = 1, \dots, m)$ in $D(e)$, we obtain holomorphic functions $b_{j,k}^{(1)}(z, w)$ and $b_{j,k}^{(2)}(z, w)$ in K_1 and K_2 such that

$$\begin{aligned} b_{j,k}(z, w) &= b_{j,k}^{(1)}(z, w) + b_{j,k}^{(2)}(z, w) \qquad \text{in } D, \\ |b_{j,k}^{(s)}(z, w)| &\le L\rho/e \quad \text{in } K_s \ (s = 1, 2). \end{aligned}$$

We set

$$B_s(z, w) := (b_{j,k}^{(s)}(z, w))_{j,k} \ \ (s = 1, 2) \qquad \text{in } K_s.$$

Thus, $B_s(z, w) \in \mathcal{M}(K_s)$ $(s = 1, 2)$ satisfies

$$B(z, w) = B_1(z, w) + B_2(z, w) \quad \text{in } D,$$
$$\|B_1\|_{K_1} \le mL\rho/e, \quad \|B_2\|_{K_2} \le mL\rho/e.$$

We also define $B^*(z,w)$ in D by the relation

$$(E - B_1(z,w))(E + B(z,w))(E - B_2(z,w)) = E + B^*(z,w) \quad \text{in } D,$$

i.e.,

$$B^* = B_1B_2 - B_1B - BB_2 + B_1BB_2;$$

hence

$$\|B^*\|_D \leq 3\,(mL\rho/e)^2 + (mL\rho/e)^3.$$

We are now ready to state and prove Cartan's lemma [**11**].

LEMMA 7.2 (Cartan's lemma). *Let $A(z,w) \in \mathcal{M}_m(D(e))$ be invertible in $D(e)$. If $A(z,\ w)$ is sufficiently close to the identity matrix E of order m in $D(e)$, then there exist $A_1(z,w) \in \mathcal{M}_m(K_1)$ and $A_2(z,w) \in \mathcal{M}_m(K_2)$ which are invertible in K_1 and K_2 and such that*

$$A(z,w) = A_1(z,w) \cdot A_2^{-1}(z,w) \quad \textit{in } D.$$

PROOF. For simplicity we omit the subscript m; e.g., $\mathcal{M}_m(E) = \mathcal{M}(E)$. To prove the lemma it suffices to find $A_1(z,w) \in \mathcal{M}_m(K_1^o)$ and $A_2(z,w) \in \mathcal{M}_m(K_2^o)$ such that $A(z,w) = A_1(z,w) \cdot A_2^{-1}(z,w)$ in D^o (where K_i^o and D^o denote the interior of K_i and D). For $n = 0, 1, \dots$, we set

$$D_n := D(e/2^n), \quad K_{n,1} := K_1(e/2^{n+1}), \quad K_{n,2} := K_2(e/2^{n+1}),$$

so that $K_{n+1,s} \subset\subset K_{n,s}$ $(s = 1,2)$, $D_{n+1} = K_{n,1} \cap K_{n,2}$ and $D^o = \lim_{n\to\infty} D_n$.

We construct sequences $B_n(z,w) \in \mathcal{M}(D_n)$, $B^{(n,1)}(z,w) \in \mathcal{M}(K_{n,1})$, and $B^{(n,2)}(z,w) \in \mathcal{M}(K_{n,2})$ inductively as follows. We define $B_0(z,w) \in \mathcal{M}(D_0)$ by the relation

$$A(z,w) = E + B_0(z,w) \quad \text{in } D_0$$

and we set $\rho_0 := \|B_0\|_{D_0} \geq 0$.

Now fix $n \geq 0$, assume that $B_n(z,w) = (b_{j,k}^{(n)}(z,w))_{j,k} \in \mathcal{M}(D_n)$ has been defined, and set $\rho_l := \|B_l\|_{D_l}$ $(l = 0, ..., n)$.

Applying Remark 7.9 following Lemma 7.1 to each $b_{j,k}^{(n)}(z,w)$ $(j,k = 1,\dots,m)$ in D_n, we obtain holomorphic functions $b_{j,k}^{(n,1)}(z,w)$ and $b_{j,k}^{(n,2)}(z,w)$ in $K_{n,1}$ and $K_{n,2}$ such that

$$\begin{aligned} b_{j,k}^{(n)}(z,w) &= b_{j,k}^{(n,1)}(z,w) + b_{j,k}^{(n,2)}(z,w) \qquad \text{in } D_{n+1}, \\ |b_{j,k}^{(n,s)}(z,w)| &\leq 2^{n+1}L\rho_n/e \quad \text{in } K_{n,s} \quad (s = 1,2). \end{aligned}$$

We write

$$B^{(n,s)}(z,w) := (b_{j,k}^{(n,s)}(z,w))_{j,k} \qquad \text{in } K_{n,s} \quad (s = 1,2).$$

Thus $B^{(n,s)}(z,w) \in \mathcal{M}(K_{n,s})$ $(s = 1,2)$ satisfies

$$\begin{aligned} B_n(z,w) &= B^{(n,1)}(z,w) + B^{(n,2)}(z,w) \quad \text{in } D_{n+1}, \\ \|B^{(n,s)}\|_{K_{n,s}} &\leq M\,2^n\rho_n \qquad (s = 1,2), \end{aligned} \tag{7.13}$$

where $M := 2mL/e > 1$ is independent of n. We then define $B_{n+1}(z,w) \in \mathcal{M}(D_{n+1})$ by

$$(E - B^{(n,1)}(z,w))(E + B_n(z,w))(E - B^{(n,2)}(z,w)) = E + B_{n+1}(z,w)) \quad \text{in } D_{n+1},$$

i.e.,

$$B_{n+1} = B^{(n,1)}B^{(n,2)} - B^{(n,1)}B_n - B_nB^{(n,2)} + B^{(n,1)}B_nB^{(n,2)}.$$

Thus if we set $\rho_{n+1} = \|B_{n+1}\|_{D_{n+1}}$, we have

$$\rho_{n+1} \leq 3M^2(2^n\rho_n)^2 + M^3(2^n\rho_n)^3. \tag{7.14}$$

This implies that if $\rho_0 > 0$ is sufficiently small, then

$$\rho_n \leq 1/4^n \ (n = 0, 1, \dots). \tag{7.15}$$

In fact, by setting $\tau_n = 4^n\rho_n$ $(n = 0, 1, \dots)$, we have, from (7.14),

$$\tau_{n+1} \leq 12M^2\tau_n^2 + 4M^3\tau_n^3.$$

Consequently, if we take a sufficiently small $\rho_0 = \tau_0$ with $0 < \rho_0 < 1$, then $\{\tau_n\}_n$ decreases to 0, so that

$$\rho_n = \tau_n/4^n \leq \tau_0/4^n \leq 1/4^n,$$

which proves (7.15).

Together with (7.13), this implies that if we take $\rho_0 > 0$ sufficiently small, i.e., if $A(z,w) = E + B_0(z,w)$ is sufficiently close to the identity matrix E in $D(e)$, then we have

$$\|B_n\|_{D_n} \leq 1/4^n, \qquad \|B^{(n,s)}\|_{K_{n,s}} \leq M/2^n \quad (s = 1, 2). \tag{7.16}$$

Now for $n = 0, 1, \dots,$ we define

$$A_{n,1}(z,w) := (E - B^{(n,1)}(z,w))(E - B^{(n-1,1)}(z,w)) \cdots (E - B^{(0,1)}(z,w)) \quad \text{in } K_1,$$
$$A_{n,2}(z,w) := (E - B^{(0,2)}(z,w))(E - B^{(1,2)}(z,w)) \cdots (E - B^{(n,2)}(z,w)) \quad \text{in } K_2,$$

so that

$$A_{n,1}(z,w)A(z,w)A_{n,2}(z,w) = E - B_{n+1}(z,w) \quad \text{in } D.$$

It follows from (7.16) and Proposition 7.6 that $A_{n,1}(z,w) \in \mathcal{M}(K_1)$ and $A_{n,2}(z,w) \in \mathcal{M}(K_2)$ are invertible in K_1 and K_2, and that the sequences $\{A_{n,1}(z,w)\}_n$ and $\{A_{n,2}(z,w)\}_n$ are uniformly convergent in K_1 and K_2. Thus,

$$A_1(z,w) := \lim_{n\to\infty} A_{n,1}(z,w) \in \mathcal{M}(K_1^o),$$
$$A_2(z,w) := \lim_{n\to\infty} A_{n,2}(z,w) \in \mathcal{M}(K_2^o),$$

which are also invertible in K_1 and K_2 with $\|A_s\|_{K_s} \leq 3$ $(s = 1, 2)$. Inequality (7.16) also implies that

$$A_1(z,w)A(z,w)A_2(z,w) = E \quad \text{on } D^o,$$

as desired. □

REMARK 7.10. 1. Since D is closed in $\mathbf{C}^{n+1}$ and $e > 0$ can be taken as small as we want in Cartan's lemma, we shall use the lemma in the following form: Let $A(z,w) \in \mathcal{M}_m(D)$ be invertible and sufficiently close to the identity matrix E on D. Then there exist $A_i(z,w)$ $(i = 1, 2)$ invertible on K_i such that $A(z,w) = A_1(z,w) \cdot A_2(z,w)$ on D.

2. Cartan's lemma holds for any $A(z,w) \in \mathcal{M}_m(D(e))$ which is invertible in $D(e)$ in the case when $G \subset \mathbf{C}_z^n$ is simply connected. For, in this case, $A(z,w)$ can be written as a product of a finite number of holomorphic matrices $A_k(z,w)$ $(k = 1, \dots, \nu)$ which are sufficiently close to E and are invertible in $D(e)$. However, we will not need this fact.

Let p and q be positive integers. We assume G (stated in (7.12)) is a *closed polydisk* in $\mathbf{C}^n_z$, and we use the same notation K_1, K_2, D, and K as before in $\mathbf{C}^{n+1}$. We consider p holomorphic vector-valued functions of rank λ in K_1 and q holomorphic vector-valued functions of rank λ in K_2:

$$f_j(z,w) \quad (j=1,\dots,p) \quad \text{in } K_1, \qquad g_j(z,w) \quad (j=1,\dots,q) \quad \text{in } K_2.$$

We let $\mathcal{J}^\lambda\{f\}$ and $\mathcal{J}^\lambda\{g\}$ denote the $\mathcal{O}$-modules generated by $\{f_j(z,w)\}_j$ and $\{g_j(z,w)\}_j$ in K_1 and K_2.

Then we obtain the following corollary.

COROLLARY 7.5. *Assume that each $f_j(z,w)$ $(j=1,\dots,p)$ belongs to $\mathcal{J}^\lambda\{g\}$ on D, and that each $g_j(z,w)$ $(j=1,\dots,q)$ belongs to $\mathcal{J}^\lambda\{f\}$ on D. Then there exist a finite number of holomorphic vector-valued functions $F_j(z,w)$ $(j=1,\dots,p+q)$ in $K:=K_1\cup K_2$ such that the $\mathcal{O}$-module $\mathcal{J}^\lambda\{F\}$ generated by $\{F_j(z,w)\}_j$ in K is equivalent to $\mathcal{J}^\lambda\{f\}$ on K_1 and to $\mathcal{J}^\lambda\{g\}$ on K_2.*

PROOF. By the hypothesis we can find $A_{q,p}(z,w)=(\alpha_{j,k}(z,w))_{j,k}\in\mathcal{M}_{q,p}(D)$ and $B(z,w)_{p,q}=(\beta_{j,k}(z,w))_{j,k}\in\mathcal{M}_{p,q}(D)$ satisfying

$$\begin{aligned}(f_1,\dots,f_p) &= (g_1,\dots,g_q)\,A_{q,p} \quad \text{in } D,\\ (g_1,\dots,g_q) &= (f_1,\dots,f_p)\,B_{p,q} \quad \text{in } D.\end{aligned}$$

On the other hand, since $D=G\times D'$, where G is a closed polydisk in $\mathbf{C}^n_z$ and D' is a rectangle in $\mathbf{C}_w$, by Runge's theorem, given $\varepsilon>0$, there exist $A'_{q,p}(z,w)=(\alpha'_{j,k}(z,w))_{j,k}\in\mathcal{M}_{q,p}(\mathbf{C}^{n+1})$ and $B'_{p,q}(z,w)=(\beta'_{j,k}(z,w))_{j,k}\in\mathcal{M}_{p,q}(\mathbf{C}^{n+1})$ such that, for each j,k,

$$\begin{aligned}|\alpha_{j,k}(z,w)-\alpha'_{j,k}(z,w)| &\le \varepsilon \quad \text{for } (z,w)\in K_1,\\ |\beta_{j,k}(z,w)-\beta'_{j,k}(z,w)| &\le \varepsilon \quad \text{for } (z,w)\in K_2.\end{aligned}$$

If we write

$$\begin{aligned}(f_1,\dots,f_p) &= (g_1,\dots,g_q)\cdot A'_{q,p}+(g_1,\dots,g_q)\cdot[A_{q,p}-A'_{q,p}]\\ &\equiv (g'_1,\dots,g'_p)+(g_1,\dots,g_q)\cdot A''_{q,p} \quad \text{in } D,\end{aligned}$$

then we see that $g'_j(z,w)\in\mathcal{J}^\lambda\{g\}$ $(j=1,\dots,p)$ on K_2, $g'_j(z,w)$ is close to $f_j(z,w)$ on D, and $A''_{q,p}(z,w)\in\mathcal{M}_{q,p}(D)$ is close to the zero matrix on D. Analogously, we have

$$\begin{aligned}(g_1,\dots,g_q) &= (f_1,\dots,f_p)\cdot B'_{p,q}+(f_1,\dots,f_p)\cdot[B_{p,q}-B'_{p,q}]\\ &\equiv (f'_1,\dots,f'_q)+(f_1,\dots,f_p)\cdot B''_{p,q} \quad \text{in } D,\end{aligned}$$

so that $f'_j(z,w)\in\mathcal{J}^\lambda\{f\}$ $(j=1,\dots,q)$ on K_1, $f'_j(z,w)$ is close to $g_j(z,w)$ on D, and $B''_{p,q}(z,w)\in\mathcal{M}_{p,q}(D)$ is close to the zero matrix on D. We then have

$$\begin{aligned}(f'_1,\dots,f'_q) &= (g_1,\dots,g_q)-[(g'_1,\dots,g'_p)+(g_1,\dots,g_q)A''_{q,p}]\cdot B''_{p,q}\\ &= (g_1,\dots,g_q)\cdot[E-A''_{q,p}B''_{p,q}]-(g'_1,\dots,g'_p)\cdot B''_{p,q}\\ &\equiv (g_1,\dots,g_q)\cdot C_{q,q}+(g'_1,\dots,g'_p)\cdot B''_{p,q} \quad \text{in } D,\end{aligned}$$

where $C_{q,q}(z,w)\in\mathcal{M}_{q,q}(D)$ is close to the identity matrix E of order q in D. Consequently,

$$\begin{aligned}(f_1,\dots,f_p,f'_1,\dots,f'_q) &= (g'_1,\dots,g'_p,g_1,\dots,g_q)\begin{pmatrix}E_{q,q} & B''_{p,q}\\ A''_{q,p} & C_{q,q}\end{pmatrix}\\ &\equiv (g'_1,\dots,g'_p,g_1,\dots,g_q)\cdot R_{p+q,p+q} \quad \text{in } D,\end{aligned}$$

where $R_{p+q,p+q}(z,w) \in \mathcal{M}_{p+q}(D)$ is close to the identity matrix E of order $p+q$ in D. Applying Cartan's lemma (see 1 of Remark 7.10), there exist $A_1(z,w) \in \mathcal{M}_{p+q}(K_1)$ and $A_2(z,w) \in \mathcal{M}_{p+q}(K_2)$ which are invertible in K_1 and K_2 and such that

$$R_{p+q}(z,w) = A_2(z,w) \cdot A_1^{-1}(z,w) \quad \text{in } D.$$

Thus, if we set

$$(F_1, \dots, F_{p+q}) = \begin{cases} (f_1, \dots, f_p, f_1', \dots, f_q') \cdot A_1 & \text{in } K_1, \\ (g_1', \dots, g_p', g_1, \dots, g_q) \cdot A_2 & \text{in } K_2, \end{cases}$$

then $F_j(z,w)$ $(j = 1, \dots, p+q)$ is a single-valued holomorphic vector-valued function of rank λ on K. Furthermore, since $A_1(z,w)$ and $A_2(z,w)$ are invertible in K_1 and K_2, it is clear that the $\mathcal{O}$-module $\mathcal{J}^\lambda\{F\}$ generated by $\{F_j(z,w)\}_{j=1,\dots,p+q}$ is equivalent to $\mathcal{J}^\lambda\{f\}$ on K_1 and to $\mathcal{J}^\lambda\{g\}$ on K_2. □

By repeating the same procedure step by step, we reach the conclusion that *if Problem C_1 is always solvable in any polydisk in $\mathbf{C}^n$, then Problem E is always solvable in any polydisk in $\mathbf{C}^n$.* Thus it remains to prove that Problem C_1 is always solvable in any polydisk in $\mathbf{C}^n$. However, we cannot verify this directly; instead, by making careful use of Corollary 7.5, of the Cousin integral, and of the main theorem (Theorem 7.1) in the following section we shall simultaneously solve Problems C_1 and Problem E in polydisks by a double induction procedure.

7.4.3. Combination Theorem. We shall prove that Problem C_1 and Problem E are always solvable in any closed polydisk in $\mathbf{C}^n$. These two problems will be solved simultaneously by a double induction procedure.

Let $\mathbf{C}^n$ have complex variables $z = (z_1, \dots, z_n)$ and write

$$z_j = t_{2j-1} + i\, t_{2j} \qquad (i^2 = -1;\ j = 1, \dots, n),$$

where t_{2j-1} and t_{2j} are real numbers. Let a_k, b_k be $2n$ real numbers with $a_k \le b_k$ for $k = 1, \dots, 2n$, and set

$$L_k \ :\ a_k \le t_k \le b_k \quad (k = 1, \dots, 2n), \qquad E := L_1 \times \cdots \times L_{2n}.$$

We call E a box in $\mathbf{C}^n$. For a fixed $l = 1, \dots, 2n$, we consider the subset in $\mathbf{C}^n$ defined by

$$E^l \ :\ a_j < t_j < b_j \ \ (j = 1, \dots, l), \quad t_j = a_j = b_j \ \ (j = l+1, \ \dots, 2n).$$

By convention, we set $E^0 = \{(a_1, a_2, \dots, a_{2n})\}$ (one point). We call E^l a **real l-dimensional open box**, and the closure $\overline{E}^l$ of E^l in $\mathbf{C}^n$ is a real l-dimensional closed box.

Given E^l as above, we call a set O^l in $\mathbf{C}^n$ of the form

$$O^l \ :\ a_j' < t_j < b_j' \ \ (j = 1, \dots, l), \quad t_j : |t_j - a_j| < \varepsilon_j \ \ (j = l+1, \ \dots, 2n),$$

where

$$a_j' < a_j,\ b_j < b_j' \ (j = 1, \dots, l) \qquad \varepsilon_j > 0 \ (j = l+1, \dots, 2n),$$

an open box neighborhood of $\overline{E}^l$ in $\mathbf{C}^n$.

Let $\overline{E}^l (l = 0, 1, \dots, 2n)$ be a real l-dimensional closed box in $\mathbf{C}^n$. We say that Problem C_1 is solvable on the real l-dimensional closed box $\overline{E}^l$ if the following condition is satisfied: Let $F_j(z)$ $(j = 1, \dots, \nu)$ and $\Phi(z)$ be holomorphic vector-valued functions of rank λ on $\overline{E}^l$ (i.e., on a neighborhood U of $\overline{E}^l$ in $\mathbf{C}^n$) such that

$\Phi(z) \in \mathcal{J}^\lambda\{F\}$ at any point z in U. Here $\mathcal{J}^\lambda\{F\}$ denotes the $\mathcal{O}$-module generated by $\{F_j(z)\}_{j=1,\dots,\nu}$ in U. Then there exist an open box neighborhood O of $\overline{E}^l$ and holomorphic functions $A_j(z)$ $(j=1,\dots,\nu)$ on O such that $\overline{E}^l \subset\subset O \subset\subset U$ and

$$\Phi(z) = A_1(z)F_1(z) + \cdots + A_\nu(z)F_\nu(z)$$

on O. In a similar fashion, we define the notion of Problem E being solvable on $\overline{E}^l$.

LEMMA 7.3. *Fix l with $0 \le l \le 2n$. If Problem C_1 and Problem E are solvable for any real l-dimensional closed box $\overline{E}^l$, then Problem E is solvable for any real $(l+1)$-dimensional closed box $\overline{E}^{l+1}$.*

PROOF. Let

$$E^{l+1}: \ a_j < t_j < b_j \ \ (j=1,\dots,l+1), \quad t_j = c_j \ (j = l+2,\ \dots, 2n)$$

be a real $(l+1)$-dimensional box in $\mathbf{C}^n$. Let G be a neighborhood of $\overline{E}^{l+1}$, and let $(\{\psi_j^{(p)}\}_{j=1,\dots,k_p}, \delta_p)_{p\in G}$ be data for Problem E on G; i.e., δ_p is a neighborhood of p in $\mathbf{C}^n$ and $\psi_j^{(p)}$ $(j=1,\dots,k_p)$ are holomorphic vector-valued functions of rank λ in δ_p such that if $\delta_p \cap \delta_q \neq \emptyset$ $(p, q \in G)$, then $\mathcal{J}^\lambda\{\psi^{(p)}\}$ and $\mathcal{J}^\lambda\{\psi^{(q)}\}$ are equivalent to each other on $\delta_p \cap \delta_q$. Here $\mathcal{J}^\lambda\{\psi^{(p)}\}$ denotes the $\mathcal{O}$-module generated by $\{\psi_j^{(p)}\}_{j=1,\dots,k_p}$ on δ_p.

Fix a point c in $[a_{l+1}, b_{l+1}]$ and set

$$\overline{E}^l(c): \ a_j \le t_j \le b_j \ \ (j=1,\dots,l), \quad t_{l+1} = c, \ \ t_j = c_j \ (j=l+2,\ \dots, 2n),$$

Since $\overline{E}^l(c) \subset \overline{E}^{l+1}$ is a real l-dimensional closed box in $\mathbf{C}^n$, it follows that Problem E is solvable on $\overline{E}^l(c)$. Thus we can find box neighborhoods $O^*(c)$ and $O(c)$ of $\overline{E}^l(c)$ in $\mathbf{C}^n$ with $O(c) \subset\subset O^*(c)$ and a finite number of holomorphic vector-valued functions $\Psi_j^{(c)}(z)$ $(j=1,\dots,\mu_c)$ of rank λ on $O^*(c)$ such that the $\mathcal{O}$-module $\mathcal{J}^\lambda\{\Psi^{(c)}\}$ generated by $\{\Psi_j^{(c)}(z)\}_{j=1,\dots,\mu_c}$ on $O^*(c)$ is equivalent to $\mathcal{J}^\lambda\{\psi^{(p)}\}$ on $O^*(c)\cap\delta_p$, $p \in G$. Since c is an arbitrary point in $[a_{l+1}, b_{l+1}]$, it follows from the Heine- Borel theorem that there is a finite cover

$$\bigcup_{i=1}^{m} O(c_i)$$

of $\overline{E}^{l+1}$, where $a_{l+1} = c_1 < c_2 < \dots < c_m = b_{l+1}$. For simplicity, we set $O(c_i) = O_i$, $O^*(c_i) = O_i^*$, $\Psi_j^{(c_i)}(z) = \Psi_j^i(z)$ $(j=1,\dots,\nu_i = \nu_{c_i})$ and $\mathcal{J}^\lambda\{\Psi^{c_i}\} = \mathcal{J}^\lambda\{\Psi^i\}$ on O_i^* $(i=1,\dots,m)$. By shrinking O_i if necessary, we may assume that each O_i is of the form

$$O_i: \quad \begin{array}{l} \alpha_j < t_j < \beta_j \ (j=1,\dots,l), \ \gamma_i < t_{l+1} < \delta_i, \\ |t_j - c_j| < \varepsilon \ (j = l+2,\dots,2n), \end{array}$$

with

$$\begin{array}{l} \alpha_j < a_j, \ b_j < \beta_j \ (j=1,\dots,l), \\ \gamma_1 < \gamma_2 < \delta_1 < \gamma_3 < \delta_2 < \gamma_4 < \dots < \gamma_m < \delta_{m-1} < \delta_m. \end{array}$$

Now we focus on the pairs $(\mathcal{J}^\lambda\{\Psi^1\}, O_1^*)$ and $(\mathcal{J}^\lambda\{\Psi^2\}, O_2^*)$, and consider the following real $(l+1)$-dimensional box $T^{l+1} \subset\subset O_1^* \cup O_2^*$:

$$T^{l+1}: \ \alpha_j < t_j < \beta_j \ (j=1,\dots,l), \ \gamma_1 < t_{l+1} < \delta_2, \ t_j = c_j \ (j=l+2,\dots,2n).$$

We prove the following assertion:

(∗) There exist a box neighborhood U^* in $\mathbf{C}^n$ with

$$\overline{T}^{l+1} \subset\subset U^* \subset\subset O_1^* \cup O_2^*$$

and a finite number of holomorphic vector-valued functions $F_j(z)$ ($j = 1, \dots, \mu$) of rank λ on U^* such that the $\mathcal{O}$-module $\mathcal{J}^\lambda\{F\}$ generated by $\{F_j(z)\}_{j=1,\dots,\mu}$ on U^* is equivalent to $\mathcal{J}^\lambda\{\psi^{(p)}\}$ on $\delta_p \cap U^*$, $p \in G$.

To prove this, we set $Q := O_1^* \cap O_2^*$, which is a real $2n$-dimensional box in $\mathbf{C}^n$. Fix a point $t_{l+1} = q \in (\gamma_2, \delta_1)$ and consider the real l-dimensional closed box

$$\overline{K}^l : \alpha_j \le t_j \le \beta_j \ (j = 1, \dots, l),\ t_{l+1} = q,\ t_j = c_j \ (j = l+2, \dots, 2n).$$

We note that $\overline{K}^l \subset\subset Q$ and that $\mathcal{J}^\lambda\{\Psi^1\}$ and $\mathcal{J}^\lambda\{\Psi^2\}$ are equivalent to each other on Q. Since Problem C_1 is solvable on $\overline{K}^l$, it follows that there exists a box neighborhood V^* in $\mathbf{C}^n$ with

$$\overline{K}^l \subset\subset V^* \subset\subset Q$$

such that each $\Psi_j^1(z)$ ($j = 1, \dots, \nu_1$) belongs to $\mathcal{J}^\lambda\{\Psi^2\}$ on V^*, and, similarly, each $\Psi_j^2(z)$ ($j = 1, \dots, \nu_2$) belongs to $\mathcal{J}^\lambda\{\Psi^1\}$ on V^*. We can take V^* of the form

$$V^* : \alpha_j^* < t_j < \beta_j^* \ (j = 1, \dots, l),\ \gamma^* < t_{l+1} < \delta^*,\ |t_j - c_j| < \varepsilon^* \ (j = l+2, \dots, 2n),$$

where

$$\gamma_2 < \gamma^* < q < \delta^* < \delta_1 \quad \text{and} \quad 0 < \varepsilon^* < \varepsilon.$$

We define the real $2n$-dimensional boxes $U_1^* \subset\subset O_1^*$ and $U_2^* \subset\subset O_2^*$ by

$$U_1^* : \alpha_j^* < t_j < \beta_j^* \ (j = 1, \dots, l),\ \gamma_1 < t_{l+1} < \delta^*,\ |t_j - c_j| < \varepsilon^* \ (j = l+2, \dots, 2n),$$
$$U_2^* : \alpha_j^* < t_j < \beta_j^* \ (j = 1, \dots, l),\ \gamma^* < t_{l+1} < \delta_2,\ |t_j - c_j| < \varepsilon^* \ (j = l+2, \dots, 2n),$$

so that $U_1^* \cap U_2^* = V^*$ and $U^* := U_1^* \cup U_2^*$ is a box neighborhood of $\overline{T}^{l+1}$ in $\mathbf{C}^n$. It follows from Corollary 7.5 that there exist a finite number of holomorphic vector-valued functions $F_j(z)$ ($j = 1, \dots, \mu$) of rank λ on U^* such that $\mathcal{J}^\lambda\{F\}$ is equivalent to $\mathcal{J}^\lambda\{\psi^1\}$ on U_1^* and to $\mathcal{J}^\lambda\{\psi^2\}$ on U_2^*. Thus assertion (∗) is proved.

We repeat the same procedure for the pairs $(\mathcal{J}^\lambda\{F\}, U^*)$ and $(\mathcal{J}^\lambda\{\Psi^3\}, O_3^*)$ as for the pairs $(\mathcal{J}^\lambda\{\Psi^1\}, O_1^*)$ and $(\mathcal{J}^\lambda\{\Psi^2\}, O_2^*)$; continuing this process, we finally obtain a box neighborhood Λ^* of $\overline{E}^{l+1}$ in $\mathbf{C}^n$ and a finite number of holomorphic vector-valued functions $\Phi_j(z)$ ($j = 1, \dots, M$) of rank λ on Λ^* such that the $\mathcal{O}$-module $\mathcal{J}^\lambda\{\Phi\}$ generated by $\{\Phi_j(z)\}_{j=1,\dots,M}$ on Λ^* is equivalent to $\mathcal{J}^\lambda\{\psi^{(p)}\}$ on $\delta_p \cap \Lambda^*$, $p \in G$. This proves that Problem E is always solvable on any real $(l+1)$-dimensional closed box in $\mathbf{C}^n$. □

LEMMA 7.4. *Fix an integer l with $1 \le l \le 2n$. Assume that Problem C_1 is solvable for any real l-dimensional closed box and that Problem E is solvable for any real $(l+1)$-dimensional closed box. Then Problem C_1 is solvable for any real $(l+1)$-dimensional closed box.*

PROOF. Let

$$\overline{K}^{l+1} : \ a_j \le t_j \le b_j \ \ (j = 1, \dots, l+1), \quad t_j = c_j \ (j = l+2, \ \dots, 2n)$$

be a real $(l+1)$-dimensional closed box in $\mathbf{C}^n$. Let $\psi_j(z)$ ($j = 1, \dots, \nu$) and $F(z)$ be holomorphic vector-valued functions on $\overline{K}^{l+1}$ (i.e., on a neighborhood U of $\overline{K}^{l+1}$

in $\mathbf{C}^n$) such that, for each point $z_0 \in U$, there exist a neighborhood δ of z_0 and ν holomorphic functions $f_j(z)$ $(j = 1, \dots, \nu)$ on δ such that

$$F(z) = f_1(z)\psi_1(x) + \dots + f_\nu(z)\psi_\nu(z) \quad \text{on } \delta.$$

Fix c in $[a_{l+1}, b_{l+1}]$ and set

$$\overline{E}^l(c): \ a_j \le t_j \le b_j \ \ (j = 1, \dots, l), \quad t_{l+1} = c, \quad t_j = c_j \ (j = l+2, \ \dots, 2n).$$

Since $\overline{E}^l(c)$ is a real l-dimensional closed box in $\mathbf{C}^n$, it follows that Problem C_1 is solvable on $\overline{E}^l(c)$. Thus we can find box neighborhoods $O^*(c)$ and $O(c)$ of $\overline{E}^l(c)$ in U with $O(c) \subset\subset O^*(c)$ and ν holomorphic functions $f_j^{(c)}(z)$ $(j = 1, \dots, \nu)$ on $O^*(c)$ such that

$$F(z) = f_1^{(c)}(z)\psi_1(z) + \dots + f_\nu^{(c)}(z)\psi_\nu(z) \quad \text{on } O^*(c).$$

Since c was an arbitrary point in the interval $[a_{l+1}, b_{l+1}]$, it follows from the Heine-Borel theorem that there exists a finite cover

$$\bigcup_{i=1}^{m} O(c_i)$$

of $\overline{E}^{l+1}$, where $a_{l+1} = c_1 < c_2 < \dots < c_m = b_{l+1}$. For simplicity, we set $O(c_i) = O_i$, $O^*(c_i) = O_i^*$, $f_j^{(c_i)}(z) = f_j^i(z)$ $(j = 1, \dots, \nu)$ on O_i^* $(i = 1, \dots, m)$. By shrinking O_i, if necessary, we may assume that each O_i is of the form

$$O_i : \alpha_j < t_j < \beta_j \ (j = 1, \dots, l), \ \gamma_i < t_{l+1} < \delta_i, \ |t_j - c_j| < \varepsilon \ (j = l+2, \dots, 2n)$$

with

$$\begin{gathered} \alpha_j < a_j, \ b_j < \beta_j \quad (j = 1, \dots, l), \\ \gamma_1 < \gamma_2 < \delta_1 < \gamma_3 < \delta_2 < \gamma_4 < \dots < \gamma_m < \delta_{m-1} < \delta_m. \end{gathered}$$

We focus on the pairs $(\{f_j^1(z)\}_j, O_1^*)$ and $(\{f_j^2(z)\}_j, O_2^*)$, and consider the following real $(l+1)$-dimensional box $T^{l+1} \subset\subset O_1^* \cup O_2^*$:

$$T^{l+1} : \ \alpha_j < t_j < \beta_j \ (j = 1, \dots, l), \ \gamma_1 < t_{l+1} < \delta_2, \ t_j = c_j \ (j = l+2, \dots, 2n).$$

We prove the following assertion:

(**) There exist a box neighborhood W^* in $\mathbf{C}^n$,

$$\overline{T}^{l+1} \subset\subset W^* \subset\subset O_1^* \cup O_2^*,$$

and ν holomorphic functions $F_j(z)$ $(j = 1, \dots, \nu)$ on W^* such that

$$F(z) = F_1(z)\psi_1(z) + \dots + F_\nu(z)\psi_\nu(z) \quad \text{on } W^*.$$

To prove this, we set $Q^* := O_1^* \cap O_2^*$ and consider the simultaneous linear equations

$$(\Omega) \qquad f_1(z)\psi_1(z) + \dots + f_\nu(z)\psi_\nu(z) = 0 \quad \text{on } Q^*$$

and the $\mathcal{O}$-module $\mathcal{L}\{\Omega\}$ with respect to the linear relation (Ω). We note that

$$\mathbf{g}(z) := (f_1^1(z) - f_1^2(z), \dots, f_\nu^1(z) - f_\nu^2(z))$$

belongs to $\mathcal{L}\{\Omega\}$ on Q^*.

By the main theorem (Theorem 7.1), we see that $\mathcal{L}\{\Omega\}$ has a finite pseudobase $\{h^{(p)}(z)\}_{j=1,\dots,m_p}$ at each point $p \in Q^*$. Fix $t_{l+1} = q \in (\gamma_2, \delta_1)$ and consider the real l-dimensional closed box

$$\overline{K}^l : \alpha_j \leq t_j \leq \beta_j \ (j = 1, \dots, l), \ t_{l+1} = q, \ t_j = c_j \ (j = l+2, \dots, 2n).$$

We note that $\overline{K}^l \subset\subset Q^*$. Since Problem E is solvable on $\overline{K}^l$, it follows that there exist a box neighborhood V^* in $\mathbf{C}^n$ with

$$\overline{K}^l \subset\subset V^* \subset\subset Q^*$$

and a finite number of holomorphic vector-valued functions $H_j(z)$ $(j = 1, \dots, s)$ of rank ν on V^* such that $\{H_j(z)\}_{j=1,\dots,s}$ is a finite pseudobase of $\mathcal{L}\{\Omega\}$ on V^*.

We again look at the real l-dimensional closed box $\overline{K}^l$ defined above. We note that $\mathbf{g}(z)$ and $H_j(z)$ $(j = 1, \dots, s)$ are defined in V^* and that $\mathbf{g}(z)$ belongs to the $\mathcal{O}$-module $\mathcal{J}^\lambda\{H\}$ generated by $\{H_j(z)\}_{j=1,\dots,M}$ on V^* at each point of V^*. Since Problem C_1 is solvable on $\overline{K}^l$ and $\overline{K}^l \subset\subset V^*$, it follows that there exist a box neighborhood W in $\mathbf{C}^n$ with

$$\overline{K}^l \subset\subset W \subset\subset V^*$$

and s holomorphic functions $A_j(z)$ $(j = 1, \dots, s)$ on W such that

$$\mathbf{g}(z) = A_1(z)H_1(z) + \cdots + A_s(z)H_s(z) \qquad \text{on } W.$$

We write

$$W : \ \alpha_j^0 < t_j < \beta_j^0 \ (j = 1, \dots, l), \ \gamma^0 < t_{l+1} < \delta^0, \ |t_j - c_j| < \varepsilon^0 \ (j = l+2, \dots, 2n)$$

and consider the following real $2n$-dimensional boxes:

$$W_1 : \alpha_j^0 < t_j < \beta_j^0 \ (j = 1, \dots, l), \ \gamma_1 < t_{l+1} < \delta^0, \ |t_j - c_j| < \varepsilon^0 \ (j = l+2, \dots, 2n),$$

$$W_2 : \alpha_j^0 < t_j < \beta_j^0 \ (j = 1, \dots, l), \ \gamma^0 < t_{l+1} < \delta_2, \ |t_j - c_j| < \varepsilon^0 \ (j = l+2, \dots, 2n).$$

Note that $W = W_1 \cap W_2$ and $W^* := W_1 \cup W_2$ is a box neighborhood of the real $(l+1)$-dimensional closed box $\overline{T}^{l+1}$ defined above. Using the Cousin integral for each $A_j(z)$ $(j = 1, \dots, s)$ on W along a segment on $t_{l+1} = q$, we can find holomorphic functions $A_j^1(z)$ and $A_j^2(z)$ on W_1 and W_2 such that

$$A_j^1(z) - A_j^2(z) = A_j(z) \qquad \text{on } W.$$

For $j = 1, \dots, \nu$ we set

$$F_j(z) := \begin{cases} f_j^1(z) - (A_1^1(z)H_1(z) + \cdots + A_s^1(z)H_s(z)) & \text{on } W_1, \\ f_j^2(z) - (A_1^2(z)H_1(z) + \cdots + A_s^2(z)H_s(z)) & \text{on } W_2. \end{cases}$$

Then $F_j(z)$ $(j = 1, \dots, \nu)$ is a single-valued holomorphic function on W^* which satisfies

$$F(z) = F_1(z)\psi_1(z) + \cdots + F_\nu(z)\psi_\nu(z) \qquad \text{on } W^*;$$

this proves assertion $(**)$.

As usual, we repeat this for the pairs $(\{F_j(z)\}_j, W^*)$ and $(\{f_j^3(z)\}_j, O_3^*)$ (as was done for the pairs $(\{f_j^1(z)\}_j, O_1^*)$ and $(\{f_j^2(z)\}_j, O_2^*)$); continuing this procedure proves the lemma. □

Observing that, by definition, Problem C_1 and Problem E are always solvable for any real 0-dimensional closed box $\overline{E}^0$ in $\mathbf{C}^n$ (here $\overline{E}^0$ is a point in $\mathbf{C}^n$), we obtain from Lemmas 7.3 and 7.4 the following result.

THEOREM 7.5 (Combination theorem). *Problem C_1, Problem C_2 and Problem E are always solvable for any closed polydisk in $\mathbf{C}^n$.*

As a simple application of this theorem we have

COROLLARY 7.6. *Let $\Phi_j(z)\,(j = 1,\dots,\nu)$ be holomorphic functions on the closed polydisk Δ in $\mathbf{C}^n$. If the functions $\Phi_j(z)\,(j = 1,\dots,\nu)$ have no common zeros on Δ, then there exist holomorphic functions $f_j(z)\,(j = 1,\dots,\nu)$ on Δ such that*

$$f_1(z)\Phi_1(z) + \cdots + f_\nu(z)\Phi_\nu(z) = 1 \quad \text{on} \ \Delta.$$

7.4.4. Completeness. Let $D \subset \mathbf{C}^n$ be a domain and let $\Phi_j(z)\ (j = 1,\dots,\nu)$ be ν holomorphic vector-valued functions of rank λ in D. We let $\mathcal{J}^\lambda\{\Phi\}$ denote the $\mathcal{O}$-module generated by $\{\Phi_j(z)\}_{j=1,\dots,\nu}$ on D. The following completeness theorem for $\mathcal{J}^\lambda\{\Phi\}$ will be useful in the next chapter.

THEOREM 7.6. *Let δ be a domain in D and let*

$$f_\iota(z) = (f_1^\iota(z),\dots,f_\lambda^\iota(z)) \qquad (\iota = 1,2,\dots)$$

be a sequence of holomorphic vector-valued functions on δ such that

(1) *each $(f_\iota(z),\delta) \in \mathcal{J}^\lambda\{\Phi\}$ $(\iota = 1,2,\dots)$, and*
(2) *$\{f_\iota(z)\}_{\iota=1,2,\dots}$ converges uniformly to a holomorphic vector-valued function $f_0(z)$ on δ.*

Then $f_0(z)$ belongs to $\mathcal{J}^\lambda\{\Phi\}$ at each point in δ.

In order to prove this theorem, we first prove a lemma about solving Problem C_1 with local estimates. Given $f(z) = (f_1(z),\dots,f_\lambda(z))$, we define $|f(z)| = \max_{j=1,\dots,\lambda}\{|f_j(z)|\}$.

LEMMA 7.5. *Let D be a polydisk centered at the origin O in $\mathbf{C}^n$. Let $F_j(z) = (F_{1,j}(z),\dots,F_{\lambda,j}(z))$ $(j = 1,\dots,\nu)$ be ν holomorphic vector-valued functions of rank λ on D. Then we can find a polydisk $\delta_0 \subset D$ centered at O and a constant $K > 0$ with the following property. Let $f(z) = (f_1(z),\dots,f_\lambda(z))$ be a holomorphic vector-valued function on D such that*

$$f(z) = a_1(z)F_1(z) + \cdots + a_\nu(z)F_\nu(z) \quad \text{on} \ D, \tag{7.17}$$

$$|f(z)| \le 1 \quad \text{on} \ D, \tag{7.18}$$

where each $a_j(z)$ $(j = 1,\dots,\nu)$ is a holomorphic function on D. Then $f(z)$ can be written in the form

$$f(z) = a_1^0(z)F_1(z) + \cdots + a_\nu^0(z)F_\nu(z) \quad \text{on} \ \delta_0,$$

$$|a_j^0(z)| \le K \quad (j = 1,\dots,\nu) \quad \text{on} \ \delta_0,$$

where each $a_j^0(z)$ $(j = 1,\dots,\nu)$ is a holomorphic function on δ_0.

PROOF. The proof will proceed by a double induction on the dimension $n \ge 1$ and the rank $\lambda \ge 1$ as in the proof of the main theorem.

First step. The lemma is true in the case $(n,\lambda) = (1,1)$.

We fix a closed disk $\delta_0 \subset\subset D$ centered at O such that we have $F_j(z) = z^{k_j}h_j(z)$ $(j = 1,\dots,\nu)$ on δ_0 with $h_j(z) \neq 0$ on δ_0. For simplicity, suppose $k_1 \le k_j$ $(j = 2,\dots,\nu)$. Let $K := \max_{z\in\partial\delta_0}\{1/|F_1(z)|\} > 0$. Then any holomorphic function $f(z)$ satisfying (7.17) and (7.18) can be written in the form

$f(z) = A_1(z)F_1(z)$ on δ_0, where $A_1(z)$ is a holomorphic function in δ_0. Hence $|A_1(z)| \le K$ on δ_0 by the maximum modulus principle, which proves the first step of the induction.

Second step. The lemma is true in the case $(n, \lambda+1)$ if the lemma is true in the cases (n,k) $(k = 1, \dots, \lambda)$.

Let

$$F_j(z) = (F_{0,j}(z), F_{1,j}(z), \dots, F_{\lambda,j}(z)) \qquad (j = 1, \dots, \nu)$$

be a holomorphic vector-valued function of rank $\lambda + 1$ in a polydisk D centered at the origin O in $\mathbf{C}^n$. Let

$$f(z) = (f_0(z), f_1(z), \dots, f_\lambda(z))$$

be a holomorphic vector-valued function of rank $\lambda + 1$ in D with

$$(\mathcal{E}) \qquad f(z) = a_1(z)F_1(z) + \cdots + a_\nu(z)F_\nu(z) \quad \text{on } D,$$

$$|f(z)| \le 1 \quad \text{on } D.$$

We fix a polydisk $D_0 \subset\subset D$ centered at O and set $M := \max\{\sum_{j=1}^{\nu} |F_j(z)| \mid z \in D_0\} < \infty$. Define the holomorphic vector-valued functions

$$\begin{aligned} F_j^0(z) &= (F_{1,j}(z), \dots, F_{\lambda,j}(z)) \ (j = 1, \dots, \nu), \\ f^0(z) &= (f_1(z), \dots, f_\lambda(z)), \end{aligned}$$

each of rank λ in D. Then

$$(\mathcal{E}^0) \qquad f^0(z) = a_1(z)F_1^0(z) + \cdots + a_\nu(z)F_\nu^0(z) \quad \text{on } D,$$

$$(\mathcal{E}_1) \qquad f_0(z) = a_1(z)F_{0,1}(z) + \cdots + a_\nu(z)F_{0,\nu}(z) \quad \text{on } D,$$

so that $(\mathcal{E}^0)$ is of type (n, λ) and $(\mathcal{E}_1)$ is of type $(n, 1)$.

By the induction assumption applied to $(\mathcal{E}_1)$ on D, we can find a closed polydisk $\delta_1 \subset D_0$ centered at O, a constant $K_1 > 0$ independent of $f_0(z)$, and holomorphic functions $a_j^0(z)$ $(j = 1, \dots, \nu)$ on δ_1 such that

$$\begin{aligned} f_0(z) &= a_1^0(z)F_{0,1}(z) + \cdots + a_\nu^0(z)F_{0,\nu}(z) \quad \text{on } \delta_1, \\ |a_j^0(z)| &\le K_1 \quad \text{on } \delta_1. \end{aligned}$$

Note that $|f_0(z)| \le K_1 M$ on δ_1.

Now we consider the single linear equation

$$(\Omega_0) \qquad b_1(z)F_{0,1}(z) + \cdots + b_\nu(z)F_{0,\nu}(z) = 0$$

for the unknown holomorphic vector-valued function

$$b(z) = (b_1(z), \dots, b_\nu(z))$$

of rank ν, and we consider the $\mathcal{O}^\nu$-module $\mathcal{L}\{\Omega_0\}$ with respect to the linear relation (Ω_0). By the main theorem (Theorem 7.1), the $\mathcal{O}^\nu$-module $\mathcal{L}\{\Omega_0\}$ has a locally finite pseudobase at the origin O; thus we can find a finite number, say μ, of holomorphic vector-valued functions

$$\Phi_k(z) = (\Phi_{1,k}(z), \dots, \Phi_{\nu,k}(z)) \qquad (k = 1, \dots, \mu)$$

of rank ν on a closed polydisk $\delta_2 \subset \delta_1$ centered at O which generate $\mathcal{L}\{\Omega_0\}$ on δ_2 (where neither δ_2 nor $\Phi_k(z)$ $(k = 1, \dots, \mu)$ depends on $f(z)$ in $(\mathcal{E})$). We set $M' = \max\{\sum_{k=1}^{\mu} |\Phi_k(z)| \mid z \in \delta_2\} < \infty$. Note that if we set

$$a(z) - a^0(z) = (a_1(z) - a_1^0(z), \dots, a_\nu(z) - a_\nu^0(z)) \quad \text{on } \delta_1,$$

then $a(z) - a^0(z)$ belongs to $\mathcal{L}\{\Omega_0\}$ on δ_2. Since Problem C_1 is solvable on the closed polydisk δ_2, we can find holomorphic functions $c_j(z)$ $(j = 1, \dots, \mu)$ on δ_2 such that

$$a(z) - a^0(z) = c_1(z)\Phi_1(z) + \cdots + c_\mu(z)\Phi_\mu(z) \quad \text{on } \delta_2 \tag{7.19}$$

(note that for any holomorphic functions $c_j(z)$ $(j = 1, \dots, \mu)$ on δ_2, the functions $a_j(z)$ $(j = 1, \dots, \nu)$ obtained by substituting the functions $c_j(z)$ into (7.19) automatically satisfy $(\mathcal{E}_1)$ on δ_2). We substitute this expression into $(\mathcal{E}^0)$ and obtain

$$\begin{aligned} f^0(z) &- (\, a_1^0(z)F_1^0(z) + \cdots + a_\nu^0(z)F_\nu^0(z)\,) \\ &= c_1(z)\left(\Phi_{1,1}(z)F_1^0(z) + \cdots + \Phi_{\nu,1}(z)F_\nu^0(z)\right) \\ &\quad + \cdots \\ &\quad + c_\mu(z)\left(\Phi_{1,\mu}(z)F_1^0(z) + \cdots + \Phi_{\nu,\mu}(z)F_\nu^0(z)\right) \quad \text{on } \delta_2 \end{aligned}$$

as holomorphic vector-valued functions of rank λ. If we set

$$\begin{aligned} g^0(z) &= f^0(z) - (\, a_1^0(z)F_1^0(z) + \cdots + a_\nu^0(z)F_\nu^0(z)\,) \quad \text{on } \delta_2, \\ G_j(z) &= \Phi_{1,j}(z)F_1^0(z) + \cdots + \Phi_{\nu,j}(z)F_\nu^0(z) \quad (j = 1, \dots, \mu) \quad \text{on } \delta_2, \end{aligned}$$

then we have

$$(\mathcal{G}) \qquad g^0(z) = c_1(z)G_1(z) + \cdots + c_\mu(z)G_\mu(z) \quad \text{on } \delta_2.$$

Again, note that for any holomorphic functions $c_j(z)$ $(j = 1, \dots, \mu)$ satisfying these λ equations $(\mathcal{G})$ on $\delta \subset \delta_2$, the functions $a_j(z)$ $(j = 1, \dots, \nu)$ obtained by substituting the functions $c_j(z)$ into (7.19) automatically satisfy $(\mathcal{E}^0)$ on δ, and hence both $(\mathcal{E}_1)$ and $(\mathcal{E}^0)$ on δ. We have that $|g^0(z)| \le 1 + K_1 M$ on δ_2. Since equation $(\mathcal{G})$ is of type (n, λ) on δ_2, the inductive hypothesis applied to $G_j(z)$ $(j = 1, \dots, \mu)$ and δ_2 implies that there exist a closed polydisk $\delta_3 \subset \delta_2$ centered at O, a constant $K_3 > 0$ independent of $g^0(z)$, and μ holomorphic functions $c_j^0(z)$ $(j = 1, \dots, \mu)$ such that

$$\begin{aligned} g^0(z) &= c_1^0(z)G_1(z) + \cdots + c_\mu^0(z)G_\mu(z) && \text{on } \delta_3, \\ |c_j^0(z)| &\le K_3(1 + K_1 M) \ (j = 1, \dots, \mu) && \text{on } \delta_3. \end{aligned}$$

Thus, if we set

$$a^*(z) := a^0(z) + c_1^0(z)\Phi_1(z) + \cdots + c_\mu^0(z)\Phi_\mu(z) \quad \text{on } \delta_3,$$

then we have

$$\begin{aligned} f(z) &= a_1^*(z)F_1(z) + \cdots + a_\nu^*(z)F_\nu(z) && \text{on } \delta_3, \\ |a_j^*(z)| &\le K_1 + K_3(1 + K_1 M)M' \equiv K \ (j = 1, \dots, \nu) && \text{on } \delta_3. \end{aligned}$$

Since δ_3, K_1, K_3, M and M' do not depend on the choice of $f(z)$ satisfying $(\mathcal{E})$ with $|f(z)| \le 1$ on D, the second step is proved (using δ_3 and $K > 0$).

Third step. The lemma is true in the case $(n+1, 1)$ if the lemma is true in the cases (n, λ) for $\lambda = 1, 2, \dots$.

Let D be a polydisk centered at the origin O in $\mathbf{C}^{n+1}$ and let $F_j(z)$ $(j = 1, \dots, \nu)$ be holomorphic functions on D.

We may assume that the z_{n+1}-direction satisfies the Weierstrass condition for each analytic hypersurface $\Sigma_j : F_j(z) = 0$ $(j = 1, \dots, \nu)$ at $z = O$. For convenience we write $z = (z_1, \ldots, z_n)$ and $w = z_{n+1}$, so that $\mathbf{C}^{n+1} = \mathbf{C}_z^n \times \mathbf{C}_w$. We can thus find a closed polydisk $\Lambda := \Delta \times \Gamma$ centered at $(z, w) = (0, 0) = O$ in $\mathbf{C}^n \times \mathbf{C}_w$,

$$\Delta : |z_j| \le \rho \quad (j = 1, \ldots, n), \qquad \Gamma : |w| \le \eta$$

such that $F_j(z, w) \neq 0 \quad (j = 1, \ldots, \nu)$ on $\Delta \times \partial\Gamma$. Thus we have

$$F_j(z, w) = \omega_j(z, w) P_j(z, w) \quad (j = 1, \ldots, \nu) \quad \text{on } \Lambda,$$

where $\omega_j(z, w) \neq 0$ at any point $(z, w) \in \Lambda$ and where $P_j(z, w)$ is a monic pseudopolynomial in w with coefficient functions that are holomorphic on Δ,

$$P_j(z, w) = w^{k_j} + A_{j,1}(z) w^{k_j - 1} + \cdots + A_{j,k_j}(z) \quad \text{on } \Lambda, \tag{7.20}$$

and such that $\Sigma_j = \{(z, w) \in \Delta \times \mathbf{C}_w \mid P_j(z, w) = 0\}$. Thus, instead of finding a polydisk $\delta_0 \subset D$ centered at O and a constant $K > 0$ for $F_j(z, w)$ $(j = 1, \ldots, \nu)$ and D to satisfy the conclusion of the third step, it suffices to find a polydisk $\Lambda^* \subset \Lambda$ centered at O and a constant $K^* > 0$ for $P_j(z, w)$ $(j = 1, \ldots, \nu)$ and Λ.

Without loss of generality, we will assume $k_\nu \ge k_j$ $(j = 1, \ldots, \nu - 1)$; i.e., the monic pseudopolynomial $P_\nu(z, w)$ has largest degree in w among all the monic pseudopolynomials $P_j(z, w)$. Let $f(z)$ be a holomorphic function on Λ satisfying

$$(\mathcal{E}) \qquad f(z, w) = a_1(z, w) P_1(z, w) + \cdots + a_\nu(z, w) P_\nu(z, w) \quad \text{on } \Lambda,$$
$$|f(z, w)| \le 1 \quad \text{on } \Lambda,$$

where each $a_j(z, w)$ $(j = 1, \ldots, \nu)$ is a holomorphic function on Λ. By the remainder theorem applied to $P_\nu(z, w)$, we have

$$f(z, w) = q(z, w) P_\nu(z, w) + r(z, w) \qquad \text{on } \Lambda, \tag{7.21}$$

where $q(z, w)$ is a holomorphic function on Λ and $r(z, w)$ is a pseudopolynomial in w of degree at most $k_\nu - 1$ with coefficient functions that are holomorphic for z in Δ,

$$r(z, w) = \beta_0(z) w^{k_\nu - 1} + \beta_1(z) w^{k_\nu - 2} + \cdots + \beta_{k_\nu - 1}(z) \quad \text{on } \Lambda.$$

Fix $\Gamma_0 : |w| \le \eta_0 < \eta$ and $\Lambda_0 := \Delta \times \Gamma_0$. From (2) of Theorem 7.2 we can find $M > 0$, independent of $f(z, w)$, such that

$$|q(z, w)|, \ |r(z, w)| \le M \quad \text{on } \Lambda_0,$$
$$|\beta_j(z)| \le M \quad (j = 0, 1, \ldots, k_\nu - 1) \quad \text{on } \Delta.$$

Similarly we have

$$a_j(z, w) = q_j(z, w) P_\nu(z, w) + r_j(z, w) \quad (j = 1, \ldots, \nu - 1) \quad \text{on } \Lambda,$$

where each $q_j(z, w)$ is a holomorphic function on Λ and each $r_j(z, w)$ is a pseudopolynomial in w of degree at most $k_\nu - 1$ with coefficient functions which are holomorphic for z in Δ,

$$r_j(z, w) = c_{j,0}(z) w^{k_\nu - 1} + c_{j,1}(z) w^{k_\nu - 2} + \cdots + c_{j,k_\nu - 1}(z) \quad \text{on } \Lambda.$$

Therefore, from $(\mathcal{E})$ we have

$$\begin{aligned} r(z, w) - (r_1(z, w) P_1(z, w) + \cdots + r_{\nu-1}(z, w) P_{\nu-1}(z, w)) \\ = r_\nu(z, w) P_\nu(z, w), \end{aligned} \tag{7.22}$$

where $r_\nu(z, w)$ is a holomorphic function on Λ. By the division theorem we see that $r_\nu(z, w)$ must be a pseudopolynomial in w of degree at most $k_\nu - 1$ with coefficient functions which are holomorphic for z in Δ,

$$r_\nu(z, w) = c_{\nu,0}(z) w^{k_\nu - 1} + c_{\nu,1}(z) w^{k_\nu - 2} + \cdots + c_{\nu,k_\nu - 1}(z) \qquad \text{on } \Lambda.$$

Thus, by comparing the coefficients of w^k on both sides of the equation (7.22) on Λ, we obtain the following $2k_\nu$ simultaneous linear equations $(\widehat{\mathcal{E}})$ on Δ:

$$(\widehat{\mathcal{E}}) \qquad \begin{cases} \beta_k(z) &= \displaystyle\sum_{i,j} c_{i,j}(z)\mathcal{A}_{i,j}^{(k)}(z) \quad (k = 0, \dots, k_\nu - 1), \\ 0 &= \displaystyle\sum_{i,j} c_{i,j}(z)\mathcal{A}_{i,j}^{(k)}(z) \quad (k = k_\nu, k_\nu + 1, \dots, 2k_\nu - 1), \end{cases}$$

where each $\mathcal{A}_{i,j}^{(k)}(z)$ is a linear combination of the functions $\{A_{l,m}(z)\}_{l,m}$ on Δ. If we define $\widetilde{\beta}(z) := (\beta_0(z), \dots, \beta_{k_\nu-1}(z), 0, \dots, 0)$ and $\widetilde{\mathcal{A}}_{i,j}(z) := (\mathcal{A}_{i,j}^{(0)}(z), \dots, \mathcal{A}_{i,j}^{(2k_\nu-1)}$ $(z))$, then the set of equations $(\widehat{\mathcal{E}})$ can be rewritten as

$$(\widetilde{\mathcal{E}}) \qquad \widetilde{\beta}(z) = \sum_{i,j} c_{i,j}(z)\widetilde{\mathcal{A}}_{i,j}(z) \quad \text{on} \quad \Delta$$

with $|\widetilde{\beta}(z)| \leq M$ on Δ. Since this $(\widetilde{\mathcal{E}})$ is a case of the form $(n, 2k_\nu)$, it follows by the inductive hypothesis applied to $\{\widetilde{\mathcal{A}}_{i,j}(z)\}_{i,j}$ (which is determined by the given $P_j(z,w)$ $(j = 1, \dots, \nu)$ in (7.20)) and Δ that we can find a polydisk $\Delta_0 \subset \Delta$ centered at O, a constant K_1 independent of $\widetilde{\beta}(z)$, and holomorphic functions $c_{i,j}^0(z)$ on Δ_0 such that $|c_{i,j}^0(z)| \leq K_1 M$ on Δ_0 and $c_{i,j}^0(z)$ (as well as $c_{i,j}(z)$) satisfy the equations $(\widetilde{\mathcal{E}})$ on Δ_0. Conversely, if we construct pseudopolynomials $r_j^0(z,w)$ $(j = 1, \dots, \nu)$ in w of degree at most $k_\nu - 1$ using $c_{i,j}^0(z)$ (as $r_j(z,w)$ $(j = 1, \dots, \nu)$ are constructed using $c_{i,j}(z)$), then by (7.22) we obtain

$$r(z,w) = r_1^0(z,w)P_1(z,w) + \cdots + r_{\nu-1}^0(z,w)P_{\nu-1}(z,w) + r_\nu^0(z,w)P_\nu(z,w)$$

on $\Delta_0 \times \Gamma \equiv \Lambda^*$ in Λ, and $|r_j^0(z,w)| \leq K_1 M \sum_{j=0}^{k_\nu - 1} \eta^j \equiv K_2$ $(j = 1, \dots, \nu)$ on Λ^*. Since (7.21) implies that

$$f = r_1^0\, P_1 + \cdots + r_{\nu-1}^0\, P_{\nu-1} + (r_\nu^0 + q)\, P_\nu \quad \text{on} \quad \Lambda^*,$$

the third step is proved (using the polydisk Λ^* and the constant $K^* := K_2 + M > 0$). This completes the proof of the lemma. □

REMARK 7.11. Now that Lemma 7.5 is established, we can use the same double induction method with respect to the real dimension of $\mathbf{R}^{2n}$ as in section 7.4.3 to extend Lemma 7.5 from a polydisk δ_0 to an arbitrary subset $D_0 \subset\subset D$ (D is a polydisk in $\mathbf{C}^n$), where the constant $K > 0$ depends on D_0. Moreover, in Theorem 8.16 in Chapter 8 we shall extend this lemma to a more general situation using another method (by use of the open mapping theorem for Fréchet spaces based on Lemma 7.5).

We now use Lemma 7.5 to prove Thorem 7.6.

PROOF OF THEOREM 7.6. By taking a subsequence of $\{f_\iota(z)\}_{\iota=1,2,\dots}$ and a smaller δ, if necessary, we may assume that $\{f_\iota(z)\}_{\iota=1,2,\dots}$ converges uniformly to $f_0(z)$ on δ with

$$\max_{z\in\delta} \{|f_{\iota+1}(z) - f_\iota(z)|\} < 1/2^\iota \quad (\iota = 1, 2, \dots),$$

and that each $f_\iota(z)$ can be written in the form

$$f_\iota(z) = a_1^{(\iota)}(z)\Phi_1(z) + \cdots + a_\nu^{(\iota)}(z)\Phi_\nu(z) \quad \text{on} \quad \delta.$$

Fix $q \in \delta$. From Lemma 7.5, it follows that there exist a neighborhood δ_0 of q in δ and a constant $K > 0$ satisfying the conditions in the lemma for the functions $\{\Phi_j(z)\}_{j=1,\dots,\nu}$ on δ. Thus, there exist holomorphic functions $c_j^{(\iota)}(z)$ ($j = 1,\dots,\nu; \iota = 1,2,\dots$) on δ_0 with

$$f_{\iota+1}(z) - f_\iota(z) = c_1^{(\iota)}(z)\Phi_1(z) + \cdots + c_\nu^{(\iota)}(z)\Phi_\nu(z) \quad \text{on } \delta_0,$$
$$\max_{z\in\delta_0} \{|c_j^{(\iota)}(z)|\} \le K/2^\iota \quad (j = 1,\dots,\nu).$$

For $j = 1,\dots,\nu$, we set

$$c_j(z) := a_j^{(1)}(z) + \sum_{\iota=1}^{\infty} c_j^{(\iota)}(z) \quad \text{on } \delta_0;$$

then $\{c_j(z)\}_j$ converges uniformly on δ_0, and we have

$$f_0(z) = c_1(z)\Phi_1(z) + \cdots + c_\nu(z)\Phi_\nu(z) \quad \text{on } \delta_0.$$

Thus $f_0(z)$ belongs to $\mathcal{J}^\lambda\{\Phi\}$ on δ_0. □

7.5. Local Finiteness Theorem

7.5.1. ℓ-ideal. Let D be a domain in $\mathbf{C}^n$ with variables $z_1,\dots,z_n$. Let $F_j(z)$ ($j = 1,\dots,\nu$) be ν holomorphic vector-valued functions of rank λ on D,

$$F_j(z) = (F_{1,j}(z),\dots,F_{\lambda,j}(z)) \quad (j = 1,\dots,\nu).$$

Consider the set of λ homogeneous linear simultaneous equations

$$(\Omega) \qquad f_1(z)F_1(z) + \cdots + f_\nu(z)F_\nu(z) = 0$$

for the unknown holomorphic vector-valued function $f(z) = (f_1(z),\dots,f_\nu(z))$ of rank ν. We refer to this system as the linear relation (Ω). We let $\mathcal{L}\{\Omega\}$ denote the $\mathcal{O}$-module with respect to the linear relation (Ω), i.e., $\mathcal{L}\{\Omega\}$ is the set of all pairs $(f(z),\delta)$ such that $f(z) = (f_1(z),\dots,f_\nu(z))$ is a holomorphic vector-valued function of rank ν on δ which satisfies (Ω) on δ. Looking at the first components of $f(z)$, we consider the set $\ell\{\Omega\}$ of all pairs $(f_1(z),\delta)$ such that there exists at least one $(f(z),\delta)$ in $\mathcal{L}\{\Omega\}$ with $f(z) = (f_1(z),\dots,f_\nu(z))$. Then $\ell\{\Omega\}$ is an $\mathcal{O}$-ideal on D which is called the **ℓ-ideal** with respect to the **linear relation (Ω)**. Since $\mathcal{L}\{\Omega\}$ has a locally finite pseudobase at each point in $\mathcal{D}$, we have the following theorem.

THEOREM 7.7. *For the linear relation (Ω) associated to the holomorphic vector-valued functions $F_j(z)$ ($j = 1,\dots,\nu$) on D, the ℓ-ideal $\ell\{\Omega\}$ has a locally finite pseudobase at each point in D.*

In this section we will often use this theorem to show that some important $\mathcal{O}$-ideals on $\mathcal{D}$ have a locally finite pseudobase at each point in D. We next prove the following corollary, due to Oka, which is a simple application of Theorem 7.7. This corollary will not be used in the remainder of this book.

Let $\mathcal{I}$ be an $\mathcal{O}$-ideal in a domain $D \subset \mathbf{C}^n$ and let Φ be a holomorphic function on D. We define $\mathcal{I}^\Phi$ to be the set of all pairs $(f + A\Phi, \delta\cap\delta')$ where $(f,\delta) \in \mathcal{I}$ and A is a holomorphic function on δ'. In addition, we define $\mathcal{I}_\Phi$ to be the set of all pairs (φ,δ) where $\varphi = f/\Phi$ is holomorphic on δ and $(f,\delta) \in \mathcal{I}$. These are both $\mathcal{O}$-ideals on D. We call $\mathcal{I}^\Phi$ and $\mathcal{I}_\Phi$ the **adjoint** and the **quotient $\mathcal{O}$-ideals** of $\mathcal{I}$ for Φ. We note that $\mathcal{I} \subset \mathcal{I}^\Phi \cap \mathcal{I}_\Phi$.

Using this notation and terminology, we have the following result.

COROLLARY 7.7. *The $\mathcal{O}$-ideal $\mathcal{I}$ on D admits a locally finite pseudobase at each point in D if and only if the same is true for both $\mathcal{I}^\Phi$ and $\mathcal{I}_\Phi$.*

PROOF. Fix $z_0 \in D$. Assume that $\mathcal{I}$ admits a locally finite pseudobase F_j ($j = 1, \dots, \nu$) on a neighborhood δ of z_0 in D. Then $\{F_j, \Phi\}_{j=1,\dots,\nu}$ forms a locally finite pseudobase of $\mathcal{I}^\Phi$ on δ. Fix $\varphi \in \mathcal{I}_\Phi$ at a point z^* in δ. Then we have

$$(\Omega) \qquad \Phi\varphi = \alpha_1 F_1 + \cdots + \alpha_\nu F_\nu \qquad \text{in } \delta^*,$$

where $\delta^* \subset \delta$ is a neighborhood of z^* and each α_j ($j = 1, \dots, \nu$) is a holomorphic function on δ^*. Thus the restriction $\mathcal{I}_\Phi$ on δ coincides with the ℓ-ideal with respect to the linear relation (Ω) in δ. By Theorem 7.7, $\mathcal{I}_\Phi$ admits a locally finite pseudobase at z_0.

Conversely, assume that $\mathcal{I}^\Phi$ and $\mathcal{I}_\Phi$ both admit a locally finite pseudobase at z_0. We denote these pseudobases as

$$F_j + A_i\Phi \ (j = 1, \dots, \nu) \qquad \text{and} \qquad \Psi_k \ (k = 1, \dots, \mu) \qquad \text{on } \delta,$$

where δ is a neighborhood of z_0 in D. Here, $(F_j, \delta) \in \mathcal{I}$ and A_j is a holomorphic function on δ; moreover, each $\Psi_k = G_k/\Phi$ is a holomorphic function on δ where $(G_k, \delta) \in \mathcal{I}$. Let $f \in \mathcal{I}$ at a point $z^* \in \delta$. Since $f \in \mathcal{I}^\Phi$ at z^*, we have

$$f = f_1(F_1 + A_1\Phi) + \cdots + f_\nu(F_\nu + A_\nu\Phi) \qquad \text{on } \delta^*,$$

where δ^* is a neighborhood of z^* in δ and each f_j ($j = 1, \dots, \nu$) is a holomorphic function on δ^*. Thus,

$$f - f_1F_1 - \cdots - f_\nu F_\nu = (f_1A_1 + \cdots + f_\nu A_\nu)\Phi \qquad \text{on } \delta^*,$$

so that $f_1A_1 + \cdots + f_\nu A_\nu$ belongs to $\mathcal{I}_\Phi$ on δ^*. Hence, we have

$$f_1A_1 + \cdots + f_\nu A_\nu = b_1\Psi_1 + \cdots + b_\mu\Psi_\mu \qquad \text{on } \delta^*,$$

where each b_k ($k = 1, \dots, \mu$) is a holomorphic function on δ^*. It follows that

$$f = f_1F_1 + \cdots + f_\nu F_\nu + b_1G_1 + \cdots + b_\mu G_\mu \qquad \text{on } \delta^*.$$

Consequently, the restriction of $\mathcal{I}$ to δ coincides with the $\mathcal{O}$-ideal generated by $\nu+\mu$ holomorphic functions $\{F_j, G_k\}$ on δ. □

EXAMPLE 7.4. Let $\Delta = (|x| < 1) \times (|y| < 1)$ and $\Delta' = (|x| < 1) \times (0 < |y| < 1)$ in $\mathbf{C}^2$. Let $\mathcal{I}$ be the set of all pairs (f, δ), $\delta \subset \Delta$ satisfying the following: if $\delta \subset \Delta'$, then f can be an arbitrary holomorphic function on δ'; if $\delta \not\subset \Delta'$, then $f = \alpha\, xy$, where α is a holomorphic function on δ. Then $\mathcal{I}$ is an $\mathcal{O}$-ideal on Δ, but $\mathcal{I}$ does not admit a locally finite pseudobase at the origin 0 in Δ. For if $\mathcal{I}$ had a pseudobase $\{\alpha_j\, xy\}$ ($j = 1, \dots, \nu$) in a neighborhood V of 0 in Δ, then their common zero set in V would contain $\{xy = 0\}$. However, at the point $(0, y) \in V$ with $y \neq 0$ the constant function 1 belongs to $\mathcal{I}$, which is a contradiction.

The adjoint $\mathcal{I}^y$ of $\mathcal{I}$ for the function y and the quotient $\mathcal{I}_x$ for the function x are generated by the function y on Δ. However, neither $\mathcal{I}_y$ nor $\mathcal{I}^x$ admit a locally finite pseudobase at the origin. For both $\mathcal{I}_y$ and $\mathcal{I}^x$ consist of the collection of all pairs (f, δ) with $\delta \subset \Delta$ satisfying the following: if $\delta \subset \Delta'$, then f can be an arbitrary holomorphic function on δ; if $\delta \not\subset \Delta'$, then $f = \alpha x$, where α is a holomorphic function on δ. Hence this collection does not admit a locally finite pseudobase at the origin.

7.5.2. *G*-**ideal.** Let D be a domain in $\mathbf{C}^n$ and let $\mathcal{I}$ be an $\mathcal{O}$-ideal on D. Fix $p \in D$. If each holomorphic function $f(z)$ belonging to $\mathcal{I}$ at the point p vanishes at p, then we say that p is a zero point of $\mathcal{I}$. We call the set $E(\mathcal{I})$ of all such p in D the **zero set of** $\mathcal{I}$. Note that for $q \in D$, we have $q \notin E(\mathcal{I})$ if and only if each holomorphic function $f(z)$ at q belongs to $\mathcal{I}$ at the point q. It is clear that $E(\mathcal{I})$ is a closed set in D. Furthermore, if $\mathcal{I}$ has a locally finite pseudobase at each point in D, then $E(\mathcal{I})$ is an analytic set in D.

Conversely, let E be a closed set in D. We consider the set $G\{E\}$ of all pairs $(f(z), \delta)$ such that $\delta \subset D$ and $f(z)$ is a holomorphic function on δ which satisfies $f(z) = 0$ on $E \cap \delta$. Then $G\{E\}$ becomes an $\mathcal{O}$-ideal on D, called the **geometric ideal** for E on D (or the *G*-**ideal** for E). We will need the following theorem concerning G-ideals.

THEOREM 7.8. *Let Σ be an analytic set in a domain D in $\mathbf{C}^n$. Then the G-ideal $G\{\Sigma\}$ on D has a locally finite pseudobase at each point in D.*

We first prove Theorem 7.8 in the special case given as Proposition 7.7 below. For the sake of convenience, we use the following notation: $\mathbf{C}^n = \mathbf{C}^r_z \times \mathbf{C}^{n-r}_w$, where $\mathbf{C}^r_z$ has variables $z_1, \dots, z_r$ and $\mathbf{C}^{n-r}_w$ has variables $w_1, \dots, w_{n-r}$. Let D be a domain in $\mathbf{C}^r_z$ and let $\Lambda = D \times \mathbf{C}^{n-r}_w \subset \mathbf{C}^n$. For each w_j ($j = 1, \dots, n-r$), we consider a monic pseudopolynomial

$$P_j(z, w_j) = w_j^{l_j} + a_{j,1}(z) w_j^{l_j - 1} + \cdots + a_{j,l_j}(z)$$

with respect to w_j, where each $a_{j,k}(z)$ ($1 \le k \le l_j$) is a holomorphic function on D and $P_j(z, w_j)$ has no multiple factors. We set

$$\tilde{\Sigma} = \bigcap_{j=1}^{n-r} \{(z, w_1, \dots, w_{n-r}) \in \Lambda \mid P_j(z, w_j) = 0\},$$

which is a pure r-dimensional analytic set in Λ.

Then we have the following proposition.

PROPOSITION 7.7. *The G-ideal $G\{\tilde{\Sigma}\}$ on Λ is generated by $n - r$ pseudopolynomials $P_j(z, w_j)$ ($j = 1, \dots, n-r$) on Λ.*

PROOF. We prove this by induction on $n - r \ge 1$ (the number of pseudopolynomials). We first assume that $n - r = 1$, i.e., $\tilde{\Sigma}$ is an analytic hypersurface in $\Lambda := D \times \mathbf{C}_w$ defined by the zero set of a single monic pseudopolynomial $P(z, w)$ with no multiple factors whose coefficients are holomorphic functions on D. Fix $p_0 \in \Lambda$. Let $f(z, w)$ be any holomorphic function at p_0 belonging to $G\{\tilde{\Sigma}\}$ at p_0. Fix a sufficiently small polydisk $\lambda := \delta \times \gamma \subset\subset D \times \mathbf{C}_w$ centered at p_0 such that $f(z, w)$ is holomorphic on $\overline{\lambda}$ and $P(z, w) \neq 0$ in $\delta \times \partial\gamma$. Then we can write $P(z, w) = P'(z, w) P''(z, w)$ in λ, where $P'(z, w)$ is a monic pseudopolynomial with respect to w and $P''(z, w) \neq 0$ in $\overline{\lambda}$. Since $f(z, w) = 0$ on $\overline{\lambda} \cap \{P'(z, w) = 0\}$ and $P'(z, w)$ has no multiple factors, it follows from the Weierstrass preparation theorem that $f(z, w) = P'(z, w)\omega(z, w)$ on λ, where $\omega(z, w)$ is a holomorphic function on λ (which may have zeros on λ). We thus have $f(z, w) = P(z, w)(\omega(z, w)/P''(z, w)) =: P(z, w)\omega_1(z, w)$ on λ, where $\omega_1(z, w)$ is a holomorphic function on λ. Consequently, $P(z, w)$ is a pseudobase of $G\{\tilde{\Sigma}\}$ on Λ.

We next assume that the proposition is true for $n - r \ge 1$, and prove it for

$n-r+1$. Let $\tilde{\Sigma}$ be the pure r-dimensional analytic set in $\Lambda := D \times \mathbf{C}_w^{n-r+1} \subset \mathbf{C}_z^r \times \mathbf{C}_w^{n-r+1} = \mathbf{C}^{n+1}$ defined by

$$\tilde{\Sigma} = \bigcap_{j=1}^{n-r+1} \{(z,w) \in \Lambda \mid P_j(z,w_j) = 0\},$$

where each $P_j(z,w_j)$ $(j = 1, \ldots, n-r+1)$ is a monic pseudopolynomial in w_j with no multiple factors whose coefficients are holomorphic functions on D. For later use we write

$$\begin{aligned} w &= (w_1, \ldots, w_{n-r}, w_{n-r+1}) = (w', w_{n-r+1}), \\ \Lambda' &= D \times \mathbf{C}_{w'}^{n-r}, \qquad \tilde{\Sigma}' = \bigcap_{j=1}^{n-r} \{(z,w) \in \Lambda' \mid P_j(z,w_j) = 0\}, \end{aligned}$$

We also let σ_{n-r+1} denote the zero set of the discriminant $d_{n-r+1}(z)$ in D of $P_{n-r+1}(z, w_{n-r+1})$ with respect to w_{n-r+1}, so that σ_{n-r+1} is an $(r-1)$-dimensional analytic hypersurface in D.

Now let $p_0 \in \Lambda$, and let $f(z,w)$ be any holomorphic function at p_0 which belongs to $G\{\tilde{\Sigma}\}$ at p_0. We claim that there exists a neighborhood λ_0 of p_0 in Λ such that

$$f(z,w) = a_1(z,w)P_1(z,w_1) + \cdots + a_{n-r+1}(z,w)P_{n-r+1}(z,w_{n-r+1}) \quad \text{on } \lambda_0, \tag{7.23}$$

where each $a_j(z,w)$ $(j = 1, \ldots, n-r+1)$ is a holomorphic function on λ_0.

To prove this, we set $p_0 = (z_0, w_0) = (z_0, w_{0,1}, \ldots, w_{0,n-r}, w_{0,n-r+1}) = (z_0, w_0', w_{0,n-r+1})$. In case $p_0 \in \Lambda \backslash \tilde{\Sigma}$, we have $P_j(z_0, w_{0,j}) \neq 0$ for some j $(1 \leq j \leq n-r+1)$. Thus, if we set $f(z,w) = (f(z,w)/P_j(z,w_j))P_j(z,w_j) =: a_j(z,w)P_j(z,w)$, then $a_j(z,w)$ is a holomorphic function in a neighborhood λ_0 of p_0 in which $P_j(z,w_j) \neq 0$. This proves our claim (7.23).

We next study the case $p_0 = (z_0, w_0) \in \tilde{\Sigma}$. We take a polydisk $\lambda := \delta \times \gamma \subset D \times \mathbf{C}_w^{n-r+1}$ centered at (z_0, w_0) in which $f(z,w)$ is holomorphic. We write

$$\begin{aligned} &\gamma := \gamma_1 \times \cdots \times \gamma_{n-r} \times \gamma_{n-r+1} \subset \mathbf{C}_w^{n-r+1}, \quad &\gamma' := \gamma_1 \times \cdots \times \gamma_{n-r} \subset \mathbf{C}_w^{n-r}, \\ &\lambda' := \delta \times \gamma' \subset D \times \mathbf{C}_w^{n-r} \subset \mathbf{C}^n, \quad &\lambda := \lambda' \times \gamma_{n-r+1} \subset \mathbf{C}^{n+1}. \end{aligned}$$

By taking a suitably smaller polydisk λ centered at (z_0, w_0) if necessary, we may assume that

$$P_{n-r+1}(z, w_{n-r+1}) \neq 0 \quad \text{on } \delta \times \partial\gamma_{n-r+1}.$$

Thus, we have

$$P_{n-r+1}(z, w_{n-r+1}) = P'(z, w_{n-r+1})P''(z, w_{n-r+1}) \quad \text{on } \delta \times \gamma_{n-r+1},$$

where both $P'(z, w_{n-r+1})$ and $P''(z, w_{n-r+1})$ are monic pseudopolynomials whose coefficients are holomorphic functions on δ such that

$$\begin{aligned} P'(z, w_{n-r+1}) &\neq 0 \text{ on } \delta \times [\mathbf{C}_{w_{n-r+1}} \setminus \gamma_{n-r+1}], \\ P''(z, w_{n-r+1}) &\neq 0 \text{ on } \delta \times \gamma_{n-r+1}; \end{aligned}$$

furthermore $P'(z, w_{n-r+1})$ has no multiple factors. We let l, l', and l'' denote the orders of P_{n-r+1}, P', and P'' with respect to w_{n-r+1}, so that $l = l' + l''$. Considering $P'(z, w_{n-r+1})$ as a monic pseudopolynomial with respect to w_{n-r+1}

whose coefficients are holomorphic functions on λ', we can apply the remainder theorem on $\lambda = \lambda' \times \gamma_{n-r+1}$ to obtain

$$f(z,w) = q(z,w)P'(z,w_{n-r+1}) + r(z,w',w_{n-r+1}) \quad \text{on } \lambda. \tag{7.24}$$

Here $q(z,w)$ is a holomorphic function on λ and $r(z,w',w_{n-r+1})$ is a pseudopolynomial with respect to w_{n-r+1} of degree at most $l'-1$; i.e.,

$$r(z,w',w_{n-r+1}) = A_0(z,w')w_{n-r+1}^{l'-1} + \cdots + A_{l'-1}(z,w') \quad \text{on } \lambda' \times \mathbf{C}_{w_{n-r+1}},$$

where $A_j(z,w')$ $(j=0,1,\dots,l'-1)$ is a holomorphic function on λ'.

We want to show that for each $j=0,1,\dots,l'-1$,

$$A_j(z,w') = 0 \quad \text{on } \tilde{\Sigma}' \cap \lambda'. \tag{7.25}$$

To see this, let $(a,b) \in \lambda' \subset D \times \mathbf{C}^{n-r}_{w'}$ be any point of $\tilde{\Sigma}' \cap \lambda'$ such that $a \in D \setminus \sigma_{n-r+1}$, so that $P_{n-r+1}(a,w_{n-r+1}) = 0$ has l distinct solutions in $\mathbf{C}_{w_{n-r+1}}$. Hence, $P'(a,w_{n-r+1}) = 0$ has l' distinct solutions in γ_{n-r+1}, say, $\zeta_1(a),\dots,\zeta_{l'}(a)$. Since $(a,b,\zeta_k(a)) \in \tilde{\Sigma} \cap \lambda$ $(k=1,\dots,l')$, it follows from (7.24) that $r(a,b,\zeta_k(a)) = 0$ $(k=1,\dots,l')$. Since $r(a,b,w_{n-r+1})$ is a polynomial with respect to w_{n-r+1} of degree at most $l'-1$, we have $r(a,b,w_{n-r+1}) \equiv 0$ on $\mathbf{C}_{w_{n-r+1}}$, and hence $A_j(a,b) = 0$ $(j=0,1,\dots,l'-1)$. By analytic continuation, $A_j(z,w') = 0$ $(j=0,1,\dots,l'-1)$ for any point $(z,w') \in \tilde{\Sigma}' \cap \lambda'$, which proves (7.25).

Since $\tilde{\Sigma}'$ is defined by $n-r$ pseudopolynomials, from the inductive hypothesis we conclude that there exists a neighborhood λ'_0 of (z_0,w'_0) in λ' such that, for each $j=0,1,\dots,l'-1$,

$$A_j(z,w') = \alpha_1^{(j)}(z,w')P_1(z,w_1) + \cdots + \alpha_{n-r}^{(j)}(z,w')P_{n-r}(z,w_{n-r}) \text{ on } \lambda'_0,$$

where each $\alpha_i^{(j)}(z,w')$ $(1 \le i \le n-r)$ is a holomorphic function on λ'_0. If we set $\lambda_0 := \lambda'_0 \times \gamma_{n-r+1}$, which is a neighborhood of (z_0,w_0) in Λ, then we have

$$\begin{aligned} f(z,w) &= \frac{q(z,w)}{P''(z,w_{n-r+1})} \cdot P_{n-r+1}(z,w_{n-r+1}) \\ &\quad + \sum_{k=1}^{n-r} \left(\sum_{j=0}^{l'-1} \alpha_k^{(j)}(z,w')w_{n-r+1}^j \right) P_k(z,w_k) \\ &\equiv \sum_{k=1}^{n-r+1} a_k(z,w)P_k(z,w_k) \qquad \text{on } \lambda_0, \end{aligned}$$

where each $a_k(z,w)$ $(k=1,\dots,n-r+1)$ is a holomorphic function on λ_0. This proves our claim in the case $p_0 = (z_0,w_0) \in \tilde{\Sigma}$ for $n-r+1$. By induction we complete the proof of the proposition. □

PROOF OF THEOREM 7.8. Let Σ be an analytic set in a domain D in $\mathbf{C}^n$. Let $z_0 \in \Sigma$. We fix a polydisk Δ centered at z_0 in D and decompose $\Delta \cap \Sigma$ into irreducible components: $\Delta \cap \Sigma = \Sigma_1 \cup \cdots \cup \Sigma_q$. We let $G\{\Sigma_j\}$ $(j=1,\dots,q)$ denote the G-ideal for Σ_j in Δ. We note that $G\{\Sigma\}\,|_\Delta$ coincides with $\bigcap_{j=1}^q G\{\Sigma_j\}$. Using Theorem 7.4, to prove Theorem 7.8 it suffices to prove that each $G\{\Sigma_j\}$ $(j=1,\dots,q)$ has a locally finite pseudobase at the point z_0. For simplicity in notation we write $\Sigma_j = \Sigma$ and assume that Σ is of dimension r $(0 \le r < n)$. By performing a coordinate change and taking a smaller polydisk Δ if necessary, we may assume that $z = (z_1,\dots,z_r,z_{r+1},\dots,z_n) = (z',z_{r+1},\dots,z_n)$ satisfies the

Weierstrass condition for Σ at any point z on Σ. Thus, if we write $\Delta = \Delta' \times \Gamma \subset \mathbf{C}^r_{z'} \times \mathbf{C}^{n-r}_{z_{r+1},\dots,z_n}$, then $\Sigma \cap (\Delta' \times \partial\Gamma) = \emptyset$. It follows from Theorem 2.2 in Chapter 2 that there exists a monic pseudopolynomial $P_j(z', z_j)$ $(j = r+1, \dots, n)$ with respect to z_j whose coefficients are holomorphic functions on Δ', such that, if we define

$$\tilde{\Sigma} := \bigcap_{j=r+1}^{n} \{(z', z_{r+1}, \dots, z_n) \in \Delta' \times \mathbf{C}^{n-r} \mid P_j(z', z_j) = 0\},$$

then Σ is one of the irreducible components of $\tilde{\Sigma}$ in Δ. We may also assume that each $P_j(z', z_j)$ $(j = r+1, \dots, n)$ has no multiple factors. We let Σ' denote the union of the remaining irreducible components of $\tilde{\Sigma}$, so that $\tilde{\Sigma} = \Sigma \cup \Sigma'$. From Remark 2.7 of Lemma 2.5, Σ' is itself an analytic set in Δ which can be written in the form

$$\Sigma' = \bigcap_{i=1}^{\lambda} \{z \in \Delta \mid \varphi_i(z) = 0\},$$

where $\varphi_i(z)$ $(i = 1, \dots, \lambda)$ is a holomorphic function on all of Δ.

Consider the following system of homogeneous linear equations:

$$(\Omega) \qquad f(z)\varphi_i(z) = f_{r+1,i}(z)P_{r+1}(z', z_{r+1}) + \cdots + f_{n,i}(z)P_n(z', z_n) \qquad (i = 1, \dots, \lambda) \quad \text{on } \Delta;$$

equivalently,

$$\begin{aligned} f \cdot (\varphi_1, \varphi_2, \dots, \varphi_\lambda) &= f_{r+1,1} \cdot (P_{r+1}, 0, \dots, 0) + \cdots + f_{n,1} \cdot (P_n, \dots, 0, 0) \\ &\quad + \cdots + f_{r+1,\lambda} \cdot (0, \dots, 0, P_{r+1}) + \dots + f_{n,\lambda} \cdot (0, \dots, 0, P_n). \end{aligned}$$

Here the functions $(\varphi_i(z), P_j(z', z_j))$ $(i = 1, \dots, \lambda;\ j = r+1, \dots, n)$ on Δ are known (given); the unknown functions are $(f(z), f_{k,i}(z))$ $(k = r+1, \dots, n;\ i = 1, \dots, \lambda)$. Thus, the linear relation (Ω) is of rank λ and the $\mathcal{O}$-module $\mathcal{L}\{\Omega\}$ is of rank $1 + \lambda(n-r)$. We consider the ℓ-ideal $\ell\{\Omega\}$ with respect to (Ω) in Δ. By Theorem 7.7, $\ell\{\Omega\} = \{(f, \delta)\}_{\delta \subset \Delta}$ has a locally finite pseudobase at each point in Δ. To prove the theorem it thus suffices to prove that $\ell\{\Omega\}$ is equivalent to $G\{\Sigma\}$ as an $\mathcal{O}$-ideal on Δ.

To verify this, fix $z^0 \in \Delta$ and let $f(z)$ be any holomorphic function at z^0 which belongs to $\ell\{\Omega\}$ at z^0. Then there exists a neighborhood δ of z^0 in Δ such that

$$f(z)\varphi_i(z) = f_{r+1,i}(z)P_{r+1}(z', z_{r+1}) + \cdots + f_{n,i}(z)P_n(z', z_n) \ (i = 1, \dots, \lambda),$$

where each $f_{k,i}(z)$ $(k = r+1, \dots, n; i = 1, \dots, \lambda)$ is a holomorphic function on δ. Take a point $\zeta = (\zeta', \zeta_{r+1}, \dots, \zeta_n) \in (\Sigma \setminus \Sigma') \cap \delta$. Then $\varphi_i(\zeta) \neq 0$ for some i $(1 \leq i \leq \lambda)$. Since $\Sigma \subset \tilde{\Sigma}$ implies that $P_k(\zeta', \zeta_k) = 0$ $(k = r+1, \dots, n)$, it follows that $f(\zeta)\varphi_i(\zeta) = 0$, and hence $f(\zeta) = 0$. By continuity, this implies $f(z) = 0$ on $\Sigma \cap \delta$ (since $\Sigma \cap \Sigma'$ is of dimension $r-1$), so that $(f, \delta) \in G\{\Sigma\}$. Thus, $f(z)$ belongs to $G\{\Sigma\}$ at z^0.

Conversely, let $z^0 \in \Delta$ and let $f(z)$ belong to $G\{\Sigma\}$ at z^0. There exists a neighborhood δ of z^0 in Δ such that $f(z) = 0$ on $\delta \cap \Sigma$. Then each function $f(z)\varphi_i(z)$ $(i = 1, \dots, \lambda)$ is a holomorphic function in δ such that $f(z)\varphi_i(z) = 0$ on $(\Sigma \bigcup \Sigma') \cap \delta$, i.e., $(f(z)\varphi_i(z), \delta) \in G\{\tilde{\Sigma}\}$. From Proposition 7.7 we have

$$f(z)\varphi_i(z) = a_{r+1,i}(z)P_{r+1}(z', z_{r+1}) + \cdots + a_{n,i}(z)P_n(z', z_n) \ (i = 1, \dots, \lambda)$$

in a neighborhood $\delta_0 \subset \delta$ of z^0, where each $a_{k,i}(z)$ $(k = r+1, \ldots, n;\ i = 1, \ldots, \lambda)$ is a holomorphic function on δ_0. This means that $(f(z), \delta_0) \in \ell\{\Omega\}$, so that $f(z)$ belongs to $\ell\{\Omega\}$ at z^0.

Consequently, $G\{\Sigma\}$ and $\ell\{\Omega\}$ are equivalent as $\mathcal{O}$-ideals on Δ. □

Theorem 7.8 combined with the solvablity of Problem E in a closed polydisk implies the following corollary.

COROLLARY 7.8. *Let $\overline{\Delta}$ be a closed polydisk in $\mathbf{C}^n$ and let Σ be an analytic set in $\overline{\Delta}$. Then there exist a finite number of holomorphic functions $f_j(z)$ $(j = 1, \ldots, \nu)$ on Δ such that $\Sigma \cap \Delta$ is equal to the common zero set of $f_j(z)$ $(j = 1, \ldots, \nu)$ in Δ.*

REMARK 7.12. Theorem 3.4 in Chapter 3 (the main theorem in Oka [**45**]) follows immediately from this fact.

We give another proof of Theorem 7.8; this is due to Oka [**50**].

REMARK 7.13. After a slight change in notation, together with the use of Theorem 7.4, we may assume that Σ is an r-dimensional irreducible analytic set in the polydisk Λ centered at the origin 0 in $\mathbf{C}^n$. Here $\Lambda = \Delta \times \Gamma \subset \mathbf{C}^r_z \times \mathbf{C}^s_w$ and $r + s = n$ with $\Sigma \cap [\Delta \times \partial\Gamma] = \emptyset$. We set $\Gamma = \Gamma_1 \times \cdots \times \Gamma_s$, where Γ_j $(j = 1, \ldots, s)$ is a disk in $\mathbf{C}_{w_j}$. We let $\mathcal{D}$ denote the projection of Σ over the polydisk Δ; this is a ramified domain over Δ without relative boundary. Finally we let m denote the number of sheets of $\mathcal{D}$ over Δ. Thus Σ can be written in the form $w_j = \xi_j(\tilde{z})$ $(j = 1, \ldots, s)$, $\tilde{z} \in \mathcal{D}$, where each $\xi_j(\tilde{z})$ is a single-valued holomorphic function on $\mathcal{D}$ with $\xi_j(\tilde{z}) \in \Gamma_j$. We let Σ_j $(j = 1, \ldots, s)$ denote the projection of Σ onto the $(r+1)$-dimensional polydisk $\Lambda_j := \Delta \times \Gamma_j$. Then Σ_j is an analytic hypersurface in Λ_j, so that Σ_j can be written as

$$\Sigma_j = \{(z, w_j) \in \Delta \times \mathbf{C}_{w_j} \mid P_j(z, w_j) = 0\},$$

where $P(z, w_j)$ is a polynomial in w_j of degree at most m whose coefficients are holomorphic functions on Δ; moreover $P(z, w_j)$ has no multiple factors. Thus, $w_j = \xi_j(\tilde{z})$ satisfies $P_j(z, w_j) = 0$, where z is the projection of $\tilde{z}$ onto Δ. We forcus on $j = 1$. By taking a coordinate transformation of $\mathbf{C}^s$ sufficiently close to the identity transformation, if necessary, we may assume that Σ and Σ_1 are in one-to-one correspondence except for an analytic set of dimension at most $r - 1$; thus the projection $\mathcal{D}_1$ of Σ_1 over Δ, which is a ramified domain over Δ without boundary, coincides with $\mathcal{D}$. In particular, $\partial P_1(z, w_1)/\partial w_1 \not\equiv 0$ on Σ_1, and hence on Σ. Thus each $\xi_j(\tilde{z})$ $(j = 2, \ldots, s)$ defines a weakly holomorphic function on Σ_1, and Σ can thus be considered as a lifting of the first kind of Σ_1 by $w_j = \xi_j(\tilde{z})$ $(j = 2, \ldots, s)$. For each $j = 2, \ldots, s$, using Remark 7.4 there exists a *linear* polynomial $\Phi^*_j(z, w_1, w_j)$ in w_j of the form:

$$\Phi^*_j(z, w_1, w_j) = w_j \frac{\partial P_1(z, w_1)}{\partial w_1} - \Phi_j(z, w_1) \quad (j = 2, \ldots, s)$$

which vanishes on Σ. Here $\Phi_j(z, w_1)$ is a polynomial in w_1 of degree at most $m - 1$ whose coefficients are holomorphic functions on Δ. We set $M = (s-1)m$.

We consider the following linear equation (Ω) defined on Λ:

$$(\Omega) \qquad f(z, w) \left(\frac{\partial P_1(z, w_1)}{\partial w_1} \right)^M$$

$$= f_1(z,w)P_1(z,w_1) + \cdots + f_s(z,w)P_s(z,w_s) + g_2(z,w)\Phi_2^*(z,w_1,w_2) + \cdots + g_s(z,w)\Phi_s^*(z,w_1,w_s),$$

where $P_j(z,w_j)$ and $\Phi_k^*(z,w_1,w_k)$ are known functions on Λ, and $(f, f_1, \dots, f_s, g_2, \dots, g_s)$ is an unknown holomorphic vector-valued function of rank $2s$ in (z,w). By the main theorem (Theorem 7.1), it remains to prove that our G-ideal $G\{\Sigma\}$ on Λ is equivalent to the l-ideal $l\{\Omega\} := \{(f(z,w),\lambda)\}$ (here $\lambda \subset \Lambda$) with respect to the linear relation (Ω) on Λ.

Fix $(f(z,w),\lambda) \in l\{\Omega\}$. We note that each Φ_k^* as well as each P_j vanishes on Σ in Λ. Since $f(z,w)$ satisfies equation (Ω) on λ for some holomorphic functions f_j, g_k on λ, it follows from the fact that $\partial P_1(z,w_1)/\partial w_1 \not\equiv 0$ on Σ that $f(z,w) = 0$ on $\Sigma \cap \lambda$; thus $(f,\lambda) \in G\{\Sigma\}$.

Conversely, let $f(z,w)$ be a holomorphic function belonging to $G\{\Sigma\}$ at a point (z_0,w_0) in Λ. By Remark 7.7, there exists a sufficiently small polydisk $\lambda = \delta \times \gamma \subset \Delta \times \Gamma$ centered at (z_0,w_0) with $\Sigma \cap (\delta \times \partial\gamma) = \emptyset$ such that

$$f(z,w) = \varphi_1(z,w)P_1(z,w) + \cdots + \varphi_s(z,w)P_s(z,w) + \sum_{|\mathbf{j}|=0}^{M} \beta_{\mathbf{j}}(z) w_1^{j_1} \cdots w_s^{j_s}$$

for $\mathbf{j} = (j_1, \dots, j_s)$, $|\mathbf{j}| = j_1 + \cdots + j_s$, $0 \le j_k \le m-1$, where each $\varphi_j(z,w)$ is a holomorphic function on λ and where each $\beta_{(\mathbf{j})}(z)$ is a holomorphic function on δ. We set $\gamma := \gamma_1 \times \cdots \gamma_s$, where γ_j $(j = 1, \dots, s)$ is a disk in Γ_j. Multiplying both sides of the above formula by $(\partial P_1/\partial w_1)^M$ and using the functions $\Phi_j^*(z,w_1,w_j)$ $(j = 2, \dots, s)$, we have

$$f(z,w)\left(\frac{\partial P_1(z,w_1)}{\partial w_1}\right)^M = \tilde{\varphi}_1(z,w)P_1 + \cdots + \tilde{\varphi}_s(z,w)P_s + \tilde{\psi}_2(z,w)\Phi_2^* + \cdots + \tilde{\psi}_s(z,w)\Phi_s^* + H(z,w_1),$$

where $\tilde{\varphi}_j(z,w)$ and $\tilde{\psi}_k(z,w)$ are holomorphic functions on λ, and where $H(z,w_1)$ is a polynomial in w_1 whose coefficients are holomorphic functions of $z \in \delta$ (independent of w_k $(k = 2, \dots, s)$). Since $f(z,w) = 0$ on $\Sigma \cap \lambda$, we have $H(z,w_1) = 0$ on $\Sigma \cap \lambda$; thus $H(z,w_1)$ vanishes on the analytic hypersurface Σ_1 in the $(r+1)$-dimensional polydisk λ_1, where $\lambda_1 = \delta \times \gamma_1$. It follows that $H(z,w_1) = P_1(z,w_1)h(z,w_1)$ in λ_1, where $h(z,w_1)$ is a holomorphic function on λ_1. Hence $(f(z,w),\lambda) \in l\{\Omega\}$. Thus $G\{\Sigma\}$ and $l\{\Omega\}$ are equivalent as $\mathcal{O}$-ideals on Λ. □

7.5.3. Projection. Let D_1 be a domain in $\mathbf{C}_z^n$ with variables $z_1, \dots, z_n$ and let D_2 be a bounded domain in $\mathbf{C}_w^m$ with variables $w_1, \dots, w_m$. We set $D = D_1 \times D_2 \subset \mathbf{C}_z^n \times \mathbf{C}_w^m$. Let $\mathcal{I}$ be an $\mathcal{O}$-ideal in D. Consider the set $\mathcal{J}$ of all pairs $(f(z),\delta)$ such that $\delta \subset D_1$ and $f(z)$ is a holomorphic function in δ with the following property: $f(z)$, regarded as a holomorphic function on $\delta \times D_2$, belongs to $\mathcal{I}$ at each point (z,w) in $\delta \times D_2$. Then $\mathcal{J}$ is an $\mathcal{O}$-ideal on D_1, which is called the **projection of $\mathcal{I}$ onto** D_1. We write $\mathcal{J} =: \mathcal{P}\{\mathcal{I}\}$. Clearly, if an $\mathcal{O}$-ideal $\widetilde{\mathcal{I}}$ in D is equivalent to $\mathcal{I}$ on $D_1 \times D_2$ as $\mathcal{O}$-ideals, then $\mathcal{P}\{\widetilde{\mathcal{I}}\}$ is equivalent to $\mathcal{P}\{\mathcal{I}\}$ on D_1. We let E and E_1 denote the zero sets of $\mathcal{I}$ and $\mathcal{P}\{\mathcal{I}\}$ in D and D_1, respectively. We also denote by $p(E)$ the projection of E onto D_1. Then $p(E) \subset E_1$. Moreover, if $E \cap (D_1 \cap \partial D_2) = \emptyset$, then $p(E) = E_1$.

We have the following theorem.

THEOREM 7.9. *Let $\mathcal{I}$ be an $\mathcal{O}$-ideal in $D = D_1 \times D_2$ such that*

(1) *$\mathcal{I}$ has a locally finite pseudobase at each point of D, and*
(2) *the zero set Σ of $\mathcal{I}$ contains no points in a neighborhood of $D_1 \times \partial D_2$ in $D_1 \times \mathbf{C}^m_w$.*

Then the projection $\mathcal{P}\{\mathcal{I}\}$ of $\mathcal{I}$ onto D_1 has a locally finite pseudobase at each point in D_1.

PROOF. Let $z_0 \in D_1$, and let us prove that $\mathcal{P}\{\mathcal{I}\}$ has a locally finite pseudobase at the point z_0. By condition (2), the section $\Sigma(z_0) = \{w \in D_2 \mid (z_0, w) \in \Sigma\}$ of Σ at $z = z_0$ consists of a finite number of points $(z_0, w^{(1)}), \dots, (z_0, w^{(q)})$. For each point $(z_0, w^{(j)})$ $(j = 1, \dots, q)$, there exists a polydisk $\lambda_j := \delta \times \Gamma_j \subset\subset D_1 \times D_2$ centered at $(z_0, w^{(j)})$ such that $\Sigma \cap (\overline{\delta} \times \partial\Gamma_j) = \emptyset$. We let $\mathcal{I}_{\lambda_j}$ $(j = 1, \dots, q)$ denote the restriction of $\mathcal{I}$ to λ_j and we set $\mathcal{P}^{(j)} = \mathcal{P}\{\mathcal{I}_{\lambda_j}\}$ (the projection of $\mathcal{I}_{\lambda_j}$ onto δ). By definition of the projection of an $\mathcal{O}$-ideal, we see that $\mathcal{P}\{\mathcal{I}\}|_\delta$ is equivalent to $\bigcap_{j=1}^q \mathcal{P}^{(j)}$ as an $\mathcal{O}$-ideal on δ. By Theorem 7.4 it suffices to prove that each $\mathcal{P}^{(j)}$ $(j = 1, \dots, q)$ has a locally finite pseudobase at each point in δ. In other words, we may assume from the beginning that $D_1 = \delta$ (a polydisk in $\mathbf{C}^n_z$); $D_2 = \Gamma$ (a polydisk in $\mathbf{C}^m_w$); $\mathcal{I}$ is an $\mathcal{O}$-ideal on the closed polydisk $\overline{\lambda} = \overline{\delta} \times \overline{\Gamma}$ satisfying condition (1) on $\overline{\lambda}$; and condition (2) becomes $\Sigma \cap (\overline{\delta} \times \partial\Gamma) = \emptyset$, where Σ is the zero set of $\mathcal{I}$ on $\overline{\lambda}$.

Moreover, we may assume $m = 1$. For assume that the theorem is true in this case and let $m > 1$. Set $\Gamma = \Gamma_1 \times \cdots \times \Gamma_m$, where Γ_j $(j = 1, \dots, m)$ is a disk in the plane $\mathbf{C}_{w_j}$, and $\delta_{m-1} := \delta \times \Gamma_1 \times \cdots \times \Gamma_{m-1} \subset \mathbf{C}^n_z \times \mathbf{C}^{m-1}_{w_1, \dots, w_{m-1}}$. Since $\Sigma \cap (\overline{\delta_{m-1} \cap \partial\Gamma_m}) = \emptyset$ from (2), it follows from the assumption for $m = 1$ that the projection $\mathcal{P}_{m-1}\{\mathcal{I}\} = \mathcal{I}_{m-1}$ of $\mathcal{I}$ onto δ_{m-1} has a locally finite pseudobase at each point of δ_{m-1}. We note that the zero set Σ_{m-1} of $\mathcal{I}_{m-1}$ in $\overline{\delta}_{m-1}$ coincides with the projection of the analytic set Σ onto $\overline{\delta}_{m-1}$ (which is an analytic set from Proposition 2.3 in Chapter 2), so that, if we define $\delta_{m-2} := \delta \times \Gamma_1 \times \cdots \times \Gamma_{m-2}$, then $(\overline{\delta_{m-2} \times \partial\Gamma_{m-1}}) \cap \Sigma_{m-1} = \emptyset$. We repeat the same procedure to obtain $\mathcal{I}_{m-2} = \mathcal{P}_{m-2}\{\mathcal{I}_{m-1}\}, \dots, \mathcal{I}_0 = \mathcal{P}_0\{\mathcal{I}_1\}$ where $\mathcal{I}_{j-1}$ $(j = 1, \dots, m-1)$ is the projection of $\mathcal{I}_j$ onto $\delta_{j-1} := \delta \times \Gamma_1 \times \cdots \times \Gamma_{j-1}$ (here $\delta_0 := \delta$) and $\mathcal{I}_{j-1}$ has a locally finite pseudobase at each point of δ_{j-1}. Thus, $\mathcal{I}_0$ has a locally finite pseudobase at each point of δ. On the other hand, we see from the definition of the projection of an $\mathcal{O}$-ideal that $\mathcal{I}_0$ is equivalent to $\mathcal{P}\{\mathcal{I}\}|_\delta$ as an $\mathcal{O}$-ideal on δ.

Thus, taking $m = 1$, we may assume $\lambda = \overline{\delta} \times \Gamma \subset \mathbf{C}^n_z \times \mathbf{C}_w$, where Γ is a disk in the plane $\mathbf{C}_w$. Let $z' \in \delta$. Since $\Sigma \cap (\overline{\delta} \times \partial\Gamma) = \emptyset$, the section $\Sigma(z')$ of Σ at $z = z'$ consists of a finite number of points $(z', w_1), \dots, (z', w_\mu)$, where $w_j \in \Gamma$ $(j = 1, \dots, \mu)$. By conditions (1) and (2), there exist a polydisk $\lambda_j := \delta' \times \gamma_j \subset \lambda$ centered at (z', w_j) $(j = 1, \dots, \mu)$ and a finite number of holomorphic functions $\Phi_1^{(j)}(z, w), \dots, \Phi_{\nu_j}^{(j)}(z, w)$ on $\overline{\lambda_j}$ such that (i) if we let $\mathcal{J}\{\Phi^{(j)}\}$ denote the $\mathcal{O}$-ideal generated by $\{\Phi_k^{(j)}(z, w)\}_{k=1,\dots,\nu_j}$ on λ_j, then $\mathcal{J}\{\Phi^{(j)}\}$ is equivalent to $\mathcal{I}|_{\lambda_j}$ as an $\mathcal{O}$-ideal on λ_j, and (ii) $\Phi_1^{(j)}(z, w) \neq 0$ on $\delta' \times (\partial\gamma_j)$ for each $j = 1, \dots, \mu$. We let $\mathcal{P}^{(j)}$ $(j = 1, \dots, \mu)$ denote the projection $\mathcal{J}\{\Phi^{(j)}\}$ onto δ'. Since $\mathcal{P}\{\mathcal{I}\}|_{\delta'}$ is equivalent to $\bigcap_{j=1}^\mu \mathcal{P}^{(j)}$ as an $\mathcal{O}$-ideal on δ', from Theorem 7.4 it suffices to prove that each $\mathcal{P}^{(j)}$ $(j = 1, \dots, \mu)$ has a locally finite pseudobase at each point of δ'.

To simplify the notation, we set $\delta' = \Delta \subset \mathbf{C}^n_z$, $\gamma_j = \Gamma \subset \mathbf{C}_w$, $\Lambda = \Delta \times \Gamma \subset \mathbf{C}^n_z \times \mathbf{C}_w$, $\nu_j = \nu$, $\Phi_k^{(j)}(z, w) = \Phi_k(z, w)$ $(k = 1, \dots, \nu)$, and $\mathcal{J}\{\Phi^{(j)}\} = \mathcal{J}\{\Phi\}$.

Since $\Phi_1(z,w) \neq 0$ on $\Delta \times \partial\Gamma = \emptyset$, we have

$$\Phi_1(z,w) = P_1(z,w)\omega_1(z,w) \quad \text{on } \Lambda,$$

where $\omega_1(z,w) \neq 0$ on Λ and $P_1(z,w)$ is a monic pseudopolynomial in w satisfying

$$\begin{gathered} P_1(z,w) = w^l + A_1^{(1)}(z)w^{l-1} + \cdots + A_l^{(1)}(z) \quad \text{in } \Delta \times \mathbf{C}_w, \\ \{(z,w) \in \Delta \times \mathbf{C}_w \mid P_1(z,w) = 0\} \subset\subset \Lambda, \end{gathered} \tag{7.26}$$

where each $A_j^{(1)}(z)$ $(j = 1, \dots, l)$ is a holomorphic function on Δ. By the remainder theorem for $P_1(z,w)$ on Λ we have

$$\Phi_j(z,w) = Q_j(z,w)P_1(z,w) + R_j(z,w) \ (j = 2, \dots, \nu) \quad \text{on } \Lambda,$$

where each $Q_j(z,w)$ is a holomorphic function on Λ and each $R_j(z,w)$ is a pseudopolynomial in w satisfying

$$R_j(z,w) = A_0^{(j)}(z)w^{l-1} + \cdots + A_{l-1}^{(j)}(z) \quad \text{on } \Delta \times \mathbf{C}_w,$$

where each $A_k^{(j)}(z)$ $(k = 0, 1, \dots, l-1)$ is a holomorphic function on Δ. Clearly $\mathcal{J}\{\Phi\}$ is equivalent to the $\mathcal{O}$-ideal $\mathcal{G}$ generated by $P_1(z,w)$, $R_2(z,w)$, $\dots$, $R_\nu(z,w)$ on Λ. Hence it suffices to prove that the projection $\mathcal{P}\{\mathcal{G}\}$ of $\mathcal{G}$ onto Δ has a locally finite pseudobase at each point in Δ.

Let $z_0 \in \Delta$ and let $f(z)$ be any holomorphic function belonging to $\mathcal{P}\{\mathcal{G}\}$ at z_0. Since $f(z)$ belongs to $\mathcal{G}$ at each point of $\{z_0\} \times \overline{\Gamma}$, we can find a polydisk $\delta \subset \Delta$ centered at z_0 such that, at each point $\eta \in \Gamma$, there exist a disk $\gamma_\eta \subset \Gamma$ centered at η and ν holomorphic functions $f_j(z,w)$ $(j = 1, \dots, \nu)$ on $v_\eta := \delta \times \gamma_\eta$ with

$$f(z) = f_1(z,w)P_1(z,w) + f_2(z,w)R_2(z,w) + \cdots + f_\nu(z,w)R_\nu(z,w) \tag{7.27}$$

on v_η. Since $f(z)$, $P_1(z,w)$, and $R_j(z,w)$ $(j = 2, \dots, \nu)$ are holomorphic functions on the polydisk $\delta \times \Gamma$ and since Problem C_1 is solvable on this polydisk, we may assume that each $f_j(z,w)$ $(j = 1, \dots, \nu)$ is a holomorphic function in $\delta \times \Gamma$ satisfying equation (7.27) on $\delta \times \Gamma$. Again using the remainder theorem for $P_1(z,w)$ on $\delta \times \Gamma$ and condition (7.26), we have, for each $j = 2, \dots, \nu$,

$$f_j(z,w) = q_j(z,w)P_1(z,w) + r_j(z,w) \quad \text{on } \delta \times \Gamma,$$

where $q_j(z,w)$ is a holomorphic function on $\delta \times \Gamma$ and $r_j(z,w)$ is a pseudopolynomial in w of degree at most $l-1$ with

$$r_j(z,w) = a_0^{(j)}(z)w^{l-1} + a_1^{(j)}(z)w^{l-2} + \cdots + a_{l-1}^{(j)}(z) \ \ (j = 2, \dots, \nu); \tag{7.28}$$

here, each $a_k^{(j)}(z)$ $(k = 0, 1, \dots, l-1)$ is a holomorphic function on δ. Substituting these into (7.27), we have

$$f(z) = r_1(z,w)P_1(z,w) + r_2(z,w)R_2(z,w) + \cdots + r_\nu(z,w)R_\nu(z,w) \quad \text{on } \delta \times \Gamma, \tag{7.29}$$

where $r_1(z,w)$ is a certain holomorphic function on $\delta \times \Gamma$. By use of the division theorem for $P_1(z,w)$, we see that $r_1(z,w)$ must, in fact, be a pseudopolynomial in w of degree at most $l-2$; i.e.,

$$r_1(z,w) = a_0^{(1)}(z)w^{l-2} + a_1^{(1)}(z)w^{l-3} + \cdots + a_{l-2}^{(1)}(z) \quad \text{in } \delta \times \mathbf{C}_w, \tag{7.30}$$

where each $a_k^{(1)}(z)$ $(k = 0, 1, \ldots, l-2)$ is a holomorphic function on δ. Therefore, comparing the coefficients of w^j $(j = 0, 1, \ldots, 2l-2)$ in equation (7.29), we obtain $2l-1$ equations on δ:

$$(\Omega)\qquad \begin{cases} f(z) = a_{l-2}^{(1)}(z)A_l^{(1)}(z) + a_{l-1}^{(2)}(z)A_{l-1}^{(2)}(z) + \cdots + a_{l-1}^{(\nu)}(z)A_{l-1}^{(\nu)}(z), \\ 0 = a_{l-2}^{(1)}(z)A_{l-1}^{(1)}(z) + a_{l-1}^{(1)}(z)A_l^{(1)}(z) + \cdots + a_{l-2}^{(\nu)}(z)A_{l-1}^{(\nu)}(z), \\ \quad\vdots \qquad\qquad\qquad \vdots \\ 0 = a_0^{(1)}(z) + a_0^{(2)}(z)A_0^{(2)}(z) + \cdots + a_0^{(\nu)}(z)A_0^{(\nu)}(z). \end{cases}$$

Or, equivalently,

$$\begin{aligned} & f\cdot(1, 0, \ldots\ldots\ldots, 0) \\ &= a_{l-2}^{(1)}\cdot(A_l^{(1)}, A_{l-1}^{(1)} \ldots, A_l^{(1)}, 1, 0, \ldots, 0) \\ &\quad + a_{l-1}^{(2)}\cdot(A_{l-1}^{(2)}, A_{l-2}^{(2)}, \ldots, A_0^{(2)}, 0, \ldots, 0) \\ &\quad + \cdots + \ a_0^{(\nu)}\cdot(0, \ldots, 0, A_{l-1}^{(\nu)}, A_{l-2}^{(\nu)}, \ldots, A_0^{(\nu)}). \end{aligned}$$

Therefore, if $f(z)$ belongs to $\mathcal{P}\{\mathcal{G}\}$ at $z_0 \in \Delta$, then we can find a polydisk $\delta \subset \Lambda$ centered at z_0 and $l\nu - 1$ holomorphic functions $a_{l-2}^{(1)}(z), a_{l-1}^{(2)}(z), \cdots, a_0^{(\nu)}(z)$ on δ which satisfy the $2l-1$ equations (Ω) on δ. In other words, $f(z)$ belongs to the ℓ-ideal $\ell\{\Omega\}$ with respect to the linear relation (Ω). Here we consider (Ω) as a linear system of $2l-1$ homogeneous equations determined by the known holomorphic functions 1, $A_l^{(1)}(z), \ldots, A_0^{(\nu)}(z)$ on Δ, where the unknown holomorphic vector-valued function $(f(z), a_{l-2}^{(1)}(z), \cdots, a_0^{(\nu)}(z))$ is of rank $\mu := \nu l$.

Conversely, let $z_0 \in \Delta$ and let $f(z)$ be any holomorphic function belonging to $\ell\{\Omega\}$ at z_0. Thus, there exist a neighborhood δ_0 of z_0 in Δ and $\mu - 1$ holomorphic functions $\{a_k^{(j)}(z)\}_{j,k}$ on δ_0 such that $f(z)$ is holomorphic in δ_0, and $(f(z), a_{l-2}^{(1)}(z), \cdots, a_0^{(\nu)}(z))$ satisfies equations (Ω) on δ_0. If we construct the pseudopolynomials $r_1(z,w)$ and $r_j(z,w)$ $(j = 2, \ldots, \nu)$ with respect to w using $\{a_k^{(1)}(z)\}_k$ and $\{a_k^{(j)}(z)\}_k$ from (7.30) and (7.28), then $\{f(z,w),\ r_i(z,w)\ (i = 1, \ldots, \nu)\}$ satisfies equation (7.29) on $\delta_0 \times \Gamma$, so that $(f(z), \delta_0)$ belongs to $\mathcal{P}\{\mathcal{G}\}$. Therefore, $\mathcal{P}\{\mathcal{G}\}$ is equivalent to $\ell\{\Omega\}$ as an $\mathcal{O}$-ideal on Δ; thus, it follows from Theorem 7.7 that $\mathcal{P}\{\mathcal{G}\}$ has a locally finite pseudobase at each point of Δ. This completes the proof of Theorem 7.9. □

This theorem combined with Theorem 7.5 implies the following corollary.

COROLLARY 7.9. *Let* $\Lambda = \Delta \times \Gamma$ *be a closed polydisk in* $\mathbf{C}_z^n \times \mathbf{C}_w^m$. *Let* $\Phi_j(z,w)$ $(j = 1, \ldots, \nu)$ *be holomorphic functions on* Λ *whose common zero set* Σ *satisfies* $\Sigma \cap (\Delta \times \partial\Gamma) = \emptyset$. *Then there exist holomorphic functions* $\varphi_k(z)$ $(k = 1, \ldots, \kappa)$ *on* Δ *such that*

(1) $\varphi_k(z) = \sum_{j=1}^{\nu} \alpha_j^{(k)}(z,w)\Phi_j(z,w)$ *on* Λ, *where the* $\alpha_j^{(k)}(z,w)$ *are holomorphic functions on* Λ*; and*
(2) *any holomorphic function* $f(z)$ *on* Δ *of the form*

$$f(z) = \sum_{j=1}^{\nu} a_j(z,w)\Phi_j(z,w)$$

on Λ, *where the* $a_j(z,w)$ *are holomorphic function on* Λ, *can be written in the form* $f(z) = \sum_{k=1}^{\kappa} b_k(z)\varphi_k(z)$ *on* Δ, *where the* $b_k(z)$ *are holomorphic functions on* Δ.

7.5.4. *Z***-ideal.** Let $D \subset \mathbf{C}^n$ be a domain. Let Σ be an analytic set in D, and let $F(z)$ be a holomorphic function on D such that $F(z) \not\equiv 0$ on each irreducible component of Σ in D. We consider the set $\mathcal{I}$ of all pairs $(f(z), \delta)$, where $\delta \subset D$ is a domain and $f(z)$ is a holomorphic function on δ satisfying

1. if $\delta \cap \Sigma = \emptyset$, then $f(z)$ is an arbitrary holomorphic function on δ;
2. if $\delta \cap \Sigma \neq \emptyset$, then $(f(z)/F(z))\,|_{\delta\cap\Sigma}$ is a weakly holomorphic function on $\delta \cap \Sigma$.

Then $\mathcal{I}$ is an $\mathcal{O}$-ideal on D. We call $\mathcal{I}$ the Z-**ideal** with respect to $F(z)$ and Σ, and we use the notation $\mathcal{I} = Z\{\Sigma, F\}$.[6] Note that the zero set of $Z\{\Sigma, F\}$ is contained in Σ.

We have the following theorem concerning Z-ideals.

THEOREM 7.10. *For any analytic set Σ in D and any holomorphic function $F(z)$ such that $F(z) \not\equiv 0$ on each irreducible component of Σ in D, the Z-ideal $Z\{F, \Sigma\}$ has a locally finite pseudobase at each point in D.*

PROOF. Fix $z_0 \in \Sigma$. We prove that $Z\{F, \Sigma\}$ has a locally finite pseudobase at z_0. Fix a sufficently small polydisk Δ centered at z_0 in D so that $\Delta \cap \Sigma$ can be decomposed into irreducible components Σ_j $(j = 1, \dots, l)$ in Δ such that each Σ_j passes through z_0. Since $F(z) \not\equiv 0$ on Σ_j, we can consider the Z-ideal $Z\{F, \Sigma_j\}$ $(j = 1, \dots, l)$ as defined on Δ. Since $Z\{F, \Sigma\}|_\Delta = \bigcap_{j=1}^{l} Z\{F, \Sigma_j\}$ as an $\mathcal{O}$-ideal in Δ, it follows from Theorem 7.4 that we need only show that each $Z\{F, \Sigma_j\}$ $(j = 1, \dots, l)$ has a locally finite pseudobase at z_0.

To prove $Z\{F, \Sigma_j\}$ has a locally finite pseudobase at z_0, as usual to simplify notation, we write $\Sigma = \Sigma_j$ in Δ and assume Σ is of dimension r. After a suitable linear coordinate transformation, we can assume the coordinate system $z = (z_1, \dots, z_r, z_{r+1}, \dots, z_n)$ satisfies the Weierstrass condition for Σ at z_0. Thus we can find a polydisk $\Delta_0 := \Delta_0^r \times \Delta_0^{n-r} \subset\subset \Delta$ centered at the point $z_0 = (z_0', z_0'')$ such that $(\Sigma \cap \Delta_0) \cap [\Delta_0^r \times (\partial \Delta_0^{n-r})] = \emptyset$ and such that $\Sigma \cap \Delta_0$ can be described as

$$z_j = \xi_j(z_1, \dots, z_r) \quad (j = r+1, \dots, n),$$

where $\tilde{z}' = (z_1, \dots, z_r)$ varies over a ramified domain $\tilde{\Delta}_0^r$ over Δ_0^r without relative boundary. By Theorem 6.4 in Chapter 6, there exists a polydisk δ in Δ_0^r centered at the point z_0' such that, upon taking $\tilde{\delta}$ to be the part of $\tilde{\Delta}_0^r$ over δ, there are a finite number of bounded holomorphic functions $\varphi_j(z_1, \dots, z_r)$ $(j = 1, \dots, m)$ on $\tilde{\delta}$, say $|\varphi_j| < M$ $(j = 1, \dots, m)$, such that, if we take the polydisk $\Gamma : |w_j| < M$ $(j = 1, \dots, m)$ in $\mathbf{C}_w^m$, then the r-dimensional irreducible analytic set $\tilde{\Sigma}$ in $\Lambda := \delta \times \Delta_0^{n-r} \times \Gamma \subset \mathbf{C}_z^n \times \mathbf{C}_w^m$ defined by

$$\tilde{\Sigma} : \begin{cases} z_j = z_j \quad (j = 1, \dots, r), \\ z_j = \xi_j(z_1, \dots, z_r) \quad (j = r+1, \dots, n), \\ w_k = \varphi_k(z_1, \dots, z_r) \quad (k = 1, \dots, m), \end{cases}$$

where $\tilde{z}' = (z_1, \dots, z_r)$ varies over $\tilde{\delta}$, has a singularity set $\tilde{\sigma}$ in Λ with $\dim \tilde{\sigma} \le r-2$, i.e., the analytic set $\tilde{\Sigma}$ in Λ is the lifting of the first kind of the analytic set Σ in $\delta \times \Delta_0^{n-r}$ with singular set of dimension at most $r-2$.

[6] Intuitively, the Z-ideal $Z\{\Sigma, F\}$ is the collection of all holomorphic functions $f(z)$ on $\delta \subset D$ such that $f(z)\,|_{\delta\cap\Sigma}$ vanishes on the given zero set of $F(z)|_\Sigma$. Thus, if we set $S := \Sigma \cap \{F = 0\}$ and denote by $G\{S\}$ the G-ideal for S in D, then $Z\{\Sigma, F\} \subset G\{S\}$.

We regard $F(z)$ as a holomorphic function on Λ (constant for w), so that $F(z) \not\equiv 0$ on $\tilde{\Sigma}$. We can thus consider the Z-ideal $Z\{F, \tilde{\Sigma}\}$ on Λ. We note that $\tilde{\Sigma} \bigcap [(\delta \times \Delta_0^{n-r}) \times \partial\Gamma] = \emptyset$, and $\Sigma \mid_{\delta \times \Delta_0^{n-r}}$ is equal to the projection of the analytic set $\tilde{\Sigma}$ onto $\delta \times \Delta_0^{n-r}$. Furthermore, since, as we have noted, the family of weakly holomorphic functions on $v \subset \Sigma$ can be identified with the family of weakly holomorphic functions on $\pi_0^{-1}(v) \subset \tilde{\Sigma}$ via $\pi_0 : \mathbf{C}_z^n \times \mathbf{C}_w^m \to \mathbf{C}_z^n$, we see from the definition of the projection of an $\mathcal{O}$-ideal that the projection $\tilde{\mathcal{P}}$ of the Z-ideal $Z\{F, \tilde{\Sigma}\}$ onto $\delta \times \Delta_0^{n-r}$ is equivalent to the Z-ideal $Z\{F, \Sigma\}|_{\delta \times \Delta_0^{n-r}}$ as an $\mathcal{O}$-ideal on $\delta \times \Delta_0^{n-r}$. Therefore, to prove that $Z\{F, \Sigma\}$ has a locally finite pseudobase at the point z_0, it suffices from Theorem 7.9 to verify that $Z\{F, \tilde{\Sigma}\}$ has a locally finite pseudobase at each point in Λ. To this end, let (z, w') be any point in Λ. By Corollary 7.2, we can find a polydisk Λ' centered at (z', w') in Λ and a finite number of universal denominators $\psi_j(z, w)$ $(j = 1, \ldots, q)$ in Λ' for $\tilde{\Sigma} \cap \Lambda'$ such that

$$ß := \bigcap_{j=1}^{q} \{(z, w) \in \Lambda' \mid \psi_j(z, w) = 0\} \subset \tilde{\sigma}.$$

We set $\tilde{\Sigma}' = \Lambda' \cap \tilde{\Sigma}$, and we consider the G-ideal $\mathcal{G}\{\tilde{\Sigma}'\}$ with respect to $\tilde{\Sigma}'$ in Λ'. By Theorem 7.8 and the solvablity of Problem E in the polydisk Λ', we can find a finite pseudobase $G_j(z, w)$ $(j = 1, \ldots, s)$ of $\mathcal{G}\{\tilde{\Sigma}'\}$ on Λ'.

Consider the following system of q homogeneous linear equations (Ω) (determined by the known holomorphic functions $\psi_j(z, w)$ $(j = 1, \ldots, q)$, $F(z)$, and $G_k(z, w)$ $(k = 1, \ldots, s)$ on Λ') for the unknown holomorphic vector-valued functions $(f(z, w), f^{(j)}(z, w), g_k^{(j)}(z, w))$ $(j = 1, \ldots, q;\ k = 1, \ldots, s)$:

$$(\Omega) \qquad \begin{aligned} f(z, w)\psi_j(z, w) &= f^{(j)}(z, w)F(z) \\ &+ g_1^{(j)}(z, w)G_1(z, w) + \cdots + g_s^{(j)}(z, w)G_s(z, w) \quad (j = 1, \ldots, q). \end{aligned}$$

We will prove that the ℓ-ideal $\ell\{\Omega\}$ (the collection of first components $(f(z, w), \lambda)$ of the $\mathcal{O}$-module $\mathcal{L}\{\Omega\}$ with respect to the linear relation (Ω) in Λ') is equivalent to the Z-ideal $Z\{F, \tilde{\Sigma}'\}$ as an $\mathcal{O}$-ideal on Λ'.

To verify this, let $(z_0, w_0) \in \Lambda'$ and let $f(z, w)$ be any holomorphic function belonging to $Z\{F, \tilde{\Sigma}'\}$ at (z_0, w_0). Thus, $f(z, w)/F(z, w)$ is a weakly holomorphic function on $\tilde{\Sigma}' \cap \lambda_0$, where λ_0 is a neighborhood of (z_0, w_0) in Λ'. For each $j = 1, \ldots, q$, the function $(f(z, w)/F(z, w)) \cdot \psi_j(z, w)$ is the restriction of a holomorphic function $f^{(j)}(z, w)$ in a neighborhood λ' of (z_0, w_0) in λ_0. Thus, there exist a neighborhood $\lambda'' \subset \lambda'$ of (z_0, w_0) and s holomorphic functions $g_k^{(j)}(z, w)$ $(k = 1, \ldots, s)$ in λ'' such that

$$f(z, w)\psi_j(z, w) = f^{(j)}(z, w)F(z) + g_1^{(j)}(z, w)G_1(z, w) + \cdots + g_s^{(j)}(z, w)G_s(z, w)$$

in λ''. Therefore, $f(z, w)$ belongs to $\ell\{\Omega\}$ at the point (z_0, w_0).

Conversely, let $(z_0, w_0) \in \Lambda'$ and let $f(z, w)$ belong to $\ell\{\Omega\}$ at (z_0, w_0). We can find a neighborhood λ_0 of (z_0, w_0) in Λ' and a holomorphic vector-valued function $(f(z, w), f^{(j)}(z, w), g_k^{(j)}(z, w))$ $(j = 1, \ldots, q; k = 1, \ldots, s)$ which satisfies equations (Ω) in λ_0. Let $(z', w') \in \lambda_0 \setminus \tilde{\sigma}$. Since $ß \subset \tilde{\sigma}$, we have $\psi_j(z', w') \neq 0$ for some $j = 1, \ldots, q$. Thus, dividing both sides of (Ω) by $\psi_j(z, w)$ in a small neighborhood $\lambda' \subset \lambda_0$ of (z', w'), we have

$$f(z, w) = \tilde{f}^{(j)}(z, w)F(z) + \tilde{g}_1^{(j)}(z, w)G_1(z, w) + \cdots + \tilde{g}_s^{(j)}(z, w)G_s(z, w)$$

in λ', where $\tilde{f}^{(j)}(z,w)$ and $\tilde{g}_k^{(j)}(z,w)$ $(k=1,\dots,s)$ are holomorphic functions in λ'. It follows that $(f(z,w)/F(z))|_{\tilde{\Sigma}'\cap\lambda'}$ is a weakly holomorphic function on $\tilde{\Sigma}'\cap\lambda'$. We thus see that $f(z,w)/F(z)$ is a weakly holomorphic function on $\tilde{\Sigma}'\cap\lambda_0$ (an analytic set of dimension r) except perhaps at points of $\tilde{\sigma}$ (an analytic set of dimension at most $r-2$). Using Remark 7.2, it follows that $f(z,w)/F(z)$ is a weakly holomorphic function on all of $\tilde{\Sigma}'\cap\lambda_0$, i.e., $f(z,w)$ belongs to $Z\{F,\tilde{\Sigma}'\}$ at the point (z_0,w_0).

By Theorem 7.7, $\ell\{\Omega\}$ has a locally finite pseudobase at any point $(z,w)\in\Lambda'$, and hence so does $Z\{F,\tilde{\Sigma}'\}=Z\{F,\Sigma\}|_{\Lambda'}$. Theorem 7.10 is proved. □

Let Σ be an analytic set in a domain $D\subset\mathbf{C}^n$. We let $\mathcal{O}_w(\Sigma)$ denote the set of all pairs $(f(z),v)$ such that $v\subset\Sigma$ is an open set in Σ and $f(z)$ is a weakly holomorphic function on v. Let $z_0\in\Sigma$. We say that $f(z)$ belongs to $\mathcal{O}_w(\Sigma)$ at the point z_0 if there exists a pair $(f(z),u)\in\mathcal{O}_w(\Sigma)$, where u is a neighborhood of z_0 in Σ. In Theorem 7.10 we consider the special case where $F(z)$ is a universal denominator $W_0(z)$ for Σ in D such that $W_0(z)\not\equiv 0$ on each irreducible component of Σ.

Then we obtain the following corollary.

COROLLARY 7.10. *Let $z_0\in\Sigma$ and let $\Phi_j(z)$ $(j=1,\dots,\nu)$ be a pseudobase of the Z-ideal $Z\{W_0,\Sigma\}$ in a neighborhood Δ of z_0 in D. Then for any $z'\in\Delta\cap\Sigma$ and for any $f(z)$ belonging to $\mathcal{O}_w(\Sigma)$ at the point z', we can find a neighborhood δ' of z' in Δ and ν holomorphic functions $\alpha_j(z)$ $(j=1,\dots,\nu)$ on δ' such that*

$$f(z)=\alpha_1(z)\frac{\Phi_1(z)}{W_0(z)}+\cdots+\alpha_\nu(z)\frac{\Phi_\nu(z)}{W_0(z)}\bigg|_{\Sigma\cap\delta'}. \tag{7.31}$$

PROOF. Let $z'\in\Delta\cap\Sigma$ and let $f(z)$ belong to $\mathcal{O}_w(\Sigma)$ at the point z'. Then $f(z)W_0(z)$ is a holomorphic function on a neighborhood $v\subset\Sigma$ of z' on Σ. That is, there exists a holomorphic function $F(z)$ on a neighborhood δ_0 of z' in Δ such that

$$f(z)W_0(z)=F(z)\,|_{\delta_0\cap\Sigma}.$$

This means that $F(z)$ belongs to $Z\{W_0,\Sigma\}$ at the point z', so that, using Theorem 7.10, we can find a neighborhood $\delta'\subset\delta_0$ of z' in Δ and ν holomorphic functions $\alpha_j(z)$ $(j=1,\dots,\nu)$ on δ' such that

$$F(z)=\alpha_1(z)\Phi_1(z)+\cdots+\alpha_\nu(z)\Phi_\nu(z)\quad\text{on }\delta',$$

which implies (7.31). □

7.5.5. W-ideal. Let D be a domain in $\mathbf{C}^n$ and let Σ be an analytic set in D. We consider the set $\mathcal{I}$ of all pairs $(f(z),\delta)$ such that $\delta\subset D$ is an open set and $f(z)$ is a universal denominator of Σ in δ. (If $\delta\cap\Sigma=\emptyset$, then all holomorphic functions on δ belong to $\mathcal{I}$ on δ.) Then $\mathcal{I}$ becomes an $\mathcal{O}$-ideal in D. We call it the **W-ideal** with respect to Σ, and we write $\mathcal{I}=W\{\Sigma\}$.

Then we have the following theorem.

THEOREM 7.11. *For any analytic set in Σ in D, the W-ideal $W\{\Sigma\}$ has a locally finite pseudobase at each point in D.*

PROOF. Let $z_0\in D$. From Proposition 7.4 we take a polydisk Δ centered at z_0 in D and a universal denominator $W_0(z)$ on Δ for Σ such that $W_0(z)\not\equiv 0$ on each irreducible component of $\Sigma\cap\Delta$ in Δ. We can thus construct the Z-ideal $Z\{W_0,\Sigma\}$ in Δ. By Theorem 7.10 and Theorem 7.8, we can find a polydisk Δ_0 in Δ, centered at z_0, such that there exist a finite pseudobase $\Phi_j(z)$ $(j=1,\dots,\nu)$ of the Z-ideal

$Z\{W_0, \Sigma\}$ in Δ_0 and a finite pseudobase $G_k(z)$ $(k = 1, \dots, s)$ of the G-ideal $G\{\Sigma\}$ in Δ_0.

Consider the following linear system of ν homogeneous equations determined by the known holomorphic functions $W_0(z)$, $\Phi_j(z)$ $(j = 1, \dots, \nu)$, and $G_k(z)$ $(k = 1, \dots, s)$ in Δ_0 for the unknown holomorphic vector-valued function $(f(z), f^{(j)}(z), g_k^{(j)}(z))$ $(j = 1, \dots, \nu; k = 1, \dots, s)$ of rank $1 + \nu + s$:

$$(\Omega) \qquad f(z)\Phi_j(z) = f^{(j)}(z)W_0(z) + g_1^{(j)}(z)G_1(z) + \cdots + g_s^{(j)}(z)G_s(z).$$

We let $\ell\{\Omega\}$ denote the ℓ-ideal with respect to (Ω), i.e., for $\delta \subset \Delta_0$, the set of the first components $(f(z), \delta)$ of the $\mathcal{O}$-module $\mathcal{L}\{\Omega\}$ with respect to the linear relation (Ω) in Δ_0. Then the W-ideal $W\{\Sigma\}|_{\Delta_0}$ is equivalent to $\ell\{\Omega\}$ as $\mathcal{O}$-ideals in Δ_0.

To see this, let $z' \in \Delta_0$. In case $z' \in \Delta_0 \setminus \Sigma$, an arbitrary holomorphic function $f(z)$ at z' belongs to both $W\{\Sigma\}$ and $\ell\{\Omega\}$ at z'. Thus we may assume $z' \in \Delta_0 \cap \Sigma$. Let $f(z)$ belong to $W\{\Sigma\}$ at the point z' and fix $j \in \{1, \dots, \nu\}$. Since $\Phi_j(z)/W_0(z)$ belongs to $\mathcal{O}_w(\Sigma)$ at z', we can find a neighborhood δ' of z' in Δ_0 and a holomorphic function $f^{(j)}(z)$ in δ' such that $(\Phi_j(z)/W_0(z)) \cdot f(z) = f^{(j)}(z)\,|_{\Sigma\cap\delta'}$. Thus, we can find a neighborhood $\delta'' \subset \delta'$ of z' and s holomorphic functions $g_k^{(j)}(z)$ $(k = 1, \dots, s)$ on δ'' such that

$$f(z)\Phi_j(z) = f^{(j)}(z)W_0(z) + g_1^{(j)}(z)G_1(z) + \cdots + g_s^{(j)}(z)G_s(z)$$

on δ''. It follows that $f(z)$ belongs to $\ell\{\Omega\}$ at z'.

Conversely, let $a \in \Delta_0$ and let $f(z)$ belong to $\ell\{\Omega\}$ at a. We can find a neighborhood δ_0 of a in Δ_0 and $\nu + s$ holomorphic functions $f^{(j)}(z)$, $g_k^{(j)}(z)$ $(j = 1, \dots, \nu;\ k = 1, \dots, s)$ on δ_0 which satisfy (Ω) on δ_0. Thus,

$$f(z)\frac{\Phi_j(z)}{W_0(z)} = f^{(j)}(z)\Bigg|_{\Sigma\cap\delta_0} \qquad (j = 1, \dots, \nu).$$

In order to prove that $f(z) \in W\{\Sigma\}$ on δ_0, let $z' \in \Sigma \cap \delta_0$ and let $h(z)$ belong to $\mathcal{O}_w(\Sigma)$ at z'. We see from Corollary 7.10 that there exist a neighborhood δ' of z' in δ_0 and ν holomorphic functions $\alpha_j(z)$ $(j = 1, \dots, s)$ on δ' such that

$$h(z) = \alpha_1(z)\frac{\Phi_1(z)}{W_0(z)} + \cdots + \alpha_\nu(z)\frac{\Phi_\nu(z)}{W_0(z)}\Bigg|_{\Sigma\cap\delta'}.$$

It follows that

$$h(z)f(z) = \alpha_1(z)f^{(1)}(z) + \cdots + \alpha_\nu(z)f^{(\nu)}(z)|_{\Sigma\cap\delta'}.$$

Since the right-hand side is a holomorphic function in δ', it follows that $f(z)$ belongs to $W\{\Sigma\}$ on δ_0 and hence at the point a. Thus, $\ell\{\Sigma\}$ and $W\{\Sigma\}$ are equivalent as $\mathcal{O}$-ideals on Δ_0, and Theorem 7.7 again yields Theorem 7.11. □

COROLLARY 7.11. *Let Σ be an r-dimensional analytic set in a domain D in $\mathbf{C}^n$ and let σ be the singular set of Σ. Then the common zero set τ of the W-ideal $W\{\Sigma\}$ with respect to Σ in D is an analytic set in D of dimension at most $r - 1$ and $\tau \subset \sigma$.*

CHAPTER 8

Analytic Spaces

8.1. Analytic Spaces

We begin by defining an analytic space of dimension n. Fix an integer $n \geq 1$ and let $\mathcal{V}$ be a connected Hausdorff space such that for each point $p \in \mathcal{V}$, there exists a neighborhood δ_p of p in $\mathcal{V}$ satisfying the following conditions:

(i) there exists a homeomorphism ϕ_p from δ_p onto a ramified domain λ_p over $\mathbf{C}^n$;

(ii) for any distinct points p, q in $\mathcal{V}$, the mapping $\phi_q \circ \phi_p^{-1}$ is an analytic mapping from $\phi_p(\delta_p \cap \delta_q)$ onto $\phi_q(\delta_p \cap \delta_q)$. Precisely, if
$$\phi_q \circ \phi_p^{-1} : \phi_p(\delta_p \cap \delta_q) \to \phi_q(\delta_p \cap \delta_q)$$
via
$$w = (\psi_1(z), \ldots, \psi_n(z)) := \phi_q \circ \phi_p^{-1}(z_1, \ldots, z_n),$$
then each $\psi_j(z)$ $(j = 1, \ldots, n)$ is a holomorphic function on the ramified domain $\phi_p(\delta_p \cap \delta_q) \subset \lambda_p$ over $\mathbf{C}^n$.

We call $\mathcal{V}$ an **analytic space of dimension** n. The triple $(\delta_p, \lambda_p, \phi_p)$ is called a local coordinate neighborhood of p in $\mathcal{V}$. Furthermore, if we can take λ_p to be a univalent domain in $\mathbf{C}^n$ for each $p \in \mathcal{V}$, then we call $\mathcal{V}$ a **complex manifold of dimension** n. In the case $n = 1$, $\mathcal{V}$ is a Riemann surface of one complex variable.

Let $\mathcal{V}$ be an analytic space of dimension n. A connected open set in $\mathcal{V}$ is called a domain in $\mathcal{V}$. Occasionally we omit the connectivity condition for a domain. Let D be a domain in $\mathcal{V}$ and let $f(p)$ be a complex-valued function on D. If for any point p in D with local coordinate neighborhood $(\delta_p, \lambda_p, \phi_p)$ the function $f \circ \phi_p^{-1}$ is holomorphic on the ramified domain $\phi_p(\delta_p \cap D) \subset \lambda_p$ over $\mathbf{C}^n$, then we say that $f(p)$ is a **holomorphic function** on D. Let $K \subset \mathcal{V}$ be a closed set. We say that a complex-valued function $f(p)$ is holomorphic on K if there exists an open neighborhood D of K in $\mathcal{V}$ such that $f(p)$ is defined and holomorphic on D.

Let $\mathcal{V}_1$ and $\mathcal{V}_2$ be analytic spaces of dimensions n and m. Let $\varphi : \mathcal{V}_1 \to \mathcal{V}_2$ be a mapping from $\mathcal{V}_1$ into $\mathcal{V}_2$. If for any open set $v \subset \mathcal{V}_2$ and for any holomorphic function $f(p)$ on v, the function $\widetilde{f} := f \circ \varphi$ is a holomorphic function on $\varphi^{-1}(\varphi(\mathcal{V}_1) \cap v) \subset \mathcal{V}_1$, then we say that $\varphi(p)$ is an **analytic mapping** from $\mathcal{V}_1$ into $\mathcal{V}_2$. Furthermore, if $m = n$ and if there exists a one-to-one analytic mapping from $\mathcal{V}_1$ onto $\mathcal{V}_2$, then we say that $\mathcal{V}_1$ and $\mathcal{V}_2$ are **analytically equivalent**.

8.1.1. Examples of Analytic Spaces.

An analytic space of dimension $n \geq 2$ is a canonical generalization of a Riemann surface of one complex variable. However, an analytic space of dimension $n \geq 2$ is not always a complex manifold (as shown in Example 6.3), in contrast to the fact that a Riemann surface of one complex variable is locally uniformizable at each point. We present some other examples of analytic spaces of dimension $n \geq 2$ which illustrate differences with the Riemann surface case.

1. T. Radó [**61**] showed that any Riemann surface $\mathcal{R}$ satisfies the second axiom of countability; i.e., there exist a countable number of open sets U_n ($n = 1, 2, \dots$) in $\mathcal{R}$ such that for any point p in $\mathcal{R}$, the collection $\{U_n\}_n$ contains a fundamental neighborhood basis of p in $\mathcal{R}$. This axiom is not necessarily satisfied by an analytic space of dimension $n \geq 2$.

EXAMPLE 8.1. [1] Let $\mathbf{C}^2 = \mathbf{C}_x \times \mathbf{C}_y$ with variables x and y, and let $\mathbf{P}_z$ be the Riemann sphere with variable z. Fix $a \in \mathbf{C}$. In the product space $\mathbf{C}^2 \times \mathbf{P}_z$, we consider the analytic hypersurface

$$\Sigma_a \ : \ yz - x + a = 0. \tag{8.1}$$

Since Σ_a is nonsingular in $\mathbf{C}^2 \times \mathbf{P}_z$, it follows that Σ_a can be considered as a 2-dimensional complex submanifold in $\mathbf{C}^2 \times \mathbf{P}_z$.

Let

$$\pi_a : (x, y, z) \in \Sigma_a \to (x, y) \in \mathbf{C}^2$$

be the projection from Σ_a to $\mathbf{C}^2$. We let L denote the complex line $y = 0$ in $\mathbf{C}^2$ and consider the inverse image $\pi_a^{-1}(L)$ in Σ_a. Thus, $\pi_a^{-1}(L)$ consists of two irreducible components

$$L_a = \mathbf{C}_x \times \{0\} \times \{\infty\} \quad \text{and} \quad L_a^* = \{(a, 0)\} \times \mathbf{P}_z,$$

so that $L_a \cap L_a^* = \{(a, 0, \infty)\}$. We set

$$\Sigma_a^* := \Sigma_a \setminus L_a;$$

then $\pi_a(\Sigma_a^*) \cap L = \{(a, 0)\}$. Now let E be an arbitrary subset of $\mathbf{C}$. We set

$$M := \bigcup_{a \in E} \Sigma_a^*$$

and we will define an identification in M to form a new space M_E. This identification is defined as follows. Let $p \in \Sigma_a^*$ and $q \in \Sigma_b^*$, where $a, b \in E$. If $a \neq b$ and $\pi_a(p) = \pi_b(q)$, then we identify p with q. The space M_E obtained by this identification in M canonically becomes a 2-dimensional complex manifold. We have an analytic mapping π from M_E into $\mathbf{C}^2$ such that $\pi_{|\Sigma_a^*} = \pi_{a|\Sigma_a^*}$ for $a \in E$, and hence $\pi(M_E) \cap L = E$. We put $\tilde{L}_a^* = L_a^* \setminus \{(a, 0, \infty)\}$. Then, for any $a \in E$, the points of $\tilde{L}_a^*$ are not identified with any other points.

Suppose we take a set E which contains an uncountable number of points in $\mathbf{C}$. Then the complex manifold M_E does not satisfy the second axiom of countability. To verify this, first note from above that for each $a \in E$, the set $\tilde{L}_a^*$ is a subset of M_E. Furthermore, for each $a \in E$, there uniquely exists a smallest open neighborhood $\tilde{\Sigma}_a^*$ of $\tilde{L}_a^*$ in M_E such that $\pi_{|\tilde{\Sigma}_a^*} = \pi_{a|\Sigma_a^*}$. Hence $\tilde{\Sigma}_a^* \cap \tilde{L}_b^* = \emptyset$ for all $b \neq a$ ($a, b \in E$). Since $M_E = \bigcup_{a \in E} \tilde{\Sigma}_a^*$ and since E is uncountable, M_E does not satisfy the second axiom of countability.

2. A Riemann surface admits a non-constant meromorphic function. This is not always true for complex manifolds of dimension $n \geq 2$.

[1] This example is due to E. Calabi and M. Rosenlicht [**5**].

EXAMPLE 8.2. [2] In $\mathbf{C}^2$ with variables x, y we set $M := \mathbf{C}^2 \setminus \{(0,0)\}$. Let α, β be complex numbers with $|\alpha|,\ |\beta| > 1$. We consider the following analytic automorphism:

$$T : (x, y) \in M \to (x', y') = (\alpha x, \beta y) \in M,$$

and we let Γ denote the automorphism subgroup of M generated by T; i.e., $\Gamma = \{T^n : n = 0, \pm 1, \dots\}$. Since T has no fixed points in M and since, given $(a, b) \in M$, the orbit $\{T^n(a, b) | n = 0, \pm 1, \dots\}$ has no accumulation point in M, it follows that $\mathcal{M} := M/\Gamma$ (the quotient space of M modulo Γ) is a compact, complex manifold of dimension 2. We note that one of the fundamental regions of M for Γ is

$$(\{|x| < \alpha\} \times \{1 \le |y| < \beta\}) \cup (\{1 \le |x| < \alpha\} \times \{|y| < \beta\}).$$

Assume that there is no pair of integers $(h, k) \neq (0, 0)$ such that $\alpha^h = \beta^k$. Then $\mathcal{M}$ does not admit a non-constant meromorphic function.

PROOF. Let $\pi : M \to \mathcal{M}$ be the canonical mapping such that $\pi \circ T^n = \pi$ $(n = 0, \pm 1, \dots)$ on M. Assume that there exists a non-constant meromorphic function $g(p)$ on $\mathcal{M}$. If we set $G(x, y) := g(\pi(x, y))$ on M, then $G(x, y)$ is a non-constant meromorphic function in $M = \mathbf{C}^2 \setminus \{(0, 0)\}$. By Levi's theorem (Theorem 4.2), $G(x, y)$ has a meromorphic extension to $(0, 0)$. Since $g(p)$ has a pole in $\mathcal{M}$, $G(x, y)$ should have a pole S at $(0, 0)$ in $\mathbf{C}^2$. To see this, let $p = (x, y)$ be a pole of $G(x, y)$. Then each point $p_n = T^{-n}(p)$ $(n = 1, 2, \dots)$ is a pole of $G(x, y)$. Since $\{p_n\}_n$ tends to the origin $(0, 0)$, $G(x, y)$ cannot be holomorphic at the origin. Hence $G(x, y)$ has a pole at the origin.

Thus S determines an analytic hypersurface Σ in $\mathbf{C}^2$ passing through $(0, 0)$. We fix a small polydisk Δ centered at $(0, 0)$ and let $\Sigma_1, \dots, \Sigma_\nu$ be the irreducible components of $\Sigma \cap \Delta$. Since $T^k(\Sigma) = \Sigma$ $(k = 0, \pm 1, \pm 2, \dots)$ and $T(0, 0) = (0, 0)$, for some l with $1 \le l \le \nu$ and some j with $1 \le j \le \nu$ we have $T^l(\Sigma_j) = \Sigma_j \cap \Delta^l$ (here $\Delta^l = T^l(\Delta)$). We may assume that Σ_j can be written in the form

$$\Sigma_j : \ y = a_h x^{\frac{h}{p}} + a_{h+1} x^{\frac{h+1}{p}} + \cdots \qquad (a_h \neq 0)$$

in a neighborhood of $(0, 0)$ in Δ, where $p \ge 1$ and h are integers. Thus, $T^l(\Sigma_j)$ is of the form

$$T^l(\Sigma_j) : \ \beta^l y = a_h (\alpha^l x)^{\frac{h}{p}} + a_{h+1} (\alpha^l x)^{\frac{h+1}{p}} + \cdots$$

in a neighborhood of $(0, 0)$ in Δ. It follows from the uniqueness of the Puiseux series expansion that $\beta^l = \alpha^{\frac{lh}{p}}$; i.e., $\alpha^{lh} = \beta^{lp}$, which contradicts our assumption. □

EXAMPLE 8.3. In $\mathbf{C}^n$ with $n \ge 2$ variables $z_1, \dots, z_n$, we consider $2n$ vectors

$$w_k = (\omega_k^1, \dots, \omega_k^n) \qquad (k = 1, \dots, 2n)$$

which are linearly independent over $\mathbf{R}$. Let

$$g_k : z = (z_1, \dots, z_n) \in \mathbf{C}^n \to z' = z + \omega_k \qquad (k = 1, \dots, 2n)$$

be a parallel translation of $\mathbf{C}^n$. We let Γ denote the automorphism subgroup of $\mathbf{C}^n$ generated by the g_k $(k = 1, \dots, 2n)$. The quotient space $\mathcal{M}_\omega := \mathbf{C}^2/\Gamma$ canonically becomes a compact, complex manifold of dimension n. We call $\mathcal{M}_\omega$ an **n-dimensional complex torus**. We let $\pi : \mathbf{C}^2 \to \mathcal{M}_\omega$ denote the canonical projection such that $\pi \circ g = \pi$ for $g \in \Gamma$. If there exists a non-constant meromorphic

[2] This example is due to H. Hopf [**31**].

function $g(p)$ on $\mathcal{M}_\omega$, then $G(z) := g(\pi(z))$ is a non-constant meromorphic function on $\mathbf{C}^n$ which has periods ω_k $(k = 1, \ldots, 2n)$, i.e., $G(z)$ is a so-called Abel function on $\mathbf{C}^n$ with $2n$ periods ω_k $(k = 1, \ldots, 2n)$. It is known in this case[3] that if we consider the $(n, 2n)$-matrix

$$C = \begin{pmatrix} \omega_1^1 & \cdots & \omega_{2n}^1 \\ \vdots & \ddots & \vdots \\ \omega_1^n & \cdots & \omega_{2n}^n \end{pmatrix},$$

then there exists an invertible $(2n, 2n)$-matrix A with integer coefficients such that

$$(*) \qquad CA^{-1}C' = 0, \qquad i\overline{C}A^{-1}C' > 0 \quad (i^2 = -1),$$

where A^{-1} denotes the inverse matrix of A and C' is the transpose matrix of C. Thus if we take $2n$ vectors ω_k $(k = 1, \ldots, 2n)$ which do not satisfy condition $(*)$, then the complex torus $\mathcal{M}_\omega$ does not admit a non-constant meromorphic function.

8.2. Analytic Polyhedra

8.2.1. Extension Theorem. We next consider analytic polyhedra in an analytic space. First of all, we mention that a non-compact complex analytic space $\mathcal{V}$ does not necessarily admit enough global holomorphic functions f to separate points; i.e., it is not necessarily true that for $p, q \in \mathcal{V}$ with $p \neq q$, there exists f holomorphic in $\mathcal{V}$ with $f(p) \neq f(q)$. Thus we introduce the following notion of separability. Let $\mathcal{V}$ be an analytic space of dimension n and let E be a subset of $\mathcal{V}$. If for distinct points $p, q \in E$ there exists a holomorphic function f on an open set D in $\mathcal{V}$ containing E such that $f(p) \neq f(q)$, then we say that E satisfies the **separation condition**.

Let $\mathcal{P}$ be a compact set in $\mathcal{V}$ which can be described in the following manner: there exist an open set D with $\mathcal{P} \subset\subset D \subset \mathcal{V}$ and finitely many holomorphic functions $\varphi_j(p)$ $(j = 1, \ldots, m)$ on D such that $\mathcal{P}$ consists of a finite number of (closed) connected components of the set

$$D_\varphi := \bigcap_{j=1}^m \{p \in D \mid |\varphi_j(p)| \leq 1\}.$$

Then we call $\mathcal{P}$ a **generalized analytic polyhedron** in $\mathcal{V}$; the functions φ_j $(j = 1, \ldots, m)$ are defining functions of $\mathcal{P}$. Of course, some connected components of D_φ may not be relatively compact in D.

If a generalized analytic polyhedron $\mathcal{P}$ in $\mathcal{V}$ satisfies the separation condition, then we say that $\mathcal{P}$ is an **analytic polyhedron** in $\mathcal{V}$. When we want to emphasize the domain D where the defining functions $\varphi_j(p)$ $(j = 1, \ldots, m)$ of $\mathcal{P}$ are holomorphic, we will say that $\mathcal{P}$ is an analytic polyhedron in $\mathcal{V}$ with defining functions on D.[4]

Let $\mathcal{P}$ be an analytic polyhedron in an analytic space $\mathcal{V}$ of dimension n and let φ_j $(j = 1, \ldots, m)$ be defining functions of $\mathcal{P}$. Let

$$\overline{\Delta} : |z_j| \leq 1 \ (j = 1, \ldots, m)$$

[3]See the textbook of C. L. Siegel [**65**], Vol. III, p. 72.

[4]In the 2-dimensional complex manifold Σ_a studied in Example 8.1, we consider the subset

$$\mathcal{P} = \Sigma_a \cap \{(x, y, z) \in \mathbf{C}^2 \times \mathbf{P}_z \mid |x - a| \leq 1, \ |y| \leq 1\}$$

of Σ_a. Then $\mathcal{P}$ is a generalized analytic polyhedron in Σ_a, but it is not an analytic polyhedron since $\mathcal{P}$ contains the compact 1-dimensional analytic set L_a^*.

be the closed unit polydisk in $\mathbf{C}^m$. We consider the following analytic mapping from $\mathcal{P}$ into $\overline{\Delta}$:

$$\Phi : \ z_j = \varphi_j(p) \ \ (j = 1, \dots , m), \quad p \in \mathcal{P}.$$

Then the image $\Sigma := \Phi(\mathcal{P})$ of $\mathcal{P}$ is an n-dimensional analytic set in $\overline{\Delta}$ such that $\Phi(\partial\mathcal{P}) \subset \partial\Delta$. By adding more holomorphic functions on $\mathcal{P}$, if necessary, we may assume that $\mathcal{P}$ satisfies the separation condition and that the points of $\mathcal{P}$ and Σ are in one-to-one correspondence via Φ. We call Σ a **model** for $\mathcal{P}$ on $\overline{\Delta}$. Furthermore, if Σ is normal in $\overline{\Delta}$, i.e., if there exists an open polydisk

$$\Delta^{(\epsilon)} : |z_j| < 1 + \epsilon \quad (j = 1, \dots , m)$$

sufficiently close to $\overline{\Delta}$ such that Σ can be analytically extended to be an analytic set $\tilde{\Sigma}$ in $\Delta^{(\epsilon)}$ with $\tilde{\Sigma}$ normal in $\Delta^{(\epsilon)}$, then we say that Σ is a **normal model** on $\overline{\Delta}$.

We have the following theorem concerning analytic polyhedra in an analytic space.

THEOREM 8.1 (Normalization Theorem). *An analytic polyhedron $\mathcal{P}$ in an analytic space $\mathcal{V}$ of dimension n always has a normal model on a closed polydisk in $\mathbf{C}^\nu$, where $\nu \geq n$ is an integer depending on $\mathcal{P}$.*

PROOF. First we construct a model Σ of $\mathcal{P}$ on the closed unit polydisk $\overline{\Delta}$ in $\mathbf{C}^m$,

$$\Sigma : z_j = \varphi_j(p) \quad (j = 1, \dots , m), \quad p \in \mathcal{P}.$$

Then we consider the W-ideal $W\{\Sigma\}$ with respect to Σ on $\overline{\Delta}$. Since $W\{\Sigma\}$ has a locally finite pseudobase at each point in $\overline{\Delta}$ by Theorem 7.11, and since Problem E is always solvable in $\overline{\Delta}$, we can find a finite number of universal denominators $W_i(z)$ $(i = 1, \dots , q)$ for Σ on $\overline{\Delta}$ such that $\{W_i(z)\}_{i=1,\dots,q}$ forms a pseudobase of $W\{\Sigma\}$ at each point in $\overline{\Delta}$. Since the common zero set of the $W_i(z)$ $(i = 1, \dots , q)$ on $\overline{\Delta}$ is contained in the set of singularities σ of Σ by Corollary 7.2, we obtain a universal denominator $W(z)$ (a linear combination of the $W_i(z)$ $(i = 1, \dots , q)$ with constant coefficients) for Σ on $\overline{\Delta}$ such that $W(z) \not\equiv 0$ on any irreducible component of Σ on $\overline{\Delta}$.

Next we consider the Z-ideal $Z\{W, \Sigma\}$ with respect to this universal denominator $W(z)$ and Σ on $\overline{\Delta}$. Since $Z\{W, \Sigma\}$ has a locally finite pseudobase at each point in $\overline{\Delta}$ by Theorem 7.10 and since Problem E is solvable on $\overline{\Delta}$, it follows that we can find a finite number of holomorphic functions

$$Z_k(z) \quad (k = 1, \dots , l) \quad \text{on } \overline{\Delta}$$

such that $\{Z_k(z)\}_{k=1,\dots,l}$ forms a pseudobase of $Z\{W, \Sigma\}$ at each point in $\overline{\Delta}$. Therefore, each quotient $Z_k(z)/W(z)$ $(k = 1, \dots , l)$ is a weakly holomorphic function on Σ, and these become holomorphic functions $\psi_k(p)$ on the analytic polyhedron $\mathcal{P}$; i.e.,

$$\psi_k(p) = \frac{Z_k}{W}(\Phi(p)) \ \ (k = 1, \dots , l) \quad \text{on } \mathcal{P}.$$

We may assume $|\psi_k(p)| < 1$ $(k = 1, \dots , l)$ on $\mathcal{P}$. Then we construct the n-dimensional analytic set

$$\tilde{\Sigma} : z_j = \varphi_j(p) \ (j = 1, \dots , m), \ w_k = \psi_k(p) \ \ (k = 1, \dots , l), \quad p \in \mathcal{P},$$

in the closed unit polydisk

$$\overline{\Lambda} : |z_j| \leq 1 \ (j = 1, \dots , m), \ |w_k| \leq 1 \ (k = 1, \dots , l)$$

in $\mathbf{C}^{m+l} = \mathbf{C}^m_z \times \mathbf{C}^l_w$, i.e., $\tilde{\Sigma}$ is a lifting of Σ of the first kind through ψ_k $(k = 1, \dots, l)$.

We shall show that $\tilde{\Sigma}$ is a normal model of $\mathcal{P}$ in $\overline{\Lambda}$.

First of all, since Σ is a model of $\mathcal{P}$ in $\overline{\Delta}$, it follows that $\tilde{\Sigma}$ is a model of $\mathcal{P}$ in $\overline{\Lambda}$. Let $\tilde{v} \subset \tilde{\Sigma}$ be an open set and let $f(z, w)$ be a weakly holomorphic function for $\tilde{\Sigma}$ on $\tilde{v}$. Since $\tilde{\Sigma}$ and Σ are analytically isomorphic via the mapping π_0 induced by the projection from $\mathbf{C}^{n+l}$ to $\mathbf{C}^n$, $f(z, w)$ can be considered as a weakly holomorphic function $\tilde{f}(z)$ for Σ on $v \subset \Sigma$ such that $f(z, w) = \tilde{f}(z)$ for $z \in v = \pi_0^{-1}(\tilde{v})$. Let $(z', w') \in \tilde{v}$ with $z' \in v$. As shown in Corollary 7.10, there exist a neighborhood δ' of z' in $\overline{\Delta}$ such that $\delta' \cap \Sigma \subset v$ and l holomorphic functions $\alpha_k(z)$ $(k = 1, \dots, l)$ in δ' such that

$$\tilde{f}(z) = \alpha_1(z)\frac{Z_1(z)}{W(z)} + \cdots + \alpha_l(z)\frac{Z_l(z)}{W(z)}\bigg|_{\Sigma \cap \delta'}.$$

In other words,

$$f(z, w) = \alpha_1(z)w_1 + \cdots + \alpha_l(z)w_l \,|_{(\delta' \times \mathbf{C}^l) \cap \tilde{\Sigma}},$$

so that $f(z, w)$ is a holomorphic function for $\tilde{\Sigma}$ in a neighborhood of (z', w'). Thus, $\tilde{\Sigma}$ is normal on $\overline{\Lambda}$. □

We next prove an extension theorem concerning a normal model of an analytic polyhedron in an analytic space.

THEOREM 8.2 (Extension Theorem). *Let $\mathcal{P}$ be an analytic polyhedron in an analytic space $\mathcal{V}$ of dimension n. Let Σ be a normal model of $\mathcal{P}$ in the closed unit polydisk $\overline{\Delta}$ in $\mathbf{C}^m$ with variables $z_1, \dots, z_m$,*

$$\Phi : p \in \mathcal{P} \to z = \Phi(p) = (\varphi_1(p), \dots, \varphi_m(p)) \in \Sigma.$$

Given a holomorphic function $f(p)$ on $\mathcal{P}$, there exists a holomorphic function $F(z)$ on $\overline{\Delta}$ such that

$$f(p) = F(\Phi(p)), \qquad p \in \mathcal{P}.$$

PROOF. We consider the G-ideal $G\{\Sigma\}$ of Σ on $\overline{\Delta}$. Since $G\{\Sigma\}$ has a locally finite pseudobase at each point in $\overline{\Delta}$ and since Problem E is solvable on $\overline{\Delta}$, there exist a finite number of holomorphic functions $G_k(z)$ $(k = 1, \dots, s)$ on $\overline{\Delta}$ such that $\{G_k(z)\}_{k=1,\dots,s}$ forms a pseudobase at each point of $\overline{\Delta}$. Since $\mathcal{P}$ is a closed analytic polyhedron in $\mathcal{V}$ and $\overline{\Delta}$ is a closed polydisk in $\mathbf{C}^m$, we can find an open analytic polyhedron $\mathcal{P}^{(\epsilon)}$ with $\mathcal{P} \subset\subset \mathcal{P}^{(\epsilon)} \subset\subset \mathcal{V}$ and an open polydisk $\Delta^{(\epsilon)}$ with $\overline{\Delta} \subset\subset \Delta^{(\epsilon)}$ in $\mathbf{C}^m$ such that $f(p)$ and $\varphi_j(p)$ $(j = 1, \dots, m)$ are holomorphic in $\mathcal{P}^{(\epsilon)}$, $\Sigma^{(\epsilon)}$ is a normal model of $\mathcal{P}^{(\epsilon)}$ on $\Delta^{(\epsilon)}$,

$$\Phi : p \in \mathcal{P}^{(\epsilon)} \to z = (\varphi_1(p), \dots, \varphi_m(p)) \in \Sigma^{(\epsilon)},$$

and such that each $G_k(z)$ $(k = 1, \dots, s)$ is a holomorphic function on $\Delta^{(\epsilon)}$ with the property that $\{G_k(z)\}_{k=1,\dots,s}$ is a pseudobase of the G-ideal $G\{\Sigma^{(\epsilon)}\}$ at each point of $\Delta^{(\epsilon)}$. We consider the following collection $\mathcal{C}$ of pairs $(f_\zeta(z), \delta_\zeta)$, where $\zeta \in \Delta^{(\epsilon)}$ and δ_ζ is a neighborhood of ζ in $\Delta^{(\epsilon)}$:

1. If $\zeta \notin \Sigma^{(\epsilon)}$, then we take a neighborhood δ_ζ of ζ in $\Delta^{(\epsilon)}$ such that $\delta_\zeta \cap \Delta^{(\epsilon)} = \emptyset$, and we take $f_\zeta(z) \equiv 1$ on δ_ζ.

2. If $\zeta \in \Sigma^{(\epsilon)}$, we take a neighborhood δ_ζ of ζ in $\Delta^{(\epsilon)}$ and a holomorphic function $f_\zeta(z)$ in δ_ζ such that $f_\zeta(\Phi(p)) = f(p)$ for $p \in \Phi^{-1}(\delta_p \cap \Sigma^{(\epsilon)})$. This can be done by Theorem 8.1.

Then $\mathcal{C}$ is a C_2-distribution in $\Delta^{(\epsilon)}$ with respect to $G_k(z)$ $(k = 1, \dots, s)$. To see this, let ζ_1, ζ_2 be distinct points in $\Delta^{(\epsilon)}$. If at least one of these points is not contained in $\Sigma^{(\epsilon)}$, there is nothing to prove since the common zero set of $G_k(z)$ $(k = 1, \dots, s)$ coincides with $\Sigma^{(\epsilon)}$. Thus we may assume that $\zeta_1, \zeta_2 \in \Sigma^{(\epsilon)}$. Since $f_{\zeta_1}(z) - f_{\zeta_2}(z) = 0$ on $\delta_{\zeta_1} \cap \delta_{\zeta_2}$, it follows that $f_{\zeta_1}(z) - f_{\zeta_2}(z)$ belongs to $G\{\Sigma^{(\epsilon)}\}$ at each point of $\delta_{\zeta_1} \cap \delta_{\zeta_2}$. Thus, $\mathcal{C}$ is a C_2-distribution in $\Delta^{(\epsilon)}$ with respect to $G_k(z)$ $(k = 1, \dots, s)$.

Since Problem C_2 is solvable in the closed disk $\overline{\Delta}$, we can find a holomorphic function $F(z)$ on $\overline{\Delta}$ such that, for any $\zeta \in \overline{\Delta}$, $F(z) - f_\zeta(z)$ belongs to $G\{\Sigma^{(\epsilon)}\}$ at each point of δ_ζ. Consequently, $f(p) = F(\Phi(p))$ for all $p \in \mathcal{P}$, which proves the theorem. □

We call $F(z)$ a **holomorphic extension** of $f(p)$ to $\overline{\Delta}$. Similarly, given a holomorphic vector-valued function $f(p) = (f_1(p), \dots, f_\nu(p))$ on $\mathcal{P}$, if each $F_j(z)$ $(j = 1, \dots, \nu)$ is a holomorphic extension of $f_j(p)$ on $\overline{\Delta}$, then we call the holomorphic vector-valued function $F(z) = (F_1(z), \dots, F_\nu(z))$ on $\overline{\Delta}$ a **holomorphic extension of the vector-valued function** $f(p)$ to $\overline{\Delta}$.

Theorem 8.1 and Theorem 8.2 are the **lifting principles in analytic spaces**, which are perhaps the most important results in this book.[5]

REMARK 8.1. Let $\mathcal{D}$ be a ramified domain over $\mathbf{C}^n$. Assume that $\mathcal{D}$ has a normal model with defining functions $\varphi_j(p)$ $(j = 1, \dots, m)$. Let $p_0 \in \mathcal{D}$ and let $f(p)$ be a holomorphic function at p_0. From Theorem 8.2, we see that

$$f(p) = \sum_{j_1, \dots, j_m} a_{j_1, \dots, j_m} (\varphi_1(p))^{j_1} \cdots (\varphi_m(p))^{j_m} \tag{8.2}$$

in a neighborhood of $p_0 \in \mathcal{D}$. This expansion may be considered as a generalization of the Puiseux expansion in the case of one complex variable. We remark that for one-variable Puiseux expansions we have uniqueness of coefficients, but this is not necessarily the case in several complex variables. As a simple example of this phenomenon, define the analytic polyhedron $\mathcal{P} \subset \mathbf{C}^2$ as $\{(x, y) \in \mathbf{C}^2 : |x| < 1,\ |y| < 1,\ |x + y| < 1\}$. The function $xy^2 + x^2y = xy(x + y)$ yields an example of nonuniqueness.

REMARK 8.2. Let $\mathcal{P}$ be a generalized analytic polyhedron in an analytic space $\mathcal{V}$ of dimension n. Let $\varphi_j(p)$ $(j = 1, \dots, m)$ be defining functions for $\mathcal{P}$ in a domain $D \subset \mathcal{V}$. Let $\overline{\Delta}^m : |z_j| \le 1$ $(j = 1, \dots, m)$ denote the closed unit polydisk in $\mathbf{C}^m$. We consider the analytic mapping

$$\Phi : p \in \mathcal{P} \to z = (\varphi_1(p), \dots, \varphi_m(p)) \in \overline{\Delta}^m$$

and its image Σ in $\overline{\Delta}^m$,

$$\Sigma : \ z_j = \varphi_j(p), \quad (j = 1, \dots, m).$$

Consider the following condition:

[5]These principles were first established by K. Oka [**51**]. He states them in a slightly different form (see Remark 8.2).

(∗) $\mathcal{P}$ and Σ are in one-to-one correspondence except perhaps for points lying on an analytic set σ of dimension at most $n-1$.

Equivalently, for each irreducible component Σ_k ($k=1,\dots,s$) of Σ in $\overline{\Delta}^m$, there exists a point $z^0 \in \Sigma_k$ such that $\Phi^{-1}(z^0)$ consists of one point. Then *the normalization theorem holds for* $\mathcal{P}$. Precisely, there exist defining functions $\psi_k(p)$ ($k=1,\dots,l$) of $\mathcal{P}$ in a domain $D_0 \subset D$ such that

$$\tilde{\Sigma}: \quad w_k = \psi_k(p) \quad (k=1,\dots,l), \quad p \in \mathcal{P},$$

is a normal model of $\mathcal{P}$ in the closed polydisk $\overline{\Delta}^l$ in $\mathbf{C}^l$ (so that $\mathcal{P}$ and $\tilde{\Sigma}$ are necessarily in one-to-one correspondence), and hence $\mathcal{P}$ is necessarily an analytic polyhedron in $\mathcal{V}$ with defining functions $\psi_k(p)$ ($k=1,\dots,l$) on $D_0 \subset \mathcal{V}$.

In fact, replacing the condition in Theorem 8.1 that $\mathcal{P}$ and Σ are in one-to-one correspondence via Φ by this weaker condition (∗), the family of all holomorphic functions $f(p)$ on $\mathcal{P}$ and the family of all weakly holomorphic functions $F(z)$ on Σ are still in one-to-one correspondence via $F(\Phi) = f$ on $\mathcal{P}$. Since the remaining arguments in the proof of Theorem 8.1 are unchanged, we obtain the normalization theorem for generalized analytic polyhedra $\mathcal{P}$ satisfying condition (∗).

Once we have this normalization result, we see by following the proof of Theorem 8.2 that *the extension theorem also remains valid for such generalized analytic polyhedra* $\mathcal{P}$.

This remark, combined with Theorem 6.1, implies the following corollary.

COROLLARY 8.1. *Any ramified domain over* $\mathbf{C}^n$ *locally has a normal model.*

The type of generalized analytic polyhedra $\mathcal{P}$ satisfying condition (∗) originally studied by Oka [**51**] were the following. Let $\mathcal{V}$ be a ramified domain over $\mathbf{C}^n_z$ with branch set S. Let $\mathcal{P}$ be a generalized analytic polyhedron in $\mathcal{V}$ with defining functions on D (here, $\mathcal{P} \subset\subset D \subset \mathcal{V}$). If there exists a holomorphic function $\psi(\tilde{z})$ on D such that $\psi(\tilde{z})$ has different function elements over each point $z \in \underline{D} \setminus \underline{S}$ (where $\underline{D}$ and $\underline{S}$ denote the projections of D and S onto $\mathbf{C}^n_z$), then the normalization theorem and the extension theorem hold for $\mathcal{P}$.

REMARK 8.3. Let Σ be an analytic set of pure dimension r in the closed polydisk $\overline{\Delta}$ in $\mathbf{C}^n$. In the beginning of the proof of Theorem 8.1, we proved the existence of a global universal denominator $W(z)$ on all of $\overline{\Delta}$ for Σ such that $W(z) \not\equiv 0$ on each irreducible component of Σ by introducing the notion of a W-ideal $W\{\Sigma\}$ (this ideal was not discussed in Oka's papers). Oka proved the existence of such a universal denominator $W(z)$ in the following manner. Let σ be the set of singularities of Σ in $\overline{\Delta}$; thus σ is an analytic set in $\overline{\Delta}$ of dimension at most $r-1$. Let $\varphi_j(z)$ ($j=1,\dots,m$) be a pseudobase of the G-ideal $G\{\sigma\}$ on $\overline{\Delta}$, so that the common zero set of $\varphi_j(z)$ ($j=1,\dots,m$) is equal to σ. Fix any point $z_0 \in \overline{\Delta}$. By Corollary 7.2, there exist a polydisk δ centered at z_0 and a finite number of universal denominators $\psi_k(z)$ ($k=1,\dots,\nu$) in δ such that $T_{z_0} := \bigcap_{k=1}^{\nu}\{z \in \delta \mid \psi_k(z) = 0\} \subset \sigma \cap \delta$. Since $\varphi_j(z) = 0$ ($j=1,\dots,m$) on T_{z_0}, it follows from the Hilbert-Rückert Nullstellensatz for holomorphic functions (see Appendix in this chapter) that there exist a polydisk $\delta_0 \subset\subset \delta$ centered at z_0 and an integer $\alpha_j \geq 1$ such that

$$\varphi_j(z)^{\alpha_j} = a_1^{(j)}(z)\psi_1(z) + \cdots + a_\nu^{(j)}(z)\psi_\nu(z) \quad \text{on} \quad \delta_0,$$

where each $a_k(z)$ ($k=1,\dots,\nu$) is a holomorphic function in δ_0. Thus, $\varphi_j(z)^{\alpha_j}|_{\delta_0}$ ($j=1,\dots,m$) is a universal denominator for Σ on δ_0. Since $\overline{\Delta}$ is a closed polydisk, we can find sufficiently large integers A_j $(j = 1, \ldots, m)$ so that each $\varphi_j(z)^{A_j}$ is a universal denominator for Σ on the whole $\overline{\Delta}$. Since $\bigcap_{j=1}^m \{z \in \overline{\Delta} \mid \varphi_j(z)^{A_j} = 0\} = \sigma$, it follows that some linear combination $W(z)$ of $\varphi_j(z)^{A_j}$ $(j = 1, \ldots, m)$ is a universal denominator for Σ on $\overline{\Delta}$ with $W(z) \not\equiv 0$ on each irreducible component of Σ.

REMARK 8.4. To prove the extension theorem (Theorem 8.2) for analytic polyhedra in a univalent domain in $\mathbf{C}^n$, we do not need the fact that a Z-ideal has a locally finite pseudobase at each point. We only require the fact that a G-ideal has a locally finite pseudobase at each point and that Problems C_2 and E are solvable on polydisks in $\mathbf{C}^n$.

We studied $\mathcal{O}^\lambda$-modules ($\mathcal{O}$-modules of rank λ) on a domain in $\mathbf{C}^n$; we can define $\mathcal{O}^\lambda$-modules on a domain in an analytic space $\mathcal{V}$ in the same manner.

Let $\mathcal{P}$ be an analytic polyhedron in a domain D in an analytic space $\mathcal{V}$ and let Σ be a normal model of $\mathcal{P}$ on the closed unit polydisk $\overline{\Delta}$ via the one-to-one mapping

$$\Phi : \ p \in \mathcal{P} \to z = (\varphi_1(p), \ldots, \varphi_m(p)) \in \Sigma \subset \overline{\Delta},$$

where $\varphi_j(p)$ $(j = 1, \ldots, m)$ is a holomorphic function on D (here $\mathcal{P} \subset\subset D$) and $\overline{\Delta}$: $|z_j| \leq 1$. By Theorem 8.2, any holomorphic function $f(p)$ on $\mathcal{P}$ has a holomorphic extension $F(z)$ on $\overline{\Delta}$. We shall show that this kind of extension theorem holds for any $\mathcal{O}^\lambda$-module on $\mathcal{P}$.

Let $\mathcal{I}^\lambda$ be an $\mathcal{O}^\lambda$-module on $\mathcal{P}$. Let $\zeta \in \overline{\Delta}$. We define pairs $(f_\zeta(z), \delta_\zeta)$ where δ_ζ is a neighborhood of ζ in $\overline{\Delta}$ and $f_\zeta(z)$ is a holomorphic vector-valued function of rank λ on δ_ζ, as follows:

1. If $\zeta \notin \Sigma$, we take δ_ζ and $f(z)$ such that $\delta_\zeta \cap \Sigma = \emptyset$ and $f_\zeta(z)$ is any holomorphic function on δ_ζ.
2. If $\zeta \in \Sigma$, we take δ_ζ and $f(z)$ such that $\delta_\zeta \subset \overline{\Delta}$ and $f(\Phi(p))$ belongs to $\mathcal{I}^\lambda$ at each point of $\Phi^{-1}(\delta_\zeta \cap \Sigma)$.

Since Σ is a normal model of $\mathcal{P}$ on $\overline{\Delta}$, the set $\{(f_\zeta(z), \delta_\zeta)\}_{\zeta \in \overline{\Delta}}$ of all such pairs is an $\mathcal{O}^\lambda$-module on $\overline{\Delta}$. We let $\tilde{\mathcal{I}}^\lambda$ denote this $\mathcal{O}^\lambda$-module and we call it the **extension $\mathcal{O}^\lambda$-module** of $\mathcal{I}^\lambda$ on $\overline{\Delta}$.

We have the following lemma.

LEMMA 8.1. *$\mathcal{I}^\lambda$ is generated by a finite number of holomorphic vector-valued functions of rank λ on $\mathcal{P}$ if and only if the same is true for $\tilde{\mathcal{I}}^\lambda$ on $\overline{\Delta}$.*

To be precise, let $G_k(z)$ $(k = 1, \ldots, \mu)$ be a pseudobase of the G-ideal $G\{\Sigma\}$ on $\overline{\Delta}$. For any h $(h = 1, \ldots, \lambda)$ and k $(k = 1, \ldots, \mu)$ we consider the following $\lambda\mu$ holomorphic vector-valued functions of rank λ on $\overline{\Delta}$:

$$\psi_{h,k}(z) = (\underbrace{\overbrace{0, \ldots, 0, G_k(z)}^{h}, 0, \ldots, 0}_{\lambda}). \tag{8.3}$$

From the definition of $\tilde{\mathcal{I}}^\lambda$ for any $\mathcal{O}^\lambda$-module $\mathcal{I}^\lambda$, we always have $\psi_{h,k}(z) \in \tilde{\mathcal{I}}^\lambda$ for any h, k.

Then the following statements are valid.

1. Assume that $\mathcal{I}^\lambda$ is equivalent to an $\mathcal{O}^\lambda$-module $\mathcal{J}^\lambda\{F\}$ generated by a finite number of holomorphic vector-valued functions $F_j(p)$ $(j = 1, \ldots, \nu)$ on $\mathcal{P}$:

$$F_j(p) = (F_{1,j}(p), \ldots, F_{\lambda,j}(p)).$$

We let $\tilde{F}_j(z)$ $(j = 1, \ldots, \nu)$ denote a vector-valued function on $\overline{\Delta}$ which is a holomorphic extension of $F_j(p)$. Then $\tilde{\mathcal{I}}^\lambda$ is equivalent to the $\mathcal{O}^\lambda$-module $\mathcal{J}^\lambda\{\tilde{F}, \psi\}$ generated by $\tilde{F}_j(z)$ $(j = 1, \ldots, \nu)$ and $\psi_{h,k}(z)$ $(h = 1, \ldots, \lambda;\ k = 1, \ldots, \mu)$ on $\overline{\Delta}$.

2. Conversely, assume that $\tilde{\mathcal{I}}^\lambda$ is equivalent to an $\mathcal{O}^\lambda$-module $\mathcal{J}^\lambda\{\tilde{F}\}$ generated by a finite number of holomorphic vector-valued functions $\tilde{F}_j(z)$ $(j = 1, \ldots, s)$ of rank λ on $\overline{\Delta}$. Then $\mathcal{I}^\lambda$ is equivalent to the $\mathcal{O}^\lambda$-module $\mathcal{J}^\lambda\{F\}$ generated by s holomorphic vector-valued functions $F_j(p)$ $(j = 1, \ldots, s)$ such that $F_j(p) = \tilde{F}_j(\Phi(p))$ on $\mathcal{P}$.

PROOF. For the proof of 1, let $\zeta_0 \in \overline{\Delta}$ and let $f(z) = (f_1(z), \ldots, f_\lambda(z))$ be any holomorphic vector-valued function belonging to $\tilde{\mathcal{I}}^\lambda$ at the point ζ_0. If $\zeta_0 \notin \Sigma$, then some $G_k(z) \neq 0$ at ζ_0. Thus we can write

$$f(z) = \frac{f_1(z)}{G_k(z)}\psi_{1,k}(z) + \cdots + \frac{f_\lambda(z)}{G_k(z)}\psi_{\lambda,k}(z)$$

in a neighborhood of ζ_0 in $\overline{\Delta}$, so that $f(z) \in \mathcal{J}^\lambda\{\tilde{F}, \psi\}$ at the point ζ_0. If $\zeta_0 \in \Sigma$, then $f(\Phi(p))$ belongs to $\mathcal{I}^\lambda$ at the point $p_0 = \Phi^{-1}(\zeta_0)$. Thus

$$f(\Phi(p)) = a_1(p)F_1(p) + \cdots + a_\nu(p)F_\nu(p)$$

in a neighborhood v of p_0 on $\mathcal{P}$, where $a_j(p)$ $(j = 1, \ldots, \nu)$ is a holomorphic function on v. There exists a holomorphic extension $\tilde{a}_j(z)$ $(j = 1, \ldots, \nu)$ of $a_j(p)$ in a neighborhood δ_0 of ζ_0, i.e., $a_j(p) = \tilde{a}_j(\Phi(p))$ for $p \in v \cap (\Phi^{-1}(\delta_0))$. We thus have

$$f(z) = \tilde{a}_1(z)\tilde{F}_1(z) + \cdots + \tilde{a}_\nu(z)\tilde{F}_\nu(z) \qquad \text{on } \delta_1 \cap \Sigma,$$

where $\delta_1 \subset \delta_0$ is a neighborhood of ζ_0. It follows that $f(z) \in \mathcal{J}^\lambda\{\tilde{F}, \psi\}$ at the point ζ_0. On the other hand, since any $f(z)$ belonging to $\mathcal{J}^\lambda\{\tilde{F}, \psi\}$ at $\zeta_0 \in \overline{\Delta}$ clearly belongs to $\tilde{\mathcal{I}}^\lambda$ at ζ_0, we obtain 1.

For the proof of 2, let $p_0 \in \mathcal{P}$ and let $f(p)$ be any holomorphic vector-valued function of rank λ belonging to $\mathcal{I}^\lambda$ at p_0. Let $\zeta_0 = \Phi(p_0)$. Since Σ is a normal model of $\mathcal{P}$, there exists a holomorphic vector-valued extension $\tilde{f}(z)$ of $f(p)$ in a neighborhood δ_0 of ζ_0 in $\overline{\Delta}$ so that $(\tilde{f}(z), \delta_0) \in \mathcal{I}^\lambda\{\tilde{F}\}$. Thus, we can find a holomorphic function $\tilde{a}_j(z)$ $(j = 1, \ldots, s)$ on a neighborhood $\delta_1 \subset \delta_0$ of ζ_0 such that

$$\tilde{f}(z) = \tilde{a}_1(z)\tilde{F}_1(z) + \cdots + \tilde{a}_s(z)\tilde{F}_s(z) \quad \text{on } \delta_1.$$

If we set $a_j(p) = \tilde{a_j}(\Phi(p))$ $(j = 1, \ldots, s)$, then we have $f(p) = a_1(p)F_1(p) + \cdots + a_s(p)F_s(p)$ on $\Phi^{-1}(\delta_1) \cap \Sigma$, so that $f \in \mathcal{J}^\lambda\{F\}$ at the point p_0. □

8.2.2. Various Problems on Analytic Polyhedra. Using Lemma 8.1, various problems which were solved on a closed polydisk in $\mathbf{C}^n$ in Chapter 7 may be solved on an analytic polyhedron in an analytic space $\mathcal{V}$.

Let $\mathcal{V}$ be an analytic space of dimension n and let $\mathcal{P}$ be an analytic polyhedron in $\mathcal{V}$. Let Σ be a normal model of $\mathcal{P}$ in the closed unit polydisk $\overline{\Delta}$ in $\mathbf{C}^m$. Let $\varphi_j(p)$ $(j = 1, \ldots, m)$ be defining functions of $\mathcal{P}$ in $D \subset \mathcal{V}$ with respect to Σ, so that, if we set

$$\Phi : p \in \mathcal{P} \to z = (z_1, \ldots, z_m) = (\varphi_1(p), \ldots, \varphi_m(p)) \in \overline{\Delta},$$

then Φ is one-to-one from $\mathcal{P}$ onto Σ with $\Sigma = \Phi(\mathcal{P})$ and $\Phi(\partial\mathcal{P}) \subset \partial\overline{\Delta}$.

We shall consider the $\mathcal{O}^\nu$-module defined as follows. Let $F_j(p)$ $(j = 1, \ldots, \nu)$ be a holomorphic vector-valued function of rank λ on $\mathcal{P}$. We consider the following system of λ homogeneous linear equations:

$$(\Omega) \qquad f_1(p)F_1(p) + \cdots + f_\nu(p)F_\nu(p) = 0,$$

determined by the given holomorphic vector-valued functions $F_j(p)$ $(j = 1, \ldots, \nu)$ of rank λ on $\mathcal{P}$ for the unknown holomorphic vector-valued function $f(p) = (f_1(p), \ldots, f_\nu(p))$ of rank ν on a domain $\delta \subset \mathcal{P}$. The set of all solutions $\{(f(p), \delta)\}_{\delta \subset \mathcal{P}}$ determines an $\mathcal{O}^\nu$-module on $\mathcal{P}$. We call this the **$\mathcal{O}$-module with respect to the linear relation** (Ω), and denote it by $\mathcal{L}\{\Omega\}$.

We have the following theorem.

THEOREM 8.3 (Oka). *The $\mathcal{O}$-module $\mathcal{L}\{\Omega\}$ with respect to any linear relation (Ω) has a locally finite pseudobase at each point of $\mathcal{P}$.*

PROOF. We let $\tilde{F}_j(z)$ $(j = 1, \ldots, \nu)$ denote a holomorphic, vector-valued extension of $F_j(p)$ to $\overline{\Delta}$. We use the functions $\psi_{h,k}(z)$ $(h = 1, \ldots, \lambda;\ k = 1, \ldots, \mu)$ defined in (8.3). Consider the following system of λ homogeneous linear equations in $\overline{\Delta}$:

$$(\tilde{\Omega}) \qquad \tilde{f}_1(z)\tilde{F}_1(z) + \cdots + \tilde{f}_\nu(z)\tilde{F}_\nu(z) + \sum_{h,k} \tilde{g}_{h,k}(z)\psi_{h,k}(z) = 0,$$

determined by the given holomorphic vector-valued functions $\tilde{F}_j(z)$ $(j = 1, \ldots, \nu)$ and $\psi_{h,k}(z)$ $(h = 1, \ldots, \lambda;\ k = 1, \ldots, \mu)$ of rank λ on $\overline{\Delta}$ for the unknown holomorphic vector-valued function $\tilde{f}(z) = (\tilde{f}_j(z), \tilde{g}_{h,k}(z))$ of rank $\kappa := \nu + \lambda\mu$ on a domain $\delta \subset \overline{\Delta}$. We let $\mathcal{L}\{\tilde{\Omega}\}$ denote the $\mathcal{O}$-module with respect to the linear relation $(\tilde{\Omega})$ on $\overline{\Delta}$. By Theorem 7.1, $\mathcal{L}\{\tilde{\Omega}\}$ has a locally finite pseudobase $\tilde{H}_j(z)$ $(j = 1, \ldots, s)$ on $\overline{\Delta}$, i.e., $\tilde{H}_j(z)$ is a holomorphic vector-valued function of rank κ on $\overline{\Delta}$ such that the $\mathcal{O}$-module $\mathcal{J}^\kappa\{\tilde{H}\}$ generated by $\tilde{H}_j(z)$ $(j = 1, \ldots, s)$ is equivalent to $\mathcal{L}\{\tilde{\Omega}\}$ on $\overline{\Delta}$. We set $H_j(p) = \tilde{H}_j(\Phi(p))$ $(j = 1, \ldots, s)$, so that $H_j(p)$ is of the form

$$H_j(p) = (H_{1,j}(p), \ldots, H_{\nu,j}(p), \overbrace{0, \ldots, 0}^{\lambda\mu}) \quad \text{on } \mathcal{P}.$$

If we define $H_j^0(p) := (H_{1,j}(p), \ldots, H_{\nu,j}(p))$ $(j = 1, \ldots, s)$, which is a holomorphic vector-valued function of rank ν on $\mathcal{P}$, then we will show that $\mathcal{L}\{\Omega\}$ is equivalent to the $\mathcal{O}$-module $\mathcal{J}^\nu\{H^0\}$ generated by the $H_j^0(p)$ $(j = 1, \ldots, s)$ on $\mathcal{P}$.

To see this, we first note that $H_j^0(p) \in \mathcal{L}\{\Omega\}$ $(j = 1, \ldots, s)$ on $\mathcal{P}$. Next, we take $p_0 \in \mathcal{P}$ and $f(p) = (f_1(p), \ldots, f_\nu(p)) \in \mathcal{L}\{\Omega\}$ at the point p_0 and set $z_0 = \Phi(p_0) \in \overline{\Delta}$. We let $\tilde{f}(z) = (\tilde{f}_1(z), \ldots, \tilde{f}_\nu(z))$ denote a holomorphic vector-valued extension of $f(p)$ in a neighborhood δ of z_0 on $\overline{\Delta}$. Then we have

$$\tilde{f}_1(z)\tilde{F}_1(z) + \cdots + \tilde{f}_\nu(z)\tilde{F}_\nu(z) = 0 \quad \text{on } \delta \cap \Sigma.$$

Thus there exist holomorphic functions $g_{h,k}(z)$ $(h = 1, \ldots, \lambda;\ k = 1, \ldots, \mu)$ in a neighborhood $\delta_1 \subset \delta$ of z_0 such that

$$\tilde{f}_1(z)\tilde{F}_1(z) + \cdots + \tilde{f}_\nu(z)\tilde{F}_\nu(z) + \sum_{h,k} g_{h,k}(z)\psi_{h,k}(z) = 0 \quad \text{on } \delta_1,$$

so that $\tilde{f}(z) := (\tilde{f}_j(z), g_{h,k}(z)) \in \mathcal{L}\{\tilde{\Omega}\}$ on δ_1. Thus,

$$\tilde{f}(z) = \tilde{\alpha}_1(z)\tilde{H}_1(z) + \cdots + \tilde{\alpha}_s(z)\tilde{H}_s(z)$$

in a neighborhood $\delta_2 \subset \delta_1$ of z_0. If we set $\alpha_j(z) = \tilde{\alpha}_j(\Phi(p))$ $(j = 1, \ldots, s)$ on $v := \Phi^{-1}(\delta_2 \cap \Sigma)$, then by taking the first ν components we have

$$f(p) = \alpha_1(p)H_1^0(p) + \cdots + \alpha_s(p)H_s^0(p)$$

on the neighborhood v of p_0 in $\mathcal{P}$; i.e., $f(p) \in \mathcal{J}^\nu\{H^0\}$ at p_0. □

We also have the following two theorems.

THEOREM 8.4 (Problem C_1). *Problem C_1 is solvable for any analytic polyhedron $\mathcal{P}$ in an analytic space $\mathcal{V}$.*

PROOF. Let $H(p)$ and $F_j(p)$ $(j = 1, \ldots, \nu)$ be holomorphic vector-valued functions of rank λ on $\mathcal{P}$. We let $\mathcal{J}^\nu\{F\}$ denote the $\mathcal{O}$-module on $\mathcal{P}$ generated by the $F_j(p)$ $(j = 1, \ldots, \nu)$. Assume that $H(p)$ belongs to $\mathcal{J}^\nu\{F\}$ at each point in $\mathcal{P}$. We claim that there exist ν holomorphic functions $\alpha_j(p)$ $(j = 1, \ldots, \nu)$ on $\mathcal{P}$ such that

$$H(p) = \alpha_1(p)F_1(p) + \cdots + \alpha_\nu(p)F_\nu(p) \quad \text{on } \mathcal{P}. \tag{8.4}$$

To prove this, we take a normal model Σ of $\mathcal{P}$ in the polydisk $\overline{\Delta}$ in $\mathbf{C}^m$ defined by use of the defining functions $\varphi_j(p)$ $(j = 1, \ldots, m)$ of $\mathcal{P}$ in $D \subset \mathcal{V}$, and we set

$$\Phi : p \in \mathcal{P} \to z = (\varphi_1(p), \ldots, \varphi_m(p)) \in \Sigma \subset \overline{\Delta}.$$

Let $\tilde{H}(z)$ and $\tilde{F}_j(z)$ $(j = 1, \ldots, \nu)$ be holomorphic vector-valued extensions of $H(p)$ and $F_j(p)$ on $\overline{\Delta}$. Let $\psi_{h,k}(z)$ $(h = 1, \ldots, \lambda;\ k = 1, \ldots, \mu)$ be the holomorphic vector-valued functions of rank λ on $\mathcal{P}$ defined by (8.3). We let $\mathcal{J}^\lambda\{\tilde{F}, \psi\}$ denote the $\mathcal{O}^\lambda$-module generated by $\tilde{F}_j(z)$ $(j = 1, \ldots, \nu)$ and $\psi_{h,k}(z)$ $(h = 1, \ldots, \lambda;\ k = 1, \ldots, \mu)$ on $\overline{\Delta}$. Since Σ is a normal model of $\mathcal{P}$ in $\overline{\Delta}$, we deduce from the fact that $H(p) \in \mathcal{J}^\lambda\{F\}$ at each point in $\mathcal{P}$ that $\tilde{H}(p) \in \tilde{\mathcal{J}}^\lambda\{\tilde{F}, \psi\}$ at each point in $\overline{\Delta}$. Since Problem C_1 is solvable in the closed polydisk $\overline{\Delta}$, there exist $\kappa := \nu + \lambda\mu$ holomorphic functions $\tilde{\alpha}_j(z)$ $(j = 1, \ldots, \nu)$, $\tilde{\beta}_{h,k}(z)$ $(k = 1, \ldots, \lambda; h = 1, \ldots, \mu)$ such that

$$\tilde{H}(z) = \tilde{\alpha}_1(z)\tilde{F}_1(z) + \cdots + \tilde{\alpha}_\nu(z)\tilde{F}_\nu(z) + \sum_{h,k} \tilde{\beta}_{h,k}(z)\psi_{h,k}(z) \quad \text{on } \overline{\Delta}.$$

Setting $\alpha_j(p) = \tilde{\alpha}_j(\Phi(p))$ $(j = 1, \ldots, \nu)$ on $\mathcal{P}$ proves (8.4). □

THEOREM 8.5 (Problem C_2). *Problem C_2 is solvable for any analytic polyhedron $\mathcal{P}$ in an analytic space $\mathcal{V}$.*

PROOF. We use the same notation Σ, $\overline{\Delta} \subset \mathbf{C}^m$, $\varphi_j(p)$ $(j = 1, \ldots, m)$ and $\Phi : \mathcal{P} \to \Sigma \subset \overline{\Delta}$ as in the proof of the previous theorem. Let $F_j(p)$ $(j = 1, \ldots, \nu)$ be a holomorphic vector-valued function of rank λ on $\mathcal{P}$ and let $\mathcal{J}^\lambda\{F\}$ denote the $\mathcal{O}$-module generated by $F_j(p)$ $(j = 1, \ldots, \nu)$ on $\mathcal{P}$. Let $\mathcal{C} = \{h_q(p), \delta_q\}_{q \in \mathcal{P}}$ be any C_2-distribution with respect to $\mathcal{J}^\lambda\{F\}$, i.e., if $\delta_{q_1} \cap \delta_{q_2} \neq \emptyset$, then $h_{q_1}(p) - h_{q_2}(p) \in \mathcal{J}^\lambda\{F\}$ at each point of $\delta_{q_1} \cap \delta_{q_2}$. We claim that there exists a holomorphic vector-valued function $H(p)$ of rank λ on $\mathcal{P}$ such that, for each $q \in \mathcal{P}$,

$$H(p) - h_q(p) \in \mathcal{J}^\lambda\{F\} \quad \text{at each point in } \delta_q. \tag{8.5}$$

To prove this, we consider the same $\mathcal{O}$-module $\mathcal{J}^\lambda\{\tilde{F}, \psi\}$ on $\overline{\Delta}$ as in the proof of the previous theorem, and we form the following distribution $\tilde{\mathcal{C}} := \{(\tilde{f}_\zeta(z), \tilde{\delta}_\zeta)\}_{\zeta \in \overline{\Delta}}$ on $\overline{\Delta}$, where $\tilde{\delta}_\zeta$ is a neighborhood of ζ in $\overline{\Delta}$ and $\tilde{f}_\zeta(z)$ is a holomorphic vector-valued function of rank λ on $\tilde{\delta}_\zeta$:

1. If $\zeta \notin \Sigma$, we take $\tilde{\delta}_\zeta$ such that $\tilde{\delta}_\zeta \cap \Sigma = \emptyset$ and set $\tilde{f}_\zeta(z) \equiv 0$.
2. If $\zeta \in \Sigma$, we set $q = \Phi^{-1}(\zeta)$ and take $\tilde{\delta}_\zeta = \Phi^{-1}(\delta_q)$; and $\tilde{f}_\zeta(\Phi(p)) = h_q(p)$ on δ_q.

Since Σ is a normal model of $\mathcal{P}$ on $\overline{\Delta}$, we can form such a distribution $\tilde{\mathcal{C}}$ at each point of $\overline{\Delta}$. Also, since $\mathcal{C}$ is a C_2-distribution on $\mathcal{P}$ with respect to $\mathcal{J}^\lambda\{F\}$, we see that $\tilde{\mathcal{C}}$ is a C_2-distribution on $\overline{\Delta}$ with respect to $\mathcal{J}^\lambda\{\tilde{F}, \psi\}$. Since Problem C_2 is solvable on the closed polydisk $\overline{\Delta}$, there exists a holomorphic vector-valued function $\tilde{H}(z)$ on $\overline{\Delta}$ such that, for each $\zeta \in \overline{\Delta}$, $\tilde{H}(z) - \tilde{f}_\zeta(z) \in \mathcal{J}^\lambda\{\tilde{F}, \psi\}$ at each point in $\tilde{\delta}_\zeta$. If we set $H(p) = \tilde{H}(\Phi(p))$ for $p \in \mathcal{P}$, then $H(p)$ satisfies (8.5). □

Finally, we prove the following theorem.

THEOREM 8.6 (Problem E). *Problem E is solvable for any analytic polyhedron $\mathcal{P}$ in an analytic space $\mathcal{V}$.*

PROOF. Let $\mathcal{J}^\lambda$ be an $\mathcal{O}^\lambda$-module on $\mathcal{P}$ which has a locally finite pseudobase at each point in $\mathcal{P}$. We use the same notation, Σ, Φ, and $\overline{\Delta} \subset \mathbf{C}^m$, as in the previous theorem. We let $\tilde{\mathcal{J}}^\lambda$ denote the extended $\mathcal{O}^\lambda$-module of $\mathcal{J}^\lambda$ to the polydisk $\overline{\Delta}$. By statement 1 of Lemma 8.1, $\tilde{\mathcal{J}}^\lambda$ has a locally finite pseudobase at each point in $\overline{\Delta}$. Since Problem E is solvable for the polydisk $\overline{\Delta}$, $\tilde{\mathcal{J}}^\lambda$ is equivalent to the $\mathcal{O}^\lambda$-module $\mathcal{J}^\lambda\{\tilde{F}\}$ generated by a finite number of holomorphic vector-valued functions $\tilde{F}_j(z)$ $(j = 1, \dots, \nu)$ of rank λ on $\overline{\Delta}$. If we set $F_j(p) = \tilde{F}_j(\Phi(p))$ $(j = 1, \dots, \nu)$ for $p \in \mathcal{P}$, then statement 2 of Lemma 8.1 implies that the $\mathcal{O}$-module $\mathcal{J}^\lambda\{F\}$ generated by $F_j(p)$ $(j = 1, \dots, \nu)$ is equivalent to $\mathcal{J}^\lambda$ on $\mathcal{P}$. □

8.2.3. Runge Problem. Let $\mathcal{V}$ be an analytic space of dimension n. Let $\mathcal{P}_1$ and $\mathcal{P}_2$ be analytic polyhedra in $\mathcal{V}$ such that the defining functions of both $\mathcal{P}_1$ and $\mathcal{P}_2$ are defined on the same set U: $\mathcal{P}_1, \mathcal{P}_2 \subset\subset U \subset \mathcal{V}$. Assume that $\mathcal{P}_1 \subset\subset \mathcal{P}_2^\circ$ (the interior of $\mathcal{P}_2$). Then we have the following lemma.

LEMMA 8.2. *The pair $(\mathcal{P}_1, \mathcal{P}_2)$ satisfies the Runge theorem.*

PROOF. Let Σ_1 (resp. Σ_2) be a normal model in the unit polydisk $\overline{\Delta}_1 \subset \mathbf{C}^m_z$ (resp. $\overline{\Delta}_2 \subset \mathbf{C}^l_w$) of $\mathcal{P}_1$ (resp. $\mathcal{P}_2$) whose defining functions are $\varphi_j(p)$ $(j = 1, \dots, m)$ (resp. $\psi_k(p)$ $(k = 1, \dots, l)$) defined on U. We assume that $|\varphi_j(p)| \le M$ $(j = 1, \dots, m)$ on $\mathcal{P}_2$. We set $\overline{\Delta}_1^* := |z_j| \le M$ $(j = 1, \dots, m)$, so that $\overline{\Delta}_1 \subset\subset \overline{\Delta}_1^* \subset \mathbf{C}^m_z$. We consider the holomorphic mapping

$$\Phi : p \in \mathcal{P}_2 \to (z, w) = (\varphi_1(p), \dots, \varphi_m(p), \psi_1(p), \dots, \psi_l(p)) \in \overline{\Delta}_1^* \times \overline{\Delta}_2$$

and define $\Sigma = \Phi(\mathcal{P}_2)$. Since Σ_1 (resp. Σ_2) is a normal model on $\overline{\Delta}_1$ (resp. $\overline{\Delta}_2$) of $\mathcal{P}_1$ (resp. $\mathcal{P}_2$), it follows from the hypothesis $\mathcal{P}_1 \subset\subset \mathcal{P}_2^\circ$ that $\Sigma \cap (\overline{\Delta}_1 \times \overline{\Delta}_2)$ (resp. Σ) is a normal model of $\mathcal{P}_1$ (resp. $\mathcal{P}_2$) in $\overline{\Delta}_1 \times \overline{\Delta}_2$ (resp. $\overline{\Delta}_1^* \times \overline{\Delta}_2$).

Let $f(p)$ be a holomorphic function on $\mathcal{P}_1$. Let $K \subset\subset \mathcal{P}_1^\circ$ and let $\epsilon > 0$. We fix a holomorphic extension $F(z, w)$ of $f(p)$ on $\overline{\Delta}_1 \times \overline{\Delta}_2$ and set $\tilde{K} = \Phi(K) \subset\subset \overline{\Delta}_1 \times \overline{\Delta}_2$. From the Taylor expansion of $F(z, w)$ in $\overline{\Delta}_1 \times \overline{\Delta}_2$, we can find a polynomial $\tilde{Q}(z, w)$ of (z, w) such that $|F(z, w) - \tilde{Q}(z, w)| < \epsilon$ on $\tilde{K}$. If we set $Q(p) = \tilde{Q}(\Phi(p))$ for $p \in \mathcal{P}_2$, then $Q(p)$ is a holomorphic function on $\mathcal{P}_2$ such that $|f(p) - Q(p)| < \epsilon$ on K. □

Using the same notation $\mathcal{P}_1 \subset\subset \mathcal{P}_2^\circ \subset\subset U \subset \mathcal{V}$ as in the previous lemma, we let $\mathcal{J}^\lambda$ be an $\mathcal{O}^\lambda$-module on U. We have the following lemma.

LEMMA 8.3. *Assume that $\mathcal{J}^\lambda$ has a locally finite pseudobase at each point in U. Let $f(p)$ be a vector-valued holomorphic function of rank λ on $\mathcal{P}_1$ such that $f(p) \in \mathcal{J}^\lambda$ at each point of $\mathcal{P}_1$. Let $\epsilon > 0$. Then there exists a holomorphic vector-valued function $F(p)$ of rank λ on $\mathcal{P}_2$ such that*

1. *$F(p) \in \mathcal{J}^\lambda$ at each point of $\mathcal{P}_2$;*
2. *$||F(p) - f(p)|| < \epsilon$ for $p \in \mathcal{P}_1$.*

PROOF. We fix an analytic polyhedron $\mathcal{P}$ in $\mathcal{V}$ with defining functions on U such that $\mathcal{P}_1 \subset\subset \mathcal{P}^\circ \subset\subset \mathcal{P}_2^\circ$, and $\mathcal{P}$ is so close to $\mathcal{P}_1$ that the given function $f(p)$ is defined on $\mathcal{P}$ and $f(p) \in \mathcal{J}^\lambda$ at each point of $\mathcal{P}$. Since $\mathcal{J}^\lambda$ has a locally finite pseudobase at each point in U, it follows from Theorem 8.6 that there exist a finite number of holomorphic vector-valued functions $\psi_j(p)$ $(j = 1, \ldots, \nu)$ of rank λ on $\mathcal{P}_2$ such that the $\mathcal{O}^\lambda$-module $\mathcal{J}\{\psi\}$ generated by $\psi_j(p)$ $(j = 1, \ldots, \nu)$ on $\mathcal{P}_2$ is equivalent to $\mathcal{J}^\lambda$ on $\mathcal{P}_2$. Thus, the function $f(p)$ on $\mathcal{P}$ belongs to $\mathcal{J}\{\psi\}$ at each point of $\mathcal{P}$. By Theorem 8.4, there exist ν holomorphic functions $\alpha_j(p)$ $(j = 1, \ldots, \nu)$ on $\mathcal{P}$ such that

$$f(p) = \alpha_1(p)\psi_1(p) + \cdots + \alpha_\nu(p)\psi_\nu(p) \quad \text{on } \mathcal{P}.$$

Since the pair $(\mathcal{P}, \mathcal{P}_2)$ satisfies Runge's theorem (by Lemma 8.2), we can find a holomorphic function $A_j(p)$ $(j = 1, \ldots, \nu)$ on $\mathcal{P}_2$ such that $|A_j(p) - \alpha_j(p)| < \epsilon'$ on $\mathcal{P}_1$, where $0 < \epsilon' < \epsilon/(\|\psi_1\|_{\mathcal{P}_2} + \cdots + \|\psi_\nu\|_{\mathcal{P}_2})$ (here $\|\psi_i\|_{\mathcal{P}_2} = \max\{||\psi_i(p)|| \mid p \in \mathcal{P}_2\}$). Consequently, if we define $F(p) = A_1(p)\psi_1(p) + \cdots + A_\nu(p)\psi_\nu(p)$ on $\mathcal{P}_2$, then $F(p) \in \mathcal{J}^\lambda$ at each point in $\mathcal{P}_2$ and

$$\|F(p) - f(p)\| \le \epsilon'(\|\psi_1\|_{\mathcal{P}_2} + \cdots + \|\psi_\nu\|_{\mathcal{P}_2}) < \epsilon \quad \text{for } p \in \mathcal{P}_1,$$

as desired. □

8.3. Stein Spaces

8.3.1. Definition of Stein Spaces. Let $\mathcal{V}$ be an analytic space of dimension n. Let $U \subset \mathcal{V}$ be a domain and let $K \subset\subset U$ be a compact set. We let $\mathcal{H}(U)$ denote the family of all holomorphic functions $f(p)$ on U. We define

$$\widehat{K}_U := \bigcap_{f \in \mathcal{H}(U)} \{q \in U \mid |f(q)| \le \max_{p \in K} |f(p)|\ \}.$$

We call $\widehat{K}_U$ the **holomorphic hull of** K with respect to U. If $\widehat{K}_U = K$, then we say that K is **holomorphically convex** with respect to U. Let V be an open set in $\mathcal{V}$ which contains U. If for any compact set $K \subset\subset U$ we have $\widehat{K}_V \subset U$, then we say that U is holomorphically convex in V. In case $U = V$, we say that U is a **holomorphically convex domain**.

If a domain U in $\mathcal{V}$ satisfies the following three conditions:

1. U satisfies the second axiom of countability;
2. U is holomorphically convex;
3. U satisfies the separation condition;

then we say that U is a **holomorphically complete domain** in $\mathcal{V}$. In case $U = \mathcal{V}$, we say that $\mathcal{V}$ is a holomorphically complete space, or a **Stein space**.[6]

For example, the interior $\mathcal{P}^\circ$ of an analytic polyhedron $\mathcal{P}$ in an analytic space $\mathcal{V}$ is always holomorphically complete. To see this, we fix a model Σ of $\mathcal{P}$ in

[6]It is known that conditions 2 and 3 imply condition 1 (see H. Grauert [**21**]).

the unit polydisk Δ centered at the origin in $\mathbf{C}^m$: $\Phi : p \in \mathcal{P} \to z = \Phi(p) = (\varphi_1(p), \dots, \varphi_m(p)) \in \Sigma$. There exist holomorphic functions $G_k(z)$ $(k = 1, \dots, \nu)$ in Δ such that $\Sigma = \bigcap_{k=1}^{\nu} \{z \in \Delta \mid G_k(z) = 0\}$. For $0 < \eta < 1$, we define $K(\eta) := \{p \in \mathcal{P} \mid |G_k(\Phi(p))| < \eta \ (k = 1, \dots, \nu), \ |\varphi_i(p)| < 1 - \eta \ (i = 1, \dots, m)\}$. Then $K(\eta) \subset\subset \mathcal{P}^o$ and $\lim_{\eta \to 0} K(\eta) = \mathcal{P}^o$, yielding the result.

REMARK 8.5. The notion of Stein space in the case of complex manifolds was introduced by K. Stein **[67]** in order to study more general spaces in which the Cousin problems, Runge theorems, expansion theorems, etc., hold as in the case of a (univalent) domain of holomorphy in $\mathbf{C}^n$. However, as noted in Example 6.8, even in the case of a ramified domain D over $\mathbf{C}^n$, a domain of holomorphy (which is a complex manifold) is not necessarily a Stein space, unlike the case of a univalent domain in $\mathbf{C}^n$.

From the definition of a holomorphically complete domain, we immediately obtain the following proposition.

PROPOSITION 8.1. *Let U be a holomorphically complete domain in an analytic space $\mathcal{V}$. Then there exists a sequence of analytic polyhedra $\mathcal{P}_\nu$ $(\nu = 1, 2, \dots)$ in $\mathcal{V}$ with defining functions on U such that*

$$\mathcal{P}_\nu \subset\subset \mathcal{P}^o_{\nu+1} \quad (\nu = 1, 2, \dots), \qquad U = \lim_{\nu \to \infty} \mathcal{P}_\nu.$$

Using the notation of the proposition, each pair $(\mathcal{P}_\nu, \mathcal{P}_{\nu+1})$ satisfies the Runge theorem according to Lemma 8.2. We thus obtain the following corollary by applying the usual techniques.

COROLLARY 8.2. *Let U be a holomorphically complete domain in an analytic space $\mathcal{V}$ and let $\mathcal{P}$ be an analytic polyhedron in $\mathcal{V}$ with defining functions on U. Then the pair $(\mathcal{P}, U)$ satisfies the Runge theorem.*

8.3.2. Approximation Condition. Let $\mathcal{V}$ be an analytic space of dimension n. If there exists a sequence of holomorphically complete domains U_j $(j = 1, 2, \dots)$ in $\mathcal{V}$ such that

1. $U_j \subset\subset U_{j+1}$ $(j = 1, 2, \dots)$, $\quad \mathcal{V} = \lim_{j \to \infty} U_j$, and
2. for each $j = 1, 2, \dots$, the domain U_j is holomorphically convex in U_{j+1},

then we say that the analytic space $\mathcal{V}$ satisfies the **approximation condition**.

It is clear from Lemma 8.2 and Proposition 8.1 that any Stein space $\mathcal{V}$ admits such a sequence U_j $(j = 1, 2, \dots)$. The converse is also true; to prove it, we first demonstrate the following theorem.

THEOREM 8.7 (Runge theorem). *If the analytic space $\mathcal{V}$ satisfies the approximation condition, i.e., if there exists a sequence of holomorphically complete domains U_j $(j = 1, 2, \dots)$ satisfying conditions* 1 *and* 2*, then the pair $(U_1, \mathcal{V})$ satisfies the Runge theorem.*

PROOF. Let $f(p)$ be a holomorphic function on U_1, let $K \subset\subset U_1$, and let $\epsilon > 0$. By condition 1, we have $\widehat{K}_{U_1} \subset\subset U_1$. Since U_1 is holomorphically convex in U_2, we have $\widehat{K}_{U_2} = \widehat{K}_{U_1}$, so that we can find an analytic polyhedron $\mathcal{P}_2$ with defining functions on U_2 such that $K \subset\subset \mathcal{P}_2 \subset\subset U_1$. Since $f(p)$ is holomorphic on $\mathcal{P}_2$, it follows from Corollary 8.2 that there exists a holomorphic function $f_2(p)$ on U_2 such that $|f_2(p) - f(p)| < \epsilon/2$ on $\mathcal{P}_2$. By repeating the same procedure for $f_2(p)$

on $\mathcal{P}_2 \subset\subset U_2$ as for $f(p)$ on $K \subset\subset U_1$, we obtain a sequence of analytic polyhedra $\mathcal{P}_{j+1}$ $(j = 1, 2, \dots)$ in U_{j+1},

$$\mathcal{P}_{j+1} \subset\subset U_j \subset\subset \mathcal{P}_{j+2} \subset\subset U_{j+1} \ (j = 1, 2, \dots),$$

and a sequence of holomorphic functions $f_{j+1}(p)$ $(j = 1, 2, \dots)$ on U_{j+1} such that

$$|f_{j+1}(p) - f_j(p)| < \epsilon/2^j \quad (j = 0, 1, \dots) \qquad \text{on } \mathcal{P}_{j+1}$$

(we define $f_1(p) := f(p)$ on U_1). If we set

$$F(p) = f(p) + \sum_{j=1}^{\infty}(f_{j+1}(p) - f_j(p)), \qquad p \in \mathcal{V},$$

then this series converges uniformly on any compact set in $\mathcal{V}$, so that $F(p)$ is a holomorphic function on $\mathcal{V}$ which satisfies

$$|F(p) - f(p)| \leq \sum_{j=1}^{\infty} |f_{j+1}(p) - f_j(p)| < \sum_{j=1}^{\infty} \epsilon/2^j = \epsilon \qquad \text{on } K.$$

Thus the pair $(U_1, \mathcal{V})$ satisfies the Runge theorem. □

We now obtain the following theorem.

THEOREM 8.8 (Approximation theorem). *If the analytic space $\mathcal{V}$ satisfies the approximation condition, then $\mathcal{V}$ is a Stein space.*

PROOF. Let U_j $(j = 1, 2, \dots)$ be a sequence of holomorphically complete domains in $\mathcal{V}$ which satisfies the approximation conditions 1 and 2. Since each U_j $(j = 1, 2, \dots)$ satisfies the second axiom of countability, and $\mathcal{V} = \bigcup_{j=1}^{\infty} U_j$, it follows that $\mathcal{V}$ satisfies the second axiom of countability.

Let $p_1, p_2 \in \mathcal{V}$ be distinct points, and take a sufficiently large integer j_0 so that $p_1, p_2 \in U_{j_0}$. Since U_{j_0} satisfies the separation condition, there exists a holomorphic function $f(p)$ on U_{j_0} such that $f(p_1) \neq f(p_2)$. By Theorem 8.7, the pair $(U_{j_0}, \mathcal{V})$ satisfies Runge's theorem; thus we can approximate $f(p)$ by a holomorphic function $F(p)$ on $\mathcal{V}$ with the property that $F(p_1) \neq F(p_2)$. Thus, $\mathcal{V}$ satisfies the separation condition.

Let $K \subset\subset \mathcal{V}$ be a compact set. We fix U_{j_0} with $K \subset\subset U_{j_0}$. Then $\widehat{K}_{U_{j_0}} \subset\subset U_{j_0}$. It follows from condition 2 that $\widehat{K}_{U_{j_0}} = \widehat{K}_{U_j}$ for all $j \geq j_0$; we denote this set by $\widehat{K}$. Let $p_0 \in \mathcal{V} \setminus \widehat{K}$ and fix j sufficiently large so that $\widehat{K} \bigcup \{p_0\} \subset\subset U_j$. Then there exists a holomorphic function $f(p)$ on U_j such that $|f(p_0)| > \max\{|f(q)| \mid q \in K\}$. Since the pair $(U_j, \mathcal{V})$ satisfies Runge's theorem, there exists a holomorphic function $F(p)$ on $\mathcal{V}$ such that $|F(p_0)| > \max\{|F(q)| \mid q \in K\}$; i.e., $p_0 \notin \widehat{K}_{\mathcal{V}}$. Hence, $\widehat{K}_{\mathcal{V}} = \widehat{K} \subset\subset \mathcal{V}$, so that $\mathcal{V}$ is holomorphically complete. Consequently, $\mathcal{V}$ is a Stein space. □

REMARK 8.6. In the case when $\mathcal{V}$ is a bounded domain in $\mathbf{C}^n$, H. Behnke and K. Stein [**2**] proved that the approximation theorem holds without the approximation condition 2; i.e., if D is a bounded domain in $\mathbf{C}^n$ with the property that there exists a sequence of holomorphically convex domains D_j $(j = 1, 2, \dots)$ in D such that $D_j \subset\subset D_{j+1}$ $(j = 1, 2, \dots)$ and $D = \bigcup_{j=1}^{\infty} D_j$, then D is a holomorphically convex domain.

PROOF. Let $r > 0$. For $j = 1, 2, \dots$, we set $D_j^{(r)} = \{z \in D_j \mid d_{D_j}(z) > r\}$, where $d_{D_j}(z)$ is the Euclidean distance from z to ∂D_j. Using Corollary 1.2 in section 1.5.3, we first note that, for a set G with $D_j^{(r)} \subset\subset G \subset\subset D_j$, there exists an analytic polyhedron $\mathcal{P}$ with defining functions on D_j such that $D_j^{(r)} \subset\subset \mathcal{P} \subset\subset G$. We next choose a subsequence $\{D_{j_k}\}_k$ of $\{D_j\}_j$ and a sequence $r_k > 0$ $(k = 1, 2, \dots)$ such that

$$D_{j_k} \subset\subset D_{j_{k+2}}^{(r_{k+2})} \subset\subset D_{j_{k+1}} \quad (k = 1, 2, \dots). \tag{8.6}$$

To verify this last inclusion, for $1 \le j < \nu$ we set

$$\begin{aligned} m_j &= \min\{d(p,q) \mid p \in \partial D_j,\ q \in \partial D\}, \\ m_{j,\nu} &= \min\{d(p,q) \mid p \in \partial D_j,\ q \in \partial D_\nu\}, \end{aligned}$$

where $d(p, q)$ denotes the Euclidean distance between p and q in $\mathbf{C}^n$. Similarly, we define

$$\begin{aligned} M_j &= \max\{d(p,q) \mid p \in \partial D_j,\ q \in \partial D\} \quad \text{and} \\ M_{j,\nu} &= \max\{d(p,q) \mid p \in \partial D_j,\ q \in \partial D_\nu\}. \end{aligned}$$

Since D is bounded in $\mathbf{C}^n$, we have $0 < m_{j,\nu} < m_j$; $0 < M_{j,\nu} < M_j$; $m_j, M_j \to 0$ as $j \to \infty$; and $M_{j,\nu} \to M_j$, $m_{j,\nu} \to m_j$ as $\nu \to \infty$. We let $D_{j_1} = D_1$. We choose $j_2 > j_1$ such that $m_{j_1} > M_{j_2}$. Then we take $j_3 > j_2$ such that

$$m_{j_1,j_3} > M_{j_2,j_3} \quad \text{and} \quad m_{j_2} > M_{j_3}.$$

If we take $r_3 > 0$ with $m_{j_1,j_3} > r_3 > M_{j_2,j_3}$, it follows that $D_{j_1} \subset\subset D_{j_3}^{(r_3)} \subset\subset D_{j_2}$. Similarly, since $m_{j_2} > M_{j_3}$, we can take $j_4 > j_3$ such that $m_{j_2,j_4} > M_{j_3,j_4}$ and $m_{j_3} > M_{j_4}$, and then we can take $r_4 > 0$ with $m_{j_2,j_4} > r_4 > M_{j_3,j_4}$ to obtain $D_{j_2} \subset\subset D_{j_4}^{(r_4)} \subset\subset D_{j_3}$. Repeating this procedure, we obtain (8.6).

We now set $j_k = k$ in (8.6), i.e., $D_k \subset\subset D_{k+2}^{(r_{k+2})} \subset\subset D_{k+1}$ $(k = 1, 2, \dots)$. From the first statement, there exists a sequence of analytic polyhedra $\mathcal{P}^{(k)}$ $(k = 1, 2, \dots)$ with defining functions in D_k such that $D_k \subset\subset \mathcal{P}^{(k+2)} \subset\subset D_{k+1}$. We let $\mathcal{H}_{k,k+2}$ (resp. $\mathcal{H}_k$) denote the holomorphic hull of $\overline{D_k}$ relative to D_{k+2} (resp. D). Using Theorem 3.5, it follows that $\mathcal{H}_{k,k+2} \subset\subset D_{k+1}$, and hence that any holomorphic function $f(z)$ on D_{k+1} can be uniformly approximated on D_k by a sequence of holomorphic functions $f_\nu(z)$ $(\nu = 1, 2, \dots)$ on D_{k+2}. Thus, by standard techniques, we conclude that $\mathcal{H}_k = \mathcal{H}_{k,k+2} \subset\subset D$, as desired. □

In the case when D is an *unbounded* domain in $\mathbf{C}^n$, the approximation theorem also holds without the approximation condition 2. To see this, let D be an unbounded domain in $\mathbf{C}^n$ having the same property as the bounded domain D in Remark 8.6. Fix $z_0 \in D$. For $r > 0$, we let B_r denote the ball centered at z_0 with radius r and we let $D^{(r)}$ denote the connected component of the open set $D \cap B_r$. Then using Remark 8.6, we see that each $D^{(p)}$ $(p = 1, 2, \dots)$ is a holomorphically convex domain. Moreover, $D^{(p)}$ is holomorphically convex in $D^{(p+1)}$. For take $K \subset\subset D^{(p)}$. Then $K \subset\subset B_r$ for $r < p$ sufficiently close to p. It follows that $\widehat{K}_{D^{(p+1)}} \subset\subset B_r \cap D^{(p+1)} \subset D^{(p)}$. Thus, condition 2 is satisfied.

In the case of a general analytic space $\mathcal{V}$, we cannot drop the approximation condition 2 in order to verify the approximation theorem.

EXAMPLE 8.4. [7] We recall the Calabi-Rosenlicht example (Example 8.1). Let ν be a positive integer and set $E_\nu = \{1/2^j \mid j = 1, \dots, \nu\} \subset \mathbf{C}$. Let $M_\nu := M_{E_\nu}$ be the 2-dimensional complex manifold associated to E_ν defined in the example. If we set $\mathbf{C}_y^* = \mathbf{C}_y \setminus \{0\}$, then

$$\begin{aligned} M_\nu &= (\mathbf{C}_x \times \mathbf{C}_y^* \times \{0\}) \cup \left(\bigcup_{j=1}^{\nu} \{(1/2^j, 0, \mathbf{C}_z)\right) \\ &\equiv (\mathbf{C}_x \times \mathbf{C}_y^* \times \{0\}) \cup \left(\bigcup_{j=1}^{\nu} \tilde{L}_j^*\right). \end{aligned}$$

The topology on M_ν can be described as follows: For a sequence $(x_n, y_n) \in \mathbf{C}_x \times \mathbf{C}_y^*$, we have $(x_n, y_n, 0) \in \mathbf{C}_x \times \mathbf{C}_y^* \times \{0\} \to (1/2^j, 0, z)$ as $n \to \infty$ if and only if $(x_n, y_n) \to (1/2^j, 0)$ and $(x_n - 2^{-j})/y_n \to z$ as $n \to \infty$. Note that $(1/2^j, \mathbf{C}_y, 0) \subset M_\nu$ $(j = 1, 2, \dots, \nu)$. Thus, we have $M_\nu \subset M_{\nu+1}$ $(\nu = 1, 2, \dots)$, and hence $M := \bigcup_{\nu=1}^{\infty} M_\nu$ is a 2-dimensional complex manifold. We will prove that each M_ν $(\nu = 1, 2, \dots)$ is a Stein manifold but M is not a Stein manifold.

PROOF. We consider the following non-singular analytic hypersurface Σ_ν in $\mathbf{C}^3$:

$$\Sigma_\nu : \quad yz = \prod_{j=1}^{\nu} \left(x - \frac{1}{2^j}\right).$$

Thus Σ_ν is a Stein manifold. Furthermore, Σ_ν is holomorphically equivalent to M_ν. To see this, we can form a one-to-one holomorphic mapping Φ_ν from M_ν onto Σ_ν, where

$$\Phi_\nu : \begin{array}{ll} (x, y, 0) \in \mathbf{C}_x \times \mathbf{C}_y^* \times \{0\} & \to (x, y, y^{-1} \prod_{j=1}^{\nu} (x - 1/2^j)) \in \Sigma_\nu \\ (1/2^j, 0, z) \in \tilde{L}_j^* \ (j = 1, \dots, \nu) & \to (1/2^j, 0, z \prod_{k=1, k \neq \nu}^{n} (1/2^j - 1/2^k)) \in \Sigma_\nu. \end{array}$$

Thus, M_ν is a Stein manifold. To prove that M is not Stein, let $K = \{|x| \leq 1\} \times \{1/2 \leq |y| \leq 1\} \times \{0\} \subset\subset M_1 \subset M$. Let $f(p)$ be holomorphic function on M. Fix $j = 1, 2, \dots$ and let $\overline{\Delta}_j$ denote the disk

$$\overline{\Delta}_j := \{1/2^j\} \times \{|y| \leq 1\} \times \{0\} \subset\subset M.$$

By the maximum principle we have

$$|f(1/2^j, 0, 0)| \leq \max\{|f(1/2^j, y, 0)| \mid |y| = 1\} \leq \max\{|f(p)| \mid p \in K\},$$

so that $(1/2^j, 0, 0) \in \widehat{K}_M$ $(j = 1, 2, \dots)$. Since $\{(1/2^j, 0, 0) \mid j = 1, 2, \dots\}$ is not relatively compact in M, it follows that M is not holomorphically convex.

In this construction, $M_\nu \subset M_{\nu+1}$ but M_ν is not relatively compact in $M_{\nu+1}$. However, it is easy to see that we can construct $M'_\nu \subset\subset M_\nu$ such that $M'_\nu \subset\subset M'_{\nu+1}$; M'_ν is a Stein manifold; and $M = \bigcup_{\nu=1}^{\infty} M'_\nu$. □

REMARK 8.7. Let $\mathcal{V}$ be a Stein space of dimension n. Let $p_0 \in \mathcal{V}$. Then there exists a local coordinate chart $(\delta_0, \lambda_0, \phi_0|_{\delta_0})$ for p_0 in $\mathcal{V}$ such that ϕ_0 is a holomorphic mapping defined on all of $\mathcal{V}$ into $\mathbf{C}^n$.

---

[7] This example is due to T. Ueda [**75**].

PROOF. Let $p_0 \in \mathcal{V}$ and let $(\delta, \lambda, \varphi)$ be a local coordinate chart for p_0 in $\mathcal{V}$. Since $\mathcal{V}$ satisfies the separation condition, the holomorphically convex hull of the one-point set $\{p_0\}$ in the Stein space $\mathcal{V}$ is $\{p_0\}$ itself. There thus exists an analytic polyhedron $\mathcal{P}$ with defining functions on $\mathcal{V}$ such that $p_0 \in \mathcal{P}^\circ$ (the interior of $\mathcal{P}$) and $\mathcal{P} \subset\subset \delta$. Thus, φ is holomorphic on $\mathcal{P}$. Since the pair $(\mathcal{P}, \mathcal{V})$ satisfies Runge's theorem, there exists a holomorphic mapping φ_0 on $\mathcal{V}$ into $\mathbf{C}^n$ which is uniformly close to φ on $\mathcal{P}$. Set $\delta_0 = \mathcal{P}^\circ$ and $\lambda_0 = \varphi_0(\mathcal{P}^\circ)$, which is a ramified domain over $\mathbf{C}^n$. Then the triple $(\delta_0, \lambda_0, \varphi_0|_{\delta_0})$ satisfies the necessary requirements. If λ is univalent in $\mathbf{C}^n$, so is λ_0 for φ_0 suffciently close to φ. □

8.3.3. Various Problems in a Stein Space. In a Stein space, many of the theorems which hold in a domain of holomorphy in $\mathbf{C}^n$ hold without any change. In this section we consider a few of these theorems.

1. Cousin problems

Cousin problems I and II in an analytic space are posed just as in a univalent domain in $\mathbf{C}^n$. We have the following two theorems.

THEOREM 8.9 (Cousin I problem). *A Cousin I problem is always solvable in a Stein space $\mathcal{V}$.*

PROOF. As shown in Chapter 3, a Cousin I problem is solvable in a domain D in $\mathbf{C}^n$ if there exists a sequence of domains D_j $(j = 1, 2, \dots)$ in D such that $D_j \subset\subset D_{j+1}$ $(j = 1, 2, \dots)$, $D = \lim_{j\to\infty} D_j$, and

(1) the Cousin I problem is solvable on each $\overline{D_j}$ $(j = 1, 2, \dots)$;
(2) the pair (D_j, D_{j+1}) $(j = 1, 2, \dots)$ satisfies Runge's theorem.

The same fact holds in an analytic space. In a Stein space $\mathcal{V}$, there exists a sequence of analytic polyhedra $\mathcal{P}_j$ $(j = 1, 2, \dots)$ in $\mathcal{V}$ with defining functions on $\mathcal{V}$ such that $\mathcal{P}_j \subset\subset \mathcal{P}_{j+1}^o$ and $\lim_{j\to\infty} \mathcal{P}_j = \mathcal{V}$. Since each $\mathcal{P}_j$ $(j = 1, 2, \dots)$ has a normal model Σ_j in the polydisk $\overline{\Delta}_j$ in $\mathbf{C}^{m_j}$ for which Oka's lifting principles (Theorems 8.1 and 8.2) hold, we can show by arguments used in Lemmas 3.3 and 3.4 that the Cousin I problem is solvable on each $\mathcal{P}_j$. Since $(\mathcal{P}_j, \mathcal{P}_{j+1})$ satisfies Runge's theorem, it follows from (1) and (2) that the Cousin I problem is also solvable in the Stein space $\mathcal{V}$. □

THEOREM 8.10 (Cousin II problem). *Let $\mathcal{C} = \{(f_p, \delta_p)\}_{p\in\mathcal{V}}$ be a Cousin II distribution in a Stein space $\mathcal{V}$. If $\mathcal{C}$ has a continuous solution in $\mathcal{V}$, then $\mathcal{C}$ has an analytic solution in $\mathcal{V}$.*

PROOF. The same proof of Theorem 3.8 yields that if a Cousin II distribution $\mathcal{C}$ has a continuous solution in the space $\mathcal{V}$, then $\mathcal{C}$ has an analytic solution in $\mathcal{V}$ under the assumption that the Cousin I problem is always solvable in $\mathcal{V}$. This assumption is guaranteed by Theorem 8.9; thus $\mathcal{C}$ has an analytic solution in $\mathcal{V}$. □

2. Problem C_1 and Problem C_2.

Let $\mathcal{V}$ be a Stein space. Let $F_j(p)$ $(j = 1, \dots, \nu)$ be ν holomorphic vector-valued functions of rank λ in $\mathcal{V}$: $F_j(p) = (F_{1,j}(p), \dots, F_{\lambda,j}(p))$ $(j = 1, \dots, \nu)$, $p \in \mathcal{V}$. We let $\mathcal{J}^\lambda\{F\}$ denote the $\mathcal{O}^\lambda$-module generated by $F_j(p)$ $(j = 1, \dots, \nu)$ in $\mathcal{V}$. We also let $\mathcal{L}\{\Omega\}$ denote the $\mathcal{O}^\nu$-module with respect to the linear relation

$$(\Omega) \qquad f_1(p)F_1(p) + \cdots + f_\nu(p)F_\nu(p) = 0,$$

i.e., $\mathcal{L}\{\Omega\} = \{(f(p), \delta)\}_{\delta \subset \mathcal{V}}$, where the holomorphic vector-valued function $f(p) = (f_1(p), \ldots, f_\nu(p))$ of rank ν in δ satisfies the λ equations (Ω) in δ.

We have the following two theorems.

THEOREM 8.11 (Problem C_1). *Problem C_1 is always solvable in a Stein space $\mathcal{V}$.*

PROOF. Let $H(p)$ be a holomorphic vector-valued function of rank λ in $\mathcal{V}$ such that $H(p)$ belongs to $\mathcal{J}^\lambda\{F\}$ at each point p in $\mathcal{V}$. We want to find a holomorphic vector-valued function $A(p) = (A_1(p), \ldots, A_\nu(p))$ on $\mathcal{V}$ such that

$$H(p) = F_1(p)A_1(p) + \cdots + F_\nu(p)A_\nu(p), \quad p \in \mathcal{V}.$$

Let $\mathcal{P}_k$ $(k = 1, 2, \ldots)$ be a sequence of analytic polyhedra in $\mathcal{V}$ such that

$$\mathcal{P}_j \subset\subset \mathcal{P}^o_{j+1} \quad (j = 1, 2, \ldots), \qquad \mathcal{V} = \lim_{k \to \infty} \mathcal{P}_k,$$

and where the defining functions of each $\mathcal{P}_k$ are defined in $\mathcal{V}$.

Choose $\epsilon_k > 0$ $(k = 1, 2, \ldots)$ such that $\sum_{k=1}^{\infty} \epsilon_k < \infty$. As usual, it suffices to find, for each $k = 1, 2, \ldots$, a holomorphic vector-valued function $A^k(p) = (A^k_1(p), \ldots, A^k_\nu(p))$ of rank ν on $\mathcal{P}_k$ such that

(i) $H(p) = F_1(p)A^k_1(p) + \cdots + F_\nu(p)A^k_\nu(p), \quad p \in \mathcal{P}_k$, and
(ii) $||A^{k+1}(p) - A^k(p)|| < \epsilon_k, \quad p \in \mathcal{P}_k$.

Then $A(p) := \lim_{k \to \infty} A^k(p)$ is uniformly convergent on any $K \subset\subset \mathcal{V}$, which proves our result.

For $k = 1$, we can find a holomorphic vector-valued function $A^1(p)$ of rank ν on $\mathcal{P}_1$ which satisfies condition (i) by Theorem 8.4. Assuming that there exists an $A^k(p)$ on $\mathcal{P}_k$ which satisfies condition (i), we now construct $A^{k+1}(p)$ on $\mathcal{P}_{k+1}$ which satisfies condition (i) on $\mathcal{P}_{k+1}$ and which also satisfies (ii) together with $A^k(p)$ on $\mathcal{P}_k$. To do this, using Theorem 8.4, we first find a holomorphic vector-valued function $\tilde{A}^{k+1}(p)$ of rank ν on $\mathcal{P}_{k+1}$ which satisfies condition (i) on $\mathcal{P}_{k+1}$. Then $\tilde{A}^{k+1}(p) - A^k(p)$ belongs to $\mathcal{L}\{\Omega\}$ on $\mathcal{P}_k$. Next we find a pseudobase $\Phi_l(p)$ $(l = 1, \ldots, s)$ of $\mathcal{L}\{\Omega\}$ on $\mathcal{P}_{k+1}$,

$$\Phi_l(p) = (\Phi_{1,l}(p), \ldots, \Phi_{\nu,l}(p)), \quad p \in \mathcal{P}_{k+1},$$

by combining Theorems 8.3 and 8.6. Furthermore, from Theorem 8.4 there exist s holomorphic functions $a_l(p)$ $(l = 1, \ldots, s)$ on $\mathcal{P}_k$ such that

$$\tilde{A}^{k+1}(p) - A^k(p) = a_1(p)\Phi_1(p) + \cdots + a_s(p)\Phi_s(p), \quad p \in \mathcal{P}_k.$$

Since the pair $(\mathcal{P}_k, \mathcal{V})$ satisfies Runge's theorem (by Corollary 8.2), for each $l = 1, \ldots, s$ there exists a holomorphic function $\tilde{a}_l(p)$ on $\mathcal{V}$ such that

$$|\tilde{a}_l(p) - a_l(p)| < \epsilon'_k \quad (l = 1, \ldots, s), \quad p \in \mathcal{P}_k,$$

where $\epsilon'_k < \epsilon_k / (\sum_{l=1}^{s} \|\Phi_l(p)\|_{\mathcal{P}_{k+1}})$. Setting

$$A^{k+1}(p) = \tilde{A}^{k+1}(p) + \tilde{a}_1(p)\Phi_1(p) + \cdots + \tilde{a}_s(p)\Phi_s(p), \quad p \in \mathcal{P}_{k+1},$$

it follows easily that $A^{k+1}(p)$ satisfies condition (i) on $\mathcal{P}_{k+1}$. Moreover, for $p \in \mathcal{P}_k$,

$$\begin{aligned} ||A^{k+1}(p) - A^k(p)|| &= ||(A^{k+1}(p) - \tilde{A}^{k+1}(p)) + (\tilde{A}^{k+1}(p) - A^k(p))|| \\ &= ||(\tilde{a}_1(p) - a_1(p))\Phi_1(p) + \cdots + (\tilde{a}_s(p) - a_s(p))\Phi_s(p)|| \\ &\le \epsilon'_k \, (\sum_{l=1}^{s} ||\Phi_l||_{\mathcal{P}_{k+1}}) < \epsilon_k, \end{aligned}$$

so that $A^{k+1}(p)$, together with $A^k(p)$, satisfies condition (ii) on $\mathcal{P}_k$. Thus we have constructed $A^k(p)$ $(k = 1, 2, \dots)$ on $\mathcal{P}_k$ by induction, proving the result. $\square$

THEOREM 8.12 (Problem C_2). *Problem C_2 is always solvable in a Stein space $\mathcal{V}$.*

PROOF. We prove this under the assumption that the completeness property of the same type as in Theorem 7.6 in $\mathbf{C}^n$ holds in the analytic space $\mathcal{V}$. This fact will be proved later in Proposition 8.3. Let $\mathcal{C} = \{(h_q(p), \delta_p)\}_{q\in\mathcal{V}}$ be a C_2 distribution with respect to $\mathcal{J}^\lambda\{F\}$. Choose $\epsilon_k > 0$ $(k = 1, 2, \dots)$ such that $\sum_{k=1}^{\infty} \epsilon_k < \infty$. Utilizing the completeness property, it suffices to construct, for each $k = 1, 2, \dots$, a holomorphic vector-valued function $H^k(p)$ of rank λ on $\mathcal{P}_k$ such that

(i) $H^k(p) - h_q(p) \in \mathcal{J}^\lambda\{F\}$ at each point in $\mathcal{P}_k \cap \delta_q$ for each $q \in \mathcal{P}_k$, and
(ii) $\|H^{k+1}(p) - H^k(p)\| < \epsilon_k, \quad p \in \mathcal{P}_k$.

For again, if such a sequence $H^k(p)$ $(k = 1, 2, \dots)$ on $\mathcal{P}_k$ exists, then $H(p) := \lim_{k\to\infty} H^k(p)$ is uniformly convergent on each $K \subset\subset \mathcal{V}$ and $H(p)$ is a solution to Problem C_2 on $\mathcal{V}$ for the given C_2 distribution $\mathcal{C}$ (under the above completeness assumption).

To begin, by Theorem 8.4, we can find an $H^1(p)$ on $\mathcal{P}_1$ which satisfies condition (i) on $\mathcal{P}_1$. Assuming that there exists an $H^k(p)$ on $\mathcal{P}_k$ which satisfies condition (i) on $\mathcal{P}_k$, we shall construct $H^{k+1}(p)$ on $\mathcal{P}_{k+1}$ which satisfies condition (i) on $\mathcal{P}_{k+1}$ and satisfies condition (ii) together with this $H^k(p)$ on $\mathcal{P}_k$.

We first find, by Theorem 8.5, a vector-valued function $\tilde{H}^{k+1}(p)$ on $\mathcal{P}_{k+1}$ which satisfies condition (i) on $\mathcal{P}_{k+1}$. Since $\tilde{H}^{k+1}(p) - H^k(p)$ belongs to $\mathcal{L}\{\Omega\}$ on $\mathcal{P}_k$, we thus can find s holomorphic functions $\beta_l(p)$ $(l = 1, \dots, s)$ on $\mathcal{P}_k$ such that

$$\tilde{H}^{k+1}(p) - H^k(p) = \beta_1(p)\Phi_1(p) + \cdots + \beta_s(p)\Phi_s(p), \quad p \in \mathcal{P}_k,$$

where the $\Phi_l(p)$ $(l = 1, \dots, s)$ constitute a pseudobase of $\mathcal{L}\{\Omega\}$ on $\mathcal{P}_{k+1}$. Since the pair $(\mathcal{P}_k, \mathcal{V})$ satisfies Runge's theorem, for each $l = 1, \dots, s$ there exists a holomorphic function $\tilde{\beta}_l(p)$ in $\mathcal{V}$ such that

$$|\tilde{\beta}_l(p) - \beta_l(p)| < \epsilon'_k, \quad p \in \mathcal{P}_k,$$

where $0 < \epsilon'_k < \epsilon_k / (\sum_{l=1}^{s} \|\Phi_l(p)\|_{\mathcal{P}_{k+1}})$. If we set

$$H^{k+1}(p) = \tilde{H}^{k+1}(p) + \tilde{\beta}_1(p)\Phi_1(p) + \cdots + \tilde{\beta}_s(p)\Phi_s(p), \quad p \in \mathcal{P}_{k+1},$$

then $H^{k+1}(p)$ satisfies condition (i) on $\mathcal{P}_{k+1}$ and satisfies condition (ii) together with $H^k(p)$ on $\mathcal{P}_k$. Thus we have constructed $H^k(p)$ $(k = 1, 2, \dots)$ on $\mathcal{P}_k$ by induction, proving the result. $\square$

3. Problem E

Let $\mathcal{V}$ be a Stein space and let $\mathcal{J}^\lambda$ be an $\mathcal{O}^\lambda$-module on $\mathcal{V}$ such that $\mathcal{J}^\lambda$ has a locally finite pseudobase at each point in $\mathcal{V}$. It is *not* necessarily true that $\mathcal{J}^\lambda$ has a finite pseudobase on all of $\mathcal{V}$.

Oka's counterexample for the pseudobase of Problem E.[8] We consider $\mathbf{C}^4$ with variables x_1, x_2, y_1, y_2. Let $\nu \geq 3$ be an integer. We consider the following four polynomials in $\mathbf{C}^4$:

$$F_1 = y_1^\nu - x_1^{\nu-1}, \ F_2 = y_2^\nu - x_2^\nu x_1, \ F_3 = y_1 y_2 - x_1 x_2, \ F_4 = x_1^{\nu-2} + x_2^\nu,$$

[8]This example is due to K. Oka [**52**]

and we let Σ denote the analytic set in $\mathbf{C}^4$ defined by $F_1 = F_2 = F_3 = F_4 = 0$. We will show that Σ is a 1-dimensional analytic set in $\mathbf{C}^4$. We consider the G-ideal $G\{\Sigma\}$ with respect to Σ in $\mathbf{C}^4$.

Then we have the following lemma.

LEMMA 8.4. *If $G_k(p)$ $(k = 1, \ldots, s)$ is a locally finite pseudobase of $G\{\Sigma\}$ at the origin O in $\mathbf{C}^4$, then $s \geq \nu - 1$.*

PROOF. We consider the ramified domain $\mathcal{R}$ over $\mathbf{C}^2_{x_1,x_2}$ defined by the function $\sqrt[\nu]{x_1}$, so that $\mathcal{R}$ is (holomorphically) isomorphic to $\mathbf{C}^2_{t,x_2}$ with variables (t, x_2) via the mapping

$$T: \quad (t, x_2) \in \mathbf{C}^2_{t,x_2} \to (\tilde{x}_1, x_2) = (t^\nu, x_2) \in \mathcal{R}.$$

We consider the analytic set $\mathcal{S}$ in $\mathbf{C}^4$ defined by $F_1 = F_2 = F_3 = 0$. Then $\mathcal{S}$ is isomorphic to $\mathcal{R}$.

Indeed, we have

$$\begin{aligned} y_1 &= (x_1^{1/\nu})^{\nu-1} \quad \text{since } F_1 = 0, \\ y_2 &= x_2(\epsilon x_1^{1/\nu}) \quad \text{since } F_2 = 0, \end{aligned}$$

where ϵ is a ν-th root of unity; i.e., $\epsilon^\nu = 1$. From $F_3 = 0$ we have $(x_1^{1/\nu})^{\nu-1} x_2 \epsilon x_1^{1/\nu} = x_1 x_2$, so that $\epsilon = 1$. Thus, $\mathcal{S}$ is the 2-dimensional irreducible analytic set in $\mathbf{C}^4$ defined by

$$\mathcal{S}: \quad y_1 = (x_1^{1/\nu})^{\nu-1}, \quad y_2 = x_2 x_1^{1/\nu},$$

where (x_1, x_2) varies over $\mathcal{R}$. Thus, $\mathcal{S}$ is isomorphic to $\mathbf{C}^2_{t,x_2}$ via

$$\pi : (t, x_2) \in \mathbf{C}^2_{t,x_2} \to (x_1, x_2, y_1, y_2) = (t^\nu, x_2, t^{\nu-1}, x_2 t) \in \mathcal{S}.$$

We remark that $F_4 = x_1^{\nu-2} + x_2^\nu$ depends only on the variables x_1 and x_2. Thus, if we let σ denote the analytic set in $\mathbf{C}^2_{x_1,x_2}$ defined by $F_4 = 0$, then we have $\Sigma = \mathcal{S} \cap [\sigma \times \mathbf{C}^2_{y_1,y_2}]$. Setting

$$\tilde{\sigma} = \{(t, x_2) \in \mathbf{C}^2_{t,x_2} \mid t^{\nu(\nu-2)} + x_2^\nu = 0\},$$

which is an analytic hypersurface in $\mathbf{C}^2_{t,x_2}$, we see that π gives a bijection from $\tilde{\sigma}$ onto Σ. Thus Σ consists of ν irreducible 1-dimensional analytic sets in $\mathbf{C}^4$. Let $F(x_1, x_2, y_1, y_2)$ be a holomorphic function defined in a neighborhood δ of the origin O in $\mathbf{C}^4$. Then $F\,|_{\mathcal{S}\cap\delta}$ can be written in the form

$$f(t, x_2) := F(t^\nu, x_2, t^{\nu-1}, x_2 t), \quad (t, x_2) \in \gamma,$$

where γ is a neighborhood of $(t, x_2) = (0, 0)$ in $\mathbf{C}^2_{t,x_2}$, so that $f(t, 0)$ is of the form

$$f(t, 0) = a + a_{\nu-1} t^{\nu-1} + a_\nu t^\nu + \cdots, \tag{8.7}$$

where $a, a_{\nu-1}, a_\nu, \ldots$ are constants.

Now assume that $F(x_1, x_2, y_1, y_2) \in G\{\Sigma\}$ on δ. We remark that $\tilde{\sigma} \cap \gamma$ is an *analytic hypersurface* in γ which is the zero set of the function $t^{\nu(\nu-2)} + x_2^\nu$; this function has no multiple factors. Since $f(t, x_2) = 0$ on $\gamma \cap \tilde{\sigma}$, it follows that

$$f(t, x_2) = (t^{\nu(\nu-2)} + x_2^\nu)\, h(t, x_2),$$

where $h(t, x_2)$ is a holomorphic function on a neighborhood $\gamma_0 \subset \gamma$ of $(0, 0)$, so that $f(t, 0)$ is of the form

$$\begin{aligned} f(t, 0) &= t^{\nu(\nu-2)}(b_0 + b_1 t + b_2 t^2 + \cdots + b_{\nu-2} t^{\nu-2}) \\ &\quad + \text{terms of order higher than } \nu(\nu-2) + \nu - 1, \end{aligned} \tag{8.8}$$

where $b_0, b_1, \ldots$ are constants.

Conversely, let $f(t, x_2)$ be a holomorphic function in a neighborhood of $(0,0)$ in $\mathbf{C}^2_{t,x_2}$ of the form

$$f(t, x_2) = (t^{\nu(\nu-2)} + x_2^\nu)\, h(t, x_2).$$

Then we can find an $F(x_1, x_2, y_1, y_2) \in G\{\Sigma\}$ on a neighborhood δ_0 of O in $\mathbf{C}^4$ such that $F\,|_{\delta_0 \cap \mathcal{S}} = f$.

Indeed, we note that $f(p) := f(\pi(p))$ is a weakly holomorphic function on $\mathcal{S}$ in a neighborhood δ of the origin O in $\mathbf{C}^4$ with $f\,|_{\delta\cap\Sigma} = 0$. Since $\Sigma \subset \mathcal{S}$, if we could holomorphically extend $f(p)$ to a holomorphic function $F(x_1, x_2, y_1, y_2)$ in a neighborhood δ_0 of O in $\mathbf{C}^4$, then necessarily $F(x_1, x_2, y_1, y_2) \in G\{\Sigma\}$ on δ_0.

Since $t^{\nu-1} = y_1$ and $h(t, x_2) = \sum_{n,m=0}^{\infty} a_{mn} t^m x_2^n$, it suffices to prove that the weakly holomorphic functions

$$f_i(t, x_2) = (t^{\nu(\nu-2)} + x_2^\nu)\, t^i \quad (i = 0, 1, \ldots, \nu - 2) \tag{8.9}$$

on $\mathcal{S}$ have holomorphic extensions $\Phi_i(x_1, x_2, y_1, y_2)$ in $\mathbf{C}^4$. To this end, for $i = 0$, since $x_1 = t^\nu$ we can take

$$\Phi_0(x_1, x_2, y_1, y_2) = x_1^{\nu-2} + x_2^\nu.$$

For $i = 1, \ldots, \nu - 2$, since $x_1 = t^\nu$, $y_1 = t^{\nu-1}$, and $y_2 = t x_2$, we can take

$$\Phi_i(x_1, x_2, y_1, y_2) = x_1^{i-1} y_1^{\nu-i} + x_2^{\nu-i} y_2^i \quad (i = 1, \ldots, \nu - 2) \tag{8.10}$$

in $\mathbf{C}^4$, so that the converse is true.

We proceed to prove the lemma by contradiction. Assume that there exist $\nu - 2$ holomorphic functions

$$G_k(x_1, x_2, y_1, y_2) \quad (k = 1, \ldots, \nu - 2)$$

on a neighborhood Δ of the origin O in $\mathbf{C}^4$ such that the $\mathcal{O}$-ideal $\mathcal{J}\{G\}$ generated by $G_k(x_1, x_2, y_1, y_2)$ $(k = 1, \ldots, \nu - 2)$ on Δ is equivalent to $G\{\Sigma\}$ on Δ. By (8.8) we have

$$\begin{aligned} g_k(t, x_2) &:= G_k\,|_{\Sigma\cap\Delta}, \\ g_k(t, 0) &= t^{\nu(\nu-2)}(b_{k,0} + b_{k,1} t + \cdots + b_{k,\nu-2} t^{\nu-2}) \\ &\quad + \text{terms of order higher than } \nu(\nu-2) + \nu - 1. \end{aligned}$$

Since each $\Phi_i(x_1, x_2, y_1, y_2)$ $(i = 0, 1, \ldots, \nu - 2)$ defined by (8.10) belongs to $G\{\Sigma\}$ in $\mathbf{C}^4$, there exist $\nu - 2$ holomorphic functions $C_k^i(x_1, x_2, y_1, y_2)$ $(k = 1, \ldots, \nu - 2)$ defined on a neighborhood $\Delta_0 \subset \Delta$ of the origin O in $\mathbf{C}^4$ such that

$$\begin{aligned} \Phi_i(x_1, x_2, y_1, y_2) &= C_1^i(x_1, x_2, y_1, y_2) G_1(x_1, x_2, y_1, y_2) \\ &\quad + \cdots + C_{\nu-2}^i(x_1, x_2, y_1, y_2) G_{\nu-2}(x_1, x_2, y_1, y_2), \end{aligned}$$

$$(i = 0, 1, \ldots, \nu - 2), \quad (x_1, x_2, y_1, y_2) \in \Delta_0.$$

We restrict $\Phi_i(x_1, x_2, y_1, y_2)$ to $\mathcal{S} \cap \{x_2 = 0\}$ and set

$$f_i(t, 0) = \Phi_i(t^\nu, 0, t^{\nu-1}, 0), \quad c_k^i(t, 0) = C_k^i(t^\nu, 0, t^{\nu-1}, 0) \quad (i = 0, 1, \ldots, \nu - 2)$$

in a neighborhood e_0 of $t = 0$ in $\mathbf{C}_t$. From (8.9) and (8.7) we have

$$f_i(t, 0) = t^{\nu(\nu-2)+i}, \quad c_k^i(t, 0) = a_k^i + a_{k,\nu-1}^i t^{\nu-1} + a_{k,\nu}^i t^\nu + \cdots,$$

where $a_k^i, a_{k,\nu-1}^i, \dots$ are constants. Since

$$f_i(t,0) = \sum_{k=1}^{\nu-2} c_k^i(t,0) g_k(t,0) \quad (i = 0, 1, \dots, \nu-2),$$

it follows that

$$\begin{aligned} t^{\nu(\nu-2)+i} &= \sum_{k=1}^{\nu-2} a_k^i \, t^{\nu(\nu-2)}(b_{k,0} + b_{k,1}t + \cdots + b_{k,\nu-2}t^{\nu-2}) \\ &\quad + \text{terms of order higher than } \nu(\nu-2)+\nu-1, \\ &\quad (i = 0, 1, \dots, \nu-2) \quad \text{on } e_0. \end{aligned}$$

Consequently,

$$\begin{aligned} t^i &= \sum_{k=1}^{\nu-2} a_k^i \, b_{k,0} + t \sum_{k=1}^{\nu-2} a_k^i \, b_{k,1} + \cdots + t^{\nu-2} \sum_{k=1}^{\nu-2} a_k^i \, b_{k,\nu-2}, \\ &\quad (i = 0, 1, \dots, \nu-2) \quad \text{on } e_0, \end{aligned}$$

so that

$$\begin{pmatrix} a_1^0 & a_2^0 & \cdots & a_{\nu-2}^0 \\ a_1^1 & a_2^1 & \cdots & a_{\nu-2}^1 \\ & \vdots & & \vdots \\ a_1^{\nu-2} & a_2^{\nu-2} & \cdots & a_{\nu-2}^{\nu-2} \end{pmatrix} \begin{pmatrix} b_{1,0} & b_{1,1} & \cdots & b_{1,\nu-2} \\ b_{2,0} & b_{2,1} & \cdots & b_{2,\nu-2} \\ & \vdots & & \vdots \\ b_{\nu-2,0} & b_{\nu-2,1} & \cdots & b_{\nu-2,\nu-2} \end{pmatrix} = E_{\nu-1},$$

where $E_{\nu-1}$ is the $(\nu-1, \nu-1)$ identity matrix. Since $(a_j^i)_{i,j}$ is a $(\nu-1, \nu-2)$ matrix and $(b_{i,j})_{i,j}$ is a $(\nu-2, \nu-1)$ matrix, such an equality is impossible. Thus, Lemma 8.4 is proved. □

For each integer $\nu \geq 3$ we let $F_k^\nu(x_1, x_2, y_1, y_2)$ $(k = 1, 2, 3, 4)$ denote the four polynomials in $\mathbf{C}^4$ defined above. Let a_ν $(\nu = 1, 2, \dots)$ be a sequence of complex numbers such that $\lim_{\nu\to\infty} a_\nu = \infty$. In $\mathbf{C}^5$ with variables x_1, x_2, y_1, y_2, y_3 we consider the analytic set of dimension 1 given by

$$\Sigma_\nu : \quad F_1^\nu = \cdots = F_4^\nu = 0, \quad y_3 = a_\nu,$$

and we set $\Sigma = \bigcup_{\nu=1}^\infty \Sigma_\nu$ in $\mathbf{C}^5$. We consider the G-ideal $G\{\Sigma\}$ in $\mathbf{C}^5$. Then $G\{\Sigma\}$ has a locally finite pseudobase on all of $\mathbf{C}^5$. However, Lemma 8.4 implies that there is no finitely generated $\mathcal{O}$-ideal $\mathcal{J}$ on all of $\mathbf{C}^5$ which is equivalent to the G-ideal $G\{\Sigma\}$ at each point of $\mathbf{C}^5$.

REMARK 8.8. We see from the proof of Lemma 8.4 that (1) $F_4 = x_1^{\nu-2} + x_2^\nu$ is a universal denominator of the 2-dimensional analytic set $\mathcal{S}$ in $\mathbf{C}^4$ with $F_4 \not\equiv 0$ on $\mathcal{S}$; (2) the Z-ideal $Z\{F_4, \mathcal{S}\}$ is equivalent to the G-ideal $G\{\Sigma\}$ at the origin O in $\mathbf{C}^4$; and (3) $G\{\Sigma\}$ is equivalent to the $\mathcal{O}$-ideal $\mathcal{O}\{\Phi\}$ generated by Φ_j $(j = 0, 1, \dots, \nu-2)$ at O in $\mathbf{C}^4$.

In fact, statements (2) and (3) follow immediately from the proof. To see (1), let $p_0 = (x_1^0, x_2^0, y_1^0, y_2^0) \in \mathcal{S}$. The singular set τ of $\mathcal{S}$ is contained in $x_1 = 0$. Since $\mathcal{S} \cap \{x_1 = 0\} = \{(0, x_2, 0, 0) \mid x_2 \in \mathbf{C}\}$ and since $F_4 \neq 0$ at $(0, x_2)$ if $x_2 \neq 0$, it suffices to prove that, for any weakly holomorphic function $f(p)$ at the origin O on $\mathcal{S}$, the function $F_4 \cdot f$ is holomorphic at O on $\mathcal{S}$. Since $\Sigma = \mathcal{S} \cap \{F_4 = 0\}$, $F_4 \cdot f$ is a weakly holomorphic function at O on $\mathcal{S}$ which vanishes on Σ in a neighborhood of

O. Under this condition we have shown that there exists a holomorphic function $F(x_1, x_2, y_1, y_2)$ in a neighborhood of O in $\mathbf{C}^4$ such that $F|_{\mathcal{S}} = F_4 \cdot f|_{\mathcal{S}}$, as desired.

8.4. Quantitative Estimates

In this section we extend Theorem 8.2 (extension theorem) and Theorem 8.4 (Problem C_1) to quantitative results with estimates (see Chapter I in Oka [**52**]). Our proofs will be done by a combination of Oka's theorems (which have already been proved) and the open mapping theorem in a Fréchet space. These theorems with estimates will be applied to obtain a subglobal pseudobase of an $\mathcal{O}^\lambda$-module on a Stein space $\mathcal{V}$ which has a locally finite pseudobase at each point in $\mathcal{V}$. In addition, we will use these results in Chapter 9 to show that any analytic space admitting a strictly pseudoconvex exhaustion function is a Stein space.

8.4.1. Open Mapping Theorem. Let $\mathcal{E}$ be a vector space over $\mathbf{C}$ equipped with a metric $d(x,y)$ such that $d(x,y) = d(x-y,0)$, where 0 denotes the zero vector in $\mathcal{E}$. Assume that:

(i) $\mathcal{E}$ has a fundamental system of convex and circled neighborhoods V_n ($n = 1, 2, \dots$) of 0 in $\mathcal{E}$. Here circled means that $\lambda V_n \subset V_n$ for any $\lambda \in \mathbf{C}$ with $|\lambda| \le 1$. (Note that V_n is not, in general, relatively compact).
(ii) $\mathcal{E}$ is complete with respect to the metric $d(x,y)$.
(iii) For any $\alpha, \beta \in \mathbf{C}$, the mapping $\mathcal{S} : (x,y) \in \mathcal{E} \times \mathcal{E} \to \alpha x + \beta y \in \mathcal{E}$ is continuous.
(iv) For any $x \in \mathcal{E}$, $\lim_{n\to\infty} x/n = 0$.

Then we call $\mathcal{E}$ a **Fréchet space**.

The following theorem will be useful in this section.

THEOREM 8.13 (Open mapping theorem). *Let $\mathcal{E}_1$ and $\mathcal{E}_2$ be Fréchet spaces equipped with metrics $d_1(x,y)$ and $d_2(u,v)$. Let $\varphi : \mathcal{E}_1 \to \mathcal{E}_2$ be a continuous linear mapping from $\mathcal{E}_1$* **onto** *$\mathcal{E}_2$. Then φ is an open mapping.*

PROOF. (cf. [**28**]) From (iii) it suffices to verify that for any neighborhood V of the zero vector 0 in $\mathcal{E}_1$, $\varphi(V)$ is a neighborhood of the zero vector 0 in $\mathcal{E}_2$. We first prove that for any neighborhood V of 0 in $\mathcal{E}_1$, $\overline{\varphi(V)}$ is a neighborhood of 0 in $\mathcal{E}_2$. To this end, using (i) we may assume that V is a convex and circled neighborhood of 0 in $\mathcal{E}_1$. Set $W = \varphi(V)$. By linearity of φ, W is convex and circled in $\mathcal{E}_2$, so that $\overline{W}$ is a convex and circled closed set in $\mathcal{E}_2$. Since φ is surjective, it follows from (iv) that $\mathcal{E}_2 = \bigcup_{n=1}^\infty nW = \bigcup_{n=1}^\infty n\overline{W}$. Since $n\overline{W}$ ($n = 1, 2, \dots$) is closed in the complete metric space $\mathcal{E}_2$, it follows from the Baire category theorem that for some integer n, $n\overline{W}$ contains an interior point u_0 in $\mathcal{E}_2$. Thus we can find a convex and circled neighborhood G of 0 in $\mathcal{E}_2$ such that $u_0 + G \subset n\overline{W}$ ($n \ge n_0$). Consequently, $u_0/n + G/n \subset \overline{W}$. In particular, $u_0/n \subset \overline{W}$, so that $-u_0/n \subset \overline{W}$. Since $\overline{W}$ is convex, we have

$$\frac{G}{2n} = \frac{1}{2}\left\{\left(-\frac{u_0}{n}\right) + \left(\frac{u_0}{n} + \frac{G}{n}\right)\right\} \subset \overline{W},$$

which proves $\overline{W}$ is a neighborhood of 0 in $\mathcal{E}_2$.

Let V be any convex and circled neighborhood of 0 in $\mathcal{E}_1$. We now show that $\overline{\varphi(V)} \subset \varphi(2V)$. Using (i) and (iv), we can find a sequence of convex and circled neighborhoods V_j ($j = 1, 2, \dots$) of 0 in $\mathcal{E}_1$ such that $V_j \subset \{x \in \mathcal{E}_1 \mid d_1(x, 0) < 1/2^j\}$

$(j = 1, 2, \dots)$, $2\overline{V}_1 \subset V$, and $2\overline{V}_{j+1} \subset V_j$ $(j = 1, 2, \dots)$. Let $y_0 \in \overline{\varphi(V)}$. Since $\overline{\varphi(V_1)}$ was shown to be a neighborhood of 0 in $\mathcal{E}_2$, there exist $x_0 \in V$ and $y_1 \in \overline{\varphi(V_1)}$ such that $y_0 - y_1 = \varphi(x_0)$. In a similar manner, we can find $x_1 \in V_1$ and $y_2 \in \overline{\varphi(V_2)}$ such that $y_1 - y_2 = \varphi(x_1)$. We inductively choose a sequence of points $x_n \in V_n$ and $y_n \in \overline{\varphi(V_n)}$ $(n = 1, 2, \dots)$ such that $y_n - y_{n+1} = \varphi(x_n)$ $(n = 1, 2, \dots)$. Since $V_n \to 0$ as $n \to 0$, it follows that $\varphi(V_n)$ and hence $\overline{\varphi(V_n)} \to 0$ as $n \to \infty$, so that $\lim_{n\to\infty} y_n = 0$. If we set $a_n = x_0 + x_1 + \cdots + x_n$ $(n = 1, 2, \dots)$, we see from $d(x_n, 0) < 1/2^n$ that $\{a_n\}_n$ is a Cauchy sequence in $\mathcal{E}_1$, so that the limit $a = \lim_{n\to\infty} a_n$ exists in $\mathcal{E}_1$. Since $a_n \in V + V_1 + \cdots + V_n \subset 2V$ $(n = 1, 2, \dots)$, we have $a \in 2\overline{V}$. Also, $\varphi(a) = \lim_{n\to\infty} \varphi(a_n) = \lim_{n\to\infty}((y_0 - y_1) + \cdots + (y_n - y_{n+1}))$ $= \lim_{n\to\infty}(y_0 - y_n) = y_0$, so that $\overline{\varphi(V)} \subset \varphi(2\overline{V})$.

For any convex and circled neighborhood V of 0 in $\mathcal{E}_1$, $\varphi(2\overline{V})$ is a neighborhood of the origin 0 in $\mathcal{E}_2$. The collection of these sets $2\overline{V}$ is a fundamental neighborhood system of 0 in $\mathcal{E}_1$, proving Theorem 8.13. □

Let $\mathcal{V}$ be an analytic space and let $U \subset \mathcal{V}$ be a domain. Let $\lambda \geq 1$ be an integer and let $\mathcal{O}^\lambda(U)$ denote the set of all holomorphic vector-valued functions $f(p) = (f_1(p), \dots, f_\lambda(p))$ of rank λ on U. Thus, $\mathcal{O}^\lambda(U)$ is a vector space over $\mathbf{C}$. In case $\lambda = 1$, we write $\mathcal{O}^1(U) = \mathcal{O}(U)$.

Let U_j $(j = 1, 2, \dots)$ be a sequence of domains in U such that

$$U_j \subset\subset U_{j+1} \quad (j = 1, 2, \dots), \qquad U = \lim_{j\to\infty} U_j.$$

For any $f(p) \in \mathcal{O}^\lambda(U)$, we set

$$m_j(f) = \sup_{p \in U_j} ||f(p)|| \quad (j = 1, 2, \dots),$$

so that $m_j(f) \leq m_{j+1}(f) < \infty$. In general, we can have $\lim_{j\to\infty} m_j(f) = +\infty$. For $f(p), g(p) \in \mathcal{O}^\lambda(U)$, we define

$$d^\lambda(f, g) = \sum_{j=1}^{\infty} \frac{1}{2^j} \frac{m_j(f - g)}{1 + m_j(f - g)} < 1.$$

Since $h(r) := r/(1 + r)$ is a concave increasing function on $[0, \infty)$ with $h(0) = 0$, it follows that $d^\lambda(f, g)$ is a metric on $\mathcal{O}^\lambda(U)$ with $d^\lambda(f, g) = d^\lambda(f - g, 0)$. We call $d^\lambda(f, g)$ the **canonical metric** on $\mathcal{O}^\lambda(U)$ (relative to $\{U_j\}_j$). We shall prove that $\mathcal{O}^\lambda(U)$ is a Fréchet space with respect to this metric $d^\lambda(f, g)$. Indeed, let $f_n(p)$ $(n = 1, 2, \dots)$ and $f(p)$ belong to $\mathcal{O}^\lambda(U)$. Then we see that $\lim_{n\to\infty} d^\lambda(f_n, f) = 0$ if and only if $\lim_{n\to\infty} f_n(p) = f(p)$ uniformly on any compact $K \subset\subset U$. We thus see that conditions (ii), (iii), and (iv) are satisfied. For condition (i) it suffices to set

$$V_j = \{f(p) \in \mathcal{O}^\lambda(U) \mid m_j(f) < 1/j\} \quad (j = 1, 2, \dots).$$

We have the following proposition.

PROPOSITION 8.2. *Let $M > 0$. Then there exists K with $0 < K < 1$ such that if $\|f(p)\| \leq M$ on U then $d^\lambda(f, 0) \leq K$. Conversely, fix K with $0 < K < 1$. Then there exists an $M > 0$ such that $d^\lambda(f, 0) \leq K$ implies $||f(p)|| \leq M$ on U_1.*

PROOF. To prove the first assertion, take $K = M/(M+1) < 1$. For the second one, since

$$K \geq \sum_{j=1}^{\infty} \frac{1}{2^j} \frac{m_j(f)}{1+m_j(f)} \geq \sum_{j=1}^{\infty} \frac{1}{2^j} \frac{m_1(f)}{1+m_1(f)} = \frac{m_1(f)}{1+m_1(f)},$$

we can take $M = K/(1-K) > 0$. □

8.4.2. Quantitative Estimates in the Existence Theorem. Let $\mathcal{V}$ be an analytic space of dimension n and let $\mathcal{P}$ be a (closed) analytic polyhedron in $\mathcal{V}$ with defining functions on $D : \mathcal{P} \subset\subset D \subset \mathcal{V}$. Let Σ be a normal model of $\mathcal{P}$ in the closed unit polydisk $\overline{\Delta} \subset \mathbf{C}^m$ and let

$$\Phi : p \in \mathcal{P} \to z = (z_1, \dots, z_m) = (\varphi_1(p), \dots, \varphi_m(p)) \in \Sigma$$

denote the normalization mapping of $\mathcal{P}$ into $\mathbf{C}^m$; here each $\varphi_j(p)$ $(j = 1, 2, \dots)$ is a holomorphic function on D. We let $\mathcal{P}^\circ$ and Δ denote the interiors of $\mathcal{P}$ in $\mathcal{V}$ and of $\overline{\Delta}$ in $\mathbf{C}^m$.

The following theorem, which will be proved without using the theory of Fréchet spaces, is an essential ingredient in proving the main theorems in this section.

THEOREM 8.14 (Interior extension theorem). *Let $f(p)$ be a holomorphic function on $\mathcal{P}^\circ$. There exists a holomorphic function $F(z)$ on Δ with*

$$f(p) = F(\Phi(p)), \quad p \in \mathcal{P}^\circ.$$

PROOF. Choose r_k, $0 < r_k < 1$ $(k = 1, 2, \dots)$, such that $r_k < r_{k+1}$ $(k = 1, 2, \dots)$ and $\lim_{k\to\infty} r_k = 1$. For each $k = 1, 2, \dots$, set

$$\overline{\Delta}_k : |z_j| \leq r_k \quad (j = 1, \dots, m), \qquad \Sigma_k = \Sigma \cap \overline{\Delta}_k.$$

Let $\mathcal{P}_k = \Phi^{-1}(\Sigma_k) \subset\subset \mathcal{P}^\circ$. Choose ϵ_k, $0 < \epsilon_k < 1$, so that $\sum_{k=1}^{\infty} \epsilon_k < \infty$. We would like to construct a sequence of holomorphic functions $F_k(z)$ on $\overline{\Delta}_k$ $(k = 1, 2, \dots)$ such that, for each $k = 1, 2, \dots$,

$$\begin{aligned} &f(p) = F_k(\Phi(p)), \quad p \in \mathcal{P}_k, \\ &|F_{k+1}(z) - F_k(z)| < \epsilon_k, \quad z \in \overline{\Delta}_k; \end{aligned} \tag{8.11}$$

for then $F(z) = \lim_{k\to\infty} F_k(z)$ converges uniformly on any compact $K \subset\subset \mathcal{P}^\circ$ and $F(z)$ satisfies $F(\Phi(p)) = f(p)$, $p \in \mathcal{P}^\circ$. We construct such a sequence $F_k(z)$ $(k = 1, 2, \dots)$ on $\mathcal{P}_k$ satisfying condition (8.11) by induction.

By Theorem 8.2, there exists a holomorphic function $F_1(z)$ on $\overline{\Delta}_1$ such that $f(p) = F_1(\Phi(p))$ on $\mathcal{P}_1$. Fix $k \geq 1$ and assume that we have constructed holomorphic functions $F_j(z)$ on $\overline{\Delta}_j$ $(j = 1, \dots, k)$ such that $F_j(\Phi(p)) = f(p)$ on $\mathcal{P}_j$ $(j = 1, \dots, k)$ and $|F_{j+1}(z) - F_j(z)| < \epsilon_j$ on $\overline{\Delta}_j$ $(j = 1, \dots, k-1)$. By Theorem 8.2, there exists a holomorphic function $\tilde{F}_{k+1}(z)$ on $\overline{\Delta}_{k+1}$ such that $f(p) = \tilde{F}_{k+1}(\Phi(p))$ on $\mathcal{P}_{k+1}$. Consider the G-ideal $G\{\Sigma_{k+1}\}$ on $\overline{\Delta}_{k+1}$. Since $G\{\Sigma_{k+1}\}$ has a locally finite pseudobase at each point of $\overline{\Delta}_{k+1}$, it follows from Theorem 8.6 (Problem E for a closed polydisk) that there exist a finite number of holomorphic functions $G_l(z)$ $(l = 1, \dots, s)$ on $\overline{\Delta}_{k+1}$ such that the $\mathcal{O}$-ideal $\mathcal{J}\{G\}$ generated by $G_l(z)$ $(l = 1, \dots, s)$ on $\overline{\Delta}_{k+1}$ is equivalent to $G\{\Sigma_{k+1}\}$ on $\overline{\Delta}_{k+1}$. Since $\tilde{F}_{k+1}(z) - F_k(z) = 0$ on Σ_k, it follows that $\tilde{F}_{k+1}(z) - F_k(z) \in G\{\Sigma_k\}$ on

$\overline{\Delta}_k$. By Theorem 8.4 (Problem C_1 for a closed polydisk), there exist s holomorphic functions $\tilde{\alpha}_l(z)$ $(l = 1, \dots, s)$ on $\overline{\Delta}_k$ such that

$$\tilde{F}_{k+1}(z) - F_k(z) = \tilde{\alpha}_1(z)G_1(z) + \cdots + \tilde{\alpha}_s(z)G_s(z), \quad z \in \overline{\Delta}_k.$$

Since the pair of polydisks $(\overline{\Delta}_k, \overline{\Delta}_{k+1})$ satisfies Runge's theorem, we can find a holomorphic function $\alpha_j(z)$ $(l = 1, \dots, s)$ on $\overline{\Delta}_{k+1}$ such that

$$|\alpha_l(z) - \tilde{\alpha}_l(z)| < \epsilon'_k \qquad (l = 1, \dots, s), \quad z \in \overline{\Delta}_k$$

where $0 < \epsilon'_k < \epsilon_k / [\max_{z \in \overline{\Delta}_{k+1}} \{|G_1(z)| + \cdots + |G_s(z)|\}]$. If we set

$$F_{k+1}(z) = \tilde{F}_{k+1}(z) + \alpha_1(z)G_1(z) + \cdots + \alpha_s(z)G_s(z), \quad z \in \overline{\Delta}_{k+1},$$

then $F_{k+1}(z)$ is a holomorphic function on $\overline{\Delta}_{k+1}$ with $F_{k+1}(\Phi(p)) = f(p)$ for $p \in \mathcal{P}_{k+1}$ and $|F_{k+1}(z) - F_k(z)| < \sum_{l=1}^{s} |\alpha_l(z) - \tilde{\alpha}_l(z)||G_l(z)| < \epsilon_k$ for $z \in \overline{\Delta}_k$. Thus we have inductively constructed $F_k(z)$ $(k = 1, 2, \dots)$ on $\overline{\Delta}_k$ satisfying condition (8.11). □

Using the same notation $\mathcal{P}$, $\overline{\Delta}$, $\mathcal{P}^\circ$, Δ, Σ, $\mathcal{P}_k$, $\overline{\Delta}_k$, and Σ_k, recall that Δ_k : $|z_j| < r_j$ $(j = 1, \dots, m)$ and we thus have $\Delta_k \subset\subset \Delta_{k+1}$ $(k = 1, 2, \dots)$ and $\Delta = \lim_{k \to \infty} \Delta_k$. We consider the set $\mathcal{O}(\Delta)$ of all holomorphic functions $F(z)$ on Δ. By the method mentioned in the previous section, the vector space $\mathcal{O}(\Delta)$ with the canonical metric $d_1(F, G)$ relative to $\{\Delta_k\}_k$ becomes a Fréchet space. Similarly, using $\mathcal{P}_k^\circ \subset\subset \mathcal{P}_{k+1}^\circ$ $(k = 1, 2, \dots)$ and $\mathcal{P}^\circ = \lim_{n \to \infty} \mathcal{P}_k^\circ$, we consider the Fréchet space $\mathcal{O}(\mathcal{P}^\circ)$ of all holomorphic functions $f(p)$ on $\mathcal{P}^\circ$ with the canonical metric $d_2(f, g)$ relative to $\{\mathcal{P}_k^\circ\}_k$.

Consider the following linear mapping φ from $\mathcal{O}(\Delta)$ to $\mathcal{O}(\mathcal{P}^\circ)$:

$$\varphi : \ F(z) \to f(p) = F(\Phi(p)), \quad p \in \mathcal{P}^\circ.$$

Since the topology for $\mathcal{O}(\Delta)$ and for $\mathcal{O}(\mathcal{P}^\circ)$ is uniform convergence on each compact set in Δ and in $\mathcal{P}^\circ$, it follows that φ is a continuous mapping on $\mathcal{O}(\Delta)$. By Theorem 8.14, φ is *surjective*. Thus, the open mapping theorem can be applied to φ.

We have the following theorem.

THEOREM 8.15 (Extension theorem with estimates). *There exists a constant $K > 0$ such that for any $f(p) \in \mathcal{O}(\mathcal{P}^\circ)$, there exists $F(z) \in \mathcal{O}(\Delta)$ with*

$$\begin{aligned} F(\Phi(p)) &= f(p), \quad p \in \mathcal{P}^\circ, \\ \max_{z \in U_1}\{|F(z)|\} &\leq K \max_{p \in \mathcal{P}^\circ}\{|f(p)|\}. \end{aligned}$$

PROOF. It suffices to prove the existence of such a constant $K > 0$ for $f(p) \in \mathcal{O}(\mathcal{P}^\circ)$ with $\max_{p \in \mathcal{P}^\circ}\{|f(p)|\} \leq 1$. Fix $0 < \rho < 1$ and let $B_\rho = \{F(z) \in \mathcal{O}(\Delta) \mid d_1(F, 0) < \rho\}$ and $M_\rho = \rho/(1 - \rho) > 0$, which satisfies Proposition 8.2. By the open mapping theorem, $\varphi(B_\rho)$ contains a neighborhood $A_\eta := \{f(p) \in \mathcal{O}(\mathcal{P}^\circ) \mid d_2(f, 0) \leq \eta\}$ of the origin O in $\mathcal{O}(\mathcal{P}^\circ)$. Take $f(p) \in \mathcal{O}(\mathcal{P}^\circ)$ with $\max_{p \in \mathcal{P}^\circ}\{|f(p)|\} \leq 1$. Since $\eta f(p) \in A_\eta$, there exists $F_0(z) \in \mathcal{O}(\Delta)$ with $d_1(F_0, 0) < \rho$ such that $F_0(\Phi(p)) = \eta f(p)$, $p \in \mathcal{P}^\circ$. By Proposition 8.2 we have $|F_0(z)| \leq M_\rho := \rho/(1 - \rho)$ on U_1. If we define $F(z) = F_0(z)/\eta$ on Δ, then $F(\Phi(p)) = f(p)$, $p \in \mathcal{P}^\circ$, and $|F(z)| \leq M_\rho/\eta$ on U_1. Thus $K := M_\rho/\eta > 0$ satisfies the conclusion of the theorem. □

8.4.3. Completeness. Using the previous theorem we can extend the completeness theorem (Theorem 7.6) for $\mathcal{O}^\lambda$-modules in a domain in $\mathbf{C}^n$ to the case of an analytic space $\mathcal{V}$. Let $\mathcal{V}$ be an analytic space of dimension n. Let $W \subset \mathcal{V}$ be a domain and let $F_j(p)$ $(j = 1, \ldots, \nu)$ be ν holomorphic vector-valued functions of rank λ on W. We let $\mathcal{J}^\lambda\{F\}$ denote the $\mathcal{O}^\lambda$-module generated by $F_j(p)$ $(j = 1, \ldots, \nu)$ on W.

We have the following result.

PROPOSITION 8.3. *$\mathcal{J}^\lambda\{F\}$ is complete in the topology of uniform convergence on compact sets in W.*

To be precise, completeness in this sense means the following. Let $f_i(p)$ $(i = 1, 2, \ldots)$ be a sequence of holomorphic vector-valued functions of rank λ on the common domain $U \subset W$. Assume that (1) $\lim_{i\to\infty} f_i(p) = f(p)$ is uniformly convergent on any $K \subset\subset U$, and (2) each $f_i(p)$ $(i = 1, 2, \ldots)$ belongs to $\mathcal{J}^\lambda\{F\}$ at each point of U. Then $f(p)$ belongs to $\mathcal{J}^\lambda\{F\}$ at each point of U.

PROOF. Let $p_0 \in U$. We can take a sufficiently small analytic polyhedron $\mathcal{P}$ in a domain $D \subset U$ such that $p_0 \in \mathcal{P}^\circ$ (Corollary 8.1). Fix a normal model Σ of $\mathcal{P}$ in the closed unit polydisk $\overline{\Delta}$ in $\mathbf{C}^n$,

$$\Phi : p \in \mathcal{P} \to z = (\varphi_1(p), \ldots, \varphi_m(p)) \in \Sigma,$$

where each $\varphi_j(p)$ $(j = 1, \ldots, m)$ is a holomorphic function on D. We take holomorphic extensions $\tilde{F}_j(z)$ $(j = 1, \ldots, \nu)$ of $F_j(p)$ on $\overline{\Delta}$; thus $\tilde{F}_j(\Phi(p)) = F_j(p)$ in $\mathcal{P}$. Fix a closed polydisk $\overline{\Delta}_1 \subset\subset \Delta$ such that $\mathcal{P}_1^\circ = \Phi^{-1}(\Sigma \cap \Delta_1)$ contains the point p_0. Since $\lim_{i\to\infty} f_i(p) = f(p)$ uniformly on $\mathcal{P}$, there exists $M > 0$ such that $||f_i(p)|| \leq M$ $(i = 1, 2, \ldots)$ on $\mathcal{P}$. By Theorem 8.15, for each $i = 1, 2, \ldots$, there exists a holomorphic extension $\tilde{f}_i(z)$ in Δ such that

$$f_i(p) = \tilde{f}_i(\Phi(p)), \quad p \in \mathcal{P}^\circ,$$
$$||\tilde{f}_i(z)|| \leq KM \qquad (i = 1, 2, \ldots), \quad z \in \Delta_1,$$

where $K > 0$ is a constant independent of $i = 1, 2, \ldots$. Thus, $\{\tilde{f}_i(z)\}_i$ is a normal family on Δ_1. Let $\delta \subset\subset \Delta_1$ be a neighborhood of the point $z_0 = \Phi(p_0)$. Then there exists a subsequence $\{\tilde{f}_{i_k}(z)\}_k$ of $\{\tilde{f}_i(z)\}_i$ which converges uniformly on δ, say $\tilde{f}(z) = \lim_{k\to\infty} \tilde{f}_{i_k}(z)$ on δ, so that $\tilde{f}(z)$ is a holomorphic vector-valued function of rank λ on δ satisfying $\tilde{f}(\Phi(p)) = f(p)$ for $p \in \Phi^{-1}(\Sigma \cap \delta)$.

Recall the holomorphic vector-valued functions $\psi_{k,l}(z)$ $(k = 1, \ldots, \lambda;\ l = 1, \ldots, s)$ of rank λ on $\overline{\Delta}$ which were constructed using the pseudobase $G_l(z)$ $(l = 1, \ldots, s)$ of the G-ideal $G\{\Sigma\}$ on $\overline{\Delta}$ defined by (8.3). Let $\mathcal{J}^\lambda\{\tilde{F}, \Psi\}$ denote the $\mathcal{O}^\lambda$-module generated by $\tilde{F}_j(z)$ $(j = 1, \ldots, \nu)$ and $\psi_{k,l}(z)$ $(k = 1, \ldots, \lambda;\ l = 1, \ldots, s)$ on $\overline{\Delta}$. Since $f_i(p) \in \mathcal{J}^\lambda\{F\}$ $(i = 1, 2, \ldots)$ at each point of $\mathcal{P}$, it follows that $\tilde{f}_i(z) \in \mathcal{J}^\lambda\{\tilde{F}, \Psi\}$ $(i = 1, 2, \ldots)$ at each point of $\overline{\Delta}$. Since $\lim_{k\to\infty} \tilde{f}_{i_k}(z) = \tilde{f}(z)$ uniformly on δ, it follows from Theorem 7.6 that $\tilde{f}(z) \in \mathcal{J}^\lambda\{\tilde{F}, \Psi\}$ at each point of δ. Since $\tilde{f}(\Phi(p)) = f(p)$ on $v := \Phi^{-1}(\Sigma \cap \delta)$, which is a neighborhood of the point p_0 in W, we see that $f(p) \in \mathcal{J}^\lambda\{F\}$ at each point of v. □

Using this completeness result, we generalize Lemma 8.3.

LEMMA 8.5. *Let $\mathcal{V}$ be a Stein space and let $\mathcal{J}^\lambda$ be an $\mathcal{O}^\lambda$-module on $\mathcal{V}$ which has a locally finite pseudobase at each point in $\mathcal{V}$. Let $\mathcal{P}$ be an analytic polyhedron in $\mathcal{V}$ with defining functions on $\mathcal{V}$ and let $f(p)$ be a holomorphic vector-valued function of rank λ on $\mathcal{P}$ such that $f(p) \in \mathcal{J}^\lambda$ at each point of $\mathcal{P}$. Given $\varepsilon > 0$, there exists a holomorphic vector-valued function $F(p)$ of rank λ on $\mathcal{V}$ such that*

1. *$F(p) \in \mathcal{J}^\lambda$ at each point in $\mathcal{V}$, and*
2. *$||F(p) - f(p)|| < \epsilon$ for each $p \in \mathcal{P}$.*

PROOF. Let $\mathcal{P}_j$ $(j = 1, 2, \dots)$ be a sequence of analytic polyhedra in $\mathcal{V}$ with defining functions on $\mathcal{V}$ such that $\mathcal{P} \subset\subset \mathcal{P}_1^\circ$, $\mathcal{P}_j \subset\subset \mathcal{P}_{j+1}^\circ$ $(j = 1, 2, \dots)$, and $\mathcal{V} = \lim_{j\to\infty} \mathcal{P}_j$. Choose $\epsilon_k > 0$ $(k = 1, 2, \dots)$ such that $\sum_{j=1}^\infty \epsilon_k < \epsilon$. By Theorem 8.6 (Problem E), there exist a finite number of holomorphic vector-valued functions $\varphi_j(p)$ $(j = 1, \dots, s)$ of rank λ defined in a neighborhood U of $\mathcal{P}_1$ such that the $\mathcal{O}^\lambda$-module $\mathcal{J}^\lambda\{\varphi\}$ generated by $\varphi_j(p)$ $(j = 1, \dots, s)$ on $\mathcal{P}_1$ is equivalent to $\mathcal{J}^\lambda$ on $\mathcal{P}_1$.

Using Theorem 8.6, we can find a holomorphic vector-valued function $a(p) = (a_1(p), \dots, a_s(p))$ of rank s on $\mathcal{P}$ such that

$$f(p) = a_1(p)\varphi_1(p) + \cdots + a_s(p)\varphi_s(p), \quad p \in \mathcal{P}.$$

Since the pair $(\mathcal{P}, \mathcal{P}_1)$ satisfies Runge's theorem (Lemma 8.2), there exists a holomorphic vector-valued function $A(p) = (A_1(p), \dots, A_s(p))$ of rank s on $\mathcal{P}_1$ such that

$$||A(p) - a(p)|| < \epsilon_1' \qquad \text{for } p \in \mathcal{P},$$

where $0 < \epsilon_1' < \epsilon_1/(||\varphi_1||_{\mathcal{P}_1} + \cdots + ||\varphi_s||_{\mathcal{P}_1})$. If we set

$$F_1(p) = A_1(p)\varphi_1(p) + \cdots + A_s(p)\varphi_s(p), \quad p \in \mathcal{P}_1,$$

then $F_1(p)$ is a holomorphic vector-valued function of rank λ on $\mathcal{P}_1$ such that $F_1(p) \in \mathcal{J}^\lambda$ at each point of $\mathcal{P}_1$ and

$$||F_1(p) - f(p)|| \le \epsilon_1' \, (||\varphi_1(p)|| + \cdots + ||\varphi_s(p)||) < \epsilon_1, \quad p \in \mathcal{P}.$$

Similarly, there exists $F_2(p)$ of rank λ on $\mathcal{P}_2$ such that $F_2(p) \in \mathcal{J}^\lambda$ at each point of $\mathcal{P}_2$ and $||F_2(p) - F_1(p)|| < \epsilon_2$ for $p \in \mathcal{P}_1$. Thus, inductively we construct a vector-valued function $F_j(p)$ $(j = 1, 2, \dots)$ of rank λ on $\mathcal{P}_j$ such that $F_j(p) \in \mathcal{J}^\lambda$ at each point of $\mathcal{P}_j$ and $||F_j(p) - F_{j-1}(p)|| < \epsilon_j$ on $\mathcal{P}_{j-1}$, where $F_0(p) = f(p)$ and $\mathcal{P}_0 = \mathcal{P}$. It follows that $F(p) := \lim_{j\to\infty} F_j(p)$ converges uniformly on any compact set in $\mathcal{V}$. Thus, $F(p)$ is a holomorphic vector-valued function of rank λ on $\mathcal{V}$ which belongs to $\mathcal{J}^\lambda$ at each point of $\mathcal{V}$ by Proposition 8.3. We also have $||F(p) - f(p)|| \le \sum_{j=1}^\infty ||F_j(p) - F_{j-1}(p)|| < \sum_{j=1}^\infty \epsilon_j < \epsilon$ on $\mathcal{P}$, which proves the lemma. □

8.4.4. Quantitative Estimates for Problem C_1. We return to the situation in 8.4.2. Let $\mathcal{V}$ be an analytic space and let $\mathcal{P}$ be an analytic polyhedron in $\mathcal{V}$ with defining functions on D, $\mathcal{P} \subset\subset D \subset \mathcal{V}$. Let $\mathcal{P}^\circ$ denote the interior of $\mathcal{P}$ in $\mathcal{V}$. We let $\mathcal{O}^\lambda(\mathcal{P}^\circ)$ and $\mathcal{O}^\nu(\mathcal{P}^\circ)$ denote the spaces of all holomorphic vector-valued functions on $\mathcal{P}^\circ$ of rank λ and ν. Let F_j $(j = 1, \dots, \nu)$ be ν holomorphic vector-valued functions of rank λ on the closed analytic polyhedron $\mathcal{P}$ and let $\mathcal{J}^\lambda\{F\}$ be the $\mathcal{O}^\lambda$-module generated by F_j $(j = 1, \dots, \nu)$ on $\mathcal{P}$.

We have the following theorem.

THEOREM 8.16 (Estimates for Problem C_1). *Let $U \subset\subset \mathcal{P}^\circ$. Then there exists a constant $K > 0$ such that for any $H(p) \in \mathcal{O}^\lambda(\mathcal{P}^\circ)$ with $H(p) \in \mathcal{J}^\lambda\{F\}$ at each point in $\mathcal{P}^\circ$, there exists $A(p) = (A_1(p), \ldots, A_\nu(p)) \in \mathcal{O}^\nu(\mathcal{P}^\circ)$ such that*

$$H(p) = A_1(p)F_1(p) + \cdots + A_\nu(p)F_\nu(p), \quad p \in \mathcal{P}^\circ,$$
$$\max_{p\in U}\{||A(p)||\} \leq K \max_{p\in\mathcal{P}^\circ}\{||H(p)||\}.$$

PROOF. Let U_j $(j = 1, 2, \ldots)$ be a sequence of domains in $\mathcal{P}^\circ$ such that $U = U_1$, $U_j \subset\subset U_{j+1}$ $(j = 1, 2, \ldots)$, and $\mathcal{P}^\circ = \lim_{j\to\infty} U_j$. We let $d^\lambda(f, g)$ and $d^\nu(f, g)$ denote the canonical distances with respect to $\{U_j\}_j$ on $\mathcal{O}^\lambda(\mathcal{P}^\circ)$ and on $\mathcal{O}^\nu(\mathcal{P}^\circ)$. Then $\mathcal{O}^\lambda(\mathcal{P}^\circ)$ and $\mathcal{O}^\nu(\mathcal{P}^\circ)$ are Fréchet spaces with respect to $d^\lambda(f, g)$ and $d^\nu(f, g)$. We let $\mathcal{F}$ denote the set of all holomorphic vector-valued functions $f(p)$ of rank λ on $\mathcal{P}^\circ$ such that $f(p) \in \mathcal{J}^\lambda\{F\}$ at each point of $\mathcal{P}^\circ$; thus $\mathcal{F}$ is a linear subspace of $\mathcal{O}^\lambda(\mathcal{P}^\circ)$. By Proposition 8.3, $\mathcal{F}$ is complete with respect to the metric $d^\lambda(f, g)$, so that $\mathcal{F}$ is a Fréchet space.

Next we consider the continuous linear mapping φ from $\mathcal{O}^\nu(\mathcal{P}^\circ)$ to $\mathcal{F}$ given by

$$\varphi : A(p) = (A_1(p), \ldots, A_\nu(p)) \to H(p) = A_1(p)F_1(p) + \cdots + A_\nu(p)F_\nu(p).$$

By Theorem 8.11, φ is *surjective*. Thus, the open mapping theorem can be applied to φ.

We fix ρ, $0 < \rho < 1$, and let $\delta_\rho = \{A(p) \in \mathcal{O}^\nu(\mathcal{P}^\circ) \mid d^\nu(A, 0) < \rho\}$. By Proposition 8.2, there exists $M_\rho > 0$ such that $||A(p)|| < M_\rho$ on U for all $A(p) \in \delta_\rho$. Since $\varphi(\delta_\rho)$ is an open neighborhood of the zero vector in $\mathcal{F}$, there exists η, $0 < \eta < 1$, such that $V_\eta = \{H(p) \in \mathcal{F} \mid ||H(p)|| < \eta\} \subset \varphi(\delta_\rho)$. We show that $K := M_\rho/\eta > 0$ satisfies the conclusion of the theorem. To prove this, we may assume that $H(p)$ satisfies $||H(p)|| \leq 1$ on $\mathcal{P}^\circ$. We then have $d^\lambda(\eta H, 0) \leq \eta/(1 + \eta) < \eta$, so that we can find $A(p) = (A_1(p), \ldots, A_\nu(p)) \in \delta_\rho$ such that $\varphi(A) = \eta H$ on $\mathcal{P}^\circ$. Consequently,

$$H(p) = \frac{A_1(p)}{\eta}F_1(p) + \cdots + \frac{A_\nu(p)}{\eta}F_\nu(p), \quad p \in \mathcal{P}^\circ;$$
$$||A(p)/\eta|| \leq M_\rho/\eta = K, \quad p \in U,$$

which proves the theorem. □

8.4.5. Applications of Quantitative Estimates. We give some applications of Theorem 8.16 (Problem C_1 with quantitative estimates) concerning the existence of a subglobal normal model of a Stein space $\mathcal{V}$ and of a subglobal pseudobase of an $\mathcal{O}^\lambda$-module on $\mathcal{V}$ having a locally finite pseudobase at each point in $\mathcal{V}$.

1. Subglobal normal model

Let $\mathcal{V}$ be an analytic space. Let $\mathcal{P}$ be an analytic polyhedron in $\mathcal{V}$ with defining functions on all of $\mathcal{V}$. We showed that $\mathcal{P}$ has a normal model Σ in a polydisk $\overline{\Delta}$ in $\mathbf{C}^m$ via the mapping $\Phi : p \in \mathcal{P} \to z = \Phi(p) = (\varphi(p_1), \ldots, \varphi_m(p)) \in \Sigma$, where $\varphi_j(p)$ $(j = 1, \ldots, m)$ is a holomorphic function in a domain G in $\mathcal{V}$. In general, we cannot assume that $\varphi_j(p)$ is holomorphic on all of $\mathcal{V}$. However, if $\mathcal{V}$ is a Stein space, this is possible.

Let $\mathcal{V}$ be an analytic space of dimension n. Let $\mathcal{P} : |\varphi_j(p)| \leq 1$ $(j = 1, \ldots, m)$ be an analytic polyhedron in $\mathcal{V}$ where $\varphi_j(p)$ $(j = 1, \ldots, m)$ is a holomorphic function on a domain G with $\mathcal{P} \subset\subset G \subset \mathcal{V}$ such that

$$\Phi : z_j = \varphi_j(p) \ (j = 1, \dots, m), \quad p \in \mathcal{P},$$

is a normal model $\Sigma = \Phi(p)$ in the closed unit polydisk $\overline{\Delta}$ in $\mathbf{C}^m$.

Let $0 < \alpha < 1$ and $0 < \epsilon < 1$. Let $\psi_j(p)$ $(j = 1, \dots, m)$ be holomorphic functions in a domain G_1, $\mathcal{P} \subset\subset G_1 \subset G$, such that

$$|\varphi_j(p) - \psi_j(p)| < \epsilon, \quad p \in \mathcal{P}. \tag{8.12}$$

If $\epsilon > 0$ is sufficiently small relative to α, then the set

$$\mathcal{P}^* := \{p \in G_1 \mid |\psi_j(p)| \leq 1 - \alpha, \ j = 1, ..., m\}$$

is an analytic polyhedron in $\mathcal{V}$ such that $\mathcal{P}^* \subset\subset \mathcal{P}^\circ$. We consider the image

$$\Sigma^* : w_j = \psi_j(p) \ (j = 1, \dots, m), \quad p \in \mathcal{P}^*$$

in the polydisk $\overline{\Delta}^* : |w_j| \leq 1 - \alpha$ in $\mathbf{C}^m_w$, and we set

$$\Psi : p \in \mathcal{P}^* \to w = \Psi(p) = (\psi_1(p), \dots, \psi_m(p)) \in \Sigma^*.$$

We obtain the following stability result concerning the normal model.

LEMMA 8.6. *For sufficiently small $\epsilon > 0$, Σ^* (as well as Σ in $\overline{\Delta}$) is a normal model in $\overline{\Delta}^*$.*

PROOF. Take a polydisk $\overline{\Delta}_1 \subset\subset \Delta$ such that $\mathcal{P}^* \subset\subset \Phi^{-1}(\Sigma \cap \overline{\Delta}_1)$. By Theorem 8.15 and (8.12), there exist a constant $K > 0$ (depending on $\overline{\Delta}_1$) and a holomorphic function $F_j(z)$ $(j = 1, \dots, m)$ in $\overline{\Delta}$ such that

$$\begin{aligned} F_j(\Phi(p)) &= \psi_j(p) - \varphi_j(p), \quad p \in \mathcal{P}^\circ, \\ |F_j(z)| &\leq K\epsilon, \quad z \in \overline{\Delta}_1. \end{aligned} \tag{8.13}$$

We consider the following analytic mapping from Δ into $\mathbf{C}^m_w$:

$$T: \ w_j = z_j + F_j(z) \quad (j = 1, \dots, m).$$

For $\epsilon > 0$ sufficiently small, it follows from (8.13) that T is injective on $\overline{\Delta}_1$ with $\overline{\Delta}^* \subset T(\overline{\Delta}_1)$, and $T(\Phi(p)) = \Psi(p)$ on $\mathcal{P}^*$; i.e., $T(\Sigma) \,|_{T^{-1}(\Sigma^*)} = \Sigma^*$.

Now let $f(p)$ be a weakly holomorphic function at a point p_0 on Σ^*. We set $\tilde{p}_0 = T^{-1}(p_0)$ and $\tilde{f} = f \circ T$, which is a weakly holomorphic function at the point $\tilde{p}_0$ on Σ. Since Σ is normal at $\tilde{p}_0$ on Σ, we can find a holomorphic function $F(z)$ in a neighborhood δ of $\tilde{p}_0$ in $\overline{\Delta}$ such that $F \,|_{\Sigma \cap \delta} = \tilde{f} \,|_{\Sigma \cap \delta}$. If we set $H(w) = F(T^{-1}(w))$ for $w \in \delta^* := T(\delta)$ (so that δ^* is a neighborhood of p_0 in $\overline{\Delta}^*$), then $H(w)$ is a holomorphic function on δ^* with $H \,|_{\Sigma^* \cap \delta^*} = F \,|_{\Sigma \cap \delta} = \tilde{f} \,|_{\Sigma \cap \delta} = f \,|_{\Sigma^* \cap \delta^*}$. Thus, $f(p)$ is holomorphic at the point p_0. Therefore, Σ^* is a normal model of $\mathcal{P}^*$ in $\overline{\Delta}^*$. □

This result, combined with Runge's theorem in a Stein space, yields the following theorem.

THEOREM 8.17 (Subglobal normalization). *Let $\mathcal{V}$ be a Stein space and let $\mathcal{P}$ be an analytic polyhedron in $\mathcal{V}$ with defining functions on all of $\mathcal{V}$. Then $\mathcal{P}$ has a normal model $\widehat{\Sigma} : w_j = \psi_j(p)$ $(j = 1, \dots, \mu)$ in a polydisk $\overline{\Delta}$ in $\mathbf{C}^\mu$, where $\psi_j(p)$ $(j = 1, \dots, \mu)$ is a holomorphic function on $\mathcal{V}$.*

PROOF. By Theorem 8.1, $\mathcal{P}$ has a normal model $\Sigma : z_j = \varphi_j(p)$ $(j = 1, \ldots, m)$, $p \in \mathcal{P}$ in the closed unit polydisk $\overline{\Delta}^m$ in $\mathbf{C}^m$, where $\varphi_j(p)$ $(j = 1, \ldots, m)$ are holomorphic functions in a domain G, $\mathcal{P} \subset\subset G \subset \mathcal{V}$. By taking a smaller domain, if necessary, we can take $\eta > 0$ sufficiently small so that, if we set $\mathcal{P}_\eta : |\varphi_j(p)| \le 1 + \eta$ $(j = 1, \ldots, m)$, then $\mathcal{P} \subset\subset \mathcal{P}_\eta \subset\subset G$ and $\Sigma_\eta : z_j = \varphi_j(p)$ $(j = 1, \ldots, m)$ is a normal model of $\mathcal{P}_\eta$ in $\overline{\Delta}_\eta : |z_j| \le 1 + \eta$ $(j = 1, \ldots, m)$. Fix $\epsilon > 0$. Since the pair $(\mathcal{P}_\eta, \mathcal{V})$ satisfies Runge's theorem (Corollary 8.2), we can find a holomorphic function $\psi_j(p)$ $(j = 1, \ldots m)$ on $\mathcal{V}$ such that

$$|\psi_j(p) - \varphi_j(p)| < \epsilon \quad \text{on } \mathcal{P}_\eta.$$

If $\epsilon > 0$ is sufficiently small, then $\mathcal{P}^* : |\psi_j(p)| \le 1 + \eta/2$ $(j = 1, \ldots, m)$ is an analytic polyhedron in G with $\mathcal{P} \subset \mathcal{P}^*$. Furthermore, Lemma 8.6 implies that $\Sigma^* : w_j = \psi_j(p)$ $(j = 1, \ldots, m)$ is a normal model of $\mathcal{P}^*$ in the polydisk $\overline{\Delta}_{\eta/2} :=$ $|w_j| \le 1 + \eta/2$ $(j = 1, \ldots, m)$ in $\mathbf{C}^m_w$. The theorem is proved by setting

$$\widehat{\Sigma} : \quad w_j = \psi_j(p) \ (j = 1, \ldots, m)$$

in the polydisk $\overline{\Delta}_{\eta/2}$ in $\mathbf{C}^m$. □

2. Subglobal finite pseudobase

Let $\mathcal{V}$ be an analytic space and let $\mathcal{J}^\lambda$ be an $\mathcal{O}^\lambda$-module on $\mathcal{V}$. We say that $\mathcal{J}^\lambda$ has a **subglobal finite pseudobase** in $\mathcal{V}$ if $\mathcal{J}^\lambda$ satisfies the following condition. Let E be an arbitrary compact set in $\mathcal{V}$. Then there exist a finite number of holomorphic vector-valued functions $F_k(p)$ $(k = 1, \ldots, \nu)$ of rank λ on $\mathcal{V}$ such that:

1. Each $F_k(p)$ $(k = 1, \ldots, \nu)$ belongs to $\mathcal{J}^\lambda$ at each point of $\mathcal{V}$.
2. $\mathcal{J}^\lambda$ is generated by $F_k(p)$ $(k = 1, \ldots, \nu)$ on E, i.e., if we let $\mathcal{J}^\lambda\{F\}$ denote the $\mathcal{O}^\lambda$-module generated by $F_k(p)$ $(k = 1, \ldots, \nu)$, then $\mathcal{J}^\lambda\{F\}$ is equivalent to $\mathcal{J}^\lambda$ on E.

Then we have the following theorem.

THEOREM 8.18. *Let $\mathcal{V}$ be a Stein space and let $\mathcal{J}^\lambda$ be an $\mathcal{O}^\lambda$-module on $\mathcal{V}$ which has a locally finite pseudobase at each point in $\mathcal{V}$. Then $\mathcal{J}^\lambda$ has a subglobal finite pseudobase in $\mathcal{V}$.*

To prove this we prove the following lemma on the stability of a pseudobase.

LEMMA 8.7. *Let $\mathcal{P}$ be a closed analytic polyhedron in $\mathcal{V}$ with defining functions in a domain $U \subset \mathcal{V}$. Let $F_j(p)$ $(j = 1, \ldots, \nu)$ be a holomorphic vector-valued function of rank λ on $\mathcal{P}$ and let $\mathcal{J}^\lambda\{F\}$ denote the $\mathcal{O}^\lambda$-module generated by $F_j(p)$ $(j = 1, \ldots, \nu)$ on $\mathcal{P}$. Let $\epsilon > 0$ and let $F_j^*(p)$ $(j = 1, \ldots, \nu)$ be a holomorphic vector-valued function of rank λ on $\mathcal{P}$ such that $F_j^*(p) \in \mathcal{J}^\lambda\{F\}$ at each point of $\mathcal{P}$ and*

$$||F_j^*(p) - F_j(p)|| < \epsilon \quad (j = 1, \ldots, \nu) \qquad \textit{for } p \in \mathcal{P}. \tag{8.14}$$

Then for $\epsilon > 0$ sufficiently small, the $\mathcal{O}^\lambda$-module $\mathcal{J}^\lambda\{F^*\}$ generated by $F^*(p)$ $(j = 1, \ldots, \nu)$ is equivalent to $\mathcal{J}^\lambda\{F\}$ on $\mathcal{P}$.

PROOF. Let $\mathcal{P} : |\varphi_j(p)| \le 1$ $(j = 1, \ldots, m)$ with $\mathcal{P} \subset\subset U$, where $\varphi_j(p)$ $(j = 1, \ldots, m)$ is a holomorphic function on U. We take $\eta > 0$ sufficiently small so that $\mathcal{P}_\eta \subset\subset U$, where $\mathcal{P}_\eta : |\varphi_j(p)| \le 1 + \eta$ $(j = 1, \ldots, m)$ and $F_j^*(p) \in \mathcal{J}^\lambda\{F\}$ $(j = 1, \ldots, \nu)$ at each point of $\mathcal{P}_\eta$. By Theorem 8.4, there exists a holomorphic vector-valued function $A^{(j)}(p) = (A_1^{(j)}(p), \dots, A_\nu^{(j)}(p))$ $(j = 1, \dots, m)$ of rank ν on $\mathcal{P}_\eta$ such that, for $j = 1, \dots, \nu$,

$$F_j^*(p) - F_j(p) = A_1^{(j)}(p)F_1(p) + \dots + A_\nu^{(j)}(p)F_\nu(p), \quad p \in \mathcal{P}_\eta.$$

We may assume that each $A^{(j)}(p)$ on $\mathcal{P}_\eta$ satisfies $||A^{(j)}(p)|| < K\epsilon$ on $\mathcal{P}$ for some constant $K > 0$ (depending on $\mathcal{P}_\eta$ and $\mathcal{P} \subset\subset \mathcal{P}_\eta^\circ$ but not on ϵ) by Theorem 8.16 and (8.14). Therefore, by taking a smaller $\epsilon > 0$ if necessary, we can write $F_j(p)$ $(j = 1, \dots, \nu)$ in the form

$$\begin{aligned} F_j(p) &= B_1^{(j)}(p)F_1^*(p) + \dots + (1 + B_j^{(j)}(p))F_j^*(p) + \dots + B_\nu^{(j)}(p)F_\nu^*(p) \\ &\qquad (j = 1, \dots, \nu), \qquad p \in \mathcal{P}, \end{aligned}$$

where each $B_k^{(j)}(p)$ $(j, k = 1, \dots, \nu)$ is a uniformly small holomorphic function on $\mathcal{P}$. It follows that $\mathcal{J}^\lambda\{F^*\}$ is equivalent to $\mathcal{J}^\lambda\{F\}$ on $\mathcal{P}$. □

PROOF OF THEOREM 8.18. Let $E \subset\subset \mathcal{V}$ be given. We take an analytic polyhedron $\mathcal{P}$ in $\mathcal{V}$ with defining functions on $\mathcal{V}$ such that $E \subset\subset \mathcal{P}^\circ$. By Theorem 8.6, we can find a finite number of holomorphic vector-valued functions $F_j(p)$ $(j = 1, \dots, \nu)$ of rank λ on $\mathcal{P}$ such that the $\mathcal{O}^\lambda$-module $\mathcal{J}^\lambda\{F\}$ generated by the $F_j(p)$ $(j = 1, \dots, \nu)$ is equivalent to $\mathcal{J}^\lambda$ on $\mathcal{P}$. Given $\epsilon > 0$, by Lemma 8.5 we can find a holomorphic vector-valued function $F_j^*(p)$ $(j = 1, \dots, \nu)$ of rank λ on $\mathcal{V}$ such that $F_j^*(p) \in \mathcal{J}^\lambda$ at each point of $\mathcal{P}$ and $||F_j^*(p) - F_j(p)|| < \epsilon$ on $\mathcal{P}$. By Lemma 8.7, for sufficiently small $\epsilon > 0$, the $\mathcal{O}^\lambda$-module $\mathcal{J}^\lambda\{F^*\}$ generated by $F_j^*(p)$ $(j = 1, \dots, \nu)$ on $\mathcal{V}$ is equivalent to $\mathcal{J}^\lambda\{F\}$ on $\mathcal{P}$. It follows that $\mathcal{J}^\lambda$ has a subglobal finite pseudobase in $\mathcal{V}$. □

3. Representation of meromorphic functions

Let $\mathcal{V}$ be a Stein space and let $g(p)$ be a meromorphic function on $\mathcal{V}$. To be precise, $g(p)$ is a single-valued holomorphic function on $\mathcal{V}$ except for at most an analytic hypersurface Σ; and, at any point $q \in \mathcal{V}$, there exist two holomorphic functions $h_q(p)$ and $k_q(p)$ on a neighborhood δ_q of q in $\mathcal{V}$ such that $h_q(p)$ and $k_q(p)$ are relatively prime on δ_q and $g(p) = h_q(p)/k_q(p)$ on δ_q. To be precise, this means that for $q_1, q_2 \in \mathcal{V}$ with $\delta_{q_1} \cap \delta_{q_2} \neq \emptyset$, $h_{q_1}(p)$ $(k_{q_1}(p))$ has the same zero set, counted with multiplicity, as $h_{q_2}(p)$ $(k_{q_2}(p))$ in the sense that both $h_{q_1}(p)/h_{q_2}(p)$ and $k_{q_1}(p)/k_{q_2}(p)$ can be holomorphically extended to non-zero holomorphic functions on $\delta_{q_1} \cap \delta_{q_2}$. Hence the data determined by the denominators $\{(k_q(p), \delta_q)\}_{q \in \mathcal{V}}$ defines a Cousin II distribution $\mathcal{C}$ on $\mathcal{V}$. If the distribution $\mathcal{C}$ admits a solution $K(p)$ of the Cousin II problem on $\mathcal{V}$, then $H(p) = K(p) \cdot g(p)$ is a single-valued holomorphic function on $\mathcal{V}$. It follows that $g(p) = H(p)/K(p)$ on $\mathcal{V}$, where $H(p)$ and $K(p)$ are relatively prime at each point in $\mathcal{V}$ (i.e., this is a solution of the Poincaré problem for $g(p)$). As shown in Chapter 3, the Cousin II problem cannot always be solved, even in a product domain in $\mathbf{C}^2$. However, using Theorem 8.18 regarding $\mathcal{O}$-ideals, we have the following theorem.

THEOREM 8.19. *Any meromorphic function $g(p)$ on a Stein space $\mathcal{V}$ can be represented in the form $g(p) = H(p)/K(p)$ on $\mathcal{V}$, where $H(p)$ and $K(p)$ are holomorphic functions on $\mathcal{V}$ (which are not necessarily relatively prime at each point of $\mathcal{V}$).*

PROOF. We use the notation $h_q(p)$, $k_q(p)$, δ_q $(q \in \mathcal{V})$ associated with $g(p)$. For a fixed point $q \in \mathcal{V}$, we consider the $\mathcal{O}$-ideal $\mathcal{I}_q$ generated by the function $k_q(p)$ on δ_q. If $q_1, q_2 \in \mathcal{V}$ with $\delta_{q_1} \cap \delta_{q_2} \neq \emptyset$, then $\mathcal{I}_{q_1}$ and $\mathcal{I}_{q_2}$ are equivalent on $\delta_{q_1} \cap \delta_{q_2}$ (since $k_{q_2}(p) = (h_{q_2}(p)/h_{q_1}(p))k_{q_1}(p)$ on $\delta_{q_1} \cap \delta_{q_2}$, where $h_{q_2}(p)/h_{q_1}(p)$ is a non-zero holomorphic function on $\delta_{q_1} \cap \delta_{q_2}$). Thus, the collection $\mathcal{I}$ of the the $\mathcal{O}$-ideals $\{\mathcal{I}_q\}_{q \in \mathcal{V}}$ becomes an $\mathcal{O}$-ideal on $\mathcal{V}$ which has a locally finite pseudobase (indeed, one element) at each point in $\mathcal{V}$. We let Σ denote the zero set of the $\mathcal{O}$-ideal $\mathcal{I}$ on $\mathcal{V}$, i.e., Σ consists of the pole set together with the points of indeterminacy of $g(p)$ in $\mathcal{V}$. If $\Sigma = \emptyset$, there is nothing to prove. If $\Sigma \neq \emptyset$, then Theorem 8.18 implies that there exists a holomorphic function $K(p)$ on $\mathcal{V}$ such that $K(p) \not\equiv 0$ on $\mathcal{V}$ and $K(p) \in \mathcal{I}$ at each point of $\mathcal{V}$. Thus, $K(p) = c_q(p) \cdot k_q(p)$ near a point $q \in \mathcal{V}$, where $c_q(p)$ is a holomorphic function in a neighborhood δ'_q of q (where $c_q(p)$ may have zeros in δ'_q). If we set $H(p) = g(p) \cdot K(p)$ on $\mathcal{V}$, then $H(p)$ is a single-valued holomorphic function on $\mathcal{V}$, so that $g(p) = H(p)/K(p)$ on $\mathcal{V}$. □

8.5. Representation of a Stein space

In this section we show that a Stein space $\mathcal{V}$ of dimension n can be realized as an analytic set in $\mathbf{C}^{2n+1}$, and as a distinguished ramified domain over $\mathbf{C}^n$ (this will be defined in 8.5.2). The results in this section are due to E. Bishop [**3**].

8.5.1. Distinguished Analytic Polyhedra. Let $\mathcal{V}$ be an analytic space of dimension n and let $U \subset \mathcal{V}$ be a domain. Let $\mathcal{P}$ be an analytic polyhedron in $\mathcal{V}$ whose defining functions are defined in U; i.e., there exist a finite number of holomorphic functions $\varphi_j(p)$ $(j = 1, \ldots, \nu)$ in U such that $\mathcal{P}$ consists of a finite number of compact, connected components of the set $U_\varphi := \bigcap_{j=1}^{\nu} \{p \in U \mid |\varphi_j(p)| \leq 1\}$. We consider the closed unit polydisk $\overline{\Delta}$ in $\mathbf{C}^\nu$,

$$\overline{\Delta} : \ |z_j| \leq 1 \ (j = 1, \ldots, \nu),$$

and the mapping

$$\Phi : p \in \mathcal{P} \to z = (\varphi_1(p), \ldots, \varphi_\nu(p)) \in \overline{\Delta}.$$

We set $\Sigma = \Phi(\mathcal{P})$, which is an analytic set in $\overline{\Delta}$ with $\partial\Sigma \subset \partial\Delta$. Since $\mathcal{P}$ satisfies the separation condition, $\mathcal{P}$ does not contain any compact analytic set of positive dimension. Thus, $\nu \geq n$ and Σ is of dimension n. Moreover, for each $z \in \Sigma$, $\Phi^{-1}(z)$ consists of a finite number d of points in $\mathcal{P}$, where d is the same for all $z \in \Sigma$ except perhaps for an analytic set of dimension at most $n-1$. If $\nu = n$, we say that $\mathcal{P}$ is a **distinguished analytic polyhedron** in $\mathcal{V}$ (whose defining functions are defined on U). Then $\Sigma = \Delta$, and $\mathcal{P}$ is mapped in a one-to-one fashion onto a finitely sheeted, ramified domain $\mathcal{D}$ over $\overline{\Delta}$ without relative boundary.

By definition, at any point of the analytic space $\mathcal{V}$, there exists a distinguished analytic polyhedron neighborhood V of p in $\mathcal{V}$.

We have the following proposition, which is of fundamental importance in this section.

PROPOSITION 8.4. *Let $\mathcal{V}$ be an analytic space of dimension n and let $U \subset \mathcal{V}$ be a domain. Let $\mathcal{P}$ be an analytic polyhedron in $\mathcal{V}$ whose defining functions are defined on U. Let K be a compact set in $\mathcal{V}$ such that $K \subset\subset \mathcal{P}^o$ (the interior of $\mathcal{P}$ in $\mathcal{V}$) and let W be a domain in $\mathcal{V}$ such that $\mathcal{P} \subset W \subset\subset U$. Then there exists a distinguished analytic polyhedron $\mathcal{Q}$ in $\mathcal{V}$, whose defining functions are defined in U, such that $K \subset\subset \mathcal{Q}^o \subset\subset W$.*

To prove this we need the following two lemmas.

LEMMA 8.8. *Under the same notation as in Proposition 8.4, we write the analytic polyhedron $\mathcal{P}$ as a finite union of compact, connected components of the set*

$$U_\varphi := \bigcap_{j=1}^{\nu} \{p \in U \mid |\varphi_j(p)| \leq 1\},$$

where $\varphi_j(p)$ $(j = 1, \ldots, \nu)$ is a nonconstant holomorphic function in U. We set

$$\sigma = \{p \in U \mid \varphi_1(p) = 0\},$$

which is an analytic hypersurface in U. Assume $\nu \geq n+1$. Then for any $\varepsilon > 0$, there exists a holomorphic function $\psi_j(p)$ $(j = 2, \ldots, \nu)$ on U such that

(i) $|\varphi_j(p) - \psi_j(p)| < \varepsilon \quad (j = 2, \ldots, \nu)$ *on W, and*

(ii) *for any $a = (a_1, \ldots, a_{\nu-1}) \in \mathbf{C}^{\nu-1}$, the set*

$$S_a := \left\{p \in W \setminus \sigma \mid \frac{\psi_{k+1}(p)}{\varphi_1(p)} = a_k \ (k = 1, \ldots, \nu - 1)\right\}$$

consists of at most a finite number of points in W.

PROOF. We consider $\mathbf{C}^{\nu+n}$ with variables $z_1, \ldots, z_\nu$ and $w_1, \ldots, w_n$. Noting that $\nu \geq n+1$, we consider the following set in $\mathbf{C}^{\nu+n}$:

$$\Sigma^* : z_j = \varphi_j(p) \ (j = 1, \ldots, \nu), \ w_k = \frac{\varphi_{k+1}(p)}{\varphi_1(p)} \ (k = 1, \ldots, n), \quad p \in U \setminus \sigma;$$

this is an n-dimensional analytic set in the domain $D \setminus \{z_1 = 0\}$, where

$$D = \{(z, w) \in \mathbf{C}^{\nu+n} : |z_j| < 1 \ (j = 1, ..., \nu), \ w \in \mathbf{C}^n\}.$$ [9]

By Theorem 2.3, there exists a coordinate system $Z' = (z'_1, \ldots, z'_\nu, w'_1, \ldots, w'_n)$ sufficiently close to the original coordinate system $Z = (z_1, \ldots, z_\nu, w_1, \ldots, w_n)$ such that Σ^* satisfies the Weierstrass condition for $(w'_1, \ldots, w'_n)$ at each point of Σ^*; i.e., the projection π_n of Σ^* onto $\mathbf{C}_{w'_1, \ldots, w'_n}$ has the property that for any $a = (a_1, \ldots, a_n) \in \mathbf{C}^n$, the set $\pi_n^{-1}(a)$ is isolated in Σ^*. Here, $Z' = AZ$, where A is a $(\nu+n, \nu+n)$ matrix sufficiently close to the unit matrix $E_{\nu+n,\nu+n}$. This means that if we set $A = (\delta_{i,j} + \varepsilon_{i,j})_{i,j}$, where $\delta_{i,j}$ is the Kronecker delta and $|\varepsilon_{i,j}| << 1$, then the set of points p in U with

$$(*) \begin{cases} \varepsilon_{1,\nu+1}\varphi_1(p) + \cdots + \varepsilon_{\nu,\nu+1}\varphi_\nu(p) + (1 + \varepsilon_{\nu+1,\nu+1})\frac{\varphi_2(p)}{\varphi_1(p)} \\ \qquad + \cdots + \varepsilon_{\nu+n,\nu+n}\frac{\varphi_{n+1}(p)}{\varphi_1(p)} = a_1 \\ \qquad \vdots \qquad \vdots \qquad \vdots \\ \varepsilon_{1,\nu+n}\varphi_1(p) + \cdots + \varepsilon_{\nu,\nu+n}\varphi_\nu(p) + \varepsilon_{\nu+1,\nu+n}\frac{\varphi_2(p)}{\varphi_1(p)} \\ \qquad + \cdots + (1 + \varepsilon_{\nu+n,\nu+n})\frac{\varphi_{n+1}(p)}{\varphi_1(p)} = a_n \end{cases}$$

is isolated in U. Therefore, if we define, for $p \in U$,

$$\psi_j(p) := \varphi_j(p) + \varphi_1(p) \sum_{k=1}^{\nu} \epsilon_{k,\nu+j}\varphi_k(p) + \sum_{l=1}^{n} \epsilon_{\nu+l,\nu+j}\varphi_{l+1}(p) \ (j = 2, \ldots, n+1),$$

$$\psi_j(p) := \varphi_j(p) \ (j = n+2, \ldots, \nu),$$

[9] If we set $\overline{\Sigma}^* : z_j = \varphi_j(p) \ (j = 1, \ldots, \nu), \ \varphi_1(p)w_k = \varphi_{k+1}(p)(k = 1, \ldots, n), \ p \in U$, then $\overline{\Sigma}^*$ ia an n-dimensional analytic set in D and is equal to the closure of Σ^* in D.

then $|\psi_j(p) - \varphi_j(p)| < \varepsilon$ $(j = 2, \dots, \nu)$ on W (for we can choose $\varepsilon_{i,j}$ sufficiently small relative to $\varepsilon > 0$). Equation $(*)$ and the condition $\nu - 1 \geq n$ imply that, given $a = (a_1, \dots, a_n, \dots, a_{\nu-1}) \in \mathbf{C}^{\nu-1}$, the set of points p in U such that $\frac{\psi_{j+1}(p)}{\varphi_1(p)} = a_j$ $(j = 1, \dots, \nu - 1)$ is isolated in U. $\square$

To prove Proposition 8.4, using Lemma 8.8 we may assume that for any $(a_2, \dots, a_\nu) \in \mathbf{C}^{\nu-1} \setminus \sigma$, the analytic set in U defined by

$$\frac{\varphi_j(p)}{\varphi_1(p)} = a_j \quad (j = 2, \dots, \nu) \tag{8.15}$$

has dimension 0.

Given a number $r > 1$ and an integer $N \geq 1$, we set

$$F_k(p) := (r\varphi_1(p))^N - (r\varphi_k(p))^N \quad (k = 2, \dots, \nu),$$

which is a holomorphic function on U, and we set

$$E_{r,N} := \{\, p \in U \ : \ |F_k(p)| \leq 1 \quad (k = 2, \dots, \nu) \,\},$$

so that $E_{r,N}$ is a closed subset of U defined by $\nu - 1$ holomorphic functions in U.

We have the following lemma.

LEMMA 8.9. *Under the same notation as in Proposition 8.4, if $r > 1$ is sufficiently close to 1 and $N = N(r) \geq 1$ is sufficiently large, then there exist a finite number of connected components Q_j of $E_{r,N}$ whose union $\mathcal{Q} = \bigcup Q_j$ satisfies*

$$K \subset\subset \mathcal{Q}^o \subset\subset W, \tag{8.16}$$

where $\mathcal{Q}^o$ denotes the interior of $\mathcal{Q}$ in $\mathcal{V}$. Then $\mathcal{Q}$ is an analytic polyhedron in U which satisfies condition (8.16) and is defined by $\nu - 1$ holomorphic functions $F_k(p)$ $(k = 2, \dots, \nu)$ in U.

PROOF. We fix a domain V with smooth boundary ∂V in W such that

$$\mathcal{P} \subset\subset V \subset\subset W.$$

Since K is a compact set in $\mathcal{P}^o$, we can choose $r > 1$ sufficiently close to 1 so that $|r\varphi_j(p)| < 1$ $(j = 1, \dots, \nu)$ on K. Therefore, there exists an integer N_0 such that $K \subset\subset E^o_{r,N}$ (the interior of $E_{r,N}$) for all $N \geq N_0$. We let $\mathcal{Q}_{r,N}$ denote the smallest union of connected components of $E_{r,N}$ which contains K. To prove the lemma, it suffices to show that

$$\mathcal{Q}_{r,N} \subset V \quad \text{for sufficiently large } N. \tag{8.17}$$

We prove this by contradiction; thus we assume there exist an infinite number of integers $N \geq N_0$ such that $\mathcal{Q}_{r,N} \not\subset V$. For simplicity we write $E_{r,N} = E_N$ and $\mathcal{Q}_{r,N} = \mathcal{Q}_N$. For such N, since $K \subset \mathcal{Q}^o_N \cap \mathcal{P}^o$ and $\mathcal{P} \subset V$, there exists a (connected) real 1-dimensional arc γ in $(\mathcal{Q}_N \cap V) \setminus \mathcal{P}^o$ which connects a point $p'_N \in \partial\mathcal{P}$ to a point $p''_N \in \partial V$.

Fix $q \in \gamma$. Since $q \notin \mathcal{P}^o$, we have $|\varphi_j(q)| \geq 1$ for some j $(1 \leq j \leq \nu)$. Using the fact that $\gamma \subset \mathcal{Q}_N$, it follows that

$$|r\varphi_1(q)|^N \geq |r\varphi_j(q)|^N - 1 \geq r^N - 1 > 8\pi N$$

(for the last inequality is true if $N = N(r)$ is sufficiently large). Since $|F_k(q)| \leq 1$ $(k = 2, \dots, \nu)$, we obtain

$$\left| 1 - \left(\frac{\varphi_k(q)}{\varphi_1(q)} \right)^N \right| = \frac{|F_k(q)|}{|r\varphi_1(q)|^N} \leq \frac{1}{|r\varphi_1(q)|^N} \leq \frac{1}{8\pi N}. \tag{8.18}$$

In particular, setting $\delta = \{p \in U \mid |\varphi_1| < 1/2\}$, which contains σ, we have $q \in U \setminus \delta$ (for $|\varphi_j(p)| \geq 1$). We define

$$\Omega_N := \{t \in \mathbf{C}_t \mid |1 - t^N| < 1/8\pi N\},$$

which consists of N mutually disjoint sets ω_l^N $(l = 1, \ldots, N)$ about each Nth-root ε_l^N of unity. Fix $k \in \{2, \ldots, \nu\}$. Then inequality (8.18) implies that $\frac{\varphi_k(q)}{\varphi_1(q)} \in \omega_l^N$ for some l $(1 \leq l \leq N)$. Since $q \in \gamma$ is arbitrary and $\frac{\varphi_k}{\varphi_1}(\gamma)$ is connected, it follows that $\frac{\varphi_k}{\varphi_1}(\gamma) \subset \omega_l^N$ and $\varphi_1(\gamma) \subset U \setminus \delta$, where l depends only on γ and k. By taking a subsequence of such N, if necessary, we can assume that ω_l^N approaches a point t_k with $|t_k| = 1$ as $N \to \infty$. Consequently, there exist infinitely many connected real 1-dimensional arcs γ_N (independent of $k = 2, \ldots, \nu$) which connect a point $p'_N \in \partial\mathcal{P}$ and a point $p''_N \in \partial V$ in $V \setminus \mathcal{P}^0$ such that $\frac{\varphi_k}{\varphi_1}(\gamma_N) \to t_k$ $(k = 2, \ldots, \nu)$ in $\mathbf{C}_t$ as $N \to \infty$. Thus we can find a continuum Γ in $(\overline{V} \setminus \mathcal{P}^0) \cap (U \setminus \delta)$ which connects a point of $\partial\mathcal{P}$ and a point of ∂V such that $\frac{\varphi_k}{\varphi_1}(\Gamma) = t_k$ $(k = 2, \ldots, \nu)$. This contradicts (8.15), and (8.17) is proved. □

PROOF OF PROPOSITION 8.4. If we repeat Lemmas 8.8 and 8.9 $(\nu - n)$ times, then we obtain Proposition 8.4. □

Using Proposition 8.4 we obtain the following proposition.

PROPOSITION 8.5. *Let $\mathcal{V}$ be a Stein space. Then there exists a sequence of distinguished analytic polyhedra $\mathcal{P}_n$ $(n = 1, 2, \ldots)$ in $\mathcal{V}$ whose defining functions are defined in $\mathcal{V}$ and such that*

$$\mathcal{P}_k \subset\subset \mathcal{P}_{k+1}^o \quad (k = 1, 2, \ldots), \qquad \mathcal{V} = \lim_{k \to \infty} \mathcal{P}_k. \tag{8.19}$$

PROOF. We first take a sequence of analytic polyhedra $\mathcal{Q}_k$ $(k = 1, 2, \ldots)$ in $\mathcal{V}$ satisfying condition (8.19) whose defining functions are defined in $\mathcal{V}$. By Proposition 8.4, there exists a distinguished analytic polyhedron $\mathcal{R}_k$ $(k = 1, 2, \ldots)$ in $\mathcal{V}$ whose defining functions are defined in $\mathcal{Q}_{k+1}$ and such that $\mathcal{Q}_k \subset\subset \mathcal{R}_k^o \subset\subset \mathcal{Q}_{k+1}$. Since each pair $(\mathcal{R}_k, \mathcal{Q}_{k+1}^o)$ and $(\mathcal{Q}_{k+1}, \mathcal{V})$ satisfies the Runge theorem, we can find a distinguished analytic polyhedron $\mathcal{P}_k$ in $\mathcal{V}$ whose defining functions are defined in $\mathcal{V}$ and such that $\mathcal{Q}_k \subset\subset \mathcal{P}_k^o \subset\subset \mathcal{Q}_{k+1}$. Thus $\mathcal{P}_k$ $(k = 1, 2, \ldots)$ satisfies the conclusion of the proposition. □

8.5.2. Distinguished Ramified Domains. Let $\mathcal{D}$ be a ramified domain over $\mathbf{C}^n$ and let $\pi : \mathcal{D} \to \mathbf{C}^n$ be the projection map. If, for any compact set K in $\mathbf{C}^n$, each connected component of $\pi^{-1}(K)$ is compact in $\mathcal{D}$, then we say that $\mathcal{D}$ is a **distinguished ramified domain** over $\mathbf{C}^n$. It is clear that a distinguished ramified domain $\mathcal{D}$ over $\mathbf{C}^n$ is a Stein space if $\mathcal{D}$ satisfies the separation condition. Indeed, from Theorem 9.3 in Chapter 9 we shall see that any distinguished ramified domain is a Stein space. Conversely, we have the following theorem.

THEOREM 8.20. *Let $\mathcal{V}$ be a Stein space of dimension n. Then $\mathcal{V}$ is holomorphically isomorphic to a distinguished ramified domain $\mathcal{D}$ over $\mathbf{C}^n$.*

PROOF. By Proposition 8.19 we can find a sequence of distinguished analytic polyhedra $\mathcal{P}_k$ $(k = 0, 1, \ldots)$ in $\mathcal{V}$ such that

$$\mathcal{P}_k \subset\subset \mathcal{P}_{k+1}^o \quad (k = 0, 1, \ldots), \qquad \mathcal{V} = \lim_{k \to \infty} \mathcal{P}_k,$$

where each $\mathcal{P}_k$ $(k = 0, 1, \dots)$ can be described as a finite union of compact connected components in $\mathcal{V}$ of the set

$$\mathcal{V}_{\varphi_k} := \bigcap_{j=1}^{n} \{p \in \mathcal{V} \mid |\varphi_j^{(k)}(p)| \leq 1\},$$

where $\varphi_j^{(k)}(p)$ $(j = 1, \dots, n)$ is a nonconstant holomorphic function on $\mathcal{V}$.

Choose $\epsilon_k > 0$ $(k = 1, 2, \dots)$ so that $\sum_{k=1}^{\infty} \epsilon_k < 1$ and p_k $(k = 1, 2, \dots)$ so that $p_1 < p_2 < \dots$ and $\lim_{n\to\infty} p_n = \infty$. We set

$$M_0 = \max_{j=1,\dots,n} \{|\varphi_j^{(0)}(p)| \mid p \in \mathcal{P}_1\} > 0,$$

and we set $c_1 = p_1 + M_0 > 0$. We can then choose an integer N_1 such that

$$\left| c_1 \left(\varphi_j^{(1)}(p)\right)^{N_1} \right| < \epsilon_1 \quad (j = 1, \dots, n) \quad \text{on } \mathcal{P}_0,$$

since $|\varphi_j^{(1)}(p)| < 1$ on the compact set $\mathcal{P}_0$ in $\mathcal{P}_1^o$.

Consequently, if for $j = 1, \dots, n$ we define

$$\psi_j^{(1)}(p) := \varphi_j^{(0)}(p) + c_1 \left(\varphi_j^{(1)}(p)\right)^{N_1} \qquad \text{on } \mathcal{V},$$

then $\psi_j^{(1)}(p)$ is a holomorphic function on $\mathcal{V}$ which satisfies

$$|\psi_j^{(1)}(p) - \varphi_j^{(0)}(p)| < \epsilon_1 \qquad (j = 1, \dots, n) \quad \text{on } \mathcal{P}_0.$$

Furthermore,

$$|\psi_1^{(1)}(p)| + \dots + |\psi_n^{(1)}(p)| \geq p_1 \quad \text{on } \partial\mathcal{P}_1.$$

To see this, let $q \in \partial\mathcal{P}_1$. Then $|\varphi_j^{(1)}(q)| = 1$ for some j $(1 \leq j \leq n)$, so that

$$|\psi_j^{(1)}(q)| \;\; \geq \;\; \left| c_1 \left(\varphi_j^{(1)}(q)\right)^{N_1} \right| \; - \; |\varphi_j^{(0)}(q)| \geq c_1 - M_0 = p_1,$$

which proves the above inequality on $\partial\mathcal{P}_1$.

We repeat the same procedure for $\psi_j^{(1)}(p)$ $(j = 1, \dots, n)$ that we used for $\varphi_j^{(0)}(p)$ $(j = 1, \dots, n)$ to obtain a holomorphic function $\psi_j^{(2)}(p)$ $(j = 1, \dots, n)$ of the form $\psi_j^{(1)}(p) + c_2 \left(\varphi_j^{(2)}(p)\right)^{N_2}$ such that

$$\begin{aligned} &|\psi_j^{(2)}(p) - \psi_j^{(1)}(p)| < \epsilon_2 \qquad (j = 1, \dots, n) \qquad &&\text{on } \mathcal{P}_1, \\ &|\psi_1^{(2)}(p)| + \dots + |\psi_n^{(2)}(p)| \geq p_2 &&\text{on } \partial\mathcal{P}_2. \end{aligned}$$

We thus inductively obtain a sequence of holomorphic functions $\{\psi_j^{(k)}(p)\}_{k=0,1,\dots}$ $(j = 1, \dots, n)$ (where we set $\psi_j^{(0)}(p) = \varphi_j^{(0)}(p)$ $(j = 1, \dots, n)$) of the form $\psi_j^{(k+1)}(p)$ $= \psi_j^{(k)}(p) + c_{k+1} \left(\varphi_j^{(k+1)}(p)\right)^{N_{k+1}}$ $(j = 1, \dots, n)$ and such that

$$\begin{aligned} &|\psi_j^{(k+1)}(p) - \psi_j^{(k)}(p)| < \epsilon_{k+1} \qquad (j = 1, \dots, n) \qquad &&\text{on } \mathcal{P}_k \\ &|\psi_1^{(k+1)}(p)| + \dots + |\psi_n^{(k+1)}(p)| \geq p_{k+1} &&\text{on } \partial\mathcal{P}_{k+1}. \end{aligned}$$

We define

$$H_j(p) = \varphi_j^{(0)}(p) + \sum_{k=0}^{\infty} (\psi_j^{(k+1)}(p) - \psi_j^{(k)}(p)) \;\; (j = 1, \dots, n) \quad \text{on } \mathcal{V}.$$

Since this sum converges uniformly on each compact set in $\mathcal{V}$, it follows that $H_j(p)$ $(j = 1, \ldots, n)$ is a holomorphic function on $\mathcal{V}$. Moreover, if we fix $p \in \partial\mathcal{P}_l$ $(l = 1, 2, \ldots)$, then we have

$$\begin{aligned}
&|H_1(p)| + \cdots + |H_n(p)| \\
&\geq |\psi_1^{(l)}(p)| + \cdots + |\psi_n^{(l)}(p)| - \sum_{j=1}^{n}\left(\sum_{k=l}^{\infty} |\psi_j^{(k+1)}(p) - \psi_j^{(k)}(p)|\right) \\
&\geq p_l - \sum_{j=1}^{n}\left(\sum_{k=l}^{\infty} \epsilon_{k+1}\right) \geq p_l - n.
\end{aligned}$$

Thus

$$|H_j(p)| \geq \frac{p_l}{n} - 1 \quad \text{for some } j \ (1 \leq j \leq n), \tag{8.20}$$

where j depends on $p \in \partial\mathcal{P}_l$.

Consider the holomorphic mapping

$$\Phi : \ p \in \mathcal{V} \to z = (H_1(p), \ldots, H_n(p)).$$

Then Φ maps $\mathcal{V}$ bijectively onto a ramified domain $\mathcal{D} = \Phi(\mathcal{V})$ over $\mathbf{C}^n$. We shall show that $\mathcal{D}$ is a distinguished ramified domain over $\mathbf{C}^n$.

To see this, let K be a compact set in $\mathbf{C}^n$ and fix a polydisk $Q: \ |z_j| < R$ $(j = 1, \ldots, n)$ such that $K \subset\subset Q$. We choose an integer $l_0 \geq 1$ such that $p_{l_0}/n - 1 > R$. Then (8.20) implies that $\Phi^{-1}(K) \cap \partial\mathcal{P}_l = \emptyset$ for $l \geq l_0$. Hence, each component $\tilde{K}$ of $\Phi^{-1}(K)$ in $\mathcal{V}$ is contained in $\mathcal{P}_{l_0}^o$ or in $\mathcal{P}_{l'+1} \setminus \mathcal{P}_{l'}^o$ for some $l' \geq l_0$ (which depends on $\tilde{K}$). Thus $\tilde{K}$ is compact in $\mathcal{V}$. Hence, $\mathcal{D}$ is a distinguished ramified domain over $\mathbf{C}^n$. □

8.5.3. Imbedding of a Stein Space. Any n-dimensional analytic set Σ in $\mathbf{C}^N$ $(N \geq n)$ can be regarded in a canonical manner as a Stein space $\mathcal{V}$ on which the holomorphic functions correspond to the weakly holomorphic functions on Σ. Conversely, any Stein space of dimension n can be represented as an n-dimensional irreducible analytic set in $\mathbf{C}^N$, where $N = 2n + 1$.[10] We prove this by first using Theorem 8.20 to prove the following theorem.

THEOREM 8.21. *Any Stein space $\mathcal{V}$ of dimension n can be mapped holomorphically onto an n-dimensional analytic set Σ in $\mathbf{C}^{n+1}$ in a one-to-one manner, except perhaps for an at most $(n-1)$-dimensional analytic set in $\mathcal{V}$.*

PROOF. Using Theorem 8.20, we can find n holomorphic functions $\phi_j(p)$ $(j = 1, \ldots, n)$ on $\mathcal{V}$ such that the mapping

$$\Phi : \ z_j = \phi_j(p) \qquad (j = 1, \ldots, n), \ \ p \in \mathcal{V},$$

gives a bijection from $\mathcal{V}$ onto a distinguished ramified domain $\mathcal{D}$ over $\mathbf{C}^n$. We let π denote the projection from $\mathcal{D}$ onto $\mathbf{C}^n$ and we write O for the origin in $\mathbf{C}^n$. We set

$$\pi^{-1}(O) = \{p_i\}_{i=1,2,\ldots} \subset \mathcal{D}. \tag{8.21}$$

[10]The imbedding of a Stein space was first studied by R. Remmert (see R. Narasimhan [**36**]).

Let r_k $(k = 1, 2, \dots)$ be a sequence of positive numbers such that $r_k < r_{k+1}$ $(k = 1, 2, \dots)$ and $\lim_{k\to\infty} r_k = \infty$, and consider the sequence of polydisks Δ_k in $\mathbf{C}^n$ defined as

$$\Delta_k : \ |z_j| \leq r_k \quad (j = 1, \dots, n;\ k = 1, 2, \dots).$$

We set $\tilde{W}_k = \pi^{-1}(\Delta_k) \subset \mathcal{V}$; in general, this set consists of an infinite number of compact, connected components. We let W_k denote the connected component of $\tilde{W}_k$ which contains the point p_1. Then $W_{k+1} \cap \tilde{W}_k$ consists of a finite number of connected components of $\tilde{W}_k$ which includes W_k. We set $H_k = W_{k+1} \cap (\tilde{W}_k \setminus W_k)$ and $R_k = W_{k+1} \setminus \tilde{W}_k$, so that

$$W_{k+1} = W_k \cup H_k \cup R_k \quad \text{and} \quad \pi(R_k) \cap \overline{\Delta}_k = \emptyset.$$

We note that W_k, H_k and the disjoint union $W_k \cup H_k$ are analytic polyhedra in $\mathcal{V}$ whose defining functions are defined in $\mathcal{V}$. We also set $\pi^{-1}(O) \cap W_k = \{p_j^{(k)}\}_{j=1,\dots,l_k}$.

Choose $\epsilon_k > 0$ $(k = 1, 2, \dots)$ with $\sum_{k=1}^{\infty} \epsilon_k < 1$ and choose $a_k > 3$ $(k = 1, 2, \dots)$ so that $a_{k+1} > a_k$ and $\lim_{k\to\infty} a_k = \infty$. Since $\mathcal{V}$ satisfies the separation condition, there exists a holomorphic function $f_1(p)$ on $\mathcal{V}$ such that

$$f_1(p_i^{(1)}) \neq f_1(p_j^{(1)}) \qquad (i, j = 1, \dots, l_1;\ i \neq j).$$

We set

$$m_1 = \min\{|f_1(p_i^{(1)}) - f_1(p_j^{(1)})| \mid i, j = 1, \dots, l_1;\ i \neq j\} > 0.$$

Next we construct a holomorphic function $f_2(p)$ on $\mathcal{V}$ such that

1. $|f_2(p) - f_1(p)| < \epsilon_1 \min\{1/2, m_1/2\}$ on W_1;
2. $|f_2(p)| > a_1 + \max_{p\in W_1}\{|f_1(p)|\}$ on H_1;
3. $f_2(p_i^{(2)}) \neq f_2(p_j^{(2)})$ $(i, j = 1, \dots, l_2;\ i \neq j)$.

To do this, we fix positive numbers M and δ with $\max_{p\in W_1}\{|f_1(p)|\} + a_1 + 1 < M$ and $0 < \delta < 1$. Since the union $U_1 := W_1 \cup H_1$ is an analytic polyhedron in $\mathcal{V}$ whose defining functions are defined in $\mathcal{V}$, it follows that the pair $(U_1, \mathcal{V})$ satisfies Runge's theorem. Noting that W_1 and H_1 are closed sets in $\mathcal{V}$ such that $W_1 \cap H_1 = \emptyset$, we can find a holomorphic function $\tilde{f}_2(p)$ on $\mathcal{V}$ such that $|\tilde{f}_2(p) - f_1(p)| < \delta$ on W_1 and $|\tilde{f}_2(p) - M| < \delta$ on H_1. Hence, $|\tilde{f}_2(p)| > \max_{p\in W_1}\{|f_1(p)|\} + a_1$ on H_1. By taking $\delta > 0$ sufficiently small, we see that $\tilde{f}_2(p)$ satisfies conditions 1 and 2. Since $\mathcal{V}$ satisfies the separation condition, we can find a holomorphic function $k(p)$ on $\mathcal{V}$ such that $k(p_i^{(2)}) \neq k(p_j^{(2)})$ $(i, j = 1, \dots, l_2;\ i \neq j)$. Hence, for $\epsilon > 0$ sufficiently small, $f_2(p) := \tilde{f}_2(p) + \epsilon k(p)$ on $\mathcal{V}$ satisfies conditions 1, 2, and 3.

We inductively construct a sequence of holomorphic functions $f_k(p)$ $(k = 1, 2, \dots)$ on $\mathcal{V}$ and a sequence of positive numbers m_k $(k = 1, 2, \dots)$,

$$m_k = \min\{|f_k(p_i^{(k)}) - f_k(p_j^{(k)})| \mid i, j = 1, \dots, l_k;\ i \neq j\ \} > 0,$$

such that

1. $|f_{k+1}(p) - f_k(p)| < \epsilon_k \min\{1/2, m_1/2, \dots, m_k/2\ \}$ on W_k;
2. $|f_{k+1}(p)| > a_k + \max_{p\in W_k}\{|f_k(p)|\}$ on H_k;
3. $f_{k+1}(p_i^{(k+1)}) \neq f_{k+1}(p_j^{(k+1)})$ $(i, j = 1, \dots, l_{k+1};\ i \neq j)$.

We set

$$F(p) = f_1(p) + \sum_{k=1}^{\infty}(f_{k+1}(p) - f_k(p)), \quad p \in \mathcal{V}.$$

Condition 1 implies that this sum converges uniformly on each compact set in $\mathcal{V}$, so that $F(p)$ is a holomorphic function on $\mathcal{V}$. On W_{k+1} ($k = 1, 2, \dots$), we have

$$|F(p) - f_{k+1}(p)| \leq \sum_{\mu=k+1}^{\infty} |f_{\mu+1}(p) - f_\mu(p)| < \sum_{\mu=k+1}^{\infty} \epsilon_\mu < 1. \tag{8.22}$$

In particular,

$$\begin{aligned} |F(p)| &\geq |f_{k+1}(p)| - 1 \quad \text{on } W_{k+1}, \\ |F(p)| &\geq a_k + \max_{p \in W_k}\{|f_k(p)|\} - 1 \quad \text{on } H_k. \end{aligned} \tag{8.23}$$

For $k = 1, 2, \dots$, we also have

$$\begin{aligned} &|F(p_i^{(k)}) - F(p_j^{(k)})| \\ &\geq |f_k(p_i^{(k)}) - f_k(p_j^{(k)})| - \sum_{\mu=k}^{\infty}(|f_{\mu+1}(p_i^{(k)}) - f_\mu(p_i^{(k)})| + |f_{\mu+1}(p_j^{(k)}) - f_\mu(p_j^{(k)})|) \\ &\geq m_k \, (1 - \sum_{\mu=k}^{\infty} \epsilon_\mu) > 0 \end{aligned}$$

for $i, j = 1, \dots, l_k$; $i \neq j$. It follows that

$$F(p_i) \neq F(p_j) \qquad (i, j = 0, 1, \dots ; \ i \neq j). \tag{8.24}$$

Now consider the following holomorphic mapping $\mathbf{F}$ from $\mathcal{V}$ into $\mathbf{C}^{n+1} = \mathbf{C}_z^n \times \mathbf{C}_w$:

$$\mathbf{F}: \ p \to (z_1, \dots, z_n, w) = (\phi_1(p), \dots, \phi_n(p), F(p)) \in \mathbf{C}^{n+1},$$

and set $\Sigma = \mathbf{F}(\mathcal{V})$ in $\mathbf{C}^{n+1}$. We shall show that $\mathbf{F}$ and Σ satisfy the conclusion of the theorem.

To this end, using (8.24) and (8.21) it suffices to show that Σ is an analytic set in $\mathbf{C}^{n+1}$; i.e., Σ is closed in $\mathbf{C}^{n+1}$. Equivalently, if we set

$$L_k = \min\left\{ \sum_{j=1}^{n} |\phi_j(p)| \ + \ |F(p)| \mid p \in W_{k+1} \setminus W_k \right\} \quad (k = 1, 2, \dots),$$

then it suffices to show that $\lim_{k \to \infty} L_k = +\infty$.

To see this, fix $p \in R_k$. Then there exists j with $1 \leq j \leq n$ such that $|\phi_j(p)| \geq r_k$. Furthermore, (8.23) implies that $|F(p)| \geq a_k - 1$ on H_k. Since $W_{k+1} \setminus W_k = R_k \cup H_k$, it follows that $L_k \geq \min\{r_k, a_k - 1\} \to +\infty$ as $k \to +\infty$. Hence, Σ is an n-dimensional analytic set in $\mathbf{C}^{n+1}$. □

REMARK 8.9. The analytic set $\Sigma = \mathbf{F}(\mathcal{V})$ in $\mathbf{C}^{n+1}$ from the above proof has the following property: Let $M_k = \max_{p \in W_k} \{|F(p)|\}$, $\Gamma_k = \{|w| < M_k\} \subset \mathbf{C}_w$, and $\Lambda_k = \overline{\Delta}_k \times \Gamma_k \ \subset \mathbf{C}^{n+1}$. Then $\Sigma \cap \Lambda_k = \mathbf{F}(W_k)$.

Indeed, for each $l \geq k$, we have, from (8.23) and condition 1 on the sequence $\{f_k\}_k$,

$$\begin{aligned} |F(p)| &> a_l + \max_{p\in W_l}\{|f_l(p)|\} - 1 \quad \text{on } H_l \\ &> M_l + 1 > M_k. \end{aligned}$$

Since $\pi^{-1}(\overline{\Delta}_k) \setminus W_k \subset \bigcup_{l=k}^{\infty} H_l$, it follows that

$$\min\ \{|F(p)| \mid p \in \pi^{-1}(\overline{\Delta}_k) \setminus W_k\ \} > M_k,$$

which proves the remark.

We also obtain the following theorems.

THEOREM 8.22. *Any Stein space $\mathcal{V}$ of dimension n is holomorphically isomorphic to an n-dimensional analytic set Σ in $\mathbf{C}^{2n+1}$.*

PROOF. We use Theorem 8.21 and maintain the notation from its proof. Let $\mathcal{S}$ be the set of points $p \in \mathcal{V}$ such that there exists at least one point $q \in \mathcal{V}$ with $p \neq q$ and $\mathbf{F}(p) = \mathbf{F}(q)$. The set $\mathcal{S}$ in $\mathcal{V}$ is an analytic set of dimension at most $n-1$. To see this, fix $p_0 \in \mathcal{S}$ and let $(z^0, w^0) = F(p_0) \in \Sigma$. From Theorem 8.21 there exist only a finite number of points $p_j \in \mathcal{V}$ $(j = 1, \dots, m)$ such that $F(p_j) = (z^0, w^0)$ and $p_j \neq p_0$. If we take a small neighborhood δ of (z^0, w^0) in $\mathbf{C}^{n+1}$, then the open set $F^{-1}(\delta \cap \Sigma)$ in $\mathcal{V}$ consists of $(m+1)$ connected components $v_i \subset \mathcal{V}$ $(i = 0, 1, \dots, m)$ such that $p_i \in v_i$. Since $\sigma := \bigcup_{j=1}^{m}[F(v_0) \cap F(v_j)]$ is an analytic set in δ whose dimension is at most $n-1$, the same is true of the analytic set $\tau := F^{-1}(\sigma) \cap v_0$ in v_0. Since $\tau = \mathcal{S} \cap v_0$, we have our desired conclusion.

We set $\Sigma^{(n-1)} = \mathbf{F}(\mathcal{S})$, which is an analytic set of dimension at most $n-1$ in $\mathbf{C}^{n+1}$. We let $\mathcal{S}^{(n-1)}$ denote the family of $(n-1)$-dimensional irreducible analytic sets in $\mathcal{S}$, say $\mathcal{S}^{(n-1)} = \{S_j\}_{j=1,2,\dots}$. We let $\mathcal{S}_1$ denote the collection of S_j such that $W_1 \cap S_j \neq \emptyset$; this is a finite collection of sets. We inductively define $\mathcal{S}_{k+1}$ $(k = 1, 2, \dots)$ as the collection of all S_j such that $W_{k+1} \cap S_j \neq \emptyset$, so that $\mathcal{S}_k \subset \mathcal{S}_{k+1}$ and $\mathcal{S}^{(n-1)} = \lim_{k\to\infty} \mathcal{S}_k$. Note that each $\mathcal{S}_k$ is a finite collection of sets S_j.

For the sake of convenience, we rename the collection of sets S_j in $\mathcal{S}_k$:

$$\mathcal{S}_k = \{S_{k,j}\}_{j=1,\dots,l_k};$$

here $0 \leq l_k < \infty$ and $S_{k+1,j} = S_{k,j}$ for $j = 1, \dots, l_k$.

On each $\mathcal{S}_k$ $(k = 1, 2, \dots)$, we fix a point $p_j^{(k)} \in S_{k,j} \cap W_k$ $(j = 1, \dots, l_k)$ such that $p_j^{(k)} = p_j^{(k+1)}$ $(j = 1, \dots, l_k)$ and such that $\Sigma^{(n-1)}$ is nonsingular at the point $\mathbf{F}(p_j^{(k)})$. We consider all points $q_{j,s}^{(k)} \in \mathcal{V}$ $(s = 1, 2, \dots)$ such that $p_j^{(k)} \neq q_{j,s}^{(k)}$ and $\mathbf{F}(p_j^{(k)}) = \mathbf{F}(q_{j,s}^{(k)})$. Since $p_j^{(k)} \in W_k$, it follows from Remark 8.9 that all points $q_{j,s}^{(k)}$ are contained in W_k and hence there are only finitely many such points, say $q_{j,s}^{(k)} \in W_k$ $(s = 1, \dots, s_j^{(k)})$, where $s_j^{(k)} < \infty$.

Let $\epsilon_k > 0$ $(k = 1, 2, \dots)$ with $\epsilon_k > \epsilon_{k+1}$ and $\sum_{k=1}^{\infty} \epsilon_k < 1$. Since $\mathcal{V}$ satisfies the separation condition, we can find a holomorphic function $g_1(p)$ on $\mathcal{V}$ such that

$$g_1(p_j^{(1)}) \neq g_1(q_{j,s}^{(1)}) \quad (j = 1, \dots, l_1;\ s = 1, \dots, s_j^{(1)}).$$

We set

$$m_1 = \min_{\substack{j=1,\dots,l_1 \\ s=1,\dots,s_j^{(1)}}} \{|g_1(p_j^{(1)}) - g_1(q_{j,s}^{(1)})|\} > 0.$$

We next construct a holomorphic function $g_2(p)$ on $\mathcal{V}$ such that

1. $|g_2(p) - g_1(p)| < \epsilon_1 \min\{1, m_1/2\}$ on W_1;
2. $g_2(p_j^{(2)}) \neq g_2(q_{j,s}^{(2)})$ $(j = 1, \ldots, l_2;\ s = 1, \ldots, s_j^{(2)})$.

To do this, since $\mathcal{V}$ satisfies the separation condition, we first find a holomorphic function $h_2(p)$ on $\mathcal{V}$ such that $h_2(p_j^{(2)}) \neq h_2(q_{j,s}^{(2)})$ $(j = 1, \ldots, l_2;\ s = 1, \ldots, s_j^{(2)})$. If we set $g_2(p) = g_1(p) + \epsilon h_2(p)$ on $\mathcal{V}$, then $g_2(p)$ satisfies conditions 1 and 2 provided ϵ is sufficiently small.

We inductively obtain a sequence of holomorphic functions $g_k(p)$ $(k = 1, 2, \ldots)$ on $\mathcal{V}$ and a sequence of positive numbers m_k $(k = 1, 2, \ldots)$ such that

1. $|g_{k+1}(p) - g_k(p)| < \epsilon_k \min\{1/2, m_1/2, \ldots, m_k/2\}$ on W_k;
2. $g_{k+1}(p_j^{(k+1)}) \neq g_{k+1}(q_{j,s}^{(k+1)})$ $(j = 1, \ldots, l_{k+1};\ s = 1, \ldots, s_j^{(k+1)})$;
3. $m_{k+1} = \min\limits_{\substack{j=1,\ldots,l_{k+1} \\ s=1,\ldots,s_j^{(k+1)}}} \{|g_{k+1}(p_j^{(k+1)}) - g_{k+1}(q_{j,s}^{(k+1)})|\} > 0.$

Next we set

$$G_1(p) = g_1(p) + \sum_{k=1}^{\infty} (g_{k+1}(p) - g_k(p)), \quad p \in \mathcal{V}.$$

By condition 1, $G_1(p)$ is a holomorphic function on $\mathcal{V}$. Furthermore, by condition 1 we have

$$\begin{aligned}
&|G_1(p_j^{(k)}) - G_1(q_{j,s}^{(k)})| \\
&\geq |g_k(p_j^{(k)}) - g_k(q_{j,s}^{(k)})| - \sum_{\mu=k}^{\infty}(|g_{\mu+1}(p_j^{(k)}) - g_\mu(p_j^{(k)})| + |g_{\mu+1}(q_{j,s}^{(k)}) - g_\mu(q_{j,s}^{(k)})|) \\
&\geq m_k\ (1 - \sum_{\mu=k}^{\infty} \epsilon_\mu) > 0 \qquad (8.25)
\end{aligned}$$

for all k, j, s. We consider the holomorphic mapping

$$\begin{aligned}
\mathbf{G}_1 : p \in \mathcal{V} \quad &\to \quad (z_1, \ldots, z_n, w_1, w_2) \\
&= \quad (\phi_1(p), \ldots, \phi_n(p), F(p), G_1(p)) \in \mathbf{C}^{n+2},
\end{aligned}$$

and we let $\Sigma_1 := \mathbf{G}_1(\mathcal{V})$, which is an n-dimensional analytic set in $\mathbf{C}^{n+2}$. Then $\mathcal{V}$ and Σ_1 are in one-to-one correspondence except perhaps for the analytic sets $\mathcal{S}^{(n-2)}$ of dimension at most $n-2$ in $\mathcal{V}$. To see this, note that (8.25) implies $p_j^{(k)} \notin \mathcal{S}^{(n-2)}$ $(k = 1, 2, \ldots;\ j = 1, \ldots, l_k)$, so that $\mathcal{S}^{(n-2)}$ does not contain the irreducible component $S_{k,j}$. Since $\mathcal{S}^{(n-1)} = \bigcup_{k,j} S_{k,j}$, it follows that dim $\mathcal{S}^{(n-2)} \leq n-2$.

We repeat the same procedure on $\mathcal{S}^{(n-2)}$ as we performed on $\mathcal{S}^{(n-1)}$ to obtain a holomorphic function $G_2(p)$ on $\mathcal{V}$ such that the mapping

$$\begin{aligned}
\mathbf{G}_2 : p \in \mathcal{V} \quad &\to \quad (z_1, \ldots, z_n, w_1, w_2, w_3) \\
&= \quad (\phi_1(p), \ldots, \phi_n(p), F(p), G_1(p), G_2(p)) \in \mathbf{C}^{n+3}
\end{aligned}$$

gives a one-to-one correspondence between $\mathcal{V}$ and the analytic set $\Sigma_2 = \mathbf{G}_2(\mathcal{V})$ in $\mathbf{C}^{n+3}$ except perhaps for the analytic set $\mathcal{S}^{(n-3)}$ in $\mathcal{V}$ which has dimension at most $n-3$.

After n repetitions of this procedure, we finally achieve the conclusion of the theorem. □

For the next theorem concerning Stein manifolds we need the following lemma.

LEMMA 8.10. *Let D be a domain in $\mathbf{C}^n$ and let*

$$\mathbf{G}: \ z \in D \to w = (g_1(z), \dots, g_{n+l}(z)) \in \mathbf{C}^{n+l}$$

(where $l \geq 1$) be a holomorphic mapping from D into $\mathbf{C}^{n+l}$. For an integer i with $0 \leq i \leq n$, we consider the following analytic set $\mathcal{E}^{(i)}$ in D:

$$\mathcal{E}^{(i)} = \left\{ z_0 \in D \mid \operatorname{rank} \left(\frac{\partial(g_1, \dots, g_{n+l})}{\partial(z_1, \dots, z_n)} \right)_{z_0} \leq i \right\}.$$

Assume that for any $a \in \mathbf{C}^{n+l}$, the set $\mathbf{G}^{-1}(a)$ is either empty or is an isolated set in D. Then $\dim\ \mathcal{E}^{(i)} \leq i$.

PROOF. We set $k = \dim\ \mathcal{E}^{(i)}$ and we need to show that $k \leq i$. Let p_0 be a nonsingular point of $\mathcal{E}^{(i)}$. We may assume that $D = \Delta^n : \ |z_j| < 1 \ (j = 1, \dots, n)$ and $\mathcal{E}^{(i)} = \Delta^k \times \{O\}$ where $\Delta^k := |z_j| < 1 \ (j = 1, \dots, k)$ and O is the origin in $\mathbf{C}^{n-k}_{z_{k+1}, \dots, z_n}$. We set $g_j|_{\mathcal{E}^{(i)}} = G_j(z_1, \dots, z_k)$ $(j = 1, \dots, n+l)$. Since for any $z = (z_1, \dots, z_k, 0, \dots, 0) = (z', 0, \dots, 0) \in \mathcal{E}^{(i)}$ we have

$$\operatorname{rank} \left(\frac{\partial(g_1, \dots, g_{n+l})}{\partial(z_1, \dots, z_n)} \right)_z \geq \operatorname{rank} \left(\frac{\partial(G_1, \dots, G_{n+l})}{\partial(z_1, \dots, z_k)} \right)_{z'},$$

it suffices to prove that $\partial(G_1, \dots, G_{n+l})/\partial(z_1, \dots, z_k)$ is of rank k at some point $z' \in \Delta^k$.

We consider the holomorphic mapping

$$\tilde{\mathbf{G}}: \ z' \in \Delta^k \to w = (G_1(z'), \dots, G_{n+l}(z')) \in \mathbf{C}^{n+l}$$

and set $\Sigma = \tilde{\mathbf{G}}(\Delta^k)$. For each $a \in \mathbf{C}^{n+l}$, since the set $\mathbf{G}^{-1}(a)$ is either empty or is an isolated set in Δ^n, the same is true of $\tilde{\mathbf{G}}^{-1}(a)$ in Δ^k. Hence Σ is a k-dimensional analytic set in a domain in $\mathbf{C}^{n+l}$. If we project Σ to a suitable k-dimensional hyperplane L in $\mathbf{C}^{n+l}$, say $\pi : \Sigma \to L$ and $L : w_j = 0$ $(j = k+1, \dots, n+l)$, then $\pi(\Sigma)$ is a ramified domain over $\mathbf{C}^k_{w_1, \dots, w_k}$ with branch set $\mathcal{S}$ of dimension at most $k-1$. Thus, $\det\left(\partial(G_{j_1}, \dots, G_{j_k})/\partial(z_1, \dots, z_k)\right) \not\equiv 0$ on Δ^k for some $(j_1, \dots, j_k)$, which proves the result. □

We study the mapping

$$\mathbf{F}: \ \mathcal{V} \to (z_1, \dots, z_n, w) = (\phi_1(p), \dots, \phi_n(p), F(p)) \in \mathbf{C}^{n+1}$$

defined in the proof of Theorem 8.21 in the case when $\mathcal{V}$ is a complex manifold of dimension n. By Remark 8.9, $\mathbf{F}$ in each Λ_k satisfies the condition in Lemma 8.10. The contrapositive of the lemma yields the following fact: on an analytic set σ of dimension r $(0 \leq r \leq n)$ in $\mathcal{V}$, the matrix $\partial(\phi_1, \dots, \phi_n, F)/\partial(\zeta_1, \dots, \zeta_n)$ is of rank at least r on σ except for an analytic set $\sigma' \subset \sigma$ of dimension $r-1$, where $(\zeta_1, \dots, \zeta_n)$ are local coordinates on the complex manifold $\mathcal{V}$.[11]

Then we have the following imbedding theorem for Stein manifolds.

THEOREM 8.23. *A Stein manifold $\mathcal{V}$ of dimension n is holomorphically isomorphic to an n-dimensional non-singular analytic set Σ in $\mathbf{C}^{2n+1}$.*

[11] By convention, an analytic set of negative dimension is the empty set.

PROOF. We maintain the notation used in the proof of Theorems 8.21 and 8.22. The proof is similar to that of Theorem 8.22; we add the rank condition to the separation condition as follows.

We set $\phi_{n+1}(p) = F(p)$ in $\mathcal{V}$. For $\alpha = 0, 1, \ldots, n-1$, we set

$$\mathcal{E}_0^{(\alpha)} = \left\{ p \in \mathcal{V} \mid \operatorname{rank} \left(\frac{\partial(\phi_1, \ldots, \phi_{n+1})}{\partial(\zeta_1, \ldots, \zeta_n)} \right)_p = \alpha \right\},$$

and we let $\mathcal{E}^{(\alpha)} = Cl\,[\mathcal{E}_0^{\alpha}]$ denote the closure of $\mathcal{E}_0^{(\alpha)}$ in $\mathcal{V}$. Then $\mathcal{E}^{(\alpha)}$ is an analytic set in $\mathcal{V}$. We note by the definition of rank that $\mathcal{E}_0^{(0)} = \mathcal{E}^{(0)}$ and $\partial\mathcal{E}^{(\alpha)} \subset \bigcup_{j=0}^{\alpha-1} \mathcal{E}^j$ ($\alpha = 1, \ldots, n-1$), so that $\mathcal{E}^{(\alpha)} \subset \bigcup_{k=0}^{\alpha} \mathcal{E}_0^{(k)}$. Lemma 8.10 implies that

$$d_\alpha := \dim \mathcal{E}^{(\alpha)} \le \alpha. \tag{8.26}$$

First step. There exists a holomorphic function $\phi_{n+2}(p)$ on $\mathcal{V}$ such that the mapping

$$\mathbf{G}: \ p \in \mathcal{V} \to (z_1, \ldots, z_n, w_1, w_2) = (\phi_1(p), \ldots, \phi_n(p), \phi_{n+1}(p), \phi_{n+2}(p)) \in \mathbf{C}^{n+2}$$

from $\mathcal{V}$ into $\mathbf{C}^{n+2}$ satisfies the following conditions:

(1) If we set $\Sigma = \mathbf{G}(\mathcal{V})$, then $\mathcal{V}$ and Σ are one-to-one except for perhaps an analytic set of dimension at most $n-2$ in $\mathcal{V}$.
(2) If we put, for each $\alpha = 0, 1, \ldots, n-1$,

$$\begin{aligned} \mathcal{F}_0^{(\alpha)} &= \left\{ p \in \mathcal{V} \mid \operatorname{rank} \left(\frac{\partial(\phi_1, \ldots, \phi_{n+1}, \phi_{n+2})}{\partial(\zeta_1, \ldots, \zeta_n)} \right)_p = \alpha \right\} \\ \mathcal{F}^{(\alpha)} &= Cl\,[\mathcal{F}_0^{(\alpha)}], \end{aligned}$$

then $\dim \mathcal{F}^{(\alpha)} \le \alpha - 1$.

To prove this, for each $\alpha = 0, 1, \ldots, n-1$, we begin by setting $\mathcal{E}^{(\alpha)} = \bigcup_{l=1}^{\infty} E_l^{(\alpha)}$, where $E_l^{(\alpha)}$ ($l = 1, 2, \ldots$) is an irreducible component of $\mathcal{E}^{(\alpha)}$. We let $\mathcal{E}^{(\alpha,1)}$ denote the collection of sets $E_l^{(\alpha)}$ such that $W_1 \cap E_l^{(\alpha)} \neq \emptyset$, where W_1 is defined in the proof of Theorem 8.21. This is a finite collection of sets.

We inductively define $\mathcal{E}^{(\alpha,k+1)}$ ($k = 1, 2, \ldots$) to be the collection of sets $E_l^{(\alpha)}$ such that $W_{k+1} \cap E_l^{(\alpha)} \neq \emptyset$ and $E_l^{(\alpha)} \notin \mathcal{E}^{(\alpha,1)} \cup \ldots \cup \mathcal{E}^{(\alpha,k)}$. Again, this is a finite collection of sets. Thus, $\mathcal{E}^{(\alpha)} = \bigcup_{k=1}^{\infty} \mathcal{E}^{(\alpha,k)}$, and this is a disjoint union except for analytic sets of dimension at most $\alpha - 1$. For convenience, we rename the collection of sets $E_l^{(\alpha)}$ in $\mathcal{E}^{(\alpha,k)}$,

$$\mathcal{E}^{(\alpha,k)} = \{E_i^{(\alpha,k)}\}_{i=1,\ldots,m^{(\alpha,k)}},$$

where $0 \le m^{(\alpha,k)} < \infty$.

We first take $k = 1$. In the proof of Theorem 8.22, we chose a point $p_j^{(1)} \in S_{1,j} \cap W_1$ ($j = 1, \ldots, l_1$) and $q_{j,s}^{(1)} \in S_{1,j} \cap W_1$ ($j = 1, \ldots, l_1$; $s = 1, \ldots, s_j^{(1)}$). Now choose a point $\tilde{p}_i^{(\alpha,1)} \in E_i^{(\alpha,1)} \cap W_1$ ($i = 1, \ldots, m^{(\alpha,1)}$) for each $\alpha = 0, 1, \ldots, n-1$

such that

$$p_j^{(1)} \neq \tilde{p}_i^{(\alpha,1)} \quad (j=1,\dots,l_1;\ i=1,\dots,m^{(\alpha,1)}),$$
$$\tilde{p}_i^{(\alpha,1)} \neq \tilde{p}_j^{(\alpha,1)} \quad (i,j=1,\dots,m^{(\alpha,1)},\ i\neq j),$$
$$\operatorname{rank}\left(\frac{\partial(\phi_1,\dots,\phi_{n+1})}{\partial(\zeta_1,\dots,\zeta_n)}\right)_{\tilde{p}_i^{(\alpha,1)}} = \alpha \quad (i=1,\dots,m^{(\alpha,1)}).$$

Fix $\epsilon_k > 0$ $(k=1,2,\dots)$ with $\epsilon_k > \epsilon_{k+1}$ and $\sum_{k=1}^{\infty}\epsilon_k < 1$. Since $\mathcal{V}$ satisfies the separation condition, using Remark 8.7 we obtain a holomorphic function $g_1(p)$ on $\mathcal{V}$ such that

$$g_1(p_j^{(1)}) \neq g_1(q_{j,s}^{(1)}) \qquad (j=1,\dots,l_1;\ s_1=1,\dots,s_j^{(1)}),$$
$$\operatorname{rank}\left(\frac{\partial(\phi_1,\dots,\phi_{n+1},g_1)}{\partial(\zeta_1,\dots,\zeta_n)}\right)_{\tilde{p}_i^{(\alpha,1)}} = \alpha+1 \quad (\alpha=0,1,\dots,n-1;\ i=1,\dots,m^{(\alpha,1)}).$$

We thus have

$$\tilde{a}_i^{(\alpha,1)} := \det\left(\frac{\partial(\phi_{i_1},\dots,\phi_{i_\alpha},g_1)}{\partial(\zeta_{j_1},\dots,\zeta_{j_{\alpha+1}})}\right)_{\tilde{p}_i^{(\alpha,1)}} \neq 0 \quad (i=1,\dots,m^{(\alpha,1)}),$$

where $(i_1,\dots,i_\alpha)$ are α distinct numbers in $(1,\dots,n)$ and $(j_1,\dots,j_{\alpha+1})$ are $\alpha+1$ distinct numbers in $(1,\dots,n)$ which depend on $\tilde{p}_i^{(\alpha,1)}$. We set

$$m_1 = \min_{\substack{j=1,\dots,l_1\\ s=1,\dots,s_j^{(1)}\\ \alpha=0,1,\dots,n-1\\ i=1,\dots,m^{(\alpha,1)}}} \{|g_1(p_j^{(1)}) - g_1(q_{j,s}^{(1)})|,\ |\tilde{a}_i^{(\alpha,1)}|\ \} > 0.$$

We next construct a holomorphic function $g_2(p)$ on $\mathcal{V}$ such that

1. $|g_1(p) - g_2(p)| < \epsilon_1 \min\{1, m_1/2\}$ on W_1;
2. if we set
$$\delta^{(1)} = \max_{\substack{(\mu_1,\dots,\mu_\alpha),\\ (i_1,\dots,i_{\alpha+1}),}} \left\{ \left|\det\left(\frac{\partial(\phi_{\mu_1},\dots,\phi_{\mu_\alpha},g_1-g_2)}{\partial(\zeta_{i_1},\dots,\zeta_{i_{\alpha+1}})}\right)_{\tilde{p}_i^{(\alpha,1)}}\right| \right\} \geq 0,$$
where $(\mu_1,\dots,\mu_\alpha)$ runs over all increasing α-tuples in $(1,\dots,n+1)$; $(i_1,\dots,i_{\alpha+1})$ runs over all increasing $(\alpha+1)$-tuples in $(1,\dots,n)$; $\alpha=0,1,\dots,n-1$; $i=1,\dots,m^{(\alpha,1)}$, then we have $\delta^{(1)} < \epsilon_1\min\{1,m_1\}$;
3. $g_2(p_j^{(\nu)}) \neq g_2(q_{j,s}^{(\nu)})$ $(\nu=1,2;\ j=1,\dots,l_2;\ s=1,\dots,s_j^{(2)})$;

$$\operatorname{rank}\left(\frac{\partial(\phi_1,\dots,\phi_{n+1},g_2)}{\partial(\zeta_1,\dots,\zeta_n)}\right)_{\tilde{p}_i^{(\alpha,2)}} = \alpha+1 \quad (\alpha=0,1,\dots,n-1;\ i=1,\dots,m^{(\alpha,2)}).$$

To do this, just as we constructed $g_1(p)$ on $\mathcal{V}$, we construct a holomorphic function $h_2(p)$ on $\mathcal{V}$ which satisfies condition 3. If we set $g_2(p) = g_1(p) + \epsilon h_2(p)$ on $\mathcal{V}$, then $g_2(p)$ satisfies conditions 1, 2, and 3 for sufficiently small ϵ.

By condition 3 we have

$$\tilde{a}_i^{(\alpha,2)} := \det\left(\frac{\partial(\phi_1,\dots,\phi_{i_\alpha},g_2)}{\partial(\zeta_{j_1},\dots,\zeta_{j_\alpha+1})}\right)_{\tilde{p}_i^{(\alpha,2)}} \neq 0 \quad (\alpha=0,1,\dots,n-1;\ i=1,\dots,m^{(\alpha,2)}),$$

where $(i_1, \dots, i_\alpha)$ is an increasing α-tuple in $(1, \dots, n+1)$ and $(j_1, \dots, j_{\alpha+1})$ is an increasing $(\alpha+1)$-tuple in $(1, \dots, n)$ which depends on $\tilde{p}_i^{(\alpha,2)}$ $(\alpha = 0, 1, \dots, n-1;\ i = 1, \dots, m^{(\alpha,2)})$. We set

$$m_2 = \min\{\ |g_2(p_j^{(2)}) - g_2(q_{j,s}^{(2)})|,\ |\tilde{a}_i^{(\alpha,2)}|\ \} > 0,$$

where $j = 1, \dots, l_2;\ s = 1, \dots, s_j^{(2)};\ \alpha = 0, 1, \dots, n-1;$ and $i = 1, \dots, m^{(\alpha,2)}$.

We inductively obtain a sequence of holomorphic functions $g_k(p)$ $(k = 1, 2, \dots)$ on $\mathcal{V}$ and a sequence of positive numbers m_k $(k = 1, 2, \dots)$ such that

1. $|g_k(p) - g_{k+1}(p)| < \epsilon_k \min\{1, m_1/2, \dots, m_k/2\}$ on W_k;
2. if we set
$$\delta^{(k)} = \max\left\{ \left| \det\left(\frac{\partial(\phi_{\mu_1}, \dots, \phi_{\mu_\alpha}, g_k - g_{k+1})}{\partial(\zeta_{i_1}, \dots, \zeta_{i_\alpha+1})} \right)_{\tilde{p}_i^{(\alpha,k)}} \right| \right\} \geq 0,$$
where $(\mu_1, \dots, \mu_\alpha)$ runs over all increasing α-tuples in $(1, \dots, n+1)$; $(i_1, \dots, i_{\alpha+1})$ runs over all increasing $(\alpha+1)$-tuples in $(1, \dots, n)$; $\alpha = 0, 1, \dots, n-1;\ i = 1, \dots, m^{(\alpha,k)}$, then we have $\delta^{(k)} < \epsilon_k \min\{1, m_1, \dots, m_k\}$;
3. $g_{k+1}(p_j^{(\nu)}) \neq g_{k+1}(q_{j,s}^{(\nu)})$ $(\nu = 1, \dots, k+1;\ j = 1, \dots, l_\nu;\ s = 1, \dots, s_j^{(\nu)})$,
$$\operatorname{rank}\left(\frac{\partial(\phi_1, \dots, \phi_{n+1}, g_{k+1})}{\partial(\zeta_1, \dots, \zeta_n)} \right)_{p_i^{(\alpha,k+1)}} = \alpha + 1$$
$$(\alpha = 0, 1, \dots, n-1;\ i = 1, \dots, m^{(\alpha,k+1)});$$
4. by condition 3, if we set
$$\tilde{a}_i^{(\alpha,k+1)} := \det\left(\frac{\partial(\phi_{i_1}, \dots, \phi_{i_\alpha}, g_{k+1})}{\partial(\zeta_{j_1}, \dots, \zeta_{j_{\alpha+1}})} \right)_{\tilde{p}_i^{(\alpha,k+1)}} \neq 0 \qquad (i = 1, \dots, m^{(\alpha,k+1)}),$$
where $(i_1, \dots, i_\alpha)$ is some increasing α-tuple in $(1, \dots, n+1)$ and $(j_1, \dots, j_{\alpha+1})$ is some increasing $(\alpha+1)$-tuple in $(1, \dots, n)$ which depends on $\tilde{p}_i^{(\alpha,k+1)}$, then
$$m_{k+1} := \min\{|g_{k+1}(p_j^{(k+1)}) - g_{k+1}(q_{j,s}^{(k+1)})|,\ |\tilde{a}_i^{(\alpha,k+1)}|\ \} > 0,$$
where $j = 1, \dots, l_{k+1};\ s = 1, \dots, s_j^{(k+1)};\ \alpha = 0, 1, \dots, n-1;$ and $i = 1, \dots, m^{(\alpha,k+1)}$.

We define

$$G_1(p) = g_1(p) + \sum_{\mu=1}^{\infty} (g_{\mu+1}(p) - g_\mu(p)), \quad p \in \mathcal{V}.$$

Then condition 1 implies that $G_1(p)$ is a holomorphic function on $\mathcal{V}$. As already shown in the proof of Theorem 8.22, we have

$$G_1(p_j^{(k)}) \neq G_1(q_{j,s}^{(k)}) \quad \text{for all } k, j, s. \tag{8.27}$$

Furthermore, for any point $\tilde{p}_i^{(\alpha,k)}$,

$$\begin{aligned}
&\left|\det\left(\frac{\partial(\phi_{i_1},\dots,\phi_{i_\alpha},G_1)}{\partial(\zeta_{j_1},\dots,\zeta_{j_{\alpha+1}})}\right)_{\tilde{p}_i^{(\alpha,k)}}\right| \\
&\quad\geq \left|\det\left(\frac{\partial(\phi_{i_1},\dots,\phi_{i_\alpha},g_k)}{\partial(\zeta_{j_1},\dots,\zeta_{j_{\alpha+1}})}\right)_{\tilde{p}_i^{(\alpha,k)}}\right| \\
&\quad\quad -\sum_{\mu=k}^{\infty}\left|\det\left(\frac{\partial(\phi_{i_1},\dots,\phi_{i_\alpha},g_{\mu+1}-g_\mu)}{\partial(\zeta_{j_1},\dots,\zeta_{j_{\alpha+1}})}\right)_{\tilde{p}_i^{(\alpha,k)}}\right| \\
&\quad\geq m_k\left(1-\sum_{\mu=k}^{\infty}\epsilon_\mu\right)>0,
\end{aligned}$$

so that

$$\operatorname{rank}\left(\frac{\partial(\phi_1,\dots,\phi_{n+1},G_1)}{\partial(\zeta_1,\dots,\zeta_n)}\right)=\alpha+1 \quad \text{at all points } \tilde{p}_i^{(\alpha,k)}, \tag{8.28}$$

where $\alpha=0,1,\dots,n-1$; $k=1,2,\dots$; $i=1,\dots,m^{(\alpha,k)}$.

Next we consider the holomorphic mapping

$$\mathbf{G}_1:\ p\in\mathcal{V}\to(z_1,\dots,z_n,w_1,w_2)=(\phi_1(p),\dots,\phi_n(p),\phi_{n+1}(p),G_1(p))\in\mathbf{C}^{n+2},$$

and set $\Sigma_1=\mathbf{G}_1(\mathcal{V})$ in $\mathbf{C}^{n+2}$. Formula (8.27) implies that $\mathcal{V}$ and Σ_1 are in one-to-one correspondence via $\mathbf{G}_1$ except for an analytic set $\mathcal{S}^{(n-2)}$ of dimension at most $n-2$. Further, for each $\alpha=0,1,\dots,n-1$, if we set

$$\begin{aligned}
\mathcal{F}_0^{(\alpha)}:&=\left\{p\in\mathcal{V}\mid \operatorname{rank}\left(\frac{\partial(\phi_1,\dots,\phi_{n+1},G_1)}{\partial(\zeta_1,\dots,\zeta_n)}\right)_p=\alpha\right\},\\
\mathcal{F}^{(\alpha)}&=Cl\,[\mathcal{F}_0^{(\alpha)}]\quad\text{in } \mathcal{V},
\end{aligned}$$

then $\mathcal{F}^{(\alpha)}$ is an analytic set in $\mathcal{V}$ of dimension at most $\alpha-1$. In fact, it is clear that $\mathcal{F}^{(\alpha)}\subset\mathcal{E}^{(\alpha)}$. Let $\mathcal{E}_j^{(\alpha)}$ be any irreducible component of $\mathcal{E}^{(\alpha)}$ and let $d_{\alpha,j}=\dim\mathcal{E}_j^{(\alpha)}$, so that $d_{\alpha,j}\leq\alpha$ by (8.26). On the other hand, formula (8.28) implies that

$$\operatorname{rank}\left(\frac{\partial(\phi_1,\dots,\phi_{n+1},G_1)}{\partial(\zeta_1,\dots,\zeta_n)}\right)_p=\alpha+1$$

for all $p\in\mathcal{E}_j^{(\alpha)}$ except for an analytic set $e_j^{(\alpha)}$ of dimension at most $d_{\alpha,j}-1$ ($\leq\alpha-1$). Therefore, $\mathcal{F}^{(\alpha)}\subset\bigcup_{j=1}^{\infty}e_j^{(\alpha)}$, and $\dim\mathcal{F}^{(\alpha)}\leq\alpha-1$.

Thus, by setting $\phi_{n+2}(p)=G_1(p)$, we complete the first step.

We have dim $\mathcal{F}^{(\alpha)}\leq\alpha-1$. In particular, if we set $\alpha=0$, then $\mathcal{F}^{(0)}=\emptyset$, i.e.,

$$\left\{p\in\mathcal{V}\mid \operatorname{rank}\left(\frac{\partial(\phi_1,\dots,\phi_{n+1},\phi_{n+2})}{\partial(\zeta_1,\dots,\zeta_n)}\right)_p=0\right\}=\emptyset. \tag{8.29}$$

Second step. If we repeat the same procedure for $\mathcal{F}^{(\alpha)}$ ($\alpha=1,2,\dots,n-1$) as we performed for $\mathcal{E}^{(\alpha)}$ ($\alpha=0,1,\dots,n-1$), then we obtain a holomorphic function $\phi_{n+3}(p)$ on $\mathcal{V}$ with the following property: if we set

$$\mathbf{G}_2:\ p\in\mathcal{V}\to(z_1,\dots,z_n,w_1,w_2,w_3)=(\phi_1(p),\dots,\phi_{n+3})\in\mathbf{C}^{n+3}$$

and $\Sigma_2 = \mathbf{G}_2(\mathcal{V})$ in $\mathbf{C}^{n+3}$, then $\mathcal{V}$ and Σ_2 are in one-to-one correspondence via $\mathbf{G}_2$ except for an analytic set $\mathcal{S}^{(n-3)}$ of dimension at most $n-3$. Moreover, for each $\alpha = 1, \ldots, n-1$, if we set

$$\mathcal{G}_0^{(\alpha)} := \left\{ p \in \mathcal{V} \mid \operatorname{rank} \left(\frac{\partial(\phi_1, \ldots, \phi_{n+3})}{\partial(\zeta_1, \ldots, \zeta_n)} \right)_p = \alpha \right\},$$
$$\mathcal{G}^{(\alpha)} = Cl\,[\mathcal{G}_0^{(\alpha)}] \quad \text{in } \mathcal{V},$$

then $\mathcal{G}^{(\alpha)}$ is an analytic set in $\mathcal{V}$ of dimension at most $\alpha - 2$. From (2) in the first step we have dim $\mathcal{F}^{(\alpha)} \leq \alpha - 2$. In particular, if we set $\alpha = 1$, then this inequality and (8.29) imply that

$$\left\{ p \in \mathcal{V} \mid \operatorname{rank} \left(\frac{\partial(\phi_1, \ldots, \phi_{n+1}, \phi_{n+2}, \phi_{n+3})}{\partial(\zeta_1, \ldots, \zeta_n)} \right)_p = 0 \text{ or } 1 \right\} = \emptyset.$$

Third step. We repeat the same procedure n times to obtain n holomorphic functions $\phi_{n+2}, \ldots, \phi_{2n+1}(p)$ on $\mathcal{V}$ such that, if we put

$$\mathbf{G}_n \;:\; p \in \mathcal{V} \to (z_1, \ldots, z_n, w_1, \ldots, w_{n+1}) = (\phi_1(p), \ldots, \phi_{2n+1}(p)) \in \mathbf{C}^{2n+1}$$

and $\Sigma_n = \mathbf{G}_n(\mathcal{V})$ in $\mathbf{C}^{2n+1}$, then $\mathcal{V}$ and Σ_n are in one-to-one correspondance via $\mathbf{G}_n$ and

$$\bigcup_{\alpha=0}^{n-1} \left\{ p \in \mathcal{V} \mid \operatorname{rank} \left(\frac{\partial(\phi_1, \ldots, \phi_{2n+1})}{\partial(\zeta_1, \ldots, \zeta_n)} \right)_p = \alpha \right\} = \emptyset.$$

This completes the proof of the theorem. □

8.6. Appendix

We shall prove the Hilbert-Rückert Nullstellensatz for holomorphic functions. This proof follows Oka [**51**].

THEOREM 8.24 (Nullstellensatz). *Let $F_j(z)$ $(j = 1, \ldots, \nu)$ be holomorphic functions defined on a neighborhood Δ of the origin O in $\mathbf{C}_z^n$ and let Σ denote the common zero set of the F_j $(j = 1, \ldots, \nu)$ in Δ. Let $\mathcal{I}\{\mathbf{F}\}$ denote the $\mathcal{O}$-ideal generated by F_j $(j = 1, \ldots, \nu)$ in Δ. If $f(z)$ is a holomorphic function defined in a neighborhood $\delta \subset \Delta$ of O in $\mathbf{C}_z^n$ with $f(z) = 0$ on $\delta \cap \Sigma$, then there exists a positive integer ρ with $f^\rho \in \mathcal{I}\{\mathbf{F}\}$ at O.*

PROOF. We let $r(\Sigma) \geq 0$ denote the dimension of the analytic set Σ at O, i.e., $r(\Sigma)$ is the maximum dimension at O of the irreducible components of Σ passing through O. The proof will be by induction on $r(\Sigma)$ (and is independent of the dimension n of $\mathbf{C}_z^n$).

We first prove that the theorem is valid if $r(\Sigma) = 0$. Let F_j, Δ, Σ and f be given as in the statement of the theorem with $r(\Sigma) = 0$. By taking a smaller neighborhood δ, if necessary, we may assume that $\delta = \delta_1 \times \cdots \times \delta_n \subset \Delta$ where δ_i $(i = 1, \ldots, n)$ is a disk in the complex plane $\mathbf{C}_{z_i}$ centered at the origin $z_i = 0$ and that $\delta \cap \Sigma = \{O\}$. Fix $i \in \{1, \ldots, n\}$ and set $\delta^* := \delta_1 \times \cdots \times \widehat{\delta_i} \times \cdots \times \delta_n$, where $\widehat{\delta_i}$ means that δ_i is omitted. We consider the projection ideal $\mathcal{P}_i$ of $\mathcal{I}\{\mathbf{F}\}$ on δ onto the disk δ_i in $\mathbf{C}_{z_i}$. Since $\Sigma \cap (\delta^* \times \partial\delta_i) = \emptyset$, it follows from Theorem 7.9 that $\mathcal{P}_i$ has a locally finite pseudobase $\varphi^{(k)}(z_i)$ $(k = 1, \ldots, \nu_i)$ on a neighborhood $\delta_i' \subset \delta_i$ of the origin $z_i = 0$ in $\mathbf{C}_{z_i}$. We note that the projection of $\Sigma \cap \delta$ onto δ_i consists of the origin $z_i = 0$ and that the common zero set of $\varphi^{(k)}(z_i)$ $(k = 1, \ldots, \nu_i)$ on δ_i' equals

$\{0\}$ in $\mathbf{C}_{z_i}$. We thus have $\varphi^{(k)}(z_i) = \alpha^{(k)}(z_i) z_i^{l_{k,i}}$ $(k = 1, \dots, \nu_i)$, where $\alpha^{(k)}(z_i)$ is a holomorphic function on δ'_i with $\alpha^{(k)}(0) \neq 0$ and $l_{k,i}$ is a positive integer. Setting $l_i := \min_{k=1,\dots,\nu_i} l_{k,i}$, we see that $\mathcal{P}_i$ is generated by $z_i^{l_i}$ in δ'_i. By the definition of the projection ideal $\mathcal{P}_i$ of $\mathcal{I}\{\mathbf{F}\}$ we see that

$$z_i^{l_i} \in \mathcal{I}\{\mathbf{F}\} \qquad \text{at each point in } \delta^* \times \{z_i = 0\} \subset \mathbf{C}^n.$$

We set $l = \max_{i=1,\dots,n}\{l_i\} \geq 1$, so that $z_i^l \in \mathcal{I}\{\mathbf{F}\}$ $(i = r+1, \dots, n)$ at the origin O in $\mathbf{C}^n$. On the other hand, since $f(O) = 0$, we can write

$$f(z) = A_1(z) z_1 + \cdots + A_n(z) z_n \qquad \text{on } \delta.$$

It follows that $f(z)^{nl} \in \mathcal{I}\{\mathbf{F}\}$ at O, which proves the theorem if $r(\Sigma) = 0$.

Now let $r \geq 1$ be an integer and assume that the theorem holds for Σ with $r(\Sigma) \leq r - 1$. Let F_j, Δ, Σ and f be given as in the statement of the theorem with $r(\Sigma) = r$. We claim that it suffices to prove the result under the assumption that Σ is of *pure* dimension r at the origin O in $\mathbf{C}^n$.

For assume the theorem is valid for any Σ with pure dimension r at O. Given a general Σ with $r(\Sigma) = r$, we have a decomposition of Σ in Δ of the form

$$\Sigma = \Sigma_0 \cup \cdots \cup \Sigma_r,$$

where Σ_j $(j = 0, 1, \dots, r)$ is a pure j-dimensional analytic set in Δ (possibly empty). As usual, we may need to take a smaller neighborhood Δ about O in $\mathbf{C}^n$ to achieve this. For each $j = 0, \dots, r-1$, we can find holomorphic functions $F_k^{(j)}(z)$ $(k = 1, \dots, \nu_j)$ in Δ whose common zero set in Δ equals $\bigcup_{k=0}^{j} \Sigma_k$. We introduce r new variables $y_1, \dots, y_r$ and consider the common zero set $\widetilde{\Sigma}$ in $\widetilde{\Delta} := \Delta \times \mathbf{C}_y^r$ of the $\nu + \nu_{r-1} + \ldots + \nu_0$ holomorphic functions

$$F_1, \dots, F_\nu,\ y_1 F_1^{(r-1)}, \dots, y_1 F_{\nu_{r-1}}^{(r-1)}, \dots, y_r F_1^{(0)}, \dots, y_r F_{\nu_0}^{(0)}.$$

Then the analytic set $\widetilde{\Sigma}$ in $\widetilde{\Delta}$ is identical with the lifting of the second kind of the analytic set Σ in Δ, and is of pure dimension r in $\widetilde{\Delta}$. We let $\widetilde{\mathcal{J}}$ denote the $\mathcal{O}$-ideal generated by these functions in $\widetilde{\Delta}$. Since $f(z) = 0$ on $\widetilde{\Sigma}$ (we regard f as being independent of the variables $y_1, \dots, y_r$), from the hypothesis that $f \in \widetilde{\mathcal{J}}$ at the origin O in $\mathbf{C}_z^n \times \mathbf{C}_y^r$, we can write

$$f = \sum_{i=1}^{\nu} \alpha_j F_i + \sum_{k=1}^{r} \sum_{j=1}^{\nu_j} \beta_j^{(k)} y_k F_j^{(r-k)},$$

where $\alpha_i, \beta_j^{(k)}$ are holomorphic functions in a neighborhood of O in $\mathbf{C}_z^n \times \mathbf{C}_y^r$. Restricting this equation to $y_1 = \cdots = y_r = 0$, we see that $f \in \mathcal{I}\{\mathbf{F}\}$ at O in $\mathbf{C}_z^n$. Thus, the theorem is valid if Σ is not necessarily pure r-dimensional at the origin O in $\mathbf{C}^n$.

Thus we can now assume that $\Sigma : F_j(z) = 0$ $(j = 1, \dots, \nu)$ is of pure dimension r at O in $\mathbf{C}^n$. We can choose coordinates

$$(z', z'') := (z_1, \dots, z_r, z_{r+1}, \dots, z_n),$$

where $z' = (z_1, \dots, z_r)$ and $z'' = (z_{r+1}, \dots, z_n)$, and a polydisk $\Delta = \Delta^{(r)} \times \Gamma \subset \mathbf{C}_{z'}^r \times \mathbf{C}_{z''}^{n-r}$ centered at the origin O such that $\Sigma \cap [\Delta^{(r)} \times \partial\Gamma] = \emptyset$. We set $\Gamma = \Gamma_{r+1} \times \cdots \times \Gamma_n$ where $\Gamma_i :\ |z_i| \leq r_i$ $(i = r+1, \dots, n)$. Fix $i \in \{r+1, \dots, n\}$. We set

$$\Lambda_i = \Delta^{(r)} \times \Gamma_i \qquad \text{and} \qquad \Gamma^* = \Gamma_{r+1} \times \cdots \times \widehat{\Gamma_i} \times \cdots \times \Gamma_n.$$

Since $\Sigma \cap [\Lambda_i \times (\partial \Gamma^*)] = \emptyset$, we see from Theorem 7.9 that the projection ideal $\mathcal{P}_i$ of $\mathcal{I}\{\mathbf{F}\}$ onto Λ_i has a finite pseudobase

$$\varphi_k^{(i)}(z', z_i) \quad (k = 1, \ldots, \mu_i) \qquad \text{on } \Lambda_i$$

(again, we take a smaller polydisk Λ_i, if necessary). We let $\mathcal{I}\{\varphi^{(i)}\}$ denote the $\mathcal{O}$-ideal generated by $\varphi_k^{(i)}(z', z_i)$ $(k = 1, \ldots, \mu_i)$ on Λ_i; thus $\mathcal{I}\{\varphi^{(i)}\}$ is equivalent to $\mathcal{P}_i$ on Λ_i. We note that the projection of the analytic set Σ onto Λ_i is an analytic set $\Sigma^{(i)}$ in Λ_i which equals the common zero set of $\varphi_k(z', z_i)$ $(k = 1, \ldots, \mu_i)$ in Λ_i. Also, we note that $\Sigma^{(i)} \cap [\Delta^{(r)} \cap \partial \Gamma_i] = \emptyset$. Since $\Sigma^{(i)}$ is a pure r-dimensional analytic set in the $(r+1)$-dimensional polydisk Λ_i, i.e., $\Sigma^{(i)}$ is an analytic hypersurface in Λ_i, it follows that $\Sigma^{(i)}$ can be described as

$$\Sigma^{(i)} : \ P_i(z', z_i) = 0 \qquad \text{in } \Lambda_i,$$

where $P_i(z', z_i)$ is a distinguished pseudopolynomial in z_i whose coefficients are holomorphic functions of z' in $\Delta^{(r)}$. From the arguments in Chapter 2, the original analytic set Σ in Λ consists of certain irreducible components of the complete algebraic analytic set defined by

$$S := \bigcap_{i=r+1}^{n} \{z = (z', z'') \in \Delta^{(r)} \times \mathbf{C}_{z''}^{n-r} \mid P_i(z', z_i) = 0\}.$$

We let Σ' denote the collection of the remaining irreducible comoponents of S, other than Σ, so that $S = \Sigma \cup \Sigma'$.

We claim that for each $i = r+1, \ldots n$, there exists a positive integer q_i such that

$$(*) \quad P_i(z', z_i)^{q_i} \in \mathcal{I}\{\varphi^{(i)}\} \qquad \text{at the origin } O \text{ in } \mathbf{C}_{z'}^r \times \mathbf{C}_{z_i}. \tag{8.30}$$

If the claim is proved, then we complete the proof of the theorem as follows. By the definition of the projection ideal $\mathcal{P}_i$ of $\mathcal{I}\{\mathbf{F}\}$ onto Λ_i, we see that each $\varphi_k^{(i)}(z', z_i) \in \mathcal{I}\{\mathbf{F}\}$ $(k = 1, \ldots, \mu_i)$ at each point in $\Lambda \subset \mathbf{C}_z^n$. Here we regard $\varphi_k^{(i)}(z', z_i)$ as being independent of the $n-r-1$ variables $z_{r+1}, \ldots, \widehat{z_i}, \ldots, z_n$. This observation, combined with claim $(*)$, implies that

$$P_i(z', z_i)^{q_i} \in \mathcal{I}\{\mathbf{F}\} \qquad \text{at the origin } O \text{ in } \mathbf{C}_z^n.$$

We set $q = \max_{i=r+1,\ldots,n} q_i \geq 1$. We now let $H(z)$ be a holomorphic function in Λ such that $H(z) = 0$ on Σ' and $H(z) \neq 0$ on each irreducible component of Σ. Since $f(z) = 0$ on Σ, we have $fH = 0$ on S. From Proposition 7.7 it follows that

$$f(z)H(z) \in \mathcal{I}\{P_{r+1}, \ldots, P_n\} \qquad \text{at } O \text{ in } \mathbf{C}_z^n.$$

If we let σ denote the common zero set of the $n - r + 1$ holomorphic functions $\{H(z), P_{r+1}(z', z_{r+1}), \ldots, P_n(z', z_n)\}$ in Λ, then the conditions imposed on $H(z)$ imply that the dimension of σ at O is less than or equal to $r-1$. Since $f = 0$ on $\sigma \subset \Sigma$, it follows from the induction hypothesis that there exists a positive integer ρ with

$$f(z)^\rho \in \mathcal{I}\{H, F_1, \ldots, F_\nu\} \qquad \text{at } O \text{ in } \mathbf{C}_z^n.$$

Hence, the above relations imply that

$$f(z)^{(\rho+1)\cdot nq} \in \mathcal{I}\{\mathbf{F}\} \qquad \text{at } O \text{ in } \mathbf{C}_z^n.$$

Thus the theorem is proved, assuming the claim $(*)$.

It remains to prove claim $(*)$ for each $i = r+1, \ldots, n$. For simplicity, we write $z_i = w$, $\Sigma^{(i)} = \Sigma$, $P_i(z', z_i) = P(z', w)$, $\varphi_k^{(i)}(z', z_i) = \varphi_k(z', w)$ $(k = 1, \ldots, \mu_i = \mu)$, $\mathcal{I}\{\varphi^{(i)}\} = \mathcal{I}\{\varphi\}$, $\Gamma_i = \Gamma \subset \mathbf{C}_w$, and $\Lambda = \Delta^{(r)} \times \Gamma \subset \mathbf{C}^r_{z'} \times \mathbf{C}_w$. We let

$$\Sigma = \Sigma_1 \cup \ldots \cup \Sigma_p$$

denote the irreducible decomposition of Σ in Λ. Since $\Sigma_j \cap [\Delta^{(r)} \times \partial\Gamma] = \emptyset$, each Σ_j $(j = 1, \ldots p)$ can be represented in the form

$$\Sigma_j : \; Q_j(z', w) = 0 \qquad \text{in } \Delta^{(r)} \times \mathbf{C}_w,$$

where each $Q_j(z', w)$ is an irreducible distinguished pseudopolynomial in w whose coefficients are holomorphic functions of z' in $\Delta^{(r)}$. We note that

$$P = Q_1 \times \cdots \times Q_p \qquad \text{on } \Delta^{(r)} \times \mathbf{C}_w.$$

Since $\varphi_1(z', w) = 0$ on Σ, it follows from the Weierstrass preparation theorem that

$$\varphi_1 = A_1 Q_1^{m_1} \cdots Q_p^{m_p} \qquad \text{in } \Lambda,$$

where A_1 is a holomorphic function on Λ with $A_1 \not\equiv 0$ on each Σ_j $(j = 1, \ldots, p)$, and m_j $(j = 1, \ldots, p)$ is a positive integer. Setting $m = \max_{j=1,\ldots,p} m_j$ and $m'_j = m - m_j \geq 0$ $(j = 1, \ldots, p)$, we have

$$Q_1^{m'_1} \cdots Q_p^{m'_p} \varphi_1 = A_1 P^m \qquad \text{in } \Lambda.$$

If $A_1(O) \neq 0$, we have $P \in \mathcal{I}\{\varphi_1\} \subset \mathcal{I}\{\varphi\}$ at the origin O in $\mathbf{C}^r_{z'} \times \mathbf{C}_w$; hence the claim $(*)$ is proved. If $A_1(O) = 0$, the common zero set σ of the $\mu + 1$ holomorphic functions $\{A_1, \varphi_1, \ldots, \varphi_\mu\}$ in Λ is of dimension $r - 1$ at O. Since $P = 0$ on $\sigma \subset \Sigma$, it follows from the induction hypothesis that there exists a positive integer ρ with

$$P^\rho = \alpha_1 A_1 + \beta_1 \varphi_1 + \cdots + \beta_\mu \varphi_\mu \qquad \text{at } O \text{ in } \mathbf{C}^r_{z'} \times \mathbf{C}_w,$$

where α_1, β_j $(j = 1, \ldots, \mu)$ are holomorphic functions at O in $\mathbf{C}^r_{z'} \times \mathbf{C}_w$. Thus

$$P^{\rho+m} = (Q_1^{m'_1} \cdots Q_p^{m'_p} + \beta_1)\varphi_1 + \beta_2 \varphi_2 + \cdots + \beta_\mu \varphi_\mu \qquad \text{at } O \text{ in } \mathbf{C}^r_{z'} \times \mathbf{C}_w,$$

which proves the claim $(*)$. □

CHAPTER 9

Normal Pseudoconvex Spaces

9.1. Normal Pseudoconvex Spaces

The main purpose of this chapter is to prove Oka's theorem that any pseudoconvex domain in $\mathbf{C}^n$ is a domain of holomorphy. We shall prove this theorem in a more general setting. The essential part of the proof of this generalization, the use of an integral equation to solve the Cousin I problem, is the same as in Oka's original work [**53**]. We first define a **normal pseudoconvex space** as an analytic space with a strictly pseudoconvex exhaustion function.[1] We shall then show that a normal pseudoconvex space is a Stein space; this statement contains Oka's theorem as a special case. In this chapter we will always assume that an analytic space satisfies the second countability axiom of Hausdorff.

9.1.1. Pseudoconvex Functions. We will define a pseudoconvex domain in an analytic space. Let $\mathcal{V}$ be an analytic space of dimension n. Let $U \subset \mathcal{V}$ be a domain and let ∂U denote the boundary of U in $\mathcal{V}$. Let $q \in \partial U$ and let $(\delta_q, \lambda_q, \phi_q)$ be a local coordinate chart for q in $\mathcal{V}$. If there exists a neighborhood $v \subset \delta_q$ such that $\phi_q(v \cap U) \subset \lambda_q$ is a ramified pseudoconvex domain over $\mathbf{C}^n$ (defined in 6.1.6), then we say that U is pseudoconvex at the boundary point q. If U is pseudoconvex at each boundary point, then we say that U is a **pseudoconvex domain** in $\mathcal{V}$. Immediately from the definition we obtain the following properties.

1. If U_1 and U_2 are pseudoconvex domains in $\mathcal{V}$, then so is $U_1 \cap U_2$.
2. Let U_k $(k = 1, 2, \dots)$ be pseudoconvex domains in $\mathcal{V}$ with $U_k \subset U_{k+1}$ $(k = 1, 2, \dots)$, $\lim_{k\to\infty} U_k = U_0$, and $U_0 \subset\subset \mathcal{V}$. Then U_0 is a pseudoconvex domain in $\mathcal{V}$.

Now let D be a domain in $\mathcal{V}$ and let $\ell(p)$ be a real-valued continuous function on D; we allow ℓ to admit the value $-\infty$. If, for any point $q \in D$, the domain $\{p \in D \mid \ell(p) < \ell(q)\} \subset D$ is pseudoconvex, then we say that $\ell(p)$ is a **pseudoconvex function** on D. As we have already shown, any continuous plurisubharmonic function in a univalent domain D in $\mathbf{C}^n$ is a pseudoconvex function on D.[2]

---

[1] We use the word "normal" for the following reasons:

(i) We showed in Chapter 8 that an analytic space can locally be mapped to a normal analytic set in a one-to-one manner.

(ii) A pseudoconvex domain in an analytic space is not always a holomorphically complete domain (Stein space). In Theorem 9.3 we shall prove that a domain which admits a strictly pseudoconvex exhaustion function is holomorphically convex; we will call such a domain in an analytic space a **normal pseudoconvex domain**.

[2] Since the notion of pseudoconvexity is local, we can define the notion of a plurisubharmonic function on a domain in a complex manifold without any ambiguity. However, there is no unique definition of plurisubharmonicity (or of pseudoconvexity) in a ramified domain over $\mathbf{C}^n$. We use the terminology "pseudoconvex function" on domains in an analytic space.

In section 3.4.1 we defined what it meant for a family of analytic hypersurfaces $\{\sigma\}_{t\in[0,1]}$ in $\mathbf{C}^n$ to satisfy Oka's condition in order to find a useful criterion for a point to be in a polynomially convex hull. We now introduce a similar type of family of analytic hypersurfaces in an analytic space $\mathcal{V}$.

Let $q \in \mathcal{V}$ and let $(\delta_q, \lambda_q, \phi_q)$ give local coordinates for q in $\mathcal{V}$. Let $I = [0,1]$ be the unit interval of the complex t-plane and let $g(p,t)$ be a complex-valued function in $\delta_q \times I$ such that

1. $g(p,t)$ is a continuous function on $\delta_q \times I$, and
2. for any fixed $t \in I$, $g(p,t)$ is a non-constant holomorphic function on δ_q.

Given $t \in I$, we consider the analytic hypersurface in δ_q defined by

$$\sigma_t : \quad g(p,t) = 0. \tag{9.1}$$

We say that $\{\sigma_t\}_{t\in I}$ is a continuous family of analytic hypersurfaces in δ_q at the point q.

Now let $\ell(p)$ be a finite real-valued continuous function defined on a domain D in $\mathcal{V}$. Fix $q \in D$ and let $\{\sigma_t\}_{t\in I}$ be a continuous family of analytic hypersurfaces in $\delta_q \subset D$ at the point q. We use the same notation as in (9.1). If $\{\sigma_t\}_{t\in I}$ satisfies the following two conditions:

1. σ_0 passes through q and $\sigma_0 \setminus \{q\}$ lies in $\{p \in \delta_q \mid \ell(p) > \ell(q)\}$;
2. for each $t > 0$, $\sigma_t \subset \{p \in \delta_q \mid \ell(p) > \ell(q)\}$,

then we say that $\{\sigma_t\}_{t\in I}$ is a **family of analytic hypersurfaces touching the domain** $\{p \in D \mid \ell(p) < \ell(q)\}$ **from outside at the point** q.

If $\ell(p)$ admits at least one such continuous family $\{\sigma_t\}_{t\in I}$, then we say that $\ell(p)$ is **strictly pseudoconvex at the point** q. If $\ell(p)$ is strictly pseudoconvex at each point q in D, then we say that $\ell(p)$ is a **strictly pseudoconvex function** on D.

By definition, any strictly pseudoconvex function on $D \subset \mathcal{V}$ is a pseudoconvex function on D. Any piecewise smooth, strictly plurisubharmonic function on a univalent domain D in $\mathbf{C}^n$ is a strictly pseudoconvex function on D. However a strictly pseudoconvex function of class C^2 on a univalent domain D in $\mathbf{C}^n$ is not always a strictly plurisubharmonic function on D. The following properties of strictly pseudoconvex functions on a domain $D \subset \mathcal{V}$ are easily verified.

1. Let $\ell(p)$ be a strictly pseudoconvex function on D and let $h(x)$ be a finite, real-valued increasing function on $[-\infty, \infty)$. Then $\ell_0(p) := h(\ell(p))$ is a strictly pseudoconvex function on D.
2. Let $\ell_i(p)$ $(i = 1, 2)$ be strictly pseudoconvex functions on D and let $\ell_0(p) = \max\{\ell_1(p), \ell_2(p)\}$. Then $\ell_0(p)$ is a strictly pseudoconvex function on D.

We have the following relationship between strictly pseudoconvex functions and pseudoconvex domains.

1. Let $\mathcal{D}$ be a ramified domain over a polydisk Δ in $\mathbf{C}^n$ and let $\pi : \mathcal{D} \to \Delta$ be the canonical projection. For any strictly plurisubharmonic function $s(z)$ on Δ, the function $\ell(p) := s(\pi(p))$ is a strictly pseudoconvex function on $\mathcal{D}$.
2. Let $\mathcal{P}$ be an analytic polyhedron in an analytic space $\mathcal{V}$ of dimension n. Let Σ be a model of $\mathcal{P}$ in the polydisk Δ in $\mathbf{C}^m$, i.e.,

$$\Phi : p \in \mathcal{P} \to z = (\varphi_1(p), \dots, \varphi_m(p)) \in \Delta,$$

where each $\varphi_j(p)$ $(j = 1, \dots, m)$ is a holomorphic function on a domain G containing $\mathcal{P}$ with $\Sigma = \Phi(\mathcal{P})$; Σ is an analytic set in Δ; and Σ and $\mathcal{P}$ are bijective except for an analytic set of dimension at most $n-1$. Then for any strictly plurisubharmonic function $s(z)$ on Δ, the function $\ell(p) := s(\Phi(p))$ is a strictly pseudoconvex function on $\mathcal{P}$.

Now let $\ell(p)$ be a real-valued continuous function on a domain U in $\mathcal{V}$. We allow ℓ to admit the value $-\infty$. For a real number a, we set[3]

$$U_a := \{p \in U \mid \ell(p) < a\}.$$

If $U_a \subset\subset U$ for each a, we say that $\ell(p)$ is an **exhaustion function** for U.

9.1.2. Normal Pseudoconvex Spaces. Let $\mathcal{V}$ be an analytic space of dimension n and let $U \subset \mathcal{V}$ be a domain. If there exists a strictly pseudoconvex exhaustion function $\ell(p)$ on U, then we say that U is a **normal pseudoconvex domain** in $\mathcal{V}$, and we call $\ell(p)$ an **associated function** on U. In the case of $U = \mathcal{V}$, we call $\mathcal{V}$ a **normal pseudoconvex space**.

We have the following theorem relating Stein spaces and normal pseudoconvex spaces.

THEOREM 9.1. *A Stein space $\mathcal{V}$ is a normal pseudoconvex space.*

PROOF. Let $n = \dim \mathcal{V}$. Theorem 8.22 implies that $\mathcal{V}$ is bijective to an analytic set Σ in $\mathbf{C}^{2n+1}$; we let

$$\Phi : \quad p \in \mathcal{V} \to z = (\varphi_1(p), \ldots, \varphi_{2n+1}(p)) \in \Sigma$$

denote this bijection. We set $s(z) := \sum_{j=1}^{2n+1} |z_j|^2$ in $\mathbf{C}^{2n+1}$ and define $\ell(p) := s(\Phi(p))$ for $p \in \mathcal{V}$. Since $s(z)$ is a strictly plurisubharmonic exhaustion function on $\mathbf{C}^{2n+1}$, it follows that $\ell(p)$ is a strictly pseudoconvex exhaustion function on $\mathcal{V}$. □

We showed in Theorem 4.6 that any univalent pseudoconvex domain D in $\mathbf{C}^n$ admits a piecewise smooth, strictly plurisubharmonic exhaustion function; hence D is a normal pseudoconvex domain. In the case of an analytic space (even in the case of a complex manifold), this is not necessarily true. Before giving some counterexamples, we verify the following proposition.

PROPOSITION 9.1. *A normal pseudoconvex domain D in an analytic space $\mathcal{V}$ cannot contain a compact analytic set τ of positive dimension.*

PROOF. Let D be a normal pseudoconvex domain with associated function $\ell(p)$. Assume that there exists a compact analytic set τ in D having positive dimension. Let $a = \max\{\ell(p) \mid p \in \tau\} < \infty$ and fix $q_0 \in \tau$ with $\ell(q_0) = a$. There exists a family of analytic hypersurfaces

$$\sigma_t : \quad g(p,t) = 0 \quad (t \in I)$$

in a neighborhood δ of q_0 in D with $q_0 \in \sigma_0$, $\sigma_0 \setminus \{q_0\} \subset \delta_a := \{p \in \delta \mid \ell(p) > a\}$, and $\sigma_t \subset \delta_a$ for each $t > 0$. Fix a one-dimensional analytic set $\tau_0 \subset \tau$ in δ passing through q_0; we may assume τ_0 is conformally equivalent to a disk Δ_0. Identifying τ_0 with Δ_0, we have

$$g(p,t) \,|_{\Delta_0} \neq 0 \quad \text{for all } t > 0,$$

while $g(q_0, 0) = 0$ and $g(p,0) \,|_{\Delta_0} \not\equiv 0$. Since $g(p,t) \to g(p,0)$ as $t \to 0$ uniformly on Δ_0, this contradicts the classical Hurwitz theorem. □

[3] U_a may be the empty set.

There exist many pseudoconvex domains in an analytic space which are not normal pseudoconvex domains.

EXAMPLE 9.1. Let $\Omega = \mathbf{C}_z \times \mathbf{P}^m$ with $m \geq 1$. This is a complex manifold, and $D := \{|z| < 1\} \times \mathbf{P}^m$ is a pseudoconvex domain which contains the compact analytic set $\{0\} \times \mathbf{P}^m$ of dimension m. Thus D is not a normal pseudoconvex domain.

EXAMPLE 9.2. Let $\mathbf{C}^n$ have variables $z_1, \dots, z_n$ and let $\mathbf{P}^{n-1}$ have homogeneous coordinates $[w_1 : w_2 : \dots : w_n]$. We consider the product space $\Omega^* := \mathbf{C}^n \times \mathbf{P}^{n-1}$ and the n-dimensional analytic set Σ in Ω^* defined by

$$\Sigma : \quad z_1 w_j - z_j w_1 = 0 \quad (j = 2, \dots, n).$$

Since Σ is non-singular in Ω^*, it follows that Σ is an n-dimensional complex manifold. If we consider the subset of Σ given by $\Sigma_1 := \Sigma \cap \{\sum_{j=1}^n |z_j|^2 < 1\}$, then Σ_1 is a strictly pseudoconvex domain in Σ. Since Σ_1 contains the $(n-1)$-dimensional compact analytic set $\{0\} \times \mathbf{P}^{n-1}$, it is not a normal pseudoconvex domain.

EXAMPLE 9.3. [4] Let $\mathbf{C}^2$ have variables $z = x + iy$ and $w = u + iv$. We consider the lattice group Γ generated by the following four linearly independent vectors (in $\mathbf{C}^2$) over $\mathbf{R}$:

$$(1, 0), \ (i, 0), \ (0, 1), \ (i\alpha, i),$$

where $\alpha > 0$ is an irrational number. We let $\mathcal{M}$ denote the quotient space $\mathbf{C}^2/\Gamma$. Then $\mathcal{M}$ is a 2-dimensional, compact, complex torus with canonical projection $\pi : \mathbf{C}^2 \to \mathcal{M}$. Let a and b be real numbers with $0 < a < b < 1$ and let

$$\Delta = \{(z, w) \in \mathbf{C}^2 \mid a < \operatorname{Re} z < b\}, \quad U = \pi(\Delta).$$

Then U is a Levi flat domain in $\mathcal{M}$. We list the following properties of U.

1. U cannot contain any compact analytic set of dimension 1.

PROOF. Let $0 < c < 1$ and define $H_c := \{(z, w) \in \mathbf{C}^2 \mid \operatorname{Re} z = c\}$, $\mathcal{H}_c := \pi(H_c)$. Then $\mathcal{H}_c$ is a real 3-dimensional compact hypersurface in $\mathcal{M}$. Fix $z_0 = c_0 + ic'_0$ with $0 < c_0 < 1$ and let

$$s_{z_0} := \{(z, w) \in \mathbf{C}^2 \mid z = z_0\}, \quad \mathcal{S}_{z_0} := \pi(s_{z_0}).$$

Then we have $\mathcal{S}_{z_0} = \{z_0\} \times (\mathbf{C}_w/[1])$, so that $\mathcal{S}_{z_0}$ is conformally equivalent to $\mathbf{C}^*$ as a Riemann surface. We note that $\mathcal{S}_{z_0} \neq \mathcal{H}_{c_0}$ and that

(*) $\mathcal{S}_{z_0}$ is dense in $\mathcal{H}_{c_0}$.

To verify (*), let $z_1 = c_1 + ic'_1$, where $0 < c_1 < 1$. Then $\mathcal{S}_{z_0} = \mathcal{S}_{z_1}$ if and only if $c_0 = c_1$ and $c'_0 - c'_1 = m\alpha + n$, where n and m are integers. Since $\mathcal{H}_{c_0} = \bigcup_{y \in \mathbf{R}_y} \mathcal{S}_{c_0 + iy}$ and since α is irrational, (*) follows.

We now prove 1 by contradiction. Thus we assume that there exists a one-dimensional compact analytic set $\mathcal{S}$ in U. Set $\mathbf{s} = \pi^{-1}(\mathcal{S})$ in Δ, which is a non-compact analytic set in Δ. Let $(z, w), (z', w') \in \mathbf{s}$. Then $\pi(z, w) = \pi(z', w')$ in $\mathcal{S}$ implies $\operatorname{Re} z = \operatorname{Re} z'$. Since $\mathcal{S}$ is compact, the single-valued harmonic function $\operatorname{Re} z$ on $\mathcal{S}$ attains its maximum on $\mathcal{S}$. Therefore, $\operatorname{Re} z = c$ (constant) on $\mathbf{s}$, and hence $\mathbf{s} = \{c + ic'\} \times \mathbf{C}_w$ where c' is a constant. Consequently, $\pi(\mathbf{s}) = \mathcal{S}_{c+ic'}$, which is not compact from (*). This contradicts the assumption that $\pi(\mathbf{s}) = \mathcal{S}$ is compact. □

[4] This example is due to H. Grauert.

2. Any holomorphic function on U is a constant.

PROOF. Let $f(p)$ be a holomorphic function on U. Let $z_0 = c + ic'$ where $a < c < b$. Since $\mathcal{S}_{z_0}$ is conformally equivalent to $\mathbf{C}^*$, it follows from the fact that $\overline{\mathcal{S}}_{z_0} = \mathcal{H}_c \subset\subset U$ that $f(p)$ must be constant on $\mathcal{S}_{z_0}$, and hence in $\mathcal{H}_c$. Consequently, $f(p)$ is a constant on U. □

3. U is not a normal pseudoconvex domain.

PROOF. We prove this by contradiction. Thus we assume that there exists an associated function $\varphi(p)$ on U. Then by the same reasoning as in 2 we see that $\varphi(p)$ is constant on each set $\mathcal{H}_c$, $a < c < b$. Therefore, for suffciently large $A > 0$, $U_A := \{p \in U \mid \varphi(p) < A\}$ is a non-empty Levi flat domain in U, which contradicts the fact that $\varphi(p)$ is an associated function on U. □

From 1 and 3 we see that U is a pseudoconvex domain in $\mathcal{M}$ containing no compact curves, but U is not a normal pseudoconvex domain.

Another proof of $(*)$: We first remark that $\mathcal{M}$ is homeomorphic to the product $T_1 \times T_2$ of two real compact tori T_1 and T_2, where

$$\begin{aligned} T_1 &= \mathbf{R}_x \times \mathbf{R}_u/[(1,0),(0,1)], \\ T_2 &= \mathbf{R}_y \times \mathbf{R}_v/[(1,0),(\alpha,1)]. \end{aligned}$$

We write $\mathcal{M} \approx T_1 \times T_2$. Since T_1 is homeomorphic to the product $\gamma_1 \times \gamma_2$ of two unit circles, we have $\mathcal{M} \approx \gamma_1 \times \gamma_2 \times T_2$. We let π_2 denote the canonical projection from $\mathbf{R}_y \times \mathbf{R}_v$ onto T_2. Since α is irrational, for any fixed c_0' with $0 < c_0' < 1$, we have that $\sigma_{c_0'} := \pi_2(\{c_0'\} \times \mathbf{R}_v)$ is dense in T_2. Fix $0 < c_0 < 1$ and set $z_0 = c_0 + ic_0'$. Since

$$\mathcal{S}_{z_0} \approx \{c_0\} \times \gamma_2 \times \sigma_{c_0'}, \quad \mathcal{H}_{c_0} \approx \{c_0\} \times \gamma_2 \times T_2,$$

it follows that $\mathcal{S}_{z_0}$ is dense in $\mathcal{H}_{c_0}$.

We remark that the domains in these three examples are domains of holomorphy, but they are not Stein spaces.

9.1.3. Local Holomorphic Completeness. We shall extend Lemma 3.5 (Oka's lemma) in $\mathbf{C}^n$ to an analytic space. Let $\mathcal{V}$ be an analytic space of dimension n. Let E and A be compact sets in $\mathcal{V}$ with $E \subset A$. Let $p \in A$ and let $\{\sigma_t\}_{t \in I}$ be a continuous family of analytic hypersurfaces in δ_p, where $I = [0,1]$ and δ_p is an open neighborhood of p in $\mathcal{V}$. If $\{\sigma_t\}_{t\in I}$ satisfies the following three conditions:

1. for $t \in I$, $\sigma_t \cap E = \emptyset$;
2. $\sigma_0 \cap A \neq \emptyset$ and $\sigma_1 \cap A = \emptyset$;
3. for $t \in I$, $(\partial \sigma_t) \cap A = \emptyset$,

then we say that $\{\sigma_t\}_{t\in I}$ satisfies **Oka's condition** at p with respect to the pair (E, A).

We have the following generalization of Lemma 3.5.

LEMMA 9.1. *Let U be a holomorphically complete domain in $\mathcal{V}$, i.e., U is a Stein space. Let $K \subset\subset U$ and let $\widehat{K} = \widehat{K}_U$ denote a holomorphically convex hull of K with respect to the holomorphic functions in U. Then for each $p \in \widehat{K}$, there* *does* **not** *exist a continuous family of analytic hypersurfaces* $\{\sigma_t\}_{t\in I}$ *in* δ_p *which satisfies Oka's condition at* p *with respect to the pair* $(K, \widehat{K})$.

PROOF. The essential part of the proof is similar to that of Lemma 3.5. In Lemma 3.5 we used the fact that $I \times \widehat{K}$ is the holomorphic hull of $I \times K$ in $V \times U \subset \mathbf{C}_t \times \mathbf{C}^n_z$, and we used the solvability of the Cousin I problem in the analytic polyhedron in the $(n+1)$-dimensional domain $V \times U$. Here we will use the solvability of the Cousin I problem in an analytic polyhedron in the n-dimensional domain U. We will prove the lemma by contradiction; hence we assume that there exists a continuous family of hypersurfaces $\{\sigma_t\}_{t\in I}$ which satisfies Oka's condition with respect to the pair $(K, \widehat{K})$ at some point $p_0 \in \widehat{K}$. Let

$$\sigma_t : \ g(t,p) = 0, \quad (t,p) \in I \times \delta,$$

where δ is a neighborhood of p_0 in U and $g(t,p)$ is a continuous function of $(t,p) \in I \times \delta$ which is a nonconstant holomorphic function in $p \in \delta$ for each fixed $t \in I$. Since $\widehat{K}$ is the holomorphic hull of K in the Stein space U, it follows that there exists an analytic polyhedron $\mathcal{P}$ in U with defining functions that are defined in U such that

$$\widehat{K} \subset \mathcal{P}^o; \ \sigma_0 \cap \mathcal{P} \neq \emptyset; \ \sigma_1 \cap \mathcal{P} = \emptyset; \text{ and } (\partial\sigma_t) \cap \mathcal{P} = \emptyset, \ t \in I.$$

Thus for each fixed $t \in I$, the meromorphic function $1/g(t,p)$ in $\delta \cap \mathcal{P}$ canonically defines a Cousin I distribution in $\mathcal{P}$. Hence for each $t \in I$ we can find a meromorphic function $H(t,p)$ in $\mathcal{P}$ whose only poles in $\delta \cap \mathcal{P}$ are given by $1/g(t,p)$. We remark that although $H(t,p)$ is not uniquely determined by $1/g(t,p)$, the proof in Theorem 8.9 of the construction of meromorphic functions with prescribed Cousin I data implies that we can take $H(t,p)$ in $I \times \mathcal{P}$ to be continuous for $(t,p) \in I \times \delta$ if $g(t,p)$ is continuous in $I \times \delta$. Setting $t_0 := \inf\{t \in I \mid \sigma_t \cap \widehat{K} = \emptyset\}$, we can choose t^* with $0 < t_0 < t^*$ sufficiently close to t_0 to insure that $H(t^*, p)$ satisfies the condition that there exists a point $q \in \widehat{K}$ with $|H(t^*, q)| > \max_{p\in K}\{|H(t^*, p)|\}$. Since the pair $(\mathcal{P}, U)$ satisfies the Runge theorem, we can find a holomorphic function $H(p)$ in U such that $|H(q)| > \max_{p\in K}\{|H(p)|\}$, which gives a contradiction to the fact that $\widehat{K}$ is the holomorphic hull of K in U. □

Let $q \in \mathcal{V}$. Using Remark 8.2, there exists a neighborhood δ_q of q in $\mathcal{V}$ which has a normal model in a polydisk Δ. Thus the set δ_q is holomorphically complete. We say that an analytic space $\mathcal{V}$ is **locally holomorphically complete** at each point.

Let U be a domain in an analytic space $\mathcal{V}$. Let $q \in \partial U$. If there exists a neighborhood δ_q of q in $\mathcal{V}$ such that $U \cap \delta_q$ is holomorphically complete, i.e., $U \cap \delta_q$ is itself a Stein space, then we say that U is locally holomorphically complete at the boundary point q. If U is locally holomorphically complete at each point q of ∂U, then we say that U is a **locally holomorphically complete domain** in $\mathcal{V}$.[5]

In the following lemmas and propositions in this section we will always assume that $\mathcal{V}$ is a normal pseudoconvex space with associated function $\ell(p)$. For a real number a, we set

$$\mathcal{V}_a := \{p \in \mathcal{V} \mid \ell(p) < a\} \subset\subset \mathcal{V}, \quad \mathcal{V}_\infty = \mathcal{V}.$$

[5] J.-E. Fornaess [**20**] gave an example of an analytic space $\mathcal{V}$ and a domain $U \subset\subset \mathcal{V}$ such that U is locally holomorphically complete in $\mathcal{V}$ but U is not holomorphically complete.

LEMMA 9.2. *Let U be a holomorphically complete domain in $\mathcal{V}$. Then for any real number a, the subset $U_a = U \cap \mathcal{V}_a$ is holomorphically convex with respect to the holomorphic functions in U.*

PROOF. Let $K \subset\subset U_a$ and let $\widehat{K}$ be the holomorphically convex hull of K with respect to U, so that $\widehat{K} \subset\subset U$. We set $b := \max_{p\in\widehat{K}}\{\ell(p)\} < \infty$. We prove the lemma by contradiction; hence we assume that $b \geq a$. Fix a point $p_0 \in \widehat{K}$ such that $\ell(p_0) = b$. Since $\ell(p)$ is strictly pseudoconvex in $\mathcal{V}$, there exists a continuous family of analytic hypersurfaces $\{\sigma_t\}_{t\in I}$ in a neighborhood δ_{p_0} of p_0 which touches $\mathcal{V}_b$ from outside at the point p_0. Since $K \subset\subset \mathcal{V}_a \subset \mathcal{V}_b$, it follows that the continuous family $\{\sigma_t\}_{t\in I}$ satisfies Oka's condition at p_0 for the pair $(K, \widehat{K})$. This contradicts Lemma 9.1. □

This lemma implies the following proposition.

PROPOSITION 9.2. *For any real number a, the domain $\mathcal{V}_a$ is a locally holomorphically complete domain in $\mathcal{V}$.*

PROOF. Let $q \in \partial\mathcal{V}_a$. Since $\mathcal{V}$ is locally holomorphically complete at each point, we can find a holomorphically complete neighborhood δ_q of q in $\mathcal{V}$. By Lemma 9.2, $\delta_q \cap \mathcal{V}_a$ is holomorphically complete; this proves the proposition. □

PROPOSITION 9.3. *Let a be a real number or $a = +\infty$. If $\mathcal{V}_a$ is holomorphically complete, then for each $c < a$, $\mathcal{V}_c$ is also holomorphically complete.*

PROOF. Since $\mathcal{V}_a$ satisfies conditions 1 and 3 in the definition of holomorphic completeness (stated in 8.3.1), so does the set $\mathcal{V}_c$. By Lemma 9.2, condition 2 for $\mathcal{V}_a$ implies condition 2 for $\mathcal{V}_c$. Thus $\mathcal{V}_c$ is holomorphically complete. □

We obtain the converse of Proposition 9.3.

PROPOSITION 9.4. *Let a be a real number or $a = +\infty$. If for each $c < a$ the set $\mathcal{V}_c$ is holomorphically complete, then $\mathcal{V}_a$ is holomorphically complete.*

PROOF. Let c_j $(j = 1, 2, \dots)$ be an increasing sequence of real numbers with $\lim_{j\to\infty} c_j = a$. Then,

$$\mathcal{V}_{c_j} \subset\subset \mathcal{V}_{c_{j+1}} \quad (j = 1, 2, \dots), \quad \mathcal{V}_a = \lim_{j\to\infty} \mathcal{V}_{c_j}.$$

Using Lemma 9.2, we conclude that $\mathcal{V}_{c_j}$ is holomorphically convex with respect to $\mathcal{V}_{c_{j+1}}$, so that $\mathcal{V}_{c_j}$ $(j = 1, 2, \dots)$ satisfies the approximation condition stated in 8.3.2. It follows from Theorem 8.8 that $\mathcal{V}_a$ is holomorphically complete. □

Let D be a relatively compact domain in an analytic space $\mathcal{V}$. We say that the Cousin I problem is solvable on the closure $\overline{D}$ of D if for any Cousin I distribution $\mathcal{C} = \{(g_q(p), \delta_q)\}_{q\in G}$ where $\overline{D} \subset\subset G$, there exists a meromorphic function $F(p)$ on a domain G^* with $\overline{D} \subset\subset G^* \subset G$ such that $F(p) - g_q(p)$ is holomorphic on $\delta_q \cap G^*$. For use in the next section, we prove the following two lemmas.

LEMMA 9.3. *Let D and D' be domains in an analytic space $\mathcal{V}$ with $D' \subset\subset D \subset\subset \mathcal{V}$. Assume that the Cousin I problem is solvable on $\overline{D}$. Let $\xi \in \partial D'$ and let $f_\xi(p)$ be a holomorphic function in a neighborhood δ_ξ of ξ in $\mathcal{V}$ satisfying the following conditions: if we let S denote the analytic hypersurface in δ_ξ determined by $f_\xi(p) = 0$ in δ_ξ, then $\xi \in S$, $S \cap [(\partial D') \setminus \{\xi\}] = \emptyset$, and $\partial S \cap \overline{D} = \emptyset$. Then there* *exist a holomorphic function* $F(p)$ *on* D' *and two neighborhoods* $\delta'_\xi, \delta''_\xi$ *of* ξ *in* D *such that* $\delta'_\xi \subset \delta''_\xi \subset \delta_\xi$ *and*

$$D' \cap \delta'_\xi \subset \{p \in D' \mid |F(p)| > 1\},$$

$$\sup\{|F(p)| \mid p \in D' \setminus \delta''_\xi\} < 1.$$

PROOF. Fix a domain G in $\mathcal{V}$ such that $\overline{D} \subset G$ and $\partial S \cap G = \emptyset$. Consider the following Cousin I distribution $\mathcal{C} = \{(g_q(p), \delta_q)\}_{q \in G}$:

1. for $q \in G \cap \delta_\xi$, we set $\delta_q = \delta_\xi \cap G$ and $g_q(p) = 1/f_\xi(p)$ on δ_q;
2. for $q \notin G \setminus \delta_\xi$, we choose δ_ξ so that $\delta_q \cap \delta_\xi = \emptyset$ and set $g_q(p) = 0$ on δ_q.

Since Cousin I is solvable in $\overline{D}$, there exists a meromorphic function $F^*(p)$ on G^*, where $\overline{D} \subset\subset G^* \subset G$, such that $F^*(p) - g_q(p)$ is holomorphic on $\delta_q \cap G^*$. Thus $F^*(p)$ is holomorphic on $G^* \setminus S$; in particular, it is holomorphic on $\overline{D'} \setminus \{\xi\}$, which contains D', and it has poles along $S \cap G^*$; i.e., $|F^*(p)| = +\infty$ on $S \cap G^*$. Since $M := \max_{p \in \overline{D} \setminus \delta_\xi}\{|F^*(p)|\} < \infty$ and $\xi \in S$, it follows that there exist two neighborhoods δ'_ξ and δ''_ξ of ξ in G^* with $\delta'_\xi \subset \delta''_\xi \subset \delta_\xi$ such that $\min_{p \in D' \cap \delta'_\xi}\{|F^*(p)|\} > M + 1$ and $\max_{p \in D' \setminus \delta''_\xi}\{|F^*(p)|\} < M + 1/2$. Setting $F(p) = F^*(p)/(M+1)$ on D' completes the proof of the lemma. □

LEMMA 9.4. *Let* D *be a domain in an analytic space* $\mathcal{V}$ *with* $D \subset\subset \mathcal{V}$. *Assume that the Cousin I problem is solvable on* $\overline{D}$. *Let* $f(p)$ *be a holomorphic function on* $\overline{D}$ *and let* S *denote the analytic hypersurface in* $\overline{D}$ *determined by* $f(p) = 0$ *on* $\overline{D}$. *Let* $p_1, p_2 \in S$. *Assume that there exists a holomorphic function* $\varphi(p)$ *in a neighborhood* V *of* S *in* $\overline{D}$ *such that* $\varphi(p_1) \neq \varphi(p_2)$. *Then there exists a holomorphic function* $\Phi(p)$ *defined on all of* $\overline{D}$ *such that* $\Phi(p_1) \neq \Phi(p_2)$.

PROOF. We fix a domain G with $\overline{D} \subset\subset G \subset\subset \mathcal{V}$ such that $f(p)$ is holomorphic in G and S is an analytic hypersurface in G. We may assume that $\varphi(p)$ is defined and holomorphic in a neighborhood V of S in G. We consider the following Cousin I distribution $\mathcal{C} = \{(g_q(p), \delta_q)\}_{q \in G}$ on G:

1. for $q \in V$, we take $\delta_q \subset V$ and set $g_q(p) = \varphi(p)/f(p)$ on δ_q;
2. for $q \in G \setminus V$, we take δ_q such that $\delta_q \cap S = \emptyset$ and set $g_q(p) \equiv 0$ on δ_q.

Since the Cousin I problem is solvable on $\overline{D}$, there exists a meromorphic function $F(p)$ on a domain G^* with $\overline{D} \subset\subset G^* \subset G$ such that $F(p) - g_q(p)$ is holomorphic on each δ_q, $q \in G^*$. If we set $\Phi(p) = F(p) \cdot f(p)$ on G^*, then $\Phi(p)$ is a holomorphic function on G^* satisfying $\Phi(p_i) = \varphi(p_i)$ $(i = 1, 2)$. Thus, $\Phi(p)$ satisfies the conclusion of the lemma. □

9.2. Linking Problem

9.2.1. Linking Condition. Let $\mathcal{V}$ be an analytic space of dimension n. Let $\mathcal{D}_1$ and $\mathcal{D}_2$ be relatively compact domains in $\mathcal{V}$ such that if we set

$$\mathcal{D} = \mathcal{D}_1 \cup \mathcal{D}_2, \quad \mathcal{D}_0 = \mathcal{D}_1 \cap \mathcal{D}_2, \tag{9.2}$$

then $\mathcal{D}$ and $\mathcal{D}_0$ satisfy the following conditions:

(L1) $\mathcal{D}$ is a normal pseudoconvex domain.
(L2) Both $\mathcal{D}_1$ and $\mathcal{D}_2$ are holomorphically complete domains in $\mathcal{V}$.

(L3) There exists a bounded holomorphic function $\varphi_0(p) = u(p) + iv(p)$ on a domain G in $\mathcal{V}$ such that $\mathcal{D}_0 \subset\subset G$ and $\mathcal{D}_0$ can be described as

$$\mathcal{D}_0 = \{p \in G \cap \mathcal{D} \mid a_1 < u(p) < a_2\},$$

where a_1 and a_2 are real numbers with $a_1 < 0 < a_2$.

We say that $\mathcal{D}$ satisfies the **linking condition**, or, more precisely, $\mathcal{D}$ satisfies the linking condition with respect to $\varphi_0(p)$ and a_i $(i = 1, 2)$.

For real numbers b_1 and b_2 with $a_1 < b_1 < 0 < b_2 < a_2$, such a domain $\mathcal{D}$ satisfies the linking condition with respect to this same function $\varphi_0(p)$ and the numbers b_i $(i = 1, 2)$, since we can write

$$\mathcal{D} = \mathcal{D}_1' \cup \mathcal{D}_2', \quad \mathcal{D}_0' = \mathcal{D}_1' \cap \mathcal{D}_2' = \{p \in G \cap \mathcal{D} \mid b_1 < u(p) < b_2\}, \tag{9.3}$$

where $\mathcal{D}_i' \subset \mathcal{D}_i$ $(i = 1, 2)$ are holomorphically complete.

We shall prove later (Theorem 9.2) that a domain $\mathcal{D} \subset\subset \mathcal{V}$ satisfying the linking condition is a holomorphically complete domain.

Define

$$\begin{aligned} \mathcal{H}_0 &= \{p \in \mathcal{D}_0 \mid u(p) = 0\}, \\ \mathcal{H}_i &= \{p \in \mathcal{D}_0 \mid u(p) = a_i\} \quad (i = 1, 2). \end{aligned}$$

We may assume from (9.2) that

$$\mathcal{H}_2 \subset \partial\mathcal{D}_1, \quad \mathcal{H}_1 \subset \partial\mathcal{D}_2. \tag{9.4}$$

Although $u(p)$ is not defined on all of $\mathcal{D}$, we use the terminology that the direction for $\mathcal{D}$ in which $u(p)$ increases (decreases) is to the right (left) – see Figure 1.

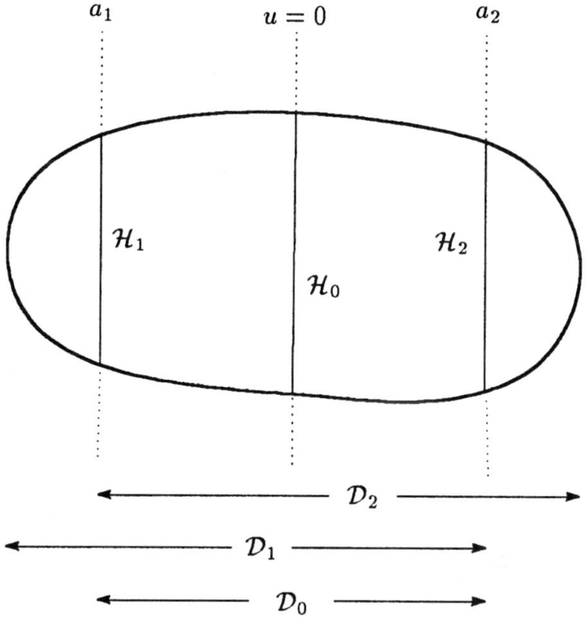

FIGURE 1. Linking condition for $\mathcal{D}$

We note from condition (L2) that $\mathcal{D}_0$ is holomorphically complete. By condition (L1) there exists an associated function $\ell(p)$ on $\mathcal{D}$. Fix a real number α and set

$$\begin{aligned} \mathcal{D}^{(\alpha)} &:= \{p \in \mathcal{D} \mid \ell(p) < \alpha\} \subset\subset \mathcal{D}, \\ \mathcal{D}_j^{(\alpha)} &:= \mathcal{D}^{(\alpha)} \cap \mathcal{D}_j \subset \mathcal{D}_j \quad (j = 0, 1, 2). \end{aligned}$$

By Lemma 9.2, each $\mathcal{D}_j^{(\alpha)}$ $(j = 0, 1, 2)$ is holomorphically complete.

We have the following lemma.

LEMMA 9.5. *$\mathcal{D}_0^{(\alpha)}$ is holomorphically convex with respect to the holomorphic functions in $\mathcal{D}_1^{(\alpha)}$ and in $\mathcal{D}_2^{(\alpha)}$. Similarly, $\mathcal{D}_0$ is holomorphically convex with respect to the holomorphic functions in $\mathcal{D}_1$ and in $\mathcal{D}_2$.*

PROOF. Since the proofs are similar, we will only prove that $\mathcal{D}_0^{(\alpha)}$ is holomorphically convex with respect to $\mathcal{D}_1^{(\alpha)}$. Let $K \subset\subset \mathcal{D}_0^{(\alpha)}$ be compact. We let $\widehat{K}$ denote the holomorphically convex hull of K with respect to $\mathcal{D}_1^{(\alpha)}$; i.e., $\widehat{K} = \widehat{K}_{\mathcal{D}_1^{(\alpha)}}$. Our goal is to show that $\widehat{K} \subset\subset \mathcal{D}_0^{(\alpha)}$. Since $\widehat{K} \subset\subset \mathcal{D}_1^{(\alpha)}$, using (9.4) it suffices to prove that $\widehat{K} \cap \mathcal{H}_1 = \emptyset$. We prove this by contradiction; thus we assume that $\widehat{K} \cap \mathcal{H}_1 \neq \emptyset$. Set

$$c = \max\{v(p) \mid p \in \widehat{K} \cap \mathcal{H}_1\} < \infty;$$

thus there exists a point $q_0 \in \widehat{K} \cap \mathcal{H}_1$ such that $\varphi_0(q_0) = a_1 + ic$. Consider the following family of analytic hypersurfaces in G:

$$\tau_t : \quad \varphi_0(p) = a_1 + (c + t)i \quad (0 \leq t < \infty).$$

From the definition of the number c we have $q_0 \in \tau_0 \cap \widehat{K}$ and $\tau_t \cap (\widehat{K} \cap \mathcal{H}_1) = \emptyset$ for all $t > 0$, so that $\tau_t \cap \widehat{K} = \emptyset$ for all $t > 0$. Moreover $\partial\tau_t \subset \partial G$ for all $t \geq 0$, so that $(\partial\tau_t) \cap \widehat{K} = \emptyset$ for all $t \geq 0$. It follows that $\{\tau_t\}_{t \in [0,\infty)}$ satisfies Oka's condition at q_0 for the pair $(K, \widehat{K})$. This contradicts Lemma 9.1. □

9.2.2. Oka's Fundamental Lemma. Let $\mathcal{D} \subset\subset \mathcal{V}$ be a domain which satisfies the linking condition. We use the same notation $\mathcal{D}_j$ $(j = 0, 1, 2)$, $\varphi_0(p) = u(p) + iv(p)$ on G where $\mathcal{D}_0 \subset\subset G$, a_i $(i = 1, 2)$, and $\mathcal{H}_j$ $(j = 0, 1, 2)$ as in the previous section. We also use the associated function $\ell(p)$ on $\mathcal{D}$ and the notation $\mathcal{D}^{(\alpha)} = \{p \in \mathcal{D} \mid \ell(p) < \alpha\}$ for each real number α. For future use, we fix a positive number ρ_0 such that

$$\rho_0 > \max\{|\varphi_0(p)| \mid p \in \overline{\mathcal{D}_0}\}. \tag{9.5}$$

We fix real numbers b_i $(i = 1, 2)$ sufficiently close to a_i with $a_1 < b_1 < 0 < b_2 < a_2$. As in (9.3) we construct domains $\mathcal{D}'_j$ $(j = 0, 1, 2)$ associated to b_i $(i = 1, 2)$ such that

$$\mathcal{D} = \mathcal{D}'_1 \cup \mathcal{D}'_2, \quad \mathcal{D}'_0 = \mathcal{D}'_1 \cap \mathcal{D}'_2 = \{p \in G \cap \mathcal{D} \mid b_1 < u(p) < b_2\}.$$

For simplicity, given $\alpha > 0$, we use the notation

$$D := \mathcal{D}^{(\alpha)}, \quad D_j := \mathcal{D}'_j \cap \mathcal{D}^{(\alpha)} \quad (j = 0, 1, 2) \tag{9.6}$$

(note the slight difference in notation between $\mathcal{D}_i$ and D_i $(i = 0, 1, 2)$). Fix a real number $\gamma < \alpha$ and consider the following set:

$$\mathbf{b} = \{p \in \mathcal{D}^{(\gamma)} \cap G \mid b_1 < u(p) < b_2\} \subset\subset \mathcal{D}_0^{(\alpha)}. \tag{9.7}$$

We remark that this set will be used in 9.2.3.

We assume that there exist a finite number of holomorphic functions $\varphi_j(p)$ $(j = 1, \dots, m)$ on D_0 which satisfy the following conditions:

1°. There exists a positive number $\delta > 0$ such that the subset
$$\mathcal{A} = \{p \in D_0 \mid -\delta \le u(p) \le \delta,\ |\varphi_j(p)| \le 1 \quad (j = 1, \dots, m)\}$$
satisfies $\mathcal{A} \subset\subset D_0$, so that $\mathcal{A}$ is an analytic polyhedron in D_0.

2°. In $\mathbf{C}^{m+1}$ with variables $z_0 = u + iv,\ z_1, \dots, z_m$, define the product domain
$$\Lambda = U \times \overline{\Delta} \subset \mathbf{C}_{z_0} \times \mathbf{C}^m_{z_1, \dots, z_m}$$
where
$$U : |z_0| \le 2\rho_0, \quad -\delta \le u \le \delta, \qquad \overline{\Delta} : |z_j| \le 1 \quad (j = 1, \dots, m),$$
and $\rho_0 > 0$ is defined in (9.5). Then the mapping
$$\Phi(p) : \quad z_i = \varphi_i(p) \quad (i = 0, 1, \dots, m)$$
gives a normal model $\Sigma = \Phi(\mathcal{A})$ of $\mathcal{A}$ in Λ.

3°. There exist positive numbers ϵ_0, ϵ_1 with $\epsilon_0, \epsilon_1 < 1$ such that, if we set
$$E = \{p \in D_0 \mid u(p) < b_1 + \epsilon_1\} \bigcup \{p \in D_0 \mid u(p) > b_2 - \epsilon_1\},$$
then we have
$$|\varphi_j(p)| < 1 - \epsilon_0 \quad (j = 1, \dots, m) \quad \text{on } E \cup \mathbf{b}.$$

The existence of such functions $\varphi_j(p)$ $(j = 1, \dots, m)$ on D_0 will be proved in the next section.[6]

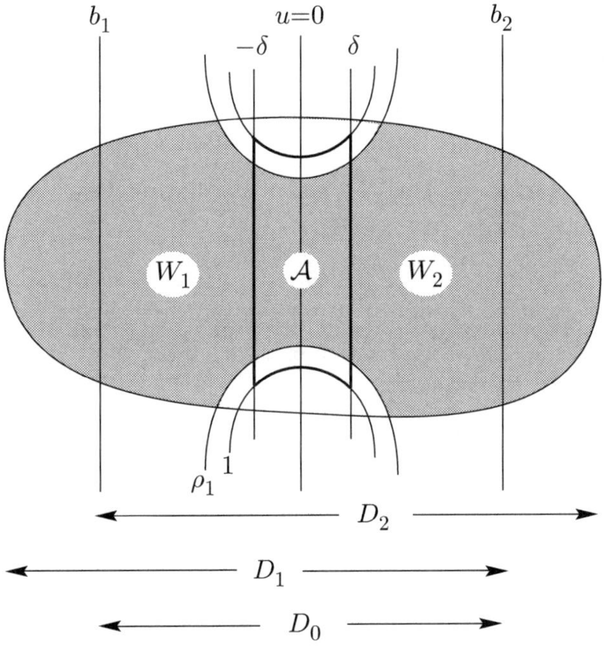

FIGURE 2. Linking condition for D

[6] Condition 1° says that a neighborhood V of $\mathcal{H}_0 \cap \partial D_0$ in D_0 is excluded from D_0 whenever $|\varphi_j(p)| > 1$ for some j $(1 \le j \le m)$. Further, condition 3° says that this neighborhood V can be chosen so that V does not intersect either of the sets $u(p) = b_1$ or $u(p) = b_2$ (see Figure 2).

We fix a positive number ρ_1 such that

$$1 - \epsilon_0 < \rho_1 < 1, \tag{9.8}$$

and we define

$$W = \{p \in D_0 \mid |\varphi_j(p)| \le \rho_1 \quad (j = 1, \dots, m)\} \bigcup (D \setminus D_0), \tag{9.9}$$

so that, by condition 3°,

$$D \cap (\mathcal{H}_1 \bigcup \mathcal{H}_2) \subset W, \qquad \mathcal{D}^{(\gamma)} \subset\subset W. \tag{9.10}$$

We divide W into two parts W_1 and W_2 using the real hypersurface $\mathcal{H}_0$ so that W_1 (W_2) is the left (right) part of W; thus $W_1 \subset D_1$ and $W_2 \subset D_2$.

In this geometric situation, we have the following fundamental lemma of Oka.

LEMMA 9.6 (Oka). [7] *Let $f(p)$ be a holomorphic function on $\mathcal{A}$. Then there exist holomorphic functions $f_1(p)$ and $f_2(p)$ on W_1 and W_2 such that both $f_1(p)$ and $f_2(p)$ can be holomorphically extended beyond $\mathcal{H}_0 \cap W$ and*

$$f(p) = f_1(p) - f_2(p) \quad \text{on} \quad \mathcal{H}_0 \cap W. \tag{9.11}$$

PROOF. We divide the proof into four steps. Afterwards, we will make a few remarks to clarify some of the details.

First step. Let δ, ϵ_0, ϵ_1 be as in 1° and 3°. We fix $\delta' > 0$ with $0 < \delta' < \delta$ and we also fix $\rho_1' > 0$ with $1 - \epsilon_0 < \rho_1 < \rho_1' < 1$. Consider the polydisk

$$\Lambda' = U' \times \overline{\Delta}' \subset \mathbf{C}_{z_0} \times \mathbf{C}^m_{z_1, \dots, z_m},$$

where

$$U' : |z_0| \le \rho_0, \quad -\delta' \le u \le \delta', \qquad \overline{\Delta}' : |z_j| \le \rho_1' \quad (j = 1, \dots, m),$$

with $\rho_0 > 0$ having been defined in (9.5). We have $\Lambda' \subset\subset \Lambda$. Using condition 2°, Theorem 8.15 implies that if $g(p)$ is a holomorphic function on $\mathcal{A}$, then $g(p)$ has a holomorphic extension $G(z_0, z_1, \dots, z_m) = G(z)$ on Λ,

$$G(\varphi_0(p), \varphi_1(p), \dots, \varphi_m(p)) = g(p) \quad \text{on } \mathcal{A},$$

with

$$\max_{z \in \Lambda'} \{|G(z)|\} \le K \max_{p \in \mathcal{A}} \{|g(p)|\},$$

where K is a constant which does not depend on the function $g(p)$ on $\mathcal{A}$.

Second step. Let $f(p)$ be a holomorphic function on $\mathcal{A}$. Since $\mathcal{A}$ is a compact set, there exists $M > 0$ such that

$$|f(p)| \le M \quad \text{on} \quad \mathcal{A}.$$

From the first step, there exists a holomorphic extension $F(z_0, z_1, \dots, z_m) = F(z)$ on Λ of $f(p)$ (i.e., $F(\varphi_0(p), \varphi_1(p), \dots, \varphi_m(p)) = f(p)$ on $\mathcal{A}$) such that

$$|F(z)| \le KM \quad \text{on} \quad \Lambda'.$$

Fix a segment $L = [-\rho_0 i, \rho_0 i]$ on the imaginary axis of the $z_0 = u + iv$-plane $\mathbf{C}_{z_0}$, and let $\mathbf{C}^+_{z_0}$ ($\mathbf{C}^-_{z_0}$) denote the right (left) half-plane divided by the imaginary axis

[7] This result was first proved in 1942 by Oka [**49**] in $\mathbf{C}^2$. In that paper, Oka used the Weil integral formula. The proof was rather complicated to understand (although the essential point – using an integral equation technique – is the same as presented here). The simpler proof given here using the lifting principle was published in 1953 by Oka [**52**]. However, the original idea of the simpler proof had been written in 1943 in Japanese (see Oka's posthumous work No. 1 in [**55**]).

in $\mathbf{C}_{z_0}$. Since $F(z)$ is holomorphic on Λ, which contains $L\times\overline{\Delta}$, we can consider the Cousin integral of $F(z)$ with respect to z_0:

$$\Psi(z) := \Psi(z_0,z_1,\dots,z_m) = \frac{1}{2\pi i}\int_L \frac{F(\zeta_0,z_1,\dots,z_m)}{\zeta_0 - z_0}d\zeta_0$$

for $(z_0,z_1,\dots,z_m)\in(\mathbf{C}_{z_0}\setminus L)\times\overline{\Delta}$. This defines a holomorphic function $\Psi_1(z)$ ($\Psi_2(z)$) on $\mathbf{C}^+_{z_0}\times\overline{\Delta}$ ($\mathbf{C}^-_{z_0}\times\overline{\Delta}$) such that both $\Psi_1(z)$ and $\Psi_2(z)$ can be holomorphically extended beyond $L\times\overline{\Delta}$ and satisfy

$$\Psi_1(z)-\Psi_2(z) = F(z) \quad \text{on} \quad L\times\overline{\Delta}. \tag{9.12}$$

We consider the polydisk $\Delta' : |z_j| < \rho_1'$ $(j=1,\dots,m)$ and its distinguished boundary

$$\Gamma : |\zeta_j| = \rho_1' \qquad (j=1,\dots,m) \quad \text{in } \mathbf{C}^m.$$

Since $\Delta'\subset\subset\Delta$, it follows from Cauchy's formula that

$$F(\zeta_0,z_1,\dots,z_m) = \frac{1}{(2\pi i)^m}\int_\Gamma \frac{F(\zeta_0,\zeta_1,\dots,\zeta_m)}{(\zeta_1-z_1)\cdots(\zeta_m-z_m)}d\zeta_1\cdots d\zeta_m$$

for $(\zeta_0,z_1,\dots,z_m)\in L\times(\Delta')^o$, where $(\Delta')^o$ is the interior of Δ'. Therefore, we have

$$\begin{aligned}&\Psi_1(z_0,z_1,\dots,z_m)\ (\Psi_2(z_0,z_1,\dots,z_m)) \\ &= \frac{1}{(2\pi i)^{m+1}}\int_{L\times\Gamma}\frac{F(\zeta_0,\zeta_1,\dots,\zeta_m)}{(\zeta_0-z_0)(\zeta_1-z_1)\cdots(\zeta_m-z_m)}d\zeta_0 d\zeta_1\cdots d\zeta_m\end{aligned} \tag{9.13}$$

for $(z_0,z_1,\dots,z_m)\in\mathbf{C}^-_{z_0}\times(\Delta')^o$ $(\mathbf{C}^+_{z_0}\times(\Delta')^o)$.

Now let $p\in D_0\cap W_1$ $(p\in D_0\cap W_2)$. Then $u(p)=\operatorname{Re}\varphi_0(p)<0$ $(u(p)>0)$, and $|\varphi_j(p)|<\rho_1<\rho_1'$ $(j=1,\dots,m)$. It follows that

$$\psi_i(p) := \Psi_i(\varphi_0(p),\varphi_1(p),\dots,\varphi_m(p)) \quad (i=1,2)$$

is a well-defined holomorphic function on $D_0\cap W_i$. Furthermore, since $|\varphi_0(p)|<\rho_0$ on $\overline{D_0}$, it follows that the $\psi_i(p)$ $(i=1,2)$ can be holomorphically extended beyond $\mathcal{H}_0\cap W_i$ and satisfy (from (9.12))

$$\psi_1(p)-\psi_2(p) = F(\varphi_0(p),\varphi_1(p),\dots,\varphi_m(p)) = f(p), \quad p\in W\cap\mathcal{H}_0. \tag{9.14}$$

We remark that from (9.13), the functions $\psi_i(p)$ on $D_0\cap W_i$ $(i=1,2)$ can be written in the form

$$\psi_i(p) = \int_{L\times\Gamma}\chi(\zeta,p)F(\zeta)d\zeta_0 d\zeta_1\cdots d\zeta_m, \quad p\in D_0\cap W_i, \tag{9.15}$$

where $\zeta=(\zeta_0,\zeta_1,\dots,\zeta_m)\in L\times\Gamma$ and

$$\chi(\zeta,p) = \frac{1}{(2\pi i)^{m+1}}\,\frac{1}{(\zeta_0-\varphi_0(p))(\zeta_1-\varphi_1(p))\cdots(\zeta_m-\varphi_m(p))}$$

for $(\zeta,p)\in(L\times\Gamma)\times D_0$. We note from (9.14) that the $\psi_i(p)$ $(i=1,2)$ can be holomorphically extended beyond $W\cap\mathcal{H}_0$ to $(D_0\bigcap W_i)\cup(W\bigcap\mathcal{A})$ and satisfy $\psi_1(p)-\psi_2(p)=f(p)$ on $W\cap\mathcal{A}$. Furthermore, there exists a constant $\tilde{K}>0$ (independent of $f(p)$ on $\mathcal{A}$) such that

$$|\psi_i(p)| \le \tilde{K}M, \quad p\in W\cap\mathcal{A}. \tag{9.16}$$

To verify this last statement, we set $U'_+ = \{z_0 \in U' \mid u \geq 0\}$; note that the boundary of this set contains L, and set $L' = (\partial U'_+) \setminus L$ (which consists of two circular arcs and one line segment). Fix $\rho_0^* > 0$ with $\rho_0 > \rho_0^* > \max\{|\varphi_0(p)| \mid p \in \overline{D_0}\}$. We then have

$$\eta := \min\{\delta', \rho_0 - \rho_0^*\} \quad \leq \quad |\zeta_0 - \varphi_0(p)|$$

for $\zeta_0 \in L'$ and $p \in W_1 \cap \mathcal{A}$. Let $p \in W_1 \cap \mathcal{A}$ and $(\zeta_1, \dots, \zeta_m) \in \Gamma$ be fixed. Then $\chi(\zeta, p)$ is holomorphic as a function of ζ_0 on U'_+. Using Cauchy's formula, we can replace L by L' and obtain

$$\begin{aligned} |\psi_1(p)| &= \left| \int_{L' \times \Gamma} \chi(\zeta, p) F(\zeta) d\zeta_0 d\zeta_1 \cdots d\zeta_m \right| \\ &\leq \frac{KM}{\eta(\rho_1' - \rho_1)^m} \cdot (\pi \rho_0) \cdot (2\pi\rho_1)^m =: K'M, \end{aligned}$$

where $K' > 0$ does not depend on $f(p)$. Similarly we have $|\psi_2(p)| \leq K'M$ on $W_2 \cap \mathcal{A}$. It follows from (9.14) that

$$|\psi_1(p)| \leq |\psi_2(p)| + |f(p)| \leq (K' + 1)M \quad \text{on } W_2 \cap \mathcal{A}.$$

Therefore, $\tilde{K} = K' + 1 > 0$ satisfies (9.16).

Third step. We fix a small rectangular neighborhood l of L in $\mathbf{C}_{z_0}$ and we form the product set $\gamma = \gamma_1 \times \cdots \times \gamma_m$, where each γ_j $(j = 1, \dots, m)$ is a thin annular neighborhood of the circle $|\zeta_j| = \rho_1'$ in $\mathbf{C}_{z_j}$. We set $\tau = l \times \gamma$, which is a neighborhood of $L \times \Gamma$ in $\mathbf{C}^{m+1}$. We consider $\chi(\zeta, p)$ as a meromorphic function on $\tau \times D_0$. From condition 3° and the relation $\rho_1' > \rho_1 > 1 - \epsilon_0$, we can choose such a neighborhood τ of $L \times \Gamma$ sufficiently small so that the pole set of $\chi(\zeta, p)$ does not intersect $(\partial D_0) \cap D \subset \{p \in D \mid u_0(p) = b_1 \text{ or } b_2\}$. Therefore, if we define

$$\mathcal{C} = \begin{cases} \chi(\zeta, p) & \text{on } \tau \times D_0, \\ 0 & \text{on } \tau \times (D \setminus D_0), \end{cases}$$

then $\mathcal{C}$ is a Cousin I distribution on $\tau \times D$, and hence on $\tau \times D_1$. Since $\tau \times D_1$ is a holomorphically complete domain, from Theorem 8.9 we conclude that there exists a solution $\chi_1(\zeta, p)$ of the Cousin I problem for $\mathcal{C}$ on $\tau \times D_1$. Thus, $\chi_1(\zeta, p)$ is a meromorphic function on $\tau \times D_1$ with $\chi_1(\zeta, p) - \chi(\zeta, p)$ holomorphic on $\tau \times D_0$; moreover, $\chi_1(\zeta, p)$ itself is holomorphic on $\tau \times (D_1 \setminus D_0)$.

Fix $\epsilon > 0$. Since $\tau \times D_0$ is holomorphically convex in $\tau \times D_1$ by Lemma 9.5, it follows from the inclusion $(L \times \Gamma) \times \mathcal{A} \subset\subset \tau \times D_0$ that there exists a holomorphic function $H_1(\zeta, p)$ on $\tau \times D_1$ such that

$$|(\chi_1(\zeta, p) - \chi(\zeta, p)) - H_1(\zeta, p)| < \epsilon \quad \text{on} \quad (L \times \Gamma) \times \mathcal{A}.$$

We set

$$\begin{aligned} h_1(\zeta, p) &= \chi_1(\zeta, p) - \chi(\zeta, p) - H_1(\zeta, p) \quad \text{on } \tau \times D_0, \\ K_1(\zeta, p) &= \chi_1(\zeta, p) - H_1(\zeta, p) \quad \text{on } \tau \times D_1, \end{aligned}$$

so that $K_1(\zeta, p)$ is a meromorphic function on $\tau \times D_1$ with the same pole set as $\chi(\zeta, p)$ and

$$\begin{aligned} K_1(\zeta, p) &= \chi(\zeta, p) + h_1(\zeta, p) \quad \text{on } \tau \times D_0, \\ |h_1(\zeta, p)| &< \epsilon \qquad \text{on } (L \times \Gamma) \times \mathcal{A}. \end{aligned}$$

In a similar fashion, we can construct a meromorphic function $K_2(\zeta,p)$ on $\tau\times D_2$ with the same pole set as $\chi(\zeta,p)$ and a holomorphic function $h_2(\zeta,p)$ on $\tau\times D_0$ such that

$$\begin{aligned} K_2(\zeta,p) &= \chi(\zeta,p)+h_2(\zeta,p) \quad \text{on } \tau\times D_0, \\ |h_2(\zeta,p)| &< \epsilon \quad \text{on } (L\times\Gamma)\times\mathcal{A}. \end{aligned}$$

Since $K_i(\zeta,p)$ $(i=1,2)$ as well as $\chi(\zeta,p)$ has no poles in $(L\times\Gamma)\times W_i$, we can form the integral

$$I_if(p)=\int_{L\times\Gamma} K_i(\zeta,p)F(\zeta)d\zeta_0d\zeta_1\cdots d\zeta_m \quad \text{for } p\in W_i,$$

which is a holomorphic function on W_i. On the other hand, we have

$$\begin{aligned} I_if(p) &= \int_{L\times\Gamma}\chi(\zeta,p)F(\zeta)d\zeta_0d\zeta_1\cdots d\zeta_m \qquad (9.17) \\ &+ \int_{L\times\Gamma} h_i(\zeta,p)F(\zeta)d\zeta_0d\zeta_1\cdots d\zeta_m \quad \text{for } p\in W_i\cap D_0. \end{aligned}$$

The second term on the right-hand side is a holomorphic function on D_0. It follows from (9.14) and (9.15) that $I_if(p)$ can be holomorphically extended beyond $W\cap\mathcal{H}_0$ and satisfies

$$\begin{aligned} I_1f(p)-I_2f(p) &= f(p)+\int_{L\times\Gamma}(\,h_1(\zeta,p)-h_2(\zeta,p)\,)F(\zeta)d\zeta_0d\zeta_1\cdots d\zeta_m \\ &\qquad \text{for } p\in W\cap\mathcal{H}_0. \end{aligned}$$

Consider the second term on the right-hand side:

$$f^{(1)}(p):=\int_{L\times\Gamma}(\,h_2(\zeta,p)-h_1(\zeta,p)\,)F(\zeta)d\zeta_0d\zeta_1\cdots d\zeta_m \quad \text{for } p\in D_0;$$

$f^{(1)}(p)$ is a holomorphic function on D_0. Let $p\in\mathcal{A}\subset\subset D_0$. Since $|h_i(\zeta,p)|<\epsilon$ on $(L\times\Gamma)\times\mathcal{A}$ and $|F(\zeta)|\le KM$ on Λ' (which contains $L\times\Gamma$), we have

$$|f^{(1)}(p)|\le(2\epsilon)\cdot KM\cdot(2\rho_0)\cdot(2\pi\rho_1')^m=:\lambda M,$$

where $\lambda=2^{m+2}K\rho_0(\pi\rho_1')^m\epsilon>0$. We assume we have chosen $\epsilon>0$ sufficiently small so that $0<\lambda<1$. Since K,ρ_0,ρ_0' are independent of $f(p)$ on $\mathcal{A}$, so is λ.

We have constructed the integral kernel $K_i(\zeta,p)$ on $(L\times\Gamma)\times D_i$ $(i=1,2)$ with the following property: given a holomorphic function $f(p)$ on $\mathcal{A}$ such that $|f(p)|\le M$ on $\mathcal{A}$, take a holomorphic extension $F(z)$ of $f(p)$ on Λ such that $|F(z)|\le KM$ on Λ', and construct

$$I_if(p)=\int_{L\times\Gamma}K_i(\zeta,p)F(\zeta)d\zeta_0d\zeta_1\cdots d\zeta_m \quad (i=1,2) \quad \text{for } p\in W_i.$$

Then $I_if(p)$ is a holomorphic function on W_i which can be holomorphically extended beyond $W_i\cap\mathcal{H}_0$ and satisfies

$$I_1f(p)-I_2f(p)=f(p)-f^{(1)}(p) \quad \text{on } W\cap\mathcal{H}_0,$$

where $f^{(1)}(p)$ is a holomorphic function on D_0 with

$$|f^{(1)}(p)|\le\lambda M \quad \text{on } \mathcal{A}.$$

Therefore $I_i f(p)$ $(i = 1, 2)$ can be holomorphically extended to $W_i' = W_i \cup (\mathcal{A} \bigcap W)$ and satisfies

$$I_1 f(p) - I_2 f(p) = f(p) - f^{(1)}(p) \quad \text{on } \mathcal{A} \cap W. \tag{9.18}$$

Furthermore, (9.16) and (9.17) imply

$$I_i f(p) = \psi_i(p) + \int_{L \times \Gamma} h_i(\zeta, p) F(\zeta) d\zeta_0 d\zeta_1 \cdots d\zeta_m \quad \text{on } \mathcal{A} \cap W$$

and

$$|I_i f(p)| \leq \tilde{K} M + \epsilon K M \, 2\rho_0 (2\pi \rho_1')^m =: K^* M \quad \text{on } \mathcal{A} \cap W, \tag{9.19}$$

where $K^* > 0$ does not depend on the function $f(p)$ on $\mathcal{A}$.

Fourth step. We repeat the same procedure for $f^{(1)}(p)$ on $\mathcal{A}$ satisfying $|f^{(1)}(p)| \leq \lambda M$ on $\mathcal{A}$ as we used for $f(p)$ on $\mathcal{A}$ satisfying $|f(p)| \leq M$ on $\mathcal{A}$; thus we use an extension function $F^{(1)}(z)$ of $f^{(1)}(p)$ on Λ such that $|F^{(1)}(z)| \leq K\lambda M$ on Λ', and we obtain $I_i f^{(1)}(p)$ on W_i $(i = 1, 2)$ such that

$$I_1 f^{(1)}(p) - I_2 f^{(1)}(p) = f^{(1)}(p) - f^{(2)}(p) \quad \text{on } \mathcal{A} \cap W,$$

where $f^{(2)}(p)$ is a holomorphic function on D_0 such that $|f^{(2)}(p)| \leq \lambda^2 M$ on $\mathcal{A}$ and $|I_i f^{(1)}(p)| \leq K^* \lambda M$ $(i = 1, 2)$ on $\mathcal{A} \cap W$.

We thus inductively construct sequences of holomorphic functions $f^{(j)}(p)$ $(j = 1, 2, \dots)$ on D_0, $F^j(z)$ $(j = 1, 2, \dots)$ on Λ, and $I_i f^{(j)}(p)$ $(i = 1, 2;\ j = 1, 2, \dots)$ on W_i' such that

$$\begin{array}{llll} |f^{(j)}(p)| & \leq \lambda^j M & (j = 1, 2, \dots) & \text{on } \mathcal{A}, \\ |F^{(j)}(p)| & \leq K\lambda^j M & (j = 1, 2, \dots) & \text{on } \Lambda', \\ |I_i^{(j)} f(p)| & \leq K^* \lambda^j M & (j = 1, 2, \dots) & \text{on } \mathcal{A} \cap W. \end{array} \tag{9.20}$$

In order to solve equation (9.11), we set

$$\begin{aligned} \tilde{f}(p) &= f(p) + \sum_{j=1}^{\infty} f^{(j)}(p) \quad \text{on } \mathcal{A}, \\ \tilde{F}(z) &= F(z) + \sum_{j=1}^{\infty} F^{(j)}(z) \quad \text{on } \Lambda'. \end{aligned}$$

Using (9.20), and the fact that $0 < \lambda < 1$, we see that $\tilde{f}(p)$ is continuous on $\mathcal{A}$ and holomorphic in $\mathcal{A}^o$ (the interior of $\mathcal{A}$) and that $\tilde{F}(z)$ is continuous on Λ' and holomorphic in $(\Lambda')^o$, with

$$\tilde{F}(\varphi_0(p), \varphi_1(p), \dots, \varphi_m(p)) = \tilde{f}(p) \quad \text{on } \mathcal{A} \cap W.$$

We construct

$$I_i \tilde{f}(p) = \int_{L \times \Gamma} K_i(\zeta, p) \tilde{F}(\zeta) d\zeta \quad (i = 1, 2) \quad \text{for } p \in W_i,$$

so that $I_i \tilde{f}(p)$ is holomorphic on W_i. We shall show that $I_i \tilde{f}(p)$ can be holomorphically extended beyond $\mathcal{H}_0 \cap W$ and satisfies

$$I_1 \tilde{f}(p) - I_2 \tilde{f}(p) = f(p) \quad \text{on } \mathcal{H}_0 \cap W.$$

Indeed, fix $p \in W_i$ $(i = 1, 2)$. Since $\sum_{j=1}^{\infty} F^{(j)}(\zeta)$ is uniformly convergent on Λ' (which contains $L \times \Gamma$), we have

$$\begin{aligned} I_i\tilde{f}(p) &= \int_{L\times\Gamma} K_i(\zeta,p)\left(F(\zeta)+\sum_{j=1}^{\infty} F^{(j)}(\zeta)\right) d\zeta_0 d\zeta_1\cdots d\zeta_m \\ &= I_i f(p) + \sum_{j=1}^{\infty} I_i f^{(j)}(p). \end{aligned}$$

Using (9.20), we see that the right-hand side is a holomorphic function in $(\mathcal{A}\cap W)^o$. Therefore $I_i\tilde{f}(p)$ can be holomorphically extended beyond $\mathcal{H}_0\cap W$ to $W_i\cup(\mathcal{A}\bigcap W)^o$. Moreover, for any $p \in (\mathcal{A}\cap W)^o$,

$$\begin{aligned} &I_1\tilde{f}(p) - I_2\tilde{f}(p) \\ &= I_1 f(p) - I_2 f(p) + \sum_{j=1}^{\infty}(I_1 f^{(j)}(p) - I_2 f^{(j)}(p)) \\ &= f(p) - f^{(1)}(p) + \sum_{j=1}^{\infty}(f^{(j)}(p) - f^{(j+1)}(p)) \\ &= f(p) \quad \text{by (9.20)}. \end{aligned}$$

Consequently, $f_i(p) := I_i\tilde{f}(p)$ $(i = 1, 2)$ on W_i solves (9.11). Lemma 9.6 is completely proved $\square$

We make the following remarks concerning this proof.

1. In the second step of the proof, formulas (9.14) and (9.15) obtained from the kernel $K(\zeta,p)$ on $(L\times\Gamma)\times D_0$ imply that the functions $\psi_i(p)$ on $W_i\cap D_0$ $(i = 1, 2)$ give the solution of equation (9.11) on the holomorphically complete domain $D_0\cap W \subset W$.
2. In the third step, we modified the kernel $\chi(\zeta,p)$ defined in $(L\times\Gamma)\times D_0$ to form the kernel $K_i(\zeta,p)$ defined in $(L\times\Gamma)\times D_i$ $(i = 1, 2)$ in order to obtain a solution $f_i(p)$ of equation (9.11) on W_i.
3. Given a holomorphic function $f(p)$ on $\mathcal{A}$, the functions $I_i f(p)$ $(i = 1, 2)$ are defined on W_i. Although $W_i \cap \mathcal{A} \subset\subset \mathcal{A}$, the difference $f^{(1)}(p) = I_1 f(p) - I_2 f(p)$ is also holomorphic on $\mathcal{A}$; this allows us to repeat the same procedure for $f^{(1)}(p)$ as for $f(p)$.

9.2.3. Examination of the Conditions. In this section, we shall construct a finite number of holomorphic functions $\varphi_j(p)$ $(j = 1, \ldots, m)$ on D_0 which satisfy conditions 1°, 2°, and 3° in the previous section.

Fix $\xi \in \mathcal{H}_0\cap\partial D_0$. Since ∂D_0 near ξ is contained in $\{p \in \mathcal{D} \mid l(p) = \alpha\}$, there exists a continuous family of analytic hypersurfaces in a neighborhood δ_ξ of ξ in $\mathcal{D}$:

$$\sigma_t : g(t,p) = 0 \quad (p \in \delta_\xi,\ t \in I)$$

which touches the domain D_0 from outside at the point ξ. We may assume that $\delta_\xi \cap (\{p \in D_0 \mid u(p) = b_1 \text{ or } b_2\}\bigcup \mathbf{b}) = \emptyset$ (recall $\mathbf{b}$ was defined in (9.7)). We fix a real number $\beta > \alpha$ sufficiently close to α so that

$$\mathcal{D}^{(\beta)} \cap (\partial\sigma_0) = \emptyset.$$

By Lemma 9.5, the domain

$$\mathcal{D}_0^{(\beta)} := \{p \in \mathcal{D}^{(\beta)} \cap G \mid a_1 < u(p) < a_2\}$$

(with $D_0 \subset\subset \mathcal{D}_0^{(\beta)}$) is holomorphically complete; hence the Cousin I problem is solvable on $\mathcal{D}_0^{(\beta)}$. Applying Lemma 9.3 for D_0, $\mathcal{D}_0^{(\beta)}$, σ_0, and $g(0,p)$ (corresponding to D', D, S and $f_\xi(p)$ in the lemma), we obtain a holomorphic function $\varphi_\xi(p)$ on D_0 and two neighborhoods $\delta'_\xi \subset \delta''_\xi$ of ξ in $\mathcal{D}_0^{(\beta)} \cap \delta_\xi$ such that

$$\delta'_\xi \cap D_0 \subset \{p \in D_0 \mid |\varphi_\xi(p)| > 1\},$$
$$\sup \{|\varphi_\xi(p)| \mid p \in D_0 \setminus \delta''_\xi\} < 1.$$

Since $\mathcal{H}_0 \cap \partial D_0$ is compact, there exist a finite number of points $\xi_j \in \mathcal{H}_0 \cap \partial D_0$ $(j = 1, \dots, \nu)$ such that the corresponding functions $\varphi_{\xi_j}(p)$ and neighborhoods δ'_{ξ_j} satisfy the condition

$$\mathcal{H}_0 \cap \partial D_0 \subset \bigcup_{j=1}^{\nu} \delta'_{\xi_j};$$

thus $\mathcal{H}_0 \cap \partial D_0 \subset \bigcup_{j=1}^{\nu}\{p \in D_0 \mid |\varphi_{\xi_j}(p)| > 1\}$. For $\epsilon_1 > 0$ sufficiently small, we have

$$\max_{j=1,\dots,\nu} \{|\varphi_{\xi_j}(p)|\} < 1 \tag{9.21}$$
$$\text{for any } p \in D_0 \text{ such that } u(p) \le b_1 + \epsilon_1 \text{ or } \ge b_2 - \epsilon_1.$$

Consequently, if we fix $\delta > 0$ sufficiently small, then the subset $\mathcal{A}$ in D_0 defined by

$$\mathcal{A} = \{p \in D_0 \mid |u(p)| \le \delta,\ |\varphi_{\xi_j}(p)| \le 1 \quad (j = 1, \dots, \nu)\}$$

satisfies condition 1°. Furthermore, using (9.21), we see that condition 3° is satisfied.

In order to verify condition 2°, i.e., in order to prove that $\mathcal{A}$ has a normal model, we first note that $\mathcal{D}_0^{(\beta)}$ $(\beta > \alpha)$ is holomorphically complete. Since $D_0 \subset\subset \mathcal{D}_0^{(\beta)}$, there exists an analytic polyhedron $\mathcal{P}$ in $\mathcal{D}_0^{(\beta)}$ with defining functions $\psi_k(p)$ $(k = 1, \dots, \mu)$ in $\mathcal{D}_0^{(\beta)}$ such that $D_0 \subset\subset \mathcal{P} \subset\subset \mathcal{D}_0^{(\beta)}$ and $\Sigma : w_k = \psi_k(p)$ $(k = 1, \dots, \mu)$ is a normal model of $\mathcal{P}$ in the polydisk $\Delta^\mu : |w_k| < 1$ $(k = 1, \dots, \mu)$. Since $\mathcal{A} \subset \mathcal{P}$, if we set $\varphi_j(p) = \varphi_{\xi_j}(p)$ $(j = 1, \dots, \nu)$ and $\varphi_{\nu+k}(p) = \psi_k(p)$ $(k = 1, \dots, \mu)$ on D_0, then for $m = \nu + \mu$, the m holomorphic functions $\varphi_j(p)$ $(j = 1, \dots, m)$ on D_0 satisfy all the conditions 1°-3°.

9.2.4. Cousin I Problem. From Lemma 9.6 we obtain the following result.

LEMMA 9.7. *Let $\mathcal{D} \subset\subset \mathcal{V}$ be a domain which satisfies the linking condition. Let $\ell(p)$ be a strictly pseudoconvex exhaustion function on $\mathcal{D}$, and for a real number γ, let $\mathcal{D}^{(\gamma)} = \{p \in \mathcal{D} \mid \ell(p) < \gamma\} \subset\subset \mathcal{D}$. Then the Cousin I problem is solvable on $\overline{\mathcal{D}^{(\gamma)}}$.*

PROOF. We use the same notation $\mathcal{D}_j$ $(j = 0, 1, 2)$, $\varphi_0(p) = u(p) + iv(p)$ on $\mathcal{D}_0 \subset\subset G$, a_i $(i = 1, 2)$, and $\mathcal{H}_j$ $(j = 0, 1, 2)$ as in 9.2.1. Let $\mathcal{C} = \{(g_q(p), \delta_q)\}_{q \in U}$ be a Cousin I distribution on a domain U, $\overline{\mathcal{D}^{(\gamma)}} \subset U \subset \mathcal{D}$. We take a real number $\beta > \gamma$ sufficiently close to γ so that $\mathcal{D}^{(\beta)} \subset\subset U$. We also take b_i $(i = 1, 2)$ with $a_1 < b_1 < 0 < b_2 < a_2$, and, for fixed $\alpha > \beta$, we write

$$D := \mathcal{D}^{(\alpha)}, \quad D_j := \mathcal{D}'_j \cap \mathcal{D}^{(\alpha)} \quad (j = 0, 1, 2)$$

(as in (9.6)). We consider the following set:

$$\mathbf{b} = \mathcal{D}^{(\gamma)} \cap \mathcal{D}_0^{(\alpha)} = \{p \in \mathcal{D}^{(\gamma)} \cap G \mid b_1 < u(p) < b_1\}$$

(similar to (9.7)). Recall that in 9.2.3 we constructed a finite number of holomorphic functions $\varphi_j(p)$ $(j = 1, \dots, m)$ on D_0 satisfying conditions 1°, 2°, and 3°.

We continue to use the same notation $\mathcal{A}$, W, W_i $(i = 1, 2)$ as in Lemma 9.6. As noted in (9.10), we have $\mathcal{D}^{(\gamma)} \subset\subset W$. Each D_i $(i = 1, 2)$ is holomorphically complete; hence the Cousin I problem is solvable in D_i. Since $D_i \subset \mathcal{D}^{(\beta)} \subset U$, there exists a meromorphic function $G_i(p)$ in D_i with the same pole set as $g_q(p)$ on each $\delta_q \cap D_i$. Thus, $G_1(p) - G_2(p)$ is holomorphic in D_0, and hence in $\mathcal{A}$. By Lemma 9.6, there exists a holomorphic function $f_i(p)$ in W_i $(i = 1, 2)$ which can be holomorphically extended beyond $\mathcal{H}_0 \cap W$ and which satisfies $f_1(p) - f_2(p) = G_1(p) - G_2(p)$ on $\mathcal{H}_0 \cap W$. We set

$$F(p) = \begin{cases} G_1(p) + f_1(p), & p \in W_1, \\ G_2(p) + f_2(p), & p \in W_2. \end{cases}$$

Then $F(p)$ is a single-valued meromorphic function on W with the same pole set as $g_q(p)$ on each $\delta_q \cap W$. Since $\mathcal{D}^{(\gamma)} \subset\subset W$, the proof is complete. □

9.3. Principal Theorem

9.3.1. Linking Theorem. Let $\mathcal{V}$ be an analytic space of dimension n and let $\mathcal{D} \subset\subset \mathcal{V}$ be a domain which satisfies the linking condition in 9.2.1. Let $\ell(p)$ be an associated function on $\mathcal{D}$, and for a real number α, set $\mathcal{D}^{(\alpha)} = \{p \in \mathcal{D} \mid \ell(p) < \alpha\}$.

We have the following theorem.

THEOREM 9.2 (Linking theorem). *A domain $\mathcal{D}$ satisfying the linking condition is holomorphically complete.*

PROOF. From Proposition 9.4 and Lemma 9.2, it suffices to prove that for any real number α, there exists an analytic polyhedron $\mathcal{P}$ in $\mathcal{D}$ with defining functions on $G \subset \mathcal{D}$ satisfying

$$\mathcal{D}^{(\alpha)} \subset\subset \mathcal{P} \subset\subset \mathcal{D}.$$

We first prove that there exists a generalized analytic polyhedron $\mathcal{P}$ in $\mathcal{D}$ such that $\mathcal{D}^{(\alpha)} \subset\subset \mathcal{P} \subset\subset \mathcal{D}$. To do this, we fix a real number β with $\alpha < \beta < \infty$ and consider the domain $\mathcal{D}^{(\beta)}$. Fix $\xi \in \partial\mathcal{D}^{(\beta)}$. There exists a continuous family of analytic hypersurfaces in a neighborhood δ_ξ in $\mathcal{D}$:

$$\sigma_t : g_\xi(t, p) = 0 \quad (p \in \delta_\xi,\ t \in I),$$

which touches $\mathcal{D}^{(\beta)}$ from outside at the point ξ. Since the analytic space $\mathcal{V}$ is locally holomorphically complete at each point, we may assume that δ_ξ is holomorphically complete and that $\overline{\delta_\xi} \cap \mathcal{D}^{(\alpha)} = \emptyset$. Choose a real number $\gamma > \beta$ sufficiently close to β so that $(\partial\sigma_0) \cap \mathcal{D}^{(\gamma)} = \emptyset$. Since the Cousin I problem is solvable on $\overline{\mathcal{D}^{(\gamma)}}$ (Lemma 9.7), it follows from Lemma 9.3 that there exist a holomorphic function $\phi_\xi(p)$ on $\mathcal{D}^{(\beta)}$ and two neighborhoods $\delta'_\xi, \delta''_\xi$ of ξ in $\mathcal{D}^{(\gamma)}$ with $\delta'_\xi \subset \delta''_\xi \subset \delta_\xi$, satisfying

$$\begin{aligned} &\mathcal{D}^{(\beta)} \cap \delta'_\xi \subset \{p \in \mathcal{D}^{(\beta)} \mid |\varphi_\xi(p)| > 1\}, \\ &\sup\{|\varphi_\xi(p)| \mid p \in \mathcal{D}^{(\beta)} \setminus \delta''_\xi\} < 1. \end{aligned} \tag{9.22}$$

Since $\partial\mathcal{D}^{(\beta)}$ is compact, we can find a finite number of points $\xi_j \in \partial\mathcal{D}^{(\beta)}$ $(j = 1, \dots, \mu)$ such that

$$\partial\mathcal{D}^{(\beta)} \subset \bigcup_{j=1}^{\mu} \delta'_{\xi_j}. \tag{9.23}$$

If we set

$$\mathcal{P} := \{p \in \mathcal{D}^{(\beta)} \mid |\varphi_{\xi_j}(p)| < 1 \ (j = 1, \dots, \mu)\},$$

then $\mathcal{P}$ is a generalized analytic polyhedron in $\mathcal{D}$ with defining functions in $\mathcal{D}^{(\beta)}$, and

$$\mathcal{D}^{(\alpha)} \subset\subset \mathcal{P} \subset\subset \mathcal{D}^{(\beta)}.$$

We next prove that $\mathcal{P}$ is an analytic polyhedron, i.e., $\mathcal{P}$ satisfies the separation condition. To prove this, it suffices to show that $\mathcal{P}$ has a normal model in a unit polydisk $\overline{\Delta}$ in $\mathbf{C}^\nu$, where $\nu \geq \mu$.

In $\mathbf{C}^\mu$ with variables $z_1, \dots, z_\mu$, let $\overline{\Delta} : |z_j| \leq 1$ $(j = 1, \dots, \mu)$ be the unit polydisk; consider the analytic mapping

$$\Phi : p \in \mathcal{P} \to z = (\varphi_{\xi_1}(p), \dots, \varphi_{\xi_\mu}(p)) \in \overline{\Delta}$$

and set $\Sigma = \Phi(\mathcal{P}) \subset \overline{\Delta}$. Then Σ is an n-dimensional analytic set in $\overline{\Delta}$ with $\partial\Sigma \subset \partial\Delta$. Let $Q \in \Sigma \setminus \partial\Sigma$. Then $\Phi^{-1}(Q)$ is an analytic set in $\mathcal{P}$. Since $\partial\Sigma \subset \partial\Delta$ and $\mathcal{P}$ contains no compact analytic sets of positive dimension, it follows that $\Phi^{-1}(Q_0)$ consists of a finite number of points in $\mathcal{P}$. Assume that there exists a point Q of Σ such that $\Phi^{-1}(Q)$ consists of more than one point. Then $\mathcal{P}$ is mapped via Φ to a ramified domain $\tilde{\Sigma}$ over the analytic set Σ without relative boundary. We let $d \geq 2$ denote the number of sheets of $\tilde{\Sigma}$ over Σ. There exists a point $Q_0 \in \partial\Sigma$ such that $\Phi^{-1}(Q_0)$ consists of d distinct points $\zeta_1^0, \dots, \zeta_d^0 \in \partial\mathcal{P}$. Let

$$Q_0 = (\alpha_1, \dots, \alpha_\mu) \in \partial\Sigma;$$

thus some α_k $(k = 1, \dots, \mu)$ satisfies $|\alpha_k| = 1$. Therefore, $\varphi_{\xi_k}(\zeta_l^0) = \alpha_k$ $(l = 1, \dots, d)$. We set

$$S = \{p \in \mathcal{D}^{(\beta)} \mid \varphi_{\xi_k}(p) = \alpha_k\},$$

so that $S \subset \delta_{\xi_k}$ by (9.22), $\zeta_l^0 \in S$ $(l = 1, \dots, d)$, and $(\partial S) \cap \mathcal{D}^{(\beta)} = \emptyset$ (since $\partial S \subset \partial\mathcal{D}^\beta$). Since $\mathcal{D}^{(\beta)} \cap \delta_{\xi_k}$ as well as δ_{ξ_k} is holomorphically complete, we can find a holomorphic function $f(p)$ in $\mathcal{D}^{(\beta)} \cap \delta_{\xi_k}$ such that $f(\zeta_l^0) \neq f(\zeta_{l'}^0)$ $(l \neq l',\ 1 \leq l, l' \leq d)$. Fix a number $\beta' < \beta$ sufficiently close to β so that $\mathcal{P} \subset\subset \mathcal{D}^{(\beta')}$. Since the Cousin I problem is always solvable on $\overline{\mathcal{D}^{(\beta')}}$ and $\varphi_{\xi_k}(p)$ (which defines S) is holomorphic in $\overline{\mathcal{D}^{(\beta')}}$, we can apply Lemma 9.4 to obtain a holomorphic function $\varphi_0(p)$ on $\mathcal{D}^{(\beta')}$ such that $\varphi_0(\zeta_l^0) \neq \varphi_0(\zeta_{l'}^0)$ $(l \neq l',\ 1 \leq l, l' \leq d)$. We may assume $|\varphi_0(p)| < 1$ on $\mathcal{P}$.

In $\mathbf{C}^{\mu+1}$ with variables $\tilde{z} = (z_0, z_1, \dots z_\mu)$, we consider the unit polydisk

$$\tilde{\Delta} : |z_j| \leq 1 \qquad (j = 0, 1, \dots, \mu)$$

and the analytic mapping

$$\tilde{\Phi} : p \in \mathcal{P} \to \tilde{z} = (\varphi_0(p), \varphi_{\xi_1}(p), \dots, \varphi_{\xi_\mu}(p)) \in \tilde{\Delta},$$

and we set $\tilde{\Phi}(\mathcal{P}) = \tilde{\Sigma}$. For any point $\zeta \in \partial\tilde{\Delta}$ of the form $\tilde{\zeta} = (z_0, \alpha_1, \dots, \alpha_\mu) = (z_0, Q_0)$, the set $\tilde{\Phi}^{-1}(\tilde{\zeta})$ in $\mathcal{P}$ consists of at most one point. Thus, $\mathcal{P}$ and $\tilde{\Sigma}$ are in one-to-one correspondence via the mapping $\tilde{\Phi}$ except perhaps on an at most

$(n-1)$-dimensional analytic set in $\mathcal{P}$. It follows from Remark 8.2 that $\mathcal{P}$ has a normal model. $\square$

9.3.2. Principal Theorem. Let $\mathcal{V}$ be a normal pseudoconvex space with associated function $\ell(p)$. For a real number a, we set $\mathcal{V}_a := \{p \in \mathcal{V} \mid \ell(p) < a\}$. We now state and prove the main lemma in this section.

LEMMA 9.8. *If $\mathcal{V}_a$ is holomorphically complete, then there exists a real number b such that $b > a$ and $\mathcal{V}_b$ is also holomorphically complete.*

PROOF. Let $\zeta \in \partial\mathcal{V}_a$. There exists a continuous family of analytic hypersurfaces in a neighborhood δ_ζ of ζ in $\mathcal{V}$,

$$\sigma_{\zeta,t} : g_\zeta(p,t) = 0 \quad (p \in \delta_\zeta,\ t \in I = [0,1]),$$

which touches the domain $\mathcal{V}_a$ from outside at the point ζ. Since the analytic space is locally holomorphically complete (Corollary 8.1), we can assume that the neighborhood δ_ζ is holomorphically complete. Further we may assume that δ satisfies the condition in Corollary 6.4 in Chapter 6. By taking a smaller neighborhood δ_ζ, if necessary, we may also assume that $g_\zeta(p,t)$ is continuous for $(p,t) \in \delta_\zeta \times I$ and $g_\zeta(p,1) \neq 0$ for any $p \in \delta_\zeta$, i.e., $\sigma_{\zeta,1} = \emptyset$. Let ϵ_ζ and ϵ'_ζ be positive numbers with $\epsilon_\zeta > \epsilon'_\zeta > 0$, and set

$$\begin{aligned} \gamma_\zeta &= \{p \in \delta_\zeta \mid |g_\zeta(p,0)| < \epsilon_\zeta\}, \\ \gamma'_\zeta &= \{p \in \delta_\zeta \mid |g_\zeta(p,0)| < \epsilon'_\zeta\}. \end{aligned}$$

Thus, γ_ζ and γ'_ζ are neighborhoods of ζ in $\mathcal{V}$ with $\gamma'_\zeta \subset \gamma_\zeta \subset \delta_\zeta$. We may assume that $\epsilon_\zeta > 0$ is sufficiently small so that $[(\partial\gamma_\zeta) \cap (\partial\delta_\zeta)] \cap \mathcal{V}_a = \emptyset$. Since $\partial\mathcal{V}_a$ is a compact set, we can find a finite number of points ζ_j $(j = 1, \dots, N)$ such that, writing γ_j and γ'_j for γ_{ζ_j} and γ'_{ζ_j}, and writing $\sigma_{j,t}$ and $g_j(p,t)$ for $\sigma_{\zeta_j,t}$ and $g_{\zeta_j}(p,t)$, we have $\partial\mathcal{V}_a \subset \bigcup_{j=1}^N \gamma'_j$. We also let ϵ_j $(j = 1, ..., N)$ denote the corresponding numbers ϵ_{ζ_j}. It follows that we can find a real number $b > a$ sufficiently close to a so that

(1) $\partial\mathcal{V}_b \subset \bigcup_{j=1}^N \gamma'_j$, and

(2) if we set

$$\tilde{\gamma}_j = \gamma_j \cap \mathcal{V}_b, \quad \tilde{\gamma}'_j = \gamma'_j \cap \mathcal{V}_b \quad (j = 1, \dots, N),$$

then $[\tilde{\gamma}_j \setminus \tilde{\gamma}'_j] \cap \{\sigma_{j,t}\}_{t\in I} = \emptyset$.

We have the following fact:

($*$) the holomorphic function $\log g_j(p,0)$ has a single-valued branch on $\tilde{\gamma}_j \setminus \tilde{\gamma}'_j$.

PROOF. Let l be a closed curve in $\tilde{\gamma}_j \setminus \tilde{\gamma}'_j$. Let l^* be the image of l under the function $\tau = g_j(p,0)$; thus l^* is a closed curve in the complex plane $\mathbf{C}_\tau$ which does not pass through the origin 0. To verify ($*$), it suffices to show that the winding number $N(0)$ of l^* about 0 is zero. We note from (2) that $g_j(p,t)$ is a continuous function for $(p,t) \in \delta_j \times I$ with $g_j(p,t) \neq 0$ on $l \times I$. For each $t \in I$ we let $l^*(t)$ denote the image of l under the function $\tau = g_j(p,t)$; thus $l^*(t)$ is a closed curve in $\mathbf{C}_\tau \setminus \{0\}$, which varies continuously with $t \in I$. Thus, if we let $N(t)$ denote the winding number of $l^*(t)$ about 0, then $N(0) = N(t)$ for all $t \in I$. On the other hand, since $g_j(p,1) \neq 0$ for each $p \in \delta_j$, it follows from Corollary 6.4 that $\log\, g_j(p,1)$ is single-valued in δ_j. Hence $N(1) = 0$, and ($*$) is proved. $\square$

We set

$$\mathcal{D}_k = \mathcal{V}_b - \bigcup_{j=k+1}^{N} \tilde{\gamma}'_j \qquad (k = 0, 1, \ldots, N),$$

where $\mathcal{D}_N = \mathcal{V}_b$, so that $\mathcal{D}_0 \subset \mathcal{D}_1 \subset \cdots \subset \mathcal{D}_{N-1} \subset \mathcal{V}_b$ and $\mathcal{D}_0 \subset\subset \mathcal{V}_a$. In addition, we set

$$\gamma_j^0 = \tilde{\gamma}_j - \left[\bigcup_{h=j+1}^{N} \tilde{\gamma}'_h\right] \qquad (j = 1, \ldots, N),$$

where $\gamma_N^0 = \tilde{\gamma}_N$. Then we have

$$\mathcal{D}_{k+1} = \mathcal{D}_k \cup \gamma_{k+1}^0 \quad (k = 0, 1, \ldots, N-1) \tag{9.24}$$

(see Figure 3).

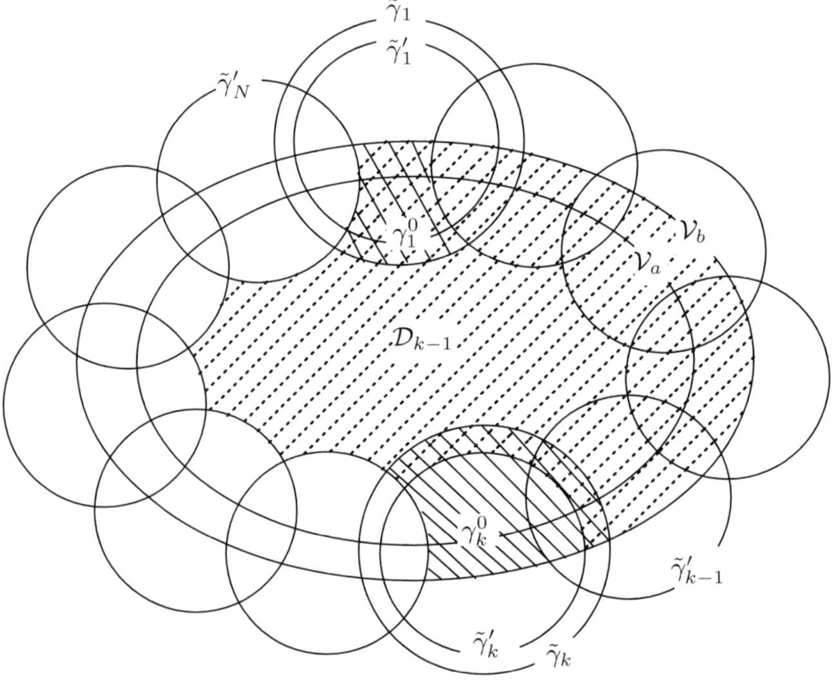

FIGURE 3. Representation of $\mathcal{D}_k$

Since $\mathcal{V}_a$ is holomorphically complete, it follows from Lemma 9.1 that $\mathcal{D}_0$ is holomorphically complete. Similarly, since δ_k is holomorphically complete and $g_k(p, 0)$ is holomorphic on δ_k, it follows again from Lemma 9.1 that each γ_k^0 $(k = 1, \ldots, N)$ is holomorphically complete.

We shall show that

(**) each $\mathcal{D}_k$ $(k = 0, 1, \ldots, N)$ is a normal pseudoconvex space.

PROOF. We prove this by reverse induction. We first note that $\mathcal{D}_N = \mathcal{V}_b$ is a normal pseudoconvex space, since $\ell_N(p) = 1/(b - \ell(p))$ is a strictly pseudoconvex exhaustion function on $\mathcal{V}_b$. We next assume that $\mathcal{D}_{k+1}$ is a normal pseudoconvex space with associated function $\ell_{k+1}(p) > 0$ on $\mathcal{D}_{k+1}$. We will construct a strictly pseudoconvex exhaustion function for $\mathcal{D}_k$. Using $g_{k+1}(p, 0)$ in δ_{k+1} we can construct a strictly pseudoconvex function $\eta_{k+1}(p)$ in $\gamma_{k+1} \setminus \gamma'_{k+1} \subset \delta_{k+1}$, with

$$\eta_{k+1}(p) \begin{cases} < 0, & p \in \partial\gamma_{k+1}, \\ = +\infty, & p \in \partial\gamma'_{k+1}. \end{cases}$$

We set

$$\ell_k(p) = \begin{cases} \max\{\ell_{k+1}(p), \eta_{k+1}(p)\}, & p \in \mathcal{D}_k \cap \gamma^0_{k+1}, \\ \ell_{k+1}(p), & p \in \mathcal{D}_k \setminus \gamma^0_{k+1}. \end{cases}$$

Then $\ell_k(p)$ is a strictly pseudoconvex exhaustion function in $\mathcal{D}_k$; this completes the proof of $(**)$. □

Finally, we show that

$(***)$ each $\mathcal{D}_k$ $(k = 0, 1, \dots, N)$ is holomorphically complete.

PROOF. We prove this by induction. We already noted that $\mathcal{D}_0$ is holomorphically complete. Assume that $\mathcal{D}_k$ is holomorphically complete. We will prove that $\mathcal{D}_{k+1}$ is holomorphically complete. By Theorem 9.2, it suffices to show that $\mathcal{D}_{k+1}$ satisfies the linking conditions (L1), (L2), and (L3) with $\mathcal{D}_k$ and γ^0_{k+1}.

We showed that $\mathcal{D}_{k+1} = \mathcal{D}_k \cup \gamma^0_{k+1}$; γ^0_{k+1} is holomorphically complete; and $\mathcal{D}_{k+1}$ is a normal pseudoconvex space. Thus, $\mathcal{D}_{k+1}$ satisfies conditions (L1) and (L2). To verify (L3), we set

$$\varphi(p) = \log g_{k+1}(p, 0) = u(p) + iv(p) \quad \text{on } \tilde{\gamma}_{k+1} \setminus \tilde{\gamma}'_{k+1},$$

which is a single-valued holomorphic function. Then

$$\mathcal{D}_k \cap \gamma^0_{k+1} = \{p \in (\tilde{\gamma}_{k+1} \setminus \tilde{\gamma}'_{k+1}) \cap \mathcal{D}_{k+1} \mid \log \epsilon'_{k+1} < u(p) < \log \epsilon_{k+1}\},$$

so that $\mathcal{D}_{k+1}$ satisfies condition (L3). Thus $(***)$ is verified. □

Consequently, $\mathcal{D}_N = \mathcal{V}_b$ is holomorphically complete, and Lemma 9.8 is completely proved. □

From Lemma 9.8 we obtain the following theorem.

THEOREM 9.3 (Nishino [**41**]). *Any normal pseudoconvex space is a Stein space.*

PROOF. Let $\mathcal{V}$ be a normal pseudoconvex space with associated function $\ell(p)$. Let $\alpha = \min_{p \in \mathcal{V}}\{\ell(p)\}$, so that $-\infty < \alpha < \infty$, and $\mathcal{V}_\alpha$ is a compact set in $\mathcal{V}$ without interior points. Then by reasoning similar to that used in Lemma 9.8 (replacing $\mathcal{V}_a$ by $\mathcal{V}_\alpha$ with $\partial\mathcal{V}_\alpha = \mathcal{V}_\alpha$), we see that there exists a real number $\beta > \alpha$ such that $\mathcal{V}_\beta$ is holomorphically complete. Thus, we can define

$$a_0 := \sup\{a \mid \mathcal{V}_a \text{ is holomorphically complete}\},$$

so that $\alpha < a_0 \leq +\infty$.

If $a_0 < +\infty$, then $\mathcal{V}_{a_0}$ is holomorphically complete from Proposition 9.4. This contradicts Lemma 9.8. Therefore $a_0 = +\infty$. Proposition 9.4 now yields that $\mathcal{V}$ itself is holomorphically complete. □

Notice that in establishing this theorem, which is one of the main goals of the book, almost all of the results in Chapters 7 and 8 have been used. In conjuction with Theorem 4.6, we obtain the following corollary.

COROLLARY 9.1. *Any pseudoconvex univalent domain in* $\mathbf{C}^n$ *is a domain of holomorphy.*

This corollary was proved by Oka. The case $n = 2$ is in [**49**] and the case $n \geq 2$ is in [**52**].

REMARK 9.1. As shown on p. 35 in [**49**], the linking theorem (Theorem 9.2) in section 9.3.1 is needed to prove the corollary in $\mathbf{C}^n$, but we do not need the arguments in section 9.3.2.

For let D be a pseudoconvex domain in $\mathbf{C}^n$ with associated function $l(z)$. Let $D_a = \{z \in \mathbf{C}^n \mid l(z) < a\}$ for a real number a. From Proposition 9.4 it suffices to prove that each $D_a\,(a < +\infty)$ is holomorphically complete. To show this, noting that D_a is locally holomorphically complete, we divide the z_i-plane $\mathbf{C}_{z_i}$ $(i = 1, \dots, n)$ into equal rectangles $\delta_i^{(j)}$ $(j = 1, 2, \dots)$ by two systems of straight lines parallel to the x_i- and y_i-axis (where $z_i = x_i + \sqrt{-1}\,y_i$) and we set $\Delta^{(\mathbf{j})} = \delta_1^{j_1} \times \cdots \times \delta_n^{j_n}$ (where $\mathbf{j} = (j_1, \dots, j_n)$), which is a box in $\mathbf{C}^n$. We set $D^{(\mathbf{j})} = D \cap \Delta^{(\mathbf{j})}$; then D_a is a finite collection of these sets $D^{(\mathbf{j})}$. If $\Delta^{(\mathbf{j})}$ is sufficiently small, then $D^{(\mathbf{j})}$ is a holomorphically complete domain. In this situation, we can apply the linking theorem step by step to conclude that D_a is holomorphically complete.

This method can be applied to a ramified domain $\mathcal{D}$ over $\mathbf{C}^n$ with associated function $l(p)$. The arguments in section 9.3.2 were needed to prove the holomorphic completeness for a general normal pseudoconvex space.

REMARK 9.2. In a ramified domain $\mathcal{D}$ over $\mathbf{C}^n$ any generalized analytic polyhedron $\mathcal{P}$ is an analytic polyhedron, i.e., $\mathcal{P}$ satisfies the separation condition.

To see this, let $\mathcal{P} : |\varphi_j(p)| \leq 1$ $(j = 1, \dots, m)$, so that $m \geq n$; let $\Sigma : w_j = \varphi_j(p)$ $(j = 1, \dots, m)$ in $\mathbf{C}^m$; and let $\Phi : p \in \mathcal{P} \to \Sigma$. Thus Σ is an n-dimensional analytic set in the unit polydisk Δ^m in $\mathbf{C}^m$. Let $u(w)$ be a strictly plurisubharmonic exhaustion function on Δ^m. Then $s(p) := u(\Phi(p))$ is a strictly pseudoconvex exhaustion function on $\mathcal{P}$. By Theorem 9.3, $\mathcal{P}$ is a Stein space.

REMARK 9.3. H. Grauert showed in [**22**] that any relatively compact *strongly* pseudoconvex domain with piecewise smooth boundary in a complex manifold is holomorphically convex, and that a complex manifold admitting a piecewise smooth strongly plurisubharmonic exhaustion function is a Stein manifold. R. Narasimhan showed in [**37**] that an analytic space with a strongly plurisubharmonic exhaustion function is a Stein space. Here, we say that a real-valued function $s(p)$ on an analytic space $\mathcal{V}$ is strongly plurisubharmonic on $\mathcal{V}$ if any point $p \in \mathcal{V}$ has a neighborhood v in $\mathcal{V}$ which is isomorphic to an analytic set σ in a domain δ in some $\mathbf{C}^\nu$; and, if we denote this isomorphism by T, then we require that $s \circ T^{-1}$ is the restriction to σ of some strictly plurisubharmonic function in δ.

9.4. Unramified Domains Over $\mathbf{C}^n$

We shall show that any unramified pseudoconvex domain D over $\mathbf{C}^n$ is holomorphically complete and hence is a domain of holomorphy.[8] Using Theorem 9.3, it suffices to construct a strictly plurisubharmonic, piecewise smooth exhaustion function on D.

---

[8]This fact was published in 1952 by Oka in [**52**]. However, he already proved it in 1943 (see Oka's posthumous work No. 6 in [**55**]).

9.4.1. Unramified Domains over $\mathbf{C}^n$. Let $\mathcal{R}$ be an unramified domain over $\mathbf{C}^n$ with variables $z_1, \dots, z_n$ and let $\pi : \mathcal{R} \to \mathbf{C}^n$ be the canonical projection. We write $\pi(p) = \underline{p}$ for $p \in \mathcal{R}$. We let $q_r(a)$ denote the ball in $\mathbf{C}^n$ centered at a with radius r, and we let $\gamma_r(a)$ denote the polydisk in $\mathbf{C}^n$ centered at a with radius r. Let $p \in \mathcal{R}$, and let v be a univalent subset of $\mathcal{R}$ such that $p \in v$ and $\pi(v) = q_r(\underline{p})$. We call the supremum $D_{\mathcal{R}}(p)$ of such $r > 0$ the **(Euclidean) boundary distance** of $\mathcal{R}$ from the point p. If we replace $q_r(\underline{p})$ by $\gamma_r(\underline{p})$, then we call the supremum $\Delta_{\mathcal{R}}(p)$ of r with $\pi(v) = \gamma_r(\underline{p})$ the **cylindrical boundary distance** of $\mathcal{R}$ from the point p.

Let $E \subset \mathcal{R}$. Then we call $\inf\{D_{\mathcal{R}}(p) \mid p \in E\}$ the **(Euclidean) boundary distance of $\mathcal{R}$ from the set** E. Given $\rho > 0$, we also define

$$\mathcal{R}^{(\rho)} = \{p \in \mathcal{R} \mid \Delta_{\mathcal{R}}(p) > \rho\}. \tag{9.25}$$

We set

$$\delta(p) = -\log D_{\mathcal{R}}(p), \quad p \in \mathcal{R}, \tag{9.26}$$

which we call the **logarithmic boundary distance function** on $\mathcal{R}$.

We have the following theorem.

THEOREM 9.4. *If $\mathcal{R}$ is an unramified pseudoconvex domain over $\mathbf{C}^n$, then $\delta(p)$ is a plurisubharmonic function on $\mathcal{R}$.*

PROOF. This theorem was proved in the case when $\mathcal{R}$ is a univalent pseudoconvex domain D in $\mathbf{C}^n$ in Lemma 4.6. In the proof we did not need D to be univalent, i.e., the proof is valid for an unramified pseudoconvex domain $\mathcal{R}$ over $\mathbf{C}^n$. Thus Theorem 9.4 is true. □

In the case of a bounded univalent domain D in $\mathbf{C}^n$, we have $D^{(\rho)} \subset\subset D$ for each $\rho > 0$. Using this fact, we constructed a piecewise smooth, strictly plurisubharmonic exhaustion function on a univalent pseudoconvex domain in $\mathbf{C}^n$. However, in the case of an (infinitely sheeted) bounded unramified domain $\mathcal{R}$ over $\mathbf{C}^n$, it is no longer true that $\mathcal{R}^{(\rho)} \subset\subset \mathcal{R}$ for all $\rho > 0$. For example, let $\mathcal{R}$ denote the portion of the Riemann surface of $\log z$ lying over the disk $\{|z| < 1\}$. Then $\mathcal{R}^{(\rho)}$ for $0 < \rho < 1/2$ coincides with the portion of $\mathcal{R}$ lying over $\{\rho < |z| < 1 - \rho\}$, which is not relatively compact in $\mathcal{R}$. Thus, we need further analysis, which we will carry out in the following section, to construct a piecewise smooth, strictly plurisubharmonic exhaustion function on an unramified pseudoconvex domain over $\mathbf{C}^n$.

9.4.2. Family of Continuous Curves. Let l be a rectifiable curve in $\mathbf{C}^n$, i.e.,

$$l : t \in I \to z = (l_1(t), \dots, l_n(t)) \in \mathbf{C}^n,$$

where $I = [0, 1]$, $l_j(t)$ $(j = 1, \dots, n)$ is a complex-valued continuous function on I, and the Euclidean length $L(l)$ of the curve l in $\mathbf{C}^n$ is finite. We call the vector-valued function $l(t) = (l_1(t), \dots, l_n(t))$ on I a parameterization for the curve l.

We consider a sequence of rectifiable curves l_k $(k = 1, 2, \dots)$ in $\mathbf{C}^n$. If we can find a sequence of parameterizations $l^k(t)$ on I of l_k such that $l^k(t)$ $(k = 1, 2, \dots)$ converges uniformly to $l^0(t)$ on I, then we say that l_k $(k = 1, 2, \dots)$ converges uniformly to the curve

$$l^0 : z = l^0(t), \quad t \in I,$$

and we call l^0 the limiting curve of l_k $(k = 1, 2, \dots)$.

Let $\{l_\iota\}_\iota$ be a family of rectifable curves in $\mathbf{C}^n$. If for any sequence $\{l_k\}_{k=1,2,\dots}$ which is contained in $\{l_\iota\}_\iota$ we can choose a subsequence $\{l_{k_j}\}_{j=1,2,\dots}$ of $\{l_k\}_{k=1,2,\dots}$ such that $\{l_{k_j}\}_{j=1,2,\dots}$ converges uniformly to a curve l, then we say that $\{l_\iota\}_\iota$ is a **normal family**.

We have the following proposition.

PROPOSITION 9.5 (Oka). *Let $\{l_\iota\}_{\iota\in\mathcal{I}}$ be a family of rectifiable curves in $\mathbf{C}^n$ such that the set of initial points of l_ι $(\iota \in \mathcal{I})$ is bounded in $\mathbf{C}^n$ and the Euclidean length $L(l_\iota)$ of the curves l_ι $(\iota \in \mathcal{I})$ is uniformly bounded. Then $\{l_\iota\}_{\iota\in\mathcal{I}}$ is a normal family.*

PROOF. We prove this by use of the arc length parameter τ of the curve l_ι. Let $l_\iota(t)$ $(\iota \in \mathcal{I})$ be a parameterization on I of the curve l_ι and let $L_\iota(t)$ denote the length of l_ι from $l_\iota(0)$ to $l_\iota(t)$. By assumption there exists an $M > 0$ such that $L_\iota(1) \le M$ $(\iota \in \mathcal{I})$. We may assume each $L_\iota(1) > 0$. We set $\tau_\iota : t \in I \to \tau = L_\iota(t)/L_\iota(1) \in [0, 1]$ $(\iota \in \mathcal{I})$, and we let $t = \sigma_\iota(\tau)$ denote the inverse function of τ_ι. Then $l_\iota^*(\tau) := l_\iota(\sigma_\iota(\tau))$ is a parameterization on I of l_ι. We have

$$|l_\iota^*(\tau') - l_\iota^*(\tau'')| \le M\,|\tau' - \tau''| \qquad \text{for all } \tau', \tau'' \in I,$$

so that $l_\iota^*(\tau)$ $(\iota \in \mathcal{I})$ is equicontinuous on I. Since $\{l_\iota(0)\}_{\iota\in\mathcal{I}}$ is bounded in $\mathbf{C}^n$, it follows from the Arzelà-Ascoli theorem that $\{l_\iota^*(\tau)\}_{\iota\in\mathcal{I}}$ is a normal family of functions on I. Thus, $\{l_\iota\}_{\iota\in\mathcal{I}}$ is normal. □

9.4.3. Distance Function. Let $\mathcal{R}$ be an unramified domain over $\mathbf{C}^n$. Let $p_1, p_2 \in \mathcal{R}$ and let γ be a curve which connects p_1 and p_2 in $\mathcal{R}$. We let $L(\gamma)$ denote the Euclidean length of the curve $\underline{\gamma} = \pi(\gamma)$ in $\mathbf{C}^n$. We set

$$d_{\mathcal{R}}(p_1, p_2) = \inf\{L(\gamma) \mid \gamma \text{ connects } p_1 \text{ and } p_2 \text{ in } \mathcal{R}\}.$$

Let $\mathcal{R}_0$ be a connected region in $\mathcal{R}$ such that the boundary distance of $\mathcal{R}$ from $\mathcal{R}_0$ is positive, i.e.,

$$m = \inf\{D_{\mathcal{R}}(p) \mid p \in \mathcal{R}_0\} > 0.$$

Fix a point p_0 in $\mathcal{R}_0$. Let $p \in \mathcal{R}_0$ and set

$$d_{p_0}(p) = d_{\mathcal{R}_0}(p_0, p), \tag{9.27}$$

which is called the **distance function** on $\mathcal{R}_0$ with initial point p_0.

We have the following lemma.

LEMMA 9.9. *For each $M > 0$, the subset*

$$E_M := \{p \in \mathcal{R}_0 \mid d_{p_0}(p) < M\}$$

of $\mathcal{R}_0$ is relatively compact in $\mathcal{R}$.

PROOF. We prove this by contradiction; thus we assume that there exists a sequence of points p_k $(k = 1, 2, \dots)$ in E_M such that $\{p_k\}_{k=1,2,\dots}$ has no accumulation points in $\mathcal{R}$. For each $k = 1, 2, \dots$, we can find a continuous curve l_k in $\mathcal{R}_0$ which connects p_0 and p_k such that $L(l_k) < M$. We write $\underline{l_k} = \pi(l_k)$ and $\underline{p_0} = \pi(p_0)$. Then $\underline{l_k}$ is a rectifiable curve in $\mathbf{C}^n$ with initial point $\underline{p_0}$ and length $L(\underline{l_k}) = L(l_k) < M$. It follows from Proposition 9.5 that we can find a subsequence $\{\underline{l_{k_j}}\}_{j=1,2,\dots}$ of $\{\underline{l_k}\}_{k=1,2,\dots}$ which converges uniformly to a curve $\underline{l_0}$. Let $\underline{l_{k_j}}(t)$ and $\underline{l_0}(t)$ be parameterizations on I of $\underline{l_{k_j}}$ and $\underline{l_0}$ such that $\lim_{j\to\infty} \underline{l_{k_j}}(t) = \underline{l_0}(t)$ uniformly on I. We fix r with $0 < r < m/2$ and consider the band $\mathcal{B}$ along $\underline{l_0}$ with radius r; i.e., the collection of balls $q_r(\underline{l_0}(t))$, $t \in I$, in $\mathbf{C}^n$. Then we have $\underline{l_{k_j}}(t) \subset \mathcal{B}$ ($t \in I$) for sufficiently large j. Fix such a j. Then $\underline{l_0}(t)$ ($t \in I$) is contained in the band $\mathcal{B}_j$ along $\underline{l_{k_j}}$ with radius $2r < m$ and $\mathcal{B} \subset \mathcal{B}_j$. Since $D_{\mathcal{R}}(p) \geq m$ for all $p \in l_{k_j}(t)$, $t \in I$, it follows that $\mathcal{B}_j \subset\subset \mathcal{R}$, and hence $\mathcal{B} \subset\subset \mathcal{R}$. We thus have $p_{k_i} = \underline{l_{k_i}}(1) \in \mathcal{B}$ for all sufficiently large i. This is a contradiction. □

9.4.4. Modification of $d_{p_0}(p)$. Let $\mathcal{U}$ be a domain in an analytic space $\mathcal{V}$. Let $h(p)$ and $k(p)$ be two real-valued functions on $\mathcal{U}$. If, given a real number a, there exists a real number $b > 0$ such that

$$\{p \in \mathcal{U} \mid h(p) < a\} \subset \{p \in \mathcal{U} \mid k(p) < b\},$$
$$\{p \in \mathcal{U} \mid k(p) < a\} \subset \{p \in \mathcal{U} \mid h(p) < b\},$$

then we say that $h(p)$ and $k(p)$ are of **weakly bounded difference** in $\mathcal{U}$. Furthermore, if $h(p) - k(p)$ is a bounded function in $\mathcal{U}$, then we say that $h(p)$ and $k(p)$ are of **bounded difference** in $\mathcal{U}$.

Let $\mathcal{R}$ be an unramified domain over $\mathbf{C}^n$ and for $m > 0$ let

$$\mathcal{R}_0 = \mathcal{R}_{0,m} = \{p \in \mathcal{R} \mid D_{\mathcal{R}}(p) > m\} \subset \mathcal{R}. \tag{9.28}$$

We note that if $\mathcal{R}$ is finitely sheeted and bounded over $\mathbf{C}^n$, then $\mathcal{R}_0 \subset\subset \mathcal{R}$. However, if $\mathcal{R}$ is infinitely sheeted, this is not necessarily the case. We defined $d_{p_0}(p) = d_{\mathcal{R}_0}(p_0, p)$ on $\mathcal{R}_0$ where p_0 is a fixed point in $\mathcal{R}_0$. Let $\rho > 0$ and let

$$\mathcal{R}_0^{(\rho)} = \{p \in \mathcal{R}_0 \mid \Delta_{\mathcal{R}_0}(p) > \rho\}, \tag{9.29}$$

so that the polydisk $\gamma_\rho(p) \subset\subset \mathcal{R}$ for each $p \in \mathcal{R}_0^{(\rho)}$ (recall the notation from (9.25)). We shall construct a strictly plurisubharmonic function $u(p)$ on $\mathcal{R}_0^{(\rho)}$ such that $u(p)$ and $d_{p_0}(p)$ are of weakly bounded difference in $\mathcal{R}_0^{(\rho)}$.

To this end, we recall the following mean-value integral of a real-valued, continuous function $\varphi(z)$ which was studied in Chapter 4:

$$\varphi_1(z) := A_r\varphi(z) = \frac{1}{(\pi r^2)^n} \int_{\gamma_r(z)} \varphi(\zeta) dv_\zeta. \tag{9.30}$$

We have the following lemma.

LEMMA 9.10. *Let $\varphi(z)$ be a real-valued continuous function on a univalent domain D in $\mathbf{C}^n$. Let $\rho > 0$ and let $D^{(\rho)} = \{p \in D \mid \Delta_D(p) > \rho\}$. For $0 < r < \rho$ define $\varphi_1(z) = A_r\varphi(z)$ on $D^{(\rho)}$ (which is of class C^1 in $D^{(\rho)}$).*

1. *If there exists a constant $c > 0$ such that for any two points z^1, z^2 in D,*
$$|\varphi(z^1) - \varphi(z^2)| \leq c\,\|z^1 - z^2\|,$$
then we have
$$\left|\frac{\partial \varphi_1(z)}{\partial \xi}\right| \leq c, \quad z \in D^{(\rho)}.$$
Here $\partial/\partial\xi$ denotes any of the partial derivatives $\partial/\partial x_j$ or $\partial/\partial y_j$, where $z_j = x_j + \sqrt{-1}y_j$ ($j = 1, \dots, n$).[9]
2. *If $|\varphi(z)| \leq M$ on D, then*
$$\left|\frac{\partial \varphi_1(z)}{\partial \xi}\right| \leq \frac{4M}{\pi r}, \quad z \in D^{(\rho)}.$$

[9] $\|z^1 - z^2\|$ denotes the Euclidean distance between z^1 and z^2 in $\mathbf{C}^n$.

PROOF. We prove this for $\xi = x_1$. Let $a = (a_1, \dots, a_n) \in D^{(\rho)}$ and let $\triangle\xi$ be a real number such that $a' = (a_1 + \triangle\xi, a_2, \dots, a_n) \in D^{(\rho)}$. Under the assumption in 1, we have

$$\begin{aligned}
&\left| \frac{1}{\triangle\xi} \left(\int_{\gamma_r(a')} \varphi(z) dv_z - \int_{\gamma_r(a)} \varphi(z) dv_z \right) \right| \\
&\leq \frac{1}{|\triangle\xi|} \left| \int_{\gamma_r(0)} (\varphi(a' + \zeta) - \varphi(a + \zeta)) dv_\zeta \right| \\
&\leq \frac{1}{|\triangle\xi|} \int_{\gamma_r(0)} c \, \|a' - a\| dv_\zeta = c(\pi r^2)^n,
\end{aligned}$$

so that $|(\partial\varphi_1/\partial\xi)(a)| \leq c$. Thus 1 is proved.

Under the assumption in 2, we have

$$\begin{aligned}
&\left| \frac{1}{\triangle\xi} (\int_{\gamma_r(a')} \varphi(z) dv_z - \int_{\gamma_r(a)} \varphi(z) dv_z) \right| \\
&\leq \frac{M}{|\triangle\xi|} \text{ vol } (\gamma_r(a') \setminus \gamma_r(a)) \\
&\leq 4Mr \cdot (\pi r^2)^{n-1},
\end{aligned}$$

so that $|(\partial\varphi_1/\partial\xi)(a)| \leq 4M/\pi r$. Hence, 2 is proved. □

Since the mean-value integral $\varphi_1(z) = A_r\varphi(z)$ of $\varphi(z)$ is defined locally on $D^{(\rho)}$, this integration can be defined on $\mathcal{D}^{(\rho)}$ for a continuous function $\varphi(z)$ on an unramified domain $\mathcal{D}$ over $\mathbf{C}^n$.

We return to the situation (9.29). The distance function $d_{p_0}(p) = d_{\mathcal{R}_0}(p_0, p)$ on $\mathcal{R}_0$ is a real-valued continuous function in $\mathcal{R}_0$ with the property that

$$|d_{p_0}(p') - d_{p_0}(p'')| \leq \|\underline{p}' - \underline{p}''\| \tag{9.31}$$

for any two points p', p'' in $\mathcal{R}_0$ such that p', p'' are contained in a (univalent) ball in $\mathcal{R}_0$. Fix r with $0 < r < \rho$. Then we can construct the mean-value integral defined by (9.30):

$$\begin{aligned}
\varphi_1(p) &:= A_r d_{p_0}(p), \quad p \in \mathcal{R}_0^{(\rho)}, \\
\varphi_2(p) &:= A_r \varphi_1(p), \quad p \in \mathcal{R}_0^{(2\rho)}.
\end{aligned}$$

Then $\varphi_1(p)$ is of class C^1 in $\mathcal{R}_0^{(\rho)}$ and

$$\left| \frac{\partial\varphi_1}{\partial\xi}(p) \right| \leq 1, \quad p \in \mathcal{R}_0^{(\rho)}; \tag{9.32}$$

in addition, $\varphi_2(p)$ is of class C^2 in $\mathcal{R}_0^{(2\rho)}$ with

$$\left| \frac{\partial\varphi_2}{\partial\eta\partial\xi}(p) \right| \leq \left| \frac{\partial}{\partial\eta} \left(A_r \frac{\partial\varphi_1}{\partial\xi} \right) (p) \right| \leq \frac{4}{\pi r}, \quad p \in \mathcal{R}_0^{(2\rho)}. \tag{9.33}$$

We note from (9.31) and (9.32) that

$$|\varphi_2(p) - d_{p_0}(p)| \leq 2r, \quad p \in \mathcal{R}_0^{(2\rho)}. \tag{9.34}$$

Next we define the following function on $\mathcal{R}$:

$$\zeta(p) = |z_1|^2 + \cdots + |z_n|^2, \quad p \in \mathcal{R},$$

where $p = (z_1, \dots, z_n) \in \mathbf{C}^n$. Then $\zeta(p)$ is a strictly plurisubharmonic function on $\mathcal{R}$. Given a constant $K > 0$, we set

$$\lambda_K(p) = \varphi_2(p) + K\zeta(p), \quad p \in \mathcal{R}_0^{(2\rho)},$$

which is a positive-valued function of class C^2 on $\mathcal{R}_0^{(2\rho)}$.

We have the following lemma.

LEMMA 9.11. *There exists a strictly plurisubharmonic function $\varphi(z)$ on $\mathcal{R}_0^{(2\rho)}$ such that $\varphi(p)$ and $d_{p_0}(p)$ are of weakly bounded difference on $\mathcal{R}_0^{(2\rho)}$.*

PROOF. We fix a constant $K > 0$ such that $K > 4n^2/(\pi r)$. We shall show that $\varphi(p) = \lambda_K(p)$ on $\mathcal{R}_0^{(2\rho)}$ satisfies the conclusion of the lemma.

Indeed, from (9.33) we see that for any $c = (c_1, \dots, c_n) \in \mathbf{C}^n$ with $\|c\| = 1$, we have

$$\sum_{j,k=1}^{n} \frac{\partial^2 \lambda_K(p)}{\partial z_i \partial \overline{z}_j} c_i \overline{c_j} \geq -\frac{4n^2}{\pi r} + K > 0,$$

so that $\lambda_K(p)$ is a strictly plurisubharmonic function on $\mathcal{R}_0^{(2\rho)}$. It is clear from (9.34) that $\varphi_2(p)$ and $d_{p_0}(p)$ are of bounded difference on $\mathcal{R}_0^{(2\rho)}$. Let $c > 0$ and let $A_c = \{p \in \mathcal{R}_0^{(2\rho)} \mid d_{p_0}(p) < c\}$. Then the projection $\underline{A_c}$ of A_c to $\mathbf{C}^n$ is a bounded subset in $\mathbf{C}^n$, so that $\zeta(p)$ is bounded on A_c. It follows that $\lambda_K(p)$ and $d_{p_0}(p)$ are of weakly bounded difference on $\mathcal{R}_0^{(2\rho)}$. □

In the case when the projection $\underline{\mathcal{R}_0}$ of $\mathcal{R}_0$ to $\mathbf{C}^n$ is a bounded domain in $\mathbf{C}^n$, $\lambda_K(p)$ and $d_{p_0}(p)$ are of bounded difference on $\mathcal{R}_0^{(2\rho)}$.

Given $\varepsilon > 0$, by taking smaller $m > 0$ and $\rho > 0$ such that $m + 2\rho < \varepsilon$ we have from this lemma the following corollary.

COROLLARY 9.2. *Let $\mathcal{R}$ be an unramified pseudoconvex domain over $\mathbf{C}^n$, and for $\varepsilon > 0$, let*

$$\mathcal{D} = \{p \in \mathcal{R} \mid D_{\mathcal{R}}(p) > \varepsilon\}.$$

Then there exists a strictly plurisubharmonic function $\varphi(p)$ on $\mathcal{D}$ such that, for any real number a,

$$\mathcal{D}_a = \{p \in \mathcal{D} \mid \varphi(p) < a\} \subset\subset \mathcal{R}.$$

This corollary says that, if we let $\partial\mathcal{D}_a$ denote the boundary of $\mathcal{D}_a$ in $\mathcal{R}$, then $\mathcal{D}_a$ is finitely sheeted over $\mathbf{C}^n$ and

$$\partial\mathcal{D}_a \subset \{p \in \mathcal{R} \mid D_{\mathcal{R}}(p) = \varepsilon\} \cup \{p \in \mathcal{D} \mid \varphi(p) = a\}. \tag{9.35}$$

9.4.5. Construction of an Associated Function on $\mathcal{R}$. Let $\mathcal{R}$ be an unramified pseudoconvex domain over $\mathbf{C}^n$. We defined the logarithmic boundary distance function $\delta(p) = -\log D_{\mathcal{R}}(p)$ on $\mathcal{R}$ by (9.26). Since $\mathcal{R}$ is pseudoconvex, we have that $\delta(p)$ is continuous and plurisubharmonic on $\mathcal{R}$. Let a_j $(j = 1, 2, \dots)$ be a sequence of positive numbers such that

$$a_j < a_{j+1} \quad (j = 1, 2, \dots), \qquad \lim_{j\to\infty} a_j = +\infty.$$

We set

$$\mathcal{R}_j = \{p \in \mathcal{R} \mid \delta(p) < a_j\} \quad (j = 1, 2, \dots).$$

Equivalently, setting $\varepsilon_j = e^{-a_j} > 0$ $(j = 1, 2, \dots)$ (so that $\varepsilon_j > \varepsilon_{j+1}$ and $\lim_{j\to\infty} \varepsilon_j = 0$), we have $\mathcal{R}_j = \{p \in \mathcal{R} \mid D_{\mathcal{R}}(p) > \varepsilon_j\}$. Then

$$\mathcal{R}_j \subset \mathcal{R}_{j+1} \quad (j = 1, 2, \dots), \qquad \lim_{j\to\infty} \mathcal{R}_j = \mathcal{R}.$$

By Corollary 9.2, there exists a strictly plurisubharmonic function $\varphi_j(p)$ on $\mathcal{R}_j$ $(j = 1, 2, \dots)$ such that, for any real number b,

$$(\mathcal{R}_j)_b := \{p \in \mathcal{R}_j \mid \varphi_j(p) < b\} \subset\subset \mathcal{R}.$$

We note from (9.35) that $(\mathcal{R}_j)_b \subset\subset \mathcal{R}_{j+1}$. Next, let b_j $(j = 1, 2, \dots)$ be a sequence of positive real numbers such that

(1) $b_j < b_{j+1}$ $(j = 1, 2, \dots)$ and $\lim_{j\to\infty} b_j = +\infty$;

(2) if we set

$$\Delta_j = \{p \in \mathcal{R}_j \mid \varphi_{j+3}(p) < b_j\},$$

then

$$\Delta_j \subset\subset \Delta_{j+1} \ (j = 1, 2, \dots), \quad \lim_{j\to\infty} \Delta_j = \mathcal{R}.$$

This is possible by taking b_{j+1} sufficiently greater than b_j. We note that $\Delta_{j+2} \subset\subset \mathcal{R}_{j+3}$ and

$$\partial\Delta_j \subset \{p \in \mathcal{R} \mid \delta(p) = a_j\} \bigcup \{p \in \mathcal{R}_{j+3} \mid \varphi_{j+3}(p) = b_j\}.$$

We set

$$\psi_j(p) = \max\{\delta(p) - a_j, \varphi_{j+3}(p) - b_j\}, \quad p \in \Delta_{j+2} \setminus \Delta_{j-1} \quad (j = 2, 3, \dots).$$

Then $\psi_j(p)$ is a plurisubharmonic function on $\Delta_{j+2} \setminus \Delta_{j-1}$ satisfying

$$\psi_j(p) > 0 \quad (\text{resp. } = 0, < 0) \quad \text{on } \overline{\Delta_{j+2}} \setminus \overline{\Delta_j} \ (\text{resp. } \partial\Delta_j, \ \Delta_j \setminus \Delta_{j-1}).$$

Using the sequence of functions $\psi_j(p)$ on $\Delta_{j+2} \setminus \Delta_{j-1}$ $(j = 1, 2, \dots)$, we can apply standard techniques to construct a plurisubharmonic exhaustion function $\psi(p)$ on $\mathcal{R}$.

To be precise, we fix a plurisubharmonic function $\tilde{\psi}_2(p)$ on $\overline{\Delta_3}$ such that $\tilde{\psi}_2(p) > 0$ on $\overline{\Delta_3}$. We take $k_2 > 0$ sufficiently large so that

$$\min_{p\in\overline{\Delta_4}\setminus\Delta_3} \{k_2\, \psi_2(p)\} > \max\{3, \max_{p\in\overline{\Delta_3}} \{\tilde{\psi}_2(p)\}\},$$

and define

$$\tilde{\psi}_3(p) = \begin{cases} \tilde{\psi}_2(p) & \text{on } \overline{\Delta_2}, \\ \max\{\tilde{\psi}_2(p),\ k_2\, \psi_2(p)\} & \text{on } \overline{\Delta_3} \setminus \Delta_1, \\ k_2\, \psi_2(p) & \text{on } \overline{\Delta_4} \setminus \Delta_3. \end{cases}$$

Then $\tilde{\psi}_3(p) > 0$ is a plurisubharmonic function on $\overline{\Delta_4}$ such that $\tilde{\psi}_3(p) = \tilde{\psi}_2(p)$ on $\overline{\Delta_2}$ and $\tilde{\psi}_3(p) \geq 3$ on $\overline{\Delta_4} \setminus \Delta_3$. In a similar fashion, using $\tilde{\psi}_3(p)$ on $\overline{\Delta_4}$ and $\psi_3(p)$ on $\overline{\Delta_5} \setminus \Delta_2$, we obtain a plurisubharmonic function $\tilde{\psi}_4(p) > 0$ on $\overline{\Delta_5}$ such that $\tilde{\psi}_4(p) = \tilde{\psi}_3(p)$ on $\overline{\Delta_3}$ and $\tilde{\psi}_4(p) \geq 4$ on $\overline{\Delta_5} \setminus \Delta_4$. We repeat this procedure inductively and obtain a continuous plurisubharmonic exhaustion function $\tilde{\psi}(p)$ on $\mathcal{R}$.

Using the same argument in the case of a univalent pseudoconvex domain in $\mathbf{C}^n$ via the mean-value integral of $\tilde{\psi}(p)$, we modify $\tilde{\psi}(p)$ to obtain a piecewise smooth, strictly plurisubharmonic exhaustion function $\Phi(p)$ on $\mathcal{R}$.

Thus we have proved the following.

PROPOSITION 9.6. *Any unramified pseudoconvex domain over $\mathbf{C}^n$ is a normal pseudoconvex space.*

This proposition, together with the main theorem (Theorem 9.3), yields the following.

THEOREM 9.5. *Any unramified pseudoconvex domain over $\mathbf{C}^n$ is holomorphically complete, and hence is a domain of holomorphy.*

9.4.6. Unramified Covers. Let $\mathcal{V}$ be an analytic space of dimension n. Let $\tilde{\mathcal{V}}$ be another analytic space of dimension n. If $\tilde{\mathcal{V}}$ satisfies the following two conditions:

(1) there exists an analytic mapping $\tilde{\pi}$ from $\tilde{\mathcal{V}}$ onto $\mathcal{V}$; and
(2) for any point p in $\mathcal{V}$, there exists a neighborhood δ_p of p in $\mathcal{V}$ such that $\tilde{\pi}$ maps each connected component of $\tilde{\pi}^{-1}(\delta_p)$ in $\tilde{\mathcal{V}}$ in a one-to-one fashion to δ_p,

then we say that $\tilde{\mathcal{V}}$ is an **unramified cover** of $\mathcal{V}$ without relative boundary; the mapping $\tilde{\pi}$ is called the **canonical projection**.

We shall prove the following theorem.

THEOREM 9.6. [10] *Any unramified cover of a Stein space is also a Stein space.*

We devote the rest of this section to the proof of this theorem. Thus we always assume that $\mathcal{V}$ is a Stein space and that $\tilde{\mathcal{V}}$ is an unramified cover of $\mathcal{V}$ with canonical projection $\tilde{\pi}$. We first verify the following proposition.

PROPOSITION 9.7. *Suppose that there exists a sequence of analytic polyhedra $\mathcal{P}_j$ $(j = 1, 2, \dots)$ in $\mathcal{V}$ with defining functions on $\mathcal{V}$ such that*

(1) *$\mathcal{P}_j \subset\subset \mathcal{P}^0_{j+1}$ $(j = 1, 2, \dots)$ and $\lim_{j\to\infty} \mathcal{P}_j = \mathcal{V}$; and*
(2) *each connected component of $\tilde{\mathcal{P}}_j := \tilde{\pi}^{-1}(\mathcal{P}_j)$ $(j = 1, 2, \dots)$ in $\tilde{\mathcal{V}}$ is holomorphically complete.*

Then $\tilde{\mathcal{V}}$ is a Stein space.

PROOF. For each $j = 1, 2, \dots,$ we can find a connected component $\tilde{\mathcal{P}}^0_j$ of $\tilde{\mathcal{P}}_j$ in $\tilde{\mathcal{V}}$ such that

$$\tilde{\mathcal{P}}^0_j \subset \tilde{\mathcal{P}}^0_{j+1} \quad (j = 1, 2, \dots), \qquad \lim_{j\to\infty} \tilde{\mathcal{P}}^0_j = \tilde{\mathcal{V}}.$$

In general, $\tilde{\mathcal{P}}^0_j$ is infinitely sheeted over $\mathcal{P}_j$. In order to prove that $\tilde{\mathcal{V}}$ is a Stein space, using Theorem 8.8 it suffices to show that each $\tilde{\mathcal{P}}^0_j$ is holomorphically convex in $\tilde{\mathcal{P}}^0_{j+1}$. Recall that in Theorem 8.8 we assumed $\tilde{\mathcal{P}}^0_j \subset\subset \tilde{\mathcal{P}}^0_{j+1}$; however, we only used the fact that $\tilde{\mathcal{P}}^0_j \subset \tilde{\mathcal{P}}^0_{j+1}$ in the proof.

To this end, let $K \subset\subset \tilde{\mathcal{P}}^0_j$ and let $\widehat{K}$ denote the holomorphically convex hull of K with respect to $\tilde{\mathcal{P}}^0_{j+1}$. It suffices to show that

$$\text{(i)}\ \widehat{K} \subset\subset \tilde{\mathcal{P}}^0_{j+1}; \qquad \text{(ii)}\ \tilde{\pi}(\widehat{K}) \subset\subset \mathcal{P}_j.$$

Claim (i) follows from our assumption that $\tilde{\mathcal{P}}^0_{j+1}$ is holomorphically complete. To verify (ii), let $k = \tilde{\pi}(K)$, so that $k \subset\subset \mathcal{P}_j$. We let $\widehat{k}$ denote the holomorphically convex hull of k with respect to $\mathcal{P}_{j+1}$. Since any holomorphic function φ on $\mathcal{P}_{j+1}$ gives rise to the holomorphic function $\tilde{\varphi} := \varphi \circ \pi$ in $\tilde{\mathcal{P}}^0_{j+1}$, it follows easily that

[10] This theorem was first proved by K. Stein in [**68**]. The proof given here is due to the author.

$\tilde{\pi}(\widehat{K}) \subset \widehat{k}$. Since $\mathcal{P}_j$ is holomorphically convex in $\mathcal{P}_{j+1}$, we have $\widehat{k} \subset\subset \mathcal{P}_j$, proving (ii). □

9.4.7. Preparation Lemma. By Theorem 8.20, a Stein space $\mathcal{V}$ can be realized as a distinguished ramified domain $\mathcal{D}$ over $\mathbf{C}^n$ with projection π. Precisely, fixing $\rho > 0$, we use variables $z_1, \dots, z_n$ in $\mathbf{C}^n$ and we let

$$\Gamma_\rho : |z_j| < \rho \quad (j = 1, \dots, n)$$

be the polydisk centered at the origin O with radius ρ. Then each connected component of $\pi^{-1}(\Gamma_\rho)$ in $\mathcal{D}$ is a finitely sheeted ramified domain over Γ_ρ without relative boundary. To verify Theorem 9.6, it thus suffices to prove that the unramified cover $\tilde{\mathcal{D}}$ with canonical projection $\tilde{\pi}$ of the distinguished ramified domain $\mathcal{D}$ over $\mathbf{C}^n$ is a Stein space.

We fix ρ_0 and ρ_1 with $\rho_0 > \rho_1 > 0$ and we take a connected component $\mathcal{D}_1$ (resp. $\mathcal{D}_0$) of $\pi^{-1}(\Gamma_{\rho_1})$ (resp. $\pi^{-1}(\Gamma_{\rho_0})$) in $\mathcal{D}$; then $\mathcal{D}_1$ (resp. $\mathcal{D}_0$) is a finitely sheeted ramified domain over Γ_{ρ_0} (resp. Γ_{ρ_1}) without relative boundary such that $\mathcal{D}_1 \subset\subset \mathcal{D}_0$.

Let $\tilde{\mathcal{D}}_0$ be any connected domain over $\tilde{\mathcal{D}}$, so that $\tilde{\mathcal{D}}$ is an infinitely or finitely sheeted unramified cover of $\mathcal{D}_0$ without relative boundary with projection $\tilde{\pi}$, and let $\widehat{\pi} = \pi \circ \tilde{\pi}$; this maps $\tilde{\mathcal{D}}_0$ onto Γ_{ρ_0}. Let $\tilde{\mathcal{D}}_1$ be the part of $\tilde{\mathcal{D}}_0$ over Γ_{ρ_1}; this is an unramified cover of $\mathcal{D}_1$ without relative boundary. To prove Theorem 9.6, using Proposition 9.7 it suffices to show that $\tilde{\mathcal{D}}_1$ is holomorphically complete. Moreover, using Theorem 9.3 it suffices to verify the following claim:

Claim. There exists a strictly pseudoconvex exhaustion function $\tilde{\varphi}(p)$ on $\tilde{\mathcal{D}}_1$. (9.36)

In the construction of $\tilde{\varphi}(p)$ we use the following notation. Let $E \subset \mathcal{D}_0$. Then

$$\underline{E} := \pi(E) \subset \Gamma_{\rho_0} \quad \text{and} \quad \tilde{E} := \tilde{\pi}^{-1}(E) \subset \tilde{\mathcal{D}}_0,$$

so that $\underline{\mathcal{D}_1} = \Gamma_{\rho_1}$ and $\tilde{\mathcal{D}}_1 = \tilde{\pi}^{-1}(\mathcal{D}_1)$. Consider the function

$$\eta(z) = \max_{j=1,\dots,n} \left\{ \frac{1}{\rho_1 - |z_j|} \right\} + \sum_{j=1}^{n} |z_j|^2, \quad z \in \Gamma_{\rho_1},$$

and set

$$\tilde{\eta}(p) = \eta(\widehat{\pi}(p)), \quad p \in \tilde{\mathcal{D}}_1.$$

Then $\tilde{\eta}(p)$ is a strictly pseudoconvex function on $\tilde{\mathcal{D}}_1$ such that $\lim_{p \to p_0} \tilde{\eta}(p) = +\infty$ for any $p_0 \in \partial \tilde{\mathcal{D}}_1$ with $\widehat{\pi}(p_0) \in \partial \Gamma_{\rho_1}$. However, $\tilde{\eta}(p)$ is not necessarily an exhaustion function on $\tilde{\mathcal{D}}_1$ if $\tilde{\mathcal{D}}_1$ is infinitely sheeted over $\mathcal{D}_1$ (or equivalently over Γ_{ρ_1}).

On $\tilde{\mathcal{D}}_0$, we define the usual metric $d_{\tilde{\mathcal{D}}_0}(p_1, p_2)$ in the following manner. Let $p_1, p_2 \in \tilde{\mathcal{D}}_0$. We connect p_1 and p_2 by a curve $\tilde{\gamma}$ in $\tilde{\mathcal{D}}_0$ and we let $L(\tilde{\gamma})$ denote the Euclidean length of the curve $\underline{\tilde{\gamma}} = \widehat{\pi}(\tilde{\gamma})$ in $\mathbf{C}^n$. We define

$$d_{\tilde{\mathcal{D}}_0}(p_1, p_2) = \inf \{ L(\tilde{\gamma}) \mid \tilde{\gamma} \text{ joins } p_1 \text{ and } p_2 \text{ in } \tilde{\mathcal{D}}_0 \}.$$

Fix a point p_0 in $\tilde{\mathcal{D}}_0$ and define

$$\tilde{d}_{p_0}(p) := d_{\tilde{\mathcal{D}}_0}(p_0, p), \quad p \in \tilde{\mathcal{D}}_0. \tag{9.37}$$

This is a nonnegative, continuous function on $\tilde{\mathcal{D}}_0$ such that

$$|\tilde{d}_{p_0}(p) - \tilde{d}_{p_0}(q)| \le d_{\tilde{\mathcal{D}}_0}(p, q) \quad \text{for } p, q \in \tilde{\mathcal{D}}_0.$$

We call $\tilde{d}_{p_0}(p)$ the distance function on $\tilde{\mathcal{D}}_0$. Using the fact that $\mathcal{D}_0$ is a finitely sheeted ramified domain over Γ_{ρ_0} without relative boundary, we see via the method used in the proof of Lemma 9.9 that for any real number a, the set

$$(\tilde{\mathcal{D}}_0)_a := \{p \in \tilde{\mathcal{D}}_0 \mid \tilde{d}_{p_0}(p) < a\}$$

is finitely sheeted over $\mathcal{D}_0$, and hence over Γ_{ρ_0}. The set $(\tilde{\mathcal{D}}_0)_a$ has relative boundary in $\tilde{\mathcal{D}}_0$ in the case when $\tilde{\mathcal{D}}_0$ is infinitely sheeted over $\mathcal{D}_0$. Consequently, if we set

$$\tilde{\lambda}(p) = \max\{\tilde{d}_{p_0}(p),\ \tilde{\eta}(p)\}, \quad p \in \tilde{\mathcal{D}}_1,$$

then $\tilde{\lambda}(p)$ is an exhaustion function on $\tilde{\mathcal{D}}_1$, but it is not necessarily a strictly pseudoconvex function. To finish the proof of our claim, it thus suffices to construct a strictly pseudoconvex function $\tilde{\varphi}(p)$ on $\tilde{\mathcal{D}}_1$ such that $\tilde{\varphi}(p)$ and $\tilde{\lambda}(p)$ are of bounded difference on $\tilde{\mathcal{D}}_1$.

9.4.8. Canonical Coordinates. Let r be an integer with $1 \le r \le n$ and let

$$p_r : (z_1, \dots, z_n) \in \mathbf{C}^n \to (z_1, \dots, z_r) \in \mathbf{C}^r$$

be a projection from $\mathbf{C}^n$ onto $\mathbf{C}^r$. We set

$$\sigma_r = p_r \circ \pi \quad \text{on } \mathcal{D}_0.$$

Then we have the following proposition.

PROPOSITION 9.8. *After a preliminary coordinate transformation of $\mathbf{C}^n$, if necessary, there exists a sequence of analytic sets S^r $(r = 0, 1, \dots, n-1)$ in $\mathcal{D}_0$ such that:*

1. *S^{n-1} is a pure $(n-1)$-dimensional analytic set in $\mathcal{D}_0$. Each S^r $(r = 0, 1, \dots, n-2)$ is a pure r-dimensional analytic set in $\mathcal{D}_0$ with $S^r \subset S^{r+1}$.*
2. *The coordinate system $(z_1, \dots, z_n)$ satisfies the Weierstrass condition for each analytic set $\underline{S}^r$ $(r = 0, 1, \dots, n-1)$ in Γ_{ρ_0} at each point of $\underline{S}^r$.*
3. *If we set $s_0^r = S^r \setminus S^{r-1}$, then the projection σ_r from s_0^r over $\mathbf{C}^r$ is locally one-to-one, i.e., the image $\sigma_r(s_0^r)$ is an unramified domain over $\mathbf{C}^r$.*

PROOF. For $r = n-1$, we can take S^{n-1} to be the branch set $\mathcal{S}_{n-1}$ of $\mathcal{D}_0$ over $\mathbf{C}^n$. Since $\underline{S^{n-1}}$ is an analytic hypersurface in Γ_{ρ_0}, we may assume that the coordinates $(z_1, \dots, z_n)$ of $\mathbf{C}^n$ satisfy the Weierstrass condition for $\underline{S^{n-1}}$ at each point of $\underline{S^{n-1}}$. Thus we can find a finitely sheeted ramified domain D_{n-1} over $\mathbf{C}^{n-1}$ such that $\underline{S^{n-1}}$ and D_{n-1} are in one-to-one correspondence via the projection p_{n-1} except perhaps for an at most $(n-2)$-dimensional analytic set in Γ_{ρ_0}. Consequently, under the projection $\sigma_{n-1} = p_{n-1} \circ \pi$, there exists a ramified domain $\mathcal{D}^{n-1}$ over $\mathbf{C}^{n-1}$ such that S^{n-1} and $\mathcal{D}^{n-1}$ are in one-to-one correspondence except perhaps for an $(n-2)$-dimensional analytic set $(S')^{n-2}$ in $\mathcal{D}_0$; i.e.,

$$S^{n-1} : z_n = \xi_n(z_1, \dots, z_{n-1}), \tag{9.38}$$

where $(z_1, \dots, z_{n-1})$ runs over the ramified domain $\mathcal{D}^{n-1}$ over $\mathbf{C}^{n-1}$. We let τ_{n-1} denote this mapping from S^{n-1} to $\mathcal{D}^{n-1}$. Consider the branch set $\mathcal{S}_{n-2}$ of $\mathcal{D}^{n-1}$ over $\mathbf{C}^{n-1}$ and let $(S'')^{n-2}$ denote the $(n-2)$-dimensional analytic subset of S^{n-1} which corresponds to $\mathcal{S}_{n-2}$ via τ_{n-1}. We then define

$$S^{n-2} = (S')^{n-2} \cup (S'')^{n-2},$$

which is an $(n-2)$-dimensional analytic set in $\mathcal{D}_0$ with $S^{n-2} \subset S^{n-1}$.

Since $\underline{S^{n-2}}$ is an analytic set in Γ_{ρ_0}, it follows from (9.38) that after taking a suitable linear transformation of $(z_1, \ldots, z_{n-1})$ in $\mathbf{C}^{n-1}$, if necessary, the coordinates $(z_1, \ldots, z_{n-1})$ satisfy the Weierstrass condition for $\underline{S^{n-2}}$ as well as for $\underline{S^{n-1}}$. By applying the same method to $\underline{S^{n-2}}$ as was done to $\underline{S^{n-1}}$, under the projection σ_{n-2}, there exists a ramified domain $\mathcal{D}^{n-2}$ over $\mathbf{C}^{n-2}$ such that S^{n-2} and $\mathcal{D}^{n-2}$ are in one-to-one correspondence except for an at most $(n-3)$-dimensional analytic set $(S')^{n-3}$ in $\mathcal{D}_0$; i.e.,

$$S^{n-2} : z_k = \eta_k(z_1, \ldots, z_{n-2}) \quad (k = n-1, n),$$

where $(z_1, \ldots, z_{n-2})$ runs over the ramified domain $\mathcal{D}^{n-2}$ over $\mathbf{C}^{n-2}$ and

$$\eta_n(z_1, \ldots, z_{n-2}) = \xi_n(z_1, \ldots, z_{n-2}, \eta_{n-1}(z_1, \ldots, z_{n-2})).$$

We let τ_{n-2} denote the mapping from S^{n-2} onto $\mathcal{D}^{n-2}$. Next, we consider the branch set $\mathcal{S}_{n-3}$ of $\mathcal{D}^{n-2}$ over $\mathbf{C}^{n-2}$ and we let $(S'')^{n-3}$ denote the analytic subset of S^{n-2} which corresponds to $\mathcal{S}_{n-3}$ via τ_{n-2}. We then define

$$S^{n-3} = (S')^{n-3} \cup (S'')^{n-3},$$

which is an $(n-3)$-dimensional analytic subset in $\mathcal{D}_0$ with $S^{n-2} \subset S^{n-3}$.

We thus inductively obtain a pure i-dimensional analytic set S^i $(i = n-1, \ldots, 1, 0)$ in $\mathcal{D}_0$ and a ramified domain $\mathcal{D}^i$ $(i = n-1, \ldots, 1, 0)$ over $\mathbf{C}^i$ such that $S^{i-1} \subset S^i$ $(i = n-1, \ldots, 1)$; S^i and $\mathcal{D}^i$ are in one-to-one correspondence via the transformation τ_i except for an at most $(i-1)$-dimensional analytic set (which is contained in S^{i-1}) in $\mathcal{D}_0$,

$$\tau_i : S^i \to \mathcal{D}^i ;$$

and the coordinates $(z_1, \ldots, z_n)$ satisfy the Weierstrass condition for each $\underline{S^i}$. Thus, this sequence S^i $(i = n-1, \ldots, 1, 0)$ satisfies conditions 1 and 2. If we set $s_0^r = S^r \setminus S^{r-1}$ $(r = 0, 1, \ldots, n)$, where $S^n = \mathcal{D}_0$, the construction also yields that s_0^r corresponds to the ramified domain $\mathcal{D}^r$ over $\mathbf{C}^r$ after the branch set S_r of $\mathcal{D}^r$ over $\mathbf{C}^r$ and some $(r-1)$-dimensional analytic set determined by the mapping τ_r are deleted. Thus, $\mathcal{R}^r := \tau_r(s_0^r)$ is a finitely sheeted, *unramified* domain over $\mathbf{C}^r$; hence condition 3 is also satisfied locally by $\tau_r = \sigma_r$. □

We set $\tilde{\tau}_r = \tau_r \circ \tilde{\pi}$ $(r = 0, 1, \ldots, n-1)$; thus $\tilde{\tau}_r$ is a one-to-one mapping from $\tilde{s}_0^r = \tilde{\pi}^{-1}(s_0^r) \subset \tilde{\mathcal{D}}_0$ onto an unramified domain $\tilde{\mathcal{R}}^r$ over $\mathcal{R}^r$ without relative boundary. Generally this domain is infinitely sheeted. We let $\tilde{\eta}_r : \tilde{\mathcal{R}}^r \to \mathcal{R}^r$ denote the mapping which corresponds to $\tilde{\pi} : \tilde{s}_0^r \to s_0^r$ via $\tilde{\tau}_r$. Since $\tilde{\mathcal{R}}^r$ is a domain over $\mathbf{C}^r$, we have the usual distance function $d_{\tilde{\mathcal{R}}^r}(\zeta', \zeta'')$ for $\zeta', \zeta'' \in \tilde{\mathcal{R}}^r$ (just as we have $d_{\tilde{\mathcal{D}}_0}(p', p'')$ in $\tilde{\mathcal{D}}_0$ over $\mathbf{C}^n$). Since $\tilde{s}_0^r$ and $\tilde{\mathcal{R}}^r$ are in one-to-one correspondence via $\tilde{\tau}_r$, we can define the distance function $d_r(p', p'')$ for p', p'' in $\tilde{s}_0^r$ by

$$d_r(p', p'') = d_{\tilde{\mathcal{R}}^r}(\zeta', \zeta''),$$

where $\zeta' = \tilde{\tau}_r(p')$ and $\zeta'' = \tilde{\tau}_r(p'')$.

We make the following observation about the distance function $\tilde{d}_{p_0}(p)$ defined in (9.37).

REMARK 9.4. Let K be a compact set in s_0^r (equivalently, $\tau_r(K)$ is a compact set in $\mathcal{R}^r$). Then there exists a constant $c_K > 0$ such that

$$|\tilde{d}_{p_0}(p') - \tilde{d}_{p_0}(p'')| \leq c_K \, d_r(p', p'') \tag{9.39}$$

for all points p' and p'' in $\tilde{K} = \tilde{\pi}^{-1}(K) \subset \tilde{s}_0^r$ which are sufficiently close to each other so that $\tilde{\tau}_r(p')$ and $\tilde{\tau}_r(p'')$ are contained in a univalent closed ball $\tilde{B}$ in $\tilde{\mathcal{R}}^r$ over $\mathbf{C}^r$.

PROOF. We may assume that the given set K is a compact set in s_0^r so that $\tau_r(K)$ is a closed univalent ball B in the unramified domain $\mathcal{R}^r$ over $\mathbf{C}^r$. For simplicity we set $K = B$. Thus the set $\tilde{B} = \tilde{\pi}_r^{-1}(B) \subset \tilde{\mathcal{R}}^r$ satisfies $\tilde{\eta}_r(\tilde{B}) = B$. We set $\kappa = (\tilde{\tau}_r)^{-1}(\tilde{B}) \subset \tilde{s}_0^r$. Then κ can be written in the form

$$\kappa : z_k = \xi_k(z_1, \dots , z_r) \quad (k = r+1, \dots , n),$$

where each $\xi_k(z_1, \dots , z_r)$ $(k = r+1, \dots , n)$ is a single-valued holomorphic function on the closed ball B in $\mathbf{C}^r$. Thus, we can find a constant $A_K > 1$ such that

$$|\frac{\partial \xi_k}{\partial z_i}(z_1, \dots , z_r)| \leq A_K$$

for any point $(z_1, \dots , z_r) \in B$. Let $p', p'' \in \tilde{s}_0^r$ with $\tilde{\tau}_r(p'), \tilde{\tau}_r(p'') \in \tilde{B}$. We set $\widehat{\pi}(p') = (z_1', \dots , z_n')$ and $\widehat{\pi}(p'') = (z_1'', \dots , z_n'')$ in $\mathbf{C}^n$. We consider the arc $\tilde{\gamma}$ on $\tilde{s}_0^r$ connecting the points p' and p'':

$$\tilde{\gamma} : t \in [0,1] \to (z_1(t), \dots , z_n(t)),$$

where

$$\begin{aligned} z_i(t) &= z_i' + (z_i'' - z_i')t && (i = 1, \dots , r), \\ z_k(t) &= \xi_k(z_1(t), \dots , z_r(t)) && (k = r+1, \dots , n). \end{aligned}$$

Since $d_r(p', p'') = (\sum_{i=1}^r |z_i' - z_i''|^2)^{1/2}$ and $\tilde{s}_0^r \subset \tilde{\mathcal{D}}_0$, it follows that

$$\begin{aligned} |\tilde{d}_{p_0}(p') - \tilde{d}_{p_0}(p'')| &\leq d_{\tilde{\mathcal{D}}_0}(p', p'') \leq L(\tilde{\gamma}) := \int_0^1 (\sum_{j=1}^n |\frac{dz_j(t)}{dt}|^2)^{1/2} dt \\ &\leq \left(\sum_{i=1}^r |z_i' - z_i''|^2 \right)^{1/2} \left(1 + \sum_{k=r+1}^n \sum_{i=1}^r |\frac{\partial \xi_k}{\partial z_i}(z_1(t), \dots , z_r(t))|^2 \right)^{1/2} \\ &\leq d_r(p', p'')\,(1 + (n-r) r A_K^2)^{1/2}. \end{aligned}$$

Setting $c_K := (1 + (n-r) r A_K^2)^{1/2} > 0$, we obtain (9.39). □

9.4.9. Lemmas. The ramified domain $\mathcal{D}_0$ over Γ_{ρ_0} is an analytic polyhedron in a ramified domain G, where $\mathcal{D}_0 \subset\subset G$ (Remark 9.2). We can thus find a normal model Σ in a polydisk $\Gamma = \Gamma_{\rho_0} \times \Gamma^0$ in $\mathbf{C}^{n+m} = \mathbf{C}^n \times \mathbf{C}^m$, where $\mathbf{C}^m$ has variables $w_1, \dots , w_m$ and

$$\Gamma^0 : \quad |w_k| < 1 \quad (k = 1, \dots , m).$$

To be precise, we can find a one-to-one holomorphic mapping Φ from $\mathcal{D}_0$ onto Σ such that Σ is an n-dimensional analytic set in Γ:

$$\Phi : p \in \mathcal{D}_0 \to (z, w) = (\pi(p), \Phi_1(p), \dots , \Phi_m(p)) \in \Sigma. \tag{9.40}$$

We set

$$\chi(z, w) = \sum_{i=1}^n |z_i|^2 + \sum_{k=1}^m |w_k|^2 \quad \text{in } \mathbf{C}^{n+m}.$$

Using the projection $\tilde{\pi} : \tilde{\mathcal{D}}_0 \to \mathcal{D}_0$, we define

$$\tilde{\chi}(p) = \chi(\Phi \circ \tilde{\pi}(p)), \quad p \in \tilde{\mathcal{D}}_1,$$

so that $\tilde{\chi}(p)$ is a bounded, strictly pseudoconvex function on $\tilde{\mathcal{D}}_1$.

We prove the following two lemmas.

LEMMA 9.12. *Let e be an open set in s_0^r with $e \subset\subset s_0^r$. There exist a neighborhood* $\mathbf{V}$ *of e in $\mathcal{D}_0$ and a strictly pseudoconvex function $f(p)$ on $\tilde{\mathbf{V}} = \tilde{\pi}^{-1}(\mathbf{V})$ such that $f(p)$ and $\tilde{d}_{p_0}(p)$ are of bounded difference on $\tilde{\mathbf{V}}$.*

PROOF. We set $E = \tau_r(e) \subset \mathcal{R}^r$, $\tilde{e} = \tilde{\pi}^{-1}(e) \subset \tilde{s}_0^r$, and $\tilde{E} = \tilde{\tau}_r(\tilde{e}) \subset \tilde{\mathcal{S}}^r$, so that $E \subset\subset \mathcal{R}^r$ and so that $\tilde{E} \subset \tilde{\mathcal{R}}^r$ is an infinitely sheeted unramified cover of E without relative boundary. Thus, both E and $\tilde{E}$ are unramified domains over $\mathbf{C}^r$. For any point $\zeta \in \tilde{\mathcal{R}}^r$, there exists a unique point $p \in \tilde{s}_0^r \subset \tilde{\mathcal{D}}_0$ with $\tilde{\tau}_r(p) = \zeta$. Thus we define

$$D_r(\zeta) := \tilde{d}_{p_0}(p), \quad \zeta \in \tilde{\mathcal{R}}^r,$$

which is a nonnegative function on $\tilde{\mathcal{R}}^r$. Since $e \subset\subset s_0^r$, it follows from inequality (9.39) that there exists a constant $c_e > 0$ such that

$$|D_r(\zeta') - D_r(\zeta'')| \leq c_e \, \|\zeta' - \zeta''\|$$

for all points ζ', ζ'' in $\tilde{E}$ which are sufficiently close to each other. Here $\|\zeta' - \zeta''\|$ denotes the Euclidean distance between ζ' and ζ'' in $\mathbf{C}^r$; this only depends on the set e. Therefore, using the same method as was used in section 9.4.4, but now applied to $E \subset\subset \mathcal{R}^r$, we can construct a strictly plurisubharmonic function $G_r(\zeta)$ on the unramified domain $\tilde{E}$ over $\mathbf{C}^r$ such that $G_r(\zeta)$ and $D_r(\zeta)$ are of bounded difference in $\tilde{E}$. It follows that

$$g_r(p) = G_r(\tilde{\tau}_r(p)), \quad p \in \tilde{e},$$

is a strictly plurisubharmonic function on $\tilde{e}$ such that $g_r(p)$ and $\tilde{d}_{p_0}(p)$ are of bounded difference on $\tilde{e}$.

On the other hand, from Proposition 9.8, $\underline{s_0^r}$ is an r-dimensional, non-singular analytic set in a domain in Γ_{ρ_0} and the coordinates $(z_1, \ldots, z_n)$ satisfy the Weierstrass condition for $\underline{s_0^r}$. Thus s_0^r can be written in the form

$$z_k = \xi_k(z_1, \ldots, z_r) \quad (k = r+1, \ldots, n),$$

where $(z_1, \ldots, z_r)$ varies over the unramified domain $\mathcal{R}^r$ over $\mathbf{C}^r$ and $\xi_k(z_1, \ldots, z_r)$ is a single-valued holomorphic function on $\mathcal{R}^r$. Since $e \subset\subset s_0^r$, we can find a tubular neighborhood $\mathbf{V}$ of e in $\mathcal{D}_0$ of the form

$$\mathbf{V} = \bigcup_{(z_1, \ldots, z_r) \in E} (z_1, \ldots, z_r, V(z_1, \ldots, z_r)),$$

where $V(z_1, \ldots, z_r)$ is a polydisk in $\mathbf{C}^{n-r}$ centered at the point $(\xi_{k+1}(z_1, \ldots, z_r), \ldots, \xi_n(z_1, \ldots, z_r))$:

$$V(z_1, \ldots, z_r) : |z_k - \xi_k(z_1, \ldots, z_r)| < \delta \quad (k = r+1, \ldots, n),$$

and $\delta > 0$ is sufficiently small. Thus $\mathbf{V}$ is a unramified domain over $\mathbf{C}^n$. The projection T_r from $\mathbf{V}$ onto e such that

$$T_r(z_1, \ldots, z_r, V(z_1, \ldots, z_r)) = (z_1, \ldots, z_r, \xi_{k+1}(z_1, \ldots, z_r), \ldots, \xi_n(z_1, \ldots, z_r))$$

canonically determines a holomorphic mapping (contraction) $\tilde{T}_r : \tilde{\mathbf{V}} \to \tilde{e}$, where $\tilde{\mathbf{V}} = \tilde{\pi}^{-1}(\mathbf{V})$, via the relation $\tilde{\pi} \circ \tilde{T}_r = T_r \circ \tilde{\pi}$. Setting

$$g(p) := g_r(\tilde{T}_r(p)), \quad p \in \tilde{\mathbf{V}},$$

it follows that $g(p)$ is a pseudoconvex function on $\tilde{\mathbf{V}}$. Further, since $g_r(p)$ and $\tilde{d}_{p_0}(p)$ are of bounded difference on $\tilde{e}$, so are $g(p)$ and $\tilde{d}_{p_0}(p)$ on $\tilde{\mathbf{V}}$. Consequently, if we set $f(p) = g(p) + \tilde{\chi}(p)$ on $\tilde{\mathbf{V}}$, then $f(p)$ is a function satisfying the conclusion of the lemma. □

We use this lemma to prove the following.

LEMMA 9.13. *Let e be an open set in S^r $(r = 1, \dots, n-1)$ such that $e \subset\subset S^r$, and set $e' = e \cap S^{r-1} \subset\subset S^{r-1}$. Assume that there exists a neighborhood $\mathcal{U}$ of e' in $\mathcal{D}_0$ with the property that there exists a strictly pseudoconvex function $f(p)$ on $\tilde{\mathcal{U}} = \tilde{\pi}^{-1}(\mathcal{U})$ such that $f(p)$ and $\tilde{d}_{p_0}(p)$ are of bounded difference on $\tilde{\mathcal{U}}$. Then there exist a neighborhood $\mathcal{W}$ of e in $\mathcal{D}_0$ and a strictly pseudoconvex function $F(p)$ on $\tilde{W} = \tilde{\pi}^{-1}(\mathcal{W})$ such that $F(p)$ and $\tilde{d}_{p_0}(p)$ are of bounded difference on $\tilde{\mathcal{W}}$.*

PROOF. We begin with the normal model Σ of $\mathcal{D}_0$ in the polydisk $\Gamma \subset \mathbf{C}^{n+m}$ via the mapping Φ defined in (9.40). We set $\Sigma^{r-1} = \Phi(S^{r-1})$. Since Σ^{r-1} is an $(r-1)$-dimensional analytic set in Γ, we can find a finite number of holomorphic functions $f_j(z,w)$ $(j = 1, \dots, s)$ on Γ such that

$$\Sigma^{r-1} : f_j(z,w) = 0 \quad (j = 1, \dots, s).$$

For $0 < \epsilon << 1$, we define

$$H_\epsilon(z,w) := \max_{j=1,\dots,s} \{|f_j(z,w)|\} + \epsilon\chi(z,w), \quad (z,w) \in \Gamma,$$

and

$$h_\epsilon(p) := H_\epsilon(\Phi(p)), \quad p \in \mathcal{D}_0,$$

so that $h_\epsilon(p)$ is a strictly pseudoconvex function on $\mathcal{D}_0$. We take $\epsilon > 0$ sufficiently small so that if we set

$$\mathcal{H}_a := \{p \in \mathcal{D}_0 \mid h_\epsilon(p) < a\}$$

for a sufficiently small $a > 0$, then $\mathcal{H}_a$ is a tubular neighborhood of S^{r-1} in $\mathcal{D}_0$ such that $e' \subset \mathcal{H}_a \cap e \subset\subset \mathcal{U}$. We fix $0 < a_1 < a_2 < a_3$ such that $\mathcal{H}_{a_3} \cap e \subset\subset \mathcal{U}$. Since $e \setminus \mathcal{H}_{a_1} \subset\subset s_0^r$, it follows from Lemma 9.12 that there exist a neighborhood $\mathbf{V}$ of $e \setminus \mathcal{H}_{a_1}$ in $\mathcal{D}_0$ and a strictly pseudoconvex function $g(p)$ on $\tilde{\mathbf{V}}$ such that $g(p)$ and $\tilde{d}_{p_0}(p)$ are of bounded difference on $\tilde{\mathbf{V}}$. We set

$$\mathcal{W}_1 = \mathcal{H}_{a_1} \cap \mathcal{U}, \quad \mathcal{W}_2 = (\mathcal{H}_{a_3} \setminus \mathcal{H}_{a_1}) \cap \mathbf{V}, \quad \mathcal{W}_3 = \mathbf{V} \setminus \mathcal{H}_{a_3},$$

so that $\mathcal{W} := \mathcal{W}_1 \cup \mathcal{W}_2 \cup \mathcal{W}_3$ is a neighborhood of e in $\mathcal{D}_0$. Let $\tilde{\pi}^{-1}(\mathcal{W}_i) = \tilde{\mathcal{W}}_i$ $(i = 1,2,3)$ and $\tilde{\pi}^{-1}(\mathcal{W}) = \tilde{\mathcal{W}}$. For $K > 0$, we set

$$F(p) = \begin{cases} f(p), & p \in \tilde{\mathcal{W}}_1, \\ \max\{f(p), g(p) + K\,[h_\epsilon(\tilde{\pi}(p)) - a_2]\}, & p \in \tilde{\mathcal{W}}_2, \\ g(p) + K\,[h_\epsilon(\tilde{\pi}(p)) - a_2], & p \in \tilde{\mathcal{W}}_3. \end{cases}$$

We note that if $K > 0$ is sufficiently large, then $F(p)$ is a well-defined, single-valued function on $\tilde{\mathcal{W}}$. It is clear that $F(p)$ is a strictly pseudoconvex function on $\tilde{\mathcal{W}}$. Moreover, $f(p)$ and $g(p)$ are of bounded difference on $\tilde{\mathcal{W}}_2$, since each of them are of bounded difference with $\tilde{d}_{p_0}(p)$ on $\tilde{\mathcal{W}}$. Furthermore, since $K\,[h_\epsilon(\tilde{\pi}(p)) - a_2]$ is bounded in $\tilde{\mathcal{W}}_2 \cup \tilde{\mathcal{W}}_3$, we see that $F(p)$ and $\tilde{d}_{p_0}(p)$ are of bounded difference on $\tilde{\mathcal{W}}$. Hence, this $F(p)$ on $\tilde{\mathcal{W}}$ satisfies the conditions of the lemma. □

9.4.10. Proof of the Claim. We shall prove our claim (9.36), which then yields Theorem 9.6. We use the notation S^r $(r = 0, 1, \dots, n)$ from Proposition 9.8, where $S^n = \mathcal{D}_0$. Then S^0 consists of a finite number of points Q_k $(k = 1, \dots, M)$ in $\mathcal{D}_0$. By taking a larger ρ_1 $(0 < \rho_1 < \rho)$ than the given ρ_1 in claim (9.36), if necessary, we may assume that $Q_k \notin \partial\mathcal{D}_1$ $(k = 1, \dots, M)$ and $S^0 \cap \mathcal{D}_1 = \{Q_k\}_{k=1,\dots,M'}$ $(M' \leq M)$, where $\mathcal{D}_1$ is the subset of $\mathcal{D}_0$ over Γ_{ρ_1}.

For each Q_k $(k = 1, \dots, M')$, we can take a sufficiently small neighborhood $\mathcal{U}_k$ of Q_k in $\mathcal{D}_1$ such that $\mathcal{U}_k \cap \mathcal{U}_l = \emptyset$ $(k \neq l)$ and such that each connected component of $\tilde{\mathcal{U}}_k = \tilde{\pi}^{-1}(\mathcal{U}_k)$ in $\tilde{\mathcal{D}}_1$ is bijective to $\mathcal{U}_k$ via the projection $\tilde{\pi}$ (since $\tilde{\mathcal{D}}_1$ is an unramified cover over $\mathcal{D}_1$ without relative boundary). We set $\mathcal{U}^0 = \bigcup_{k=1}^{M'} \mathcal{U}_k$, which is a neighborhood of $S^0 \cap \mathcal{D}_1$, and we define

$$g_0(p) = \tilde{d}_{p_0}(\tilde{Q}_k) + \tilde{\chi}(p), \quad p \in \tilde{\mathcal{U}}_k,$$

where $\tilde{Q}_k$ denotes the point of $\tilde{\mathcal{U}}_k$ over Q_k. Then $g_0(p)$ is a strictly pseudoconvex function on $\tilde{\mathcal{U}}^0 = \tilde{\pi}^{-1}(\mathcal{U}^0)$, and $g_0(p)$ and $\tilde{d}_{p_0}(p)$ are clearly of bounded difference on $\tilde{\mathcal{U}}^0$.

Applying Lemma 9.13 for $r = 1$, $f(p) = g_0(p)$, $\mathcal{U} = \mathcal{U}^0$, and $e = S^1 \cap \mathcal{D}_1$ (so that $e \cap S^0 = \{Q_k\}_{k=1,\dots,M'}$), we can find a neighborhood $\mathcal{U}^1$ of $S^1 \cap \mathcal{D}_1$ in $\mathcal{D}_0$ and a strictly pseudoconvex function $g_1(p)$ on $\tilde{\mathcal{U}}^1 = \tilde{\pi}^{-1}(\mathcal{U}^1)$ such that $g_1(p)$ and $\tilde{d}_{p_0}(p)$ are of bounded difference on $\tilde{\mathcal{U}}^1$. Again applying the lemma for $r = 2$, $f(p) = g_1(p)$, $\mathcal{U} = \mathcal{U}^1$, and $e = S^2 \cap \mathcal{D}_1$ (so that $e \cap S^1 = S^1 \cap \mathcal{D}_1$), we can find a neighborhood $\mathcal{U}^2$ of $S^2 \cap \mathcal{D}_1$ in $\mathcal{D}_0$ and a strictly pseudoconvex function $g_2(p)$ on $\tilde{\mathcal{U}}^2 = \tilde{\pi}^{-1}(\mathcal{U}^2)$ such that $g_2(p)$ and $\tilde{d}_{p_0}(p)$ are of bounded difference on $\tilde{\mathcal{U}}^2$. We repeat this procedure n times to verify claim (9.36). □

Part II may be summarized briefly as follows. In Chapter 6 we showed that any ramified domain over $\mathbf{C}^n$ locally carries a simple function. In Chapter 7 we introduced the notion of $\mathcal{O}$-ideals and proved certain results concerning them. In particular, we proved the existence of a locally finite pseudobase for a G-ideal and for a Z-ideal at each point. This was established with the aid of a simple function. (Oka, on the other hand, proved the existence without utilizing simple functions; instead, he made very detailed and complicated constructions which heavily depend on the properties of ramified domains over $\mathbf{C}^n$). These results were then used to establish the lifting principle in an analytic space in the beginning of Chapter 8; this principle was used to prove many results for Stein spaces. In Chapter 9 we gave a geometric condition for an analytic space to be a Stein space (Levi's problem), and we gave examples of analytic spaces satisfying this condition. These examples include unramified pseudoconvex domains over $\mathbf{C}^n$.

Bibliography

[1] R. Baire, *Leçons sur les fonctions discontinues*, Gauthier-Villars, 1907.

[2] H. Behnke and K. Stein, Konvergentenfolgen von Regularitätsbereichen und der Meromorphiekonvexität, *Math. Ann.* 116 (1939), 204-216.

[3] E. Bishop. Mapping of partially analytic spaces, *Amer. J. Math.* 83 (1961), 209-242.

[4] A. B. Brown, On certain analytic continuations and analytic homeomorphisms, *Duke Math. J.* 2 (1936), 20-28.

[5] E. Calabi and M. Rosenlicht, Complex analytic manifolds without countable base, *Proc. Amer. Math. Soc.* 4 (1953), 335-340.

[6] E. Cartan, Sur la géométrie pseudo-conforme des hypersurfaces de l'espace de deux variables complexes, *Ann. Mat. Pura Appl.* (4) 11 (1932), 17-90.

[7] H. Cartan, Sur les fonctions de deux variables complexes, *Bull. Sci. Math.* 54 (1930), 99-116.

[8] H. Cartan, Sur les fonctions de deux variables complexes: les transformations d'un domaine borné D en un domaine intérieur à D, *Bull. Soc. Math. France* 58 (1930), 199-219.

[9] H. Cartan, Les problèmes de Poincaré et de Cousin pour les fonctions de plusieurs variables complexes, *C. R. Acad. Sci. Paris* 199 (1934),1284-1287.

[10] H. Cartan. Sur le premier problème de Cousin, *C. R. Acad. Sci. Paris* 207 (1938), 558-560.

[11] H. Cartan, Sur les matrices holomorphes de n variables complexes, *J. Math. Pura Appl.* (9) 19 (1940), 1-26.

[12] H. Cartan, Idéaux et modules de fonctions analytiques de variables complexes, *Bull. Soc. Math. France* 78 (1950), 29-64.

[13] H. Cartan and P. Thullen, Zur Theorie der Singularitäten der Funktionen mehrerer komplexen Veränderlichen: Regularitäts- und Konvergenzbereiche, *Math. Ann.* 106 (1932), 617-647.

[14] L. Chow, On compact analytic varieties, *Amer. J. Math.* 71 (1949), 893-914.

[15] P. Cousin, Sur les fonctions de n variables complexes, *Acta Math.* 19 (1895), 1-61.

[16] J. Duval, Convexité rationnelle des surfaces lagrangiennes, *Invent. Math.* 104 (1991), 581-599.

[17] E. E. Levi, Studii sui punti singolari essenziali delle funzioni analitiche di due o più variabili complesse, *Ann. Mat. Pura Appl.* (3) 17 (1910), 61-87.

[18] E. Fabry, Sur les rayons de convergente d'une série double, *C. R. Acad. Sci. Paris* 134 (1902), 1190-1192.

[19] M. Fékete, Über die Verteilung der Wurzeln bei gewissen algebraischen Gleichungen mit ganzzahligen Koeffizienten, *Math. Z.* 17 (1923), 228-249.

[20] J. E. Fornaess, A counterexample for the Levi problem for bordered Riemann domains, *Math. Ann.* 234 (1978), 275-277.

[21] H. Grauert, Charakterisierung der holomorph vollständigen komplexen Räume, *Math. Ann.* 129 (1955), 233-259.

[22] H. Grauert, On Levi's problem and the imbedding of real analytic manifolds, *Ann. of Math.* (2) 68 (1958), 460-472.

[23] H. Grauert and R. Remmert, Konvexität in der komplexen Analysis. Nicht-holomorph-konvexe Holomorphiegebiete und Anwendungen auf die Abbildungstheorie, *Comment. Math. Helvetici* 31 (1956), 152-160, 161-183.

[24] H. Grauert and R. Remmert, Komplexe Räume, *Math. Ann.* 136 (1958), 245-318.

[25] H. Grauert and R. Remmert, *Theorie der Steinschen Räume*, Springer-Verlag, 1977.

[26] H. Grauert and R. Remmert, *Coherent Analytic Sheaves*, Springer-Verlag, 1984.

[27] T. H. Gronwall, On the expressibility of a uniform function of several complex variables as the quotient of two functions of entire character, *Trans. Amer. Math. Soc.* 18 (1917), 50-64.

[28] R. C. Gunning and H. Rossi, *Analytic functions of several complex variables*, Prentice-Hall, 1965.

[29] F. Hartogs, Einige Folgerungen aus der Cauchyschen Integralformel bei Funktionen mehrerer Veränderlichen, *Sitzengsber. Math. Phys. Kl. Bayer. Akad. Wiss. München* 36 (1906), 223-242.

[30] F. Hartogs, Über die aus den singulären Stellen einer analytischen Funktion mehrerer Veränderlichen bestehenden Gebilde, *Acta Math.* 32 (1909), 57-79.

[31] H. Hopf, Zur Topologie der komplexen Mannigfaltigkeiten. *Studies and essays presented to R. Courant on his sixtieth birthday*, Interscience, 1948, pp. 167-185.

[32] G. Julia, Sur les familles de fonctions analytiques de plusieurs variables, *Acta Math.* 47 (1926), 53-115.

[33] K. Kato, Sur le théorème de P. Thullen et K. Stein. *J. Math. Soc. Japan* 18 (1966), 211-218.

[34] M. Klimek, *Pluripotential theory*, Clarendon Press, 1991.

[35] E. E. Levi, Sulle ipersuperficie dello spazio a 4 dimensioni che possono essere frontiera del campo di esistenza di una funzione analitica di due variabili complesse, *Ann. Mat. Pura Appl.* 18 (1911), 69-79.

[36] R. Narasimhan, Imbedding of holomorphically complete complex spaces, *Amer. J. Math.* 82 (1960), 917-934.

[37] R. Narasimhan, The Levi problem for complex spaces, *Math. Ann.* 142 (1961), 355-365.

[38] R. Nevanlinna, *Le théorème de Picard–Borel et la théorie des fonctions méromorphes*, Gauthier-Villars, 1929.

[39] T. Nishino, Les ensembles analytiques et les domaines, *J. Math. Kyoto Univ.* 1 (1962), 379-384.

[40] T. Nishino, Sur les ensembles pseudoconcaves, *J. Math. Kyoto Univ.* 1 (1962), 225-245.

[41] T. Nishino, Sur les espaces analytiques holomorphiquement complets, *J. Math. Kyoto Univ.* 1 (1962), 247-254.

[42] T. Nishino, Le théorème de Borel et le théorème de Picard, *C. R. Acad. Sci. Paris Sér. I Math.* 299 (1984), 667-668.

[43] K. Oka, Note sur les familles de fonctions analytiques multiformes etc, *J. Sci. Hiroshima Univ. A* 4 (1934), 93-98.

[44] K. Oka, Sur les fonctions analytiques de plusieurs variables, I: Domaines convexes par rapport aux fonctions rationnelles, *J. Sci. Hiroshima Univ. A* 6 (1936), 245-255.

[45] K. Oka, Sur les fonctions analytiques de plusieurs variables, II: Domaines d'holomorphie, *J. Sci. Hiroshima Univ. A* 7 (1937), 115-130.

[46] K. Oka, Sur les fonctions analytiques de plusieurs variables, III: Deuxième problème de Cousin, *J. Sci. Hiroshima Univ. A* 9 (1939), 7-19.

[47] K. Oka, Sur les fonctions analytiques de plusieurs variables, IV: Domaines d'holomorphie et domaines rationnellement convexes, *Japanese J. Math.* 17 (1941), 517-521.

[48] K. Oka, Sur les fonctions analytiques de plusieurs variables, V: L'intégrale de Cauchy, *Japanese J. Math.* 17 (1941), 523-531.

[49] K. Oka, Sur les fonctions analytiques de plusieurs variables, VI: Domaines pseudoconvexes, *Tôhoku Math. J.* 49 (1942), 15-52.

[50] K. Oka, Sur les fonctions analytiques de plusieurs variables, VII: Sur quelques notions arithmétiques, *Bull. Soc. Math. France* 78 (1950), 1-27.

[51] K. Oka, Sur les fonctions analytiques de plusieurs variables, VIII: Lemme fondamental, *J. Math. Soc. Japan* 3 (1951), 204-214.

[52] K. Oka, Sur les fonctions analytiques de plusieurs variables, IX: Domaines finis sans point critique intérieur, *Japanese J. Math.* 27 (1953), 97-155.

[53] K. Oka, *Sur les fonctions analytiques de plusieurs variables*, Iwanami Shoten, Tokyo, 1961.

[54] K. Oka, Sur les fonctions analytiques de plusieurs variables, X: Une mode nouvelle engendrant les domaines pseudoconvexes, *Japanese J. Math.* 32 (1962), 1-12.

[55] K. Oka, *Oka's posthumous works No.* 1 $\sim$ *No.* 7, Edited by T. Nishino and A. Takeuchi, Kyoto, 1980-1983. (Japanese)

[56] W. F. Osgood, *Lehrbuch der Funktionentheorie, Zweiter Band,*, reprint, Chelsea Publishing Company, 1965.

[57] E. Picard, Sur certaines équations fonctionnelles et sur une classe de surfaces algébriques, *C. R. Acad. Sci. Paris* 139 (1904), 5-9.

[58] E. Picard and G. Simart, *Théorie des fonctions algébraiques (Tome I)*, reprint, Chelsea Publishing Company, 1971.

[59] H. Poincaré, Sur les fonctions de deux variables, *Acta Math.* 2 (1883), 97-113.

[60] H. Poincaré, Les fonctions analytiques de deux variables et la représentation conforme, *Rend. Circ. Mat. Palermo* 23 (1907), 185-220.

[61] T. Radó, Über den Begriff der Riemannschen Fläche, *Acta Litt. Sci. Reg. Univ. Hungar. Szeged* 2 (1925), 101-121.

[62] F. Riesz, Sur les fonctions subharmoniques et leur rapport avec la théorie du potentiel, I, II, *Acta Math.* 48 (1926), 329-343; 54 (1930), 321-360.

[63] J.-P. Rosay, Sur une characterisation de la boule parmi les domaines de $\mathbf{C}^n$ par son groupe d'automorphismes, *Ann. Inst. Fourier* 29 (1979), no. 4, 91-97.

[64] W. Rothstein, Zur Theorie der analytischen Mannigfaltigkeiten im Räume von n komplexen Veränderlichen, *Math. Ann.* 129 (1955), 96-138.

[65] C. L. Siegel, *Topics in complex function theory, Vols.I,II,III*, Wiley-Interscience, 1969, 1971, 1973.

[66] K. Stein, Topologische Bedingungen für die Existenz analytischer Funktionen komplexer Veränderlichen zu vorgegebenen Nullstellenflächen, *Math. Ann.* 117 (1941), 727-757.

[67] K. Stein, Analytischen Funktionen mehrerer komplexer Veränderlichen zu vorgegebenen Periodizitätmoduln und das zweite Cousinsche Problem, *Math. Ann.* 123 (1951), 201-222.

[68] K. Stein, Überlagerungen holomorph-vollständiger komplexer Räume, *Arch. Math.* 7 (1956), 354-361.

[69] R. Remmert and K. Stein, Über die wesentlichen Singularitäten analytischer Mengen, *Math. Ann.* 126 (1953), 263-306.

[70] G. Stolzenberg, An example concerning rational convexity, *Math. Ann.* 147 (1962), 275-276.

[71] A. Takeuchi, Une remarque sur le deuxième mémoire de M. K. Oka, *J. Math. Kyoto Univ.* 9 (1969), 1-4.

[72] T. Terada, Sur une certaine condition sous laquelle une fonction de plusieurs variables complexes est holomorphe, *Publ. Res. Inst. Math. Sci. Koyto Univ. Ser. A* 2 (1967), 383-396.

[73] P. Thullen, Zur Theorie des Singularitäten der Funktionen zweier komplexen Varäderichen: Die Regularitätshüllen, *Math. Ann.* 106 (1932), 64-76.

[74] P. Thullen, Über die wesentlichen Singularitäten analytiqscher Funktionen und Flächen im Raume von n komplexen Veränderlichen, *Math. Ann.* 111 (1935), 137-157.

[75] T. Ueda, Modifications continues des variétés de Stein. *Publ. Res. Inst. Math. Sci. Kyoto Univ.* 13 (1977), 681-686.

[76] V. S. Vladimirov, *Methods of the Theory of Functions of Many Complex Variables.* The M.I.T. Press, 1966.

[77] A. Weil, Sur les séries de polynomes de deux variables complexes, *C. R. Acad. Sci. Paris* 194 (1932), 1304-1305.

[78] J. Wermer, Addendum to "An example concerning polynomial convexity", *Math. Ann.* 140 (1960), 322-323.

[79] H. Yamaguchi, Sur une uniformité des surfaces d'une fonction entière de deux variables complexes, *J. Math. Kyoto Univ.* 13 (1973), 417-433.

Index

Selected Titles in This Series

(Continued from the front of this publication)

159 **S. Ya. Khavinson,** Best approximation by linear superpositions (approximate nomography), 1997
158 **Hideki Omori,** Infinite-dimensional Lie groups, 1997
157 **V. B. Kolmanovskiĭ and L. E. Shaĭkhet,** Control of systems with aftereffect, 1996
156 **V. N. Shevchenko,** Qualitative topics in integer linear programming, 1997
155 **Yu. Safarov and D. Vassiliev,** The asymptotic distribution of eigenvalues of partial differential operators, 1997
154 **V. V. Prasolov and A. B. Sossinsky,** Knots, links, braids and 3-manifolds. An introduction to the new invariants in low-dimensional topology, 1997
153 **S. Kh. Aranson, G. R. Belitsky, and E. V. Zhuzhoma,** Introduction to the qualitative theory of dynamical systems on surfaces, 1996
152 **R. S. Ismagilov,** Representations of infinite-dimensional groups, 1996
151 **S. Yu. Slavyanov,** Asymptotic solutions of the one-dimensional Schrödinger equation, 1996
150 **B. Ya. Levin,** Lectures on entire functions, 1996
149 **Takashi Sakai,** Riemannian geometry, 1996
148 **Vladimir I. Piterbarg,** Asymptotic methods in the theory of Gaussian processes and fields, 1996
147 **S. G. Gindikin and L. R. Volevich,** Mixed problem for partial differential equations with quasihomogeneous principal part, 1996
146 **L. Ya. Adrianova,** Introduction to linear systems of differential equations, 1995
145 **A. N. Andrianov and V. G. Zhuravlev,** Modular forms and Hecke operators, 1995
144 **O. V. Troshkin,** Nontraditional methods in mathematical hydrodynamics, 1995
143 **V. A. Malyshev and R. A. Minlos,** Linear infinite-particle operators, 1995
142 **N. V. Krylov,** Introduction to the theory of diffusion processes, 1995
141 **A. A. Davydov,** Qualitative theory of control systems, 1994
140 **Aizik I. Volpert, Vitaly A. Volpert, and Vladimir A. Volpert,** Traveling wave solutions of parabolic systems, 1994
139 **I. V. Skrypnik,** Methods for analysis of nonlinear elliptic boundary value problems, 1994
138 **Yu. P. Razmyslov,** Identities of algebras and their representations, 1994
137 **F. I. Karpelevich and A. Ya. Kreinin,** Heavy traffic limits for multiphase queues, 1994
136 **Masayoshi Miyanishi,** Algebraic geometry, 1994
135 **Masaru Takeuchi,** Modern spherical functions, 1994
134 **V. V. Prasolov,** Problems and theorems in linear algebra, 1994
133 **P. I. Naumkin and I. A. Shishmarev,** Nonlinear nonlocal equations in the theory of waves, 1994
132 **Hajime Urakawa,** Calculus of variations and harmonic maps, 1993
131 **V. V. Sharko,** Functions on manifolds: Algebraic and topological aspects, 1993
130 **V. V. Vershinin,** Cobordisms and spectral sequences, 1993
129 **Mitsuo Morimoto,** An introduction to Sato's hyperfunctions, 1993
128 **V. P. Orevkov,** Complexity of proofs and their transformations in axiomatic theories, 1993
127 **F. L. Zak,** Tangents and secants of algebraic varieties, 1993
126 **M. L. Agranovskiĭ,** Invariant function spaces on homogeneous manifolds of Lie groups and applications, 1993

For a complete list of titles in this series, visit the AMS Bookstore at **www.ams.org/bookstore/**.